ENERGY

1 Btu = 1.0555056 kJ  .169 ft·lb$_f$

1 ft·lb$_f$ = 1.3558 J 10^{-4}) Btu

SPECIFIC ENERGY

1 Btu/lb$_m$ = 2.326 kJ/kg   1 kJ/kg = 0.4299 Btu/lb$_m$

1 Btu/lb$_m$·mol = 2.326 kJ/kg·mol   1 kJ/kg·mol = 0.4299 Btu/lb$_m$·mol

SPECIFIC ENTROPY, SPECIFIC HEAT, GAS CONSTANT

1 Btu/lb$_m$·°R = 4.1885 kJ/kg·K   1 kJ/kg·K = 0.2388 Btu/lb$_m$·°R

1 Btu/lb$_m$·mole·°R = 4.1885 kJ/kg·mol·K   1 kJ/kg·mol·K = 0.2388 Btu/lb$_m$·mole·°R

DENSITY, SPECIFIC VOLUME

1 lb$_m$/ft^3 = 16.018 kg/m^3 1 kg/m^3 = 0.062428 lb$_m$/ft^3

1 ft^3/lb$_m$ = 0.062428 m^3/kg   1 m^3/kg = 16.018 ft^3/lb$_m$

POWER

1 Btu/s = 1.055056 kJ/sec   1 hp = 2545 Btu/hr

1 hp = 550 ft·lb$_f$/sec   1 kW = 3413 Btu/hr

VELOCITY

1 mph = 1.467 ft/s   1 mph = 0.4470 m/s

1 ft/s = 0.3048 m/s

TEMPERATURE

T[°C] = T[K]·273.15   T[°F] = T[°R]·459.67

T[K] = $\frac{5}{9}$T[°R]

PHYSICAL CONSTANTS

Gravitational units conversion constants: $g_c = 32.174 \dfrac{\text{ft·lb}_m}{\text{lb}_f \cdot \text{s}^2}$

Standard gravitational acceleration: $g = 32.1740$ ft/s^2 = 9.80665 m/s^2

Standard atmospheric pressure: $P_\infty = 14.696$ psia = 101,325 kPa = 760 mm Hg

Thermodynamics

PWS Series in Engineering

Thermodynamics

Edward E. Anderson
Texas Tech University

PWS PUBLISHING COMPANY
Boston

PWS PUBLISHING COMPANY
20 Park Plaza, Boston, MA 02116-4324

Copyright © 1994 by PWS Publishing Company. All rights reserved. No part of this book may be reproduced, stored in a retrieval system, or transcribed in any form or by any means, electronic, mechanical, photocopying, or otherwise, without the prior written permission of the publisher, PWS Publishing Company, 20 Park Plaza, Boston, Massachusetts 02116.

PWS Publishing Company is a division of Wadsworth, Inc.

 This book is printed on acid-free, recycled paper.

International Thomsom Publishing
The trademark ITP is used under license.

Sponsoring Editor: *Jonathan Plant*
Developmental Editor: *Maureen Brooks*
Production Editor: *Kirby Lozyniak*
Editorial Assistant: *Cynthia Harris*
Manufacturing Coordinator: *Ellen Glisker*
Interior Designer: *Julie D. Gecha*
Interior Artist: *Christopher Hayden*
Chapter Opening Illustrations: *Academy Artworks, Inc.*
Cover Designer: *Kirby Lozyniak*
Cover Image: © *Steven Hix / FPG International*
Cover Printer: *Henry N. Sawyer Company, Inc.*
Text Printer and Binder: *R.R. Donnelley/Crawfordsville*

Printed and bound in the United States of America.

1 2 3 4 5 6 7 8 9 — 98 97 96 95 94 93

Library of Congress Cataloging-in-Publication Data
Anderson, E. E.
    Thermodynamics / Edward E. Anderson.
        p.   cm.
    Includes index.
    ISBN 0-534-93294-0
    1 Thermodynamics.   I. Title.
TJ265.A48 1993   93-25647
621.402'1—dc20   CIP

Contents

CHAPTER 4

Closed Systems   124

CHAPTER 5

Steady-Flow Systems   185

CHAPTER 6   **General Transient-Flow Systems   242**

CHAPTER 7   **Vapor Power and Refrigeration Cycles   276**

CHAPTER 8   **Gas Power and Refrigeration Cycles   351**

Contents

CHAPTER 9

Additional Thermodynamic Property Relations and Models   430

CHAPTER 10

Homogeneous Gaseous Mixtures   492

CHAPTER 11   Thermochemistry and Combustion   548

CHAPTER 12   Introduction to Equilibrium   592

Preface

The goal of this book is to help undergraduate students develop a practical, systems-based understanding of thermodynamics. It is intended for the sophomore/junior-level course in thermodynamics taken by most mechanical, civil, and electrical engineering majors. Organized around system types, rather than thermodynamics laws and principles, the book introduces the second law of thermodynamics and entropy at the outset, and then integrates the first and second laws of thermodynamics in the study of all systems throughout the book. The implications of the first and second laws with respect to cyclic systems are thoroughly presented (including work potential) before moving on to closed, steady-flow, and transient systems. From the very beginning, students are made aware that they must consider both laws simultaneously when studying the operation of any system. This approach differs from that taken by traditional texts, in which the first and second laws are studied separately, and are only integrated very late in the development of the subject.

In presenting new topics, the author's goal has been to begin with the qualitative, concrete and familiar, and then progress to the quantitative, abstract, and general. Students seem to learn thermodynamics more quickly and with greater interest when topics are grounded in the "real-world" value of the material. In addition, the author has tried to adopt a "just-in-time" approach to introducing complex topics. By avoiding such topics as multiple property models for the same substance, and temperature variable specific heats, in the early chapters, students can focus on the foundations of thermodynamics without being overwhelmed by complications that only add to the accuracy with which problems are solved. Both of these techniques are used by the author to help the student quickly master the subject and develop a true "feel" for thermodynamics.

Developing the subject of thermodynamics around systems rather than physical laws leads to a cumulative way of learning that builds deliberately on what students have learned before. Thus, as students work through the text, the work potential of an energy reservoir becomes "closed-system exergy," which in turn becomes "steady-flow exergy." This spiral, systems-based pattern of learning is not unlike the way that the subject of thermodynamics itself developed over time. Although those accustomed to more traditional presentations may find this approach somewhat repetitive, the author has found that it greatly assists students in reinforcing their understanding of what has historically been regarded as an abstract, difficult and unapproachable subject.

The author believes that a student only truly comprehends thermodynamics when the most basic principles are fully understood. This occurs when the student looks beyond the myriad of equations resulting from these principles to the principles themselves. This goal runs counter to the usual goal of solving as many problems as quickly as possible. To encourage the student to concentrate on principles rather than equations, the author has used the fewest equations practicable in presenting the material. The equations, definitions, principles and laws that form the source of all analysis and problem-solving are distinctively set off in the text. To further encourage the student to "keep it simple," a problem-solving methodology that applies only these equations, definitions, principles and laws is used throughout the book in all of the example problems. The author's "just-in-time" presentation of new material (only when needed, and only in the detail required to understand the topic at hand) further supports the goal of making students comfortable and grounded in the subject.

Chapter opening introductions are intended to link thermodynamics concepts to new discoveries and technologies, drawing on such fields as combustion engine design, aerospace and environmental engineering, and biomedical design. The author has also provided detailed coverage of engineering design issues throughout the book, addressing an important trend in engineering education.

The essence of thermodynamics is presented in the first six chapters of the book. These chapters cover basic concepts, simple property models, the first and second laws of thermodynamics, exergy (availability), and simple processes and devices. The latter chapters develop property models, thermodynamic property relationships and applications of thermodynamics to systems, mixtures, and chemical reactions. The author uses the first six chapters, plus selected topics in the systems applications chapters, for a one-semester introduction to thermodynamics course taught to all engineering disciplines. The rest of the material serves a second, one-semester application of thermodynamics course as typically taught in a mechanical engineering curriculum.

An IBM PC 3.5" disk containing the program *Thermodynamics Self-Help*, is available to all adopters for duplication or network installation. This software is a hypertext-based tutorial program designed to reinforce the learning of basic principles, concepts and definitions and is intended for IBM PCs using WINDOWS. A complete solutions manual is also available in two volumes, Chapters 1–6 and Chapters 7–12.

Some instructors may wish to follow a conventional, topical organization of materials. To accommodate these instructors, each section has been written to stand alone. This allows teachers to present the sections in any order that meets their course objectives. All the major topics of an introductory thermodynamics course are presented in this text.

The author is fortunate to have had the editorial staff at PWS and the following reviewers to assist him in the development of this book:

David G. Briggs
Rutgers University

Allan T. Kirkpatrick
Colorado State University

Kirby Chapman
Kansas State University

Peter E. Liley
Purdue University

Creighton A. Depew
University of Washington

Noam Lior
University of Pennsylvania

Steven K. Howell
Northern Arizona University

Timothy W. Wong
Arizona State University

These individuals have helped the author define an approach which allows beginning students to develop a comprehensive and thorough understanding of thermodynamics. The author is also indebted to his wife, Sharon, who has been a constant source of inspiration, a long time friend, and has allowed the author to concentrate on writing this text. Dr. Louis Gritzo, a former graduate student, and his son David, have assisted the author by developing many of the text examples and the solutions manual. They have also made many constructive suggestions, as former students of thermodynamics. The author would also like to thank Dong Shen of Texas Tech University for carefully checking the accuracy of the answer section and the solutions manual. The author thanks the many students who have taken thermodynamics under his instruction over the years. Their numerous suggestions, comments, discussions and guidance have been incorporated into this text, which was written with them and future students in mind. Finally, the author has been blessed by having been taught by many truly outstanding teachers. He is particularly grateful to Robert Fellinger of Iowa State University and to Raymond Viskanta of Purdue University who respectively taught him to appreciate thermodynamics and to seek the foundations upon which true knowledge is built.

Note To The Student

Thermodynamics is a body of knowledge based upon a set of definitions and physical principles which are often expressed in mathematical form. Students are encouraged to focus their attention on the definitions and principles rather than the numerous equations which can be derived from them. To encouraged this, we have distinguished important definitions as:

DEFINITION: *Thermodynamics*

The study of energy transformations as well as the devices and materials used to accomplish these transformations.

An important principle is distinguished as:

PRINCIPLE: First Law of Thermodynamics (Cyclical Devices)

$$\oint \delta Q = \oint \delta W$$

and important equations as:

$$\eta = \frac{W}{Q_h} \tag{2.1}$$

All the results presented in this text are simply applications of these important elements to specific situations. Equations numbered in other ways are only designated as such so that they may be referred to in future discussions and does not imply that the equation has any particular importance.

Many complete examples are also presented in this text. A specific format is used in the examples to assist the student with reducing general principles to specific applications. Students are encouraged to follow this format as it represents an organized approach to problem solving. When appropriate, tips have been added to the examples to give additional insight to the significance of the example. These tips are placed in the margin and appear in the second color. Students should review these examples and tips in preparation for working the section problems.

The use of system schematic diagrams, with labeled points at different locations in the system, is essential to any problem solution. These are accompanied by sketches of state diagrams showing all the system states with labels and processes. Students are encouraged to make it a habit to include these sketches with all problem solutions.

Most chapters include projects at the end of the chapter in addition to the problem set. These projects are more extensive than a typical problem and may require additional reading, clarification and decision making on the part of the student. Most do not have a single correct answer. Rather, the correct answer will depend on the manner in which the student approaches the project. They are intended to extend the student's knowledge beyond the material covered in the chapter and to introduce him or her to the real world of design decisions.

This text also includes tabulated thermodynamic property tables, located in Appendix B, to help the student solve problems. This is done so that the student is prepared to solve a problem when a personal computer is not available or if a problem involves a small number of states.

A separate supplement, *Tables and Charts for Thermodynamics*, By Peter Liley of Purdue University, is also available. It provides an excellent reference for laboratory and classroom use.

Edward E. Anderson

Thermodynamics

Introduction

Of Knowledge and Fish

Whatever be the detail with which you cram your student, the chance of his meeting in later life exactly that detail is almost infinitesimal; and if he does meet it, he will probably have forgotten what you taught him about it. Knowledge does not keep any better than a fish. The really useful training yields comprehension of a few general principles with a thorough grounding in the way they apply to a variety of concrete details.

Alfred North Whitehead

The harvesting and conversion of energy resources into more useful energy forms is a principal activity in any industrialized society. Today's highly developed societies are those whose members modify large quantities of energy into many more useful forms for a large variety of purposes. At modest cost, this energy transports people, reduces their labor, entertains them, allows them to communicate, and increases their productivity, among many other applications.

The 1990 supply of and demand for energy in the United States is summarized in Table 1.1. To put these data into perspective, in 1990 every resident used an average of 8–10 gallons (30–40 liters) of oil every day. Each resident spent from $3 to $5 per day on energy, or 5% to 15% of his or her income. Most of this energy was derived from oil, coal, or natural gas that was converted by a facility or a machine (such as an electricity generating station or an internal combustion engine) into a more usable form. In 1990 the United States imported approximately 20% of its energy.

TABLE 1.1

1990 Energy Supply and Demand, United States

	CONVENTIONAL UNITS	OIL EQUIVALENT (10^6 BARRELS PER DAY)	QUADRILLION Btu PER DAY *	QUADRILLION KILOJOULES PER DAY
Total Demand		38.5	81.4	86.0
Domestic Supply				
Oil and NGL		8.4	17.7	18.7
Natural Gas	17.7 trillion cubic feet	8.5	18.1	19.1
Coal	1034 million short tons	10.7	22.6	23.9
Nuclear	109 gigawatts	2.9	6.2	6.5
Hydro/Geo/Solar		1.5	3.1	3.3
Imports				
Natural Gas	1.4 trillion cubic feet	0.7	1.4	1.5
Oil		7.1	15.1	16.0

* Btu = British thermal unit

The large-scale consumption of energy has now begun to affect the natural environment. This impact is occurring at locations where the resources are harvested as well as where they are being used. This fairly recent change in energy development has the potential of curtailing the continued growth of these highly developed societies and represents one of our greatest challenges over the next few decades.

In view of the impact that energy resources and conversion have on humanity, it is surprising how few individuals understand energy, its conversion, the limits on its conversion, and the consequences of its conversion. Thermodynamics is the beginning of the study of these questions. Through the study of thermodynamics, we will learn that 100% conversion of energy to more useful forms is impossible and that energy conversion always involves some waste of energy. The details of how to accomplish an energy conversion, typical devices for energy conversion, and many other energy topics are also discussed throughout this book. This knowledge is invaluable to engineers who select, develop, design, operate, or maintain energy-conversion systems. But, it is also extremely useful to policy makers, planners, and decision makers.

This chapter begins to develop the concepts of thermodynamics, energy, and energy conversion by defining several terms that are specific to thermodynamics. We also begin developing an intuitive feel for the conversion of energy by describing some of the more common energy-conversion systems.

1.1   What Is Thermodynamics?

Webster's *New Collegiate Dictionary* defines *energy* as "the capacity for performing work." Work is defined by Webster as "the transference of energy by a process involving the motion of the point of application of a force." These definitions clearly demonstrate that work and energy are coupled concepts. To some extent, the circular nature of these definitions cloud the concept of energy. We must rely on our experiences to clarify this issue. For example, we are all familiar with the concept of energy being stored in a compressed or coiled spring and the release of energy by the burning of wood or coal, although we may not be able to define these concepts. The study of thermodynamics will clarify our understanding of energy and work (as well as heat exchange) and broaden our knowledge of energy conversion.

Thermodynamics is the study of energy conversion and the limitations on energy conversion. It is also the study of the properties of materials as related to energy transformations.

DEFINITION: *Thermodynamics*

The study of energy transformations as well as the devices and materials used to accomplish these transformations.

The spark-ignition internal-combustion engine (known as the *Otto cycle* in thermodynamics) is but one example of an energy-transformation device. Basically, the engine is designed to convert the chemical energy available in a fuel-air mixture into mechanical work that can be supplied to the transmission of an automobile, the cutter blade of a lawn mower, or any number of other useful purposes. This is accomplished by a device consisting of a cylinder (or cylinders) equipped with two or more valves and a piston that moves up and down inside the cylinder as shown in Figure 1.1. The valves open and close at the appropriate times to allow a fresh fuel–air charge to enter the volume formed by the piston cylinder through the intake manifold and to allow spent combustion gases to leave the piston cylinder through the exhaust manifold. When the values are not open, the piston cylinder is sealed so that mass cannot enter or leave the enclosed volume. A spark device (spark plug) is located inside the piston cylinder. At the appropriate time, a brief spark generated by this device initiates the combustion of the fuel–air charge.

FIGURE 1.1

*Typical
Internal-Combustion
Engine
Piston-Cylinder
Apparatus*

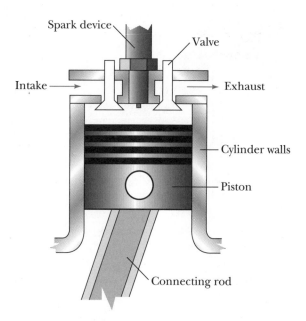

The reciprocating motion of the piston within the cylinder is generated by a crankshaft-connecting rod mechanism. The position of the piston in the cylinder is then directly related to the rotational position of the crankshaft. At its uppermost position, as shown in Figure 1.2, the piston is said to be at top dead center (TDC), and the volume of the space confined by the piston and cylinder in this position is known as the *clearance volume.* This is also the minimum volume that the substance occupying the piston-cylinder chamber can have. At its lowest position, the piston is said to be at bottom dead center (BDC). In this position, the substance filling the chamber is at its maximum volume. The difference between the confined volume at BDC and TDC (i.e., the volume swept by the piston as it moves from BDC to TDC) is known as the *cylinder displacement volume.* Engine displacement (i.e., the sum of all the engine's cylinder displacement volumes) is a commonly cited measure of engine size and, to some degree, power. For example, a 6-cylinder automobile engine with a 3-liter displacement implies that each piston displaces 0.5 liter as it moves from BCD to TDC.

FIGURE 1.2

Typical Internal-Combustion Engine with the Piston at Its Extreme Positions

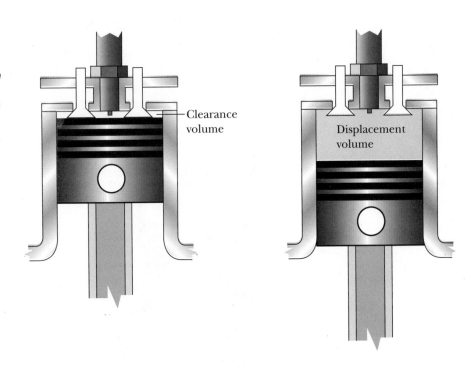

AT TOP DEAD CENTER   AT BOTTOM DEAD CENTER

In the four-stroke version of this engine, the sequence of events that form a complete cycle is:

Stroke 1   **(TDC to BDC)**   Intake of a fresh fuel–air charge while the inlet valve is open.

Stroke 2   **(BDC to TDC)**   Compression of the charge while both valves are closed with ignition occurring near the end of the stroke.

Stroke 3   (TDC to BDC)   Expansion of the combustion gases with both valves closed.

Stroke 4   (BDC to TDC)   Removal of the spent combustion gases through the open exhaust valve before the entire process is repeated.

These four processes are illustrated on the cylinder pressure–volume diagram of Figure 1.3. Each stroke ends and the next stroke begins at the solid dots; the TDC and BDC positions of the piston are indicated in this figure. Intake stroke 1 begins with the piston near TDC. At that time, the intake valve is opened and a fresh fuel–air charge is drawn in as the piston moves down to BDC. Near BDC, the intake valve is closed and the piston begins to move up. Both valves are closed during this part of the cycle, and the confined fuel–air mixture is compressed into an increasingly smaller volume during this second stroke. As a result of this compression, the pressure and temperature of the confined fuel–air mixture increases as the piston moves up. Just before the piston reaches TDC, the spark device is fired and combustion proceeds. After passing TDC, the piston begins to move down with both valves still closed as the third stroke is executed. The pressure and temperature of the gases resulting from the combustion then decrease as the gases expand and work is delivered to the crankshaft. Near BDC, the exhaust valve opens so that the spent combustion gases can be moved out of the piston cylinder as the piston again moves up during the fourth stroke. Once TDC is reached, the exhaust valve closes and the inlet valve reopens to restart the entire sequence.

FIGURE 1.3
Typical Spark-Ignition Engine Cylinder Pressure–Volume Diagram

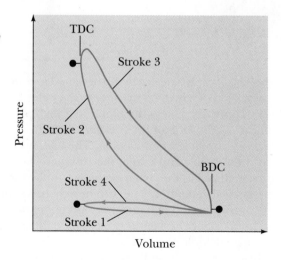

This engine illustrates three major features of a typical thermodynamic device. First, it transforms energy from one form to another (chemical to mechanical). Second, it uses working fluids whose conditions are constantly changing. Third, it periodically returns to its starting condition to begin the process once again. Most, but not all, thermodynamic devices share these features.

As a second example, consider the typical steam power plant used by electric utilities to generate large quantities of electricity (known as the *Rankine cycle* in thermodynamics). The basic plant illustrated in Figure 1.4 has four major components: boiler, turbine generator, condenser, and feedwater pump. Water in liquid and vapor forms circulates through these devices to produce mechanical work. The generator converts this work into electrical power.

FIGURE 1.4
Basic Steam-Powered Electrical Generation Station

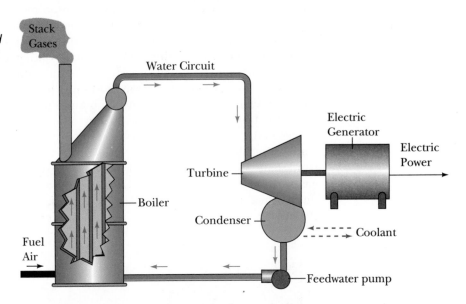

The boiler is simply a device that is designed to transfer the heat generated by the combustion of a fuel–air mixture to the water being circulated through the system. This water enters the boiler in a high-pressure liquid form. The heat from the combustion of the fuel and air converts this water into a vapor form (steam) before it leaves the boiler. The high-pressure, high-temperature steam leaves the boiler and enters the turbine, where its pressure and temperature are reduced as the steam expands and the turbine produces work to drive the electrical generator. The turbine exhausts a mixture of low-pressure steam and a small amount of liquid into the condenser, where heat is removed to convert the working fluid entirely into liquid form. This heat is removed by an environmental cooling source such as the water from a river or lake. Liquid water from the condenser is repressurized by the feedback pump before it reenters the boiler to repeat the process.

This system illustrates several additional facets of thermodynamic devices. In this case, a source of heat—the combustion gases in the boiler—is required.

Some of this heat is converted into work in the turbine, and some is rejected to the environment in the condenser. We also have only one working fluid—the circulating water. But the working fluid takes different forms (liquid and vapor) in different parts of the system.

The boiler in this system can be replaced with other energy-conversion devices such as nuclear reactors and solar concentrators. A Rankine cycle that uses a solar concentrator to heat and boil the working fluid is illustrated in Figure 1.5. In this system, the working fluid is heated and boiled in a device known as a *solar receiver.* In a receiver, concentrated solar energy is converted to heat when it is absorbed by the dark surfaces of the receiver. Solar energy converted in this way is then transferred as heat to the working fluid in much the same manner as in a conventionally fueled boiler.

FIGURE 1.5

Solar Energy–Driven Rankine Cycle

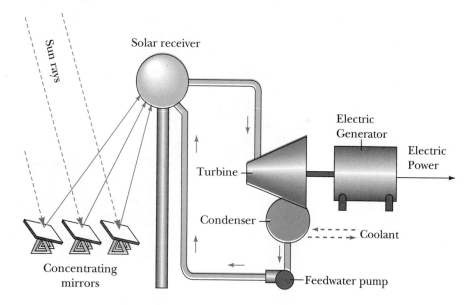

Concentration of the solar energy is required to achieve the temperatures (and consequent thermal efficiency) needed to heat the working fluid at a satisfactory rate. This can be done by using parabolic mirrors that focus the solar energy on a tube placed at the focus of the parabola. Concentration can also be done by using a large field of tilting flat mirrors that all reflect the solar energy onto the receiver. This receiver is located on a tower several hundred feet above the field of mirrors as shown in Figure 1.5. This system is thermodynamically the same as the conventionally fueled Rankine cycle discussed previously. It transfers heat to a working fluid, produces useful work, and transfers heat to its surroundings. The working fluid also passes through many states while undergoing several processes as this fluid executes its cycle. It therefore shares all the elements of other thermodynamic devices.

Our last example of a common thermodynamic device is the refrigerator, which is widely used to cool and freeze food and cool homes and automobile passenger cabins, among a variety of cooling purposes. If this device is used for heating purposes, it is known as a *heat pump*. A typical refrigerator is shown in Figure 1.6. It consists of an evaporator, condenser, compressor, and expansion device.

FIGURE 1.6
A Typical Refrigeration Device

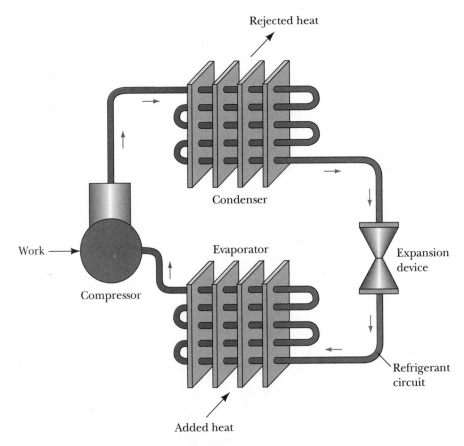

A cold, low-pressure vapor form of the working fluid enters the compressor where it is compressed to a high-temperature, high-pressure vapor. Work or power, usually from an electric motor, is applied to the compressor to accomplish this task. This vapor is next converted into a liquid in the condenser by transferring heat from the working fluid to the condenser surroundings. The liquid that results is next passed through an expansion device to reduce its pressure. This is normally done with no interactions occurring between the expansion device and its environment. On emerging from the expansion device, the working fluid is a mixture of liquid and vapor. The remainder of the liquid is converted into vapor in the evaporator by transferring heat to the working fluid mixture from the region being cooled. At this point, the working fluid is again all vapor as it reenters the compressor to begin the cycle once again.

This system, like the previous examples, uses work and heat interactions between the working fluid in the various pieces of equipment and their environments to accomplish a task (cooling in this case). Many, but not all, refrigerators maintain the refrigerated space at a low temperature by boiling the working fluid. Special fluids with low boiling points are required for this purpose. Typically, refrigerants such as ammonia and chlorofluorocarbons are used in this system.

1.2 Elementary Definitions

We begin our study of thermodynamics with several basic definitions. Most of these definitions are simply starting points through which we can share a common vocabulary. As our study develops, these defintions will take on an expanded and much more comprehensive meaning.

First, we consider the idea of a system.

> **DEFINITION: *System***
>
> A system is an arbitrarily defined fixed quantity of mass or region of space that we select to focus our attention for the purpose of analysis or understanding. A real or imaginary boundary known as a *control boundary* is always used to delineate the system from other elements in the environment.

The system plays the same role in thermodynamics as the free-body diagram plays in such mechanics courses as statics, dynamics, and strength of materials.

Notice that two types of systems exist: the fixed-mass system and the region-of-space system. The *fixed-mass system* is also known as the *closed system* or *control mass system,* and the *region-of-space system* may be called a *control volume, open system,* or *steady-flow system,* depending on the nature of the region of space selected and the accepted conventions in various disciplines. For example, aeronautical engineers and fluid mechanists prefer the *control volume* term, and thermodynamic and heat-transfer practitioners prefer the *steady-flow* or *transient system* terms.

Also observe that a system must be selected by placing a real or imaginary boundary around the substance or space selected for analysis. This boundary is known as a *control boundary* or *control surface.* We are at complete liberty to do this in any way that is convenient for the question at hand. This is *always the first step* in any thermodynamic or engineering analysis. It is also a very important step because the system selected determines the direction of analysis and the equations that can be applied. Our first system choice may not be appropriate, and we may need to start over with a better one to obtain the desired answer. If this occurs, take advantage of what you learned from your first choice to more intelligently select your second (and perhaps later) system. Just remember that starting over happens to the best thermodynamicists.

An example of a fixed-mass (closed) system is illustrated by the spring-loaded piston-cylinder device of Figure 1.7. The system is the mass confined by the control boundary (the dashed lines). The control boundary is formed by the interface between the inner piston-cylinder surfaces and the substance that fills the volume formed by the piston and cylinder. As the piston moves up or down, the control boundary changes its shape, but no mass crosses this boundary. The mass confined by this surface, also known as the *control mass,* then remains fixed regardless of the position of the piston.

FIGURE 1.7
A Typical Closed System

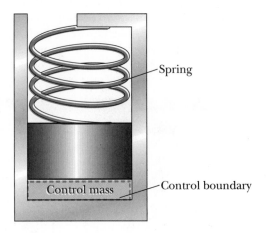

An example of an open (control-volume) system is shown in Figure 1.8. This system is a nozzle that accelerates the velocity of a fluid as it flows through the system. In this case, the control boundary consists of the interface between the inner surface of the nozzle material and the fluid that fills the nozzle,

FIGURE 1.8
A Typical Open System

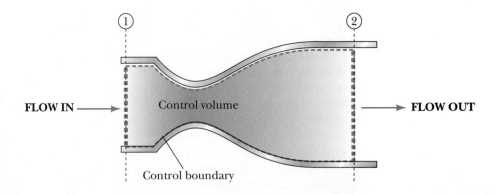

as well as the imaginary surfaces across the nozzle's entrance and exit. This control boundary is illustrated by the figure's dashed lines. The control volume formed by this control boundary does not change its shape in this case because the geometry of the nozzle passage is fixed. However, mass is permitted to enter and leave this control volume through the imaginary surfaces across the nozzle's entry and exit. Open systems can then exchange mass with their environments, while closed systems cannot.

Because our approach depends on the system selected, we must be very careful to clearly define our selection. Although everyone will eventually come to the same conclusions, the path followed to these conclusions will be determined by the system selected. Thus, we may each use a different set of equations depending on the system selected, but we will obtain the same final answer. In this book, systems are clearly defined before we begin any anaylsis. You are encouraged to do the same.

> **DEFINITION: *Surroundings***
>
> The surroundings include everything that is not part of the system being considered.

For example, suppose that for the internal-combustion engine discussed in the previous section we select as our system the region of space whose bounding surfaces are the cylinder, top of the piston, and two imaginary surfaces stretched across the valve openings. Anything outside of this control boundary is part of the surroundings. This includes the fuel–air mixture in the intake manifold waiting to enter this region, the spent combustion gases in the exhaust manifold that have left this region of space, all of the metal engine parts, the cooling water in the cooling system, the atmospheric air surrounding the engine and vehicle, and so on. The benefits of dividing a problem into a system and its surroundings is now quite clear. In this example, we only need to concentrate on the events that happen inside the volume formed by the piston and cylinder as well as the interactions that occur between the system and its surroundings. We need not become confused by anything that occurs in another part of the engine or environment.

As a second example, consider a block of solid material that is sliding down an inclined plane. If we consider the block as our system, it is a fixed-mass system as it moves down the inclined plane (if no material is worn off the portion of the block in contact with the plane). Now we only need to consider the block; we can remove the other elements (inclined plane, etc.) from the problem. Of course, we must replace the effects of the surroundings on the system in the correct manner. In this example, the inclined plane effect would be replaced with a frictional force and the gravitational effect would be replaced with a weight force; both forces act on the block.

DEFINITION: *Property*

A property of a system is any characteristic of the system that can be observed or measured. Mathematical combinations of observed or measured properties are also properties.

You are already aware of properties such as mass, weight, volume, pressure, and temperature. In fact, you have probably measured these at some time. You may not be familiar with other measurable properties, such as electrical conductivity, dielectric constant, emissivity, and several others, but each is important in various engineering disciplines.

Mathematical combinations of properties form new properties. For example, if we divide the mass of a system by its volume, a new property known as *density* is formed. Similarly, the ratio of the system volume to the system mass forms a new property known as *specific volume*. Obviously, the list of properties is open-ended, and we can define a new property any time it is convenient to deal with the problem at hand. Fortunately, only a few properties are needed for most problems.

Properties can be further subdivided into *extensive* and *intensive* properties. Extensive properties are those that depend on the *size* of the system. Volume and mass, for example, are extensive properties. Intensive properties such as pressure and temperature do not depend on system size. A good test to determine if a property is extensive or intensive is to divide the system into two equal parts and then determine which properties change. If we consider a system that contains 1 kilogram of a gas and occupies 1 cubic meter of space, then its specific volume is 1 cubic meter per kilogram. Now, divide this system into two equal parts. Each new system now contains 0.5 kilogram of gas and occupies 0.5 cubic meter of space. But the specific volume of each new system is still 1 cubic meter per kilogram. Consequently, mass and volume are extensive properties, and specific volume is an intensive property.

The ratio of any extensive property to the mass of the system forms an intensive property known as a *specific* property. We have just seen the specific volume (volume per unit mass). Several others will be introduced at later points in this book. To distinguish extensive properties from their specific counterparts, we will use capital letters for the extensive property and lowercase letters for their specific counterparts (the exception to this rule is pressure, P, and temperature, T, which are always intensive properties). Thus, volume is represented by V and specific volume is designated by v. Also note that specific properties are always intensive properties.

DEFINITION: *State*

The state of a system is simply the condition of the system as described by its properties.

The water in the steam-powered electrical generation station discussed in the previous section has different states at various places in the equipment. For example, it enters the feedwater pump as a liquid at some pressure and temperature. On leaving the boiler, it is a vapor at some higher pressure and temperature. Thus, the state of the water at the feedwater pump entry is different from the state at the boiler exit. To describe the water completely at these two states, we need to know several properties.

For the concept of states to be useful, the system must be in equilibrium or close enough that it can be modeled as if it were in equilibrium.

DEFINITION: *Equilibrium*

A system is said to be in equilibrium when its contents cannot change their condition of their own accord. Such a change can only be brought about through some action in the surroundings.

Consider the weighted piston-cylinder apparatus shown in Figure 1.9. The system is the contents of this piston cylinder, which can be a gas, liquid, or solid. A force balance on the piston tells us that the pressure of the system equals the combined weight of the piston and weights divided by the area of the piston. Thus, we can control the system pressure by adding or removing weights from the piston. Whenever a weight is added or removed, the system will change to a new state. After reaching this new state, it will no longer change its state of its own accord because all the forces will again be in balance. So it is once again in equilibrium and in a new state that will not change until more weights are added or removed.

When the forces acting on a system are all in balance, we say that the system is in *mechanical equilibrium*. *Chemical equilibrium* implies that the system will not undergo any chemical reactions of its own accord. Similarly, *thermal equilibrium* implies that the system will not change temperature of its own accord. If the system working substance consists of more than one phase (say, liquid and solid) and one phase is not being converted to the others (say, liquid to solid), then *phase equilibrium* is said to exist. For complete equilibrium to exist, all of these, plus several other forms of equilibrium (when appropriate), must exist.

FIGURE 1.9
*A Weighted
Piston-Cylinder
Apparatus*

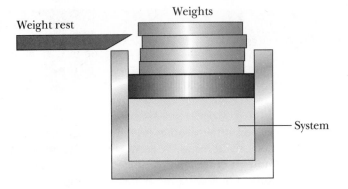

A simple system is as follows.

> ## DEFINITION: *Simple System*
>
> A system whose state is completely fixed by two independent properties that are based on the molecular behavior of the substance contained in the system.

The state of a simple system is then a point on a diagram whose two axes are the two independent properties. Each point on one of these diagrams represents a different state of the system. These diagrams are known as *state diagrams,* and are extremely useful for understanding thermodynamic devices and systems.

> ## DEFINITION: *Process*
>
> As a system transforms from one state to another, it passes through a series of intermediate states. This sequence of states is known as a *process.*

A typical process is shown in the state diagram of Figure 1.10. This process couples the initial state 1 to the final state 2. Notice that this is just one of several processes that could couple the initial and final states. During some processes, one property is held constant. These processes are usually, but not always, denoted by the *iso-* prefix. For example, a constant pressure process is known as an *isobaric process,* and a constant temperature process is known as an *isothermal process.* Processes are either *reversible* or *irreversible.*

FIGURE 1.10
*A Process on a
General State
Diagram*

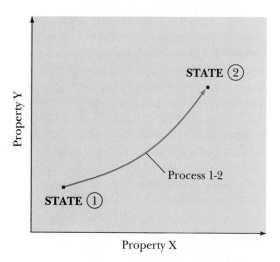

DEFINITION: *Reversible Process*

A reversible process is one in which everything involved with the
process (system and surroundings) can be returned to its original
condition after the process has been executed.

To illustrate the concept of a reversible process, consider a weight sliding
down an inclined plane with friction between the weight and plane. As the
weight moves down the plane, its elevation decreases and it consequently loses
potential energy. This energy may be stored in the surroundings for later
recovery and use. Simultaneously, the temperatures of both the weight and
plane increase as a result of the frictional effect. Nothing in the surroundings
changes as the weight moves down the plane.

The system (i.e., weight and plane) may be returned to its original condition by cooling the block and plane to their original temperatures and restoring
the weight's potential energy. As the block and plane are cooled, heat is transferred to the surroundings. This heat will certainly change the state of the
surroundings in some way. Similarly, the potential energy of the weight can
be restored by doing work on the weight to lift it to its original elevation. The
energy stored in the surroundings during the original process can be used to
restore the block's original position while that portion of the surroundings
in which this energy was stored is returned to its original condition. Thus, if
the system is returned to its original state, the surroundings will not return to
their original state as a result of the heating of the block and plane. Processes
involving friction are therefore not reversible and are said to be *irreversible*.

As an illustration of a reversible process, consider the weighted piston-cylinder apparatus of Figure 1.9. Each weight is considered to be very small. More precisely, we will consider what happens in the limit as the size of the weights approaches zero. The system is the contents of the piston-cylinder apparatus. We will force the system to undergo a process by sliding the top weight from the piston to the weight rest. Assuming this process of sliding the weight is frictionless, no work is required to move the weight to the weight rest. When the weight is moved from the piston, the system pressure decreases a small amount and the system volume increases a small amount. By removing several weights in this manner, a process will have been executed and the system will be at a new state of lower pressure and larger volume.

To determine if this process is reversible, we must ascertain if the system and its surroundings can be returned back to their original states once the process is completed. This may be done by frictionlessly sliding the weights back on the the piston one at a time. Again, because of the frictionless surface, no work is involved in sliding the weights back on to the piston. Therefore, nothing changes in the surroundings except the location of the weights, which are being repositioned on the piston. As each weight is moved back on to the piston, the system is compressed as the system pressure increases a small amount and the system volume decreases a small amount. After all the weights have been moved back on to the piston, the system will have returned to its original pressure and volume. Because both the system and surroundings have been returned to their original condition, this process is reversible.

In contrast, consider a rigid vessel divided into two chambers by a thin membrane as shown in Figure 1.11. Initially, the left chamber contains a substance at a high pressure and small volume and the right chamber is completely evacuated. When the membrane is ruptured, the substance will fill the evacuated chamber very rapidly. Considering the contents of the two chambers

FIGURE 1.11
A Partitioned Vessel System

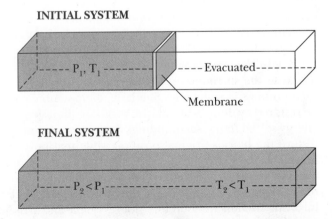

INITIAL SYSTEM

P_1, T_1 ----- Evacuated -----

Membrane

FINAL SYSTEM

$P_2 < P_1$ ----------------- $T_2 < T_1$

as the system, the rupturing of the membrane executes a change in state very similar to that of the previous weighted piston-cylinder apparatus; that is, the system (i.e., the substance) starts at a state of high pressure and small volume and ends at a state of reduced pressure and increased volume.

To determine if this change in state is reversible, we must return the system to its original condition. This may be done by replacing the membrane and then using a pump (which requires work) to move the substance in the right chamber back to the left chamber. The system has now been returned to its original state, but the surroundings have not because work has been transferred from the surroundings. This work transfer permanently changes the state of the surroundings, and the process is irreversible.

Features found in irreversible processes include:

1. friction
2. unrestrained expansion (as in the previous example)
3. heat transfer across a finite temperature difference
4. rapid mixing
5. rapid chemical reactions
6. other factors that create losses

Inspection of this list reveals that one or more of its elements are involved in any real process. The reversible process then plays the same role in thermodynamics that the frictionless plane, unstretchable cord, point mass, and other idealizations play in mechanics. Much can be learned from the study of reversible processes. At a later point, we can correct any conclusions we have reached based on the reversible process to account for irreversible effects.

DEFINITION: *Cycle*

A cycle is a series of processes that form a closed figure on a state diagram. Cycles have the characteristic that the system always returns to the same state at which it started.

A typical cycle consisting of three different processes is illustrated in the state diagram of Figure 1.12. Notice that the system always returns to the initial state 1 in this example. The minimum number of processes involved in a cycle is two; no maximum number exists.

FIGURE 1.12
A Typical Cycle

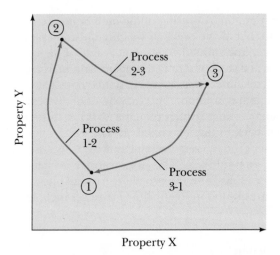

Property X

DEFINITION: *Work*

Work is an interaction between a system and its surroundings that can be reduced to the lifting of a weight in the surroundings. The magnitude of the work is given by:

$$_1W_2 \equiv \int_1^2 \mathbf{F} \cdot d\mathbf{x}$$

where $\mathbf{F}$ is the force vector exerted by the system on the surroundings at the control boundary, and $\mathbf{x}$ is the displacement vector of the system's control boundary.

This definition has three parts: (1) interaction, (2) reduction to the lifting of a weight, and (3) magnitude of work. Each is important to our understanding of the concept of work.

First, work is an interaction between a system and its surroundings; it is not a characteristic of a system or the surroundings, such as a property. Rather, it is an interaction between these two that occurs at the control boundary. For an interaction to occur, either the system or surroundings must be changing. Something must then change its state for work to occur. Because the surroundings are usually not changing their state, the system must be changing its state when work is occurring. Stated in another way, work can only occur when a system is undergoing a process. *A system that is not changing its state cannot do work.*

Of all the possible interactions between a system and its surroundings, work is the one that can be reduced to the lifting of a weight. This does not require the *lifting* of a weight, but it must be capable of being *reduced* to the lifting of a weight. For example, consider an electrical storage battery such as one used to start an automobile. Over a period of time, a current i is drawn from this battery while the voltage drop across the terminals remains at ΔV. This current could be flowing through a resistor, personal computer, light bulb, or any of several electrical devices. Selecting the storage battery as the system, an interaction occurs (i.e., the flow of current across the control boundary) between the battery and the device in the surroundings. Is this interaction work? To answer this question, we must reduce the interaction to the lifting of a weight in the environment. To this end, let us connect an electrical motor to the battery's terminals. This motor is selected to require the same ΔV and i so that the system will undergo the same process as before. The motor is attached to a drum about which a cable attached to a weight is wrapped. As the motor turns, the weight will be lifted. Hence, this interaction between the system and surroundings is work regardless of the device attached to the storage battery.

Finally, the magnitude of the work is determined by the classical mechanical definition of work as developed in classical physics and mechanics courses: the dot product of the force vector and differential displacement vector integrated over the total displacement. For work to be involved, two things must be present: a force and a displacement of the control boundary or a portion of the control boundary. *If either element is absent, work is not possible.* Note that *the force produced by the system* is used in this calculation rather than the force produced by the surroundings. This is different from the mechanics definition of work. The accepted thermodynamics work sign convention is that *work done by the system is considered positive* and *work done on the system is considered negative*. For a variety of historical reasons, this convention is just the opposite of that accepted in mechanics disciplines.

Work is measured on a total, rate, or specific (i.e., per unit of system mass) basis. We will use the symbol W (or $_iW_j$ when discussing processes) to denote the total work required or done. The symbol $\dot{W}$ will be used to denote the rate at which work is required or done (i.e., power). Finally, the symbol w is used to denote the specific work.

| EXAMPLE 1.1 | A steel ingot is pushed to the right and up along a very low friction plane by a forklift. This forklift exerts a force of 3 newton (N) acting at an angle of 30° with respect to the line of motion followed by the ingot. Determine the total work expended on the ingot to move it a distance of 2 meters (M) along the plane. |

System: The steel ingot.

Basic Equations:

$$_1W_2 = \int_1^2 \mathbf{F} \cdot d\mathbf{x}$$

Conditions: The plane is taken as frictionless, and we presume that the force exerted by the forklift remains constant through the motion of the ingot.

Solution: The system, the reaction force generated by the system, and the adopted coordinate system are shown in Figure 1.13. The choice of this coordinate system is arbitrary and will not affect the final answer. In terms of this coordinate system, the force produced by the system is:

$$\mathbf{F} = -\mathbf{i}F\cos 30° - \mathbf{j}F\sin 30°$$

and the differential displacement of the system is

$$d\mathbf{x} = \mathbf{i}\,dx + \mathbf{j}0$$

Forming the dot product, the total work is:

TIP:

In this case, the work is negative— meaning that the surroundings (i.e., forklift) are doing work on the system. Do not forget to use the force as exerted by the system on the surroundings.

$$_1W_2 = \int_{x_1}^{x_2} -F\cos 30°\,dx = -F\cos 30°\,(x_2 - x_1)$$

Substituting the information given in the problem statement,

$$_1W_2 = -3\cos 30°\,(2 - 0) = -5.20 \text{ N-m}$$

FIGURE 1.13
System of Example 1.1

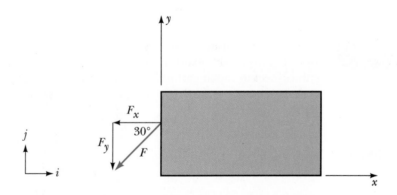

Work may be produced or consumed by a system as it changes its state. A substance that can only interact reversibly with work energy by changing its volume (or specific volume) is known as a *simple compressible substance.*

DEFINITION: *Simple Compressible Substance*

Any substance that can only reversibly perform or use work by changing its volume or specific volume.

Although substances can interact with work in other ways (e.g., ionized substances interacting with a magnetic field), our focus will be on simple compressible substances that reversibly produce or consume mechanical work through volume changes. For the most part, we will leave the treatment of other substances for other courses.

When applied to a simple compressible substance, the *state postulate of thermodynamics* (which will be discussed in more depth later) states that each state of a substance is determined by two independent properties. This very important conclusion tells us that we only need to know two independent properties of the substance that constitutes the system to determine its entire state. Thus, if the pressure and temperature of a system containing a simple compressible substance are known, the specific volume, electrical conductivity, dielectric constant, and other properties are also known. The converse is also true: If we know the pressure and electrical conductivity, we automatically know the temperature, specific volume, and so on. Certainly, equations, tables, computer models, or charts containing this property information are needed before we can do anything useful, but this result clearly specifies that any property of a system containing a simple compressible substance depends *only* on two independent properties. This considerably simplifies any equation, table, computer model, or chart needed to determine the states of a substance.

DEFINITION: *Heat Transfer*

Heat transfer (or exchange) is an interaction between a system and its surroundings that occurs at the system's control boundary and results from the system being at a temperature different from the surroundings. Heat transfer is measured in the same units as work.

Heat transfer is another form of interaction between the system and surroundings. Hence, heat transfer, like work, can only occur while the system undergoes a process. A system that is not changing its state cannot transfer heat. Only when a system changes to another state through some process will a heat transfer result. The common statement that "a gallon of gasoline contains so many units of heat" is thus incorrect. The amount of heat transfer we will realize from this gallon of gasoline depends on the process in which it is used.

Heat transfer is the interaction that is caused by a temperature difference between the system and its surroundings. This interaction occurs at the control boundary of the system. If no such temperature difference or differential exists across this boundary or the system is adequately insulated, a heat exchange interaction cannot occur. Systems having one or the other of these characteristics are known as *adiabatic systems*. Similarly, a process that is not allowed to have any heat transfer interaction is known as an *adiabatic process*. These types of systems and processes occur quite frequently in thermodynamic devices.

This definition says nothing about the magnitude of heat transfer, although it states that heat exchange units are the same as those used for work. Joules, kilojoules, and British thermal units (Btu) are commonly used for this purpose. One Btu is defined as the amount of heat transfer required to raise the temperature of a 1-pound mass of liquid water from $59.5°$ F to $60.5°$ F. This is equivalent to 778.169 foot-pounds force. The accepted sign convention for heat exchange is *heat transferred from the surroundings to the system is taken as positive* and *heat transferred from the system to the surroundings is taken as negative*. Note that this convention is the opposite of that for work.

We will use the symbol Q (or $_iQ_j$ when discussing processes) to denote the total heat transferred to or from the system. The symbol $\dot{Q}$ is used for the rate of heat transfer, and the symbol q is used for the specific heat transfer.

Thermodynamics focuses on the heat-transfer interaction and the impact of this interaction on the system and other system interactions. It does not concern itself with other factors that influence the heat transfer such as the temperature difference, materials used, surface area provided, and the movement of any fluids involved. These topics are taken up in fluid mechanics and heat-transfer courses that normally follow the study of thermodynamics.

1.3   Dimensions and Units

Certain physical quantities do not have definitions, yet we all have a very clear understanding of what they are. We all know what distance, time, and mass are, but we would be very hard pressed to express these concepts in words or other terms. Each has an associated dimension that is independent of the units needed to give the dimension a magnitude. Distance has the dimension of length (L), which is the same whether we express it in inches, meters, miles, or light-years. Similarly, time has the dimension t, mass the dimension m, temperature the dimension T, electrical current the dimension i, and light intensity the dimension I. The dimensions of all other physical quantities can be developed from these five undefined dimensions. For example, velocity has the dimension of length divided by time, acceleration the dimension of length divided by time squared, electrical charge the dimension of current times time, and specific volume the dimension of length cubed divided by mass. All of these derived dimensions result from definitions or basic equations that describe physical behavior.

We have complete freedom when assigning units to these dimensions in order to give them concrete rather than abstract meaning. Unfortunately, this has led to considerable confusion because various groups have exercised this freedom to define units in a manner convenient for them. Actually, it has also simplified matters to some degree. Imagine the size of the number that would be required to express the distance from the earth to its closest star if only centimeter or inch units were available. To reduce this confusion to some degree, we will only use the United States Customary System (USCS) and Système International d'Unités (SI) units in this book.

You may be asking, "Why two unit systems?" In fact, both systems are widely used. Although attempts have been made to eliminate the USCS system and only use the SI system, they have not been entirely successful. The modern engineer must be fluent in and capable of working in both systems. To help you acquire these skills, we will intermix the two systems throughout this book. When problems or examples are stated in USCS units, they will be solved in USCS units and the same practice will be employed for SI unit problems and examples. Do not fall into the trap of always converting to SI or USCS units. Rather, *develop the facility to work comfortably in both systems.*

The Système International d'Unités defines the units of the basic dimensions as in Table 1.2. Prefixes are attached to these units to increase or decrease their size by multiples of 10. Standard prefixes are presented in Table 1.3. Thus, a micrometer (μm) is one-millionth of a meter, and a kilometer (km) is 1000 meters. This system of prefixes and base units allows us to cover a very extensive range of sizes.

TABLE 1.2
SI Basic Dimension Units

length (L)	meter (m)
time (t)	second (s)
mass (m)	kilogram (kg)
temperature (T)	kelvin (K)
electrical current (i)	ampere (A)
light intensity (I)	candela (cd)

TABLE 1.3
Magnitude Prefixes for the SI System

Prefix	Symbol	Multiplier	Prefix	Symbol	Multiplier
atto	a	10^{-18}	deci	d	1/10
femto	f	10^{-15}	deca	da	10
pico	p	10^{-12}	hecto	h	100
nano	n	10^{-9}	kilo	k	1000
micro	μ	10^{-6}	mega	M	10^6
milli	m	1/1000	giga	G	10^9
centi	c	1/100	tera	T	10^{12}

Units derived from the basic set of dimensions and units of Table 1.2 are given new names (normally named after a famous scientist). Table 1.4 presents selected derived units that fall into this category. The same prefixes are used with these derived units. Thus, $1\ kN = 1000\ N = 1000\ kg\text{-}m/s^2$.

TABLE 1.4
Selected SI-Derived Units

newton (N)	$1\ N = 1\ kg\text{-}m/s^2$
pascal (Pa)	$1\ Pa = 1\ N/m^2$
joule (J)	$1\ J = 1\ N\text{-}m$
watt (W)	$1\ W = 1\ J/s$

The basic dimensions and units of the USCS system are presented in Table 1.5. This system includes force (F) as one basic dimension. This choice creates two complications. First, there are two pounds in this system: the force-pound and the mass-pound. These are not the same in concept and are only equal in magnitude at standard gravitational acceleration. We must then be very careful to distinguish these with the m and f subscripts when stating units.

TABLE 1.5
Units of USCS Basic Dimensions

length (L)	foot (ft)
time (t)	second (s)
mass (m)	pound-mass (lb_m)
force (F)	pound-force (lb_f)
temperature (T)	degree Rankine ($^\circ R$)
electrical current (i)	ampere (A)
light intensity (I)	candela (cd)

The second complication arises because Newton's second law of motion, $F = ma$, interrelates the force and mass dimensions. Either mass or force can then be a basic dimension, but not both. To overcome the fact that the USCS system considers both as basic units, we must introduce a unit-conversion factor into Newton's law and any results derived from this law. With the unit conversion factor, Newton's law becomes

$$F = \frac{ma}{g_c}$$

where g_c is a units conversion constant. Rearranging this equation gives

$$g_c = \frac{ma}{F}$$

One pound-force is defined as the force required to accelerate 1 pound-mass at standard gravitational acceleration ($32.174\ldots$ ft/s^2). With this definition, the g_c units conversion constant is

$$g_c = 32.174\ldots \frac{\text{ft-lb}_m}{\text{lb}_f\text{-s}^2}$$

This constant and its units may be inserted into any equation needing mass-pounds converted into force-pounds or vice versa.

EXAMPLE 1.2

A 10-pound-mass space satellite is to be accelerated at a rate of 6 feet per second squared while being launched. Determine the force that must act on this satellite to produce this acceleration.

System: The space satellite.

Basic Equations:

$$F = \frac{ma}{g_c}$$

Solution: Substituting the values from the problem statement, the force is:

$$F = \frac{10 \times 6}{32.174} \left[\text{lb}_m \frac{\text{ft}}{\text{s}^2} \frac{\text{lb}_f\text{-s}^2}{\text{ft-lb}_m} \right]$$
$$= 1.86 \text{ lb}_f$$

The USCS system also includes many traditional units such as the inch, yard, mile, gallon, and quart that are not as simply related to the base units as in the SI system. A table containing many of these conversions is printed inside the front cover of this book. These conversions may be used wherever it is appropriate to obtain the desired units.

Equations must be dimensionally correct, and the units of each term must be the same as those of the other terms. This serves as an excellent check on the correctness of an equation. For example, consider the equation

$$z = ax$$

The dimensions and units of z and ax must be the same in order for the equation to be correct. If z is a mass dimension and x is a length cubed dimension, then a must have the dimension of mass per length cubed. If z is in kilogram units and x in cubic meter units, then a must be in kilogram per cubic meter units for the equation's units to be correct. The equation

$$z = \frac{dy}{dt}$$

must also have the correct dimensions and units. If y has the dimension of length and t the dimension of time, then z must have the dimensions of length per unit of time. If y is in feet units and t is in hour units, then z must be in feet per hour units. Similarly, the equation

$$z = \int_{t_1}^{t_2} a \, dt$$

must have consistent dimensions and units. When a is an acceleration whose dimensions are length per time squared and t is a time quantity in time dimensions, z will be a velocity with the dimensions of length per unit time. If a is in m/s^2 units and t is in s units, then z will be in m/s units.

1.4 Solving Problems

It is always good practice to organize your approach to solving problems. This becomes particularly important as the problems begin to incorporate more and more physical principles and constraints. A frequently used approach to solving consists of the following steps.

1. State the problem in summary form.
2. Clearly describe the system that is to be analyzed.
3. List the principles that apply to the system and problem in a general form.
4. List the conditions and constraints that apply to the problem.
5. Proceed to solve the problem by first applying the conditions and constraints to reduce the principles to fit the system. Once this is done, substitute the information given in the problem statement to produce the solution.
6. Review the solution for errors, reasonableness, and important points made by the problem.

The purpose of restating a problem in summary form is to review the problem and become familiar with its information before attempting to formulate a solution. This may appear to be redundant when the problem is very simple and only contains one or two pieces of data, but it can very important when the problem is complex and involves many conditions, constraints, and pieces of data. The writing of this statement firmly fixes the problem and the given information in one's mind.

Next, thoroughly describe the system that will be used for solving the problem. Verbal descriptions, schematic diagrams of equipment, state diagrams, and other appropriate devices are frequently used for this purpose. This is one of the most important steps in the process because the selection of the system will dictate the form that the governing physical principles will take. For example, if the system selected is a closed system, the conservation of energy principle will have a different form than if the system were open. Everything should also be carefully labeled now so that one can clearly delineate between two system inlets, beginning and final states, and other features of the problem and system.

Basic physical principles are usually expressed in the form of equations; writing the basic equation(s) that apply to the problem is the same as stating the applicable principles. As you proceed in your study of thermodynamics, the number of principles available to you will increase and you will have to exercise judgment in selecting those that apply to the problem at hand from this set. For example, the basic principle for determining mechanical work is

$$_1W_2 \equiv \int_1^2 \mathbf{F} \cdot d\mathbf{x}$$

In addition to the physical arrangement, what distinguishes one system from another is the conditions under which each operates. For example, a system consisting of a gas trapped in a piston-cylinder device may be operated in one case with the gas pressure remaining constant and in another case with temperature remaining constant. The behavior of this system will certainly be different as these operational conditions are changed. Thus, it is very important that you *clearly state* the conditions that apply to the system before you analyze the problem. When determining mechanical work, the conditions might be that the force remains constant throughout the displacement and that the displacement occurs along a straight line, for example.

After you have carefully described the system, the conditions that apply to the problem, and the basic equations or principles, you may then proceed with solving the problem. This process is basically one of reducing the basic principles to fit the system and its conditions, performing mathematical or other manipulations that may be required to obtain the quantity sought, and then substituting the given data into this result to obtain the answer. Verbal comments, sketches, and other devices should be used throughout this step wherever they are required for clarification. In the example of determining the mechanical work, drawing a free-body diagram showing the force, writing the force and displacement in vector form, forming the dot product, and performing the integration are all part of the solution step.

Students will also find it helpful to review the whole solution to determine if the final answer is reasonable and to get an overall picture of the behavior of the system under the imposed conditions. Problems are intended to assist you in learning the material at hand. Therefore, you should carefully review each problem to be sure that you understand the material contained therein.

It is a good practice to summarize the problem at the end to ensure that you understand the problem and to simplify future reviews of the problem. We will use this procedure for solving problems in the examples throughout this textbook to help you develop your own problem-solving skills.

Students often have difficulty in finding the point at which to start a problem. When this occurs, pause and organize your approach to the problem before plunging headlong into the work. Problems are most easily organized by the following steps:

1. Clearly identify the quantity to be calculated.
2. Write the basic equation that will be used to calculate the desired quantity.
3. Identify what is known or given in each basic equation and those data that must be found by other means.
4. Write any second-level equations that are to be used to find the unknown data needed in the basic equation.
5. Identify the known and unknown quantities in the second-level equations.
6. Repeat steps 4 and 5 for any additional levels until there are no more unknown quantities.

The problem solution is then obtained by working backward from the last step to the first step as the quantities generated in each step are substituted into the preceding step. Flow charts, tree structures, and other devices can be used to organize this process, but a simple outline of the problem solution is often all that is needed. As you gain experience in solving thermodynamic problems, you will rely less on written approaches to problems and more on mentally generated approaches.

Physical units are an important consideration in thermodynamics and often a point of frustration to beginning students. I always begin a problem by converting all given data into one set of units before proceeding; I encourage you to do the same. I happen to prefer feet or meters for lengths and all length-based quantities such as area; pound-force per square inch or kilopascals for pressures; degrees Fahrenheit, Celsius, Rankine, or Kelvin for temperatures; seconds for time or time-based quantities, pound-mass, kilograms, pound-mass-moles, or kilogram-moles for mass quantities; and pound-force or newtons for force quantities. You may prefer and use another set of basic units, but always start with the same set of basic units: You will minimize problems of converting units in complicated equations and analysis. You then can concentrate on the analysis of a problem rather than the units in which the problem should be worked.

PROBLEMS

1.1 You have been asked to do a metabolism (energy) analysis of a person. How would you define the system for this purpose? What type of system is this?

1.2 You are trying to understand how a reciprocating air compressor (piston-cylinder device) works. What system would you use? What type of system is this?

1.3 How might you define a system to study the depletion of ozone in the upper layers of the earth's atmosphere?

1.4 How would you define a system to determine the rate at which an automobile adds carbon dioxide to the atmosphere?

1.5 How would you define a system to determine the temperature rise created in a lake when a portion of its water is used to cool a nearby electrical power plant?

1.6 The length of a spring can be changed by (a) applying a force to it or (b) changing its temperature (i.e., thermal expansion). What type of interaction between the system (spring) and surroundings is required to change the length of the spring in these two ways?

1.7 Is the weight of a system an extensive or intensive property?

1.8 The specific weight, γ, of a system is defined as the weight per unit volume (note that this definition violates the normal specific property–naming convention). Is the specific weight an extensive or intensive property?

1.9 Is the number of moles of a substance contained in a system an extensive or intensive property?

1.10 The molar specific volume of a system ($\bar{v}$) is defined as the ratio of the volume of the system to the number of moles of substance contained in the system. Is this an extensive or intensive property?

1.11 How would you describe the state of the air in the atmosphere? What kind of process does this air undergo from a cool morning to a warm afternoon?

1.12 How would you describe the state of the water in a bathtub? How would you describe the process that this water experiences as it cools?

1.13 A block slides down an inclined plane with friction and no restraining force. Is this process reversible or irreversible? Justify your answer.

1.14 Consider an electric refrigerator located in a room. Determine the sign of the work and heat interactions when the following are taken as the system: (a) the contents of the refrigerator, (b) all parts of the refrigerator including the contents, and (c) everything contained within the room during a winter day.

1.15 Consider an automobile traveling at a constant speed along a road. Determine the sign of the heat and work interactions, taking the following as the system: (a) the car radiator, (b) the car engine, (c) the car wheels, (d) the road, and (e) the air surrounding the car.

1.16 A personal computer is to be examined from a thermodynamic perspective. Determine the sign of the work and heat transfers when the (a) keyboard, (b) monitor, (c) processing unit, and (d) all of these are taken as the system.

1.17 A construction crane lifts a prestessed concrete beam weighing 2 tons from the ground to the top of piers that are 18 ft above the ground. Determine the amount of work considering (a) the beam and (b) the crane as the system. Express your answers in both ft-lb$_f$ and Btu.

1.18 A man weighing 180 lb$_f$ is pushing a cart that weighs 100 lb$_f$ with its contents up a ramp that is inclined at an angle of 10° from the horizontal. Determine the work needed to move along this ramp a distance of 100 ft considering (a) the man and (b) the cart and its contents as the system. Express your answers in both ft-lb$_f$ and Btu.

1.19 A man whose mass is 100 kg pushes a cart whose mass, including its contents, is 100 kg up a ramp that is inclined at an angle of 20° from the horizontal. The local gravitational acceleration is 9.8 m/s^2. Determine the work (in kJ) needed to move along this ramp a distance of 100 m considering (a) the man and (b) the cart and its contents as the system.

1.20 The force (F) required to compress a spring a distance x is given by

$$F - F_0 = kx$$

where k is the spring constant and F_0 is the preload. Determine the work required to compress a spring whose spring constant is $k = 200$ lb/in. a distance of one inch starting from its free length where $F_0 = 0$ lb$_f$. Express your answer in both ft-lb$_f$ and Btu.

1.21 Determine the work (in kJ) required to compress a spring a distance of 1 cm when its spring constant is 300 N/cm (see the previous problem) and the spring is initially compressed by a force of 100 N.

1.22 The force required to compress the gas in a gas spring a distance x is given by

$$F = \frac{\text{Constant}}{x^k}$$

where the constant is determined by the geometry of this device and k is determined by the gas used in the device. One such device has a constant of 200 lb$_f$-in.$^{1.4}$ and $k = 1.4$. Determine the work (in Btu) required to compress this device from 1 inch to 4 inches.

1.23   A gas spring is arranged to have a constant of 1000 N-m$^{1.3}$ and a $k = 1.3$ (see the previous problem). Determine the work in kilojoule required to compress this spring from 0.1 meters to 0.3 meters.

1.24   A man weighing 180 lb$_f$ pushes a block weighing 100 lb$_f$ along a horizontal plane. The dynamic coefficient of friction between the block and plane is 0.2. Assuming that the block is moving at constant speed, calculate the work required to move the block a distance of 100 ft considering (a) the man and (b) the block as the system. Express your answers in both ft-lb$_f$ and Btu.

1.25   Is g_c necessary in the SI unit system? If it is required, what are its units and magnitude?

1.26   A man weighs 180 lb$_f$ at a location where $g = 32.10$ ft/s^2. Determine his weight on the moon, where $g = 5.4$ ft/s^2.

1.27   If the mass of an object is 10 lb$_m$, what is its weight (in lb$_f$) at a location where $g = 32.0$ ft/s^2?

1.28   A lunar exploration module weighs 4000 N at a location where $g = 9.8$ m/s^2. Determine the weight of this module in newtons when it is on the moon, where $g = 1.64$ m/s^2.

1.29   What is the weight (in newtons) of an object with a mass of 200 kg at a location where $g = 9.6$ m/s^2?

1.30   The specific kinetic energy, e_k, of a moving mass is given by $V^2/2$, where V is the velocity of the mass. Determine the specific kinetic energy of a mass whose velocity is 100 ft/s in Btu/lb$_m$.

1.31   Calculate the total kinetic energy (in Btu) of an object with a mass of 15 lb$_m$ when its velocity is 100 ft/s (see the previous problem).

1.32   Determine the specific kinetic energy of a mass whose velocity is 30 m/s in kJ (See Problem 1.30).

1.33   Calculate the total kinetic energy (in kJ) of an object whose mass is 100 kg and whose velocity is 20 m/s (see Problem 1.30).

1.34   The specific potential energy of an object with respect to some datum level is given by gz, where g is the local gravitational acceleration and z is the elevation of the object above the datum. Determine the specific potential energy (in Btu/lb$_m$) of an object elevated 100 ft above a datum at a location where $g = 32.1$ ft/s^2.

1.35   Calculate the total potential energy (in Btu) of an object with a mass of 200 lb$_m$ when it is 10 ft above a datum level at a location where standard gravitational acceleration exists (see the previous problem).

1.36 Calculate the total potential energy (in Btu) of an object that is 20 ft below a datum level at a location where $g = 31.7$ ft/s^2 and that has a mass of 100 lb$_m$ (see Problem 1.34).

1.37 Determine the specific potential energy (in kJ/kg) of an object 50 m above a datum in a location where $g = 9.8$ m/s^2 (see Problem 1.34).

1.38 An object whose mass is 100 kg is located 20 m above a datum level in a location where standard gravitational acceleration exists. Determine the total potential energy (in kJ) of this object (see Problem 1.34).

1.39 Calculate the total potential energy (in kJ) of an object whose mass is 20 kg when it is located 20 m below a datum level in a location where $g = 9.5$ m/s^2 (see Problem 1.34).

1.40 Explain why the light-year has the dimension of length.

1.41 The pressure in a compressed air storage tank is 1500 kPa. What is the tank's pressure in (a) kN and m units; (b) kg, m, and s units; and (c) kg, km, and s units?

1.42 The energy stored in the spring of a railroad car is 5000 ft-lb$_f$. What is this energy in (a) lb$_m$, ft, and s units; (b) lb$_f$ and yard units; and (c) lb$_m$, mile, and hour units?

1.43 A small electrical motor produces 10 W of mechanical power. What is this power in (a) N, m, and s units; and (b) kg, m, and s units?

1.44 The pressure in a water line is 1500 kPa. What is the line pressure in (a) lb$_f$/ft^2 units and (b) lb$_f$/in.^{2} (psi) units?

1.45 As a forklift lifts a weight, it expends 5000 ft-lb$_f$ of energy. How much energy is this in (a) N-m units and (b) kN-m units?

1.46 A model aircraft internal-combustion engine produces 10 W of power. How much power is this in (a) ft-lb$_f$/s units and (b) horsepower units?

PROJECTS

The following projects are more extensive than the problems presented in this chapter. Most are open-ended and require some decisions on your part. Many can be expedited with computer-based property tables, computer spreadsheets, and student-written computer programs.

1.1 Prove that processes that use work for mixing are irreversible by considering an adiabatic system whose contents are stirred by turning a paddle wheel inside the system (e.g., stirring a cake mix with an electric mixer).

1.2 Prove that processes involving rapid chemical reactions are irreversible by considering the combustion of a natural gas (e.g., methane) and air mixture in a rigid container.

1.3 Prove that the work produced by a reversible process exceeds that produced by a equivalent irreversible process by considering a weight moving down a plane both with and without friction.

1.4 Conduct a literature survey that reviews the concept of thermal pollution and its present state of the art.

Heat Engines and Refrigerators

On the Second Law

Heat cannot of itself pass from a colder to a hotter body.

Rudolf Clausius

New energy-conversion devices are constantly being proposed and invented. Articles on the latest innovations and devices regularly appear in such magazines as *Popular Mechanics* and *Popular Science*. Innumerable patents continue to be issued for energy-conversion and energy-conservation devices by patent offices around the world. But despite all of this innovation and activity, few of these concepts and devices have been widely accepted by or affected the average human. It is natural to ask why so many of these devices are not accepted and whether any tests can be applied to a concept to determine its feasibility. In this chapter, we will begin to build the basis to answer these questions by considering general heat engines and refrigerators, which are independent of the devices and means by which they operate.

Many of the new or proposed concepts and devices are simply not competitive because of economic reasons. This was the situation with solar energy as a means of space heating, for example. Although the energy was essentially free in this case, the cost and maintenance of the equipment required to convert this energy to the desired form was so great that solar energy could not compete with other energy sources. The presence of a supporting infrastructure often gives one energy-conversion method an advantage over newer methods. This is the case with the internal-combustion engine, which has been around so long that many factories throughout the world are specifically designed to manufacture this and no other engine. The cost of converting these factories to alternate engine manufacturing is extremely high. Specialty machinery and processes have been built and installed that build only internal-combustion engines and that give them reliability and long life. All of this has taken many resources over a long period of time to put in place and will not be readily discarded.

Although economics and existing infrastructure do constrain the acceptance of new energy-conversion concepts, they do not make them impractical. The laws of nature also place limits on energy-conversion innovations and devices; in many cases, these laws make the devices impossible. Concepts that violate natural principles, of course, can never be implemented even under the best of economic and infrastucture conditions. Among the natural principles that an energy-conversion concept must satisfy are the two laws of thermodynamics. The first law is the concept that under nonrelativistic conditions energy can be neither created nor destroyed; it can only have its form altered. The second law places limits on how far the alteration of the form of energy can proceed by using the entropy property. New energy-conversion proposals must satisfy both of these laws. Most, but not all, satisfy the first law, and a large number do not satisfy the second law and thus are not possible. Once these natural laws of thermodynamics are understood, new energy-conversion

concepts may be tested against them to determine whether further consideration and development is merited.

The development of the laws of thermodynamics is begun in this chapter
as we consider and generalize our experience with everyday thermodynamic
devices such as electric-generation plants and refrigerators, which operate in
a cyclical manner while interacting with energy reservoirs. Consideration of
these general devices, which are independent of the machinery performing
the cycle, leads to the most elemental statements of the first and second laws,
which will be refined further and developed throughout the remainder of the
textbook. We also introduce here the concepts of entropy and work potential.

2.1 Heat Engines

In the previous chapter, we briefly discussed the steam-power–generation cycle,
which converts heat into mechanical work. This cycle is shown in Figure 2.1,

FIGURE 2.1

*Interactions of a
Rankine Cycle with
Its Surroundings*

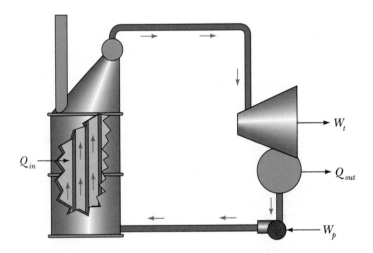

which emphasizes the working fluid system. As one or another form of this
water progresses through the various devices, it interacts with its surroundings
in several ways. While passing through the boiler, heat, Q_h, is transferred into
the water from the combustion gases, whose temperature is greater than the
water's. The temperature difference between the hotter gases and the water
causes the heat transfer. Next, the water produces work, W_t, as its pressure
and temperature are reduced in the turbine. Heat in the amount, Q_l, is next
transferred from the water to an environmental cooling fluid in the condenser.
Again, a finite temperature difference between the now hotter water and the
colder cooling fluid drives this transfer of heat. Finally, a small amount of work,

W_p, is applied to the water in the feedwater pump to increase the pressure so that the whole cycle may be repeated. A net amount of work given by $W_t - |W_p|$ is produced by this system. All devices that convert heat into work in this manner are known as *heat engines*.

In addition to converting heat to work, all of these devices share the following features:

1. a high-temperature reservoir that serves as a source of heat
2. a low-temperature reservoir that serves as a sink to which heat is rejected
3. a repeated cycle

For the steam-powered electrical-generation system, the thermal-energy source reservoir is the combustion gases inside the boiler, and the thermal-energy sink reservoir is the cooling water or air in the environment; the cycle is executed as the working fluid passes through the various pieces of equipment while returning to the boiler entrance. When a solar receiver is used in place of the conventional boiler, the thermal-energy source reservoir is provided by the solar radiation as it is absorbed on the receiver surfaces. Similar thermodynamic models can be developed for the internal-combustion and other engines. The machinery and processes used to accomplish these tasks are unimportant here. It is only necessary that the events and processes that occur in the heat engine form a continuously repeated cycle. The general heat engine is also a closed system because it does not exchange mass with its surroundings.

A general heat engine is illustrated in Figure 2.2. It consists of a thermal energy source, the device, and a thermal energy sink. The source transfers heat to the heat engine and may be characterized by its temperature, T_h. Similarly, the sink removes heat from the engine and may be characterized by its temperature, T_l. As a result of these heat transfer processes, the general engine produces work (as in the figure) as it operates in a cyclical manner.

The thermal-energy source and sink are known as *thermal-energy reservoirs*; these may be either a source or a sink, depending on whether heat is transferred from the sources or to the sinks. Thermal-energy reservoirs have the following characteristics:

1. The only interaction of interest between the reservoir and its surroundings is the heat interaction.
2. Finite amounts of heat can be removed from or added to them without changing their temperature.
3. The processes occurring inside the reservoir are reversible.

In addition to natural reservoirs such as large bodies of water and the atmospheric air, thermal-energy reservoirs can be generated by capturing solar energy, releasing nuclear energy, and combusting fuels. Many natural reservoirs vary in temperature and composition to some degree during the operation of the heat engine, and chemical and thermal pollution may result if these variations become too large. These variations also can cause errors in analysis that do not properly incorporate them.

FIGURE 2.2
*A General Heat
Engine*

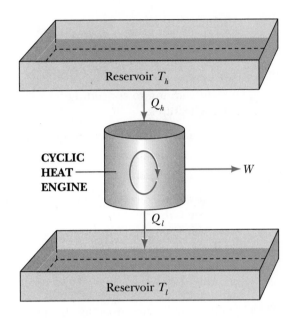

There are three interactions between the system (working fluid in the device) and its surroundings:

1. heat transfer from the high-temperature thermal-energy reservoir to the system (denoted by Q_h)
2. heat transfer from the system to the low-temperature thermal-energy reservoir (denoted by Q_l)
3. work produced by the system and delivered to the surroundings (denoted as W).

These are shown in Figure 2.2 by an arrow in the proper direction. Each occurs at the control boundary between the system and surroundings. As a result, they are considered to be external to the system. The various events that occur inside the device as the cycle is executed or repeated are internal to the system. For the time being, we will focus on the external interactions and reserve the internal cycle events for later discussion.

Inspection of any work-producing heat engine reveals that they all share these same features. Thus, careful study of a general heat engine will yield results that are applicable to a wide range of devices.

2.2   Refrigerators and Heat Pumps

Refrigerators and heat pumps are devices designed expressly for the purpose of removing heat from a low-temperature thermal-energy reservoir, which acts as a source reservoir. A schematic for a vapor-compression refrigeration system is shown in Figure 2.3. Refrigerant vapor enters the compressor where work, W, is applied to increase the pressure of the vapor. This high-pressure, high-temperature refrigerant next passes through the condenser, where it is cooled by transferring heat, Q_h, to the condenser environment, which has a temperature lower than the refrigerant's. After the refrigerant pressure is reduced in the insulated (i.e., adiabatic) throttle device, the resulting mixture of liquid and vapor is converted to a pure vapor in the evaporator as it is heated by the evaporator surroundings. These surroundings have a higher temperature than the refrigerant in the evaporator. This finite temperature difference causes the heat, Q_l, to be transferred from the evaporator surroundings to the refrigerant. Following this conversion back to a vapor, the refrigerant reenters the compressor to repeat the cycle.

FIGURE 2.3
The Interactions Between a Refrigeration System and Its Surroundings

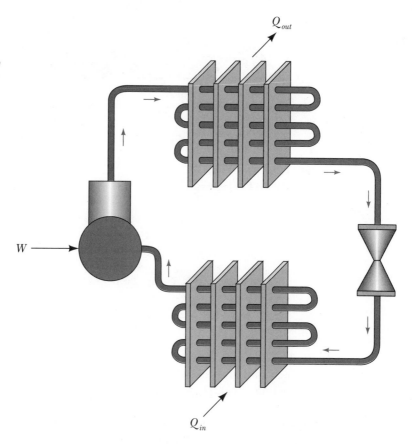

A general refrigerator or heat pump is shown in Figure 2.4. This system shares the same three components (i.e., two reservoirs and the cyclical device itself) as the heat engine. But notice that the directions of the three interactions have been reversed; the reservoir from which heat is transferred is now at the low temperature, T_l; and the reservoir to which heat is transferred is at the high temperature, T_h. Now Q_l denotes the heat transfer from the low-temperature reservoir to the system, and Q_h denotes the heat transfer from the system to the high-temperature reservoir. Also, work, W, is now done *on* rather than being produced *by* the system. When we consider an automotive air conditioner, for example, the low-temperature source reservoir is the air inside the automobile's cabin, which is being cooled to maintain passenger comfort. The high-temperature sink reservoir is the ambient air that is drawn through the condenser located in front of the automobile's radiator. The work needed to drive this system is taken from the engine and transferred to the air-conditioner compressor with the fan belt. The processes and machinery that form the general refrigerator or heat pump are temporarily unimportant. The only requirement for this device now is that it operate by repeating a regular cycle.

FIGURE 2.4
*A General
Refrigerator or Heat
Pump*

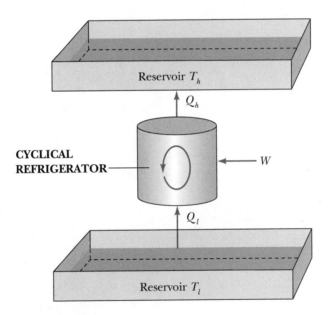

The only difference between a refrigerator and a heat pump is the manner in which they are used. A refrigerator is designed and optimized to produce the cooling effect, Q_l, while the heat pump is designed and optimized for the heating effect, Q_h. For example, a household refrigerator removes heat from the contents inside the cabinet while rejecting heat to the room in which it is located. Its primary purpose is the cooling of the food, and it is therefore considered to be a refrigerator. But during the winter months, it also heats the air in the room. If this were its primary purpose, it would be considered a heat pump. The same device can then be either a refrigerator or a heat pump, depending on its intended use. In the case of a residential heat pump, the low-temperature source reservoir is the air outside of the building. The high-temperature sink reservoir is the air inside the building, which is being heated to maintain the comfort of the building occupants. Most of these systems use an electric motor to produce the work that drives the system.

2.3 Cyclical Performance Indices

To compare one cyclical device to another, we need a *performance index* for the device. This performance index should be defined so that we can compare the index of two similar devices and decide which is better. The accepted definition for this purpose follows.

DEFINITION: *Cyclical Performance Index*

$$PI \equiv \frac{\text{Quantity of desired effect}}{\text{Quantity expended to create the desired effect}}$$

Stated another way, the *performance index* is the ratio of the desired effect to the cost of producing this effect. The larger the desired effect and the lower the cost of creating this effect, the greater the performance index. When comparing two similar devices, we would then select the one with the larger performance index. This is, of course, a general definition that can be applied to any device. We must now translate it into a form that is more directly applicable to the heat engine, refrigerator, and heat pump.

For the heat engine, the desired effect is the work produced, and the quantity expended to create this effect is the heat supplied by the high-temperature reservoir. This heat is a cost because we must purchase the fuel that is consumed to generate the heat (with the exception of a small number of "free" energy sources such as the sun and geothermal sources). The performance index for the heat engine is then

$$\eta = \frac{W}{Q_h} \qquad (2.1)$$

where W and Q_h are both in the same units. This performance index is known as the *thermal efficiency* and is given its own symbol, η. The thermal efficiency can also be calculated on a basis of power to rate of heat transfer,

$$\eta = \frac{\dot{W}}{\dot{Q}_h}$$

or a basis of specific work to specific heat addition,

$$\eta = \frac{w}{q_h}$$

A perfect heat engine would convert all the heat transferred to it into work. Its thermal efficiency would then be 100%. The second law of thermodynamics, which will be considered shortly, will have a great deal to say about this hasty conclusion.

EXAMPLE 2.1

An automobile engine produces 50 hp while its energy supply reservoir transfers $2.55(10^5)$ Btu/hr of heat to it. Determine the engine's thermal efficiency.

System: The automobile engine.

Basic Equations:
$$\eta = \frac{\dot{W}}{\dot{Q}_h}$$

Conditions: By the nature of its operation, the automobile engine continually repeats the same cycle. The energy that is released by the fuel–air combustion is modeled as an energy reservoir.

Solution: The rate of work production in Btu/hr is:

$$W = 50 \times 2545 \quad \left[\text{hp} \frac{\text{Btu/hr}}{\text{hp}} \right]$$

$$= 1.27(10^5) \ \text{Btu/hr}$$

The thermal efficiency is then

$$\eta = \frac{1.27(10^5)}{2.55(10^5)} = \underline{0.50} = 50\%$$

The desired effect for a refrigerator device is the cooling effect, Q_l, and the quantity expended to produce this effect is the work input, W. Hence, a refrigerator's performance index is

$$\text{COP}_{ref} = \frac{Q_l}{|W|} \tag{2.2}$$

where the absolute-value sign has been introduced to compensate for the work-sign convention. This index is called the *coefficient of performance* (COP) of the refrigerator. It can also be calculated on a rate or specific basis.

In the case of a heat pump, the desired effect is the heating effect, Q_h, and the quantity that produces this effect is the work input, W. The coefficient of performance of the heat pump is then

$$\text{COP}_{hp} = \frac{|Q_h|}{|W|} \tag{2.3}$$

where the additional absolute-value sign has been introduced to correct for the heat-sign convention. A rate or specific basis can also be used to calculate the heat pump's COP.

EXAMPLE 2.2   A food refrigerator produces a 1.5 kW cooling rate while using 1 kW of electrical power. Calculate this refrigerator's COP.

System: The refrigerator.

Basic Equations: $$\text{COP}_{ref} = \frac{Q_l}{|W|}$$

Conditions: A refrigerator operates by having its working fluid continually repeat the same cycle of events.

Solution: Substituting the data from the problem statement,

TIP:

If the absolute value signs confuse you, they may be dropped if you substitute only the magnitude of the heat and work interactions in the equations.

$$\text{COP}_{ref} = \frac{1.5}{|-1.0|}$$
$$= 1.5$$

2.4 The First Law of Thermodynamics (Cyclical Devices)

Uncounted observations of devices (heat engines and refrigerators) that operate in a cyclical fashion have revealed that the net amount of heat added to these devices equals the net amount of work produced by them. If we think of heat and work as "energy in transition," then we can summarize these observations by the statement "the net energy in the form of heat added (removed) to a system that operates in a cyclic manner equals the net energy in the form of work produced (consumed) by the system," or that energy must be preserved. Energy, like mass, can then change its form but must be conserved. Many experiments have been conducted to verify this observation, but none have ever found this concept to be violated. This concept must then be a basic natural law. It has been termed the *first law of thermodynamics* and is also known as the *conservation of energy principle*.

The mathematical form of this law follows.

PRINCIPLE: First Law of Thermodynamics (Cyclical Devices)

$$\oint \delta Q = \oint \delta W$$

In words, this equation states that as we go completely around the cycle, $\oint$, and sum up the various differential heat additions, δQ, we will get the same result as when we go completely around the cycle, $\oint$, and sum up the differential work productions, δW. For example, if we apply this equation to the heat engines discussed previously and illustrated in Figures 2.1 and 2.2, we get

$$\oint \delta Q = Q_h + Q_l$$
$$\oint \delta W = W$$

Therefore,

$$Q_h + Q_l = W \qquad (2.4)$$

Similarly, the first law applied to the refrigerator or heat pump of the previous section (shown in Figures 2.3 and 2.4) gives the same result:

$$Q_h + Q_l = W$$

On a rate basis, the first law applied to a cyclical device is

$$\dot{Q}_h + \dot{Q}_l = \dot{W}$$

which is useful for determining rates of heat transfer and power production or consumption. The first law on a specific basis is

$$q_h + q_l = w$$

which is convenient for calculating work and heat additions per unit of system mass.

When the first law is substituted into the thermal-efficiency expression, it takes another useful form,

$$\eta = \frac{Q_h + Q_l}{Q_h} = 1 - \frac{|Q_l|}{Q_h}$$

where the final step results from the fact that Q_l is a heat rejection. The thermal efficiency is also determined by the amounts of heat added to and rejected from the cycle. The amount of work produced has been eliminated by using the first law. This expression also can be written on a rate or specific basis. Combining the first law with the coefficients of performance gives the alternate forms

$$COP_{ref} = \frac{1}{\frac{|Q_h|}{Q_l} - 1}$$

and

$$COP_{hp} = \frac{1}{1 - \frac{Q_l}{|Q_h|}}$$

Like the heat engine, the coefficients of performance become functions of only the heat addition and rejection. The first law eliminates the work interaction from these expressions.

EXAMPLE 2.3

The heat-transfer rate to the working fluid in an electrical-generation station is $2(10^6)$ Btu/hr, and this station has a thermal efficiency of 30%. Calculate the net power produced by this station and the rate at which it rejects heat to the sink reservoir.

System: The working fluid of the electrical-generation station.

Basic Equations:

$$\eta = \frac{W}{Q_h}$$

$$\oint \delta W = \oint \delta Q$$

Conditions: As the fluid circulates among the system's components, it continually repeats a cycle.

Solution: From the thermal-efficiency expression,

TIP:

The system is the electrical-generation station working fluid, which rejects heat to the low-temperature reservoir. Thus, Q_l is negative. If we were to consider the low-temperature reservoir as the system, then Q_l would be positive because it is a heat addition to this reservoir.

$$\dot{W} = \eta \dot{Q}_h$$
$$= 0.3 \times 2(10^6)$$
$$= 0.6(10^5) \text{ Btu/hr} = 235.8 \text{ hp}$$

Rearranging the first law,

$$\dot{Q}_l = \dot{W} - \dot{Q}_h$$
$$= 0.6(10^5) - 2(10^6)$$
$$= -1.4(10^6) \text{ Btu/hr}$$

2.5 The Second Law of Thermodynamics (Cyclical Devices)

Although the first law of thermodynamics tells us that the form of energy can be changed, it gives us no idea as to which types of transformations are possible or impossible. For example, consider a block sliding along an insulated plane with friction. Because the plane is insulated, it cannot exchange heat with the block. It is therefore adiabatic. The work that we apply to this block causes its temperature to increase because of the frictional effect. Now consider the reverse process, with heat released from the block to cool it down to its original temperature. The block should now produce work, which will move it back along the plane to its original position according to the first law. We all know that this reverse process is impossible, consequently the original process is said to be *irreversible*. To answer questions of this nature, an additional law is needed. This law, like the first law, is based on many observations of the operation of cyclical and other devices.

The first scientists to state the second law of thermodynamics were Lord Kelvin[1] and Max Planck[2]. After studying several experiments, they independently concluded the following.

PRINCIPLE: Kelvin–Planck Statement of the Second Law of Thermodynamics

A cyclical heat engine cannot operate while receiving heat from a single thermal-energy reservoir and only produce work.

In other words, *heat engines must always reject heat to a thermal-energy reservoir.*

The cyclical device shown in Figure 2.5 is a direct violation of the Kelvin–Planck second law statement. If we apply the first law to this device, we conclude that $Q = W$. Substituting this into the expression for thermal efficiency, we find that this heat engine's thermal efficiency is 100%. The combination of the first and second laws proves that *no heat engine can have a thermal efficiency of 100%.*

---

[1]William Thompson Kelvin (1824–1907) was a British physicist who, from 1847 to 1851, developed the absolute-temperature scale and placed the work of his predecessors in a form that led to universal acceptance of the conservation of energy principle. As part of this work, he briefly stated the second law of thermodynamics while addressing the dissipation of energy. As a result of this statement, he is considered to be its discoverer. Lord Kelvin also invented several telegraphic and scientific instruments.

[2]Max Karl Ernst Ludwig Planck (1858–1947), a German physicist, in the 1880s clarified the concept of entropy and demonstrated its importance in thermodynamics. Following this accomplishment, he turned his attention to blackbody radiation. By applying the quantum theory initiated by Albert Einstein, he was able to resolve experimental observations of blackbody radiation. Planck's constant, which relates the quantum of energy for an oscillator to the oscillator's frequency, is named in his honor.

FIGURE 2.5
A Heat Engine that Violates the Kelvin–Planck Statement of the Second Law

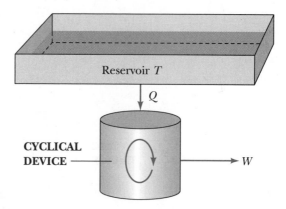

Another scientist, Rudolf Clausius[3], stated the second law in an alternate fashion.

> **PRINCIPLE: Clausius Statement of the Second Law of Thermodynamics**
>
> No cyclical device can cause heat to transfer from thermal-energy reservoirs at low temperature to reservoirs at high temperature with no other effects.

Work is therefore required to cause heat transfer from a cold to a hot reservoir.

A cyclical device that violates this statement of the second law is shown in Figure 2.6. Because this device requires no work to cause the heat transfer, its coefficient of performance is infinity. *Refrigerators or heat pumps with infinite coefficients of performance are impossible* according to this statement of the second law.

Both statements of the second law place bounds on the operation of cyclical devices. They do not tell us the exact amount of heat that must be rejected by a heat engine or the exact amount of work needed to drive a refrigerator or heat pump. Rather, they tell us the lower limit (zero) for both of these quantities. Any mathematical statement of the second law thus must be in the form of a limit or an inequality.

---

[3]Rudolf Julius Emanuel Clausius (1822–1888) was a German physicist who, by his restatement of Carnot's principle, put the theory of heat on a truer and sounder basis. In a paper presented in 1850, he stated that "heat cannot of itself pass from a colder to a hotter body." He later used this fact to stress and refine the concept of entropy. His research also made several contributions to the kinetic theory of gases.

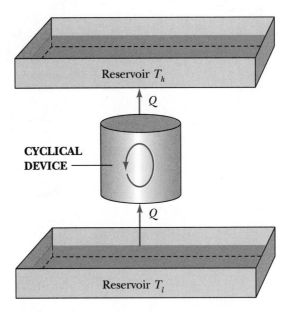

These statements are different in appearance but equivalent in concept. We may prove this by considering a heat engine that violates Kelvin–Planck; it is coupled to an actual working refrigerator as shown in Figure 2.7. The refrigerator is sized such that it requires all the work produced by the Kelvin–Planck–violating heat engine. Application of the first law to the violating engine reveals that

$$W = Q_{h,\,eng}$$

Similarly, application of the first law to the actual refrigerator gives

$$|Q_{h,\,ref}| = W + Q_l = Q_{h,\,eng} + Q_l$$

Now consider the new compound cyclical device consisting of the engine and refrigerator shown in the dotted box of Figure 2.7. This new cyclical device requires no work. Using this equation, the net heat flow between the new device formed by the previous two devices and the high-temperature reservoir is

$$|Q_h| = |Q_{h,\,ref}| - Q_{h,\,eng} = Q_{h,\,eng} + Q_l - Q_{h,\,eng} = Q_l$$

The new device then transfers an amount of heat, Q_l, from the cold reservoir to the hot reservoir while receiving no work. This is the Clausius statement of the second law. The Kelvin–Planck statement is thus equivalent to the Clausius statement. The proof that the Clausius statement is equivalent to the Kelvin–Planck statement is left to you.

FIGURE 2.7

*Systems for Proving
the Equivalence of
Second Law
Statements*

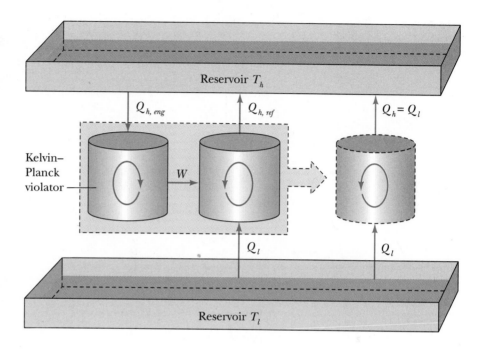

2.6 Completely Reversible Devices

Heat engines, refrigerators, and heat pumps whose cycles can be reversed such that all the work and heat interactions reverse their direction while retaining their magnitudes are known as *completely reversible devices* (or externally reversible devices). Such a device is shown in Figure 2.8. You might think of this as converting a heat engine into a refrigerator or vice versa with the magnitudes of the heat and work interactions remaining the same. When a device is reversed, we also go around the internal cycle in the opposite sense (from clockwise to counterclockwise or vice versa).

For a device to be completely reversible, every aspect of it must be reversible. This requires that the processes that constitute the device's internal cycle be reversible (as discussed in Chapter 1) so that we can go around the cycle in either direction while passing through the same series of states. In addition to the internal processes being reversible, the heat transfers between the cycle and the reservoirs must be reversible. Heat transfer can only be reversible when the temperature difference between the system and reservoirs that causes transfer is infinitesimally small. Because the reservoirs are isothermal, the portions of the cycle processes where heat is added or rejected also must be isothermal at temperatures infinitesimally close to the reservoir temperatures.

FIGURE 2.8
*A Completely
Reversible Device*

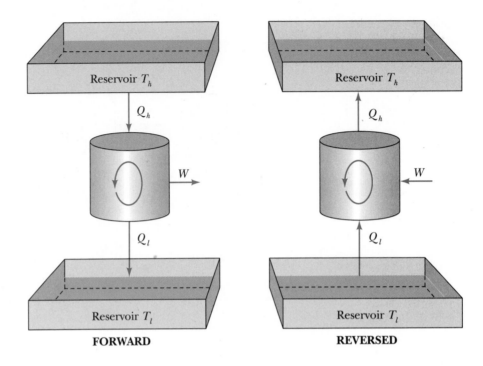

Because irreversibilities such as friction, mixing, and fast chemical reactions can never be completely eliminated, no real device can ever be completely reversible. As technology continues to improve, we get closer and closer. The completely reversible device is certainly valid in concept, but its real application is in expanding the first and second laws of thermodynamics.

One heat engine cycle that meets all of these requirements consists of the following internal processes.

1. A *reversible isothermal process* occurs at a temperature of $T_h - dT$, and heat is transferred from the hot thermal energy reservoir to the cycle working fluid. The temperature of the working fluid must be lower by an infinitesimal amount, dT, than that of the reservoir during this process to satisfy the Clausius second law statement.

2. A *reversible adiabatic process* must occur and produce an amount of work, W_2, but without exchanging heat with the surroundings.

3. A *reversible isothermal process* takes place at a temperature of $T_l + dT$, during which heat is transferred from the cycle working fluid to the low-temperature reservoir. The temperature of the working fluid must be higher by an infinitesimal amount, dT, than that of the cold reservoir during this process to satisfy the Clausius second law statement.

4. A *reversible adiabatic process*, which requires an amount of work, W_4, completes the cycle.

To satisfy the second law according to Kelvin–Planck, less work is required for the fourth process, $|W_4|$, than is produced by the second process, W_2. The net effect is that the cycle produces an amount of work $W_2 - |W_4|$ while accepting heat from the source reservoir and rejecting heat to the sink reservoir. This completely reversible heat engine was first postulated by Sadi Carnot[4] and is known as the *Carnot cycle.*

We might suspect that any completely reversible cycle produces the maximum amount of work possible because all wasteful irreversibilities have been eliminated. If this is the case, these cycles should also have the largest possible thermal efficiency. To prove this, consider two heat engines, A and B; A is a completely reversible engine, and B is any engine that may or may not be completely reversible. Both engines use the same reservoirs and are supplied the same amount of heat, Q_h, by a high-temperature reservoir, as shown in Figure 2.9.

FIGURE 2.9
*A Completely
Reversible and a
General Heat Engine*

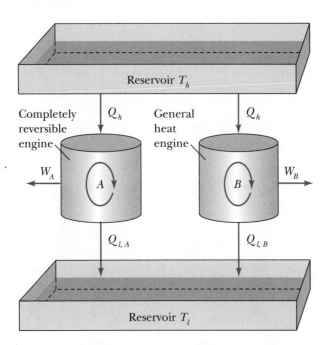

Let us first assume that the thermal efficiency of engine B is greater than that of engine A, $\eta_B > \eta_A$. Because both engines have been selected for the same heat input, Q_h, engine B will then produce more work than engine A,

$$W_B = \eta_B Q_h > W_A = \eta_A Q_h$$

[4]Sadi Carnot (1796–1832) was a French physicist and engineer who, in 1827, was one of the earliest to recognize the true nature of heat. *Carnot's principle* that the efficiency of a reversible engine depends on the temperatures between which it works is fundamental to the theories of thermodynamics.

The first law tells us that engine A will reject more heat than engine B because

$$|Q_{l,A}| = Q_h(1 - \eta_A) > |Q_{l,B}| = Q_h(1 - \eta_B)$$

Now we can reverse the completely reversible engine A and consider the new compound cyclical device enclosed by dashed lines in Figure 2.10. Because engine B produces more work than the reversed engine A requires, the compound device will produce a net amount of work, which is given by $W_B - |W_A|$ as shown in Figure 2.10. The heat rejected by the now reversed engine A is the same as the heat required by engine B. Hence, the original high-temperature reservoir can be eliminated by routing the heat rejected from the reversed engine A directly to engine B. According to the first law, the heat entering the reversed engine A is greater than that being rejected by engine B. Thus, there is a net heat addition from the low-temperature reservoir to the compound device.

FIGURE 2.10

Two Heat Engines with Engine A Reversed

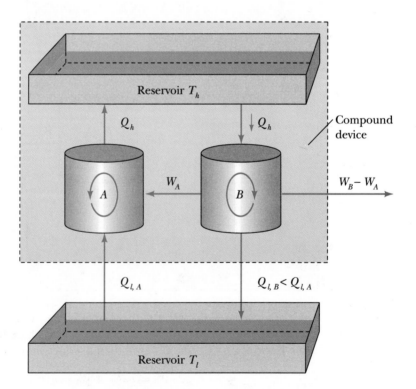

The compound device operates in a cycle, accepts heat from a single thermal-energy reservoir, and produces work. This is a direct violation of the Kelvin–Planck statement of the second law. Consequently, our assumption that $\eta_B > \eta_A$ must be incorrect. We can therefore conclude the following.

PRINCIPLE:

No heat engine can be more efficient than a completely reversible engine when operating with the same thermal-energy reservoirs.

This principle is very important because it tells us that the completely reversible heat engine is the most efficient heat engine for a given set of thermal energy reservoirs, and no other heat engine can exceed the thermal efficiency of a completely reversible engine. The thermal efficiency of a completely reversible engine is therefore the upper bound for all engines using the same reservoirs.

Next, let us consider what happens when both engine A and B are completely reversible. If we first presume that the thermal efficiency of engine B is greater than that of engine A and then reverse engine A and apply the first law, we will find that the second law is violated. If we next presume that the thermal efficiency of engine A is greater than that of engine B and then reverse engine B and apply the first law, we will find that the second law is again violated. Hence, we can conclude the following.

PRINCIPLE:

All completely reversible heat engines have the same thermal efficiency when operated with the same thermal energy reservoirs.

This principle states that the thermal efficiency of all completely reversible engines, η_{cr}, is determined by the conditions of the thermal energy reservoirs alone. Because the only difference between the reservoirs is their temperatures, η_{cr} can only depend on these two temperatures:

$$\eta_{cr} = f(T_h, T_l)$$

where the form of the function f remains to be determined.

2.7 Thermodynamic Temperature

The completely reversible engine provides a means for defining temperature independent of any special systems, property relations, or other constraints because

$$\eta_{cr} = f(T_h, T_l)$$

For this purpose, consider the three completely reversible engines shown in Figure 2.11. Note that the heat rejected by the first of these serves as the heat input to the second after passing through an intermediate reservoir at temperature T_i. The first and third of these engines are sized to operate with the same heat input Q_h. When the compound cyclical device formed by the first and second engines as well as the intermediate reservoir is considered, it must have the same heat rejection as the third engine because it has the same heat input and thermal efficiency, $f(T_h, T_l)$, as the third.

FIGURE 2.11
Three Completely Reversible Engines

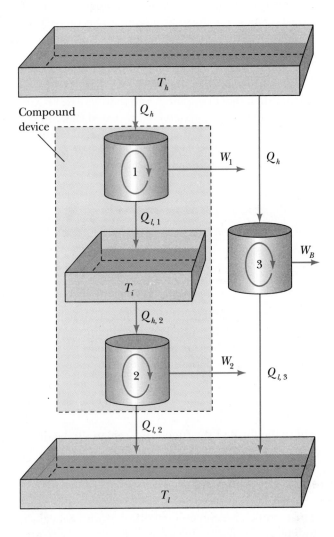

The thermal efficiency combined with the first law for the first engine tells us that

$$|Q_{l,1}| = [1 - f(T_h, T_i)]Q_h$$

This serves as the heat input to the second engine. The heat rejected by the second engine is then

$$
\begin{aligned}
|Q_{l,2}| &= [1 - f(T_i, T_l)] Q_{h,2} \\
&= [1 - f(T_i, T_l)] Q_{l,1} \\
&= [1 - f(T_i, T_l)][1 - f(T_h, T_i)] Q_h \\
&= f_1(T_i, T_l) f_1(T_h, T_i) Q_h
\end{aligned}
$$

where we have introduced a new function, $f_1(T_j, T_k) = 1 - f(T_j, T_k)$. In terms of this new function, the heat rejected by the third engine is

$$ |Q_{l,3}| = f_1(T_h, T_l) Q_h $$

Equating the heat rejected by the compound device to that rejected by engine 3 gives

$$ f_1(T_h, T_l) = f_1(T_h, T_i) f_1(T_i, T_l) $$

The only type of function that will satisfy this constraint is

$$ f_1(T_j, T_k) = \frac{f_2(T_j)}{f_2(T_k)} $$

where $f_2(T)$ is still an arbitrary function.

This is as far as deductive reasoning will take us. We still need to determine the function $f_2(T)$. Many suggestions have been made about this function. That suggested by Lord Kelvin—

$$ f_2(T) = T $$

—has been adopted by international agreement for defining thermodynamic temperature.

The temperature scale generated in this way is known as the *absolute thermodynamic temperature scale.* The SI units for this scale are Kelvin (K); the USCS units for this scale are degree Rankine ($^\circ$R). The relationship between this scale and the more familiar Celsius ($^\circ$C) and Fahrenheit ($^\circ$F) scales will be discussed in the next chapter.

The intermediate function $f_1(T_j, T_k)$ is T_j/T_k, and the thermal efficiency of a completely reversible engine operating between two thermal energy reservoirs is

$$ \eta_{cr} = 1 - \frac{T_l}{T_h} \tag{2.5} $$

As a result, the thermal efficiency of a completely reversible engine is determined entirely by the temperatures of the reservoirs, as it must be. Comparing this to the thermal efficiency combined with the first law shows the following:

$$ \left. \frac{|Q_l|}{Q_h} \right|_{cr} = \frac{T_l}{T_h} \tag{2.6} $$

EXAMPLE 2.4

A completely reversible heat engine uses a solar receiver for a high-temperature reservoir and a lake for a low-temperature reservoir. The sun maintains the receiver surfaces at 750° R, while the lake remains at 500° R. Calculate the work produced and the heat rejected by this engine when 1000 Btu are transferred from the solar receiver to the engine.

System: The working fluid of the heat engine.

Basic Equations:

$$\eta = \frac{W}{Q_h}$$

$$\eta_{cr} = 1 - \frac{T_l}{T_h}$$

$$W = Q_h + Q_l$$

Conditions: The heat engine is stated to be completely reversible in both its internal and external operations.

Solution: The thermal efficiency of a completely reversible engine is determined by the given temperatures of the thermal energy reservoirs:

$$\eta_{cr} = 1 - \frac{T_l}{T_h} = 1 - \frac{500}{750} = 0.334$$

TIP:

These results only apply to completely reversible heat engines and not to general heat engines.

From the definition of the thermal efficiency, the work produced is

$$W = \eta_{cr} Q_h = 0.334 \times 1000 = 334 \text{ Btu}$$

Rearranging the first law,

$$Q_l = W - Q_h = 334 - 1000 = -666 \text{ Btu}$$

The thermal efficiency of a completely reversible engine given by $\eta_{cr} = 1 - T_l/T_h$ is known as the *Carnot thermal efficiency* in honor of the scientist who first conceived a heat engine capable of achieving this efficiency. All of the preceding discussion can be summarized in one equation:

$$\eta \leq 1 - \frac{T_l}{T_h} = \eta_{cr} \tag{2.7}$$

where η is the thermal efficiency of any heat engine. The inequality portion of this equation applies to any engine, and the equality portion applies only to completely reversible engines.

2.8 Entropy

The thermodynamic temperature, Equation 2.6, may be rearranged as

$$\frac{|Q_l|}{T_l} = \frac{Q_h}{T_h}\bigg|_{cr}$$

This suggests that the ratio of the heat transfer to the absolute temperature at which the heat transfer occurs may have additional significance. To gain additional insight into the significance of this ratio, let us evaluate this ratio all the way around the cycle executed by the completely reversible engine shown in Figure 2.8 when it is operated in the forward manner. When the heat Q_h is added to the working substance of the cycle, the substance's temperature must remain fixed at T_h (or infinitesimally close) because the heat transfer must be reversible. The heat transfer to absolute temperature ratio for this portion of the cycle is Q_h/T_h. Similarly, during the portion of the cycle where heat is rejected by the cycle, this ratio is $-|Q_l|/T_l$. The total change in this ratio as we go completely around the cycle and return to the point in the cycle at which we started is

$$\Delta\left(\frac{Q}{T}\right) = \frac{Q_h}{T_h} - \frac{|Q_l|}{T_l}$$

According to the thermodynamic temperature, the two terms on the right-hand side of this result are equal. Thus, as we go completely around the completely reversible cycle,

$$\Delta\left(\frac{Q}{T}\right)_{cr} = 0$$

The quantity Q/T has then returned to the same value from which it started. Only quantities that are properties will return to their same value when a cycle is completed. This suggests that this ratio is a property.

To state the second law in a more general and usable form, we now introduce the concept of entropy. The change of entropy is defined as follows.

DEFINITION: *Entropy*

$$\Delta S \equiv \int_{\substack{rev \\ process}} \frac{\delta Q}{T}$$

where the integration of the heat transfer divided by the absolute thermodynamic temperature is performed along a reversible process. We will find that the entropy is extremely useful for conducting first and second law analysis of thermodynamic devices and systems. As we see more and more applications of the entropy, we will develop a better understanding of what the entropy is and how it is used. For the present, the above definition is sufficient for our purposes.

For the entropy to be a property, it must return to its original value whenever a cycle is completed; that is,

$$\Delta S = \oint_{\substack{rev \\ cycle}} \frac{\delta Q}{T} = 0$$

To prove that this is the case for all cycles, including the completely reversible cycle previously discussed, we will consider the cycle made of two general processes (not necessarily reversible) shown in Figure 2.12. We can approximate this cycle with a series of reversible isothermal and reversible adiabatic processes as shown in the figure. If we focus on the ith set of the approximating processes, we see that it forms a Carnot heat engine as presented in Section 2.6. Note that the effects caused by the right-hand reversible adiabatic process of the ith cycle are canceled by the left-hand reversible adiabatic process of the $i + 1$th cycle. Hence, the overlapping processes cancel one another and we are left with the stepwise approximation to the two actual processes. In the limit, as the number of these processes approaches infinity, the isothermal and adiabatic processes will follow the actual processes exactly.

FIGURE 2.12

A General Cycle Approximated by a Series of Carnot Cycles

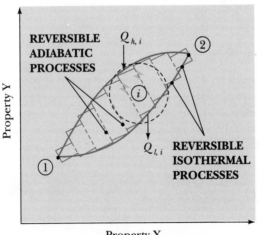
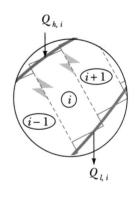

Considering the ith approximating Carnot cycle, the total entropy change, according to the entropy definition, as we traverse this Carnot cycle in a counterclockwise sense is

$$\delta(\Delta S) = \frac{\delta Q_{h,i}}{T_{h,i}} - \frac{\delta |Q_{l,i}|}{T_{l,i}}$$

But according to the absolute thermodynamic temperature,

$$\delta |Q_{l,i}| = \delta Q_{h,i} \frac{T_{l,i}}{T_{h,i}}$$

Substituting this into the previous result gives

$$\delta(\Delta S) = \delta Q_{h,i} \left[\frac{1}{T_{h,i}} - \frac{1}{T_{l,i}} \frac{T_{l,i}}{T_{h,i}} \right] = 0$$

Thus, the net entropy change for the ith Carnot cycle is zero. The total entropy change for all of the approximating Carnot cycles is then

$$\Delta S = \sum_{i=1}^{n} \frac{\delta Q}{T} = 0$$

The total entropy change of the actual processes approximated by the Carnot cycles in the limit as the number of approximating Carnot cycles becomes very large is

$$\Delta S = \lim_{n \to \infty} \sum_{i=1}^{n} \frac{\delta Q}{T} = \oint_{\substack{rev \\ cycle}} \frac{\delta Q}{T} = 0$$

Consequently, entropy must be a property of the system because it will always return to its original value whenever a cycle is completed.

2.9 Increase in Entropy Principle

To become more familiar with entropy, let us calculate the entropy change of everything involved with a general heat engine. The entropy of the engine working fluid will change as we progress from the beginning state around the cycle. Once we complete the cycle, the entropy of the working fluid will return to its original value because entropy is a property of this fluid. The net entropy change of the working fluid is consequently zero whenever the cycle is completed. There is no entropy change associated with the work interaction because the definition only concerns heat transfer.

The reservoirs are undergoing isothermal reversible processes (see Section 2.1). The entropy change of the high-temperature reservoir is

$$\Delta S_{source} = -\frac{|Q_h|}{T_h}$$

where the negative and absolute value signs have been introduced because Q_h is a heat rejection from the perspective of this reservoir. This result states that the entropy of the source reservoir decreases as the engine completes its cycle. The low-temperature reservoir's entropy changes by the amount

$$\Delta S_{sink} = \frac{Q_l}{T_l}$$

where Q_l is a heat addition from the sink reservoir's perspective. The entropy of this reservoir then increases as the engine completes its cycle.

Accounting for all of the components of the heat engine, the net entropy change is

$$\Delta S_{everything} = \Delta S_{sink} + \Delta S_{source} = \frac{Q_l}{T_l} - \frac{|Q_h|}{T_h}$$

If this engine is completely reversible, the second law and absolute temperature state that

$$Q_l = |Q_h|\frac{T_l}{T_h}$$

Combining these results, the total entropy change for a completely reversible heat engine is

$$\Delta S_{everything} = |Q_h|\left[\frac{1}{T_l}\frac{T_l}{T_h} - \frac{1}{T_h}\right] = 0$$

The sum of the entropy changes of all of the components of a completely reversible engine, including those in the surroundings with which it interacts, is zero. This is an alternate test to determine if a cyclical device is completely reversible.

When the engine is not completely reversible, the second law and absolute temperature scale tell us that

$$\frac{|Q_l|}{Q_h} > \frac{T_l}{T_h}$$

The net entropy change of all of the heat engine components (including the interacting components in the surroundings) is then

$$\Delta S_{everything} > 0$$

In words, the total entropy of the engine and all of the surrounding components that interact with the engine must increase when the heat engine is not completely reversible.

Based on these findings, we can state the following.

PRINCIPLE: Increase of Entropy Principle

$$\Delta S_{everything} \geq 0$$

This principle expresses the second law in a mathematical form. This important result states that when we calculate the entropy change for everything involved with a device or process, the total entropy must increase or remain fixed. When everything involved with a device or process is reversible, the net entropy change is zero.

It is sometimes convenient to express the increase in entropy principle as an equality rather than as an inequality. Toward this end, a second term known as the *entropy generation* can be introduced such that the increase in entropy principle becomes the following.

PRINCIPLE: Alternate Increase of Entropy Principle

$$\Delta S_{everything} - \Delta S_{gen} = 0$$

If all processes and interactions are reversible, then the entropy generation is zero. The further we deviate from reversible processes and interactions, the larger the entropy generation will become. Hence, the entropy generation serves to measure how closely we approach the ideal reversible situation. It is also a measure of the irreversibility of the processes and interactions.

EXAMPLE 2.5

A heat engine with an efficiency of 30% operates between a fossil-fuel–generated energy source at 1460° R and an environmental sink at 530° R. Compare the entropy change of the source and sink for (a) this engine and (b) a completely reversible engine using the same source and sink when both extract 1000 Btu from the source.

System: The heat engine.

Basic Equations:

$$\Delta S = \int_{\substack{rev \\ process}} \frac{\delta Q}{T}$$

$$\left.\frac{|Q_l|}{Q_h}\right|_{cr} = \frac{T_l}{T_h}$$

$$\oint \delta Q = \oint \delta W$$

$$\eta = \frac{W}{Q_h}$$

Conditions: During the second portion of the solution, the engine will be considered to be completely reversible.

Solution: The actual heat rejected to the sink by this engine is

$$\begin{aligned}
Q_{act,sink} &= (1-\eta)Q_h \\
&= (1-.3) \times 1000 \\
&= 700 \text{ Btu}
\end{aligned}$$

For this engine, the entropy change of the source is

$$\Delta S_{source} = \frac{Q_{source}}{T_{source}} = \frac{-1000}{1460} = -0.6849 \text{ Btu/}^\circ\text{R}$$

The entropy change of the sink is

$$\Delta S_{sink} = \frac{700}{530} = 1.321 \text{ Btu/}^\circ\text{R}$$

The entropy of the sink has increased by 1.321 Btu/°R, while the entropy of the source has decreased by 0.6849 Btu/°R. The net entropy generation is then 0.6361 Btu/°R, which is positive—as it must be, according to the increase in entropy principle.

    If this engine had been completely reversible, the heat removed from the source would be the same according to the problem statement, but the heat rejected to the sink would be

$$Q_{cr,sink} = \frac{T_{sink}}{T_{source}}Q_h = \frac{530}{1460}1000 = 363.0 \text{ Btu}$$

TIP:

In both cases, the entropy change of the working fluid inside the engines is zero because both engines operate using cycles that constantly return the working fluid to its original state.

and the entropy of the sink would have increased by

$$\Delta S_{sink} = \frac{363.0}{530} = 0.6849 \text{ Btu/}^\circ\text{R}$$

Because we have not changed the conditions of the source, its entropy would decrease by the same amount as before. The net entropy generation is now 0 Btu/°R, as required by the increase in entropy principle.   ◼

2.10 Work Potential of a Thermal-Energy Reservoir

Energy reservoirs such as those discussed in the previous section and Section 2.1 contain a large amount of energy (conceptually, an infinite amount). Whenever we extract some of this energy in the form of heat from these reservoirs, not all of it can be converted into the more useful work energy form according to the second law of thermodynamics. The *work potential* is that portion of this heat that can be converted into work, although it may not actually be so converted.

Whenever energy is extracted from a reservoir, some or all of it will eventually be returned to a lower temperature reservoir according to the Kelvin–Planck statement of the second law. A completely reversible engine operating on the heat received from this source reservoir and rejecting heat to a sink reservoir would produce the maximum amount of work that takes advantage of the reservoirs available to us. According to the first law and the absolute thermodynamic temperature, the maximum amount of useful work that heat transferred from the high-temperature reservoir can produce is as follows.

DEFINITION: *Work Potential of a Themal Energy Reservoir*

$$W_{ptl,s} = Q_s \left(1 - \frac{T_l}{T_s} \right)$$

where T_l is the absolute temperature of the heat-receiving, low-temperature reservoir; T_s is the temperature of the source reservoir; and the heat transfer is taken with respect to the system rather than the source reservoir.

Whenever a general heat engine operates using arbitrary reservoirs, the second law holds that it cannot produce an amount of work greater than this. Consequently, the work potential represents the upper bound on the amount of work a heat engine can produce for a given source and sink reservoir.

Because the work potential places an upper bound on the work that can be produced by an energy source, it serves as an excellent means of comparing energy sources. For example, let us consider a source that has $1(10^9)$ megajoule (MJ) of energy that can be released at 1000 K to one that has $2(10^9)$ MJ of energy that can be released at 300 K. Presuming that our sink reservoir is at 290 K (typical atmospheric air or body of water temperature), the work potential of the first source is

$$W_{ptl,s} = 1(10^9) \left(1 - \frac{290}{1000} \right) = 0.710(10^9) \text{MJ}$$

The work potential of the second source is

$$W_{ptl,s} = 2(10^9) \left(1 - \frac{290}{300} \right) = 0.067(10^9)\text{MJ}$$

The second source contains twice as much energy as the first, but can only produce about 1/10 as much work as the first source. The first source is much more valuable for the purpose of producing work.

PROBLEMS

2.1 A general heat engine has a total heat input of 1.3 kJ and thermal efficiency of 35%. How much work will it produce in kJ?

2.2 A heat engine has a heat input of $3(10^4)$ Btu/hr and a thermal efficiency of 40%. Calculate the power it will produce in horsepower.

2.3 The thermal efficiency of a general heat engine is 40%, and it produces 30 hp. At what rate is heat added to this engine in Btu/s?

2.4 A residential heat pump has a coefficient of performance of 1.4. How much heating effect in Btu/hr will result when 5 hp is supplied to this heat pump?

2.5 The coefficient of performance of a residential heat pump is 1.6. Calculate the heating effect in kilowatts this heat pump will produce when it consumes 2 kW of electrical power.

2.6 A commercial heat pump removes 10,000 Btu/hr from the source, rejects 15,090 Btu/hr to the sink, and requires 2 hp as power. What is this heat pump's coefficient of performance?

2.7 A refrigerator used for cooling food in a grocery store is to produce a 10,000 Btu/hr cooling effect, and it has a coefficient of performance (COP) of 1.35. How many kilowatts of power will this refrigerator require to operate?

2.8 A food freezer is to produce a 5-kW cooling effect, and its COP is 1.3. How many kilowatts of power will this refrigerator require for operation?

2.9 A refrigerator used to cool a computer requires 3 kW of electrical power and has a COP of 1.4. Calculate the cooling effect of this refrigerator in kilowatts.

2.10 A heat engine that propels a ship produces 500 Btu/lb_m of work while rejecting 300 Btu/lb_m of heat. What is its thermal efficiency?

2.11 A heat engine that pumps water out of an underground mine accepts 500 kJ of heat and produces 200 kJ of work. How much heat does it reject in kJ units?

2.12 A general heat engine with a thermal efficiency of 40% rejects 1000 Btu/lb_m of heat. How much heat does it consume in Btu/lb_m?

2.13 A heat pump has a COP of 1.7. Determine the heat (a) transferred to and (b) from this heat pump in kJ when 50 kJ of work are applied.

2.14 A heat pump with a COP of 1.4 is to produce a 100,000 Btu/hr heating effect. How much power does this device require in horsepower?

2.15 A food refrigerator is to provide a 15,000 Btu/hr cooling effect while rejecting 22,000 Btu/hr of heat. Calculate its COP.

2.16 An air conditioner produces a 2-kW cooling effect while rejecting 2.5 kW of heat. What is its COP?

2.17 An automotive air conditioner produces a 1-kW cooling effect while consuming 0.75 kW of power. What is the rate at which heat is rejected from this system in kW units?

2.18 Prove that a refrigerator's COP cannot exceed that of a completely reversible refrigerator that shares the same thermal-energy reservoir.

2.19 Prove that the COPs of all completely reversible refrigerators must be the same when the reservoir temperatures are the same.

2.20 Derive an expression for the COP of a completely reversible refrigerator in terms of the thermal-energy reservoir temperatures, T_l and T_h.

2.21 Derive an expression for the COP of a completely reversible heat pump in terms of the thermal-energy reservoir temperatures, T_l and T_h.

2.22 A completely reversible heat engine operates with a source reservoir at 1500° R and sink reservoir at 500° R. At what rate must heat be added to this engine (in Btu/hr) for it to produce 5 hp of power?

2.23 Plot the thermal efficiency of a completely reversible heat engine as a function of the source-reservoir temperature up to 2000° R with the sink-reservoir temperature fixed at 500° R.

2.24 A completely reversible refrigerator is driven by a 10 kW compressor and operates with thermal-energy reservoirs at 250 K and 300 K. Calculate the rate of cooling provided by this refrigerator.

2.25 A completely reversible refrigerator operates between thermal-energy reservoirs at 450° R and 540° R. How many kilowatts of power are required for this device to produce a 15,000 Btu/hr cooling effect?

2.26 Plot the COP of a completely reversible refrigerator as a function of the temperature of the sink reservoir up to 500 K with the source reservoir's temperature fixed at 250 K.

2.27 A completely reversible heat pump has a COP of 1.6 and a sink-reservoir temperature of 300 K. Calculate (a) the temperature of the source reservoir and (b) the rate of heat transfer to the sink reservoir when 1.5 kW of power is supplied to this heat pump.

2.28   An inventor claims to have devised a cyclical engine for use in space vehicles that operates with a nuclear-fuel–generated energy-source reservoir whose temperature is 1000° R and a sink reservoir that radiates heat to deep space at 550° R. He also claims that this engine produces 5 hp while rejecting heat at a rate of 15,000 Btu/hr. Is this claim valid?

2.29   You are an engineer in an electrical-generation station. You know that the flames in the boiler reach a temperature of 1200 K and that cooling water at 300 K is available from a nearby river. What is the maximum efficiency your plant will ever achieve?

2.30   As the engineer of the previous problem, you also know that the metallurgical temperature limit for the blades in the turbine is 1000 K before they will incur excessive creep. Now what is the maximum efficiency for this plant?

2.31   Supposedly, the efficiency of a completely reversible heat engine can be doubled by doubling the temperature of the energy-source reservoir. Justify the validity of this claim.

2.32   A manufacturer of ice cream freezers claims that its product has a coefficient of performance of 1.3 while freezing ice cream at 250 K when the surrounding environment is at 300 K. Is this claim valid?

2.33   A heat-pump designer claims to have a heat pump whose coefficient of performance is 1.8 when heating a building whose interior temperature is 300 K and when the atmospheric air surrounding the building is at 260 K. Is this claim valid?

2.34   Calculate the rate of entropy change (in kJ/s-K) of all the components of a refrigerator that uses 10 kW of power, rejects 14 kW of heat, and has a high-temperature energy reservoir at 400 K and a low-temperature energy reservoir at 200 K. What is the rate of cooling (in kW) produced by this refrigerator? Is this refrigerator completely reversible?

2.35   A heat engine accepts 200,000 Btu of heat from a source reservoir at 1,500° R and rejects 100,000 Btu of heat to a sink reservoir at 600° R. Calculate the entropy change of all the components of this engine and determine if it is completely reversible. How much total work does it produce?

2.36   When a system is adiabatic, what can be said about the entropy change of the substance in the system?

2.37   A completely reversible heat engine operates with a source reservoir at 1500° R and a sink reservoir at 500° R. If the entropy of the sink reservoir increases by 10 Btu/° R, how much will the entropy (in Btu/° R) of the source reservoir decrease? How much heat (in Btu) is transferred from the source reservoir?

2.38   An energy-source reservoir at 1000 K transfers heat to a completely reversible heat engine. This engine transfers heat to a sink reservoir that is at 300 K. How much heat must be transferred from the energy-source reservoir to increase the entropy of the energy-sink reservoir by 20 kJ/K?

2.39   One hundred Btu of heat are transferred directly from an energy-source reservoir at 1200° R to an energy-sink reservoir at 600° R. Calculate the entropy change of the two reservoirs and determine if the increase of entropy principle is satisfied.

2.40   In the previous problem, assume that the heat is transferred from the cold reservoir to the hot reservoir contrary to the Clausius statement of the second law. Prove that this violates the increase in entropy principle—as it must, according to Clausius.

2.41   Heat is transferred at a rate of 2 kW from a hot reservoir at 800 K to a cold reservoir at 300 K. Calculate the rate at which the entropy of the two reservoirs change and determine if the second law is satisfied.

2.42   A completely reversible air conditioner provides 36,000 Btu/hr of cooling for a space maintained at 70° F while rejecting heat to the environmental air at 110° F. Calculate the rate at which the entropies of the two reservoirs change (in Btu/° R-s) and verify that this satisfies the increase in entropy principle.

2.43   A completely reversible heat pump produces heat at a rate of 100 kW to warm a house maintained at 21° C. The exterior air, which is at 10° C, serves as the source reservoir. Calculate the rate of entropy change (in kJ/K-s) of the two reservoirs and determine if this heat pump satisfies the second law according to the increase in entropy principle.

2.44   A heat engine whose thermal efficiency is 40% uses a hot reservoir at 1300° R and a cold reservoir at 500° R. Calculate the entropy change of the two reservoirs when 1 Btu of heat is transferred from the hot reservoir to the engine. Does this engine satisfy the increase in entropy principle? If the thermal efficiency of the heat engine is increased to 70%, will the increase in entropy principle still be satisfied?

2.45   A refrigerator with a coefficient of performance of 4 transfers heat from a cold region at −20° C to a hot region at 30° C. Calculate the total entropy change (in kJ) of the regions when 1 kilojoule of heat is transferred from the cold region. Is the second law satisfied? Will this refrigerator still satisfy the second law if its coefficient of performance is 6?

2.46   A heat pump creates a heating effect of 100,000 Btu/hr for a space maintained at 530° R while using 3 kW of electrical power. What is the minimum temperature (in ° R) of the source reservoir that satisfies the second law of thermodynamics?

2.47   A proposed heat-pump design creates a heating effect of 25 kW while using 5 kW of electrical power. The thermal-energy reservoirs are at 300 K and 260 K. Is this possible according to the increase in entropy principle?

2.48   An inventor claims to have developed a heat pump that produces a 200 kW heating effect for a 293 K heated zone while only using 75 kW of power and a heat-source reservoir at 273 K. Justify the validity of this claim.

2.49   A heat engine operates with a source reservoir at 1280 K and a sink reservoir at 290 K. What is the maximum work per unit of heat that the engine can remove from the source reservoir?

2.50   A heat engine has thermal-energy reservoirs at 1260° R and 510° R. What is the maximum work per unit of heat that the engine can remove from the source reservoir?

2.51   From a work-production perspective, which is the more valuable: (a) thermal-energy reservoirs at 675 K and 325 K or (b) thermal-energy reservoirs at 625 K and 275 K?

2.52   Determine the minimum work per unit of heat transfer from the source reservoir that is required to drive a refrigerator with thermal-energy reservoirs at 273 K and 303 K.

2.53   Determine the minimum work per unit of heat transfer from the source reservoir that is required to drive a heat pump with thermal-energy reservoirs at 460° R and 535° R.

2.54   A thermodynamicist claims to have a heat engine with 50% thermal efficiency when operating with thermal-energy reservoirs at 1260° R and 510° R. Is this claim valid?

2.55   A thermodynamicist claims to have a heat pump with a COP of 1.7 when operating with thermal-energy reservoirs at 273° R and 293° R. Is this claim valid?

PROJECTS

The following projects are more extensive than the problems presented in this chapter. Most are open-ended and require decisions on the student's part. Many can be expedited with computer-based property tables, computer spreadsheets, and student-written computer programs.

2.1   The sun supplies electromagnetic energy to the earth. It appears to have an effective thermal temperature of approximately 5550 K. On a clear summer day in North America, the energy incident on a surface facing the sun is approximately 0.95 kW/m^2. The electromagnetic solar energy can be converted into thermal energy by being absorbed on a darkened surface. How might you characterize the source, entropy generation, and work potential of the sun's energy when it is to be used to produce work?

2.2   In the search to reduce thermal pollution and take advantage of renewable energy sources, some people have proposed that we take advantage of such sources as discharges from electrical power plants, geothermal energy, and ocean thermal energy. Although many of these sources contain an enormous amount of energy, the amount of work they are capable of producing is limited. How might you use the work potential to assign an "energy quality" to these proposed sources? Test your proposed "energy quality" measure by applying it to the ocean thermal source, where the temperature 100 ft below the surface is perhaps 5° F lower than at the surface. Apply it also to the geothermal source, where the temperature 1 to 2 miles below the surface is perhaps 300° F hotter than at the surface.

2.3   The maximum work that can be produced by a heat engine may be increased by lowering the temperature of its sink. Hence, it seems logical to use refrigerators to cool the sink and reduce its temperature. What is the fallacy in this logic?

2.4   Sometimes the irreversibility that is defined as $T_0 \Delta S$, where T_0 is the environment's temperature, is used in place of the potential work. Physically, what is this quantity in terms of completely reversible engines and energy reservoirs? How may it be used to assess the quality of the energy available in a thermal-energy reservoir?

2.5   At one time, it was proposed that the thermodynamic temperature function should be $\ln T$ rather than T. How would this suggestion have changed the expressions for the performance indices of completely reversible heat engines, refrigerators, and heat pumps?

Introduction to Properties

On Things

"The time has come," the Walrus

said, "To talk of many things:

Of shoes—and ships—

and sealing-wax

—Of cabbages—and kings—

And why the sea is boiling hot—

And whether pigs have wings."

Lewis Carroll

T he ultimate operation of a thermodynamic system is determined by the working fluid it uses and that fluid's properties. Which fluid is selected for a system depends on many factors, including the system's overall thermodynamic performance, the fluid's availability and cost, its compatibility with the materials used to build the system, disposal of the fluid when it is replaced or exhausted from the system, and the environmental impact of any fluid leaked or exhausted into the surroundings.

An excellent case study of a working fluid is that of the chlorofluorocarbons (CFCs). These chlorine, fluorine, and carbon compounds (there are several in the family) were specifically developed as substitutes for propane, butane, and ammonia, which were then used in refrigeration systems. The CFCs reduced the operating costs of these systems because their thermodynamic properties were superior. CFCs also made refrigeration systems much safer (propane and butane are highly flammable) and more acceptable (ammonia is toxic). Refrigeration systems then became available for household refrigerators and freezers and for building and automobile air conditioning as they became safer and more economical to operate. Simultaneously, as the CFCs became more readily available with increased manufacturing capacity, they became so inexpensive to use that they found new applications by which their vapors directly entered the atmosphere. These applications included propellants for spray cans, foaming agents for foam products such as insulation, and cleaning agents for electronic fabrication processes. CFCs accumulated in the atmosphere and now contribute to the depletion of the ozone layer. Some nations have now passed legislation to limit the production of CFCs. These compounds thus have undergone a complete transition from benefit to problem.

In this chapter, we will develop an understanding of the properties of selected substances that typify those used in many thermodynamic systems. It is not possible to treat every thermodynamic substance, but the methods and techniques developed in this chapter may readily be extended to deal with many others. Specifically, we present the properties of substances that change their phase, such as water, and gases, such as air. Our focus will be on the properties that affect the thermodynamic performance of the system. Issues of environmental impact and material compatibility, among others, will be relegated to other studies of these subjects.

3.1 Mass, Weight, and Volume

Newton conceived of the mass of a system as that property that measures the resistance of the system to a velocity change. The mass of an object is invariant and remains constant at speeds not approaching the speed of light. In the USCS unit system, the standard unit of mass is the *pound-mass*. The standard unit for mass in the SI unit system is the *kilogram*.

When dealing with chemical elements or compounds, a convenient unit of relative mass is the *mole*. The mole always contains the same number of elemental particles such as atoms, molecules, and electrons. A gram-mole is defined as $6.023(10^{23})$ elemental particles. The atomic weight, M, is the mass of an elemental particle on a scale relative to the mass of an atom of carbon-12 (it really should be called atomic mass). Atomic weights of selected elements are listed in Table 3.1 and are also available in any periodic table of the elements. The mass of an oxygen atom is $16.00/12.00$ times the mass of a carbon atom, according to this table. Molecular weight is the sum of the atomic weights of all the atoms that form a molecule. For example, diatomic oxygen, O_2, has a molecular weight of $16.00 + 16.00 = 32.00$ and is $32.00/12.00$ times more massive than a carbon-12 atom.

TABLE 3.1
Atomic Weights of Common Elements

Hydrogen	1.00797	Helium	4.0026
Carbon	12.001115	Nitrogen	14.0067
Oxygen	15.9994	Fluorine	18.9984
Neon	20.179	Sulfur	32.064
Chlorine	35.453	Argon	39.948
Aluminum	26.9815	Iron	55.847

Atomic and molecular weights are also the mass of a substance per mole of substance. One gram-mole of carbon-12 has a mass of 12.00 grams, and one pound-mass-mole of carbon-12 has a mass of 12.00 pounds-mass. The units of atomic and molecular weight are then X/X-mole, where X is any accepted unit of mass. The molecular weight of diatomic oxygen (the most common form of oxygen) is then 32.00 gm/gm-mole, 32.00 kg/kg-mole, or 32.00 lb_m/lb_m-mole.

In a chemical reaction balance, the number of atoms of each element must be the same in the products as in the reactants. In addition to conserving the number of atoms of each element, these balances are also conservation of mass statements. When we burn hydrogen in the presence of diatomic oxygen, the reaction balance is

$$4H + O_2 \rightarrow 2H_2O$$

where the quantities on the left-hand side are called the *reactants,* and those on the right-hand side are the *products.* This is read as "when four moles of hydrogen are reacted with one mole of diatomic oxygen, two moles of water

($M = 18.02$) will be produced." When we multiply the various terms in this balance by their respective molecular weights, the mass of the reactants will equal the mass of the products. In pound-mass units, there are $4 \times 1.01 + 1 \times 32.00 = 36.04$ pounds-mass of reactants and $2 \times 18.02 = 36.04$ pounds-mass of products in the preceding hydrogen–oxygen reaction.

When considering mixtures of several substances, the mole fraction (x_i) is convenient for specifying the relative amount of the various components in the mixture. The *mole fraction* is defined as follows.

DEFINITION: *Mole Fraction (y_i)*

The ratio of the number of moles of a single component in a mixture of two or more components to the total number of moles in the mixture.

For component i in a mixture, its mole fraction is

$$y_i = \frac{N_i}{\sum\limits_{i=1}^{k} N_i} \tag{3.1}$$

where N_i is the number of moles of component i. In the hydrogen–oxygen reaction of the preceding paragraph, the mole fraction of hydrogen in the reactant mixture is 0.8, while that of the diatomic oxygen is 0.2. The mole fraction of water in the products is 1.0.

Mixtures of substances also have an *apparent molecular weight*. The apparent molecular weight of a mixture is the sum of the molecular weights of each component weighted by the component's mole fraction,

$$M = \sum\limits_{i=1}^{k} y_i M_i \tag{3.2}$$

EXAMPLE 3.1

Atmospheric air is a mixture of 3.76 moles of diatomic nitrogen for every one mole of diatomic oxygen when the various trace elements are neglected. Determine the mole fraction of nitrogen in this mixture and the apparent molecular weight of the mixture.

System: One mole of O_2 and 3.76 moles of N_2.

Basic Equations:

$$y_i = \frac{N_i}{\sum\limits_{i=1}^{m} N_i}$$

$$M = \sum_{i=1}^{m} y_i M_i$$

Solution: The total number of moles of mixture is 4.76. The nitrogen mole fraction is then

$$y_{N_2} = \frac{3.76}{4.76} \left[\frac{\text{kg-mole of } N_2}{\text{kg-mole of mixture}} \right]$$
$$= 0.7899$$

The apparent molecular weight of this mixture is

$$M = 0.7899 \times 28.002 + 0.2101 \times 32.000 \left[\frac{\text{kg-mole of } i}{\text{kg-mole of mixture}} \frac{\text{kg of } i}{\text{kg-mole of } i} \right]$$
$$= 28.84 \text{ kg/kg-mole or } lb_m/lb_m\text{-mole} \quad \blacksquare$$

TIP:

When the normal trace elements of atmospheric air are included, its molecular weight is 28.97. This value should always be used as the apparent molecular weight of atmospheric air.

Mass fractions (x_i) are also useful when analyzing systems composed of mixtures of various substances. The mass fraction of a component in a mixture is defined as follows.

DEFINITION: *Mass Fraction (x_i)*

The ratio of the mass of a single component in a mixture of two or more components to the total mass of the mixture.

For component i in a mixture, the mass fraction is

$$x_i = \frac{m_i}{m_{mixture}} \qquad (3.3)$$

When a mixture consists of k components, its total mass is $\sum\limits_{i=1}^{k} m_i$. The mass fraction of component i is then

$$x_i = \frac{m_i}{\sum\limits_{i=1}^{k} m_i}$$

EXAMPLE 3.2	Determine the mass fraction of oxygen in atmospheric air; neglect the trace elements.

System: One mole of O_2 and 3.76 moles of N_2.

Basic Equations:

$$x_i = \frac{m_i}{\sum\limits_{i=1}^{n} m_i}$$

Solution: The mass of the oxygen in the mixture per kilogram-mole of oxygen is $1 \times 32.00 = 32.00$ kg, and the mass of the nitrogen in the mixture per kilogram-mole of oxygen is $3.76 \times 28.02 = 105.4$ kg. The oxygen mass fraction is then

TIP:

This answer is the same regardless of the mass units selected for the calculations.

$$x_{O_2} = \frac{32.00}{32.00 + 105.4} \left[\frac{\text{kg } O_2/\text{kg-mole } O_2}{\text{kg mix}/\text{kg-mole } O_2} \right]$$
$$= 0.233$$

System volumes are readily determined from the system's geometry. Frequently, the volume occupied by a unit of mass of the working fluid is more useful than the volume. This volume is known as the specific volume and is denoted by the symbol v.

DEFINITION: *Specific Volume (v)*

$$v = \frac{V}{m}$$

When we know the specific volume and mass of a system, the total volume is $V = mv$. As you will see, the specific volume simplifies many calculations. The reciprocal of the specific volume is called the *density*, ρ. Density is the mass per unit volume.

The weight per unit volume, known as the *specific weight*, γ, is useful when we need to know the weight force acting on a system. According to Newton's law, the weight is given by

$$w = mg$$

The weight of an object is not invariant like mass because it depends on the local gravitational acceleration, g. If we divide both sides of this equation by the volume of the system, we obtain the relation between density, specific weight, and specific volume:

$$\gamma = \rho g = \frac{g}{v}$$

Be sure to include the gravitation unit-conversion constant, g_c, when using this equation in USCS units.

EXAMPLE 3.3

A system consists of 1 liter (0.001 m^3) of a fluid whose specific volume is 0.0011 m^3/kg and 2 liters of a fluid whose specific volume is 0.00055 m^3/kg. Determine the weight of this system in newtons at standard gravitational acceleration.

System: Three liters containing two different fluids.

Basic Equations:

$$v = \frac{V}{m}$$

$$w = mg$$

Conditions: The gravitational acceleration is taken as the standard gravitational acceleration.

Solution: The total mass of the first fluid is $0.001/0.0011 = 0.9091$ kg, according to the definition of the specific volume. The total mass of the second fluid is $0.002/0.00055 = 3.636$ kg. Summing these two masses, the total system mass is 4.545 kg. Thus, the total weight of the system is

$$w = mg = 4.545 \times 9.8067 \quad \left[\text{kg} \frac{\text{m}}{\text{s}^2} \right]$$

$$= 44.58 \text{ N}$$

3.2 Pressure

When a substance is confined by the walls of a system, it exerts a force on these walls that acts perpendicular to the wall. The ratio of this normal force to the area of the wall on which it acts is the pressure.

DEFINITION: *Pressure (P)*

$$P \equiv \frac{F_n}{A}$$

where F_n is the force acting perpendicular to the area A.

From a molecular viewpoint, pressure is the result of the change in momentum experienced by moving atomic particles as they collide with and bounce off of the system walls. Rather than attempting to pursue the events experienced by all of the atomic particles in a system (a *microscopic* view), we will take a *macroscopic* view and consider the overall result of these events.

Pressures are measured with devices known as *gauges*. The scales on these gauges are normally set so that they indicate zero pressure when exposed to the local atmosphere. Consequently, *gauge pressures* are pressures relative to the local atmospheric pressure. Negative gauge pressures are termed *vacuum pressures*. Pressures measured relative to zero force acting on the walls of the system are called *absolute pressures*. Gauge and absolute pressures are related by

$$P = P_{gauge} + P_\infty$$

where P_∞ is the local atmospheric pressure. This relation is illustrated in Figure 3.1. Atmospheric pressure is influenced by factors such as altitude and weather. At standard sea-level conditions, the absolute atmospheric pressure is defined as 14.696 pounds per square inch absolute (psia) or 101.325 kPa. A *barometer* can be used to measure the local atomspheric pressure.

FIGURE 3.1

Relationship Between Gauge and Absolute Pressures

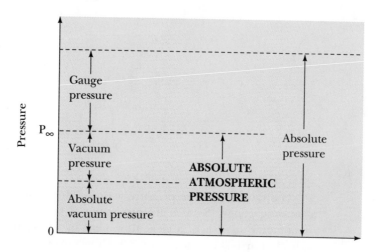

According to the definition of pressure, any convenient units of force and area may be used for pressure units. We will use the pound-force per square inch and kilopascal (one kilonewton per square meter) for the most part. Pressure units that end in "a" or the word *absolute* are used to indicate absolute pressures. When we need to clarify gauge pressures, the letter "g" or the word *gauge* will be added as a suffix to the pressure units.

The change in the pressure caused by the weight of the fluid may be found from a static force balance of the pressure and weight forces acting on a stationary fluid element (see Figure 3.2). This force balance yields

$$P - (P + dP) = \frac{\gamma V}{A}$$

$$dP = \gamma \, dz = -\rho g \, dz$$

By integrating this relationship, the difference between the pressure at two different elevations in a fluid is

$$P_2 - P_1 = - \int_{z_1}^{z_2} \rho g\, dz$$

where the negative sign indicates that the pressure decreases as we move up in the fluid. As we move deeper into a fluid, the pressure will increase because of the increasing weight of the fluid above. According to this result, the pressure at a fixed depth in the same fluid remains constant.

FIGURE 3.2

Forces Acting on a Submerged Fluid Element

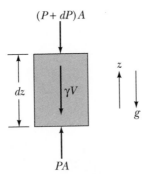

The variation of the pressure with the depth of the fluid can be used as a pressure gauge. Such a pressure gauge is known as a *manometer*. A typical manometer consists of a clear tube bent in a U-shape and is partially filled with a liquid as shown in Figure 3.3. Drawing a free-body diagram of the fluid column in the right leg of the manometer (recall that the pressure is the same at the same depth throughout a fluid) and writing the static force balance, assuming that the fluid density and the local gravitational acceleration are constant, gives

$$P_2 - P_1 = \rho g h = \frac{gh}{v}$$

The height of the column in the right leg of the manometer and the density of the manometer fluid then serve to measure the system pressure. Pressures are frequently cited on the basis of this result in terms of a fluid height. Inches of water, centimeters of mercury, and similar units are also acceptable pressure units.

FIGURE 3.3
A Manometer

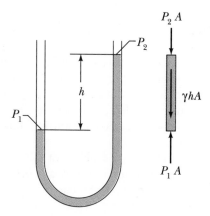

EXAMPLE 3.4

A gauge is used to measure the pressure of the air in a scuba diver's tank. This gauge indicates a pressure of 400 kPa. A barometer in the same room indicates an atmospheric pressure of 750 mm Hg (the specific weight of mercury at standard gravitational acceleration is 132.9 kN/m³). Calculate the absolute pressure of the air in the tank.

System: The air in the scuba diver's tank.

Basic Equations:

$$P = P_{gauge} + P_\infty$$
$$P_{gauge} = \rho g h$$

Solution: The atmospheric pressure is

$$P_\infty = \gamma h = 132.9 \times 0.75 \qquad \left[\frac{\text{kN}}{\text{m}^3}\text{m}\right]$$

$$= 99.68 \text{ kPa}$$

and the absolute pressure in the tank is

$$P = 99.38 + 400.0 = 499.38 \text{ kPa}$$

3.3 Temperature

In Chapter 2, we defined temperature in terms of completely reversible cyclical devices operating between isothermal thermal-energy reservoirs. Although this defines temperature, it provides no means for measuring temperatures because completely reversible devices can only be approached and never achieved. Practical means of measuring temperature are provided by the science of *thermometry*.

The basic principle of thermometry is known as the *zeroth law of thermodynamics* because it historically preceded the first and second laws. Like the first and second laws, it is based on numerous experimental observations. This law may be stated as

PRINCIPLE: Zeroth Law of Thermodynamics

When a system (say, A) is in thermal equilibrium with each of two other systems (say, B and C), these other systems must be in thermal equilibrium with one another.

Thus, if system A is in thermal equilibrium with system B (they will not exchange heat of their own accord) and system A is simultaneously in thermal equilibrium with system C, then system B must be in thermal equilibrium with system C. Heat can only be exchanged between two systems when their temperatures are not the same. The zeroth law then states that if system A has the same temperature as system B and system C, then system B has the same temperature as system C.

For practical temperature measurements, we can take system A to be a thermometer device that operates on some physical property (e.g., volume, electrical resistance, generated voltage) that changes with temperature. System B can be a standard system such as water freezing at atmospheric pressure, water boiling at atmospheric pressure, or gold melting at atmospheric pressure. The temperature of unknown systems can then be determined by comparison of the unknown system to standard systems using the thermometer according to the zeroth law; that is, we can state that the temperature of the unknown system matches some standard system. But this approach would require a very large number of standard systems to accommodate any range of temperatures. To overcome this difficulty, we use a limited number of thermometers and standard systems, along with equations to interpolate temperatures indicated by these thermometers between those of the standard systems.

Like mass, length, time, and electrical current, temperature is a fundamental dimension. We are at complete liberty to assign any convenient unit or number to it. This has led to a wide variety of temperature scales and units. The familiar USCS units of temperature are degrees Fahrenheit (°F); the common SI temperature units are degrees Celsius (°C). These are relative temperature scales chosen for convenience to describe commonly experienced temperatures. The thermodynamic temperature (see Section 2.7) uses absolute temperature scales whose USCS units are degrees Rankine (°R); its SI units are kelvin (K).

The two absolute temperature scales were selected so that their degree sizes are the same as those of the more common Celsius and Fahrenheit scales. They are related by

$$T[\mathrm{K}] = \frac{5}{9} T[^\circ \mathrm{R}]$$

Absolute scales are related to the more common scales by

$$T[^\circ \mathrm{C}] = T[\mathrm{K}] - 273.15$$

$$T[^\circ \mathrm{F}] = T[^\circ \mathrm{R}] - 459.67$$

The International Practical Temperature Scale (IPTS) was adapted to standardize temperature scales, thermometers, and standard systems, as well as to make practical temperature measurements consistent with the thermodynamic definition of temperature. The most recent modification to the IPTS was made in 1990, and the name was changed to the International Temperature Scale (ITS-90). Standard systems and their assigned temperatures on the ITS-90 scale are presented in Table 3.2. In the range from 4.2 K to 24.6 K, the standard thermometer is a *gas thermometer*, which is based on the change in a gas's pressure with temperature when its volume is kept constant. Polynomial equations are used to interpolate temperatures between the standard system points in this range. From 13.81 K to 1235 K, a *platinum resistance thermometer* serves as the standard thermometer. Resistance thermometers are based on the change in a metal's electrical resistivity with temperature. Interpolation between standard system points is done with polynomial equations in this range.

TABLE 3.2
Standard Systems of ITS-90

SYSTEM	°C	K	°F	°R
Triple point of equilibrium hydrogen	−259.34	13.81	−434.81	24.86
Triple point of equilibrium neon	−284.55	24.6	−415.39	44.28
Triple point of equilibrium oxygen	−218.75	54.4	−361.75	97.92
Triple point of equilibrium argon	−189.35	83.8	−308.83	150.84
Triple point of equilibrium water	0.01	273.16	32.02	491.69
Triple point of equilibrium gallium	29.85	303.0	85.73	545.40
Freezing point of gallium	29.77	302.92	85.59	545.26
Freezing point of tin	231.97	505.12	449.55	909.22
Freezing point of zinc	419.58	692.73	787.24	1246.91
Freezing point of aluminum	660.46	933.61	1220.83	1680.50
Freezing point of silver	961.93	1235.08	1763.47	2223.14

3.4 Internal Energy and Enthalpy

Part of the energy of a system consists of the activity of the individual atoms or molecules contained in the system. For example, consider a system that is composed of water vapor. The triatomic molecules, H_2O, are moving about this system. At a given instant of time, each of the many molecules has an individual velocity and occupies a position within this system. Consequently, each has an individual kinetic energy and potential energy. In addition, each molecule exists at a quantum energy state that is associated with the manner in which it is spinning and vibrating. The electrons of each atom of the molecules also occupy quantum energy states that are associated with the orbits that they occupy and the manner in which they spin. The net sum of all of these individual energies of the many elemental particles constitutes the macroscopic property known as *internal energy*.

DEFINITION: *Internal Energy (U)*

The energy associated with the motion, position, and internal state of the atoms or molecules of the substance contained in the system.

Internal energy is a property of a system because it is determined by the atoms or molecules that compose the system and their various energy states. Anything that alters the energy states of the particles will change the internal energy of the system. For example, when we change the temperature of the system, the velocity, position, and quantum energy state of each particle also changes. The average kinetic, potential, and quantum energy of all particles will then change, causing a change in the macroscopic internal energy. A pressure change will have a similar result. From a macroscopic perspective, we do not need to know the details about each of the many atomic particles in the system. Instead, we need to know the relationship between the internal energy and other macroscopic properties, such as pressure and temperature.

The symbol U will be used to denote the total internal energy, which may be measured in any energy units. We will use the British thermal unit (Btu) in the USCS system and the kilojoule (kJ) in the SI system. Specific internal energy will be denoted by u. According to the state postulate, the specific internal energy of a simple compressible substance depends on two other independent properties:

$$u = u(P, T)$$

$$\text{or } u = u(P, v)$$

$$\text{or } u = u(T, v)$$

In thermodynamic analysis, the property combination $u + Pv$ or $U + PV$ occurs frequently. Because this is a mathematical combination of several properties, this result is also a property. For convenience, it is given the name *enthalpy* and its own symbol, h. The specific enthalpy is defined as follows.

> **DEFINITION:** *Specific Enthalpy*
>
> $$h \equiv u + Pv$$

The total enthalpy is

$$H \equiv U + PV$$

which has the same units as internal energy.

Like all properties, the enthalpy obeys the state postulate. The specific enthalpy of a simple compressible substance then depends on two other independent properties:

$$h = h(P, T)$$
$$\text{or } h = h(P, v)$$
$$\text{or } h = h(T, v)$$

3.5 Entropy

In Chapter 2, we defined the change in entropy as

$$\Delta S \equiv \int_{\substack{rev \\ process}} \frac{\delta Q}{T}$$

and demonstrated that entropy is a property just like pressure, temperature, and internal energy, among others. It must therefore follow the state postulate. For a simple compressible substance, specific entropy only depends on two other independent properties:

$$s = s(P, T)$$
$$\text{or } s = s(P, v)$$
$$\text{or } s = s(T, v)$$

We will denote the total entropy by the S symbol and the specific entropy by the s symbol. The definition tells us that the units of entropy are those of heat divided by temperature. Typical total entropy units are Btu/$^\circ$R and kJ/K, and typical specific entropy units are Btu/lb$_m$-$^\circ$R and kJ/kg-K.

3.6 Phases

All substances can exist in several different phases, depending on the state of the substance. These phases are commonly termed *gaseous, liquid,* and *solid* (the gaseous phase is also known as the *vapor* phase), but this is a misconception because many substances have *several* liquid and solid phases. Helium, for example, has two different liquid phases, one denser than the other. Carbon has several solid phases as is apparent when one compares diamonds, common pencil lead, and carbon black. Water has only one liquid phase and several solid phases such as crystalline and amorphous ice.

Each of the various phases of a substance has the same chemical composition as the other phases but is different from the other phases in properties such as specific volume and internal energy. Solid, liquid, and gaseous water are all made up of the triatomic molecule, H_2O, while the solid phase has a smaller specific volume than the gaseous phase. In the solid phase, the molecules or atoms of the substance are very close to one another on a microscopic scale, and a large number of them occupy a unit of volume. The molecules or atoms of crystalline solids are arranged in a definite repeating geometric pattern such as a body-centered cubic or hexagonally close-packed arrangement. Amorphous solids are locally arranged in groups of fixed geometric patterns, with one group having an orientation different from its neighbors. Globally, amorphous solids have random geometric arrangements. Liquids resemble amorphous solids in that a small group of molecules or atoms may be arranged in a fixed pattern, while neighboring groups may have different arrangements or orientations. The average spacing between the molecules or atoms in a liquid is generally somewhat greater than that of a solid (water near its freezing point is an exception) because of this geometric arrangement. The specific volume of a liquid is generally somewhat larger than that of a solid, because the molecules or atoms are, on average, farther apart and fewer of them occupy the same space. In the gaseous phase, the molecules are far apart (on a microscopic scale) and do not have a regular geometric arrangement. Rather, they have greater freedom to move about and are only constrained by their collisions with other molecules or atoms and the confining walls. The gaseous phase always has the largest specific volume because of the large average distance between the molecules or atoms.

In a system that contains several different phases of a substance, we will observe definite physical boundaries that separate the various phases. A container full of liquid water and water vapor will contain either liquid droplets floating in the water vapor, a pool of liquid water at the bottom and the vapor phase above, or both, depending on the amount of time allowed for the system to settle. Whatever the configuration of the container contents, we can clearly see the interface(s) between the liquid and gaseous phases. We can define a *phase* of a chemically homogeneous system as follows.

> **DEFINITION:** *Phase*
>
> A phase of a chemically homogeneous system is any part of the system that is physically distinct and clearly separated from the other parts of the system by definite physical boundaries.

We can better appreciate the concept of phases by performing an experiment. For this purpose, we will use the weighted, frictionless piston-cylinder arrangement shown in Figure 3.4. The pressure inside the piston cylinder can be controlled by the weight of the piston. If we increase the piston weight, the pressure of the system will likewise increase. Heat will be added to the contents of the piston cylinder *at a constant rate* from the source shown in the figure. Initially, we will place a substance in liquid form in the piston cylinder at an initial temperature and then begin adding heat to the system at a constant rate while keeping the weight of the piston constant. Once we start adding heat, the temperature of the substance will begin to change. While this is happening, the pressure of the substance will not change because we are not changing the weight of the piston.

FIGURE 3.4
Weighted
Piston-Cylinder
Apparatus

Weight

Working
fluid

Heat
source

When we plot the temperature of the substance against the time from when we start adding heat, we will obtain the results shown in Figure 3.5. This figure actually contains the results of four such experiments conducted at different pressures, with P_1 being the smallest pressure and P_4 the largest pressure. At the three lowest pressures, we observe that the temperature first increases with time, then remains fixed for awhile, and then begins to increase with time at a faster rate than before. During the time that the temperature remains fixed, the system's contents are being converted from the liquid phase to the gaseous phase, and we can actually observe physical boundaries separating the two phases. At the beginning and end of the experiment, when the temperature is changing with time, only one phase occupies the system. At the earliest times, only the liquid phase occupies the system. Later, only the vapor phase occupies the system. When the pressure is increased to P_4 or more, notice that the temperature never remains constant. Rather, only the *rate* of temperature increase changes with time.

During the time when the phase is being transformed and the temperature remains constant, the substance is said to be *saturated*. The properties of the substance when this occurs are called *saturation properties,* and the state of the substance is known as a *saturation state*. Saturation then refers to conditions under which a phase transformation can be or is occurring.

FIGURE 3.5
Weighted Piston-Cylinder Device: Temperature Versus Time

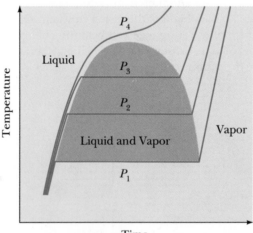

When a substance is saturated, a unique relationship exists between the pressure and temperature; that is, for a given pressure, there is only one temperature at which a phase transformation will occur. A *phase-state diagram,* such as that for water and carbon dioxide shown in Figure 3.6, is a plot of this unique relationship for the various phase transformations that each substance can undergo. In the phase-state diagram, note that the various regions correspond to single phases, while the lines indicate the conditions under which two phases will be present.

FIGURE 3.6

Phase-State Diagrams for Water and Carbon Dioxide (Distorted Scales)

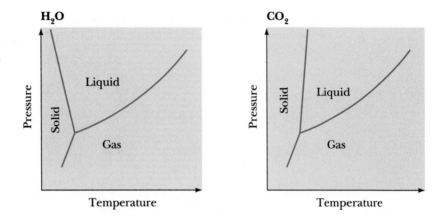

A general phase-state diagram that illustrates various terms is in Figure 3.7. Solid–liquid saturation states fall on the solid–liquid line. As we traverse this line from left to right, the substance undergoes a *melting* process. If we traverse it from right to left, the substance undergoes a *freezing* process. Solid–vapor saturation states occur along the solid–vapor line. When we traverse this line from left to right, the substance undergoes a *sublimation* process. Naphthalene (moth balls) and dry ice (CO_2) are common examples of substances that undergo sublimation. Liquid–vapor saturation states fall on the liquid–vapor line. A *boiling* process occurs when we cross this line from left to right. A *condensation* process results as we cross this line from right to left.

FIGURE 3.7

A General Phase-State Diagram

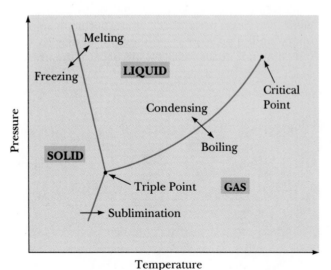

Two unique points occur on a phase-state diagram. The first is termed the *triple point.* Triple points of several substances are presented in Table 3.3. At this one combination of pressure and temperature, all three phases—solid, liquid, and vapor—can coexist simultaneously in equilibrium in varying proportions.

Triple points are so unique and repeatable that they serve as standard systems of the International Temperature Scale (ITS-90).

TABLE 3.3
Triple-Point States

SUBSTANCE	PRESSURE		TEMPERATURE	
	kPa	psia	°C	°F
Ammonia	6.1	0.89	−78	−108
Carbon Dioxide	517.	75.	−57	−70
Helium	5.1	0.75	−271	−456
Hydrogen	7.0	1.0	−259	−434
Methane	11.7	1.70	−182.5	−296.4
Nitrogen	12.5	1.81	−210	−346
Oxygen	0.15	0.022	−219	−361
Propane	3.0E-6	0.44E-6	−187.7	−305.8
Water	0.611	0.0886	0.01	32.02

The second unique point terminates the liquid–vapor line and is known as the *critical point.* Critical-point states of several substances are presented in Table 3.4. At pressures above the critical-point pressure, there is no clear demarcation between the liquid and vapor phases as is observed at pressures below the critical pressure. In the experiment summarized in Figure 3.5, we observe this at pressure P_4. Although you may have never observed this phenomenon, it is no less real and will occur if the pressure is great enough.

TABLE 3.4
Critical-Point States

SUBSTANCE	PRESSURE		TEMPERATURE		MOLAR SPECIFIC VOLUME	
	kPa	psia	°C	°F	m^3/kg-mole	ft^3/lb$_m$-mole
Ammonia	11,300	1640.	132	270	0.0723	1.16
Carbon Dioxide	7390	1071.	31	88	0.0941	1.51
Helium	230	33.	−268	−450	0.0579	0.93
Hydrogen	1300	188	−213	−400	0.0648	1.04
Methane	4599	670.7	−128.5	−199.4	0.0991	1.59
Nitrogen	3400	493	−147	−233	0.0897	1.44
Oxygen	5000	731.	−119	−182	0.0741	1.19
Propane	4242	615.3	96.7	206.0	0.195	3.13
Water	22,100	3206.	374	705	0.0558	0.896

Although the temperature–time results of Figure 3.5 and phase-state diagrams of Figures 3.6 and 3.7 contain a lot of information and give us some insight into the behavior of substances, they do not tell a complete story. If we repeat the previous experiment and plot the volume of the system against time, then we will obtain the results shown in Figure 3.8. We have repeated the experiment at the same pressures $P_1, \ldots, P_4$ as before. Recall that P_4 is at or above the critical pressure.

FIGURE 3.8
Weighted Piston-Cylinder Device: Volume Versus Time

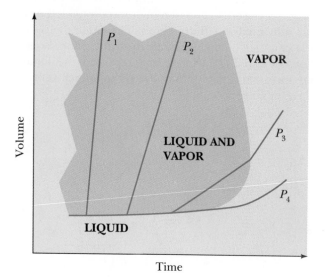

As we first begin to heat the substance that is initially a liquid, its volume is very small and changes very little with time. When the liquid's temperature reaches the saturation temperature, the volume begins to change rapidly as the liquid is being transformed into a vapor. The rate of volume change is quite high at the lower pressures—P_1, P_2, and P_3 of Figure 3.8. The volume change that occurs as the liquid is being converted to a vapor is so large that we can only see the end of the transformation on the P_3 curve with the scales selected for the figure. For water, the volume of the system at the end of the vapor conversion is 20,000 times larger than what it was at the beginning when the pressure was P_1. On the P_2 curve, the volume of water vapor at the end of the phase transformation is 250 times what it was as a liquid at the beginning. Now we focus on the P_3 curve: When the transformation is completed, the rate of volume increase is even greater than when the transformation was occurring. At or above the critical pressure, the volume increase undergoes a smooth, continuous change without any radical changes in the slope of the curve to indicate a phase transformation.

Below the critical pressure, these results show that the pressure has very little effect on the system's volume as long as the substance remains in liquid form. Substances that display this behavior are said to be *incompressible*. Properties of incompressible substances then *depend only on the temperature* and not the pressure. This is a very accurate approximation that we will make considerable use of later.

The preceding results become even more useful when we eliminate the time factor by plotting the system's pressure versus its specific volume on the property diagram shown in Figure 3.9. Note the distortion of the specific volume scale. When plotted this way, the liquid–vapor line of the phase-state diagram becomes a region where the various states represent mixtures of liquid and vapor. This region is known as the *saturation* or *mixture region*. On this diagram, the critical point remains a point, but the saturation liquid–vapor line has split into two different lines. The left branch of this bifurcation constitutes the saturated liquid line, while the right branch constitutes the saturated vapor line. States along the left branch represent the liquid portion of a liquid–vapor mixture, while states along the right branch represent the vapor portion of these mixtures.

FIGURE 3.9
Pressure-Specific Volume State Diagram for the Liquid and Vapor Phases

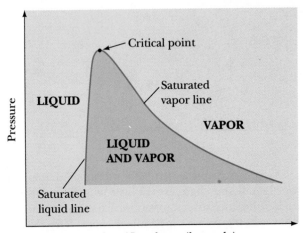

Specific volume (log scale)

Adjectives are sometimes used to clarify generic phase descriptions. The adjective *subcooled* (i.e., *below* the saturation temperature) or *compressed* (i.e., *above* the saturation pressure) are used to distinguish the liquid phase states from the saturated liquid states. The adjective *superheated* (i.e., *above* the saturation temperature) is used to distinguish vapor states from saturated vapor states. We will not use these modifiers except when required for clarity.

A general pressure-specific volume-state diagram for a phase-changing substance, including all three phases, is shown in Figure 3.10. Note that we have distorted the specific-volume scale by plotting the log of the specific volume so that solid and liquid states may be more readily seen. This distortion will be used in future P–v state diagrams. The solid–liquid and solid–vapor saturation regions are now easily seen. Note also that the triple point has become a line that represents the many ways in which the solid, liquid, and vapor phases may be proportioned in a triphase mixture.

FIGURE 3.10

Pressure-Specific Volume State Diagram for Solid, Liquid, and Vapor Phases

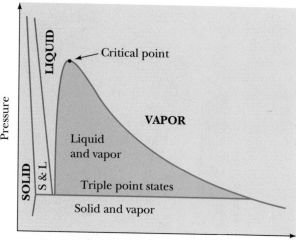

3.7 Other State Diagrams

The *P–v* and phase-state diagrams of the preceding section are special projections of a more general, three-dimensional *P–v–T* surface such as that shown in Figure 3.11. The relationship between pressure, specific volume, and temperature is known as an *equation of state*. Figure 3.11 is a graphical equation of state for a substance that has several different phases. Simpler equations of state for single-phase substances will be discussed later.

The pressure-specific volume diagram showing only the liquid and vapor phases as well as the mixture region is useful when we analyze systems such as the Rankine power-generation and vapor-compression refrigeration cycles. A general *P–v* state diagram for this purpose is shown in Figure 3.12. Several isothermal and constant-entropy processes are also shown on this state diagram. As we move upward from one isothermal process to another, the temperature increases. Hence, T_4 is greater than T_1. Similarly, as we move downward and to the left from one constant-entropy process to another, the entropy increases. The entropy of process s_3 is greater than the entropy of process s_1.

Temperature-specific entropy-state diagrams that show only the liquid and vapor phases, as well as the mixture region, are also very useful for visualizing processes and states. A general *T–s* state diagram for the liquid and vapor phases is shown in Figure 3.13. Several isobaric and constant specific-volume processes are also shown on this diagram. These are numbered such that $P_4 > P_1$ and $v_4 > v_1$. Because we will be making considerable use of this form of the *P–v* and *T–s* state diagrams, you should study them carefully, become familiar with the various process lines as they appear on these diagrams, and learn how to locate states on these diagrams.

FIGURE 3.11

*General P–v–T
Surface*

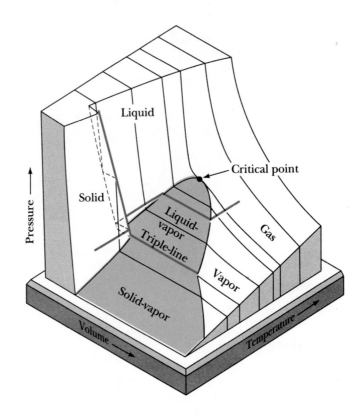

FIGURE 3.12

*A General P–v State
Diagram for the
Liquid and Vapor
Phases*

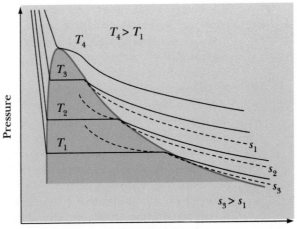

FIGURE 3.13

A General T–s State Diagram for the Liquid and Vapor Phases

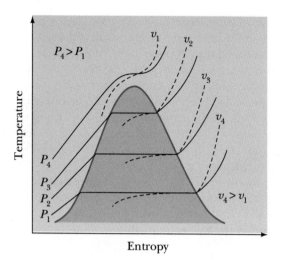

3.8 Property Tables

State diagrams and P–v–T surfaces are useful for visualizing system processes and how phases are changing, but they do not provide sufficient quantitative information for calculations. Tables that list the values of the properties at various states as determined by experimental observation and theoretical considerations are required for this purpose. These tables are known as *property tables.* Property tables for water, ammonia, and R-12 (a commonly used chlorofluorocarbon refrigerant) are contained in Appendix B. These examples illustrate the organization typical of most property tables.

Note that each table in Appendix B is broken into a group of subtables. Each subtable corresponds to a single phase or saturated mixture of phases. Table B1 is for water. It is broken into Table B1.1 ("Saturated Liquid–Vapor"), with temperature as the primary argument; Table B1.2 ("Saturated Liquid–Vapor"), with pressure as the principal argument; Table B1.3 ("Vapor"); and Table B1.4 ("Liquid"). Table B2 lists the properties of R-12. It is broken into Table B2.1 ("Saturated Liquid–Vapor"), with temperature as the principal argument, and Table B2.2 ("Vapor"). Property tables for other substances are typically divided into similar subtables.

3.8.1   Saturation Tables

The unique relationship between temperature and pressure when a substance is changing its phase is presented in the pressure and temperature columns of a saturation table such as Tables B1.1, B1.2, B2.1, and B3.1 of the appendix. Thus, we can readily determine that water boils at standard atmospheric pressure at 212° F, at 65 psia at 298° F, and at 400 kPa at 143.63° C. Do not forget that under saturation conditions, pressure and temperature are not independent of one another. Hence, pressure and temperature alone do not uniquely determine the state under saturation conditions.

In a saturated liquid–vapor table, the notation f is used as a subscript to denote the saturated liquid state, and the notation g as a subscript denotes the saturated vapor state.[1] These correspond to the left and right branches, respectively, of the bifurcated saturation line of Figure 3.9. The conditions of the liquid portion of a saturated mixture of liquid and vapor are denoted by the f subscript, while the conditions of the gaseous portion are denoted by g. The i (ice) subscript is used to denote the solid phase in solid–vapor and solid–liquid saturation tables.

When we consider a saturated mixture of a liquid and vapor phase, a certain fraction of the mixture will be in saturated vapor form, while the balance will be in the saturated liquid form. The relative proportion of the two phases is determined by the following quality.

DEFINITION: *Quality (x)*

The vapor mass fraction in a saturated liquid–vapor mixture.

According to this definition, 65% of the mass of a saturated mixture of liquid and vapor whose quality is 65% is in the saturated vapor form. The remainder of this mixture (i.e., 35% of the mass) must then be in the saturated liquid form. Hence, quality specifies the relative amounts of liquid and vapor in a saturated liquid–vapor mixture. Note that this definition only applies to mixtures of phases and not to systems that contain a single phase.

The quality is useful for determining specific properties of saturated mixtures. Consider a saturated mixture consisting of m_f units of saturated liquid mass and m_g units of saturated vapor mass. The total mass of this mixture is then $m_f + m_g$. An extensive property Z (such as volume, total internal energy, total enthalpy, or total entropy of this mixture), then has the specific counterpart $Z/(m_f + m_g)$. The liquid phase of the mixture has the specific property z_f, and the vapor phase of the mixture has the specific property z_g. The total amount of Z for the entire mixture is then $m_f z_f + m_g z_g$. Consequently, the specific value of Z for the mixture is

$$z = \frac{m_f z_f + m_g z_g}{m_f + m_g}$$

$$= \left(\frac{m_f + m_g}{m_f + m_g} - \frac{m_g}{m_f + m_g}\right) z_f + \left(\frac{m_g}{m_f + m_g}\right) z_g$$

$$= (1 - x)z_f + x z_g$$

[1]From the German $f = flussig$ = liquid, and $g = gaz$ = gas

Useful alternate forms of this result are:

$$z = z_f + xz_{fg}$$
$$z = z_g - (1-x)z_{fg} \tag{3.4}$$

where $z_{fg} = z_g - z_f$ is the change in the specific property z as the mixture is converted from a saturated liquid ($x = 0\%$) to a saturated vapor ($x = 100\%$).

EXAMPLE 3.5   A saturated mixture of water in a 5.01 ft^3 container is at 200° F. Five cubic feet of the container volume is filled with water vapor, and the remainder is filled with liquid water. Determine the mass and quality of the water in this container.

System: 5.01 ft^3 of liquid water and water vapor.

Basic Equations:
$$v = \frac{V}{m}$$

$$x = \frac{\text{mass of vapor}}{\text{total mass of mixture}}$$

160	4.745	0.01640	77.2	127.94	1062.3	127.96	1002.2	1130.1	0.2313	1.8484
170	5.996	0.01645	62.0	137.95	1065.4	137.97	996.2	1134.2	0.2473	1.8293
180	7.515	0.01651	50.2	147.97	1068.3	147.99	990.2	1138.2	0.2631	1.8109
190	9.343	0.01657	41.0	158.00	1071.3	158.03	984.1	1142.1	0.2787	1.7932
200	11.529	0.01663	33.6	168.04	1074.2	168.07	977.9	1145.9	0.2940	1.7762

Conditions: Thermodynamic equilibrium is presumed to exist.

Solution: The liquid portion of the mixture has a specific volume v_f that, according to Table B1.1 of Appendix B, is 0.01663 ft^3/lb$_m$. The vapor portion of the mixture has a specific volume v_g that, according to Table B1.1, is 33.6 ft^3/lb$_m$. The mass of the liquid portion is then

$$m_f = \frac{0.01}{0.01663} = 0.601 \text{ lb}_m$$

The mass of the vapor portion is

$$m_g = \frac{5}{33.6} = 0.149 \text{ lb}_m$$

The total mass is then $0.750\ lb_m$, and the quality is

$$x = \frac{0.149}{0.750}$$

$$= 0.199$$

The location of this state on the P–v and T–s state diagrams is shown in Figure 3.14. ▬

FIGURE 3.14
*P–v and T–s State
Diagrams for
Example 3.5*

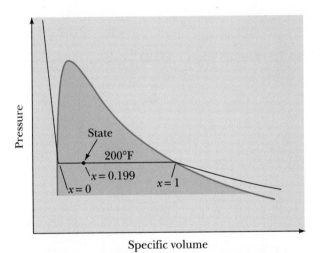

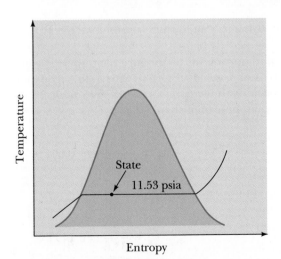

3.8.2   Single-Phase Tables

Property tables for conditions when only one phase is involved are typically organized as appendix Tables B1.3, Table B1.4, Table B2.2, and Table B3.2. Because pressure and temperature are independent properties when one phase is present, they are used as the principal arguments for organizing single-phase property tables. Additional properties such as specific volume, specific internal energy, and specific entropy are listed under each temperature–pressure pair. *Interpolation* (see Appendix A) is necessary for temperature–pressure combinations not listed in a single-phase property table.

EXAMPLE 3.6   The pressure and temperature at the outlet of a steam turbine are 5 psia and $200°$ F, respectively. What is the specific volume (in ft^3/lb_m) of the steam at the turbine outlet?

System: The water vapor at the turbine outlet.

Basic Equations: Water-property tables of Appendix B.

162.2	73.53	1063.0	1131.0	1.8441
200	**78.15**	1076.3	1148.6	1.8715
250	84.21	1093.8	1171.7	1.9052
300	90.24	1111.3	1194.8	1.9367
400	102.24	1146.6	1241.2	1.9941

Conditions: We presume that thermodynamic equilibrium exists.

Solution: According to appendix Table B1.3, the specific volume is 78.15 ft^3/lb_m. The location of this state on the P–v and T–s state diagrams is shown in Figure 3.15.

FIGURE 3.15
P–v and T–s State Diagrams for Example 3.6

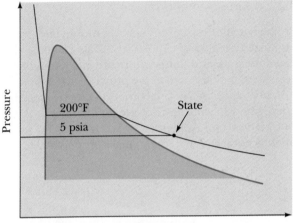

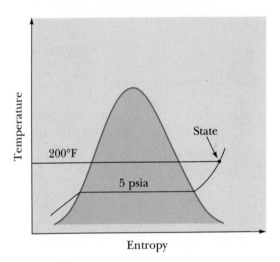

3.9 Selecting Appropriate Property Tables

With so many property tables to work with, we need an organized approach to selecting the correct table. In most situations, we know a pair of properties such as pressure and temperature, pressure and another property other than temperature, or temperature and another property other than pressure. Our goal is to use the known pair of properties and the appropriate table to determine the remaining properties of the state. In this section, we develop a means of determining the remaining properties of a state with the property tables organized in the manner of Tables B1, B2, and B3 of the appendix.

To begin with, we need a set of property tables for the substance in the system. Tables for many substances are readily available from many sources, including printed tables as well as computer magnetic and optical disks. Empirical equations that closely match experimental data are also available. Many of these empirical equations have been programmed for computers.

Selecting the appropriate table from the set is readily envisioned by considering the general-pressure–specific-volume state diagram shown in Figure 3.16. This figure illustrates the liquid, saturation, and vapor regions. Recall that, as long as equilibrium exists, a state is represented by a point on this diagram. Also recall that the left portion of the line bordering the saturation region represents saturated-liquid states, which are normally denoted by the f subscript in property tables. The right position of the line bounding the saturation region represents saturated-vapor states, which are normally denoted by the g subscript in property tables. Our task is to determine in which of the three regions we can find the known state; this will determine our choice of table to use.

FIGURE 3.16
General-Pressure–Specific-Volume Diagram

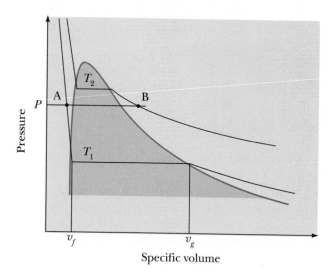

When the pressure and temperature of the state are known, we can determine the state completely only if a single phase exists. This is true because under saturation conditions the *pressure and temperature are not independent of one another.* To determine if the substance is a liquid or vapor, we begin by finding the saturation temperature for the given pressure in a saturation table. If the actual temperature is less than the saturation temperature, then the substance is a liquid. If the actual temperature is greater than the saturation temperature, then the substance is a gas.

When the known pressure is P and the known temperature is T_1 as shown in Figure 3.16, the state is point A, which is clearly in the liquid region. One then turns to the liquid table (if one is available) and uses the known pressure and temperature to determine the remaining properties (i.e., v, u, h, or s). When the known pressure is P and known temperature is T_2 as illustrated in Figure 3.16, the state is point B, which is in the vapor region. A vapor table can then be used with the known pressure and temperature to determine the remaining properties of the state.

EXAMPLE 3.7

A coal-fired boiler produces steam at 500 psia and 600° F. What is the specific internal energy of the steam produced by this boiler?

System: The water at the boiler outlet.

Basic Equations: Property tables for water in Appendix B.

450	456.39	0.01955	1.033	435.7	1119.6	437.4	768.2	1205.6	0.6360	1.4746
500	**467.13**	0.01975	0.928	447.7	1119.4	449.5	755.8	1205.3	0.6490	1.4645
550	477.07	0.01994	0.842	458.9	1119.1	460.9	743.9	1204.8	0.6611	1.4551
600	486.33	0.02013	0.770	469.4	1118.6	471.7	732.4	1204.1	0.6723	1.4464
700	503.23	0.02051	0.656	488.9	1117.0	491.5	710.5	1202.0	0.6927	1.4305

500	0.992	1139.7	1231.5	1.4923
550	1.079	1166.7	1266.6	1.5279
600	1.158	**1191.1**	1298.3	1.5585
700	1.304	1236.0	1356.7	1.6112
800	1.441	1278.8	1412.1	1.6571

Conditions: Thermodynamic equilibrium is presumed to exist.

Solution: According to appendix Table B1.2, the saturation temperature for 500 psia is 467.13° F. Because the actual temperature is greater than this, the water is gaseous (i.e., a vapor). Turning to Table B1.3 for the vapor region, we see that the specific internal energy is 1191.1 Btu/lb$_m$.

When we know either the pressure or temperature and another property, we can determine which region the state is in by comparing the known property to the corresponding saturated liquid (f subscript) and saturated vapor (g subscript) properties of a saturation table. If the known property is less than its f-subscripted counterpart, then the substance is a liquid and the state is located to the left of the saturated-liquid line of Figure 3.16. We would then use a liquid table, if one is available, to determine other properties of the state. If the known property is greater than its g-subscripted counterpart, then the substance is gaseous and the state is to the right of the saturated vapor line of Figure 3.16. A vapor table would then be used to determine the other properties of the state. When the known property is greater than or equal to its f counterpart and less than or equal to its g counterpart, then the state lies in the saturation region of Figure 3.16. A saturation table can then be used to determine the quality and other properties of the state.

EXAMPLE 3.8

Steam in the heating system pipes of a large building is at 300° F, and its specific internal energy is 1108.7 Btu/lb$_m$. What is the specific volume (in ft^3/lb$_m$) of the steam in these pipes?

System: The water in the heating pipes.

Basic Equations: Water property tables of Appendix B.

290	57.53	0.01735	7.47	259.3	1097.7	259.4	917.8	1177.2	0.4236	1.6477
300	66.98	0.01745	6.472	**269.5**	**1100.0**	269.7	910.4	1180.2	0.4372	1.6356
310	77.64	0.01755	5.632	279.8	1102.1	280.1	903.0	1183.0	0.4507	1.6238
320	89.60	0.01765	4.919	290.1	1104.2	290.4	895.3	1185.8	0.4640	1.6123
330	103.00	0.01776	4.312	300.5	1106.2	300.8	887.5	1188.4	0.4772	1.6010

228.0	20.09	1082.0	1156.4	1.7320
250	20.79	1090.3	1167.2	1.7475
300	**22.36**	1108.7	1191.5	1.7805
350	23.90	1126.9	1215.4	1.8110
400	25.43	1145.1	1239.2	1.8395

Conditions: We presume that thermodynamic equilibrium exists.

Solution: From Table B1.1, $u_g = 1100.0$ Btu/lb$_m$ and $u_f = 269.5$ Btu/lb$_m$ at 300° F. Because the known u is greater than u_g, the water is in a gaseous form, and Table B1.3 tells us that the specific volume is 22.36 ft^3/lb$_m$. ▬

EXAMPLE 3.9

Steam at 10° C enters a condenser with a specific enthalpy of 950 kJ/kg. What is the specific volume of this steam (in m^3/kg) when entering the condenser?

System: The water entering the condenser.

Basic Equations: Water property tables of Appendix B and

$$z = z_f + xz_{fg}$$

6	0.935	0.001000	137.7	25.19	2383.6	25.20	2487.2	2512.4	0.0912	9.0003
8	1.072	0.001000	120.9	33.59	2386.4	33.60	2482.5	2516.1	0.1212	8.9501
10	1.228	**0.001000**	**106.4**	42.00	2389.2	**42.01**	**2477.7**	**2519.8**	0.1510	8.9008
11	1.312	0.001000	99.86	46.20	2390.5	46.20	2475.4	2521.6	0.1658	8.8765
12	1.402	0.001001	93.78	50.41	2391.9	50.41	2473.0	2523.4	0.1806	8.8524

Conditions: Thermodynamic equilibrium is assumed.

Solution: Because the given h is between h_f and h_g, according to Table B1.1, this is a mixture of saturated liquid and vapor, and we will need the quality to determine the state. According to the quality equation,

$$x = \frac{h - h_f}{h_{fg}}$$

$$= \frac{950.0 - 42.01}{2477.7}$$

$$= 0.366.$$

The specific volume is then

$$v = v_f + x(v_g - v_f)$$

$$= 0.001 + 0.366(106.4 - 0.001)$$

$$= 38.93 \ \text{m}^3/\text{kg}$$

3.10 Incompressible-Liquid Approximation

Inspection of Table B1.4 reveals that the properties of the liquid phase do not change significantly as the pressure changes. For example, the specific volume of water at 30,000 kPa and 100° C is only 1% smaller than that at 2500 kPa and 100° C. But, the specific volume of water at 2500 kPa and 220° C is 14% greater than that at 2500 kPa and 100° C. Thus, the temperature has a larger effect on the properties of a liquid than does the pressure.

A commonly used approximation for liquids is to use the *the properties of a saturated liquid at the same temperature.* Saturated-liquid tables such as Tables B1.1, B1.2, B2.1, and B3.1 are then used in lieu of a liquid table to obtain the properties of liquids. This is an approximation that can lead to errors of a few percent, which is acceptable in many situations. Remember to read the saturation tables at the *temperature of the system,* not its pressure, when using this approximation.

Because this approximation is quite accurate in most cases, compressed-liquid tables are not always made available. A liquid table is not included for R-12 or ammonia in Appendix B for this reason. If a high degree of accuracy is required, a liquid-phase property table must be obtained.

EXAMPLE 3.10

Compare the specific internal energy of water at 2500 kPa and 200° C to that given by the incompressible-liquid approximation.

System: The water.

Basic Equations: Property tables of water in Appendix B.

140	0.001078	587.82	590.52	1.7369
180	0.001126	761.16	763.97	2.1375
200	0.001156	**849.9**	852.8	2.3294
220	0.001190	940.7	943.7	2.5174
223.99	0.001197	959.1	962.1	2.5546

Conditions: Thermodynamic equilibrium.

Solution: According to Table B1.4 for the liquid phase, the actual specific internal energy is 849.9 kJ/kg. At 200° C, the specific internal energy of saturated-liquid water is 850.65 kJ/kg, according to Table B1.2. Using 850.65 kJ/kg instead of the more accurate 849.9 kJ/kg is in error by less than 0.1%. ◼

3.11   Ideal Gases

The kinetic theory of gases considers a gas as a collection of particles that are not subject to interatomic forces, that move about randomly, and that undergo elastic collisions among themselves and the confining walls of the system. This model of a gas gives the equation of state:

$$PV = N\Re T \qquad\qquad (3.5)$$

This is known as the *ideal gas equation of state*. In this equation, N is the number of moles of gas; P and T are the absolute pressure and temperature; and $\Re$ is the *universal gas constant*

$$\Re = 8.314\frac{kJ}{kg\text{-mole K}} = 1545\frac{ft\text{-}lb_f}{lb_m\text{-mole}°R}$$

The universal gas constant is the same for all gases.

Comparison of this result to experimental measurements indicates that the ideal gas equation of state is accurate when the pressure is low and the temperature is high (i.e., the specific volume is large enough that intermolecular forces are negligible). In general, the ideal gas equation of state is accurate as long as the pressure is less than one-third of the gas's critical pressure and the temperature is two or more times greater than the gas's critical temperature. At very low pressures (approximately 0.01 of the critical pressure or less) this approximation is quite accurate at all temperatures. Because many of the substances that are gaseous under everyday conditions (air, oxygen, nitrogen, etc.,) meet this criterion, we can confidently use the ideal gas equation of state with these substances. This equation does not apply to substances whose state is near conditions where a phase change occurs. Hence, it is incorrect for substances such as water and R-12 under normal conditions.

By multiplying and dividing the right-hand side of the above molar form of the ideal gas equation by the gas molecular weight, we arrive at a more convenient form of the ideal gas equation of state,

$$PV = mRT$$

where m is the mass of the gas and R is the gas constant

$$R = \frac{\mathfrak{R}}{M}$$

Note that the gas constant applies to a specific gas, while the universal gas constant applies to all ideal gases. This result can be further rearranged by dividing by the mass into the form

$$Pv = RT$$

which incorporates the specific volume. This last form is widely used in engineering calculations. The other forms of the ideal gas equation of state are useful in specific applications.

EXAMPLE 3.11

Calculate the specific volume of dry air, $M = 28.97$, that is contained in a compressed air tank at an absolute pressure of 200 kPa and temperature of $100°$ C.

System: The air in the tank.

Basic Equations:
$$Pv = RT$$

Conditions: The air will be treated as an ideal gas.

Solution: The gas constant for air is $R = \mathfrak{R}/M = 8.314/28.97 = 0.2870$ kN-m/kg-K, and the absolute temperature is $100 + 273.15 = 373.2$ K. Rearranging the ideal gas equation of state gives

TIP:

When using the ideal gas equation of state, always convert temperatures and pressures to absolute temperatures and pressures.

$$v = \frac{RT}{P} = \frac{0.2870 \times 373.2}{200} \left[\frac{\text{kN-m}}{\text{kg-K}} \text{K} \frac{\text{m}^2}{\text{kN}} \right]$$

$$= 0.5355 \text{m}^3/\text{kg}$$

On a P–v state diagram, the isotherms of an ideal gas are lines along which $Pv = $ constant, with the constant being determined by the gas and its temperature. These are shown in Figure 3.17. Several constant-entropy lines are also shown in this figure. A general T–s state diagram for an ideal gas is shown in Figure 3.18, which also includes constant-pressure and specific-volume lines.

FIGURE 3.17
*General Ideal Gas P–v
State Diagram*

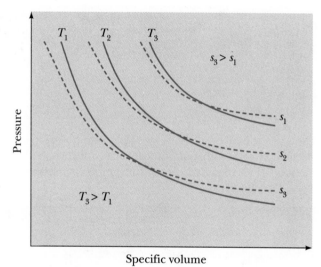

FIGURE 3.18
*General Ideal Gas T–s
State Diagram*

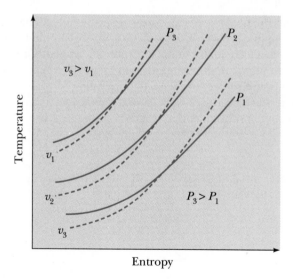

3.12 Specific Heats

Changes in the specific internal energy and enthalpy may be related to temperature changes by another property known as *specific heat.* Actually, two principal specific heats exist: the specific heat at constant specific volume and the specific heat at constant pressure. The specific heat at constant specific volume is defined as follows.

DEFINITION: *Specific Heat at Constant Specific Volume*

$$c_v \equiv \left. \frac{\partial u}{\partial T} \right|_v$$

The specific heat at constant specific volume is the rate of change of the specific internal energy with temperature when the system's specific volume is held constant. The specific heat at constant pressure is defined as follows.

DEFINITION: *Specific Heat at Constant Pressure*

$$c_P \equiv \left. \frac{\partial h}{\partial T} \right|_P$$

The specific heat at constant pressure is the rate of change of specific enthalpy with temperature when the system's pressure is held constant. Because both specific heats are mathematical combinations of other properties, they also are properties that are determined by the state of the system under consideration; thus, they are subject to the state postulate.

The state postulate tells us that the specific internal energy of a simple compressible substance is a function of temperature and specific volume:

$$u = u(T, v)$$

The differential of the specific internal energy is then

$$du = \left. \frac{\partial u}{\partial T} \right|_v dT + \left. \frac{\partial u}{\partial v} \right|_T dv$$

$$= c_v dT + \left. \frac{\partial u}{\partial v} \right|_T dv$$

During an infinitesimal process, the change in the specific internal energy is given by the product of the specific heat at constant volume and the change in temperature plus another term related to the change in the specific volume of the system.

Similarly, the specific enthalpy can be taken to be a function of temperature and pressure:

$$h = h(T, P)$$

Taking the differential of the specific enthalpy gives

$$dh = \left.\frac{\partial h}{\partial T}\right|_P dT + \left.\frac{\partial h}{\partial P}\right|_T dP$$

$$= c_P dT + \left.\frac{\partial h}{\partial P}\right|_T dP$$

The change in the enthalpy during an infinitesimal process is thus given by the product of the specific heat at constant pressure and the change in temperature plus a second term involving the change of the system's pressure during the process.

In Section 9.3 we will see that the specific internal energy and enthalpy of ideal gases depend only on temperature and are independent of the system's specific volume and pressure. The second term of the preceding differentials are then zero and do not contribute to the change of the specific internal energy or enthalpy. The change in specific internal energy of an ideal gas is therefore:

$$du = c_v dT \quad \text{(ideal gases only)} \tag{3.6}$$

The change in specific enthalpy is

$$dh = c_P dT \quad \text{(ideal gases only)} \tag{3.7}$$

Keep in mind that the two above results are valid for ideal gases only. They do not apply to more general substances such as water and R-12. The more general differentials must be used when dealing with general substances.

During a finite process from state 1 to state 2, the change in the specific internal energy of an ideal gas may be obtained by integrating Equation 3.6:

$$u_2 - u_1 = \int_{T_1}^{T_2} c_v dT$$

The change in the specific enthalpy of an ideal gas during a finite process is given by integrating Equation 3.7:

$$h_2 - h_1 = \int_{T_1}^{T_2} c_P dT$$

Knowledge of the dependence of the specific heats on the temperature of the ideal gas would allow us to perform these integrals and evaluate the change in the internal energy and enthalpy. Specific heats averaged over a temperature range are given by

$$\hat{c}_v = \frac{\int_{T_1}^{T_2} c_v \, dT}{T_2 - T_1}$$

$$\hat{c}_P = \frac{\int_{T_1}^{T_2} c_P \, dT}{T_2 - T_1}$$

These integrations simplify to

$$u_2 - u_1 = \hat{c}_v(T_2 - T_1)$$
$$h_2 - h_1 = \hat{c}_P(T_2 - T_1)$$

Specific heats of many gases are not strong functions of temperature and may be taken to be independent of temperature—over temperature ranges of a few hundred degrees—with a high degree of accuracy. When this is the case, the change in the internal energy is

$$u_2 - u_1 = c_v(T_2 - T_1)$$

and the change in the enthalpy is

$$h_2 - h_1 = c_P(T_2 - T_1)$$

The subscript 2 refers to the state at the end of the process, and subscript 1 refers to the state at the beginning of the process. Constant specific heats of several gases at temperatures typical of natural environments are listed in Table 3.5.

TABLE 3.5
Zero-Pressure Specific Heats of Selected Gases a 298 K (537° R)

Gas	Btu/lb$_m$-° R		kJ/kg-K		$k = c_P/c_v$
	c_v	c_P	c_v	c_P	
Air	0.1711	0.2397	0.7165	1.0035	1.40
Carbon Dioxide	0.1567	0.2018	0.6557	0.8446	1.29
Carbon Monoxide	0.1170	0.2479	0.7413	1.0381	1.40
Hydrogen	2.435	3.420	10.197	14.321	1.40
Nitrogen	0.1770	0.2479	0.7411	1.0380	1.40
Oxygen	0.1574	0.2195	0.6592	0.9190	1.39

At low pressures, the molar specific heats of all monatomic gases such as helium, neon, and argon are $\bar{c}_v = 2.98$ Btu/lb$_m$-mole °R $= 12.5$ kJ/kg-mole

K and $\bar{c}_P = 4.97$ Btu/lb$_m$-mole $^\circ$R $= 20.8$ kJ/kg-mole K. We frequently incur problems that involve ideal gases whose specific heats may be treated as if they are constant or that can be approached by using an average specific heat. More will be said about general substances and gases with variable specific heats in Chapter 9.

EXAMPLE 3.12

Air enters an air compressor at 70° C and is exhausted at 420° C. Calculate the change in the specific internal energy caused by this compressor.

System: The air.

Basic Equations:
$$u_2 - u_1 = c_v(T_2 - T_1)$$

Conditions: We will treat the air as an ideal gas with constant specific heats.

Solution: Getting the specific heat at constant volume for air from Table 3.5 and substituting the specified data gives

TIP:

As long as the gas is an ideal gas, this result is independent of the gas pressure. Because we are employing a temperature difference, it is not necessary to convert ° C to K.

$$u_2 - u_1 = 0.7165(420 - 70) \qquad \left[\frac{\text{kJ}}{\text{kg-K}}\text{K}\right]$$
$$= 250.8\text{kJ/kg}$$

The specific heats of an ideal gas are interrelated. By combining the definition of the enthalpy and the ideal gas equation of state, we get

$$h = u + Pv = u + RT$$

Taking the derivative of this result gives

$$dh = du + RdT$$

When we substitute the ideal gas specific-heat expressions, this becomes

$$c_P dT = c_v dT + RdT$$

Canceling the common dT term reduces this result to:

$$c_P - c_v = R \quad \text{(ideal gases only)} \tag{3.8}$$

The difference between the two specific heats of an ideal gas is then equal to the gas constant. Consequently, we only need to know one specific heat of an ideal gas. The second can always be determined by using the above result.

Constant-volume and constant-pressure processes are indistinguishable for incompressible liquids and solids because the pressure has no effect on their volumes. The constant-volume and constant-pressure specific heats are then one and the same:

$$c = c_P = c_v \quad \text{(incompressible substances only)} \qquad (3.9)$$

As a result, these modifiers are unnecessary for incompressible substances. When this is the case, we simply speak of the specific heat, c, without any modifiers. Specific heats of several liquids and solids are presented in Table 3.6.

TABLE 3.6

Specific Heats of Selected Solids and Liquids

c	Btu/lb$_m$-°R	kJ/kg-K
Water	1.00	4.18
R-12	0.215	0.229
Mercury	0.033	0.136
Benzene	0.440	1.85
Ice	0.480	2.00
Carbon (Graphite)	0.170	0.711
Aluminum	0.215	0.860
Iron	0.107	0.448
Copper	0.0930	0.388
Lead	0.0320	0.132

The term *specific heat* is somewhat misleading because it implies that it has something to do with heat, but its definitions tell us about changes in internal energy and enthalpy, not heat. This term was coined early in the development of thermodynamics when the concepts of heat, internal energy, and enthalpy were emerging and not completely understood. Do not be confused by their names. Simply remember that c_v is useful to detemine changes in internal energy, c_P is useful to determine changes in enthalpy, and c is useful when the substance is an incompressible liquid or solid.

PROBLEMS

Sketch the states and processes of the following problems on the appropriate P–v and T–s state diagrams.

3.1   Determine the molecular weight of methane (CH_4) and propane (C_3H_8) in kg/kg-mole.

3.2 What is the molecular weight of ammonia (NH_3) in lb_m/lb_m-mole?

3.3 What is the molecular weight of R-12 ($CClF_2$) in kg/kg-mole?

3.4 A gaseous mixture consists of 1 lb_m-mole of helium, 2 lb_m-moles of oxygen, 0.1 lb_m-mole of water vapor, and 1.5 lb_m-moles of nitrogen. Determine the mole fraction of the various constituents and the apparent molecular weight of this mixture in lb_m/lb_m-mole.

3.5 Trace amounts of sulfur in coal are burned in the presence of diatomic oxygen to form sulfur dioxide (SO_2). Determine the minimum mass (in kilograms) of oxygen required in the reactants and the mass (in kilograms) of sulfur dioxide in the products when 1 kilogram of sulfur is burned.

3.6 Methane is burned in the presence of diatomic oxygen. The combustion products consist of water vapor and carbon dioxide gas. Determine the mass (in lb_m) of water vapor generated when 1 pound-mass of methane is burned.

3.7 Propane fuel is burned in the presence of air (see Example 3.1). Assuming that the combustion is complete—that is, only nitrogen, water vapor, and carbon dioxide are present in the products—determine the mass fraction of carbon dioxide in the product mixture.

3.8 What is the mole and mass fraction of the water vapor in the products of the previous problem?

3.9 A container is filled with 1 kg of a fluid whose specific volume is 0.001 m^3/kg and 2 kg of a fluid whose specific volume is 0.008 m^3/kg. What is the volume of this container (in m^3) and the total weight (in newtons) of its contents at a location where $g = 9.6$ m/s^2?

3.10 A mixture is 70% by volume liquid water, whose density is 62.4 lb_m/ft^3, that is mixed with another fluid, whose density is 50.0 lb_m/ft^3. What is the specific weight (in lb_f/ft^3) of this mixture at a location where $g = 31.9$ ft/s^2?

3.11 One liter of a liquid whose specific volume is 0.0003 m^3/kg is mixed with 2 liters of a liquid whose specific volume is 0.00023 m^3/kg in a container whose total volume is 3 liters. What is the density of the resulting mixture in kg/m^3?

3.12 One pound-mass of a gas whose density is 0.001 lb_m/ft^3 is mixed with 2 pounds-mass of a gas whose density is 0.002 lb_m/ft^3 such that the pressure and temperature of the gases do not change. Determine the resulting mixture's volume in cubic feet and specific volume in ft^3/lb_m.

3.13 The *weight fraction* of a component in a mixture of several substances is defined as *the weight of the component alone divided by the total weight of the mixture*. What is the relationship between the weight fraction and the mass fraction?

3.14 The pressure at the exit of an air compressor is 150 psia. What is the pressure at the exit in kilopascals?

3.15   The absolute pressure in a compressed air tank is 200 kPa. What is the pressure in the tank in psia?

3.16   A manometer measures a pressure difference as 40 inches of water. What is this pressure difference in pounds-force per square inch?

3.17   The pressure of the helium inside a toy balloon is 1000 mm Hg. What is this pressure in kilopascals?

3.18   If the pressure inside the rubber balloon of the previous problem is 1500 mm Hg, what is this pressure in pounds-force per square inch?

3.19   The diameter of the pistons shown in Figure 3.19 are $D_1 = 3$ in. and $D_2 = 2$ in. Determine the pressure in chamber 3 in psia when the other pressures are $P_1 = 150$ psia and $P_2 = 200$ psia.

FIGURE 3.19

3.20   In Figure 3.19, the piston diameters are $D_1 = 10$ cm and $D_2 = 4$ cm. If $P_1 = 1,000$ kPa and $P_3 = 500$ kPa, what is the pressure in chamber 2 in kPa?

3.21   The piston diameters in Figure 3.19 are $D_1 = 10$ cm and $D_2 = 4$ cm. When the pressure in chamber 2 is 2000 kPa and the pressure in chamber 3 is 700 kPa, what is the pressure in chamber 1 in kPa?

3.22   The force generated by a spring is given by $F = kx$, where k is the spring constant and x is the deflection of the spring. In Figure 3.20, the spring has a spring constant of 200 lb$_f$/in. and has been compressed 2 inches. The piston diameters are $D_1 = 5$ in. and $D_2 = 2$ in. What is P_2 when $P_1 = 100$ psia and $P_3 = 20$ psia?

FIGURE 3.20

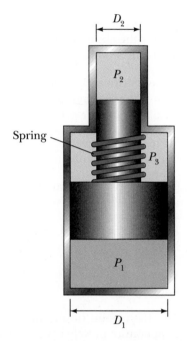

3.23 The spring of Figure 3.20 has a spring constant of 8 kN/cm. The pressures are $P_1 = 5000$ kPa, $P_2 = 10,000$ kPa, and $P_3 = 1000$ kPa. If the piston diameters are $D_1 = 8$ cm and $D_2 = 3$ cm, how far will the spring be deflected in centimeters?

3.24 In Figure 3.20, the spring has a spring constant of 50 lb$_f$/in. and has been extended 2 inches. The piston diameters are $D_1 = 3$ in. and $D_2 = 1$ in. Determine P_1 in psia when $P_2 = 800$ psia and $P_3 = 30$ psia.

3.25 Calculate the absolute pressure, P_1, of the manometer shown in Figure 3.21 in pounds-force per square inch. The local atmospheric pressure is 758 mm Hg.

FIGURE 3.21

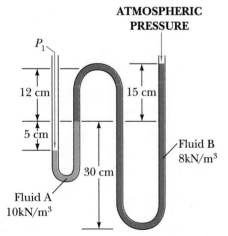

3.26   If the specific weight of fluid A in the manometer of Figure 3.21 is 100 kN/m³, what is the absolute pressure (in kPa) indicated by the manometer when the local atmospheric pressure is 90 kPa?

3.27   If the specific weight of fluid B in the manometer of Figure 3.21 is 20 kN/m³, what is the absolute pressure (in kPa) indicated by the manometer when the local atmospheric pressure is 745 mm Hg?

3.28   Steam enters a heat exchanger at 300 K. What is the temperature of this steam in degrees Fahrenheit?

3.29   The temperature of the lubricating oil in an automobile engine is measured as 150° F. What is the temperature of this oil in degrees Celsius?

3.30   Heated air is at 150° C. What is the temperature of this air in degrees Fahrenheit?

3.31   What is the temperature of the heated air of the previous problem in degrees Rankine?

3.32   The flash point of an engine oil is 363° F. What is the absolute flash-point temperature in Kelvin and degrees Rankine?

3.33   Humans are most comfortable when the temperature is between 65° F and 75° F. Express these temperature limits in degrees Celsius. Convert the size of this temperature range (10° F) to Kelvin, degrees Celsius, and degrees Rankine. Is there any difference in the size of this range as measured in relative or absolute units?

3.34   One pound-mass of water fills a 2.29 ft³ rigid container at an initial pressure of 250 psia. The container is then cooled to 100° F. Determine the initial temperature (in ° R) and final pressure (in psia) of the water.

3.35   One kilogram of R-12 fills a 0.1525 m³ rigid container at an initial temperature of −40° C. The container is then heated until the pressure is 200 kPa. Determine the final temperature (in K) and initial pressure (in kPa).

3.36   One pound-mass of water fills a 2.648 ft³ weighted piston-cylinder device at a temperature of 400° F. The piston-cylinder device is now cooled until its temperature is 100° F. Determine the final pressure (in psia) and volume (in ft³) of the water.

3.37   One kilogram of R-12 fills a 0.14 m³ weighted piston-cylinder device at a temperature of −30° C. The container is now heated until the temperature is 100° C. Determine the final volume (in m³) of the R-12.

3.38   One pound-mass of water fills a 2.29 ft³ rigid container at an initial pressure of 250 psia. The container is then cooled to 100° F. Determine the initial temperature (in ° F) and the final pressure (in psia) of the water.

3.39 Ten kilograms of R-12 fill a 1.348 m^3 rigid container at an initial temperature of $-40°$ C. The container is then heated until the pressure is 200 kPa. Determine the final temperature in $°$ C and initial pressure in kPa.

3.40 Two pounds-mass of water at 500 psia initially fill the 1.5 ft^3 left chamber of the partition divided system of Figure 1.9 (page 15). The right chamber's volume is also 1.5 ft^3, and it is initially evacuated. The partition is now ruptured, and heat is transferred to the water until its temperature is 300$°$ F. Determine the final pressure in psia and the total internal energy in Btu of the water at the final state.

3.41 One kilogram of ammonia vapor at 200 kPa fills the 0.8825 m^3 left chamber of the partitioned system of Figure 1.9 (page 15). The right chamber has twice the volume of the left and is initially evacuated. Determine the pressure of the ammonia (in kPa) after the partition has been removed and enough heat has been transferred so that the temperature of the ammonia is 0$°$ C.

3.42 One pound-mass of water fills a 2.361 ft^3 weighted piston-cylinder device at a temperature of 400$°$ F. The piston-cylinder device is now cooled until its temperature is 100$°$ F. Determine the final pressure in psia and the volume in ft^3 of the water.

3.43 Ten kilograms of R-12 fill a 1.595 m^3 weighted piston-cylinder device at a temperature of $-30°$ C. The container is now heated until the temperature is 100$°$ C. Determine the final volume of the R-12 in ft^3.

3.44 What is the specific internal energy of water at 5 psia and 200$°$ F in Btu/lb$_m$?

3.45 What is the specific volume (in ft^3/lb$_m$) of water at 500 psia, 100$°$ F?

3.46 One pound-mass of water fills a container whose volume is 2 ft^3. The pressure in the container is 100 psia. Calculate the total internal energy in the container in Btu. Calculate the total entropy in the container in Btu/$°$ R.

3.47 Three kilograms of water in a container have a pressure of 100 kPa and temperature of 360$°$ C. What is the volume of this container in m^3?

3.48 What is the specific entropy of water that is at 750 kPa, 140$°$ C in kJ/kg-K?

3.49 What is the specific volume of water that is at 5000 kPa, 160$°$ C in m^3/kg?

3.50 Two pounds-mass of R-12 fill a rigid container whose volume is 0.4 ft^3. The pressure in the container is 40 psia. Determine the temperature and total enthalpy in the container. The container is now heated until the pressure is 100 psia. Determine the temperature in $°$ F and total enthalpy in Btu when the heating is completed.

3.51 R-12, whose specific volume is 0.4284 ft^3/lb$_m$, flows through a tube at 120 psia. What is the temperature (in $°$ F) in the tube?

3.52 Three hundred kilograms of R-12 occupy a 9 m³ container at 10° C. What is the specific enthalpy of the R-12 in the container in kJ/kg?

3.53 R-12 at 200 kPa, 25° C flows through a refrigeration line. What is the specific volume (in m³/kg) of the R-12?

3.54 Two pounds-mass of water fill a weighted piston-cylinder device whose volume is 2.5 ft³. The pressure is 300 psia. The device is then heated until the temperature is 500° F. Determine the resulting change in the water's total entropy in Btu/° R.

3.55 One kilogram of R-12 initially at 600 kPa, 25° C undergoes a process during which the entropy is kept constant until the pressure is 100 kPa. Determine the final temperature of the R-12 in ° C and the final specific internal energy in kJ/kg.

3.56 Saturated R-12 vapor enters the compressor shown in Figure 3.22 at 0° F. At the compressor exit, the specific entropy is the same as at the entrance, and the pressure is 60 psia. Determine the R-12 exit temperature in ° F and the difference between the specific enthalpy (in Btu/lb$_m$) of the R-12 at the exit and entrance.

FIGURE 3.22

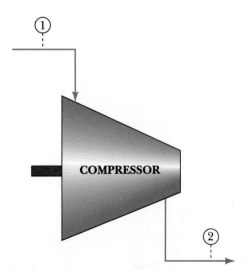

3.57 Water vapor at 35 kPa and 160° C enters the compressor shown in Figure 3.22. This water leaves the compressor at 300 kPa with the same specific entropy as at the entrance. What is the temperature in ° C and specific enthalpy in kJ/kg at the exit?

3.58 Water vapor at 6000 kPa and 400° C enters the turbine shown in Figure 3.23. This water leaves the turbine at 100 kPa with the same specific entropy as it had when it entered the turbine. Calculate the difference between the specific enthalpy (in kJ/kg) of the water at the turbine entrance and exit.

FIGURE 3.23

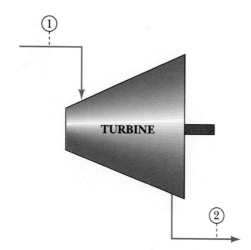

3.59 Ammonia vapor enters the turbine of Figure 3.23 at 250 psia and 175° F. The temperature of the ammonia is reduced in this turbine to 20° F; the specific entropy remains constant. Determine the change in the specific enthalpy (in Btu/lb$_m$) of the ammonia as it passes through this turbine.

3.60 One hundred kilograms of R-12 at 200 kPa are contained in a piston-cylinder device whose volume is 10.77 m³. The piston is now moved until the volume is one-half its original size. This is done such that the pressure of the R-12 does not change. Determine the final temperature (in ° C) and change in the total internal energy (in kJ/kg) of the R-12.

3.61 The spring-loaded piston-cylinder device shown in Figure 3.24 is filled with 0.5 kg of water vapor that is initially at 4000 kPa and 400° C; the spring exerts no force against the piston. The spring constant is 0.9 kN/cm and the piston diameter, D, is 20 cm (see Problem 3.22 for discussion of spring operation). The water now undergoes a process until its volume is one-half of the original volume. Calculate the final temperature in ° C and specific entropy in kJ/kg-K of the water.

3.62 In Figure 3.24, the piston-cylinder apparatus is initially filled with 0.2 lb$_m$ of an ammonia liquid-vapor mixture whose temperature is −30° F and whose quality is 80%. The spring constant is 37 lb$_f$/in. and the piston diameter, D, is 12 in. The ammonia undergoes a process that increases its volume by 40%. Calculate the final temperature in ° F and enthalpy in Btu/lb$_m$ of the ammonia.

FIGURE 3.24

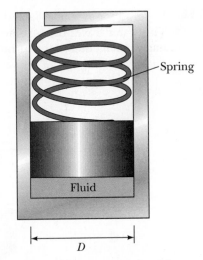

Spring

Fluid

D

3.63 Repeat Problem 3.45 using the incompressible-liquid approximation. Compare your answers to determine the accuracy of this approximation.

3.64 How much error would one expect in determining the specific enthalpy by applying the incompressible-liquid assumption to water at 1500 psia, 400° F?

3.65 How much error would result in calculating the specific entropy of water at 10,000 kPa, 100° C by using the incompressible-liquid assumption?

3.66 What is the specific volume of R-12 at 20° C, 700 kPa in m³/kg?

3.67 What is the specific entropy of R-12 at 100 psia, 60° F in kJ/kg-K?

3.68 One hundred grams of R-12 initially fill a weighted piston-cylinder device at 20 kPa, −25° C. The device is then heated until the temperature is 100° C. Determine the change in the device's volume (in m³) as a result of the heating.

3.69 What is the specific volume of oxygen at 25 psia, 80° F in m³/kg?

3.70 What is the specific volume of nitrogen at 300 kPa, 227° C in m³/kg?

3.71 One kilogram of air fills a 100-liter container at a temperature of 27° C. What is the pressure in the container in kPa?

3.72 Two pounds-mass of oxygen fill a 3 ft³ container at a pressure of 80 psia. What is the oxygen's temperature in ° R?

3.73 Two kilograms of helium are maintained at 300 kPa, 27° C in a rigid container. How large is the container in m³?

3.74 One pound-mass of argon is maintained at 200 psia, 100° F in a tank. What is the tank's volume (in ft³)?

3.75 Ten grams of oxygen fill a weighted piston-cylinder device at a pressure of 20 kPa, 100° C. The device is now cooled until the temperature is 0° C. Determine the change of the device's volume (in m³) during the cooling.

3.76   One-tenth pound-mass of helium fills a 2.5 ft³ rigid vessel at 50 psia. The vessel is heated until the pressure is 100 psia. Calculate the temperature change (in °F and °R) as a consequence of this heating.

3.77   One-tenth pound-mass of argon fills a 0.5 ft³ piston-cylinder device at 50 psia. The piston is now moved by changing the weights until the volume is twice its original size. During this process, the argon's temperature is maintained constant. Determine the final pressure in the device in psia.

3.78   In Figure 3.19, the piston diameters are $D_1 = 10$ cm and $D_2 = 4$ cm. Chamber 1 contains 1 kg of helium, chamber 2 is filled with condensing water vapor, and chamber 3 is evacuated. The entire assembly is placed in an environment whose temperature is 200° C. Determine the volume of chamber 1 in m³ when thermodynamic equilibrium has been established.

3.79   In Figure 3.19, the piston diameters are $D_1 = 3$ in. and $D_2 = 2$ in. The pressure in chamber 3 is 30 psia. Chamber 1 contains 0.5 lb_m of air, and chamber 2 is filled with condensing R-12. The entire assembly is maintained at 120° F. Determine the volume in ft³ of chamber 1.

3.80   How much will the volume of chamber 1 in the previous problem change when the pressure in chamber 3 is increased to 100 psia?

3.81   One cubic foot of air is contained in the spring-loaded piston-cylinder device shown in Figure 3.24. The spring constant is 5 lb_f/in., and the piston diameter is 10 in. (see Problem 3.22 to review the operation of a spring). When no force is exerted by the spring on the piston, the state of the air is 250 psia, 460° F. This device is now cooled until the volume is one-half its original size. Determine the change in the specific internal energy (in Btu/lb_m) and enthalpy (in Btu/lb_m) of the air.

3.82   Ten grams of nitrogen are contained in the spring-loaded piston-cylinder device shown in Figure 3.24. The spring constant is 1 kN/m, and the piston diameter is 10 cm. When the spring exerts no force against the piston, the nitrogen is at 120 kPa, 27° C. The device is now heated until its volume is 10% greater than the original volume. Determine the change in the specific internal energy (in kJ/kg) and enthalpy (in kJ/kg) of the nitrogen.

3.83   When undergoing a given temperature change, which of the two gases—air or oxygen—experiences the largest change in specific enthalpy? in specific internal energy?

3.84   What is the change in the specific internal energy (in Btu/lb_m) of air as its temperature changes from 100° F to 200° F? Is there any difference if the temperature were to change from 0° F to 100° F?

3.85   What is the change in the specific enthalpy (in kJ/kg) of oxygen as its temperature changes from 150° C to 200° C? Is there any difference if the temperature change were from 0° C to 50° C? Does the pressure at the beginning and end of this process have any effect on the specific enthalpy change?

3.86   The temperature of 2 pounds-mass of neon is increased from 70° F to 300° F. Calculate the change in the total internal energy of the neon in Btu. Would the internal energy change be any different if the neon were replaced with argon?

3.87   When argon is cooled from 400° C to 100° C, its specific enthalpy will change. Calculate the change in the specific enthalpy in kJ/kg. If neon had undergone this same change of temperature, would its change in specific enthalpy have been any different?

3.88   Air is compressed from 20 psia and 70° F to 150 psia in the compressor shown in Figure 3.22. This compressor is operated such that the air temperature remains constant. Calculate the change in the specific volume of the air in ft^3/lb_m as it passes through this compressor.

3.89   Neon is compressed from 100 kPa and 20° C to 500 kPa in an isothermal compressor such as that shown in Figure 3.22. Determine the change in the specific volume in m^3/kg and specific enthalpy in kJ/kg of the neon caused by this compression.

3.90   One pound-mass of iron is heated from 70° F to 160° F. What is the change in the iron's total internal energy and enthalpy in Btu?

3.91   The state of liquid water is changed from 50 psia, 50° F to 2000 psia, 100° F. Determine the change in the specific internal energy and enthalpy (in Btu/lb_m units) as given by the water property tables, incompressible substance model, and specific-heat model.

PROJECTS

The following projects are more extensive than the problems presented in this chapter. Most are open-ended and require some decisions on the student's part. Many can be expedited with computer-based property tables, computer spreadsheets, and student-written computer programs.

3.1   Enthalpy–entropy state diagrams (also known as *Mollier diagrams*) are more useful than temperature–entropy state diagrams for analyzing some systems. Develop an enthalpy–entropy state diagram for a substance that can undergo liquid–vapor phase transitions. Sketch isothermal, isobaric, and constant-volume processes on your diagram.

3.2   Pressure–enthalpy state diagrams are useful for studying refrigeration and like systems. Plot a pressure–enthalpy state diagram for a refrigerant that undergoes liquid–vapor phase changes. Sketch isothermal and constant-entropy processes on this diagram.

3.3 Temperature may alternatively be defined as

$$\left. \frac{\partial u}{\partial s} \right|_v$$

Prove that this definition reduces the net entropy change of two constant-volume systems filled with simple compressible substances to zero as the two systems approach thermal equilibrium.

3.4 Pressure may also be defined as

$$-\left. \frac{\partial u}{\partial v} \right|_s$$

Prove that for a simple compressible substance

$$\left. \frac{\partial s}{\partial v} \right|_u = \frac{P}{T}$$

using this definition of pressure and the definition of temperature of the previous project.

3.5 The development of the constant-pressure specific heat stated that

$$\left. \frac{\partial h}{\partial P} \right|_T = 0$$

for ideal gases. Prove this by using the preceding definitions of pressure and temperature.

3.6 The development of the constant-volume specific heat stated that

$$\left. \frac{\partial u}{\partial v} \right|_T = 0$$

for ideal gases. Prove this by using the preceding definitions of pressure and temperature.

3.7 You have a rigid tank to which a pressure gauge is attached. Describe a procedure by which you could use this tank to blend ideal gases in prescribed mole-fraction portions.

3.8 Water vapor may be treated as an ideal gas, with little error, under certain pressure and temperature conditions. Determine the pressure and temperature ranges under which water may be approximated as an ideal gas with 2% or less error.

Closed Systems

On Science

Science is eternal. It was started thousands of years ago and its progress is continuous. Principles that are deeply rooted are not likely to pass suddenly from the scene.

Theodore von Karman

We have seen that when we change a substance's properties such as pressure and temperature, we also change the energy contained therein. By creating devices that constrain and control the manner in which the properties are changed, we are capable of transforming energy and creating valuable work to light our homes and cities; propel our transportation vehicles; harvest natural resources from wells, mines, and fields; operate our computers and telecommunications; manufacture products; and perform many other tasks. The working fluids in these devices constitute systems that undergo change as they transform energy. Each device itself controls the changes experienced by its working fluid and directs the heat and work interactions between the working fluid and its surroundings. In this chapter, we focus on the closed system, which always contains the same quantity of working fluid mass.

Closed systems were the earliest to be studied by thermodynamicists because they were the first to be developed commercially. In the early 1700s, the steam engines of Savery and Newcomen took advantage of the decrease in volume that occurs when steam vapor is condensed to a liquid; they used this decrease to move water or a piston, thereby producing work. Later in the eighteenth century, James Watt[1] added the separate condenser, which increased the efficiency of these engines. Scientists and engineers who were pursuing a scientific understanding of the operation of these and like devices developed the first and second laws of thermodynamics. This knowledge led to the refinement of these engines until they reached their zenith in the early 1900s; by then they were being used to propel ships, industrial equipment, and trains.

Modern, inexpensive, and low-to-moderate power energy-transformation devices are typically closed systems, at least during some portions of their cycles. These include internal-combustion engines, air and refrigeration compressors, and Wankel engines. The cost of these devices is quite modest because they use the piston cylinder (or its equivalent, such as a Wankel engine) and heat transfer through the metal walls of the cylinder or combustion chamber to control the processes of the working fluid as the piston is moved in the cylinder.

In this chapter, we will adapt the first and second laws of thermodynamics as presented for cyclical devices in Chapter 2 to systems that have no mass crossing their control boundaries. These laws will be extended to include means of understanding the processes that closed systems undergo as they transform

---

[1]James Watt (1736–1819), was a Scottish engineer who invented the modern condensing steam engine while working as a mathematical instrument maker at the University of Glasgow. Watt's engine uses a separate condenser to keep the temperature of the work-producing cylinder high and the temperature of condensation low, thus keeping the efficiency of the engine high. His later inventions improved his basic engine and adapted it to a variety of uses.

energy. The limits on the transformation of energy that can occur in these systems will also be developed. And throughout this chapter, the properties of the working fluid as presented in Chapter 3 will be used to relate the energy transformations to other considerations, such as pressures and temperatures, that are important in designing the equipment to withstand the imposed loads and selecting materials to provide the strength and life needed by the device. The results of this chapter provide the foundation for understanding many of the applications described in later chapters.

4.1 Introduction to Closed Systems

Closed systems consist of a fixed mass of a substance that remains within the system control boundaries, regardless of the processes experienced by the system. Because of this feature, these systems are also known as *fixed-mass* or *control-mass systems*. Any property or other characteristic of the system except the mass can change. Even the chemical composition of the system can change as long the mass remains constant and within the control boundaries. For example, consider a system that consists of 1 mole of carbon and 1 mole of diatomic oxygen. If this system is ignited, combustion will occur. Once the combustion is completed, the system will consist of 1 mole of carbon dioxide. Before combustion, the system contains 12 mass units of carbon and 32 mass units of oxygen, a total of 44 mass units. After combustion, the system contains 44 mass units of carbon dioxide. The total mass before and after combustion has remained constant, while the chemical composition of this closed system has been altered.

If the chemical composition of a closed system does not change, other properties such as pressure, temperature, and specific volume can change. For example, consider a quantity of air that is sealed in a rigid vessel. As we heat this vessel, the temperature will increase, which causes the pressure, internal energy, and entropy to also increase, although the specific volume will not change. As long as the temperatures do not become exceedingly large, the vessel will always contain a fixed mass of a nitrogen and oxygen mixture that does not escape the vessel. At very high temperatures, the nitrogen and oxygen molecules will disassociate into atoms, but the quantity of mass will not change, provided that molecular speeds remain smaller than relativistic speeds.

Because no mass is allowed to enter or leave a closed system, the mass at the beginning must equal the mass at the end of the process. This allows us to define a closed system as follows.

DEFINITION: *Closed System*

Any system that, when it undergoes a process from an initial state 1 to a final state 2, obeys

$$m_1 = m_2$$

A closed system is any system whose mass remains constant (and no mass crosses its control boundary), although its composition or state may change.

Rigid and flexible vessels, and piston-cylinder devices are common examples of closed systems. The contents of the cylinder of an internal-combustion engine constitute a closed system during those portions of the cycle when the intake and exhaust valves are closed. A reciprocating air compressor is also a closed system during those portions of the cycle when the intake and outlet valves are closed. The closed system then has many practical applications, as well as being convenient for the further development of the concepts of thermodynamics.

4.2 Compression and Expansion Work Produced by a Closed System

An expression for the compression and expansion work produced by a closed system filled with a simple compressible substance may be developed by applying the definition of work from Section 1.2 to the closed system shown in Figure 4.1 as it undergoes a compression or expansion. Initially, this closed

FIGURE 4.1

Closed System Experiencing a Small Volume Change

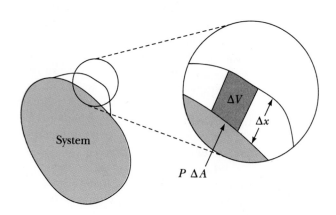

system occupies the volume indicated by the shaded area of Figure 4.1. Later, a portion of the boundary of the closed system has expanded a small amount, as indicated by the light line in the figure. Since pressure is only defined for equilibrium conditions, it is assumed that the system is very close to equilibrium during this expansion. A process of this type is known as a *quasi-equilibrium process*. As shown in the exploded view of the boundary, one small portion of the boundary, ΔA, moves a distance Δx during this expansion. The force produced by the system is $P\Delta A$, which acts in the same direction as the boundary displacement Δx. Hence, for small displacement Δx,

$$\mathbf{F} \cdot d\mathbf{x} = P\Delta x \Delta A$$

However, $\Delta x \Delta A$ is the size of the volume, ΔV, which is shown by the shading of the exploded section in Figure 4.1, that is generated by the motion of the ΔA portion of the boundary. The force-displacement product is then

$$P\Delta x \Delta A = P\Delta V$$

Substituting this result into the definition of work and taking the limit as the volume change, ΔV, approaches zero gives $\delta W = P\, dV$. When this is integrated over a process beginning at an initial state 1 and ending at a final state 2, the following result is obtained:

$$_1 W_2 = \int_{V_1}^{V_2} P\, dV \quad \text{(quasi-equilibrium processes only)} \qquad (4.1)$$

Because the system consists of a fixed mass, we can divide both sides of this equation by this mass to obtain the specific work produced:

$$_1 w_2 = \int_{v_1}^{v_2} P\, dv$$

These equations indicate two important facts. First, for a closed system filled with a simple compressible substance to produce compression and expansion work, it must change its volume or specific volume. A system that does not change its volume cannot produce compression and expansion work during a quasi-equilibrium process. Second, the amount of work produced by the system depends on how the pressure changes as the volume changes. One must know the process the system undergoes as it changes from its initial to its final state. Consequently, work is a path function rather than a point function. Statements that a system contains a certain amount of work are incorrect because the amount of work produced by the system depends on the process used to extract the work.

While developing these equations, we have assumed that the system is undergoing a quasi-equilibrium process as it changes its volume. The system is then close enough to equilibrium that the pressure of the working fluid acting on the moving control boundary is the same as the pressure throughout the rest of the system. Such a process is shown on the P–v state diagram of Figure 4.2. Recalling that integrals represent areas (see Appendix A), the shaded area of this figure is the specific work produced by the closed system as it changes its state from state 1 to state 2.

FIGURE 4.2

A Quasi-equilibrium Process Demonstrating the Definition of Compression and Expansion Work for a Closed System

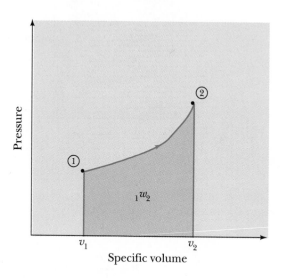

EXAMPLE 4.1

The spring-loaded piston-cylinder device shown in Figure 4.3 contains a gas that initially occupies 1 liter at 150 kPa. Its state is then changed until the gas fills 0.5 liter at 500 kPa. Determine the work done by the gas during this process. Also calculate the amount of this work that results from moving the surrounding atmospheric air (its pressure is 101 kPa) and from compression of the spring.

System: The working gas contained in the piston-cylinder device.

Basic Equations:

$$_1W_2 = \int_{V_1}^{V_2} P\,dV$$

Conditions: The spring will be treated as a linear spring with spring constant k, and a quasi-equilibrium process is presumed.

Solution: We need to begin by identifying the process generated by the spring as the gas is compressed into a smaller volume. The force exerted on the piston by the spring is directly proportional to the spring deflection:

$$F = kx$$

FIGURE 4.3
*Spring-Loaded
Piston-Cylinder
Device for
Example 4.1*

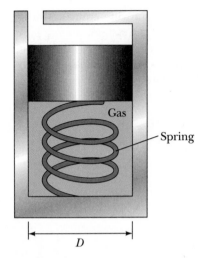

A static balance of the forces acting on the piston indicates that the pressure of the gas in the piston-cylinder device is

$$P = \frac{F}{A} = \frac{k}{A^2}xA$$

$$= \frac{k}{A^2}V$$

Consequently, the pressure is a linear function of the system volume. The process experienced by the gas can then be represented as a straight line on a P–V state diagram as shown in Figure 4.4. Because the process is always close to equilibrium, the work produced by the system is the area under the process of Figure 4.4. This area is given by

TIP:

The negative sign indicates that work has been done on the system.

$$_1W_2 = \frac{P_1 + P_2}{2}(V_2 - V_1)$$

$$= \frac{150 + 500}{2 \times 1000}(.5 - 1) \quad \left[\frac{kN}{m^2}1\frac{m^3}{1}\right]$$

$$= -162.5 \text{ J} = -0.1625 \text{ kJ}$$

The pressure of the surrounding atmosphere remains constant as it moves forward with the piston. As the piston moves a distance dx, the force exerted by the atmospheric air is $P_\infty A$, which acts in the same direction as the displacement. Then

$$
\begin{aligned}
W_\infty &= \int \mathbf{F} \cdot d\mathbf{x} \\
&= \int P_\infty A\,dx \\
&= P_\infty \int dV \\
&= P_\infty(V_{\infty,2} - V_{\infty,1}) \\
&= 101(1.0 - 0.5) \\
&= 50.5 \text{ J} = 0.0505 \text{ kJ}
\end{aligned}
$$

FIGURE 4.4
*P–V State Diagram
for Example 4.1*

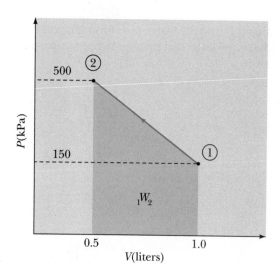

The remainder of the work needed to compress the system gas is provided by the spring as it extends its length. As it is being extended, the spring produces the amount of work

$$
\begin{aligned}
W_{sp} &= -W - W_\infty \\
&= 0.1635 - 0.0505 \\
&= 0.1120 \text{ kJ}
\end{aligned}
$$

EXAMPLE 4.2

One cubic foot of air is expanded to two cubic feet in a closed system while its temperature is maintained constant at $140°$ F. How much specific work is produced by the air as it expands?

System: The air in the closed-system device.

Basic Equations:
$$_1w_2 = \int_{v_1}^{v_2} P\,dv$$
$$Pv = RT$$

Conditions: The air will be treated as an ideal gas, and a quasi-equilibrium process will be presumed.

Solution: Because the air is being treated as an ideal gas,

$$P = \frac{RT}{v}$$

Substituting this into the expression for compression and expansion work and taking advantage of the fact that the temperature remains constant during the process gives

$$_1w_2 = RT \int_{v_1}^{v_2} \frac{dv}{v}$$

$$= RT \ln \frac{v_2}{v_1}$$

$$= RT \ln \frac{V_2}{V_1}$$

FIGURE 4.5
P–v State Diagram for Example 4.2

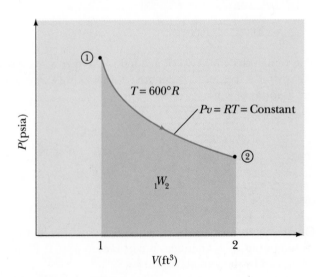

Substituting the given information into this result gives

TIP:

We do not need to convert to specific volumes because the mass is the same at states 1 and 2.

$$_1w_2 = \frac{1545}{28.97} \times \frac{600}{778} \ln\left(\frac{2}{1}\right) \quad \left[\frac{\text{ft-lb}_f}{\text{lb}_m\text{-}°\text{R}} °\text{R} \frac{\text{Btu}}{\text{ft-lb}_f}\right]$$

$$= 28.5 \text{ Btu/lb}_m$$

The graphical representation of these calculations is shown in Figure 4.5.

4.3 Generalized Work

In the preceding section we developed an expression for the compression and expansion work produced by a simple compressible substance as it reversibly changes its volume. This is but one of many modes of work. In general, we can write that work is given by

$$_1W_2 = \int_1^2 \hat{\mathbf{F}} \cdot d\hat{\mathbf{x}}$$

where $\hat{\mathbf{F}}$ is the *generalized force* and $\hat{\mathbf{x}}$ is the *generalized displacement*. For a simple compressible substance, the generalized force is the pressure, P, and the generalized displacement is the volume, V (or specific volume when calculating the specific work).

Expressions for the generalized force and displacement can be developed for other reversible modes of work. For example, consider a thin fluid film (e.g., soap film) stretched over the wire frame of Figure 4.6. One section of this frame is arranged so that it can be moved to increase or decrease the area of the fluid film. This film exerts a force at its perimeter as described by the surface tension coefficient, σ. This coefficient is defined as the surface-tension force exerted on the perimeter per unit of peripheral length. As the area of the film is increased by moving the slide wire a distance dx, the force opposing this movement is σl. The work that must be done on the film for this movement is therefore

$$\delta W = -\sigma l dx$$
$$= -\sigma dA$$

This stretching of the film is reversible because the film would produce as much work when we decrease the area by dA as it consumes when we increase the area by dA.

The work required to stretch from area A_1 to A_2 is given by

$$_1W_2 = -\int_{A_1}^{A_2} \sigma dA$$

By analogy, the generalized force and displacement for surface tension work are σ and A, respectively.

FIGURE 4.6

*Fluid Film Formed by
a Wire Frame*

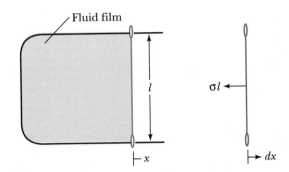

Observe that each reversible work mode involves a pair of system properties. This pair is P and V for a simple compressible substance, and σ and A for a stretching fluid film. Similar pairs can be determined for other work modes.

The generalized displacement is the only freely variable property for each work mode; that is, we can only control the volume of a simple compressible substance to produce work. The pressure will react to this volume change and cannot be controlled by using work. But the pressure can be controlled as the volume changes by adding or removing heat. For example, consider what happens when we increase the volume of a fixed mass of an ideal gas. The pressure of this gas will also change according to the ideal gas equation of state. If we wish to control this process further such that the pressure does not change as the volume increases, we must add or remove heat to control the temperature of the gas in such a manner that the pressure remains constant. Thus, we need at least one property to describe each reversible work mode plus one additional property to describe the heat interaction. This conclusion forms the basis of the state postulate that follows.

4.3.1   Other Forms of Work and Power

Chapter 1 demonstrated that electrical power is a form of work regardless of the device to which this power is applied. Physics tells us that the electrical power is given by

$$\dot{W}_{electric} = i \Delta V$$

where i is the current flowing through the device, and ΔV is the voltage drop that appears at the terminals of the device. When the current is in amperes and the voltage drop is in volts, the electrical power will be in watts. Over a period of time, the work generated by applying an electrical current is given by

$$W_{electric} = \int_{t_1}^{t_2} i(t)\Delta V(t)\, dt$$

If the current and voltage drop do not vary as time passes, this integrates to

$$W_{electric} = i\Delta V(t_2 - t_1)$$

Electrical power is independent of the substance used in the system and can occur simultaneously with and independent of compression and expansion work.

Work is also required to strain an elastic solid or spring with a force. The work required to stretch or compress an elastic spring is given by

$$W_{spring} = \frac{k}{2}(x_2^2 - x_1^2)$$

where k is the spring constant and the x's are the lengths of the spring at the beginning and end of the stretching as measured with respect to the length of the spring without force acting on it. In the case of an elastic solid undergoing a strain, the strain work is given by

$$W_{strain} = \frac{V_o E}{2}(\epsilon_2^2 - \epsilon_1^2)$$

where V_o is the original volume of the solid, E is Young's isothermal modulus, and ϵ is the strain at the beginning and ending of the process.

Work is also required to move electrical charges through electrostatic fields, magnetized entities through magnet fields, and polarized entities through electric fields. Each may be determined by applying the definition of mechanical work to the appropriate forces and displacements. Their development is left to the specialized treatises devoted to these topics.

4.4 State Postulate

Additional properties are needed to describe a system state whenever there are additional reversible modes of work. The number of independent properties needed to describe the state of a system is determined by the following.

> **DEFINITION: *State Postulate***
>
> The number of independent properties required to fix the state of a system is equal to the number of *relevant reversible work modes plus one.*

This definition is based on experimental evidence. Hence, for simple compressible substances, the only relevant work mode is compression and expansion work, and two properties are needed to fix the state of the working substance. If surface tension is the only relevant work mode, we again only need two independent properties to fix the state of the fluid film. When the surface tension is relevant and compression and expansion work is also relevant, we will need three independent properties to fix the state of the system.

Notice that we only need to concern ourselves with the relevant work modes. Surface tension is present when a simple compressible substance such as a liquid is compressed or expanded, but the work associated with the surface tension is quite small (in most cases) compared to the compression and expansion work. In this case, the surface tension work is negligible, and we can treat the liquid as a simple compressible substance.

The state postulate only determines the number of independent properties needed to describe the state of a substance. When properties become interdependent, as when a substance is undergoing a phase change, only one interdependent property may be used to describe the state. The remainder of the properties needed to specify the state must then be taken from those properties that are not interrelated in some way.

4.5 The First Law and Closed Systems

In Chapter 2 we presented the first law of thermodynamics for closed systems that operate in a cyclical manner:

$$\oint \delta Q = \oint \delta W$$

To transform this form of the first law into one that is useful for closed systems undergoing a process rather than a cycle, consider the closed system undergoing the two cycles shown in Figure 4.7. The processes constituting these cycles are general and may be reversible or irreversible.

FIGURE 4.7
Closed System Undergoing Two Different Cycles

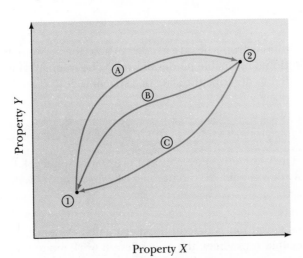

Two different cycles that the closed system can undergo are shown in the figure. In the first cycle, the system starts at state 1 and follows process A until it arrives at state 2. It then returns back to state 1 by following process B. Applying the first law to this cycle gives

$$\oint (\delta Q - \delta W) = (_1 Q_2 - {}_1 W_2)_A + (_2 Q_1 - {}_2 W_1)_B = 0$$

The first portion of the second cycle (1 – 2 along A and 2 – 1 along C) shown in the figure is identical to that of the first cycle; that is, the system starts at state 1 and progresses along process A until it reaches state 2. The second cycle differs from the first cycle in that it uses process C to return from state 2 to state 1 rather than process B. Application of the first law to the second cycle gives

$$\oint (\delta Q - \delta W) = (_1 Q_2 - {}_1 W_2)_A + (_2 Q_1 - {}_2 W_1)_C = 0$$

Equating these two results and canceling the common term involving process A yields

$$(_2 Q_1 - {}_2 W_1)_B = (_2 Q_1 - {}_2 W_1)_C$$

The quantity $Q - W$ is thereby independent of the process by which we return from state 2 to state 1, although both Q and W depend on the process used to go from state 2 to state 1. This is true whether the process between states 2 and 1 is reversible or irreversible; therefore, the quantity $Q - W$ can only depend on the beginning and ending states. Any quantity that is only determined by the states of the system is a property of the system. Hence, the quantity $Q - W$ is a property, and the quantity $_1 Q_2 - {}_1 W_2$ is the change in this property as the system follows any process from state 1 to state 2. This property is known as the *energy* of the system. We will denote the total energy with the symbol E and the specific energy with the symbol e.

With the introduction of the energy property, the first law as applied to a closed system undergoing a general process that starts at state 1 and ends at state 2 is as follows.

PRINCIPLE: The First Law for Closed Systems

$$_1 Q_2 - {}_1 W_2 = E_2 - E_1$$

Because a closed system is by definition a system of fixed mass, this result can also be written as

$$_1 q_2 - {}_1 w_2 = e_2 - e_1$$

where all quantities are now on a specific basis. We have not had to qualify the nature of the processes A, B, or C used to develop this principle. Consequently, the first law can be applied to all processes. Even irreversible processes may be analyzed by the first law.

EXAMPLE 4.3

Thirty kilojoules of heat are added to two kilograms of steam contained in a rigid, closed vessel. What is the change in the total and the specific energy of the steam as a result of this heating?

System: Two kilograms of steam contained in a closed vessel.

Basic Equations:
$$_1Q_2 - {_1W_2} = E_2 - E_1$$

Conditions: None.

Solution: Because the vessel is rigid, it cannot change its volume and produce compression and expansion work. Hence,

$$E_2 - E_1 = {_1Q_2} = 30 \text{ kJ}$$

The specific energy change is given by:

$$e_2 - e_1 = \frac{E_2 - E_1}{m} = 15 \text{ kJ/kg}$$

The effect of the heating is to increase the total and specific energy of the steam.

4.6 Forms of Energy

The first law provides an operational definition of energy, $Q - W$, but it gives us no clue about the forms of energy. Energy comes in many forms such as kinetic energy, potential energy, thermal energy, chemical energy, nuclear energy, and solar energy. To develop more useful expressions, we must apply this operational definition to closed systems that are allowed to change one form of energy at a time. For the purposes of this book, we will restrict ourselves to the mechanical forms of energy—potential, kinetic, and thermal energy.

4.6.1 Potential Energy

An expression for the potential energy may be obtained by considering an adiabatic, closed system that is only allowed to change its elevation in some gravitational field, g. No other property except the elevation is permitted to change as the system changes its state. The weight force produced by the body is mg, which acts downward as the body moves upward a small distance dz. According to our expression for work,

$$_1W_2 = \int_{z_1}^{z_2} \mathbf{F} \cdot d\mathbf{z} = -\int_{z_1}^{z_2} mg\, dz$$

The work required to move the system upward can then be determined when we know how g varies with elevation. When g does not vary with elevation, this integral gives

$$_1W_2 = -mg(z_2 - z_1)$$

By recalling that this system is adiabatic, and by substituting the preceding expression for the work into the first law, we obtain

$$mg(z_2 - z_1) = E_{p,2} - E_{p,1}$$

We have added the additional subscript p to remind ourselves that we are only considering the potential energy. The total potential energy is then

$$E_p = mgz$$

which is determined by only one property of the system: the elevation z. Similarly, the specific potential energy is

$$e_p = gz$$

The gravitational units conversion constant, g_c, must be introduced into these equations when the USCS system of units is employed.

4.6.2   Kinetic Energy

An adiabatic, closed system that is only allowed to change its velocity may be used to develop an expression for the kinetic energy. As this system experiences a small increase in velocity, dV, the system reacts with the inertial force, mdV/dt, which opposes the change in velocity. During this small change in velocity, the system will have moved forward a distance $dx = Vdt$. The total work produced by this system as it accelerates from velocity V_1 to velocity V_2 is then, assuming that relativistic velocities are not approached,

$$_1W_2 = \int_{V_1}^{V_2} \mathbf{F} \cdot d\mathbf{x} = -\int_{V_1}^{V_2} mV\,dt\,\frac{dV}{dt} = -\int_{V_1}^{V_2} mV\,dV$$

Substituting this into the first law and performing the integration, while recalling that the system is adiabatic, gives

$$m\left(\frac{V_2^2}{2} - \frac{V_1^2}{2}\right) = E_{k,2} - E_{k,1}$$

We have added the k subscript to denote kinetic energy only. The total kinetic energy is then

$$E_k = m\frac{V^2}{2}$$

and the specific kinetic energy is

$$e_k = \frac{V^2}{2}$$

Both expressions for the kinetic energy are determined by only the velocity property of the system. As before, the gravitational units conversion constant, g_c, must be included in these expressions when the USCS unit system is used.

4.6.3   Thermal Energy

The nature of the thermal energy can be determined by considering a closed system that experiences no work interactions, velocity changes, or elevation changes. As heat is added to this system, the energy of the system will certainly increase according to the first law. The only way this can happen under the stated constraints is by increasing the velocity, rotational speed, and vibration states, independently or all together, and the quantum state of the electrons of the molecules or atoms inside the system. As discussed in Chapter 3, the macroscopic effect of these changes is termed the *internal energy*.

The total thermal energy is then

$$E_t = mu = U$$

while the specific thermal energy is

$$e_t = u$$

Specific internal energy is a function of the state of the substance in the system. By the state postulate, we must know two independent properties of the system before we can determine the internal (thermal) energy of the system.

If we consider only mechanical forms of energy and substitute the preceding expressions into the first law, it becomes, on a specific basis:

$$_1q_2 - _1w_2 = (u_2 - u_1) + \left(\frac{V_2^2}{2} - \frac{V_1^2}{2}\right) + g(z_2 - z_1) \qquad (4.2)$$

This is the most commonly used form of the first law for closed systems because it relates the heat addition and work production to the observed properties of the system. Because it will be used many times, you should commit it to memory.

EXAMPLE 4.4

A rigid, closed vessel contains two pounds-mass of steam initially at 100 psia, 350° F. It is cooled to 200° F. While the vessel is being cooled, it is deaccelerated from a velocity of 100 ft/s to rest and dropped 100 ft in a location where the gravitational attraction is 32.0 ft/s². Determine the amount of heat transferred to the vessel.

System: Two pounds-mass of steam in a closed vessel.

Basic Equations:
$$_1q_2 - {_1}w_2 = (u_2 - u_1) + \left(\frac{V_2^2}{2} - \frac{V_1^2}{2}\right) + g(z_2 - z_1)$$

327.9	4.434	1105.8	1187.8	1.6034
350	**4.592**	**1115.4**	1200.4	1.6191
400	4.934	1136.2	1227.5	1.6517
450	5.265	1156.2	1253.2	1.6812
500	5.587	1175.7	1279.1	1.7085

160	4.745	0.01640	77.2	127.94	1062.3	127.96	1002.2	1130.1	0.2313	1.8484
170	5.996	0.01645	62.0	137.95	1065.4	137.97	996.2	1134.2	0.2473	1.8293
180	7.515	0.01651	50.2	147.97	1068.3	147.99	990.2	1138.2	0.2631	1.8109
190	9.343	0.01657	41.0	158.00	1071.3	158.03	984.1	1142.1	0.2787	1.7932
200	11.529	**0.01663**	**33.6**	**168.04**	**1074.2**	168.07	977.9	1145.9	0.2940	1.7762

Conditions: None.

Solution: Because the vessel is rigid, no compression or expansion work takes place, so $_1w_2 = 0$. Noting that the specific volume at the final state is the same as at the initial state ($v_1 = v_2 = 4.592$ ft³/lb$_m$) as shown in Figure 4.8, and using the steam tables, the quality at the second state is

$$x_2 = \frac{v_2 - v_{f,2}}{v_{g,2} - v_{f,2}}$$

$$= \frac{4.592 - 0.01663}{33.6 - 0.01663}$$

$$= 0.136$$

and the specific internal energy at the second state is

$$u_2 = u_{f,2} + x_2 u_{fg,2}$$
$$= 168.04 + 0.136(1074.2 - 168.04)$$
$$= 291.3 \text{ Btu/lb}_m$$

The change in specific internal energy is

$$\Delta u = u_2 - u_1$$
$$= 291.3 - 1115.4$$
$$= -824.1 \text{ Btu/lb}_m$$

The specific kinetic energy change of this system is

$$\Delta e_k = \frac{V_2^2 - V_1^2}{2g_c J}$$

$$= \frac{0^2 - 100^2}{2 \times 32.2 \times 778} \left[\frac{\text{ft}^2}{\text{s}^2} \frac{\text{lb}_f\text{-s}^2}{\text{ft-lb}_m} \frac{\text{Btu}}{\text{ft-lb}_f} \right]$$

$$= -0.200 \text{ Btu/lb}_m$$

where Joule's conversion constant, $J = 778$ ft-lb$_f$/Btu, has been introduced. The specific potential energy change is

$$\Delta e_p = \frac{g}{g_c J}(z_2 - z_1)$$

$$= \frac{32.0}{32.2 \times 778}(0 - 100) \left[\frac{\text{ft}}{\text{s}^2} \frac{\text{lb}_f\text{-s}^2}{\text{ft-lb}_m} \frac{\text{Btu}}{\text{ft-lb}_f} \right]$$

$$= -0.128 \text{ Btu/lb}_m$$

TIP:

Notice that the kinetic and potential energy changes are quite small in comparison to the internal energy change. This is often the case, and so the kinetic and potential energy terms are frequently neglected.

Substituting these results into the first law gives

$$_1 q_2 = -824.4 \text{ Btu/lb}_m$$

The total heat transfer is

$$_1 Q_2 = m_1 q_2 = -1648.8 \text{ Btu}$$

FIGURE 4.8
P–v and T–s
State Diagrams
for Example 4.4

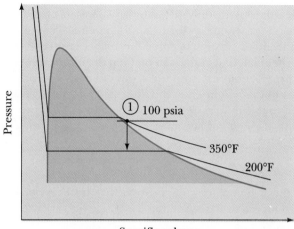

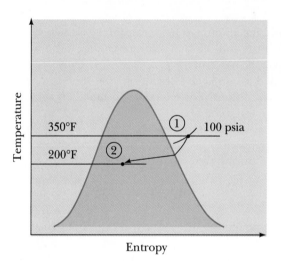

Heat and Specific Heats

In Chapter 1, we defined *heat* without much clarification. Now that we have introduced the first law, we can gain a better understanding of heat.

We begin by considering a stationary, fixed-volume system undergoing a reversible (quasi-equilibrium) process. Because the system is stationary, the kinetic and potential energy cannot change. Also, the system cannot do compression and expansion work because its volume is constant (note that this statement is only true because we are considering a reversible process). For this system, the first law tells us that

$$_1Q_2 = m(u_2 - u_1)$$

Thus, heat is the only means by which we can change the internal energy of this system (other energy forms such as chemical energy are not present). If heat is added to this system by some thermal-energy reservoir that is at a temperature higher than that of the system, then the internal energy will increase. Similarly, if heat is removed from the system to a reservoir at a temperature lower than that of the system, then the internal energy will decrease.

In the limit, as we allow the final state of this process to approach the initial state, the first law becomes

$$\delta q = du$$

If the system consists of a substance whose internal energy is only a function of temperature (see Section 3.12), the first law can be written as

$$\delta q = c_v dT$$

In words, the amount of heat transferred is the product of the constant volume specific heat and the temperature change. The word *heat* was historically included in the name of the property c_v as a consequence of this result. Although it is called a *specific heat*, it can only be used to calculate heat transfer under special conditions. For selected substances, it can always be used to calculate internal energy changes.

Reconsider the same system when it undergoes a reversible, isobaric (i.e., constant-pressure) process rather than a constant-volume process. The work produced during the isobaric process is

$$_1W_2 = P(V_2 - V_1) = mP(v_2 - v_1)$$

Substituting this into the first law and recalling the definition of enthalpy yields

$$_1Q_2 = m(u_2 - u_1) + mP(v_2 - v_1) = m(h_2 - h_1)$$

The heat transfer is then directly proportional to the change in the enthalpy. In the limit, as the final state approaches the initial state along the isobaric process, this result becomes

$$\delta q = dh$$

This reduces to

$$\delta q = c_p dT$$

if the enthalpy of the working fluid is a function of temperature only. As with c_v, c_p can be used to calculate heat transfer under special conditions. In general, it can always be used to calculate changes in enthalpy for selected substances.

EXAMPLE 4.5	Air at 100 kPa, 2° C occupies a 10-liter piston-cylinder device that is arranged to maintain constant air pressure. This device is now heated until its volume is 20 liters. Determine the heat transfer to the air and work produced by the air.

System: The air in the piston cylinder.

Basic Equations:

$$_1q_2 - {_1}w_2 = u_2 - u_1$$

$$_1w_2 = \int_1^2 p\,dv$$

$$PV = mRT$$

$$\Delta u = c_v \Delta T$$

Conditions: We will treat the air as an ideal gas with constant specific heats, presume that the process is reversible, and neglect the system's potential and kinetic energies.

Solution: Applying the equation of state to the initial state gives

$$m = \frac{P_1 V_1}{RT_1}$$

$$= \frac{100 \times 10/1000}{8.314 \times 300/28.84} \quad \left[\frac{\frac{kN}{m^2}\,m^3}{\frac{kN-m}{kg-K}K} \right]$$

$$= 0.0116 \text{ kg}$$

Because the initial mass and pressure are the same as those at the final state,

$$T_2 = T_1 \frac{V_2}{V_1}$$

$$= 300\frac{20}{10} = 600 \text{ K}$$

according to the equation of state. For a reversible, isobaric process, the work is

$$_1W_2 = \int_{V_1}^{V_2} P\,dV$$

$$= P(V_2 - V_1)$$

$$= 100\frac{20 - 10}{1000} \quad \left[\frac{kN}{m^2}\,m^3\,\frac{1}{1} \right]$$

$$= 1 \text{ kJ}$$

This process is shown in Figure 4.9. Applying the first law to the process gives

$$\begin{aligned} {}_1Q_2 &= m(u_2 - u_1) + {}_1W_2 \\ &= mc_v(T_2 - T_1) + {}_1W_2 \\ &= 0.0115 \times 0.7165(600 - 300) + 1 \\ &= 3.472 \text{ kJ} \end{aligned}$$

FIGURE 4.9

P–v and T–s State Diagrams for Example 4.5

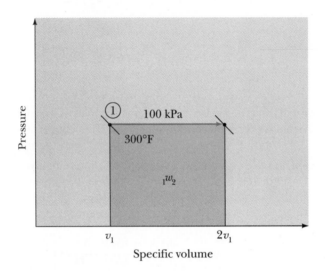

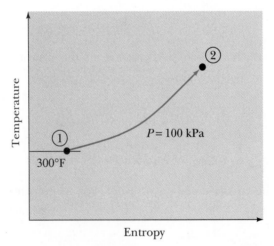

TIP:

The total heat transfer also could have been calculated using

$${}_1Q_2 = mc_p(T_2 - T_1)$$

because the pressure remains constant.

If the system's working fluid is an incompressible substance and the process is reversible, then $u = h$ and $W = 0$. The first law is then

$$\delta q = du = c\, dT$$

This form of the first law serves as the basis for defining the British thermal unit (Btu) as the amount of heat required to raise the temperature of 1 pound-mass of water from 59.5° F to 60.5° F. As a result of this definition, the USCS specific heat of liquid water at 60° F is 1 Btu/lb$_m$-° R.

4.8 Heat and Entropy

At this time, the only means we have for calculating heat transfer is the first law. Additional methods for calculating the heat transfer are required in many situations. For example, consider a rigid container equipped with a stirring device and heat source that is filled with an ideal gas. When work is done on the gas in this container with the stirring device and heat is added to the gas, the gas's temperature will increase. We can measure this temperature change with a thermometer. The first law as applied to this system and process is

$$_1q_2 - {}_1w_2 = u_2 - u_1 = c_v(T_2 - T_1)$$

The right-hand side of this result is determined by the measured temperatures, but we must know one of the two quantities on the left-hand side to complete the solution. Stirring, which involves friction, is an irreversible means of adding work to a system and consequently cannot be evaluated with

$$_1w_2 = \int_1^2 P\,dv$$

Hence, we must either find some means of measuring the work or an independent means of calculating the heat transfer. The entropy property provides us with this second alternative in many, but not all, situations.

The definition of *entropy* (see Section 2.8) tells us that heat transfer will change the entropy of the system. For a reversible process, entropy change according to this definition is

$$S_2 - S_1 = \int_1^2 \frac{\delta Q}{T} \quad \text{(reversible processes only)}$$

This can be integrated when we know the relationship between the heat transfer to the system and the system's temperature. The definition can be differentiated and rearranged to give

$$\delta Q = T\,dS \quad \text{(reversible processes only)}$$

When integrated over a reversible process, this gives:

$$_1Q_2 = \int_1^2 T\,dS \quad \text{(reversible process only)} \qquad (4.3)$$

This result tells us that the heat transfer for a *reversible* process is the area under the process path when plotted on a temperature–entropy state diagram such as that shown in Figure 4.10. This is a very important result because it gives us another means of calculating the system heat transfer besides the first law. On a specific basis, this result becomes

$$_1q_2 = \int_1^2 T\,ds \quad \text{(reversible process only)}$$

FIGURE 4.10
A Reversible Process on a Temperature–Entropy State Diagram

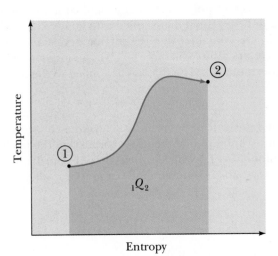

EXAMPLE 4.6

One kilogram of steam is converted from 100 kPa, 320° C to 500 kPa as it undergoes an isothermal, reversible process in a closed system. Determine the heat transfer of this process.

System: One kilogram of steam.

Basic Equations:

$$_1q_2 = \int_1^2 T\,ds \quad \text{(reversible processes only)}$$

240	2.359	2718.5	2954.5	7.9949
280	2.546	2779.6	3034.2	8.1445
320	2.732	2841.5	3114.6	**8.2849**
360	2.917	2904.2	3195.9	8.4175
400	3.103	2967.9	3278.2	8.5435

240	0.4646	2707.6	2939.9	7.2307
280	0.5034	2771.2	3022.9	7.3865
320	0.5416	2834.7	3105.6	**7.5308**
360	0.5796	2898.7	3188.4	7.6660
400	0.6173	2963.2	3271.9	7.7938

Conditions: The process is reversible.

Solution: Because the process is isothermal, the integral may be evaluated as

$$_1Q_2 = mT(s_2 - s_1)$$
$$= 1(320 + 273)(7.5308 - 8.2849) \quad \left[\text{kg K} \frac{\text{kJ}}{\text{kg-K}} \right]$$
$$= -447.2 \text{ kJ}$$

This process is illustrated in Figure 4.11.

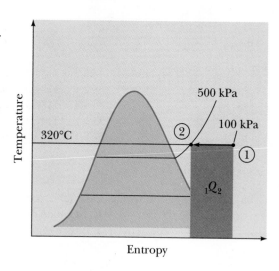

FIGURE 4.11
T–s State Diagram for Example 4.6

TIP:

Absolute thermodynamic temperatures are mandatory for making calculations with entropy.

The adiabatic, reversible process represents many practical processes including the compression and expansion processes in an automobile engine, the compression process in an air-conditioning compressor, and the compression process in an air compressor. According to the preceding results, this is also a constant-entropy process. This process is so important that it is given its own special name.

DEFINITION: *Isentropic Process*

A process in which the entropy does not change (i.e., a reversible, adiabatic process).

Although we have introduced the isentropic process in connection with closed systems, it also can be applied to other systems, as we will see in future discussions.

EXAMPLE 4.7

One pound-mass of steam is expanded isentropically from 200 psia, 400° F to 10 psia in a piston-cylinder device. Determine the work produced by this steam as it is expanded.

System: One pound-mass of steam in a piston-cylinder device.

Basic Equations:
$$_1q_2 - {_1}w_2 = u_2 - u_1$$

381.9	2.289	1114.6	1199.3	1.5464
400	2.361	**1123.5**	1210.8	**1.5600**
450	2.548	1146.4	1240.7	1.5938
500	2.724	1168.0	1268.8	1.6239
550	2.893	1188.7	1295.7	1.6512

6.0	170.03	0.01645	61.98	137.98	1065.4	138.00	996.2	1134.2	0.2474	1.8292
7.0	176.82	0.01649	53.65	144.78	1067.4	144.80	992.1	1136.9	0.2581	1.8167
8.0	182.84	0.01653	47.35	150.81	1069.2	150.84	988.4	1139.3	0.2675	1.8058
9.0	188.26	0.01656	42.41	156.25	1070.8	156.27	985.1	1141.4	0.2760	1.7963
10.0	193.19	0.01659	38.42	**161.20**	**1072.2**	161.23	982.1	1143.3	**0.2836**	**1.7877**

Conditions: The process is isentropic.

Solution: According to Table B1.3, the initial internal energy is 1123.5 Btu/lb$_m$ and the initial entropy is 1.5600 Btu/lb$_m$-° R. Because the final entropy equals the initial entropy, the final state lies in the mixture region as shown in Figure 4.12. Thus,

$$x_2 = \frac{s_2 - s_{f,2}}{s_{fg,2}}$$

$$= \frac{1.5600 - 0.2836}{1.5041}$$

$$= 0.8486$$

and the final internal energy is

$$u_2 = u_{f,2} + x_2 u_{fg,2}$$
$$= 161.2 + 0.8486 \times 911.0$$
$$= 934.3 \text{ Btu/lb}_m$$

Because the process is isentropic, there is no heat transfer, and the first law becomes

$$_1W_2 = m(u_1 - u_2)$$

$$= 1(1123.5 - 934.3) \quad \left[\text{lb}_m \frac{\text{Btu}}{\text{lb}_m} \right]$$

$$= 189.2 \text{ Btu}$$

FIGURE 4.12
T–s State Diagram for Example 4.7

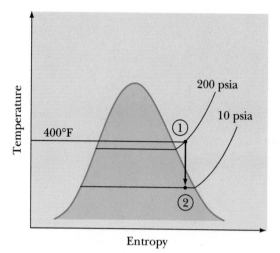

TIP:

If this example confused you, then review the use of the steam tables in Chapter 3.

4.9 Calculating Entropy Changes

Entropy is an important property for calculating heat transfer for closed and other systems, so we need to develop methods for calculating the entropy change of substances for which we do not have property tables such as the steam and refrigerant tables of Appendix B. These techniques also can be used to develop property tables for new substances.

The first technique is to simply take advantage of the definition of the specific entropy:

$$s_2 - s_1 = \int_1^2 \frac{\delta q}{T} \quad \text{(reversible processes only)}$$

To evaluate the difference between the entropies of two arbitrary states of an arbitrary substance, we need to connect these two states with one or more reversible processes and then perform the above integration over the connecting processes. The reversible, isobaric, isothermal, isentropic, and constant specific-volume processes are typically used for this purpose. Figure 4.13 illustrates the use of the isobaric and constant specific-volume reversible processes to connect two arbitrary states on a *T–s* state diagram. Notice that this produces an intermediate state denoted by *a* in the figure whose pressure is the same as that of state 2 and whose specific volume is the same as that of state 1.

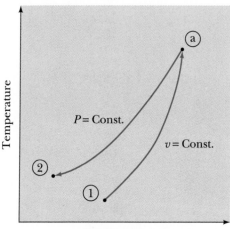

We can now use equations of state, the first law, and other known relationships to evaluate the entropy-defining integral, first between states 1 and a and then between states a and 2. The net entropy change as we progress between states 1 and 2 is then

$$s_2 - s_1 = (s_2 - s_a) + (s_a - s_1)$$

For the $1 - a$ process, the work is $\delta w = 0$ because the volume does not change. When this is substituted into the first law,

$$\delta q = du$$

Substituting this into the entropy-defining integral gives

$$s_a - s_1 = \int_1^a \frac{du}{T}$$

This can be integrated when the $u = u(T)$ relationship is known for the substance. Similarly, the first law for the $a - 2$ process is

$$\delta q = du + p\,dv = dh$$

When substituted into the entropy-defining integral, this gives

$$s_2 - s_a = \int_a^2 \frac{dh}{T}$$

This may be integrated when the $h = h(T)$ relationship is known for the substance. The net entropy change between states 1 and 2 is then

$$s_2 - s_1 = \int_1^a \frac{du}{T} + \int_a^2 \frac{dh}{T}$$

EXAMPLE 4.8

Air is compressed from 16 psia, 80° F to 100 psia, 400° F in a reciprocating piston compressor. Calculate the change in the specific entropy of the air caused by this compression.

System: An arbitrary quantity of air.

Basic Equations:

$$\Delta s = \int_1^2 \frac{\delta q}{T} \quad \text{(reversible processes only)}$$

$$\delta q - \delta w = du$$

$$Pv = RT$$

$$du = c_v \, dT$$

$$dh = c_P \, dT$$

Conditions: States 1 (16 psia, 80° F) and 2 (100 psia, 400° F) are as specified. The air will be treated as an ideal gas with constant specific heats.

Solution: According to the preceding discussion, the entropy change is

$$s_2 - s_1 = \int_1^a c_v \frac{dT}{T} + \int_a^2 c_P \frac{dT}{T}$$

$$= c_v \ln \frac{T_a}{T_1} + c_P \ln \frac{T_2}{T_a}$$

The equation of state can now be used to fix T_a. Because the specific volume at a is the same as that at 1,

$$T_a = \frac{P_a}{P_1} T_1$$

But $P_a = P_2$. Hence,

$$T_a = \frac{P_2}{P_1} T_1$$

$$= \frac{100}{16} 540$$

$$= 3375° \, \text{R}$$

Substituting the temperatures into the entropy change expression and using the specific heats of Table 3.5 gives us

$$s_2 - s_1 = 0.1711 \ln \frac{3375}{540} + 0.2397 \ln \frac{860}{3375}$$

$$= -0.0142 \; \text{Btu/lb}_m \cdot° \, \text{R}$$

Let us now consider a general differential process which is reversible. By combining the first law with the expression for work and neglecting the kinetic and potential energy changes, we get

$$\delta q = du + P\,dv$$

When this is substituted into the expression for the differential entropy change, $ds = \delta q/T$, we get

$$ds = \frac{du}{T} + \frac{P\,dv}{T}$$

This can be rearranged to read

$$T\,ds = du + P\,dv$$

Taking the derivative of the definition of the enthalpy yields

$$dh = du + P\,dv + v\,dP$$

By using this result, the preceding $T\,ds$ expression can also be written

$$T\,ds = dh - v\,dP$$

These two results are known as the *Gibbs equations*, named for J. Willard Gibbs,[2] who originally derived them:

$$
\begin{aligned}
T\,ds &= du + p\,dv \\
T\,ds &= dh - v\,dp
\end{aligned}
\tag{4.4}
$$

These equations are very important because they relate the entropy to the pressure, temperature, specific volume, internal energy, and enthalpy of the working fluid. Although we have presumed a reversible process and a closed system to obtain these results, the final results only depend on the properties of the working fluid. Consequently, the final expressions are independent of the process and system type, and they can be applied to any situation to evaluate the change in the entropy property.

---

[2]Josiah Willard Gibbs (1839–1903) was an American mathematical physicist who advanced the application of thermodynamics to chemistry. While at Yale University, he developed what are now known as the *Gibbs equations* and the *Gibbs free energy*. He was known as the first physicist to apply the second law of thermodynamics to the exhaustive discussion of the relation between chemical, electrical, and thermal energy and capacity for external work.

EXAMPLE 4.9

A weighted piston-cylinder device filled with air is arranged to maintain the air pressure at 16 psia. Calculate the change in the specific entropy of the air as it is heated from 80° F to 400° F in this device.

System: An arbitrary quantity of air.

Basic Equations:
$$T\,ds = dh - v\,dp$$
$$dh = c_p\,dT$$

Conditions: The air will be treated as an ideal gas with constant specific heats.

Solution: Because there is no change in the pressure of the air, the second Gibbs equation is the easiest to use because $v\,dP = 0$. Then

TIP:

The same answer would have been obtained if we had used the du form of the Gibbs equations or the definition of entropy. Of the three choices for evaluating entropy changes, the choice that minimizes the computational effort is the best one.

$$s_2 - s_1 = \int_1^2 \frac{dh}{T}$$

$$= c_P \int_{T_1}^{T_2} \frac{dT}{T}$$

$$= c_p \ln \frac{T_2}{T_1}$$

$$= 0.2397 \ln \frac{860}{540}$$

$$= 0.1115 \ \text{Btu/lb}_m\text{-}^\circ R$$

When the Gibbs relations are applied to an ideal gas with constant specific heats undergoing an isentropic process, some very useful results are obtained. In this case, $ds = 0$ because the entropy does not change. Also, $du = c_v\,dT$ and $dh = c_p\,dT$. With these expressions, the Gibbs relations reduce to

$$c_v\,dT = -P\,dv$$
$$c_p\,dT = v\,dP$$

Forming the ratio of these two results gives

$$\frac{dP}{P} = -\frac{c_P}{c_v}\frac{dv}{v}$$

When integrated from an initial state to a final state, this produces

$$P_1 v_1^k = P_2 v_2^k$$

where k is the ratio of the specific heats, c_p/c_v. The product of the pressure and specific volume raised to the k power then remains constant as an ideal gas with constant specific heats undergoes an isentropic process. By combining this result with the ideal gas equation of state, we get

$$\frac{T_2}{T_1} = \left(\frac{v_1}{v_2}\right)^{k-1} = \left(\frac{P_2}{P_1}\right)^{(k-1)/k}$$

The isentropic process is a special case of a more general process known as the *polytropic process*. The polytropic process is one during which Pv^n (where n is any number) remains constant throughout the process. Along this process, an arbitrary state 2 is related to the initial state 1 by

$$P_2 v_2^n = P_1 v_1^n$$

When the substance undergoing a polytropic process is an ideal gas,

$$\frac{T_2}{T_1} = \left(\frac{v_1}{v_2}\right)^{n-1} = \left(\frac{P_2}{P_1}\right)^{(n-1)/k}$$

Several ideal gas processes are represented by the polytropic process. If $n = k$, then the polytropic process is the ideal gas isentropic process. Similarly, $n = 1$ reduces the polytropic process to the ideal gas isothermal process, $n = 0$ becomes the isobaric process, and $n = \infty$ becomes the constant-volume process.

4.10 The Second Law and Closed Systems

As developed in Chapter 2, the *increase in entropy principle*

$$\Delta S_{everything} \geq 0$$

applies to closed systems as well as cyclical systems. Hence, when we calculate the entropy change of everything involved with a closed system as it undergoes a process or cycle and sum these changes, the net result must be positive or zero. The net sum of the entropy changes will be zero when all system processes and interactions are reversible. It will be positive when irreversibilities are present.

Consider a closed system that undergoes a process (say from state 1 to state 2) while exchanging heat and work with its surroundings. The net entropy change of everything is then

$$\Delta S_\infty + S_2 - S_1 \geq 0$$

Only a heat exchange between the system and surroundings will change the entropy of the surroundings according to the definition of entropy. If this system is adiabatic,

$$\Delta S_\infty = 0$$

and the entropy at the end of the process must be greater than or equal to that at the beginning. As a result of this finding, the second law is sometimes stated as *the entropy of an adiabatic system must always increase or remain fixed.*

The equality form of this principle, which includes the entropy generation, is also useful. In this form, the second law for a closed system undergoing a process is

$$\Delta S_\infty + S_2 - S_1 - S_{gen} = 0$$

where the entropy generation, S_{gen}, is always positive or zero. The entropy generation serves as a measure of the irreversibility of a system and its interactions with the surroundings. If this is zero, there are no irreversibilities present. As it becomes more positive, the irreversibilities become larger.

4.11 Forbidden States

The question often arises that if we start at some initial state 1, can a closed system proceed by some process to any arbitrary final state 2? The first law tells us that, for a closed system, any state is possible if the final internal energy satisfies

$$u_2 = u_1 + {}_1q_2 - {}_1w_2$$

In addition to the first law, the system must also satisfy the second law, which places additional constraints on the final state.

According to the *increase of entropy principle* form of the second law, as the closed system undergoes any process,

$$\Delta S_{everything} \geq 0$$

The entropy change of everything involved with the process includes the entropy change of the system itself ($S_2 - S_1$) as well as the entropy change of the system surroundings (ΔS_∞). The increase in entropy principle is then

$$S_2 - S_1 + \Delta S_\infty \geq 0$$

When rearranged, this becomes

$$S_2 \geq S_1 - \Delta S_\infty$$

In addition to satisfying the first law, the entropy of the final state must be greater than or equal to the entropy of the initial state minus the change in entropy of the surrounding components that interact with the system.

States whose entropy is less than $S_1 - \Delta S_\infty$ are known as *forbidden states* because the system cannot progress to them from state 1 irregardless of the process executed. This is true if the process is reversible or irreversible. An example of all of the forbidden states for a given initial state is shown in Figure 4.14. All those states lying in the shaded area of this T–S state diagram can never be reached by the system by any process that begins at state 1 and changes the entropy of the surroundings by ΔS_∞.

FIGURE 4.14
*T–S State Diagram
Showing the
Forbidden States*

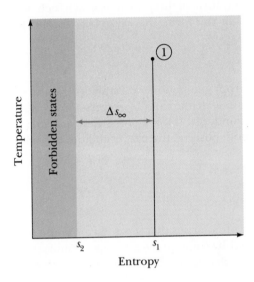

A special case is that of an adiabatic closed system. Because no heat is added to or removed from this system by any energy reservoir in the surroundings, $\Delta S_\infty = 0$ because a heat transfer is required to change the entropy of a thermal-energy reservoir. The increase of entropy principle then tells us that

$$S_2 \geq S_1$$

All states lying to the left of state 1 on a $T–S$ diagram are forbidden to the system. Only the states with the same entropy or to the right of state 1 on the $T–S$ state diagram are possible for state 2.

EXAMPLE 4.10

Determine the minimum pressure that a closed system consisting of water at 500 kPa, 200° C can achieve as it undergoes a process in which 100 kJ/kg of heat is removed from the system by a thermal-energy reservoir at 100° C.

System: Water in a closed system.

Basic Equations:

$$\Delta S_{everything} \geq 0$$

$$_1Q_2 - {_1}W_2 = m(u_1 - u_2)$$

$$\Delta S = \int_1^2 \frac{\delta Q}{T}$$

151.86	0.3749	2561.2	2748.7	6.8213
180	0.4045	2609.7	2812.0	6.9656
200	0.4249	**2642.9**	2855.4	**7.0592**
240	0.4646	2707.6	2939.9	7.2307
280	0.5034	2771.2	3022.9	7.3865

400	143.6	0.001084	0.4625	604.31	2553.6	604.74	2133.8	2738.6	1.7766	6.8959
450	147.9	0.001088	0.4140	**622.77**	**2557.6**	623.25	2120.7	2743.9	**1.8207**	**6.8565**
500	151.9	0.001093	0.3749	**639.68**	**2561.2**	640.23	2108.5	2748.7	**1.8607**	**6.8213**
600	158.9	0.001101	0.3157	669.90	2567.4	670.56	2086.3	2756.8	1.9312	6.7600
700	165.0	0.001108	0.2729	696.44	2572.5	697.22	2066.3	2763.5	1.9922	6.7080

Conditions: None.

Solution: The initial entropy of the water is 7.0592 kJ/kg-K. Because heat is being added to the thermal-energy reservoir, the entropy change of the reservoir per unit of system mass is

$$\Delta s_\infty = \frac{{}_1 q_2}{T}$$
$$= \frac{100}{373}$$
$$= 0.2681 \text{ kJ/kg-K}$$

The final specific entropy of the water can therefore not be less than 6.7911 kJ/kg-K. According to the first law, the final internal energy is

$$u_2 = u_1 + {}_1 q_2$$
$$= 2642.9 + (-100)$$
$$= 2542.9 \text{ kJ/kg}$$

Inspection of the T–S state diagram of Figure 4.15 shows that the lowest final pressure is achieved when $s_2 = 6.7911$ kJ/kg-K. Using this entropy and the final internal energy to interpolate in the steam tables, we find that the minimum final pressure is 480 kPa.

FIGURE 4.15
T–S State Diagram for Example 4.10

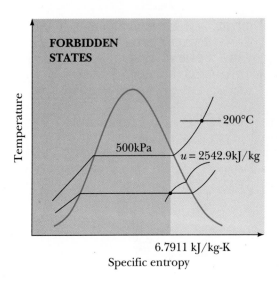

4.12 Process Efficiencies

Chapter 2 introduced thermodynamic efficiency as a means of evaluating the performance of cycles. When evaluating a process or a device that executes a process, we need a different type of efficiency. For this purpose, we define a process efficiency for a system that produces a result as follows.

DEFINITION: *Process Efficiency (η_p)*

$$\eta_p = \frac{\text{Result of actual process}}{\text{Result of an ideal process}} \quad \text{(producing system)}$$

Notice that this definition involves three things: a result, an actual process, and an ideal process. The result is the effect the process is intended to produce. It can be work production, increase of kinetic energy, heat rejection, or another effect. For example, an expansion process is often intended to produce work by reducing the pressure of the working fluid. The result is the work produced.

The actual process involves irreversible elements such as friction and mixing. These elements reduce the amount of the result produced in comparison to that which could have been produced had they not been present. The ideal process is then the same process without the irreversibilities. For example, the ideal counterpart of an adiabatic expansion is an adiabatic, reversible (i.e., isentropic) expansion. In this case, the process efficiency is the fraction of the work produced by an isentropic expansion between the same pressures.

Work is produced by an expansion process from a high pressure to a low pressure. The maximum work would be produced if the process is adiabatic, according to the first law. The ideal process is an adiabatic, reversible process between the initial high pressure and the final low pressure. Application of the first law to the actual adiabatic, work-producing expansion gives

$$w_{act} = u_1 - u_2$$

A reversible, adiabatic expansion would produce the amount of work

$$w_{ideal} = u_1 - u_{2,s}$$

where $u_{2,s}$ is the internal energy of the working fluid when its final entropy is the same as the initial entropy and its pressure is the same as the final pressure, as shown in Figure 4.16 for a phase-changing substance. The work-producing, adiabatic expansion efficiency is then

$$\eta_e = \frac{w_{act}}{w_s} = \frac{u_1 - u_2}{u_1 - u_{2,s}}$$

The process efficiency when the closed system consumes a quantity such as work or heat is defined as follows.

DEFINITION: *Process Efficiency* (η_c)

$$\eta_c = \frac{\text{Quantity consumed in an ideal process}}{\text{Actual quantity consumed}} \qquad \text{(Consuming system)}$$

FIGURE 4.16

T–S State Diagram Showing States for an Adiabatic Work-Producing Process

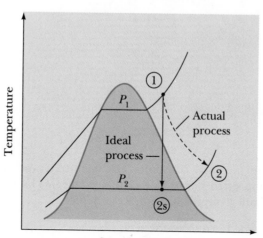

Notice that this process efficiency is the inverse of the process efficiency of a producing system. This is necessary because irreversibilities increase the amount of the quantity consumed as compared to that needed to accomplish the same effect with a reversible ideal process.

When a process consumes work during an adiabatic compression in a closed system from an initial low pressure to a final high pressure, the first law tells us that the work required is

$$-_1w_2 = u_2 - u_1$$

If this had been done reversibly while keeping the initial and final pressures the same, the work required would be

$$-_1w_{2,s} = u_{2,s} - u_1$$

where $u_{2,s}$ is the internal energy of the working fluid when its final entropy is the same as the initial entropy and its pressure is the same as the final state, as shown in Figure 4.17 for a single-phase substance. In this case, the work-consuming adiabatic compression efficiency is

$$\eta_c = \frac{-_1w_{2,s}}{-_1w_2} = \frac{u_{2,s} - u_1}{u_2 - u_1}$$

FIGURE 4.17
T–S State Diagram Showing States for an Adiabatic Work-Consuming Process

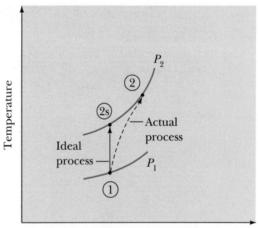

EXAMPLE 4.11

Determine the work consumed and the final temperature when air is adiabatically compressed from 14 psia, 70° F to 200 psia by a closed system device with an adiabatic compression efficiency of 85%.

System: Air contained in a closed system.

Basic Equations:

$$_1q_2 - _1w_2 = u_2 - u_1$$

$$pv^k = \text{constant}$$

$$k = \frac{c_P}{c_v}$$

Conditions: The air will be treated as an ideal gas with constant specific heats.

Solution: The ideal process is an isentropic process between P_1 and P_2. The ideal final temperature would then be

$$T_{2,s} = T_1 \left(\frac{P_2}{P_1} \right)^{(k-1)/k}$$

$$= 530 \left(\frac{200}{14} \right)^{0.4/1.4}$$

$$= 1133 \text{ K}$$

Because the process is adiabatic, the ideal work consumption according to the first law is

$$\begin{aligned}
_1w_2 &= u_1 - u_{2,s} \\
&= c_v(T_1 - T_{2,s}) \\
&= 0.1711(530 - 1133) \\
&= -103.2 \text{ Btu/lb}_m
\end{aligned}$$

TIP:

Because the entropy of the final state cannot be less than that of the initial state, the actual final temperature of an adiabatic compression will always be greater than or equal to the isentropic final temperature. Similarly, the actual final temperature in an adiabatic expansion also will always be greater than or equal to that of the isentropic expansion.

Applying the definition of the adiabatic compression efficiency gives

$$w_{act} = \frac{w_{ideal}}{\eta_c}$$

$$= \frac{-103.2}{0.85}$$

$$= -121.4 \text{ Btu/lb}_m$$

When the first law is applied to the actual process,

$$T_2 = T_1 - \frac{w_{act}}{c_v}$$

$$= 530 - \frac{-121.4}{0.174}$$

$$= 1239° \text{ R}$$

4.13 Closed-System Exergy

The preceding efficiencies are sometimes termed *first law efficiencies* because they compare the process output to that produced by an ideal process whose final state is different from the actual final state. However, they do not tell us the best possible result for the actual final state. We must turn to the second law coupled with the first law to answer this question.

For any process, engineers are interested in producing the maximum amount of work (work-producing process) or consuming the minimum amount of work (work-consuming process). If we have a closed system existing in some state (say P and T), we would like to know the maximum amount of work that this system can produce. This would occur if we used the best possible reversible processes to reduce the state of the system to that of the surrounding environment (P_0 and T_0). The system would then be in equilibrium with its surroundings. When the system comes to equilibrium with its surroundings it is said to be in its *dead state* because it will no longer change its state and consequently can produce no more work of its own accord. We have therefore extracted the maximum amount of work from the system by taking it through reversible processes to its dead state. If the system is capable of producing work by other means such as a velocity change, elevation change, or chemical reaction, then each of these would have to be brought to equilibrium with the surroundings to achieve the dead state. The same can be said for systems that involve electrical energy, magnetic energy, nuclear energy, and other energy forms. The work produced by each of these modes as it is reversibly reduced to the dead state also contributes to the maximum work produced by the system.

When the system undergoes a small expansion, dV, as its state progresses reversibly toward the dead state, it produces an amount of work, δW. Not all of this work is useful; some of it must be used to push back the environmental fluid as the system changes its volume as shown in Figure 4.18. The work used to move the environmental fluid is $P_0 \, dV$ with respect to the system. The net *useful work* resulting from this expansion is then

$$\delta W_{use} = \delta W - P_0 \, dV$$

FIGURE 4.18

Work Produced as a Closed System Reversibly Transitions to the Dead State

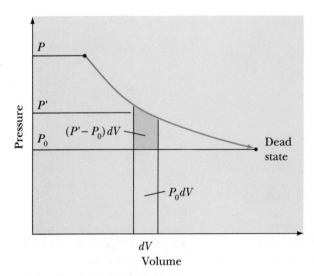

This is illustrated as the shaded area in Figure 4.18.

The system must also reject heat, δQ, to the environment as it adjusts its temperature to that of the dead state as illustrated in Figure 4.19. A portion of this heat could be converted to work by passing it through a completely reversible engine operating between the system temperature, T', and the environmental dead state temperature, T_0. The work produced by this completely reversible engine as referenced to the system would be

$$\delta W_{eng} = -\delta Q \left(1 - \frac{T_0}{T'} \right)$$

according to the discussion of Sections 2.6 and 2.7. This work is shown as the shaded area in Figure 4.19.

FIGURE 4.19
*Maximum Work
Produced by Heat
Transfer When a
Closed System
Transitions to the
Dead State*

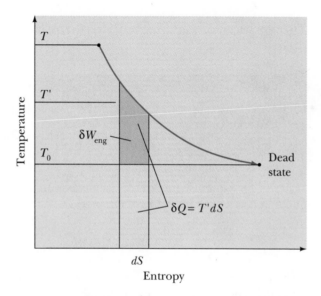

Accounting for both work productions, the maximum useful work this system could produce as it is reversibly taken to the dead state is

$$\delta W_{max,use} = \delta W - P_0 dV - \delta Q \left(1 - \frac{T_0}{T'} \right)$$

According to the first law for a closed system

$$\delta W = \delta Q - dU$$

The definition of the entropy tells us that

$$dS = \frac{\delta Q}{T'}$$

When these two results are substituted into the previous expression for the maximum useful work production, the result is

$$\delta W_{max,use} = -dU - P_0 dV + T_0 dS$$

The full amount of maximum useful work is achieved when the system is brought into equilibrium with the environment by the process of the preceding expression. Integrating this result between the actual system state and the dead state gives

$$W_{max,use} = -(U_0 - U) - P_0(V_0 - V) + T_0(S_0 - S)$$

This work is called the *closed-system exergy*, Φ (also known as the *closed-system availability*), which we can write on a specific basis as follows.

DEFINITION: *Closed System Specific Exergy (ϕ)*

$$\phi = (u + P_0 v - T_0 s) - (u_0 + P_0 v_0 - T_0 s_0)$$

In terms of the closed-system exergy, the maximum useful work that a closed system that is not exchanging heat with energy reservoirs can produce is

$$W_{max,use} = m\phi$$

The closed-system exergy is a mathematical combination of system properties and properties the system would have if it were reduced to P_0 and T_0. Although it is not a true property because it depends on the environmental conditions as well as the state of the system, it can be considered as a property because the variations in environmental conditions are normally modest. We will take the environmental conditions to be 14.7 psia (101 kPa), 77° F (25° C) throughout the remainder of this book for the purposes of determining exergies unless stated differently.

EXAMPLE 4.12

Some engineers have suggested that small motor vehicles can be propelled by expanding compressed air stored in an on-board storage tank through an air motor. What is the maximum amount of work that can be produced by such a system containing 2 kilograms of air at 1000 kPa, 327° C?

System: Air in a closed system.

Basic Equations:

$$\phi = (u + P_0 v - T_0 s) - (u_0 + P_0 v_0 - T_0 s_0)$$

$$Pv = RT$$

$$dh = c_P\, dT$$

$$T\, ds = dh - v\, dP$$

Conditions: The air will be treated as an ideal gas with constant specific heats.

Solution: According to the ideal gas equation of state,

$$v = \frac{0.2870 \times 600}{1000} = 0.1722 \text{ m}^3/\text{kg}$$

$$v_0 = \frac{0.2870 \times 298}{101} = 0.8468 \text{ m}^3/\text{kg}$$

When combined with the ideal gas equation of state and the definition of the specific heats and then integrated, the Gibbs equation gives

$$s_0 - s = c_P \ln \frac{T_0}{T} - R \ln \frac{P_0}{P}$$

$$= 1.0035 \ln \frac{298}{600} - \frac{8.314}{28.97} \ln \frac{101}{1000}$$

$$= -0.04433 \text{ kJ/kg-K}$$

Substituting these results into the expression for ϕ produces

$$\phi = c_v(T - T_0) + P_0(v - v_0) - T_0(s - s_0)$$

$$= 0.7165(600 - 298) + 101(0.1722 - 0.8468) - 298 \times (-0.04433)$$

$$= 161.5 \text{ kJ/kg}$$

TIP:

At the most, this system can only produce 322.9 kJ of work.

Thus, the total closed-system exergy of the system is

$$\Phi = m\phi = 2 \times 161.5 = 322.9 \text{ kJ}$$

When a closed system undergoes an adiabatic change of state from state 1 to state 2, the maximum amount of useful work that could have been produced is given by the change in the closed-system exergy:

$$(_1w_2)_{max} = -(\phi_2 - \phi_1)$$

If the system is not adiabatic and is receiving heat from thermal-energy reservoirs at temperatures T_j, additional work could have been produced by passing this heat through a completely reversible engine operating between the source or sink temperature and the environmental temperature as discussed in Section 2.10. Each thermal-energy reservoir interacting with the system would then produce additional work

$$w_j = q_j \left(1 - \frac{T_0}{T_j} \right)$$

where q_j is taken with regard to the system. In this case, the *work potential* for the process is

$$(_1w_2)_{ptl} = -(\phi_2 - \phi_1) + \sum_{j=1}^{k} q_j \left(1 - \frac{T_0}{T_j} \right)$$

A second law effectiveness for a work-producing process can be defined as follows.

DEFINITION: *Closed-System Second Law Effectiveness*

$$\eta_{2nd} = \frac{_1w_2}{-\Delta\phi + \sum\limits_{j=1}^{n} q_j(1 - T_0/T_j)}$$

This differs from the first law efficiency, which presumes that the best possible process is the adiabatic, reversible, or other ideal reversible process. The second law effectiveness makes no presumptions about the ideal process occurring between states 1 and 2. Rather, it compares the maximum work potential regardless of the actual process to the actual work realized. Alternate forms of the second law effectiveness can be defined for other processes and systems.

EXAMPLE 4.13

Steam at 400 psia, 600° F is expanded in a closed system to 100 psia, 350° F while producing 80 Btu/lb$_m$ of work and exchanging heat with a sink reservoir at 300° F. Determine the second law effectiveness of this expansion.

System: The steam in the closed system.

Basic Equations:

$$\phi = (u + P_0v + T_0s) - (u_0 + P_0v_0 + T_0s_0)$$
$$_1q_2 - _1w_2 = u_2 - u_1$$

500	1.284	1150.1	1245.2	1.5282
550	1.383	1174.6	1277.0	1.5605
600	**1.476**	**1197.3**	1306.6	**1.5892**
700	1.650	1240.4	1362.5	1.6397
800	1.816	1282.1	1416.6	1.6844

327.9	4.434	1105.8	1187.8	1.6034
350	**4.592**	**1115.4**	1200.4	**1.6191**
400	4.934	1136.2	1227.5	1.6517
450	5.265	1156.2	1253.2	1.6812
500	5.587	1175.7	1279.1	1.7085

Conditions: None.

Solution: Applying the first law to the actual process gives the heat rejected to the sink reservoir:

$$_1q_2 = u_2 - u_1 + {_1}w_2$$
$$= 1115.4 - 1197.3 + 80$$
$$= -1.9 \text{ Btu/lb}_m$$

The specific work potential for this process is

$$_1w_{ptl,2} = (\phi_1 - \phi_2) + q_s\left(1 - \frac{T_0}{T_s}\right)$$

$$= (u_1 - u_2) + P_0(v_1 - v_2) - T_0(s_1 - s_2) + q_s\left(1 - \frac{T_0}{T_s}\right)$$

$$= (1197.3 - 1115.4) + \frac{14.7 \times 144}{778}(1.476 - 4.592)$$

$$-537(1.5892 - 1.6191) - 1.9\left(1 - \frac{537}{760}\right)$$

$$= 88.92 \text{ Btu/lb}_m$$

Hence, the second law effectiveness is

$$\eta_{2nd} = \frac{{_1}w_2}{{_1}w_{ptl,2}}$$

$$= \frac{80}{88.92}$$

$$= 0.90$$

TIP:

This system did not produce the full 88.92 Btu/lb$_m$ of potential work because of irreversibilities associated with the internal process, heat transfer to the reservoir, and work consumed to displace the environmental fluid.

PROBLEMS

Sketch the appropriate P–V (v) and T–s state diagrams for the following problems, showing all states and processes.

4.1 The volume of 1 kg of helium in a piston-cylinder device is initially 5 m^3. Its volume is changed to 3 m^3, while its pressure is maintained at 200 kPa. Determine the helium's initial and final temperatures in K, as well as the work required to compress the helium in kJ.

4.2 Calculate the total work (in Btu) for process 1–3 shown in Figure 4.20A.

FIGURE 4.20

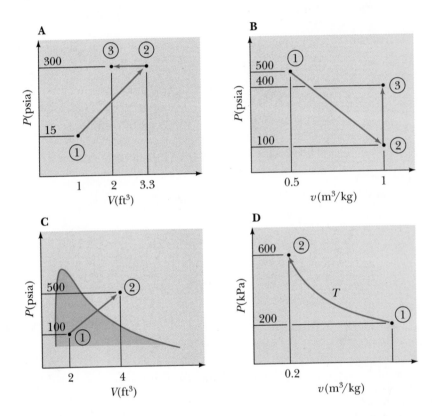

4.3 Calculate the total work (in kJ) for process 1–3 shown in Figure 4.20B when the system consists of 2 kg of nitrogen.

4.4 One cubic foot of saturated liquid water at 400° F is expanded isothermally in a closed system until its quality is 80%. Determine the total work produced by this expansion in Btu.

4.5 Calculate the total work (in Btu) produced by the process of Figure 4.20C.

4.6 Calculate the total work (in kJ) produced by the process of Figure 4.20D when the system consists of 3 kg of oxygen.

4.7 A mixture of gases consists of 1 kg-mole of carbon dioxide, 1 kg-mole of nitrogen, and $\frac{1}{3}$ kg-mole of oxygen. Determine the total amount of work in kJ units required to compress this mixture isothermally from 10 kPa, 27° C to 100 kPa.

4.8 A mixture of gases consisting of $\frac{1}{10}$ pound-mass-mole of nitrogen and $\frac{2}{10}$ pounds-mass-mole of oxygen fills a piston-cylinder device. Initially, this gas is at 300 psia and occupies 5 ft³. Its volume is then changed to 10 ft³ while its temperature is held constant. Determine the total work produced during this process in Btu.

4.9 One pound-mass of water that is initially at 200° F with a quality of 10% occupies a spring-loaded piston-cylinder device such as that in Figure 4.21. This device is now heated until the pressure is 100 psia and the temperature is 450° F. Determine the total work produced by this process in Btu.

FIGURE 4.21

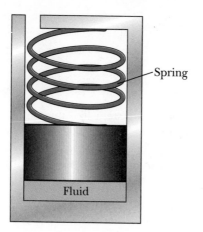

Spring

Fluid

4.10 One-half kilogram of water that is initially at 1000 kPa, 10% quality occupies a spring-loaded piston-cylinder device like that in Figure 4.21. This device is now cooled until the water is a saturated liquid at 100° C. Calculate the total work produced by this process in kJ.

4.11 A mixture of helium and nitrogen with a nitrogen mass fraction of 35% is contained in a piston-cylinder device arranged to maintain a fixed pressure of 100 psia. Determine the specific work produced (in Btu/lb_m) as this device is heated from 100° F to 500° F.

4.12 A mixture of nitrogen and carbon dioxide has a nitrogen mole fraction of 85%. Determine the specific work (in kJ/kg) required to compress this mixture isothermally in a piston-cylinder device from 100 kPa, 27° C to 500 kPa.

4.13 A closed system undergoes a process in which there is no internal energy change. During this process, the system produces 1.6×10^6 ft-lb_f of work. Calculate the heat transfer for this process in Btu.

4.14 Complete Table 4.1.

TABLE 4.1

$_1Q_2$ Btu	$_1W_2$ Btu	E_1 Btu	E_2 Btu	m lb$_m$	$e_2 - e_1$ Btu/lb$_m$
350	—	1020	860	3	—
350	130	550	—	5	—
—	260	600	—	2	150
−500	—	1400	900	7	—
—	−50	1000	—	3	−200

4.15 Complete Table 4.2.

TABLE 4.2

$_1Q_2$ kJ	$_1W_2$ kJ	E_1 kJ	E_2 kJ	m kg	$e_2 - e_1$ kJ/kg
280	—	1020	860	3	—
−350	130	550	—	5	—
—	260	300	—	2	−150
300	—	750	500	1	—
—	−200	—	300	2	−100

4.16 A closed system like that shown in Figure 4.22 is operated in an adiabatic manner. First, 15,000 ft-lb$_f$ of work are done by this system. Then work is applied to the stirring device to raise the internal energy of the system working fluid by 10.28 Btu. Calculate the net increase in the internal energy of this system in Btu.

FIGURE 4.22

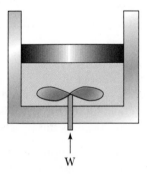

W

4.17 A substance is contained in a well-insulated, rigid container that is equipped with a stirring device similar to the piston-cylinder device of Figure 4.22. Determine the change in the internal energy (in Btu) of this substance when 15 Btu of work are applied to the stirring device. *Note:* This is not a reversible process because it involves mixing. Work consumption is then possible even if the volume does not change.

4.18 A rigid container equipped with a stirring device like that of Figure 4.22 contains 1.5 kg of motor oil. Determine the rate of specific energy increase (in watts) when 1 watt of heat is added to the oil and 1.5 watt of work are applied to the stirring device. (See the note with the previous problem.)

4.19 Nitrogen at 20 psia, $100°$ F initially occupies 1 ft^3 in a rigid container equipped with a stirring paddle wheel like that of Figure 4.22. After 5000 ft-lb$_f$ of paddle wheel work are done on the air, what is the final air temperature in$°$ F?

4.20 An adiabatic, 3 ft^3, rigid container is divided into two equal volumes by a thin membrane as illustrated in Figure 1.11. Initially, one of these chambers is filled with air at 100 psia, $100°$ F, while the second chamber is evacuated. Determine the internal energy change of the air in Btu when the membrane is ruptured. Also determine the air pressure (in psia) after the membrane is ruptured.

4.21 Two adiabatic, 2 m^3 chambers are interconnected by a valve as shown in Figure 4.23. Initially, the valve is closed, with one chamber containing oxygen at 1000 kPa, $127°$ C and the other chamber evacuated. The valve is then opened until the oxygen fills both chambers and they have the same pressure. Determine the total internal energy change in kJ and the final oxygen pressure in kPa.

FIGURE 4.23

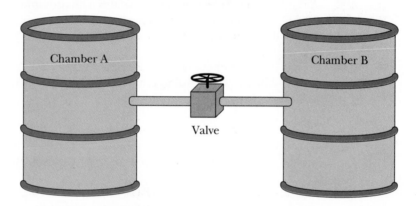

4.22 A spherical soap bubble with a surface-tension coefficient of 0.005 lb$_f$/ft is expanded from a diameter of 0.5 in. to 2.0 in. How much work (in Btu) is required to expand this bubble?

4.23 As a spherical ammonia vapor bubble rises in liquid ammonia, its diameter changes from 1 cm to 3 cm. Calculate the amount of work produced by this bubble in kilojoule when the surface-tension coefficient of ammonia is 0.02 N/m.

4.24 A 0.5-in diameter, 12-in-long steel rod with a Young's modulus of 30,000 lb $_f$/in^2 is stretched 0.125 in. How much work does this require in Btu?

4.25 A steel rod of 0.5 cm diameter and 10 m length is stretched 3 cm. Young's modulus for this steel is 21 kN/cm^2. How much work (in kJ) is required to stretch this rod?

4.26 A spring whose spring constant is 200 lb$_f$/in. has an initial 100 lb$_f$ acting on it. Determine the work (in Btu) required to compress it another 1 inch.

4.27 How much work (in kJ) can a spring whose spring constant is 3 kN/cm produce after it has been compressed 3 centimeters from its unloaded length?

4.28 Determine the specific internal energy change (in Btu/lb_m) of air in a rigid container as its temperature is increased by adding 50 Btu/lb_m of heat.

4.29 Nitrogen in a rigid vessel is cooled by releasing 100 kJ/kg of heat. Determine the specific internal energy change of the nitrogen in kJ/kg.

4.30 An adiabatic closed system is accelerated from 0 m/s to 30 m/s. Determine the specific energy change of this system in Btu/lb_m.

4.31 An adiabatic closed system is raised 100 m at a location where the gravitational attraction is 9.8 m/s². Determine the specific energy change of this system in kJ/kg.

4.32 A rigid 10-liter vessel originally contains a mixture of liquid water and vapor at 100° C, 12.3% quality. It is then heated until its temperature is 150° C. Calculate the total amount of heat required for this purpose in kJ units.

4.33 A rigid 1 ft³ vessel contains R-12 originally at –22° F, 27.7% quality. This container is then heated until its temperature is 100° F. Calculate the total amount of heat (in Btu) required to do this.

4.34 Two pounds-mass of saturated liquid water at 400° F are heated at constant pressure in a piston-cylinder device until they are a saturated vapor. Determine the total amount of heat required for this process in Btu.

4.35 Saturated water vapor at 300 kPa is isothermally cooled in a closed system until it is a saturated liquid. Calculate the amount of heat (in kJ/kg) required for this process.

4.36 Steam at 10 psia, 13% quality is contained in a spring-loaded piston-cylinder device such as that in Figure 4.21, which has an initial volume of 0.2 ft³. This device is now heated until the volume is 0.5 ft³ and the pressure is 40 psia. Determine the total heat added to and work produced by this device in Btu.

4.37 Refrigerant-12 at 600 kPa, 150° C is contained in a spring-loaded piston-cylinder device such as that shown in Figure 4.21; its initial volume is 0.3 m³. This device is now cooled until its temperature is –30° C and its volume is 0.1 m³. Determine the total heat added to and the total work produced by this device in kJ.

4.38 One pound-mass of air at 10 psia, 70° F is contained in a piston-cylinder device. Next the air is compressed reversibly to 100 psia while the temperature is maintained constant. Determine the total amount of heat (in Btu) added to the air during this compression.

4.39 One kilogram of air at 200 kPa, 127° C is contained in a piston-cylinder device. This air is now allowed to expand in a reversible, isothermal process until its pressure is 100 kPa. Determine the total amount of heat added to the air (in kJ) during this expansion.

4.40 Saturated water vapor in a closed system is condensed by cooling it at constant pressure to a saturated liquid at 40 kPa. Determine the specific heat transfer and work for this process in kJ/kg.

4.41 Nitrogen at 100 psia, 300° F in a rigid container is cooled until its pressure is 50 psia. Determine the specific work and heat transfer during this process in Btu/lb$_m$.

4.42 Nitrogen at 15 psia, 70° F in a rigid vessel is heated until its pressure is 50 psia. Calculate the specific work and heat transfer for this process in Btu.

4.43 One pound-mass of oxygen is heated from 70° F to 500° F. Determine the total amount of heat (in Btu) required when this is done in (a) a constant-volume process and (b) an isobaric process.

4.44 Ten kilograms of nitrogen are heated from 20° C to 250° C. Determine the total amount of heat (in kJ) required when this is done in (a) a constant-volume process and (b) an isobaric process.

4.45 Does the concept of a specific heat property (either at constant volume or constant pressure) have a meaning for substances that are undergoing a phase change? Why or why not?

4.46 A closed system containing 2 pounds-mass of air undergoes an isothermal process from 100 psia, 400° F to 10 psia. Determine the initial volume of this system in ft^3, the work produced by this system in Btu, and the heat transfer to this system in Btu.

4.47 A closed system containing argon undergoes an isothermal process from 200 kPa, 100° C to 50 kPa. This requires 1500 kJ of heat. Determine the mass of this system in kg and the amount of work produced by this system in kJ.

4.48 A closed piston-cylinder system containing carbon dioxide undergoes an isobaric process from 15 psia, 80° F to 200° F. Determine the specific work and heat transfer associated with this process in Btu/lb$_m$.

4.49 A closed piston-cylinder system contains nitrogen. Initially this system is at 1000 kPa, 427° C. It now undergoes an isobaric process until its temperature is 27° C. Determine the final pressure in kPa and specific-heat transfer (in kJ/kg) associated with this process.

4.50 Ten pounds-mass of helium are confined in a spring-loaded piston-cylinder device like that shown in Figure 4.21. This system is heated from 15 psia, 70° F to 100 psia, 280° F. Determine the total heat transfer to and work produced by this system in Btu.

4.51   One kilogram of carbon dioxide is confined in a spring-loaded piston-cylinder device such as that in Figure 4.21. This system is heated from 100 kPa, 25° C to 1000 kPa, 300° C. Determine the total heat transfer to and work produced by this device in kJ.

4.52   One pound-mass of air is contained in a well-insulated, rigid vessel equipped with a stirring paddle wheel similar to the device of Figure 4.22. The initial state of this air is 30 psia, 40° F. How much work (in Btu) must be added to the air with the paddle wheel to raise the air pressure to 50 psia? Also calculate the final air temperature in° F.

4.53   Air is contained in a piston-cylinder device like that in Figure 4.22. Initially the air is at 500 kPa, 27° C. The paddle wheel within the gas is turned by an external electric motor until 50 kJ/kg of work has been performed on the gas. During this process, heat is transferred to maintain a constant air temperature while allowing the gas volume to triple. Calculate the amount of heat transfer required in kJ/kg.

4.54   A mixture of nitrogen and carbon dioxide with a carbon dioxide mass fraction of 50% has a c_v of 0.792 kJ/kg-K. This mixture is heated at constant pressure in a closed system from 120 kPa, 30° C to 200° C. Calculate the specific work produced during this heating in kJ/kg.

4.55   The mixture of the preceding problem is placed in a rigid vessel and cooled from 200 kPa, 200° C until its pressure is 100 kPa. Determine the specific-heat transfer during this process in kJ/kg.

4.56   Saturated R-12 vapor at 100° F is condensed at constant pressure to a saturated liquid in a closed piston-cylinder system. Calculate the specific-heat transfer and work production for this process in Btu/lb$_m$.

4.57   Saturated water vapor at 200° C is isothermally condensed to a saturated liquid in a piston-cylinder device. Calculate the specific-heat transfer and work production for this process in kJ/kg.

4.58   Saturated water vapor at 200° C is condensed to a saturated liquid at 50° C in a spring-loaded piston-cylinder device. Determine the specific-heat transfer for this process in kJ/kg.

4.59   Two pounds-mass of saturated liquid ammonia with an initial temperature of 10° F are contained in a well-insulated, weighted piston-cylinder apparatus. This apparatus contains an electrical resistor as illustrated in Figure 4.24 to which 10 volts are applied causing a current of 2 amperes to flow through the resistor. Determine the amount of time (in hours) required for the ammonia to be converted to a saturated vapor. Also determine the ammonia's final temperature (in° F).

FIGURE 4.24

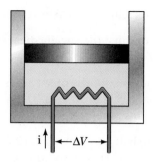

4.60 A well-insulated, rigid vessel contains 3 kg of saturated liquid water at 40° C. This vessel also contains an electrical resistor as shown in Figure 4.24 that draws 10 amperes when 50 volts are applied. Determine the final temperature (in° C) of the vessel's contents after the resistor has been operating for 30 minutes.

4.61 Determine the total heat transfer (in Btu) for process 1–3 shown in Figure 4.25A.

FIGURE 4.25

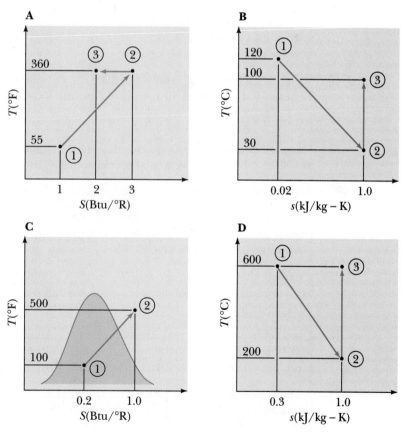

4.62 Determine on a specific basis the heat transfer (in kJ/kg) for process 1–3 shown in Figure 4.25B.

4.63 Determine the total heat transfer for process 1–2 shown in Figure 4.25C in Btu.

4.64 Determine the specific-heat transfer for process 1–3 shown in Figure 4.25D in kJ/kg.

4.65 One pound-mass of R-12 is expanded isentropically in a closed system from 100 psia, 100° F to 10 psia. Determine the total heat transfer and work production for this process in Btu.

4.66 One-half kilogram of R-12 is expanded isentropically from 600 kPa, 30° C to 140 kPa. Determine the total heat transfer and work production for this expansion in kJ.

4.67 Water at 50° F, 81.4% quality is compressed isentropically in a closed system to 500 psia. How much work does this require in Btu/lb$_m$?

4.68 Water at 70 kPa, 100° C is compressed isentropically in a closed system to 4000 kPa. Determine the final temperature of the water (in° C) and the specific work (in kJ/kg) required for this compression.

4.69 Refrigerant-12 at 240 kPa, 20° C undergoes an isothermal process in a closed system until its quality is 20%. On a specific basis how much work and heat transfer is required (in kJ/kg)?

4.70 Using the definition of *entropy*, calculate the entropy difference between air at 15 psia, 70° F and air at 40 psia, 250° F in Btu/lb$_m$-° R.

4.71 Repeat the preceding problem using a Gibbs equation.

4.72 Using the definition of *entropy*, calculate the entropy difference between oxygen at 150 kPa, 39° C and oxygen at 150 kPa, 337° C in kJ/kg-K.

4.73 Repeat the preceding problem using a Gibbs equation.

4.74 According to the definition of *entropy*, which of the two gases—helium or nitrogen—experiences the greatest entropy change as its state is changed from 2000 kPa, 427° C to 200 kPa, 27° C?

4.75 Calculate the change in the specific entropy in Btu/lb$_m$-° R as water is cooled at a constant pressure of 50 psia from a saturated vapor to a saturated liquid by using a Gibbs equation. Use the steam tables to verify your result.

4.76 Use a Gibbs equation to calculate the change in the specific entropy in kJ/kg-K as R-12 is heated at a constant pressure of 200 kPa from a saturated liquid to a saturated vapor. Use the R-12 tables to verify your answer.

4.77 Air is expanded from 200 psia, 500° F to 100 psia, 50° F. Determine the change in the air's specific entropy (in Btu/lb$_m$-° R) by using a Gibbs equation.

4.78 Air is expanded from 2000 kPa, 500° C to 100 kPa, 50° C. Determine the change in the air's specific entropy (in kJ/kg-K) by using a Gibbs equation.

4.79 Determine the final temperature in° C when air is expanded isentropically from 1000 kPa, 477° C to 100 kPa in a piston-cylinder device.

4.80 A mixture of ideal gases has a specific heat ratio (k) of 1.35 and an apparent molecular weight of 32. Determine the specific work (in Btu/lb$_m$) required to compress this mixture isentropically in a closed system from 15 psia, 70° F to 150 psia.

4.81 A mixture of hydrogen and oxygen has a hydrogen mass fraction of 0.33. Determine the difference between the specific entropy when this mixture is at 100 psia, 300° F and when it is at 20 psia, 300° F in Btu/lb$_m$-° R.

4.82 Air is expanded isentropically from 100 psia, 500° F to 20 psia in a closed system. Determine the final temperature of the air in° R.

4.83 Nitrogen is compressed isentropically from 100 kPa, 27° C to 1000 kPa in a piston-cylinder device. Determine the final temperature of the nitrogen in K.

4.84 Which of the two gases—helium or nitrogen—has the highest final temperature as it is compressed isentropically from 15 psia, 70° F to 150 psia in a closed system?

4.85 Which of the two gases—neon or air—has the lowest final temperature as it is expanded isentropically from 1000 kPa, 500° C to 100 kPa in a piston-cylinder device?

4.86 Derive a general expression for the work produced by an ideal gas as it undergoes a polytropic process in a closed system from initial state 1 to final state 2. Your result should be in terms of the initial pressure and temperature and the final pressure.

4.87 Air is expanded in a polytropic process with $n = 1.5$ from 200 pisa, 600° F to 10 psia in a piston-cylinder device. Determine the final temperature of the air in ° R.

4.88 How much specific work is produced and specific heat transferred during the expansion of the previous problem in Btu/lb$_m$?

4.89 Which of the two gases—neon or air—produces the greatest amount of work when expanded from P_1 to P_2 in a closed system polytropic process with $n = 1.2$?

4.90 Argon is compressed in a polytropic process with $n = 1.2$ from 120 kPa, 30° C to 1200 kPa in a piston-cylinder device. Determine the final temperature of the argon in K.

4.91 How much specific work is produced and specific heat transferred during the compression of the previous problem in kJ/kg?

4.92 Which of the two gases—neon or air—requires the least amount of work when compressed in a closed system from P_1 to P_2 using a polytropic process with $n = 1.5$?

4.93 Air at 15 psia, 70° F is compressed adiabatically in a closed system to 200 psia. What is the minimum temperature (in ° R) of the air following this compression?

4.94 Nitrogen at 900 kPa, 300° C is expanded adiabatically in a closed system to 100 kPa. Determine the minimum nitrogen temperature (in K) after the expansion.

4.95 What is the minimum specific enthalpy (in kJ/kg) that water can achieve as it is expanded adiabatically in a closed system from 1500 kPa, 320° C to 100 kPa?

4.96 What is the minimum specific internal energy (in Btu/lb$_m$) that R-12 can achieve as it is compressed adiabatically from 20 psia, 85% quality to 120 psia in a closed system?

4.97 Is it possible to expand water at 30 psia, 70% quality to 10 psia in a closed system undergoing an isothermal, reversible process while exchanging heat with an isothermal energy reservoir at 300° F?

4.98 Is it possible to cool and condense R-12 to a saturated liquid from 1000 kPa, 200° C in a closed system undergoing an isobaric, reversible process while exchanging heat with an isothermal energy reservoir at 100° C?

4.99 Can 2 lb$_m$ of air in a closed system at 16 psia, 100° F be compressed adiabatically to 100 psia and a volume of 4 ft^3?

4.100 What is the maximum volume (in m^3) that 3 kg of oxygen at 950 kPa, 373° C can be adiabatically expanded to in a piston-cylinder device if the final pressure is to be 100 kPa?

4.101 Can saturated water vapor at 30 psia be condensed to a saturated liquid in an isobaric, closed system process while only exchanging heat with an isothermal energy reservoir at 200° F?

4.102 A 100 lb$_m$ block of a solid material whose specific heat is 0.5 Btu/lb$_m$-° R is at 70° F. It is heated with 1 lb$_m$ of saturated water vapor that has a constant pressure of 14.7 psia. Determine the final temperature of the block and water in ° F, and the entropy change (in Btu/° R) of (a) the block, (b) the water, and (c) the entire system. Is this process possible? Why?

4.103 Ten grams of computer chips with a specific heat of 0.3 kJ/kg-K are initially at 20° C. These chips are cooled by placement in 5 grams of saturated liquid R-12 at temperature –40° C. Presuming that the pressure remains constant while the chips are being cooled, determine the entropy change (in kJ/K) of (a) the chips, (b) the R-12, and (c) the entire system. Is this process possible? Why?

4.104 It has been suggested that air at 100 kPa, 25° C can be cooled by first compressing it adiabatically in a closed system to 1000 kPa and then expanding it adiabatically back to 100 kPa. Is this possible?

4.105 One kilogram of air is in a piston-cylinder apparatus that can only exchange heat with an energy reservoir at 300 K. Initially this air is at 100 kPa, 27° C. Someone claims that the air can be compressed to 250 kPa, 37° C. Calculate the entropy change (in kJ/K) of (a) the air and (b) the reservoir. Is the claim true or false? Why?

4.106 Steam at 100 psia, 650° F is expanded adiabatically in a closed system to 10 psia. Determine the specific work produced (in Btu/lb$_m$) and the steam's final temperature (in ° F) for an adiabatic expansion efficiency of 80%.

4.107 Refrigerant-12 at 700 kPa, 40° C is expanded adiabatically in a closed system to 60 kPa. Determine the specific work produced (in kJ/kg) and final enthalpy (in kJ/kg) for an adiabatic expansion efficiency of 80%.

4.108 Two pounds-mass of air at 15 psia, 70° F are compressed adiabatically to 100 psia. How much work (in Btu) is required to compress this air if the adiabatic compression efficiency is 95%?

4.109 Three kilograms of helium at 100 kPa, 27° C are adiabatically compressed to 900 kPa. If the adiabatic compression efficiency is 80%, how much total work (in kJ) is required and what is the helium's final temperature (in K)?

4.110 Develop an expression for the efficiency of an isothermal expansion process. Clearly define the ideal process and the final state.

4.111 You are to expand a gas adiabatically from 300 psia, 400° F to 10 psia in a piston-cylinder device. Which of the two choices—air with an adiabatic expansion efficiency of 90% or neon with an adiabatic expansion efficiency of 80%—will produce the most work?

4.112 One hundred kilograms of saturated steam vapor at 100 kPa are to be adiabatically compressed to 1000 kPa. How much total work (in kJ) is required if the adiabatic compression efficiency is 90%?

4.113 Ten pounds-mass of R-12 are expanded without any heat transfer in a closed system from 120 psia, 100° F to 20 psia. If the adiabatic expansion efficiency is 95%, what is the final volume of this steam in ft^3?

4.114 Which is the more valuable for work production in a closed system—10 ft^3 of air at 100 psia, 300° F or 20 ft^3 of helium at 80 psia, 200° F?

4.115 Which has the capability to produce the most work in a closed system—1 kg of steam at 1000 kPa, 240° C or 1 kg of R-12 at 1000 kPa, 240° C?

4.116 Air is expanded in an adiabatic closed system from 150 psia, 100° F to 15 psia with an isentropic expansion efficiency of 95%. What is the second law effectiveness of this expansion?

4.117 Refrigerant-12 at 2000 kPa, 75° C is expanded in an isentropic closed system to 100 kPa with an adiabatic expansion efficiency of 85%. Determine the second law effectiveness of this expansion.

4.118   Steam is condensed in a closed system at a constant pressure of 6 psia from a saturated vapor to a saturated liquid by rejecting heat to a thermal-energy reservoir at 90° F. Determine the second law effectiveness of this process.

4.119   Refrigerant-12 is converted in a closed system using a reversible constant pressure process from a saturated liquid to a saturated vapor by heating it with a thermal-energy reservoir at 0° C. Is it more effective to do this at 100 kPa or 150 kPa?

4.120   Which is the more effective—adiabatically expanding air from 200 psia, 300° F to 10 psia, 40° F in a closed system or the same expansion while rejecting 30 Btu/lb$_m$ of heat to a thermal-energy reservoir at 40° F?

PROJECTS

The following projects are more extensive than the problems presented in this chapter. Most are open-ended and require some decisions on the student's part. Many can be expedited with computer-based property tables, computer spreadsheets, and student-written computer programs.

4.1   You are designing a closed-system, isentropic-expansion process using an ideal gas that operates between two pressures, P_1 and P_2. Which of the gases listed in Table 3.5 will produce the greatest amount of work? Which will require the least amount of work in a compression process?

4.2   What are the generalized force and displacement for two bodies governed by Newton's law of gravitational attraction?

4.3   What are the generalized force and displacement for an elastic solid when it is strained?

4.4   Compressed gases and phase-changing liquids are used to store energy in rigid containers. What are the advantages and disadvantages of each substance as a means of storing energy?

4.5   An electrical-generation utility sometimes pumps liquid water into an elevated reservoir during periods of low electrical consumption. This water is used to generate electricity during periods when the demand for electricity exceeds the utility's ability to produce electricity. Discuss this energy-storage scheme from a first and second law perspective as compared to storing a compressed phase-changing substance.

4.6   You are given the task of compressing a fixed quantity of gas from P_1, T_1 to P_2, T_1. You wish to do this using a polytropic process coupled with a constant-pressure (P_2) heat-transfer process. Determine the polytropic process that optimizes the second law effectiveness.

4.7   Some engineers have suggested that air compressed into tanks can be used to propel personal transportation vehicles. Current compressed-air tank technology permits us to compress and safely hold air at up to 4000 psia. Tanks made of composite materials require about 10 lb$_m$ of construction materials for each 1 ft^3 of stored gas. Approximately 0.01 hp is required per pound of vehicle weight to move a vehicle at a speed of 30 miles per hour. What is the maximum range that this vehicle can have? Account for the weight of the tanks only and assume perfect conversion of the energy in the compressed air.

4.8 Pressure changes across atmospheric weather fronts are typically a few centimeters of mercury, while the temperature changes are typically 2–20° C. Develop a plot of front pressure change versus front temperature change that will cause a maximum wind velocity of 10 m/s or more.

4.9 Someone has suggested that the device shown in Figure 4.26 be used to move the maximum force, F, against the spring, which has a spring constant of k. This is accomplished by changing the temperature of the liquid–vapor mixture in the container. You are to design such a device to close sun-blocking window shutters which require a maximum force of 0.5 lb$_f$. The piston must move 6 inches to close these shutters completely. You elect to use R-12 as the working fluid and arrange the liquid–vapor mixture container such that the temperature changes from 70° F when shaded from the sun to 100° F when exposed to the full sun. Select the sizes of the various components in this system to do this task. Also select the necessary spring constant and the amount of R-12 to be used.

FIGURE 4.26

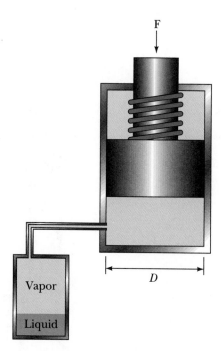

Steady-Flow Systems

The Wonder of Science

To discover that the universe is structured, and moves, according to mathematical laws is to experience one of the most profound insights into the basic order of the cosmos.

Thomas Goldstein [1]

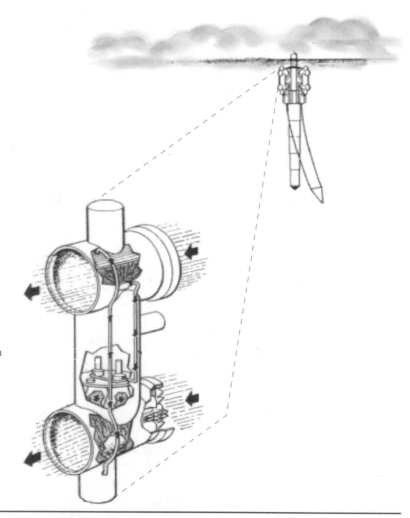

[1]*Dawn of Modern Science,* Houghton-Mifflin, Boston, MA, 1988.

F *low devices* are systems through which mass passes while exchanging heat and work with the system surroundings. A third interaction, the exchange of mass, is introduced when these devices are considered. Steady-flow systems also have the characteristic that properties and other quantities change as one changes location in the system, but they do not change as time passes. Everything about the steady-flow system is thus a function of position and not a function of time.

Exchanging mass across the system control boundary allows processes to occur in the system that alter the state of the mass as it passes through the system. A good example is the boiler in an electrical power plant. Air and fuel enter the boiler to be burned and form stack gas (flue gas), which consists primarily of nitrogen, carbon dioxide, water, and small particles of ash (when the fuel is coal or other solid). The stack gas leaves the boiler through the smoke stack, where it is dispersed by atmospheric air. In the industrialized nations of North America, Europe, and Eastern Asia, where large quantities of electricity are used, considerable amounts of carbon dioxide and ash particles are generated by these and other plants. The quantity of these gases being released has now reached the point where they are raising the level of these quantities in the atmosphere above naturally occurring levels. The world's forests convert carbon dioxide back to carbon and oxygen, but as they are being reduced at the same time that these extra gases are being emitted, the natural level of carbon dioxide in the atmospheric air is increasing. This increase in carbon dioxide and small particles in the atmosphere is changing the energy balance between the earth, sun, and space. Although it is uncertain at this time if this will lead to global warming or cooling, it is certain that our industrial activities now have the potential to change the temperature of the earth's atmosphere.

Another example of the impact of exchanging mass in an electrical power plant with that of the environment is the cooling water that is taken from a lake or river and warmed in the condenser before being returned to its source. Large quantities of this water pass through the condenser, which raises the water temperature by a few degrees Fahrenheit or Celsius. On returning to the river or lake, this water creates a zone of warm water. For example, this zone for large power plants on the Mississippi River covers as much as one-half or more of the river's width and extends for several miles downstream. These zones alter the river's ecology because some biological species prefer the warmer water while others do not. Eventually, this heat is transferred from the water to the environmental air through evaporation or other heat-transfer processes that occur at the air–water interface. This can have a similar impact on the biological species that live in this air.

In this chapter, the first and second laws as developed for the closed system will be extended to the steady-flow system. These laws will be developed in general for steady-flow systems that have several streams entering and leaving them, and then these same laws will be specialized to steady-flow systems with one inlet stream and one outlet stream. The concepts of mass flow rate, flow work, and flow exergy or availability are also developed in this chapter. These laws and concepts are applied then throughout this chapter to many of the more common steady-flow devices such as compressors, turbines, heat exchangers, separators, and mixers.

5.1 Introduction to Steady-Flow Systems

Steady-flow systems constitute a very important class of systems that is frequently encountered by engineers. Typical examples include aircraft engines, certain types of air compressors, air turbines, steam turbines, boilers, pneumatic lines and control devices, and pumps. All of these devices have one thing in common: Each has one or more fluid streams entering as well as leaving it. The characteristics of these streams do not change with time, and there is no accumulation or depletion of mass within the device itself. Devices that have this characteristic are also known as *steady-state, steady-flow systems; steady-flow control volumes;* and *flow systems;* among other names. Unlike the previous closed systems, which always had a fixed mass confined within a control boundary and which did not exchange mass with its surroundings, steady-flow systems are best analyzed by concentrating on a fixed region of space known as a *control volume* (see Figure 5.1).

FIGURE 5.1

A Typical Control Volume Used for Steady-Flow Systems

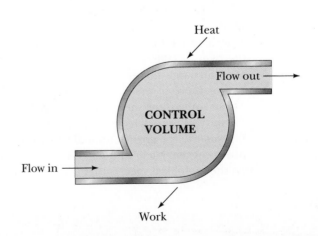

For the purposes of generality, we can define a steady-flow system as follows.

> **DEFINITION:** *Steady-Flow System*
>
> Any fixed region-of-space system through which a fluid flows and the properties of this fluid, either internal to the system or at its boundaries, do not change as time passes.

Notice that this definition only fixes the properties of the working fluid with regard to time, not location. The pressure, temperature, velocity, and other properties near the inlet of a device such as a turbine will always remain the same but will certainly be different from the same properties near the outlet, which is spatially separated from the inlet. That the fluid properties are independent of time and not location distinguishes the steady-flow system from other systems that exchange mass with their surroundings.

The first step in analyzing a steady-flow system is to identify a *control volume,* the boundaries of which (i.e., *control boundary*) separate the system from its surroundings as in Figure 5.1. Typically, the boundaries of this control volume are the same as the internal walls of the device being analyzed plus a limited number of imaginary boundaries through which mass enters and leaves the device. The control-volume boundary often cuts a rotating shaft(s) that couples internal parts such as impellers, turbine stages, and compressor stages to the environment. The torque being applied to this shaft is transmitting work to or removing work from the selected control volume. Heat also can be added to or removed from the control volume through the walls of the device because the device's contents can be hotter or colder than the surroundings. The contents of the control volume can then exchange mass, heat, and work with the surroundings.

5.2 Conservation of Mass

We now consider the events that happen to a closed system over a short period of time, Δt, as it passes through a control volume as shown in Figure 5.2. At time t, this fixed-mass system consists of the mass contained in the control volume, $m_{cv,t}$, plus the small masses, Δm_1 and Δm_2, which are about to enter the region where they will become part of the control-volume mass. Although we have shown two such masses about to enter the region, this can be an arbitrary number. The total mass of our fixed-mass system at time t is then $m_{cv,t} + \Delta m_1 + \Delta m_2$. At $t + \Delta t$, the masses Δm_1 and Δm_2 have just entered the control volume to become part of the control volume mass, while the masses Δm_3 and Δm_4, which were previously part of the control-volume mass, have just left. Our fixed-mass system now consists of the mass in the control volume, $m_{cv,t+\Delta t}$, and the two masses, Δm_3 and Δm_4. As before, the number of masses leaving the region can be arbitrary; we have only selected two for our illustration. The total mass of our fixed-mass system at time $t + \Delta t$ is $m_{cv,t+\Delta t} + \Delta m_3 + \Delta m_4$.

FIGURE 5.2

Mass Passing Through a Control Volume

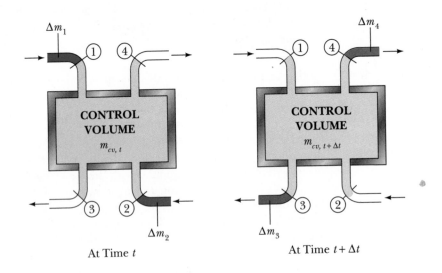

At Time t At Time $t + \Delta t$

Because the mass of the fixed mass system is the same at time $t + \Delta t$ as at time t,

$$m_{cv,t} + \Delta m_1 + \Delta m_2 = m_{cv,t+\Delta t} + \Delta m_3 + \Delta m_4$$

By the definition of a steady-flow system, the mass contained inside the control volume does not change as time passes. The preceding result thereby reduces to

$$\Delta m_1 + \Delta m_2 = \Delta m_3 + \Delta m_4$$

This states that the amount of mass entering the control volume over a time period Δt must exactly equal the amount that leaves over the same time period. This result can be put into a rate form by dividing both sides by the elapsed time and taking the limit as $\Delta t \rightarrow 0$, which yields

$$\dot{m}_1 + \dot{m}_2 = \dot{m}_3 + \dot{m}_4$$

where $\dot{m}$ is the mass flow rate, $\lim_{\Delta t \to 0} \Delta m / \Delta t$. In terms of the mass flow rates, the conservation of mass principle for steady-flow systems is as follows.

PRINCIPLE: Conservation of Mass for Steady-Flow Systems

$$\sum_{in} \dot{m} = \sum_{out} \dot{m}$$

The most frequently incurred steady-flow system is one with a single inlet and a single outlet. For this system, the conservation of mass principle reduces to

$$\dot{m}_1 = \dot{m}_2$$

where subscript 1 refers to the inlet stream and subscript 2 refers to the outlet stream.

To relate the mass flow rate at each inlet and outlet to the fluid properties at these boundaries, we need to focus on a small section of the control-volume boundary such as that shown in Figure 5.3. For the purposes of discussion, we have selected a small mass that is entering the control volume through a section of the control boundary. A small portion of the control boundary is identified by the vector, $\Delta \mathbf{A}$, whose length represents the size of the small area and which is directed outward perpendicularly from the control volume. The mass crossing this area has the velocity $\mathbf{V}$ whose magnitude is V. This mass also has the properties P, v, u, and others as it enters the control volume. The area as projected perpendicular to the velocity is $\mathbf{V} \cdot \Delta \mathbf{A}/V$, while the length of Δm measured along the direction of $\mathbf{V}$ is $V \Delta t$. Hence, the mass (given by the ratio of the volume to the specific volume) is

$$\Delta m = V \Delta t \frac{\mathbf{V} \cdot \Delta \mathbf{A}}{V v}$$

Dividing this result by the elapsed time, Δt, and taking the limit as $\Delta t \to 0$ gives the mass flow rate for this small section as

$$\Delta \dot{m} = \frac{\mathbf{V} \cdot \Delta \mathbf{A}}{v}$$

By adding the contributions of each small area where mass crosses the control boundary, we obtain

$$\dot{m} = \sum_{cb} \frac{\mathbf{V} \cdot \Delta \mathbf{A}}{v}$$

In the limit as $\Delta\mathbf{A} \to 0$, this result reduces to:

$$\dot{m} = \int_{cb} \frac{\mathbf{V} \cdot d\mathbf{A}}{v} \tag{5.1}$$

This result is general and allows the velocity and properties of the fluid to vary between different locations on the inlet and outlet control boundaries.

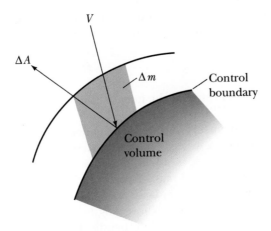

FIGURE 5.3

Small Mass Entering a Control Volume

Often the velocity is perpendicular to the control boundary. Also, the fluid properties and velocity frequently do not vary between points on each inlet and outlet segment of the control boundary. This situation is called *uniform flow*. When uniform flow exists, the preceding integral reduces to

$$\dot{m} = \frac{VA}{v}$$

for each inlet and outlet stream. The conservation of mass principle for uniform, steady flow becomes

$$\sum_{in} \frac{VA}{v} = \sum_{out} \frac{VA}{v}$$

Although this expression is less general, it is very useful in many situations.

EXAMPLE 5.1

Steam flows through a uniform, steady-flow system with a single inlet and a single outlet. The inlet area is 1 ft². At the inlet, the velocity is 20 ft/s, and the steam is at 700 psia, 700° F. The outlet area is 10 ft² and the steam leaves as a saturated vapor at 6 psia. What is the steam velocity at the outlet?

System: A steady-flow control volume with a single inlet and single outlet.

Basic Equations:

$$\sum_{in} \frac{VA}{v} = \sum_{out} \frac{VA}{v}$$

550	0.728	1149.0	1243.2	1.4723
600	0.793	1177.5	1280.2	1.5081
700	0.907	1226.9	1344.4	1.5661
800	1.011	1272.0	1402.9	1.6145
900	1.109	1315.6	1459.3	1.6576

6.0	170.03	0.01645	61.98	137.98	1065.4	138.00	996.2	1134.2	0.2474	1.8292
7.0	176.82	0.01649	53.65	144.78	1067.4	144.80	992.1	1136.9	0.2581	1.8167
8.0	182.84	0.01653	47.35	150.81	1069.2	150.84	988.4	1139.3	0.2675	1.8058
9.0	188.26	0.01656	42.41	156.25	1070.8	156.27	985.1	1141.4	0.2760	1.7963
10.0	193.19	0.01659	38.42	161.20	1072.2	161.23	982.1	1143.3	0.2836	1.7877

Conditions: It will be presumed that uniform steady-flow exists.

Solution: We will number the inlet as state 1 and the outlet as state 2. According to the steam tables, $v_1 = 0.907$ ft^3/lb$_m$ and $v_2 = 61.98$ ft^3/lb$_m$. The conservation of mass principle reduces to

$$\left(\frac{VA}{v}\right)_1 = \left(\frac{VA}{v}\right)_2$$

Solving for the outlet velocity gives

$$V_2 = \frac{A_1}{A_2}\frac{v_2}{v_1}V_1$$

$$= \frac{1}{10}\frac{61.98}{0.907}20 \quad \left[\frac{\text{ft}^2}{\text{ft}^2}\frac{\text{ft}^3/\text{lb}_m}{\text{ft}^3/\text{lb}_m}\frac{\text{ft}}{\text{s}}\right]$$

$$= 136.7 \text{ ft/s}$$

Flow rates are often quoted as *volume flow rates* for the volume of fluid passing through the control surface per unit of time. A volume flow rate is given by the product of the mass flow rate and the specific volume:

$$\dot{V} = \dot{m}v$$

For uniform, steady flow, the volume flow rate at an inlet or outlet is also given by

$$\dot{V} = AV$$

Unlike the mass flow rate, the volume flow rate is not necessarily conserved in steady-flow systems with a single inlet and single outlet because the specific volume can change.

5.3 Flow Work

Work is required to move fluid masses across the control boundary of a control volume. This work is known as *flow work*. Flow work is not available to drive devices that are external to the control volume. Its only use is to move the fluid into and out of the control volume. Either work must be supplied to the control-volume device to accomplish this movement of the fluid (e.g., a pump) or devices in other portions of the system cause the fluid to move (e.g., the feedwater pump in an electrical-generation station). Flow work is defined as follows.

DEFINITION: *Flow Work*

The work required to move the working fluid across the control boundaries of the control volume.

An expression for flow work may be developed by considering a small mass that is being pushed across the control boundary into the control volume by the fluid behind it as shown in Figure 5.4. The forces shown in this figure are those exerted by the small mass system. The pressure acting on this mass causes it to react with a force $P\Delta A$ that moves the fluid into the control volume. The amount of work required to move this mass into the control volume over the distance Δx is

$$\begin{aligned} \delta W_{fw} &= -P\Delta A\Delta x \\ &= -P\Delta V \\ &= -\Delta mPv \end{aligned}$$

The negative sign expresses the work with respect to the system (i.e., work is done on the system as the mass is moved into the system). Dividing this result by the amount of time required for the mass to enter the control volume, summing the contributions of all the masses crossing the inlet control surface, and taking the limit as $\Delta t \to 0$ gives the power needed to move the fluid into the control volume:

$$\dot{W}_{fw} = -\int_{cb} Pv\,d\dot{m}$$

When uniform flow exists, this result for one inlet control boundary reduces to

$$\dot{W}_{fw} = -Pv\dot{m}$$

Similarly, the flow power for one outlet stream is

$$\dot{W}_{fw} = Pv\dot{m}$$

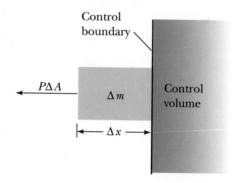

FIGURE 5.4

Force Acting on a Small Mass Entering a Control Volume

When a steady-flow system has several inlets and outlets, the net power required to move the fluid across all control-boundary segments is given by

$$\dot{W}_{fw} = \sum_{out} Pv\dot{m} - \sum_{in} Pv\dot{m}$$

If the system has a single inlet and outlet, $\dot{m}_{in} = \dot{m}_{out}$, this reduces to

$$w_{fw} = \frac{\dot{W}_{fw}}{\dot{m}} = P_2 v_2 - P_1 v_1$$

Here 1 denotes the inlet and 2 the outlet. The specific flow work, w_{fw}, is the flow work produced as one unit of mass passes through the control volume.

EXAMPLE 5.2

The vapor-separation unit in a steam power plant separates a stream containing a liquid–vapor mixture of water into a saturated liquid stream and a saturated vapor stream. Consider a plant where a mixture at 5000 kPa, 92% quality enters the separation unit and the two outlet streams exit at 5000 kPa. Determine the mass flow rates of the two outlet streams and the flow power for this unit when it processes 30 kg/s of mixture.

System: Steady-flow control volume with a single inlet and two outlets as shown in Figure 5.5.

FIGURE 5.5
Control Volume for
Example 5.2

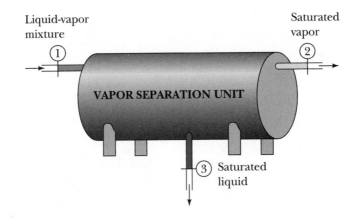

Basic Equations:

$$\sum_{in} \frac{VA}{v} = \sum_{out} \frac{VA}{v}$$

$$\dot{W}_{fw} = \sum_{out} Pv\dot{m} - \sum_{in} Pv\dot{m}$$

3500	242.6	0.001235	0.05707	1045.4	2603.7	1049.8	1753.7	2803.4	2.7253	6.1253
4000	250.4	0.001252	0.04978	1082.3	2602.3	1087.3	1714.1	2801.4	2.7964	6.0701
4500	257.5	0.001269	0.04406	1116.2	2600.1	1121.9	1676.4	2798.3	2.8610	6.0199
5000	264.0	**0.001286**	**0.03944**	1147.8	2597.1	1154.2	1640.1	2794.3	2.9202	5.9734
6000	275.6	0.001319	0.03244	1205.4	2589.7	1213.4	1571.0	2784.3	3.0267	5.8892

Conditions: Uniform steady flow is assumed to exist.

Solution: According to the steam tables, $v_2 = 0.0012859$ m^3/kg   and $v_3 = 0.03944$ m^3/kg. The specific volume at the inlet is

$$v_1 = v_f + xv_{fg}$$
$$= v_3 + x(v_2 - v_3)$$
$$= xv_2 + (1 - x)v_3$$

Because 92% of the inlet stream mass is vapor,

$$\dot{m}_2 = x\dot{m}_1 = 0.92 \times 30 = 27.6 \text{ kg/s}$$

According to the conservation of mass principle,

$$\dot{m}_3 = \dot{m}_1 - \dot{m}_2 = (1 - x)\dot{m}_1$$
$$= 30.0 - 27.6$$
$$= 2.4 \text{ kg/s}$$

The separator flow power is then

$$\frac{\dot{W}_{fw}}{\dot{m}_1} = P\left[v_2\frac{\dot{m}_2}{\dot{m}_1} + v_3\frac{\dot{m}_3}{\dot{m}_1} - v_1\right]$$

$$= P[xv_2 + (1-x)v_3 - v_1]$$

$$= 0$$

5.4 The First Law and Steady-Flow Systems

We may adapt the closed system form of the first law,

$$Q - W = E_f - E_i$$

to a steady-flow system by considering a fixed amount of mass as it passes through the control volume over a time Δt (as shown in Figure 5.6), just as we developed the conservation of mass principle. Initially at time t, this mass consists of the mass in the control volume and the two masses about to enter the control volume, $m_{cv,t} + \Delta m_1 + \Delta m_2$. At the end when the time is $t + \Delta t$, this mass consists of the mass in the control volume and the two masses that have just left the control volume, $m_{cv,t+\Delta t} + \Delta m_3 + \Delta m_4$.

FIGURE 5.6

A Mass Passing Through a Control Volume

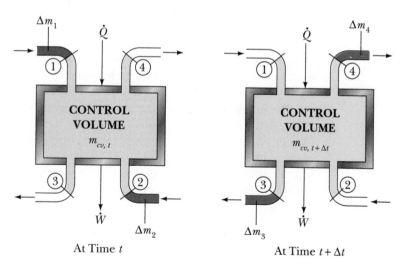

At Time t At Time $t + \Delta t$

As this mass is passing through the control volume, heat is being added to the portion of it in the control volume at the rate $\dot{Q}$, and work is being produced by the portion of it in the control volume at the rate $\dot{W}_{cv}$. The total heat added to this mass over the period Δt is then $\dot{Q}\Delta t$, and the total work produced by the mass in the control volume over the same period is $\dot{W}_{cv}\Delta t$. The closed-system first law as applied to this mass is therefore

$$\dot{Q}\Delta t - \dot{W}_{cv}\Delta t = (E_{m_{cv,t+\Delta t}} - E_{m_{cv,t}}) + E_{\Delta m_3} + E_{\Delta m_4} - E_{\Delta m_1} - E_{\Delta m_2}$$

According to the definition of steady-flow systems, the total energy contained in the control volume, like the mass, does not change as time passes. Hence, $E_{m_{cv,t+\Delta t}} = E_{m_{cv,t}}$, and the first law reduces to

$$\dot{Q}\Delta t - \dot{W}_{cv}\Delta t = E_{\Delta m_3} + E_{\Delta m_4} - E_{\Delta m_1} - E_{\Delta m_2}$$

The work produced by the contents of the control volume consists of two components, the *shaft work* and the *flow work*. The shaft work is that produced by the process that the fluid undergoes in the control volume; it may be used to drive or power devices in the control volume's surroundings. It is called shaft work because it is normally transmitted to the surroundings through a rotating shaft that has been cut by the control-boundary surfaces. It can also be transmitted in other ways such as electrical work. When uniform flow exists, the net work produced as the mass passes through the control volume is

$$\dot{W}_{cv}\Delta t = \dot{W}\Delta t + P_3 v_3 \Delta m_3 + P_4 v_4 \Delta m_4 - P_1 v_1 \Delta m_1 - P_2 v_2 \Delta m_2$$

according to the discussion of the previous section. In this result, $\dot{W}\Delta t$ is the shaft work produced by the system. When this result is substituted into the first law, the first law becomes

$$\dot{Q}\Delta t - \dot{W}\Delta t = (e + Pv)_3 \Delta m_3 + (e + Pv)_4 \Delta m_4$$
$$-(e + Pv)_1 \Delta m_1 - (e + Pv)_2 \Delta m_2$$

From this point on, the term *work* in connection with control volumes means shaft work or other forms of work that interact with the working fluid in the control volume and does not include the flow work.

The energy of the masses entering and leaving the control volume consists of internal, kinetic, and potential energy if we only consider mechanical forms of energy. All of the streams have a specific energy $u + e_k + e_p$. Each mass term of the above first law then contains the quantity $u + Pv + e_k + e_p$. The definition of enthalpy reduces each term to $h + e_k + e_p$. This further simplifies the first law as applied to a steady-flow system to

$$\dot{Q}\Delta t - \dot{W}\Delta t = (h + e_k + e_p)_3 \Delta m_3 + (h + e_k + e_p)_4 \Delta m_4$$
$$-(h + e_k + e_p)_1 \Delta m_1 - (h + e_k + e_p)_2 \Delta m_2$$

When this is divided by the elapsed time, Δt, and the limit as $\Delta t \to 0$ is taken, it becomes

$$\dot{Q} - \dot{W} = (h + e_k + e_p)_3 \dot{m}_3 + (h + e_k + e_p)_4 \dot{m}_4$$
$$-(h + e_k + e_p)_1 \dot{m}_1 - (h + e_k + e_p)_2 \dot{m}_2$$

For an arbitrary number of inlet and outlet streams, the first law can be written as follows.

PRINCIPLE: First Law for Steady-Flow Systems

$$\dot{Q} - \dot{W} = \sum_{out}(h + e_k + e_p)\dot{m} - \sum_{in}(h + e_k + e_p)\dot{m}$$

In words, this form of the first law states that the rate at which heat is added to the control volume minus the rate at which shaft or other work is produced by the control volume must equal the rate at which energy and flow work leave the control volume minus the rate at which they enter the control volume.

EXAMPLE 5.3

A chilled-water heat-exchange unit is frequently used in commercial air-conditioning systems. Chilled water passing through this unit is warmed as it cools a stream of warm air. This occurs in a well-insulated box with no shafts protruding from it, and the two streams are not allowed to intermix. Consider such a unit designed to cool 5000 liters/s of air at 100 kPa, 30° C to 100 kPa, 18° C by using chilled water at 8° C. Determine the mass flow rate of water through this unit when the water exits at 15° C.

System: Steady-flow control volume with two inlets and two outlets as shown in Figure 5.7.

FIGURE 5.7
Control Volume for
Example 5.3

Basic Equations:

$$\dot{Q} - \dot{W} = \sum_{out}(h + e_k + e_p)\dot{m} - \sum_{in}(h + e_k + e_p)\dot{m}$$

$$Pv = RT$$
$$dh = c_p\,dT$$

Conditions: There is no shaft work or heat transfer associated with this control volume. All fluid kinetic and potential energies will be neglected. The air will be treated as an ideal gas with constant specific heats, and the chilled water will be treated as an incompressible liquid.

Solution: The specific volume of the air at its inlet is

$$v_1 = \frac{RT_1}{P_1}$$

$$= \frac{8.314}{28.97} \frac{303}{100}$$

$$= 0.8696 \text{ m}^3/\text{kg}$$

The mass flow rate at the air inlet is then

$$\dot{m}_1 = \frac{\dot{V}_1}{v_1}$$

$$= \frac{5000 \times 0.001}{0.8696} \quad \left[\frac{\text{liter}}{\text{s}} \frac{\text{m}^3}{\text{liter}} \frac{\text{kg}}{\text{m}^3}\right]$$

$$= 5.750 \text{ kg/s}$$

Neglecting the kinetic and potential energy terms and noting that there is no shaft work or heat transfer for the system selected, the first law reduces to

$$\dot{m}_{water}(h_4 - h_2) = \dot{m}_{air}(h_1 - h_3)$$

Solving for the water mass flow rate, expressing the enthalpy of the water as $c\Delta T$, and expressing the enthalpy of the air as $c_P\Delta T$, we reduce this to

$$\dot{m}_{water} = \frac{c_P(T_1 - T_3)}{c(T_4 - T_2)}\dot{m}_{air}$$

$$= \frac{1.0035(30 - 18)}{4.18(15 - 8)}5.750$$

$$= 2.366 \text{ kg/s}$$

The steady-flow system with a single inlet and outlet is a commonly occurring special case. For this system, the conservation of mass principle reduces to

$$\dot{m}_1 = \dot{m}_2$$

and the first law becomes

$$_1q_2 - _1w_2 = h_2 - h_1 + \frac{1}{2}(V_2^2 - V_1^2) + g(z_2 - z_1) \qquad (5.2)$$

where subscript 1 refers to the inlet and subscript 2 the outlet. Take careful note of the similarity between this steady-flow form of the first law and the closed-system form of the first law as given by Equation 4.2 (page 145). For steady-flow systems with one inlet and one outlet, we replace the internal energy of the closed-system first law with the enthalpy, and the specific heat transfer and specific work production are interpreted as *per unit of mass passing through the control volume* rather than as *per unit of system mass.*

EXAMPLE 5.4

A nozzle is a converging–diverging passage such as that shown in Figure 5.8. Its purpose is to convert thermal energy into kinetic energy by adiabatically expanding the gas flowing through it. One such nozzle has air entering it at 200 psia, 340° F and is designed to increase the velocity of this air from 20 ft/s to 1000 ft/s. Determine the temperature of the air as it leaves this nozzle.

System: A steady-flow control volume with one inlet and one outlet as shown in Figure 5.8.

FIGURE 5.8
A Nozzle for Accelerating a Fluid

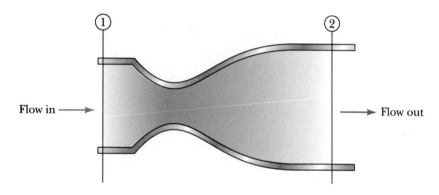

Flow in ⟶   ⟶ Flow out

Basic Equations:

$$_1q_2 - {}_1w_2 = h_2 - h_1 + \frac{1}{2}(V_2^2 - V_1^2) + g(z_2 - z_1)$$

$$dh = c_P dT$$

Conditions: There is no shaft work or heat transfer associated with this control volume. Potential energy will be neglected. The air will be treated as an ideal gas with constant specific heats.

Solution: Neglecting the potential-energy term, the first law reduces to

$$h_2 - h_1 = c_P(T_2 - T_1) = \frac{V_1^2 - V_2^2}{2}$$

Solving for the outlet temperature, T_2, and converting the units gives

$$T_2 = T_1 + \frac{V_1^2 - V_2^2}{2c_P g_c J}$$

$$= 340 + \frac{20^2 - 1000^2}{2 \times 0.2397 \times 32.174 \times 778} \left[\frac{\frac{\text{ft}^2}{\text{s}^2}}{\frac{\text{Btu}}{\text{lb}_m\text{-}°\text{F}} \frac{\text{ft-lb}_m}{\text{lb}_f\text{-s}^2} \frac{\text{ft-lb}_f}{\text{Btu}}} \right]$$

$$= 257° \text{F}$$

EXAMPLE 5.5

Water in a lake is adiabatically pumped into a storage tower located 300 ft above the lake as illustrated in Figure 5.9. The pipe between the lake and pump has an 8-inch diameter, and the pipe between the pump and storage tank has a 10-inch diameter. At the pump inlet, the water velocity is 10 ft/s. Presuming that the temperature of the lake water is 50° F and that the temperature of the water does not change, determine the specific work in (Btu/lb$_m$) required to pump this water.

System: Steady-flow control volume with one inlet and one outlet.

Basic Equations:

$$_1q_2 - _1w_2 = h_2 - h_1 + \frac{1}{2}(V_2^2 - V_1^2) + g(z_2 - z_1)$$

$$\frac{A_1 V_1}{v_1} = \frac{A_2 V_2}{v_2}$$

Conditions: The water will be treated as being incompressible. Because the temperature of the water does not change, the enthalpy of the water will not change. Standard gravitational acceleration will be used.

FIGURE 5.9
*Water Piping System
for Example 5.5*

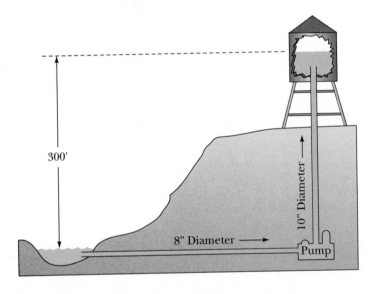

Solution: The velocity in the pipe after the pump is given by

$$V_2 = \frac{A_1}{A_2}V_1$$

$$= \left(\frac{D_1}{D_2}\right)^2 V_1$$

$$= \left(\frac{8}{10}\right)^2 10 = 6.4 \text{ ft/s}$$

From the problem specifications, the first law reduces to

$$_1w_2 = \frac{V_2^2 - V_1^2}{2g_cJ} + \frac{g}{g_cJ}(z_1 - z_2)$$

TIP:

Note that the work required is quite small even for this large elevation change and dense fluid. Potential-energy changes are often neglected because they are quite small. The kinetic-energy change is also negligible in this example.

where g_c and J have been introduced to adjust the units. Substituting the data from the problem statement,

$$_1w_2 = \frac{6.4^2 - 10^2}{2 \times 32.174 \times 778} + \frac{32.174}{32.174 \times 778}(0 - 300) \left[\frac{\frac{ft}{s^2}\,ft}{\frac{ft\text{-}lb_m}{lb_f\text{-}s^2}\,\frac{ft\text{-}lb_f}{Btu}} \right]$$

$$= -0.0012 - 0.386$$
$$= -0.387\ \text{Btu/lb}_m$$

EXAMPLE 5.6

Steam at 500 psia, 800° F enters an adiabatic steam turbine such as that shown in Figure 3.23 with a velocity of 100 ft/s. It exits this turbine as a saturated vapor at 15 psia with a velocity of 50 ft/s. Calculate the specific work (Btu/lb$_m$) produced as the steam passes through the steam turbine.

System: Steady-flow control volume with a single inlet–outlet.

Basic Equations:
$$_1q_2 - _1w_2 = h_2 - h_1 + \frac{1}{2}(V_2^2 - V_1^2) + g(z_2 - z_1)$$

600	1.158	1191.1	1298.3	1.5585
700	1.304	1236.0	1356.7	1.6112
800	1.441	1278.8	**1412.1**	1.6571
900	1.572	1321.0	1466.5	1.6987
1000	1.701	1363.3	1520.7	1.7371

14.696	211.99	0.01672	26.80	180.10	1077.6	180.15	970.4	1150.5	0.3121	1.7567
15	213.03	0.01672	26.29	181.14	1077.9	181.19	969.7	**1150.9**	0.3137	1.7551
20	227.96	0.01683	20.09	196.19	1082.0	196.26	960.1	1156.4	0.3358	1.7320
25	240.08	0.01692	16.31	208.44	1085.3	208.52	952.2	1160.7	0.3535	1.7142
30	250.34	0.01700	13.75	218.84	1088.0	218.93	945.4	1164.3	0.3682	1.6996

Conditions: The flow is taken to be uniform and steady with no heat transfer. Fluid potential energies will be neglected.

Solution: We will number the inlet as state 1 and the outlet as state 2. According to the steam tables, $h_1 = 1412.1$ Btu/lb$_m$ and $h_2 = 1150.9$ lb$_m$. The first law reduces to

TIP:

The kinetic energy is quite small as compared to the enthalpy change. The majority of the work produced results from the enthalpy change, not the kinetic-energy change. This is not atypical, and the kinetic and potential energies are frequently, but not always, neglected.

$$_1w_2 = h_1 - h_2 + \frac{V_1^2 - V_2^2}{2g_cJ}$$

$$= 1412.1 - 1150.9 + \frac{100^2 - 50^2}{2 \times 32.174 \times 778}$$

$$= 261.2 + 0.15$$
$$= 261.3 \text{ Btu/lb}_m$$

The turbine then produces 261.3 Btu/lb$_m$ of work for each pound-mass of steam that passes through it.

5.5 The Second Law and Steady-Flow Systems

Chapter 2 pointed out that the second law deals with the direction in which processes and interactions such as heat must proceed and places limits on them. According to Clausius, heat cannot transfer from cold to hot bodies of its own accord. A cup of tea sitting in a cooler environment will always decrease its temperature until the tea's temperature is the same as the environment's. In this situation, the tea will never get hotter. Thus, the direction of the process and its lower limit are set by the second law.

When the second law is applied to steady-flow systems, many similar conclusions can be reached. Consider, for example, the fluid-to-air heat exchanger that is used to keep the engine of an automobile cool by transferring heat from the engine cooling fluid to the surrounding air. This device is commonly known as a *radiator.* A schematic of this device is shown in Figure 5.10; the temperatures of the cooling fluid and air are plotted as they pass through this device. The engine cooling fluid enters this device with a temperature of $T_{h,f}$ and leaves at a lower temperature, $T_{c,f}$. The environmental air enters this device at a temperature of $T_{c,a}$ and leaves with a temperature of $T_{h,a}$. At every location inside the heat exchanger, the cooling fluid temperature must be greater than that of the air in order to transfer heat from the fluid to the air while satisfying the Clausius statement of the second law. This must also be the case for the temperatures at which the fluid and air leave this device. The second law then tells us that the transfer of heat can only continue in the heat exchanger until the temperature of the cooling fluid equals that of the air. Expansion of the increase in entropy principle will permit us to quantify these conclusions for steady-flow systems.

FIGURE 5.10

Schematic Diagram of an Automotive Radiator

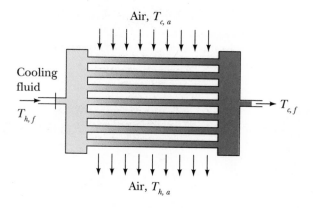

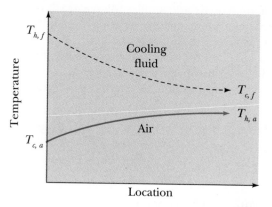

The increase in entropy principle for a fixed-mass system,

$$\Delta S_{everything} \geq 0$$

may be converted to a steady-flow system form in the same manner as the conservation of mass principle and the first law. Again we consider the events that happen to the fixed-mass system of Figure 5.6 as it passes through the control volume.

At time t, the entropy of the mass is $S_{cv,t} + \Delta m_1 s_1 + \Delta m_2 s_2$. The entropy of the mass at time $t + \Delta t$ is $S_{cv,t+\Delta t} + \Delta m_3 s_3 + \Delta m_4 s_4$. Over time period Δt, the entropy of the surroundings will have increased by $S_{\infty,t+\Delta t} - S_{\infty,t}$. The increase of the entropy of everything involved with the mass as it passes through the control volume is then

$$(S_{\infty,t+\Delta t} - S_{\infty,t}) + (S_{cv,t+\Delta t} - S_{cv,t}) + \Delta m_3 s_3 + \Delta m_4 s_4 - \Delta m_1 s_1 - \Delta m_2 s_2 \geq 0$$

Because this is a steady-flow system, $S_{cv,t} = S_{cv,t+\Delta t}$. Dividing through by the elapsed time and taking the limit as $\Delta t \to 0$ reduces the increase in entropy principle to

$$\dot{S}_\infty + \dot{m}_3 s_3 + \dot{m}_4 s_4 - \dot{m}_1 s_1 - \dot{m}_2 s_2 \geq 0$$

where $\dot{S}_\infty$ is the rate of entropy increase occurring in components external to the control volume (i.e., the surroundings' entropy generation rate) as a result of the process in the control volume. For an arbitrary number of streams, this may be written as follows.

PRINCIPLE: Second Law for Steady-Flow Systems

$$\dot{S}_\infty + \sum_{out} \dot{m}s - \sum_{in} \dot{m}s \geq 0$$

As before, the equals sign applies when everything is done reversibly and the greater-than sign applies whenever any irreversibilities are present in the system or its interactions with the surroundings.

The second law for steady-flow systems may be expressed as an equality by introducing the entropy-generation rate, $\dot{S}_{gen}$, as was done in Chapter 2. With the entropy-generation rate, the second law becomes

$$\dot{S}_\infty + \sum_{out} \dot{m}s - \sum_{in} \dot{m}s - \dot{S}_{gen} = 0 \qquad (5.3)$$

Here the entropy-generation rate must be positive or zero. It will only be zero when the process occurring in the system and the system's interactions with the surroundings are reversible. The more positive the entropy-generation rate, the greater the losses that occur with the system and the less effective the system becomes.

The forbidden states for steady-flow systems can be determined from this expression. For example, the forbidden states for one of the outlet streams (say, o') are determined by solving the inequality form of the second law for $s_{o'}$,

$$s_{o'} \geq \sum_{in} \frac{\dot{m}s}{\dot{m}_{o'}} - \sum_{out \neq o'} \frac{\dot{m}s}{\dot{m}_{o'}} - \frac{\dot{S}_\infty}{\dot{m}_{o'}}$$

As with closed systems, the entropy of a stream has a minimum value that sets the lower limit for the states that a stream can achieve. The acceptable states are determined by the surroundings' entropy-generation rate as well as the entropy and mass flow rates of all the streams.

EXAMPLE 5.7

A portion of the steam passing through a steam turbine is sometimes removed for the purposes of feedwater heating. Consider an adiabatic steam turbine with 12,000 kPa, 560° C steam entering it at a rate of 20 kg/s, steam being bled from it at a rate of 3 kg/s with a pressure of 1000 kPa, and an exit state of 100 kPa, 100° C. Determine the minimum temperature (and quality if appropriate) of the bleed steam.

System: A uniform steady-flow control volume with one inlet and two outlets as shown in Figure 5.11.

Basic Equations:

$$\dot{S}_\infty + \sum_{out} \dot{m}s - \sum_{in} \dot{m}s \geq 0$$

$$\sum_{in} \dot{m} = \sum_{out} \dot{m}$$

480	0.02576	2984.4	3293.5	6.4154
520	0.02781	3068.0	3401.8	6.5555
560	0.02977	3149.0	3506.2	**6.6840**
600	0.03164	3228.7	3608.3	6.8037
640	0.03345	3307.5	3709.0	6.9164

99.63	1.694	2506.1	2675.5	7.3594
100	1.696	2506.7	2676.2	**7.3614**
120	1.793	2537.3	2716.6	7.4668
160	1.984	2597.8	2796.2	7.6597
200	2.172	2658.1	2875.3	7.8343

1000	179.9	0.001127	0.1944	761.68	2583.6	762.81	2015.3	2778.1	**2.1387**	**6.5865**
1500	198.3	0.001154	0.1318	843.16	2594.5	844.89	1947.3	2792.2	2.3150	6.4448
2000	212.4	0.001177	0.0996	906.44	2600.3	908.79	1890.7	2799.5	2.4474	6.3409
2500	224.0	0.001197	0.07998	959.11	2603.1	962.11	1841.0	2803.1	2.5547	6.2575
3000	233.9	0.001217	0.06668	1004.8	2604.1	1008.4	1795.7	2804.2	2.6457	6.1869

Conditions: Uniform steady flow with no heat transfer. The steam kinetic and potential energies will be neglected.

Solution: According to the steam tables, $s_1 = 6.6840$ kJ/kg-K   and $s_3 = 7.3614$ kJ/kg-K. The surroundings' entropy-generation rate is zero because there is no heat exchange to alter the entropy of the surroundings. According to the second law,

$$s_2 \geq \frac{\dot{m}_1 s_1}{\dot{m}_2} - \frac{\dot{m}_3 s_3}{\dot{m}_2}$$

$$\geq \frac{\dot{m}_1 s_1}{\dot{m}_2} - \frac{(\dot{m}_1 - \dot{m}_2) s_3}{\dot{m}_2}$$

$$\geq \frac{20 \times 6.6840}{3} - \frac{(20 - 3)7.3614}{3}$$

$$\geq 2.8454 \text{ kJ/kg-K}$$

whose forbidden states are shown in Figure 5.11. Inspection of the steam tables reveals that the minimum temperature at the bleed-steam outlet would be the saturation temperature for 1000 kPa,

$$T_{2,min} = 179.9° \text{ C}$$

The minimum quality of the bleed steam is

$$x_{2,min} = \frac{s_{2,min} - s_{2,f}}{s_{2,fg}}$$

$$= \frac{2.8454 - 2.1387}{4.4476}$$

$$= 0.158$$

FIGURE 5.11

Control Volume and T–s State Diagram for Example 5.7

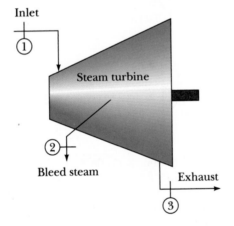

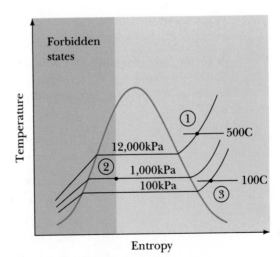

The steady-flow system with a single inlet–outlet is the most commonly occurring steady-flow system. For this case, the second law reduces to

$$\Delta s_\infty + s_2 - s_1 \geq 0 \tag{5.4}$$

where 1 refers to the inlet and 2 to the outlet, and Δs_∞ is the specific entropy change of the components in the surroundings interacting with the system. All entropies are on a per unit of mass passing through the system basis. Outlet forbidden states for this case must satisfy

$$s_2 \geq s_1 - \Delta s_\infty$$

which is illustrated in Figure 5.12. Notice the similarity between the forbidden states of a steady-flow system with a single inlet and outlet and the forbidden states of the closed system as presented in Section 4.11.

FIGURE 5.12
Forbidden States for a Single Inlet–Outlet Uniform Steady-Flow System

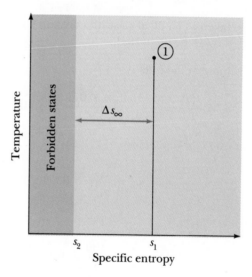

The adiabatic steady-flow system does not exchange heat with components in its surroundings, and the entropy of the surroundings is not changed by the process occurring in the system. In this event, $\dot{s}_\infty$ or s_∞ of the preceding expressions is zero, and the second law becomes

$$\sum_{out} \dot{m} s \geq \sum_{in} \dot{m} s$$

or with a single inlet and outlet,

$$s_2 \geq s_1$$

where the equals sign applies when the process occurring in the system is reversible. Consequently, the adiabatic, reversible process in a uniform steady-flow system with a single inlet–outlet is the isentropic process.

EXAMPLE 5.8

A reversible, adiabatic, uniform, steady-flow air compressor compresses air from 15 psia, 70° F to 180 psia. The air enters the compressor at a rate of 100 ft³/s through a 1 ft² opening. It exits the compressor through a 0.3 ft² pipe. Determine the power needed to drive this compressor.

System: A uniform steady-flow system that operates reversibly and adiabatically.

Basic Equations:
$$\dot{V} = \dot{m}v$$
$$\dot{m}_1 = \dot{m}_2$$

$$_1q_2 - {}_1w_2 = h_2 - h_1 + \frac{1}{2}(V_2^2 - V_1^2) + g(z_2 - z_1)$$

$$Pv = RT$$
$$dh = c_p\,dT$$

Conditions: The air will be taken as an ideal gas with constant specific heats. Fluid stream potential energies will be neglected.

Solution: The specific volume at the inlet is

$$v_1 = \frac{RT_1}{P_1} = \frac{1545 \times 530}{28.97 \times 15 \times 144} = 13.09 \text{ ft}^3/\text{lb}_m$$

The mass flow rate at the inlet is

$$\dot{m}_1 = \dot{m}_2 = \frac{\dot{V}_1}{v_1} = 7.639 \text{ lb}_m/\text{s}$$

At the inlet, the velocity is

$$V_1 = \frac{\dot{V}_1}{A_1} = \frac{100}{1} = 100 \text{ ft/s}$$

Because the process is isentropic and involves an ideal gas with constant specific heats, $P_1v_1^k = P_2v_2^k$ (see Section 4.9). The outlet temperature is then

$$T_2 = T_1\left(\frac{P_2}{P_1}\right)^{(k-1)/k} = 530\left(\frac{180}{15}\right)^{0.4/1.4}$$
$$= 1078° \text{ R} = 618° \text{ F}$$

According to the ideal gas equation of state,

$$v_2 = \frac{RT_2}{P_2} = 2.218 \text{ ft}^3/\text{lb}_m$$

The velocity at the outlet is therefore

$$V_2 = \frac{\dot{m}_2 v_2}{A_2} = \frac{7.639 \times 2.218}{0.3}$$
$$= 56.48 \ \text{ft/s}$$

Because the system is adiabatic and $\Delta h = c_P \Delta T$, the first law becomes

$$_1 w_2 = c_P(T_1 - T_2) + \frac{V_1^2 - V_2^2}{2}$$

TIP:

Study this example carefully because it incorporates almost every topic discussed to this point. You should also sketch the T–s state diagram to gain additional understanding of this example.

$$= 0.2397(70 - 618) + \frac{100^2 - 56.48^2}{2 \times 32.174 \times 778}$$
$$= -131.2 \ \text{Btu/lb}_m$$

Finally, the power required by this compressor is

$$\dot{W} = \dot{m}_1 w_2$$

$$= \frac{7.64 \times (-131.2)}{2545} \quad \left[\frac{\text{lb}_m}{\text{s}} \ \frac{\text{Btu}}{\text{lb}_m} \ \frac{\text{hp}}{\text{Btu/s}} \right]$$

$$= -0.3938 \ \text{hp}$$

5.6   Common Steady-Flow Processes

Many different processes are used to transform energy between its various forms, including heat and work. The nature of the process the working fluid undergoes is determined by the equipment through which it passes. In this section, we will examine the more frequently occurring processes and the equipment associated with them.

5.6.1   The Isobaric Process

The isobaric (i.e., constant pressure) process normally occurs in steady-flow passages and devices that are not undergoing shaft-work interactions. This process occurs as fluids flow through pipes and ducts, such as those that line the interior of a boiler or heat exchanger. Exhaust stacks, heating and ventilating ducts, condensers, and constant area flow lines are additional examples of isobaric flow-passage processes. It also occurs in chambers or vessels where fluids are mixed, separated, or undergoing chemical reactions. Typical examples of these vessels include open feedwater heaters, combustion chambers in gas turbines, liquid–vapor separators, and boiler combustion chambers.

The reversible isobaric process is an approximation (a highly useful approximation) to the actual process occurring in the system. For example, the operating pressure in the tubes of a modern steam boiler is several hundred psia. The difference in the pressure between the water inlet and the steam outlet is typically on the order of one atmosphere as a result of the fluid friction inside the tubes. As a result, treating the process as a reversible isobaric process is quite accurate in most cases.

5.6.2   The Isothermal Process

The reversible isothermal process is normally associated with devices in which a phase transition is occurring. Examples include steam boilers, refrigeration evaporators, and condensers. These processes are also isobaric as well as isothermal because of the phase transition that occurs. Although there can be a small temperature difference between the inlet and outlet of such a device (i.e., a result of the pressure drop or other factors), it is typically quite small and can often be neglected as a first approximation.

This process is an integral part of the best possible cycles (i.e., Carnot and other completely reversible cycles). As a result, several devices such as regenerators and recuperaters have been developed to simulate or approximate the reversible isothermal process. Unfortunately, they have not been entirely successful or cost-effective. Even today, development programs continue to seek equipment in which an isothermal process occurs without a phase transition.

5.6.3   The Isentropic Process

Isentropic processes are normally associated with work or kinetic energy-conversion equipment such as turbines, compressors, nozzles, and diffusers. In these devices, we seek a complete coupling between the work interaction, thermal energy, kinetic energy, and potential energy without any heat transfer. In the absence of heat transfer, the first law becomes

$$-\dot{W} = \sum_{out}(h + e_k + e_p)\dot{m} - \sum_{in}(h + e_k + e_p)\dot{m}$$

which clearly shows that the energy change is directly converted to useful work or the work is directly converted to energy change.

A process can be easily made adiabatic by simply enclosing the equipment in sufficient thermal insulation. Eliminating the irreversibilities resulting from fluid friction, unrestrained expansion, mixing, and other factors is more difficult. Hence, the isentropic process is an idealization of the process occurring in this type of equipment. As the design of this equipment continues to evolve, the actual processes approach this idealization.

5.6.4 The Polytropic Process

The *polytropic process* is a generalization of the isentropic and other processes that are used when the working fluid is an ideal gas. A polytropic process is any process for which

$$Pv^n = \text{constant}$$

applies. In this relation, n is any arbitrary number. Notice that this result includes the isentropic process (i.e., $n = k$), the isobaric process (i.e., $n = 0$), the isothermal process (i.e., $n = 1$), and the constant volume process (i.e., $n = \infty$) when the working fluid is an ideal gas. Ideal gases undergoing polytropic processes also obey

$$\frac{T_1}{T_2} = \left(\frac{P_1}{P_2}\right)^{n-1/n} = \left(\frac{v_2}{v_1}\right)^{n-1}$$

Because the polytropic process coupled with the ideal gas equation of state is very general, it is quite useful for analyzing ideal gas processes.

5.6.5 The Throttling Process

Throttling processes occur in obstructed flow passages such as valves, flow meters, capillary tubes, and other devices that reduce the pressure of the working fluid without any shaft work or heat interactions. Application of the first law to one of these devices, neglecting the fluid's kinetic and potential energy, gives

$$h_2 = h_1$$

The throttling process is one in which the enthalpy remains constant while the other properties of the working fluid change.

Reducing the pressure of the working fluid in this manner involves considerable fluid friction and mixing. As a result, this process is highly irreversible and wasteful of the energy in the working fluid. It is only used in special applications or when economically advantageous. It is primarily used as a means to control a device such as a turbine, to reduce the pressure between the condenser and evaporator of a refrigeration cycle inexpensively, or to measure mass flow rates and quality of a mixture.

EXAMPLE 5.9

A steam line at 200 psia carries wet steam. To measure the quality of this steam, the valve in the adiabatic throttling calorimeter shown in Figure 5.13 is partially opened until a steady flow is established through the calorimeter. Once steady flow is established, the pressure and temperature at the outlet of this calorimeter are measured to be 14.7 psia, 250° F. What is the quality in the steam line?

FIGURE 5.13
*Throttling
Calorimeter for
Example 5.9*

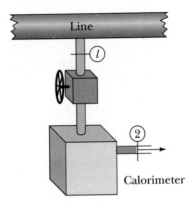

System: A uniform steady-flow system with a single inlet–outlet.

Basic Equations:

$$_1q_2 - {}_1w_2 = h_2 - h_1 + \frac{1}{2}(V_2^2 - V_1^2) + g(z_2 - z_1)$$

212.0	26.80	1077.6	1150.5	1.7567
250	28.42	1091.5	**1168.8**	1.7832
300	30.52	1109.6	1192.6	1.8157
400	34.67	1145.6	1239.9	1.8741
500	38.77	1181.8	1287.3	1.9263

170	368.47	0.01821	2.676	340.8	1112.7	341.3	855.6	1196.9	0.5270	1.5600
180	373.13	0.01827	2.533	345.7	1113.4	346.3	851.5	1197.8	0.5329	1.5553
190	377.59	0.01833	2.405	350.4	1114.0	351.0	847.5	1198.6	0.5386	1.5507
200	381.86	0.01839	2.289	354.9	1114.6	355.6	843.7	1199.3	0.5440	1.5464
250	401.04	0.01865	1.845	375.4	1116.7	376.2	825.8	1202.1	0.5680	1.5274

Conditions: The device is adiabatic and does not exchange work with its surroundings. Fluid kinetic and potential energies will be neglected.

Solution: Because no heat or work is transferred and all energies are negligible except the enthalpy, the first law becomes

$$h_2 = h_1 = h_{f,1} + x_1 h_{fg,1}$$

Hence, the enthalpy in the steam line is 1168.8 Btu/lb$_m$, according to the steam tables. Using this enthalpy and the pressure in the steam line, the quality in the steam line is

$$x_1 = \frac{h_1 - h_{f,1}}{h_{fg,1}}$$

$$= \frac{1168.8 - 355.6}{843.7}$$

$$= 0.9638$$

5.7 Other Similarities to Closed Systems

We have seen that the first and second laws for a uniform steady-flow system with a single inlet–outlet are very similar to those of a closed system. We will develop two other similarities—heat and work transfer—in this section. You should carefully note these similarities because they are very useful for remembering the various important equations of closed and steady-flow, single inlet–outlet systems.

As an element of mass passes through the control volume, it undergoes a process of some type as it transitions from inlet state 1 to exit state 2. While undergoing this process in the control volume, it may also transfer heat and produce or consume work. The specific heat added to this mass as it undergoes a reversible process while passing through the control volume is

$$_1q_2 = \int_1^2 T\,ds \quad \text{(reversible process only)} \qquad (5.5)$$

according to the definition of entropy for a fixed mass. Because heat is only transferred to this mass as it passes through the control volume, this is also the specific-heat transfer to the control volume. Consequently, the specific-heat addition for a steady-flow system with a single inlet–outlet is calculated in the same way as for a closed system; that is, the specific-heat transfer is the area under the reversible process path when plotted on a T–s state diagram, such as that of Figure 4.10 (page 148).

EXAMPLE 5.10

Carbon dioxide undergoes a reversible, isobaric process in a steady-flow device with a single inlet–outlet. The pressure is maintained at 500 kPa, the temperature at the inlet is 200° C, and the temperature at the outlet is 25° C. Calculate the heat transfer for this process.

System: A uniform steady-flow system with a single inlet–outlet.

Basic Equations:

$$_1q_2 = \int_1^2 T\,ds$$

$$T\,ds = dh - v\,dP$$
$$dh = c_p\,dT$$

Conditions: The process is stated to be reversible and isobaric. The carbon dioxide will be taken to be an ideal gas with constant specific heats.

Solution: The vdP term of the Gibbs equation is zero because the pressure is constant. The specific-heat transfer is then

TIP:

This approach to calculating the heat transfer avoids the necessity of evaluating the kinetic and potential energy changes as well as the work production of the first law.

$$\begin{aligned} {}_1q_2 &= \int_1^2 dh \\ &= h_2 - h_1 \\ &= c_P(T_2 - T_1) \\ &= 0.8446(25 - 200) \\ &= -147.8 \text{ kJ/kg} \end{aligned}$$

In differential form, the first law for a steady-flow, single inlet–outlet system is

$$\delta q - \delta w = dh + de_k + de_p$$

The *dh* form of the Gibbs equations reduces this to

$$\delta q - \delta w = T\,ds + vdP + de_k + de_p$$

When the process is reversible, $\delta q = T\,ds$, and the first law becomes

$$\delta w = -vdP - de_k - de_p$$

Thus, the total power produced per unit of mass flow rate when a reversible process occurs inside a steady-flow, single inlet–outlet system is

$$ {}_1w_2 = -\int_1^2 vdP - (e_{k,2} - e_{k,1}) - (e_{p,2} - e_{p,1}) $$

(reversible processes only)   (5.6)

When the kinetic and potential energies of the inlet and outlet stream are small, the specific work added to the control volume is the area to the left of the reversible process path when plotted on a P–v state diagram as in Figure 5.14.

FIGURE 5.14

Comparison of Closed System and Steady-Flow, Single Inlet–Outlet System Work

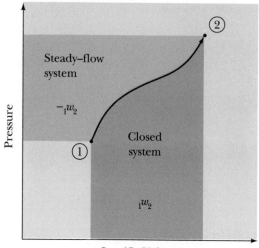

Specific Volume

| EXAMPLE 5.11 | Derive an equation for the specific work produced when an ideal gas undergoes a reversible polytropic process ($n \neq 1$) in a single inlet–outlet, steady-flow system. The final result should be in terms of the inlet and outlet pressures and the inlet temperature. |

System: An arbitrary uniform steady-flow system with a single inlet–outlet.

Basic Equations:

$$_1 w_2 = -\int_1^2 v \, dP - (e_{k,2} - e_{k,1}) - (e_{p,2} - e_{p,1})$$

$$Pv = RT$$
$$Pv^n = \text{constant}$$

Conditions: The gas is an ideal gas. Kinetic and potential energies of the gas will be neglected. The process is reversible.

Solution: The constant for this process is either $P_1 v_1^n$ or $P_2 v_2^n$. At an arbitrary point along the process path, the specific volume is

$$v = \left(P_1^{1/n} v_1 \right) P^{-1/n}$$

Substituting this into the work integral gives

$$_1 w_2 = -P_1^{1/n} v_1 \int_1^2 P^{-1/n} \, dP$$

Integrating this expression yields

$$_1w_2 = -\frac{n}{n-1}P_1^{1/n}v_1\left[P_2^{(n-1)/n} - P_1^{(n-1)/n}\right]$$

TIP:

**When n = 1, this
integration produces
the result**

$_1w_2 = RT_1 \ln \dfrac{P_1}{P_2}$

$$= -\frac{n}{n-1}P_1^{1/n}v_1 P_1^{(n-1)/n}\left[\left(\frac{P_2}{P_1}\right)^{(n-1)/n} - 1\right]$$

$$= -\frac{n}{n-1}RT_1\left[\left(\frac{P_2}{P_1}\right)^{(n-1)/n} - 1\right]$$

5.8 Steady-Flow Process Efficiencies

First law process efficiencies for steady-flow systems with a single inlet–outlet are defined in the same way as for closed systems (see Section 4.12); that is, they are the ratio of the actual result to the ideal result for a producing system and the ratio of the ideal result to the actual result for a consuming system.

Turbines operate adiabatically and are intended to produce work. They are typically designed so that the working fluid's kinetic and potential energy changes are small. The actual specific work produced by such a turbine is

$$_1w_{2,act} = h_1 - h_2$$

If all of the internal irreversibilities could be eliminated, the process through the turbine would be isentropic between the same inlet and outlet pressures. The ideal specific work produced would then be

$$_1w_{2,ideal} = h_1 - h_{2s}$$

where state 2s is the state whose pressure is the same as the actual outlet pressure and whose entropy equals that of the actual inlet state. These states are illustrated in Figure 4.16 (page 161). The efficiency of the turbine is then

$$\eta_{turbine} = \frac{h_1 - h_2}{h_1 - h_{2s}}$$

Nozzles are flow passages designed to increase the kinetic energy of a fluid stream adiabatically as its pressure is reduced without any work being added or removed. The first law tells us that the actual increase in the specific kinetic energy is

$$\Delta e_{k,actual} = h_1 - h_2$$

If this fluid acceleration is accomplished reversibly with the same inlet and outlet pressures, the ideal specific kinetic-energy increase would be

$$\Delta e_{k,ideal} = h_1 - h_{2s}$$

where state $2s$ is the same as that of an adiabatic turbine. Consequently, the efficiency of the nozzle is

$$\eta_{nozzle} = \frac{h_1 - h_2}{h_1 - h_{2s}}$$

which is the same result as for a turbine.

Compressors operate adiabatically between a fixed inlet and outlet pressure. By consuming work, they increase the pressure of the working fluid as it passes through the compressor. When the small kinetic and potential energy changes are neglected, the first law tells us that the amount of specific work consumed is

$$-_1 w_{2,actual} = h_1 - h_2$$

Ideally, the compressor would operate reversibly between the same two pressures. The specific work consumed would then be

$$-_1 w_{2,ideal} = h_1 - h_{2s}$$

where state $2s$ is illustrated in Figure 4.17 (page 167). Combining these expressions, the compressor efficiency is

$$\eta_{compressor} = \frac{h_1 - h_{2s}}{h_1 - h_2}$$

which is the reciprocal of the turbine efficiency.

A diffuser is a flow passage designed to decelerate the fluid adiabatically without any work interaction, thereby decreasing its kinetic energy while increasing the pressure. The actual specific kinetic-energy change is given by the first law as

$$\Delta e_{k,actual} = h_1 - h_2$$

Ideally, we would decelerate the fluid isentropically, in which case the change in specific kinetic energy would be

$$\Delta e_{k,ideal} = h_1 - h_{2s}$$

The efficiency of the diffuser is therefore

$$\eta_{diffuser} = \frac{h_1 - h_{2s}}{h_1 - h_2}$$

where the $2s$ state is the same as that of a compressor.

EXAMPLE 5.12   Steam enters an adiabatic nozzle at 450 psia, 500° F with a velocity of 10 ft/s. It leaves this nozzle at 300 psia. Determine the outlet temperature and velocity when the efficiency of the nozzle is 90%.

System: A uniform steady-flow nozzle.

Basic Equations:
$$_1q_2 - {_1}w_2 = h_2 - h_1 + \frac{1}{2}(V_2^2 - V_1^2) + g(z_2 - z_1)$$

Conditions: The process is adiabatic, there is no shaft work, and potential energy changes will be neglected.

456.4	1.033	1119.6	1205.6	1.4746
500	1.123	1145.1	**1238.5**	**1.5097**
550	1.215	1170.7	1271.9	1.5436
600	1.300	1194.3	1302.5	1.5732
700	1.458	1238.2	1359.6	1.6248

300	417.43	0.01890	1.544	393.0	1118.2	**394.1**	**809.8**	1203.9	**0.5883**	1.5115
350	431.82	0.01912	1.327	408.7	1119.0	409.9	795.0	1204.9	0.6060	1.4978
400	444.70	0.01934	1.162	422.8	1119.5	424.2	781.2	1205.5	0.6218	1.4856
450	456.39	0.01955	1.033	435.7	1119.6	437.4	768.2	1205.6	0.6360	1.4746
500	467.13	0.01975	0.928	447.7	1119.4	449.5	755.8	1205.3	0.6490	1.4645

Solution: The actual and ideal nozzle processes are shown in Figure 5.15. At state $2s$ the entropy is 1.5097 Btu/lb$_m$ $\cdot°$ R and the pressure is 300 psia. This state is in the mixture region and the quality at $2s$ is

$$x_{2s} = \frac{s_1 - s_{f,2s}}{s_{fg,2s}}$$

$$= \frac{1.5097 - 0.5883}{0.9232}$$

$$= 0.998$$

The enthalpy at state $2s$ is $h_{f,2s} + x_{2s}h_{fg,2s} = 394.1 + 0.998 \times 809.8 = 1202.3$ Btu/lb$_m$. Employing the nozzle efficiency, the enthalpy of the actual state 2 is

$$h_2 = h_1 - \eta_{nozzle}(h_1 - h_{2s})$$
$$= 1238.5 - 0.9(1238.5 - 1202.3)$$
$$= 1205.9 \text{ Btu/lb}_m$$

Interpolating in the steam tables (see Appendix A), the actual exhaust temperature is

$$T_2 = 419° \text{ F}$$

Solving the first law for the outlet velocity gives

$$V_2 = \sqrt{V_1^2 + 2(h_1 - h_2)}$$
$$= \sqrt{10^2 + 2 \times 32.2 \times 778(1238.5 - 1205.9)}$$
$$= 1278 \text{ ft/s}$$

FIGURE 5.15

T–s State Diagram for Example 5.10

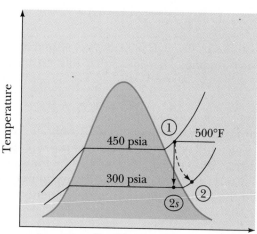

5.9 Flow Exergy

Whenever a steady-flow device alters the state of the fluid in its streams, an engineer will naturally ask if this has been done in a way that produces the greatest amount of work (or consumes the least amount). For example, when the steam that is exhausted from the turbine of an electrical-generation station is condensed in the condenser, it is condensed at the expense of producing additional work. The flow exergy will provide a means for determining the maximum amount of additional work that could have been produced for the steam states at the condenser inlets and outlets. This information may then be used to develop and assess alternative means to accomplish the same change of states in a manner that improves the system's overall performance.

The exergy of steady-flow system streams (also known as stream exergy and availability) is defined in a manner similar to that of closed systems. The *flow exergy* for one inlet or outlet stream is defined as the maximum work the fluid of this stream could produce if it were reversibily reduced to the dead state. This includes bringing the fluid to rest, changing its elevation to some common reference elevation, and bringing the pressure and temperature to P_0 and T_0. The work required to push the environment back is not included, because the control volume does not expand or contract, thereby moving the environmental fluid.

Following the derivation of Section 4.13, the maximum power a stream could produce under these stipulations as it is reduced to the dead state would be

$$\dot{W}_{max} = \dot{m}[(h - h_0) + e_k + e_p - T_0(s - s_0)]$$

The specific flow exergy, ψ, of this stream would be

$$\psi = \frac{\dot{W}_{max}}{\dot{m}} = (h - h_0) + e_k + e_p - T_0(s - s_0)$$

As a stream passes through a steady-flow system and does not exchange heat with its surroundings, the power that this stream has the potential to produce is

$$\dot{W}_{ptl} = \dot{m}(\psi_{in} - \psi_{out})$$

When a stream is exchanging heat with an energy reservoir as it passes through a steady-flow system, additional power could be produced from the heat transfer. This can be accomplished by transferring this heat through a completely reversible heat engine. The maximum useful power for any heat transfer to the control volume from components in the surroundings is (see Section 2.10).

$$\dot{W}_{max,j} = \dot{Q}_j\left(1 - \frac{T_0}{T_j}\right)$$

where T_j is the temperature of the thermal-energy reservoir exchanging heat with the control volume, and $\dot{Q}_j$ is taken with respect to the control volume.

When a control volume has several inlets and outlets and exchanges heat with several energy reservoirs in its surroundings, the power this system could potentially produce is

$$\dot{W}_{ptl} = \sum_{in} \dot{m}\psi - \sum_{out} \dot{m}\psi + \sum_{j=1}^{k} \dot{Q}_j\left(1 - \frac{T_0}{T_j}\right)$$

If the control volume has one inlet and one outlet, the preceding expression becomes

$$w_{ptl} = \psi_1 - \psi_2 + \sum_{j=1}^{k} Q_j\left(1 - \frac{T_0}{T_j}\right)$$

A second law effectiveness can be defined for a steady-flow system in the same manner as for a closed system. For a steady-flow system, the second law effectiveness is

$$\eta_{2nd} = \frac{\dot{W}_{actual}}{\dot{W}_{ptl}}$$

EXAMPLE 5.13

Steam is expanded in an adiabatic steady-flow steam turbine from 1000 kPa, 280° C to 120° C. Calculate the second law effectiveness of this turbine when its efficiency is 85%.

System: The system is an adiabatic, steady-flow steam turbine with a single inlet and outlet.

Basic Equations:

$$_1q_2 - {}_1w_2 = (h_2 - h_1) + (e_{k,2} - e_{k,1}) + (e_{p,2} - e_{p,1})$$

$$\eta_{turbine} = \frac{h_1 - h_2}{h_1 - h_{2,s}}$$

$$\eta_{2nd} = \frac{\dot{W}_{actual}}{\dot{W}_{ptl}}$$

$$\psi = (h - h_0) - T_0(s - s_0)$$

$$\dot{W}_{ptl} = \sum_{in} \dot{m}\psi - \sum_{out} \dot{m}\psi + \sum_{j=1}^{k} \dot{Q}_j \left(1 - \frac{T_0}{T_j}\right)$$

200	0.2060	2621.9	2827.9	6.6940
240	0.2275	2692.9	2920.4	6.8817
280	0.2480	2760.2	**3008.2**	**7.0465**
320	0.2678	2826.1	3093.9	7.1962
360	0.2873	2891.6	3178.9	7.3349

100	101.4	0.001044	1.673	418.94	2506.5	419.04	2257.0	2676.1	1.3069	7.3549
110	143.3	0.001052	1.210	461.14	2518.1	461.30	2230.2	2691.5	1.4185	7.2387
120	198.5	0.001060	0.8919	503.50	2529.3	503.71	2202.6	2706.3	1.5276	7.1296
130	270.1	0.001070	0.6685	546.02	2539.9	546.31	2174.2	2720.5	1.6344	7.0269
140	361.3	0.001080	0.5089	588.74	2550.0	589.13	2144.7	2733.9	1.7391	6.9299

Conditions: Fluid kinetic and potential energies will be neglected.

Solution: Working with the steam tables, $h_1 = 3008.2$ kJ/kg, $s_1 = 7.0465$ kJ/kg-K, $h_{2s} = 2673.7$ kJ/kg, $x_{2s} = 0.9852$, and $P_2 = 198.5$ kPa. The enthalpy at the actual outlet state 2 is

$$\begin{aligned} h_2 &= h_1 - \eta_{turbine}(h_1 - h_{2s}) \\ &= 3008.2 - 0.85(3008.2 - 2763.7) \\ &= 2800.4 \text{ kJ/kg} \end{aligned}$$

Entering the steam tables with P_2 and h_2 yields $T_2 = 128.3°\,C$ and $s_2 = 7.1736$ kJ/kg-K. When the first law is applied to the turbine, the actual specific work produced is

$$_1 w_{2,act} = h_1 - h_2$$
$$= 207.8 \text{ kJ/kg}$$

When the steam at the inlet and outlet is reduced to 101 kPa, 25° C, it will have condensed to a liquid. The incompressible-liquid approximation allows us to use the saturated liquid properties at 25° C of the steam tables for the dead state. Hence, $h_0 = 104.89$ kJ/kg and $s_0 = 0.3674$ kJ/kg-K. The flow exergy of the inlet is

$$\psi_1 = (h_1 - h_0) - T_0(s_1 - s_0)$$
$$= (3008.2 - 104.9) - 298(7.0465 - 0.3674)$$
$$= 912.9 \text{ kJ/kg}$$

TIP:

The first law efficiency compares the turbine performance to that of the best possible adiabatic turbine. The second law effectiveness compares the performance of the turbine to the potential power or work using the same inlet and outlet states.

and $\psi_2 = 667.3$ kJ/kg in the same manner. Using the definition of the second law effectiveness,

$$\eta_{2nd} = \frac{_1 w_{2,act}}{\psi_1 - \psi_2}$$
$$= \frac{207.8}{912.9 - 667.3}$$
$$= 0.846$$

In addition to measuring the maximum work that a steady-flow system might produce, the flow exergy can also be used to compare alternatives or identify opportunities for energy savings. For example, we might calculate the flow exergy of all the streams leaving an electrical power plant. The results of this calculation would tell us how much more work or power this plant could produce if we were to pass these streams through further work-producing devices before exhausting them to the environment. We could also calculate the increase or decrease of the work potential for each process and device in the system. Those devices with the greatest increase in work potential represent the best opportunities for improving the plant's overall energy conversion. The difference between the work potential and actual work produced is the *lost work potential*. Processes and devices with large lost work potentials also present opportunities for system improvements. This approach to analyzing a system is known as *exergy and lost work potential accounting*. Analyses of this type are illustrated in the following examples.

EXAMPLE 5.14	An electrical power plant exhausts 465 lb_m/s of stack gases (treat as air) at 450° F. It also uses 210,000 lb_m/s of cooling water from a lake at 55° F. This water is returned to the lake at 72° F. Which of the two streams leaving this plant is the more wasteful?

System: A steady-flow system.

Basic Equations:
$$\psi = (h - h_0) + e_k + e_p - T_0(s - s_0)$$
$$T\,ds = dh - v\,dP$$
$$dh = c_p\,dT$$

Conditions: Fluid kinetic and potential energies will be neglected. The stack gases will be treated as air.

Solution: Because the pressure of both streams is the same as that of the surrounding environment, the Gibbs equation intergrates to

$$s - s_0 = c_p \ln \frac{T}{T_0}$$

The rate at which flow exergy is exhausted in the stack gases is

$$\dot{\psi}_{sg} = \dot{m}_{sg}\left[c_p(T - T_0) - c_P T_0 \ln \frac{T_0}{T}\right]$$

$$= 465\left[0.2397(450 - 77) - 0.2397 \times 573\ln\frac{910}{537}\right]$$

$$= 7888 \text{ Btu/s}$$

TIP:

Although more energy is available in the cooling water, it is less valuable because its temperature is not very different from its lake surroundings. The stack gas stream has a $450 - 77 = 373°$ F temperature difference to operate some heat-engine device. This makes the energy available in this stream more valuable.

The environment and dead state for the cooling water is the lake and its state. The rate at which flow exergy is exhausted by the cooling water is

$$\dot{\psi}_{cw} = \dot{m}_{cw}\left[c(T - T_0) - cT_0 \ln \frac{T}{T_0}\right]$$

$$= 210,000\left[1.0(72 - 55) - 1.0 \times 515\ln\frac{532}{515}\right]$$

$$= 57,656 \text{ Btu/s}$$

Clearly, the cooling water stream is the more wasteful because it has the greatest potential for doing additional work.

EXAMPLE 5.15

A stream of air at 100 kPa is to be heated from $20°$ C to $80°$ C. This is to be done using condensing steam (condensing from a saturated vapor to a saturated liquid). The steam can be condensed at either 70 kPa or 150 kPa. What is the difference between these two options?

System: The air and condensing-steam streams in a system similar to that shown in Figure 5.7.

Basic Equations:

$$_1q_2 - _1w_2 = (h_2 - h_1) + (e_{k,2} - e_{k,1}) + (e_{p,2} - e_{p,1})$$
$$\psi = (h - h_0) + e_k + e_p - T_0(s - s_0)$$
$$dh = c_p\, dT$$

70	89.95	0.001036	2.365	376.63	2494.5	**376.70**	**2283.3**	2660.0	**1.1919**	**7.4797**
80	93.50	0.001039	2.087	391.58	2498.8	391.66	2274.1	2665.8	1.2329	7.4346
90	96.71	0.001041	1.869	405.06	2502.6	405.15	2265.7	2670.9	1.2695	7.3949
100	99.63	0.001043	1.694	417.36	2506.1	417.46	2258.0	2675.5	1.3026	7.3594
150	111.4	0.001053	1.159	466.94	2519.7	**467.11**	**2226.5**	2693.6	**1.4336**	**7.2233**

Conditions: Kinetic and potential energies of all fluids will be neglected. The air will be treated as an ideal gas with constant specific heats.

Solution: We will select one unit of air flow as the basis for all calculations. As one unit of air passes through this system, it will receive

$$_1q_3 = h_3 - h_1 = c_P(T_3 - T_1)$$
$$= 1.0035(80 - 20) = 60.21 \text{ kJ/kg-air}$$

units of heat from the condensing-steam stream. The flow exergy of the air stream would then be increased by

$$\Delta\psi_{air} = (h_3 - h_1) - T_0(s_3 - s_1)$$

$$= c_p(T_3 - T_1) - T_0 c_P \ln \frac{T_3}{T_1}$$

$$= 1.0035(353 - 293) - 298 \times 1.0035 \ln \frac{353}{293}$$

$$= 4.500 \text{ kJ/kg-air}$$

per unit of heated air.

When the heat is supplied by a 70 kPa condensing-steam stream, the mass of condensing steam per unit of air is

$$m_{steam} = \frac{_2q_4}{h_4 - h_2} = \frac{_2q_4}{-h_{fg}}$$

$$= \frac{-61.74}{-2283.3} = 0.02704 \text{ kg/kg-air}$$

The flow exergy of the condensing steam would decrease by

$$
\begin{aligned}
\Delta\psi_{steam} &= m_{steam}(\psi_4 - \psi_2) \\
&= m_{steam}[(h_4 - h_2) - T_0(s_4 - s_2)] = m_{steam}[-h_{fg} + T_0 s_{fg}] \\
&= 0.02704[-2283.3 + 298 \times 6.2878] = -11.07 \text{ kJ/kg-air}
\end{aligned}
$$

TIP:

The loss of a system's or a process's ability to produce work results from irreversibilities such as the transfer of heat over finite temperature differences in this example.

per unit of air flow. The condensing steam has reduced its ability to produce work by 11.07 kJ/kg-air, while the air stream has increased its ability to produce work by 4.500 kJ/kg-air. Potential work production in the amount of 6.57 kJ/kg-air has then been lost as a result of the heat transfer between the two streams.

When 150 kPa condensing steam is used, the mass of condensing steam per unit of air flow is 0.02773 kg/kg-air, and the change in the condensing steam's flow exergy is −13.90 kJ/kg-air. In this case, the lost work potential is 9.40 kJ/kg-air as a result of the heat being transferred between the two streams.

When the pressure of the condensing steam is increased, less steam is required, and the ability of the system to produce work is reduced. Which of the two options is preferred depends on the intended purpose of this system. Clearly, the lower pressure option preserves more of the system's valuable work-production capability.

PROBLEMS

Sketch the control volume with clearly labeled inlets and outlets for the following problems. Also sketch the appropriate P–v and T–s state diagrams for the following problems, showing all states and processes.

5.1 A steady-flow compressor such as that illustrated in Figure 3.22 (p. 118) compresses helium from 15 psia, 70° F at the inlet to 200 psia, 600° F at the outlet. The outlet area and velocity are 0.01 ft² and 100 ft/s. The inlet velocity is 50 ft/s. Determine the mass flow rate (lb_m/s) and inlet area (ft²).

5.2 Air is expanded from 1000 kPa, 600° C at the inlet of a steady-flow turbine to 100 kPa, 200° C at the outlet. The inlet area and velocity are 0.1 m² and 30 m/s. The outlet velocity is 10 m/s. Determine the mass flow rate (kg/s) and outlet area (m²).

5.3 Refrigerant-12 is expanded isentropically from 800 kPa, 60° C at the inlet of a steady-flow turbine to 100 kPa at the outlet. The outlet area is 1 m², and the inlet area is 0.5 m². Calculate the inlet and outlet velocities (ft/s) when the mass flow rate is 0.5 kg/s.

5.4 Water enters the constant 130-mm inside-diameter tubes of a boiler at 7,000 kPa, 65° C. This water leaves the tubes at 6,000 kPa, 440° C with a velocity of 80 m/s. Calculate the velocity (m/s) of the water at the tube inlet and the inlet volume flow rate in liters/s.

5.5 A steam pipe is to transport 200 lb$_m$/s of steam at 200 psia, 600° F. Calculate the minimum diameter (ft) this pipe can have so that the steam velocity will not exceed 59 ft/s.

5.6 A steady-flow water pump increases the water pressure from 10 psia at the inlet to 100 psia at the outlet. The water enters this pump at 60° F through a 0.03-ft-diameter opening and exits through a 0.05-ft-diameter opening. Determine the velocity (ft/s) of the water at the inlet and outlet when the mass flow rate through the pump is 1 lb$_m$/s. Do these velocities change significantly if the inlet temperature is raised to 100° F?

5.7 An open feedwater heater such as that shown in Figure 5.16 heats the feedwater by mixing it with hot steam. Consider an electrical power plant with an open feedwater heater that mixes 0.1 lb$_m$/s of steam at 10 psia, 200° F with 2.0 lb$_m$/s of feedwater at 10 psia, 100° F to produce 10 psia, 120° F feedwater at the outlet. The diameter of the outlet pipe is 0.5 ft. Determine the mass flow rate (lb$_m$/hr) and feedwater velocity (ft/s) at the outlet. Would the outlet flow rate and velocity be significantly different if the temperature at the outlet were 180° F?

FIGURE 5.16

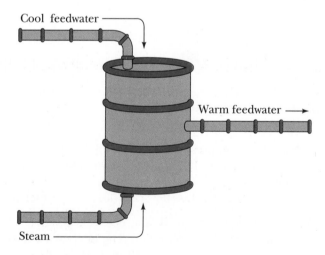

Cool feedwater

Warm feedwater →

Steam

5.8 An adiabatic open feedwater heater like that in Figure 5.16 in an electrical power plant mixes 0.2 kg/s of steam at 100 kPa, 160° C with 10 kg/s of feedwater at 100 kPa, 50° C to produce feedwater at 100 kPa, 60° C at the outlet. Determine the outlet mass flow rate (kg/s) and velocity (m/s) when the outlet pipe diameter is 0.03 m.

5.9 The fuel mixer in a natural gas burner mixes methane (CH_4) with air to form a combustible mixture at the outlet. Determine the mass flow rates (lb_m/s) at the two inlets needed to produce 0.01 lb_m/s of an ideal combustion mixture at the outlet.

5.10 Air enters a gas turbine at 150 psia, 700° F and leaves at 15 psia, 100° F. Determine the inlet and outlet volume flow rates (in ft^3/s) when the mass flow rate through this turbine is 5 lb_m/s.

5.11 A cyclone separator like that in Figure 5.17 is used to remove fine solid particles, such as fly ash, that are suspended in a gas stream. In the flue-gas system of an electrical power plant, the weight fraction of fly ash in the exhaust gases is approximately 0.001. Determine the mass flow rates (in lb_m/s) at the two outlets (flue gas and fly ash) when 10 lb_m/s of flue gas plus fly ash enters this unit.

FIGURE 5.17

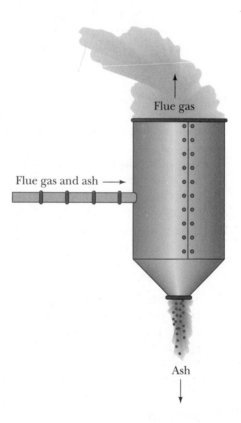

Flue gas

Flue gas and ash →

Ash

5.12 Air enters the 1 m^2 compressor inlet of an aircraft engine at 100 kPa, 20° C with a velocity of 180 m/s. Determine the volume flow rate (m^3/s) at the engine's entrance and the mass flow rate (kg/s) at the engine's exit.

5.13   A water pump increases the water pressure from 10 psia to 50 psia. Determine the flow power (hp) required to pump 1.2 ft³/s of water. Does the water temperature at the inlet have any significant effect on the required flow power?

5.14   Determine the specific flow work (Btu/lb$_m$) for the water pump of the preceding problem.

5.15   An air compressor compresses 10 liters/s of air at 120 kPa, 20° C to 1000 kPa, 300° C. Determine the specific flow work (kJ/kg) required by the compressor.

5.16   The air compressor of the preceding problem requires 4500 watts of power. How much of this power (kW) is being used to increase the pressure of the air versus the power needed to move the fluid through the compressor?

5.17   A constant-pressure R-12 vapor separation unit like that of Figure 5.5 (p. 194) separates the liquid and vapor portions of a saturated mixture into two separate outlet streams. Determine the flow power (kW) needed to pass 3 liters/s of R-12 at 200 kPa, 70% quality through this unit. What is the mass flow rate (kg/s) of the two outlet streams?

5.18   A heat exchanger like that in Figure 5.7 (p. 197) passes two separate streams through it; one is being heated by cooling the other. Consider a heat exchanger that uses hot air to warm cold water. One hundred ft³/minute of air enter this exchanger at 20 psia, 200° F and leave at 17 psia, 100° F. The water enters this unit at 20 psia, 50° F at a rate of 0.5 lb$_m$/s and exits at 17 psia, 90° F. Determine the total flow power (hp) required for this unit and the specific flow work (Btu/lb$_m$) for the air and water streams.

5.19   Determine the flow power (kW) of a compressor that compresses helium from 150 kPa, 20° C to 400 kPa, 200° C. The helium enters this compressor through a 0.1 m² pipe at a velocity of 15 m/s.

5.20   The mass flow rate of a compressed air line is divided into two equal streams by a T-fitting in the line. The compressed air enters this 1-in-diameter fitting at 200 psia, 100° F with a velocity of 150 ft/s. Each outlet has the same diameter as the inlet, and the air at these outlets has a pressure of 180 psia and a temperature of 95° F. Determine the velocity (ft/s) of the air at the outlets and the flow power (hp) for the T-fitting.

5.21   Two streams of water are mixed in an insulated container to form a third stream in a system similar to that in Figure 5.16. The first stream has a flow rate of 30 kg/s and a temperature of 90° C. The flow rate of the second stream is 200 kg/s, and the temperature of this stream is 50° C. What is the temperature of the third stream leaving the insulated container?

5.22 An adiabatic air compressor compresses 10 liters/s of air at 120 kPa, 20° C to 1000 kPa, 300° C. Determine the specific work (kJ/kg) required by the compressor.

5.23 Calculate the power (kW) required to drive the air compressor of the previous problem.

5.24 An adiabatic gas turbine like that shown in Figure 3.23 (p. 119) expands air at 1000 kPa, 500° C to 100 kPa, 150° C. The air enters the turbine through a 0.2 m² opening with a velocity of 40 m/s and exhausts through a 1 m² opening. Determine the turbine mass flow rate (kg/s) and power (kW) it produces.

5.25 Water is pumped from a 200-ft-deep well into a 100-ft-high storage tank. Determine the power (kW) that would be required to pump 200 gallons per minute.

5.26 The water behind Hoover Dam in Nevada is 675 ft higher than the Colorado River below it. At what rate (lb_m/s) must water pass through the hydraulic turbines of this dam to produce 100 MW of power if the turbines are 100% efficient?

5.27 A grist mill of the 1800s employed a water wheel that was 30 ft high; 100 gallons per minute of water flowed on to the wheel near the top. How much power (hp) could this water wheel have produced?

5.28 Windmills slow the air and cause it to fill a larger channel as it passes through the blades (as shown in Figure 5.18). Consider a circular windmill with a 7-m-diameter rotor in a 10 m/s wind on a day when the atmospheric pressure is 100 kPa and the temperature is 20° C. The wind speed behind the windmill is measured at 9 m/s, and the temperature is 20° C. Determine the diameter (m) of the wind channel downstream from the rotor and the power produced by this windmill (hp), presuming that the air is incompressible.

5.29 An automobile moving through the air causes the air velocity (measured with respect to the car) to decrease and fill a larger flow channel as illustrated in Figure 5.19. An automobile has an effective flow channel upstream of 30 ft². The car is traveling at 60 miles per hour on a day when the barometric pressure is 30 inches of mercury and the temperature is 80° F. Behind the car, the air velocity (with respect to the car) is measured at 55 miles per hour, and the temperature is 80° F. Determine the power (hp) required to move this car through the air and the area (ft²) of the effective flow channel behind the car.

FIGURE 5.18

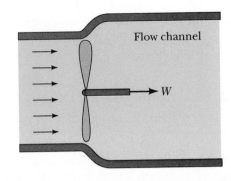

Flow channel

W

FIGURE 5.19

Flow channel

5.30 An adiabatic air compressor compresses air from 15 psia, 70° F to 300 psia, 800° F. The air enters the compressor through a 1.5 ft² opening with a velocity of 100 ft/s. It exits through a 0.8 ft² opening. Calculate the compressor mass flow rate (lb$_m$/s) and power (hp) it requires.

5.31 Calculate the specific work consumption of the previous problem in Btu/lb$_m$.

5.32 An adiabatic gas turbine expands air at 500 psia, 800° F to 60 psia, 250° F. If the volume flow rate at the exit is 50 ft³/s, the inlet area is 0.6 ft², and the outlet area is 1.2 ft², what is the power produced by this turbine in hp?

5.33 Calculate the specific work production of the previous problem in Btu/lb$_m$.

5.34 A steam turbine operates with 1500 kPa, 360° C steam at its entrance and saturated steam vapor at 30° C at its exit. The mass flow rate of the steam is 30 kg/s, and the turbine produces 9000 kW of power. Determine the rate at which heat is lost through the insulation surrounding this turbine.

5.35 Saturated liquid water is heated in a steady-flow steam boiler at a constant pressure of 800 psia. No shaft work is required for this process. Determine the amount of heat required per pound-mass of heated steam in Btu/lb$_m$ units when the outlet temperature is 600° F.

5.36 Determine the rate of heat transfer (Btu/s) when the mass flow rate through the boiler of the preceding problem is 10 lb$_m$/s.

5.37 A 110-volt electrical heater is used to warm 3 ft³/s of air at 14.7 psia, 65° F to 14.7 psia, 85° F. How much current in amperes must be supplied to this heater?

5.38 The fan on a personal computer (PC) draws 0.5 ft³/s of air at 14.7 psia, 70° F through the box containing the CPU and other circuits. This air leaves at 14.7 psia, 80° F. Calculate the electrical power (kW) dissipated by the PC circuits.

5.39 A 110-V electric hot-water heater warms 0.1 liter/s of water at 15° C to 20° C. Calculate the current in amperes that must be supplied to this heater.

5.40 Steam enters a long, insulated pipe at 1500 kPa, 360° C where the diameter of the pipe is 0.15 m. The steam velocity at the entrance is 10 m/s. It exits this pipe where its diameter is 0.1 m and the state of the water is 1000 kPa, 300° C. Calculate the mass flow rate (kg/s) of the steam and the velocity (m/s) at the pipe outlet.

5.41 Nitrogen flows through a long, constant-diameter adiabatic pipe. It enters at 100 psia, 120° F and leaves at 50 psia, 70° F. Calculate the velocity (ft/s) of the nitrogen at the pipe's inlet and outlet.

5.42 Refrigerant-12 at 600 kPa, 100° C undergoes a reversible, isothermal expansion to 200 kPa in a steady-flow device with one inlet and one outlet. Determine the power (kW) produced by this device and the rate of heat transfer (kW) when the flow rate through the device is 1 kg/s.

5.43 Carbon dioxide is compressed in a reversible isothermal process from 15 psia, 70° F to 60 psia using a steady-flow device with one inlet and one outlet. Determine the specific work (Btu/lb$_m$) required and the specific-heat transfer (Btu/lb$_m$) for this compression.

5.44 Determine the change in the specific work and heat transfer when the process of the previous problem is isentropic rather than isothermal.

5.45 A portion of the steam passing through a steam turbine is sometimes removed for the purposes of feedwater heating as shown in Figure 5.11 (p. 206). Consider an adiabatic steam turbine with 12,000 kPa, 560° C steam entering it at a mass flow rate of 20 kg/s. Steam is bled from this turbine at a mass flow rate of 1 kg/s with a state of 1000 kPa, 200° C. The remaining steam leaves the turbine at 100 kPa, 100° C. Determine the power (hp) produced by this turbine.

5.46 The stators in a gas turbine are designed to increase the kinetic energy of the gas passing though them adiabatically without any shaft work. Air enters a set of these nozzles at 300 psia, 700° F with a velocity of 80 ft/s and exits at 250 psia, 645° F. Calculate the velocity (ft/s) at the exit of the nozzles. *Note:* The kinetic energy change is quite large in this situation.

5.47 The diffuser in a jet engine is a passage like that of Figure 5.8 (p. 199), which is designed to decrease the kinetic energy of the air entering the engine compressor without any work or heat interactions. Calculate the velocity (ft/s) at the exit of a diffuser when air at 15 psia, 70° F enters it with a velocity of 850 ft/s and the exit state is 30 psia, 210° F.

5.48 Is it possible for an adiabatic liquid–vapor separator to separate wet steam at 100 psia, 90% quality, so that the pressure of the outlet streams is greater than 100 psia?

5.49 Steam is expanded in an isentropic turbine with a single inlet and outlet. At the inlet, the steam is at 2000 kPa, 360° C. The steam pressure at the outlet is 100 kPa. Calculate the specific work (kJ/kg) produced by this turbine.

5.50 Air enters an isentropic turbine at 150 psia, 900° F through a 0.5 ft^2 area with a velocity of 500 ft/s. It leaves at 15 psia with a velocity of 100 ft/s. Calculate the air temperature (° F) at the turbine exit and the power (hp) produced by this turbine.

5.51 The compressor in a refrigerator compresses saturated R-12 vapor at 0° F to 200 psia. Calculate the specific work (Btu/lb$_m$) required by this compressor when the compression process is isentropic.

5.52 An ammonia compressor compresses saturated ammonia vapor at −10° C to 800 kPa. How much specific work (kJ/kg) does this require when the process is isentropic?

5.53 An isentropic steam turbine processes 5 kg/s of steam at 4000 kPa, which is exhausted at 70 kPa, 100° C. Five percent of this flow is diverted for feedwater heating at 700 kPa. Determine the power produced by this turbine in kW.

5.54 Air at 500 psia, 700° F is expanded in an adiabatic gas turbine to 20 psia. Calculate the maximum specific work (Btu/lb$_m$) that this single inlet–outlet turbine can produce.

5.55 Nitrogen at 120 kPa, 30° C is compressed to 600 kPa in a single inlet–outlet adiabatic compressor. Calculate the minimum specific work (kJ/kg) needed for this process.

5.56 Oxygen at 50 psia, 200° F with a velocity of 10 ft/s is expanded in an adiabatic nozzle like that of Figure 5.8 with no work exchange. What is the maximum velocity (in ft/s) of the oxygen at the outlet of this nozzle when the outlet pressure is 20 psia?

5.57 Steam enters an adiabatic diffuser similar to that of Figure 5.8 at 150 kPa, 120° C with a velocity of 200 m/s. The diffuser has no work interactions. Determine the maximum velocity (m/s) that the steam can have at the outlet when the outlet pressure is 300 kPa.

5.58 Steam at 60 psia, 350° F is mixed with water at 40° F in an adiabatic steady-flow device like that in Figure 5.16 (p. 226) with no shafts protruding from it. The steam enters this device at a rate of 0.05 lb_m/s, while the water enters at a rate of 1 lb_m/s. Determine the maximum temperature (° F) of the water leaving this device when the outlet pressure is 60 psia.

5.59 An inventor claims to have invented an adiabatic steady-flow device with a single inlet–outlet that produces 100 kW when expanding 1 kg/s of air from 900 kPa, 300° C to 100 kPa. Is this claim valid?

5.60 A chilled-water heat-exchange unit like that in Figure 5.7 (p. 197) is designed to cool 5000 liters/s of air at 100 kPa, 30° C to 100 kPa, 18° C by using water at 8° C. Determine the maximum water outlet temperature (° C) when the mass flow rate of the water is 2 kg/s.

5.61 Refrigerant-12 is adiabatically expanded from 100 psia, 100° F to a saturated vapor at 10 psia. Determine the specific entropy generation (Btu/lb_m-° R) of this process.

5.62 In an ice-making plant, water is frozen at atmospheric pressure by evaporating saturated ammonia liquid at −10° C. The ammonia leaves this evaporator as a saturated vapor, and the plant is sized to produce 4000 kg/hr of ice. Determine the rate at which entropy is generated (kJ/kg-K-hr) by the evaporation of the ammonia.

5.63 The air in a large building is kept warm by heating it with steam in a heat exchange unit like that of Figure 5.7. Saturated steam vapor enters this unit at 35° C and leaves as a saturated liquid at 32° C. Air at 1-atmosphere pressure enters the unit at 20° C and leaves at 30° C. Determine the rate of entropy generation (kJ/K-hr) when this unit processes 10,000 kg/hr of steam.

5.64 Argon is expanded in an isentropic turbine from 2000 kPa, 500° C to 200 kPa. Determine the outlet temperature (° C) and specific work (kJ/kg) produced by this turbine.

5.65 Air is compressed in an isentropic compressor from 15 psia, 70° F to 200 psia. Determine the outlet temperature (° F) and specific work consumption (Btu /lb_m) of this compressor.

5.66 Oxygen enters an insulated 5-inch-diameter pipe with a velocity of 200 ft/s. At the pipe entrance, the oxygen is at 40 psia, 70° F; and at the exit, it is at 35 psia, 73° F. Calculate the rate (Btu/lb_m-° R-hr) at which the oxygen is generating entropy.

5.67 Nitrogen is compressed in an adiabatic compressor from 100 kPa, 17° C to 600 kPa, 227° C. Calculate the specific entropy generation of this process in kJ/kg-K.

5.68   Water enters a boiler at 500 psia as a saturated liquid and leaves at 600° F. Calculate the specific heat transfer (Btu/lb$_m$) to the water.

5.69   Refrigerant-12 enters a refrigeration condenser at 1200 kPa, 80° C and leaves as a saturated liquid. Determine the specific heat transfer (kJ/kg) from the R-12.

5.70   Air is expanded in an adiabatic nozzle by a polytropic process with $n = 1.3$. It enters the nozzle at 100 psia, 200° F with a velocity of 100 ft/s and exits at a pressure of 25 psia. No work is exchanged between the nozzle and its surroundings. Calculate the exit temperature (° F) and velocity (ft/s) of the air.

5.71   How would the nozzle exit temperature and velocity of the previous problem change if the polytropic exponent were 1.2?

5.72   An adiabatic capillary tube is used in some refrigeration systems to reduce the pressure of the saturated liquid in the condenser to the lower pressure in the evaporator. Determine the quality of R-12 at the exit of this tube for a system that operates the condenser at 120° F and the evaporator at 10° F.

5.73   Determine the specific entropy generation (Btu/lb$_m$-° R) of the previous problem.

5.74   Wet steam in a steam line at 2000 kPa is throttled to 100 kPa, 120° C. What is the quality in the steam line?

5.75   To control an isentropic steam turbine, a throttle valve is placed in the steam line supplying the turbine inlet as shown in Figure 5.20. Steam at 6000 kPa, 400° C is supplied to the throttle inlet, and the turbine exhaust pressure is set at 70 kPa. Compare the specific work (kJ/kg) produced by this steam turbine when the throttle valve is completely open and when it is partially closed so that the pressure at the turbine inlet is 3000 kPa.

FIGURE 5.20

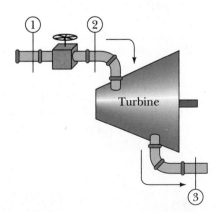

5.76 Calculate the specific heat transfer (Btu/lb$_m$) for the reversible steady-flow system (with a single inlet and outlet) process 1–3 shown in Figure 5.21A.

5.77 Calculate the specific heat transfer (kJ/kg) for the reversible steady-flow system (with a single inlet and outlet) process 1–3 shown in Figure 5.21B.

5.78 Saturated liquid water is heated at constant pressure in a reversible steady-flow device (with a single inlet and outlet) until it is a saturated vapor. Calculate the specific heat transfer (kJ/kg) when this is done at 800 kPa.

5.79 Calculate the specific heat transfer (Btu/lb$_m$) for the reversible steady-flow system (with a single inlet and outlet) process 1–2 shown in Figure 5.21C.

FIGURE 5.21

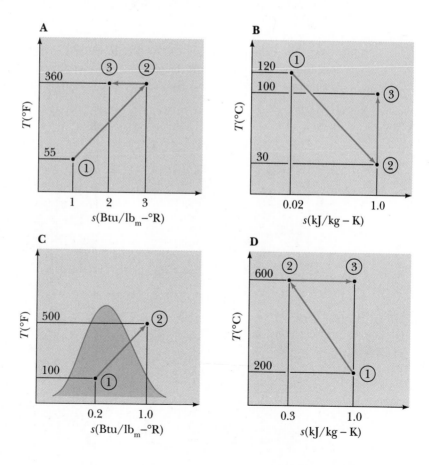

5.80 Calculate the specific-heat transfer (kJ/kg) for the reversible steady-flow system (with a single inlet and outlet) process 1–3 shown in Figure 5.21D.

5.81 Air is isothermally compressed from 16 psia, 75° F to 120 psia in a reversible steady-flow device (with a single inlet and outlet). Calculate the specific work (Btu/lb_m) required for this compression.

5.82 Calculate the specific work production (Btu/lb_m) for the reversible steady-flow system (with a single inlet and outlet) process 1–3 in Figure 5.22A.

5.83 Calculate the specific work production (kJ/kg) for the reversible steady-flow system (with a single inlet and outlet) process 1–3 illustrated in Figure 5.22B.

5.84 Calculate the specific work production (Btu/lb_m) for the reversible steady-flow system (with a single inlet and outlet) process 1–2 in Figure 5.22C.

5.85 Calculate the specific work produced (kJ/kg) by an ideal gas undergoing the reversible steady-flow system (with a single inlet and outlet) process 1–2 in Figure 5.22D.

5.86 Saturated steam vapor at 150° C is compressed to 1000 kPa, while its specific volume remains constant. Determine the specific work (kJ/kg) required to do this in a reversible steady-flow device (with a single inlet and outlet).

FIGURE 5.22

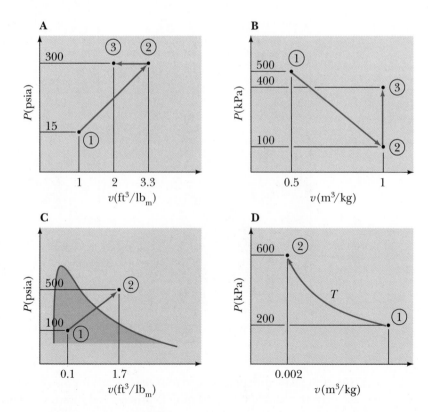

5.87   Air is expanded in an adiabatic turbine of 85% efficiency. The state at the inlet is 2200 kPa, 300° C, and the outlet pressure is 200 kPa. Calculate the air outlet temperature (° C) and the specific work produced (kJ/kg) by this turbine.

5.88   Rework the previous problem with a turbine efficiency of 90%.

5.89   Calculate the specific entropy generation (kJ/kg-K) of the previous problem.

5.90   Steam at 450 psia, 550° F is expanded to 5 psia in an adiabatic turbine with an efficiency of 92%. Determine the power (hp) produced by this turbine when the mass flow rate is 2 lb$_m$/s.

5.91   Rework the previous problem with a turbine efficiency of 90%.

5.92   Steam at 400 psia, 500° F is expanded in an adiabatic turbine to 20 psia. What is the efficiency of this turbine if the steam is exhausted as a saturated vapor?

5.93   Air is expanded from 2000 kPa, 327° C to 100 kPa in an adiabatic turbine. Determine the efficiency of this turbine when the air is exhausted at 0° C.

5.94   An adiabatic argon unit compresses argon at 200 kPa, 27° C to 2000 kPa. If the argon leaves this compressor at 550° C, what is the compressor efficiency?

5.95   An adiabatic refrigeration unit compresses saturated R-12 vapor at 0° C to 600 kPa, 50° C. What is the efficiency of this compressor?

5.96   The exhaust nozzle of a jet engine such as that in Figure 5.8 (p. 199) adiabatically expands air at 300 kPa, 180° C to 100 kPa. Determine the air velocity (m/s) at the exit when the inlet velocity is negligible and the nozzle efficiency is 96%.

5.97   A refrigeration unit compresses saturated R-12 vapor at 20° C to 1000 kPa. How much power (kW) is required to compress 0.5 kg/s of R-12 with a compressor efficiency of 85%?

5.98   An adiabatic diffuser similar to that in Figure 5.8 at the inlet of a jet engine increases the pressure of the air that enters the diffuser at 13 psia, 30° F to 20 psia. What will the air velocity (ft/s) at the diffuser exit be if the diffuser efficiency is 82% and the diffuser inlet velocity is 1000 ft/s?

5.99   A steam turbine is equipped to bleed 6% of the inlet steam for feedwater heating. It is operated with 500 psia, 600° F steam at the inlet, a bleed pressure of 100 psia, and an exhaust pressure of 5 psia. Calculate the work produced by this turbine in hp units when the efficiency between the inlet and bleed point is 97% and the efficiency between the bleed point and exhaust is 95%.

Hint: Treat this turbine as two separate turbines, with one operating between the inlet and bleed conditions and the other operating between the bleed and exhaust conditions. Do not forget that the mass flow rate through the second turbine is not the same as through the first.

5.100   What is the overall efficiency of the turbine described in the previous problem?

5.101   Air enters a compressor at 14.7 psia, 77° F and is compressed to 140 psia, 200° F. Determine the minimum specific work required for this compression in Btu/lb_m with the same inlet and outlet states. Does the minimum work require an adiabatic compressor?

5.102   Helium is expanded in a turbine from 1500 kPa, 300° C to 100 kPa, 25° C. Determine the maximum specific work this turbine can produce in kJ/kg. Does the maximum work require an adiabatic turbine?

5.103   An adiabatic refrigerator unit compresses R-12 from a saturated vapor at 30 psia to 120 psia, 120° F. What is the minimum power (in hp) required by this compressor when its mass flow rate is 0.1 lb_m/s?

5.104   An adiabatic turbine operates with air entering at 550 kPa, 425 K, and 150 m/s and leaving at 110 kPa, 325 K, and 50 m/s. Determine the actual and maximum specific work (kJ/kg) production of the turbine. Why are the maximum and actual work not the same?

5.105   Calculate the second law effectiveness of the turbine in the preceding problem.

5.106   An adiabatic steam diffuser has steam entering at 500 kPa, 200° C, and 30 m/s. This steam is exhausted as a saturated vapor at 200 kPa. Calculate the actual and maximum outlet velocity in m/s.

5.107   Calculate the second law effectiveness of the diffuser in the preceding problem.

5.108   Saturated steam is generated in a boiler by converting a saturated liquid to a saturated vapor at 200 psia. This is done by transferring heat from the combustion gases, which are at 500° F, to the water in the boiler tubes. Calculate the effect of the heat transfer on the water's work-producing capability.

5.109   Does increasing the temperature of the combustion gases in the preceding problem to 800° F alter the water's work-producing ability?

5.110   To control an isentropic steam turbine, a throttle valve is placed in the steam line supplying the turbine inlet like that in Figure 5.20 (p. 234). Steam at 6000 kPa, 700° C is supplied to the throttle inlet, and the turbine exhaust pressure is set at 70 kPa. What is the effect on the stream exergy at the turbine inlet when the throttle valve is partially closed such that the pressure at the turbine inlet is 3000 kPa. Compare the second law effectiveness of this system when the valve is partially open to when it is fully open.

5.111 A steam turbine is equipped to bleed 6% of the inlet steam for feedwater heating. It is operated with 500 psia, 600° F steam at the inlet, a bleed pressure of 100 psia, and an exhaust pressure of 5 psia. The turbine efficiency between the inlet and bleed point is 97%, and the efficiency between the bleed point and exhaust is 95%. Calculate this turbine's second law effectiveness.

PROJECTS

The following projects are more extensive than the problems presented in this chapter. Most are open-ended and require decisions on the student's part. Many can be expedited with computer-based property tables, computer spreadsheets, and student-written computer programs.

5.1 Sometimes the claim is made that work will not change the entropy of a fluid as it passes through an adiabatic steady-flow system with a single inlet–outlet. Prove this claim valid or invalid.

5.2 Heat transfer is often said to occur only when the temperature of the working fluid changes. Is this true for a steady-flow system?

5.3 You have been given the responsibility of picking a steam turbine for an electrical-generation station that is to produce 300 MW of electrical power that will sell for $0.05 per kilowatt-hour. The boiler will produce steam at 700 psia, 700° F, and the condenser is planned to operate at 80° F. The cost of generating and condensing the steam is $0.01 per kilowatt-hour of electricity produced. You have narrowed your selection to the three turbines in Table 5.1. Your criterion for selection is to pay for the equipment as quickly as possible. Which turbine should you choose?

TABLE 5.1
Turbines of Project 5.3

TURBINE	CAPACITY (MW)	η	COST ($MILLION)	OPERATING COST ($/kW-hr)
A	50	0.9	5	0.01
B	100	0.92	11	0.01
C	100	0.93	10.5	0.015

5.4 In large gas-compression stations (for example, on a natural gas pipeline), the compression is done in several stages as in Figure 5.23. At the end of each stage, the compressed gas is cooled at constant pressure back to the temperature at the inlet of the compressor. Consider a compression station that is to compress a gas (say methane) from P_1 to P_4 in N stages, where each stage is an isentropic compressor coupled to a reversible, isobaric cooling unit. Determine the $N - 1$ intermediate pressures at the outlet of each stage of compression that minimizes the total work required. How does this work compare to the work needed to do the entire compression with one isentropic compressor?

FIGURE 5.23
A Multistage Gas-Compression Facility

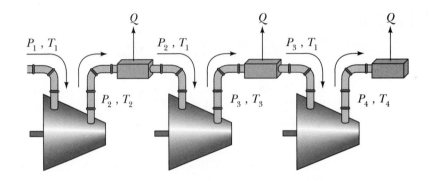

5.5   When an ideal gas is expanded from P_1, T_1 to pressure P_2 in a polytropic process, heat has to be added to the gas for some values of n and rejected for others. Determine the criterion for n when heat has to be added to the gas. For those values of n that require a heat addition, this heat is supplied by an energy reservoir maintained at T_1. Which value of n mimimizes the entropy generation in this case? When heat must be rejected, it is rejected to an energy reservoir maintained at T_2. Which value of n minimizes the entropy generation in this case?

5.6   A steam boiler may be thought of as a heat exchanger such as that in Figure 5.7 (p. 197). The combustion gases may be modeled as a stream of air because their thermodynamic properties are close to those of air. Using this model, consider a boiler that is to boil saturated liquid water at 500 psia to a saturated vapor while maintaining the water pressure constant. Determine the temperature at which the air (i.e., combustion gases) must enter this unit so that the transfer of exergy from the air to the boiling water is done at the minimum loss.

5.7   An adiabatic nozzle is designed to accelerate the speed of an ideal gas from (effectively) 0 m/s, P_1, and T_1 to V m/s. As the efficiency of this nozzle decreases, the pressure at the nozzle exit must also be decreased to maintain the speed at V. Plot the change in the flow exergy as a function of the nozzle efficiency for an ideal gas (say, air).

5.8 The temperature of the air in a building can be maintained during the winter by using different methods of heating. Compare heating this air in a heat-exchange unit with condensing steam to heating it with an electric-resistance heater. Which alternative generates the least entropy or lost work potential?

General Transient-Flow Systems

Entropy

The three statements concerning change of entropy–that it depends on the initial and final states, that it is zero for a reversible process, and that it is positive for an irreversible process–are of great importance, both for the determination of entropy and for the prediction of possible changes in systems.

Encyclopedia Britannica

ow that we have seen the first and second laws of thermodynamics applied to the special cases of cycles, closed systems, and steady-flow systems, we next extend our expression of these laws so that they apply to all types of systems. This chapter accomplishes this by considering control-volume systems that allow all quantities to depend on time and their spatial location. These systems also can interact with their surroundings by exchanging mass, heat, and work. The results developed by considering this most general of systems will have wide application, because they are not constrained in any way.

Many systems are not cyclic, closed, or steady-flow systems. For example, the automobile traveling a highway is a transient-flow system. For each meter traveled by this automobile, its mass decreases with time as the fuel in the fuel tank is burned in the engine to produce power and the products of fuel combustion are exhausted to the surrounding air. A large rocket being launched into space is even more dramatic as it exhausts tremendous quantities of combustion products and produces very large quantities of power in its very short lifetime. This power is generated at the expense of the energy, which was initially stored on the rocket as fuel and oxidizer. The filling and emptying of tanks containing gases and other substances are additional examples of transient-flow systems. These systems consequently are an important class, with many practical applications.

The most general form of the first and second laws of thermodynamics are developed in this chapter in a manner very similar to that used in Chapter 5. Once these laws are developed, we will show that the closed and steady-flow system forms of these laws are special cases of these general expressions. This chapter also emphasizes how to apply the general laws to transient-flow systems.

6.1 Introduction to Transient-Flow Systems

Transient-flow systems are the most general systems used by engineers. They can have masses that enter and leave them like the steady-flow system, but the properties of an amount of mass contained in the control volume also can change as time passes. This variation of properties with time while mass is exchanged with the surroundings distinguishes the transient-flow system from the closed and steady-flow system. We may define a transient-flow system as follows.

> ### DEFINITION: *Transient-Flow System*
>
> Any region-of-space system whose properties change with time while exchanging mass with the surroundings through the control boundaries.

A rocket such as that shown in Figure 6.1 is a typical transient-flow system. As time passes, the amount of fuel and oxidizer stored in the rocket diminish as they are burned in the engine. As a result, the rocket's mass decreases with time. The gases exhausted through the engine are a transfer of mass from the rocket to the surroundings. Other examples include bottles being filled and emptied, and energy-storage devices such as compressed air tanks, hot-water tanks, and fuel tanks. Notice that all these share the features of changing with time and mass transfer.

FIGURE 6.1
A Rocket as a Transient-Flow System

Control volume

As with steady-flow systems, the first step in analyzing a transient-flow system is identifying a control volume such as that in Figure 6.1. The control volume must be a region of space that we can clearly identify as time passes. This region is bounded by a control boundary, including imaginary surfaces through which mass is transferred. Unlike the steady-flow system, this control volume need not have a fixed geometry, and the shape of the control boundary can change. A hot-air balloon is an excellent example. If we define the control volume as the fluid contained within the balloon envelope with an imaginary surface across the opening through which hot air is introduced, then the control volume can have a variety of shapes as time passes.

6.2 Conservation of Mass

The mass of a transient-flow system must be conserved as with all nonrelativistic systems. This important principle can be adapted to the transient-flow system by tracing the events that a fixed mass experiences as it interacts with the system over a short period of time, Δt. A representation of a typical transient-flow system at an initial time, t, and a short time later, $t + \Delta t$, is presented in Figure 6.2. We will focus our attention on the fixed mass, which consists of the mass in the control volume, $m_{cv,t}$; the mass about to enter the control volume, Δm_1, at time t; the mass in the control volume, $m_{cv,t+\Delta t}$; and the mass that has just left the control volume, Δm_2, at time $t + \Delta t$. The number of streams entering and leaving the control volume is arbitrary. Only one of each is considered for the purpose of illustration.

During this elapsed time, conditions in the control volume are also changing. Both the properties of the working fluid in the control volume and the configuration of the control volume can be changing as time passes. The mass, energy, entropy, and so on contained in the control volume are then functions of time as well as location.

Equating the mass of the fixed-mass system at time $t + \Delta t$ to that at time t gives

$$m_{cv,t+\Delta t} - m_{cv,t} = \Delta m_1 - \Delta m_2$$

FIGURE 6.2
A Transient-Flow System Changing with Time

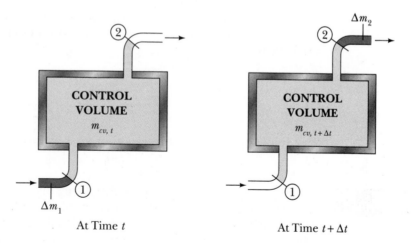

At Time t At Time $t + \Delta t$

Dividing this result by the time span, Δt, and taking the limit as $\Delta t \to 0$, reduces this result to

$$\frac{dm_{cv}}{dt} = \dot{m}_1 - \dot{m}_2$$

When there are several inlet and outlet streams, this result becomes

> **PRINCIPLE: Conservation of Mass (Transient-Flow Systems)**
>
> $$\frac{dm_{cv}}{dt} = \sum_{in} \dot{m} - \sum_{out} \dot{m}$$

In words, the general conservation of mass principle is

$$\begin{bmatrix} \text{Rate of increase} \\ \text{of control-volume} \\ \text{mass} \end{bmatrix} = \begin{bmatrix} \text{Rate at which mass} \\ \text{enters the control} \\ \text{volume} \end{bmatrix} - \begin{bmatrix} \text{Rate at which mass} \\ \text{leaves the control} \\ \text{volume} \end{bmatrix}$$

The transient-flow system statement of the conservation of mass principle is the most general form of this important principle. It contains both the closed-system and steady-flow forms of the conservation of mass principle. When the system is a closed system, there are no mass transfers and the conservation of mass principle reduces to

$$\frac{dm_{cv}}{dt} = 0$$

Therefore, the mass of the control volume remains constant as time passes when the system is closed. At the end of a process, the mass of the closed system is the same as at the beginning of the process even if the composition of the working fluid and the size or shape of the control volume changes. When the system is a steady-flow system, $dm_{cv}/dt = 0$, and the conservation of mass principle becomes

$$\sum_{in} \dot{m} = \sum_{out} \dot{m}$$

This is identical to the conservation of mass expression presented in Chapter 5.

EXAMPLE 6.1

A 6 ft^3 gas cylinder (i.e., a rigid vessel) is being filled with oxygen as shown in Figure 6.3. Oxygen is pumped into this cylinder at 500 psia, 100° F through a 0.01 ft^2 opening with a velocity of 2 ft/s. Initially, the oxygen in the cylinder is at 10 psia, 60° F. Determine the mass of oxygen in the cylinder after two minutes of pumping.

System: The oxygen contained in the gas cylinder.

Basic Equations: $$\frac{dm_{cv}}{dt} = \sum_{in} \dot{m} - \sum_{out} \dot{m}$$

FIGURE 6.3

Gas Cylinder for
Example 6.1

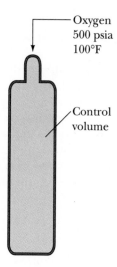

Oxygen
500 psia
100°F

Control
volume

Conditions: The volume of the oxygen in the system remains constant, and the oxygen will be treated as an ideal gas.

Solution: The specific volume of the mass entering the cylinder is

$$v = \frac{RT}{P} = \frac{1545 \times 560}{32.0 \times 500 \times 144} = 0.3755 \text{ ft}^3/\text{lb}_m$$

The mass flow rate at the inlet is

$$\dot{m}_{in} = \frac{AV}{v} = \frac{0.01 \times 2}{0.3755} = 0.05326 \text{ lb}_m/\text{s}$$

Mass within the control volume then increases at a rate of $dm_{cv}/dt = 0.05326$ lb_m/s. After 120 seconds, the mass in the control volume will have increased by 6.391 lb_m. The initial mass in the control volume is

$$m_i = \frac{V}{v} = \frac{VP}{RT}$$

$$= \frac{6 \times 10 \times 144 \times 32.0}{1545 \times 520}$$

$$= 0.344 \text{ lb}_m$$

After two minutes, the mass in the gas cylinder will be

$$m_f = m_i + \Delta m_{added} = 0.344 + 6.391 = 6.735 \text{ lb}_m$$

6.3 First Law for a Transient-Flow System

The first law of thermodynamics may also be adapted to a transient-flow system by again considering the events that happen to a fixed-mass system over a short period of time as it interacts with the transient-flow system-control volume. A fixed-mass system interacting with a transient-flow system-control volume is shown in Figure 6.4. Again, for the purpose of illustration, we have shown only one inlet and one outlet, although the number of inlets and outlets is arbitrary. As in the previous section, the fixed mass system consists of $m_{cv,t} + \Delta m_1$ at time t and $m_{cv,t+\Delta t} + \Delta m_2$ at time $t + \Delta t$.

FIGURE 6.4
A General Transient-Flow System

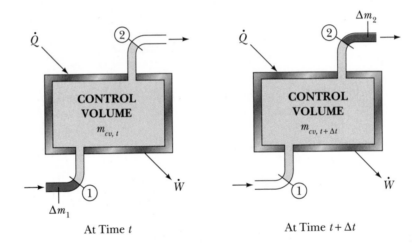

At Time t At Time $t + \Delta t$

During this time, heat is being added to the mass in the control volume at a rate of $\dot{Q}$. The total heat added to the control-volume contents during the elapsed time is $\dot{Q}\Delta t$. The power being produced by the control volume contents is $\dot{W}$. This power consists of both the flow power needed to move the working fluid in and out of the control volume and any power that crosses the control boundary or results from the motion of the control boundary, which may be used to drive some device in the surroundings. Over the elapsed time, the total work produced by the control volume contents is $\dot{W}\Delta t$.

The first law as applied to the fixed-mass system is

$$_iQ_f - {}_iW_f = E_f - E_i$$

where $_iW_f$ includes all forms of work interactions, and E includes all forms of energy. The total flow work associated with the inlet 1 over the elapsed time is $-P_1 v_1 \Delta m_1$ (see Section 5.3). Over the same time period, the total flow work for the outlet 2 is $P_2 v_2 \Delta m_2$. The total work for the fixed-mass system over the time period Δt then becomes

$$_iW_f = \dot{W}\Delta t - P_1 v_1 \Delta m_1 + P_2 v_2 \Delta m_2$$

where $\dot{W}$ is now the power exchanged between the contents of the control volume and the surroundings at the control boundary. This can include shaft power, electrical power, power created by the expansion of the control volume, and other forms of power. When these expressions for the total work and heat interactions are substituted into the first law, it reduces to

$$\dot{Q}\Delta t - \dot{W}\Delta t = E_f - E_i + P_2 v_2 \Delta m_2 - P_1 v_1 \Delta m_1$$

The initial energy of the fixed mass at time t is the combined energy of the control volume contents at time t and the energy of Δm_1,

$$E_i = E_{cv,t} + e_1 \Delta m_1$$

The final energy of the fixed mass at time $t + \Delta t$ is

$$E_f = E_{cv,t+\Delta t} + e_2 \Delta m_2$$

Substituting these expressions into the first law gives

$$\dot{Q}\Delta t - \dot{W}\Delta t = E_{cv,t+\Delta t} - E_{cv,t} + (e_2 + P_2 v_2)\Delta m_2 - (e_1 + P_1 v_1)\Delta m_1$$

When only the mechanical forms of energy are considered and uniform flow exists at the inlet, the specific energy of the mass entering the control volume is $u_1 + e_{k,1} + e_{p,1}$. Similarly, the specific energy at the outlet is $u_2 + e_{k,2} + e_{p,2}$. With these expressions, the first law reduces to

$$\dot{Q}\Delta t - \dot{W}\Delta t = E_{cv,t+\Delta t} - E_{cv,t} + (h_2 + e_{k,2} + e_{p,2})\Delta m_2 - (h_1 + e_{k,1} + e_{p,1})\Delta m_1$$

When this result is divided by Δt and the limit as $\Delta t \to 0$ is taken, it becomes

$$\dot{Q} - \dot{W} = \frac{dE_{cv}}{dt} + (h_2 + e_{k,2} + e_{p,2})\dot{m}_2 - (h_1 + e_{k,1} + e_{p,1})\dot{m}_1$$

If we include several inlets and outlets, the first law takes the following general form.

PRINCIPLE: First Law (Transient-Flow Systems)

$$\dot{Q} - \dot{W} = \frac{dE_{cv}}{dt} + \sum_{out}(h + e_k + e_p)\dot{m} - \sum_{in}(h + e_k + e_p)\dot{m}$$

When stated in words, this is:

$$\begin{bmatrix} \text{The rate at which} \\ \text{heat is added to} \\ \text{the control volume} \\ \text{contents} \end{bmatrix} - \begin{bmatrix} \text{The rate at which} \\ \text{work is produced} \\ \text{by the control} \\ \text{volume contents} \end{bmatrix} = \begin{bmatrix} \text{The rate at which} \\ \text{the energy contained} \\ \text{in the control volume} \\ \text{changes with time} \end{bmatrix}$$

$$+ \begin{bmatrix} \text{The rate at which} \\ \text{energy and flow work} \\ \text{are transported out of} \\ \text{the control volume} \end{bmatrix} - \begin{bmatrix} \text{The rate at which} \\ \text{energy and flow work} \\ \text{are transported into} \\ \text{the control volume} \end{bmatrix}$$

This very important principle states that the rate at which energy is transformed between heat, work, and energy stored in the control volume *and* energy transported by masses entering and leaving the control volume must be conserved. Students should memorize this particular form of the first law because it contains the law's specialized forms as presented in Chapters 4 and 5, as well as its transient-flow form.

When the system is closed, there are no mass exchanges, and the general first law reduces to

$$\dot{Q} - \dot{W} = \frac{dE_{cv}}{dt}$$

Integrating this result over the time it takes the system to execute a process from an initial state 1 to a final state 2 gives

$$_1Q_2 - {}_1W_2 = E_2 - E_1$$

This is the closed-system form of the first law as discussed in Chapter 4.

By definition, the energy contained in a steady-state control volume does not change as time passes. Hence, $dE_{cv}/dt = 0$. Furthermore, the conservation of mass principle, when applied to a system with a single inlet and single outlet, tells us that $\dot{m}_1 = \dot{m}_2 = \dot{m}$. The general first law then becomes

$$\frac{\dot{Q}}{\dot{m}} - \frac{\dot{W}}{\dot{m}} = (h + e_k + e_p)_2 - (h + e_k + e_p)_1$$

or

$$_1q_2 - {}_1w_2 = (h + e_k + e_p)_2 - (h + e_k + e_p)_1$$

This agrees with the form of the first law as presented in Chapter 5.

EXAMPLE 6.2

An initially evacuated, $3\,\text{m}^3$, adiabatic, rigid vessel is to be filled with steam from a steam line maintained at 2000 kPa, 400° C. Determine the temperature (and quality, if appropriate) and mass in the vessel when it has been filled to the point where the pressure is one-half that in the line.

System: The control volume illustrated in Figure 6.5.

FIGURE 6.5

Control Volume for
Example 6.2

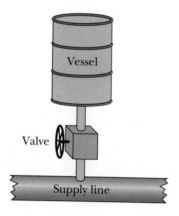

Basic Equations:

$$\frac{dm_{cv}}{dt} = \sum_{in} \dot{m} - \sum_{out} \dot{m}$$

$$\dot{Q} - \dot{W} = \frac{dE_{cv}}{dt} + \sum_{out}(h + e_k + e_p)\dot{m} - \sum_{in}(h + e_k + e_p)\dot{m}$$

320	0.1308	2807.9	3069.5	6.8452
360	0.1411	2877.0	3159.3	6.9917
400	0.1512	2945.2	**3247.6**	7.1271
440	0.1611	3013.4	3335.5	7.2540
500	0.1757	3116.2	3467.6	7.4317

440	0.3257	3023.6	3349.3	7.5883
500	0.3541	3124.4	3478.5	7.7622
540	0.3729	**3192.6**	3565.6	7.8720
600	0.4011	**3296.8**	3697.9	8.0290
640	0.4198	3367.4	3787.2	8.1290

Conditions: This is a transient system with no heat transfer. There is also no work interaction because the control volume is rigid (it cannot move anything in the
surroundings) and contains no shafts that have been cut by the control-volume
boundaries. The kinetic and potential energies will be neglected.

Solution: The energy contained in the control volume is $m_{cv}u_{cv}$ when the kinetic and
potential energies of the working fluid in the control volume are neglected.
When the above conditions are applied to the first law, it becomes

$$\frac{d(mu)_{cv}}{dt} = \dot{m}_1 h_1$$

The subscript 1 represents the conditions at the inlet to the control volume. Similarly, the conservation of mass principle becomes

$$\frac{dm_{cv}}{dt} = \dot{m}_1$$

Combining these two results gives

$$d(mu)_{cv} = h_1 \, dm_{cv}$$

Integrating this expression from the initial evacuated condition i to the final partially filled condition f, while taking advantage of the fact that h_1 is the same as that in the line, h_l, which remains constant during the filling process, yields

$$(mu)_{cv,f} - (mu)_{cv,i} = (m_{cv,f} - m_{cv,i}) h_l$$

Because $m_{cv,i} = 0$, this result simplifies to

$$u_{cv,f} = h_l$$

The final specific internal energy in the control volume is the same as the specific enthalpy in the line. According to the steam tables, this is 3247.6 kJ/kg. Interpolating in the steam tables (see Appendix A) at 1000 kPa gives the final temperature,

$$T_{cv,f} = 572^\circ \text{C}$$

The final specific volume of the steam in the vessel for the now known final pressure and temperature is 0.3878 m³/kg, according to the steam tables. The mass added to the vessel is then

$$m_{added} = m_{cv,f}$$
$$= \frac{V}{v_{cv,f}}$$
$$= \frac{3}{0.3878}$$
$$= 7.736 \text{ kg}$$

TIP:

Review this example thoroughly because it typifies many transient-flow system problems.

EXAMPLE 6.3

A 10 ft³ rigid vessel is initially filled with air at 500 psia, 140° F. As some of this air is vented to the surroundings through a small opening, the vessel is simultaneously heated so that the air temperature in the vessel remains constant. Determine the total amount of heat required when the vessel's final air pressure is 15 psia.

System: A 10 ft³ rigid vessel with a small opening through which air can flow. This opening will be denoted as 1.

Basic Equations:

$$\frac{dm_{cv}}{dt} = \sum_{in} \dot{m} - \sum_{out} \dot{m}$$

$$\dot{Q} - \dot{W} = \frac{dE_{cv}}{dt} + \sum_{out}(h + e_k + e_p)\dot{m} - \sum_{in}(h + e_k + e_p)\dot{m}$$

Conditions: No work interaction exists because the control volume is rigid and contains no shafts that have been cut by the control-volume boundaries. The kinetic and potential energies will be neglected. The air will be treated as an ideal gas with constant specific heats.

Solution: The initial mass in the vessel is

$$m_i = \frac{V}{v_i} = \frac{VP_i}{RT_i}$$

$$= \frac{10 \times 500 \times 144 \times 28.97}{1545 \times 600} = 22.50 \text{ lb}_m$$

The final mass in the vessel is

$$m_f = \frac{V}{v_f} = \frac{VP_f}{RT_f}$$

$$= \frac{10 \times 15 \times 144 \times 28.97}{1545 \times 600} = 0.6750 \text{ lb}_m$$

Because mass is only removed from the system, the conservation of mass equation reduces to

$$\frac{dm_{cv}}{dt} = -\dot{m}_1$$

Applying the stated conditions to the first law reduces it to

$$\dot{Q} = \frac{dE_{cv}}{dt} + (h\dot{m})_1$$

Now $E_{cv} = (mu)_{cv}$, and the properties of the mass leaving the control volume are the same as the properties of the air remaining in the control volume, $h_1 = h_{cv}$. When these findings are substituted into the first law, it becomes

$$\dot{Q} = \frac{d(mu)_{cv}}{dt} + h_{cv}\dot{m}_1 = \frac{d(mu)_{cv}}{dt} + (u + Pv)_{cv}\dot{m}_1$$

Combining the first law with the reduced conservation of mass equation gives

$$\dot{Q} = \frac{d(mu)_{cv}}{dt} + (u+Pv)_{cv}\dot{m}_1 = \frac{d(mu)_{cv}}{dt} - (u+Pv)_{cv}\frac{dm_{cv}}{dt}$$

When this is multiplied by dt and integrated from the initial to the final state, noting that the temperature of the air in the control volume does not change during this time, it becomes

$$_iQ_f = \int_i^f d(mu)_{cv} - (u+Pv)_{cv}\int_i^f dm_{cv}$$

$$= (mu)_{cv,f} - (mu)_{cv,i} - (u+Pv)_{cv}(m_{cv,f} - m_{cv,i})$$

$$= (u_{cv} - u_{cv} - (Pv)_{cv})(m_{cv,i} - m_{cv,f})$$

$$= (Pv)_{cv}(m_{cv,i} - m_{cv,f})$$

$$= RT_{cv}(m_{cv,i} - m_{cv,f})$$

$$= \frac{1545}{28.97 \times 778}600(22.50 - 0.6750)$$

$$= 897.6 \text{ Btu}$$

6.4 Second Law for a Transient-Flow System

The increase in entropy principle for a fixed mass system, $\Delta S_{everything} \geq 0$, will be used to develop a second law of thermodynamics expression for a transient-flow system following the same procedure as in the previous section. We can begin by breaking the entropy change of everything into the entropy change occurring in the surroundings, ΔS_∞, and the entropy change of the mass, ΔS_{sys}. Over the elapsed time Δt, the entropy of the surroundings will have changed by the amount

$$\Delta S_\infty = S_{\infty,t+\Delta t} - S_{\infty,t}$$

When calculating the entropy change of the surroundings, we need only consider those surroundings' components whose entropy changes as a result of interacting with the fixed mass.

Initially the entropy of the fixed mass system shown in Figure 6.4 consists of the entropy of the mass in the control volume and the entropy of the mass about to enter the control volume, $S_{cv,t}+\Delta m_1 s_1$. The final entropy of this mass is composed of the entropy of the mass in the control volume and the entropy of the mass that has just left the control volume, $S_{cv,t+\Delta t} + \Delta m_2 s_2$.

Substituting these results into the increase of entropy principle gives

$$S_{\infty,t+\Delta t} - S_{\infty,t} + S_{cv,t+\Delta t} - S_{cv,t} + \Delta m_2 s_2 - \Delta m_1 s_1 \geq 0$$

When this result is divided by the elapsed time, Δt, and the limit as $\Delta t \rightarrow 0$ is taken, it becomes

$$\frac{dS_\infty}{dt} + \frac{dS_{cv}}{dt} + \dot{m}_2 s_2 - \dot{m}_1 s_1 \geq 0$$

When more than one inlet and outlet are used, this expression may be generalized to the following.

PRINCIPLE: Second Law (Transient-Flow Systems)

$$\frac{dS_\infty}{dt} + \frac{dS_{cv}}{dt} + \sum_{out} \dot{m}s - \sum_{in} \dot{m}s \geq 0$$

When we rearrange this important result, it may be stated as:

$$\begin{bmatrix} \text{The rate of increase} \\ \text{of the entropy of the} \\ \text{surrounding components} \\ \text{that interact with the} \\ \text{control volume} \end{bmatrix} + \begin{bmatrix} \text{The rate of increase} \\ \text{of the entropy of the} \\ \text{control volume contents} \end{bmatrix}$$

$$+ \begin{bmatrix} \text{The rate at which} \\ \text{entropy leaves the} \\ \text{control volume} \end{bmatrix} - \begin{bmatrix} \text{The rate at which} \\ \text{entropy enters the} \\ \text{control volume} \end{bmatrix} \geq 0$$

This is the most general form of the second law, which can be applied to any system, including those discussed in the preceding chapters.

We can also express this as an equality by introducing the entropy-generation rate as we did in Chapter 2:

$$\frac{dS_\infty}{dt} + \frac{dS_{cv}}{dt} - \dot{S}_{gen} + \sum_{out} \dot{m}s - \sum_{in} \dot{m}s \geq 0$$

In this form, the general second law states that

$$\begin{bmatrix} \text{The rate of increase} \\ \text{of the entropy of the} \\ \text{surrounding components} \\ \text{that interact with the} \\ \text{control volume} \end{bmatrix} + \begin{bmatrix} \text{The rate of increase} \\ \text{of the entropy of the} \\ \text{control volume contents} \end{bmatrix} -$$

$$\begin{bmatrix} \text{The rate at which} \\ \text{entropy is generated} \\ \text{by irreversibilities} \end{bmatrix} + \begin{bmatrix} \text{The rate at which} \\ \text{entropy leaves the} \\ \text{control volume} \end{bmatrix} - \begin{bmatrix} \text{The rate at which} \\ \text{entropy enters the} \\ \text{control volume} \end{bmatrix} \geq 0$$

As in Chapter 2, the entropy-generation rate is zero when everything involved with the system's process is reversible. If any irreversibilities are associated with the process internal to the system or the heat transfer between the system and surroundings, then the entropy-generation rate will be positive and not zero. The entropy-generation rate serves to measure the irreversibilities of the system and how closely the system approaches the ideal reversible system.

When the system is closed, there are no mass transfers, and the transient-flow form of the second law reduces to

$$\frac{dS_{cv}}{dt} \geq -\frac{dS_\infty}{dt}$$

When this is integrated over the time it takes for the closed system to execute a process from an initial state 1 to a final state 2, the result is

$$S_2 - S_1 \geq S_{\infty,1} - S_{\infty,2}$$

In words, the increase of the closed system's entropy must equal or exceed the decrease of the surroundings' entropy. If the process is adiabatic, no interactions exist to alter the entropy of the surroundings. The final entropy of the system must then equal or exceed the initial system entropy.

The steady-flow, single inlet–outlet form of the second law is also a special case of the law's transient-flow form. In this case, $dS_{cv}/dt = 0$, according to the definition of a steady-flow system. The conservation of mass principle reduces to $\dot{m}_1 = \dot{m}_2 = \dot{m}$ for a single inlet–outlet. The transient-flow form of the second law then becomes

$$\frac{dS_\infty/dt}{\dot{m}} + s_2 - s_1 \geq 0$$

The subscript 1 refers to the inlet, and 2 refers to the outlet. If this steady-flow system is also adiabatic, this result simplifies to

$$s_2 \geq s_1$$

This states that the specific entropy of the outlet stream equals or exceeds the specific entropy of the inlet stream. This is the same conclusion that we reached for the adiabatic, steady-flow, single inlet–outlet system in Chapter 5.

EXAMPLE 6.4

How does the final specific entropy of a single-phase working fluid in a rigid, adiabatic vessel compare to the initial specific entropy when mass is allowed to leave the vessel through an opening?

System: A rigid, adiabatic vessel filled with a single-phase working fluid. The single outlet will be denoted as 1.

Basic Equations:

$$\frac{dm_{cv}}{dt} = \sum_{in} \dot{m} - \sum_{out} \dot{m}$$

$$\frac{dS_{\infty}}{dt} + \frac{dS_{cv}}{dt} + \sum_{out} \dot{m}s - \sum_{in} \dot{m}s \geq 0$$

Conditions: The process is adiabatic as mass escapes from the vessel.

Solution: With only a single outlet, the conservation of mass principle becomes

$$\frac{dm_{cv}}{dt} = -\dot{m}_1$$

and the second law becomes

$$\frac{dS_{cv}}{dt} + (\dot{m}s)_1 \geq 0$$

because the process is adiabatic. The properties of the fluid leaving through the outlet are the same as the properties of the fluid remaining in the vessel. Then, $s_1 = s_{cv}$, and the second law further reduces to

$$\frac{dS_{cv}}{dt} - s_{cv}\frac{dm_{cv}}{dt} \geq 0$$

Noting that $S_{cv} = m_{cv}s_{cv}$ and performing the derivative of the first term of this expression yields

$$m_{cv}\frac{ds_{cv}}{dt} + s_{cv}\frac{dm_{cv}}{dt} - s_{cv}\frac{dm_{cv}}{dt} \geq 0$$

or

$$m_{cv}\frac{ds_{cv}}{dt} \geq 0$$

TIP:

All working fluid states whose specific entropy is less than the initial specific entropy are forbidden to this system. This is only true if the system is adiabatic.

Multiplying this result by dt/m_{cv}, and integrating from the initial state i to the final state f, results in

$$s_{cv,f} - s_{cv,i} \geq 0$$

Therefore, the final specific entropy of the working fluid remaining in the vessel after some mass has been removed is greater than the initial specific entropy.   ◾

EXAMPLE 6.5

A reversible, adiabatic steam turbine is connected between two well-insulated tanks as shown in Figure 6.6. The smaller tank has a volume of 10 m³, and the larger has a volume of 100 m³. Initially, the smaller tank is filled with steam at 3000 kPa and 280° C, while the larger tank is evacuated. This steam is now allowed to slowly flow out of the small tank, through the turbine, and into the larger tank until the state of the steam in the large tank matches that of the small tank. How much work will the turbine produce during this process?

FIGURE 6.6
Steam-Turbine System of Example 6.5

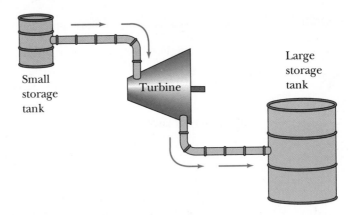

System: The water vapor that initially fills the small tank and then fills both the small and large tank at the end of the process.

233.90	0.0667	2604.1	2804.2	6.1869
240	0.0682	2619.7	2824.3	6.2265
280	**0.0771**	**2709.9**	2941.3	**6.4462**
320	0.0850	2788.4	3043.4	6.6245
360	0.0923	2861.7	3138.7	6.7801

80	93.50	0.001039	2.087	391.58	2498.8	391.66	2274.1	2665.8	1.2329	7.4346
90	96.71	0.001041	1.869	405.06	2502.6	405.15	2265.7	2670.9	1.2695	7.3949
100	99.63	0.001043	1.694	417.36	2506.1	417.46	2258.0	2675.5	1.3026	7.3594
150	111.4	**0.001053**	**1.159**	466.94	2519.7	467.11	2226.5	2693.6	**1.4336**	**7.2233**
200	120.2	**0.001061**	**0.8857**	504.49	2529.5	504.70	2201.9	2706.7	**1.5301**	**7.1271**

Basic Equations:

$$\frac{dm_{cv}}{dt} = \sum_{in} \dot{m} - \sum_{out} \dot{m}$$

$$\dot{Q} - \dot{W} = \frac{dE_{cv}}{dt} + \sum_{out}(h + e_k + e_p)\dot{m} - \sum_{in}(h + e_k + e_p)\dot{m}$$

$$\frac{dS_\infty}{dt} + \frac{dS_{cv}}{dt} + \sum_{out} \dot{m}s - \sum_{in} \dot{m}s \geq 0$$

Conditions: This process is reversible.

Solution: According to the water property tables, the relevant properties at the initial state are $v_i = 0.0771$ m^3/kg, $u_i = 2788.4$ kJ/kg, and $s_i = 6.4462$ kJ/kg-K.

By selecting the system in the above manner, we have created a closed system whose volume and shape are going to change as the process goes forward. As the steam moves from the small tank to the large tank, the system boundaries move with it, and there is no exchange of mass between the system and its surroundings. The conservation of mass principle then reduces to

$$\frac{dm_{cv}}{dt} = 0$$

From the initial conditions, the mass in the system is

$$m = \frac{V_i}{v_i}$$

$$= \frac{10}{0.0771}$$

$$= 129.9 \text{ kg}$$

At the end of the process, this mass will fill both tanks. The final specific volume is then

$$v_f = \frac{V_f}{m}$$

$$= \frac{10 + 100}{129.9}$$

$$= 0.8468 \text{ m}^3/\text{kg}$$

The second law reduces to

$$\frac{dS_\infty}{dt} + \frac{dS_{cv}}{dt} \geq 0$$

because there is no mass exchange. The system is adiabatic, and no heat transfers occur to change S_∞. The inequality becomes an equality because the process is reversible. With these two conditions, the second law further reduces to

$$\frac{dS_{cv}}{dt} = 0$$

When integrated, this produces

$$S_f = S_i$$

The final specific entropy then equals the initial specific entropy because the mass of the system does not change. Using this entropy and the final specific volume to interpolate and read the property tables of Appendix B, we find that the final state is a mixture of liquid and vapor whose pressure is 185 kPa. The specific internal energy at the final state is found to be $u_f = 2274.5$ kJ/kg.

When the first law is applied to this system, it reduces to

$$_iW_f = -m(u_f - u_i)$$

because there is no transfer of heat or exchange of mass. When the above data are substituted, the work produced by this process is

$$_1W_2 = -129.9(2274.5 - 2788.4)$$

$$= 66,760 \text{ kJ}$$

6.5 Combined First and Second Laws

The first and second laws may be combined into a form that is useful for conducting analyses of exergy, work potential, and second law effectiveness, among others. This form also provides additional insight into systems, their operation, and optimization.

We begin with the equality form of the second law, which includes the entropy-generation term. On a rate basis, this form of the second law is

$$\frac{dS_\infty}{dt} + \frac{dS_{cv}}{dt} - \dot{S}_{gen} + \sum_{out} \dot{m}s - \sum_{in} \dot{m}s = 0$$

where $\dot{S}_{gen}$ is the rate at which entropy is generated by irreversibilities. The term dS_∞/dt is the rate at which entropy is generated by transferring heat between thermal-energy reservoirs and the system. When there are several reservoirs, this term becomes

$$\frac{dS_\infty}{dt} = -\sum_{j=0}^{k} \frac{\dot{Q}_j}{T_j}$$

where the heat transfers are taken with respect to the system and the $j = 0$ reservoir is the natural environment surrounding the system. The second law may then be written as

$$\frac{dS_{cv}}{dt} - \sum_{j=0}^{k} \frac{\dot{Q}_j}{T_j} - \dot{S}_{gen} + \sum_{out} \dot{m}s - \sum_{in} \dot{m}s = 0$$

Note that, unlike the inequality form of the second law, all the terms on the left-hand side must sum to zero.

In the first law,

$$\dot{Q} - \dot{W} = \frac{dE_{cv}}{dt} + \sum_{out}(h + e_k + e_p)\dot{m} - \sum_{in}(h + e_k + e_p)\dot{m}$$

$\dot{Q}$ is simply the net sum of all the heat transfers between the system and the various thermal-energy reservoirs, $\Sigma_{j=0}^{k}\dot{Q}_j$, including any heat transfer to the natural environment. With this result, the first law can be written as

$$\sum_{j=0}^{k} \dot{Q}_j - \dot{W} = \frac{dE_{cv}}{dt} + \sum_{out}(h + e_k + e_p)\dot{m} - \sum_{in}(h + e_k + e_p)\dot{m}$$

We will now multiply the second law by T_0 and subtract the result from the right-hand side of the first law. Effectively, this subtracts nothing from the right-hand side of the first law, because the second law sums to zero when the entropy-generation term is included. Performing this step and additional algebraic manipulations yields

$$\dot{W} = \sum_{j=1}^{k} \dot{Q}_j\left(1 - \frac{T_0}{T_j}\right) - \frac{d(E - T_0 S)_{cv}}{dt} - T_0\dot{S}_{gen}$$

$$+ \sum_{in}(h - T_0 s + e_k + e_p)\dot{m} - \sum_{out}(h - T_0 s + e_k + e_p)\dot{m}$$

(6.1)

Notice that any heat transfer to the natural environment is no longer included because $1 - T_0/T_j$ is zero for this reservoir.

This important result provides us with an alternative means of determining the power a system produces as it undergoes a process. But it also gives us additional insights into system behavior and optimization, which is best demonstrated by examining each term individually. The entropy-generation rate term, $T_0\dot{S}_{gen}$, is a consequence of introducing the entropy generation as a mathematical edifice, which reduced the increase of entropy principle to an equality. According to that principle, it must always be positive and can only be zero when everything about the system is reversible. When it is zero, the system will produce the maximum possible power because there are no irreversibilities present to degrade the power produced. The work term, $\dot{W}$, is the actual power produced by the control volume. When the entropy-generation term is set to zero, the work term becomes the work the system has the potential of producing (i.e., work potential). The difference between the work potential and actual work produced was defined in Chapters 4 and 5 as the *lost work potential*. The term $T_0\dot{S}_{gen}$ thereby represents the rate at which work potential is being lost.

The term $\dot{Q}_j(1 - T_0/T_j)$ is the power that a completely reversible heat engine would produce as it removes heat at the rate $\dot{Q}_j$ (whose sign is taken with respect to the control volume) from a reservoir at temperature T_j and rejects heat to the natural environment. This power is the maximum that can be produced by any heat engine using this reservoir and the natural environment as a sink reservoir. The summation of these terms over all of the thermal-energy reservoirs interacting with the system is just the rate at which work potential is produced by all reservoirs interacting with the control volume.

When the term $d(E - T_0 S)_{cv}/dt$ is integrated from an initial time to some later final time, it becomes $(E_f - E_i) - T_0(S_f - S_i)$, which is also the change in the closed system exergy if we disregard the work required to move the environmental fluid when the control volume changes its size. The change in $E - T_0 S$ is the potential work generated by the process occurring in the control volume. Some of this potential work must be used to move the surrounding natural environmental fluid (note that this is not necessary if the system is in free space or the system executes a cycle).

The term $h - T_0 s + e_k + e_p$ is the flow exergy of each stream as measured with respect to a dead state where $h_0 = 0$ and $s_0 = 0$. The sum of the mass flow rates times these terms for all of the inlet streams then represents the rate at which work potential is added to the control volume by the streams. The same sum over the outlet streams is the rate at which work potential is removed from the control volume by these streams.

With these interpretations, Equation 6.1 can be stated in words as

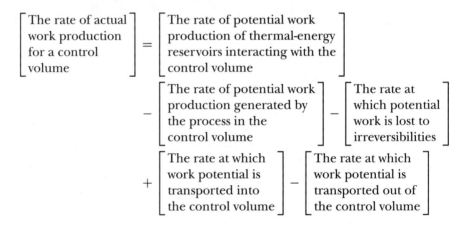

The actual work production (or rate of work production) is the difference between the potential work production (or rate of potential work production) and the lost work production (or rate of lost work production). By setting the lost work production term to zero, we can replace the word *actual* in the left-hand side of the above result with *potential* and use this result to calculate the *work potential* generated by the control volume.

EXAMPLE 6.6

What does the combined first and second law tell us about a cyclical heat engine that operates between thermal-energy reservoirs at temperatures T_h and T_l? The temperature of the low-temperature reservoir need not be the same as that of the natural environment that surrounds the heat engine.

System: The cyclical operating heat engine will be taken as the control volume.

Basic Equations:

$$\dot{Q} - \dot{W} = \frac{dE_{cv}}{dt} + \sum_{out}(h + e_k + e_p)\dot{m} - \sum_{in}(h + e_k + e_p)\dot{m}$$

$$\dot{W} = \sum_{j=1}^{k} \dot{Q}_j \left(1 - \frac{T_0}{T_j}\right) - \frac{d(E - T_0 S)_{cv}}{dt} - T_0 \dot{S}_{gen}$$

$$+ \sum_{in}(h - T_0 s + e_k + e_p)\dot{m} - \sum_{out}(h - T_0 s + e_k + e_p)\dot{m}$$

Conditions: We will consider the internal process only when we complete an integer number of cycles; that is, the beginning and final states for the internal processes are one and the same. The rate at which the engine receives heat from the high-temperature reservoir is $\dot{Q}_h$, and the rate at which the engine rejects heat to the low-temperature reservoir is $-|\dot{Q}_l|$ (the negative sign is necessary because this is a heat rejection from the perspective of the heat engine).

Solution: The various stream terms are zero because there are no streams entering or leaving the heat engine. The $d(E - T_0 S)_{cv}/dt$ is also zero because we are only considering the time required for the heat engine to complete an integer number of cycles. The heat-source summation term becomes

$$\sum_{j=1}^{k} \dot{Q}_j \left(1 - \frac{T_0}{T_j}\right) = \dot{Q}_h \left(1 - \frac{T_0}{T_h}\right) - |\dot{Q}_l| \left(1 - \frac{T_0}{T_l}\right)$$

$$= (\dot{Q}_h - |\dot{Q}_l|) - T_0 \left(\frac{\dot{Q}_h}{T_h} - \frac{|\dot{Q}_l|}{T_l}\right)$$

With these results, the combined first and second laws reduce to

$$\dot{W} = (\dot{Q}_h - |\dot{Q}_l|) - T_0 \left(\frac{\dot{Q}_h}{T_h} - \frac{|\dot{Q}_l|}{T_l}\right) - T_0 \dot{S}_{gen}$$

When the first law is applied to the heat engine, it becomes

$$\dot{W} = (\dot{Q}_h - |\dot{Q}_l|)$$

This further reduces the combined laws to

$$\dot{S}_{gen} = -\left(\frac{\dot{Q}_h}{T_h} - \frac{|\dot{Q}_l|}{T_l}\right)$$

If this engine is completely reversible, then $\dot{S}_{gen} = 0$ and the above result becomes

$$\frac{\dot{Q}_h}{|\dot{Q}_l|} = \frac{T_h}{T_l}$$

As we demonstrated in Chapter 2, this is the basis of the definition of the thermodynamic temperature scale.

EXAMPLE 6.7

Determine the work potential added to a 100-liter rigid container of compressed air like that shown in Figure 6.5 as it is adiabatically filled from a compressed air line that is maintained at 10,000 kPa and 300° C. The air in the container is originally at 100 kPa and 20° C, and the adiabatic filling process is discontinued when the pressure in the container equals that in the supply line.

System: The rigid adiabatic container with an opening for mass transfer will be taken as the control volume.

Basic Equations:

$$\frac{dm_{cv}}{dt} = \sum_{in} \dot{m} - \sum_{out} \dot{m}$$

$$\dot{Q} - \dot{W} = \frac{dE_{cv}}{dt} + \sum_{out}(h + e_k + e_p)\dot{m} - \sum_{in}(h + e_k + e_p)\dot{m}$$

$$\dot{W} = \sum_{j=1}^{k} \dot{Q}_j\left(1 - \frac{T_0}{T_j}\right) - \frac{d(E - T_0 S)_{cv}}{dt} - T_0 \dot{S}_{gen}$$

$$+ \sum_{in}(h - T_0 s + e_k + e_p)\dot{m} - \sum_{out}(h - T_0 s + e_k + e_p)\dot{m}$$

Conditions: The air will be treated as an ideal gas with constant specific heats. No work is done by the system, and no heat is transferred to the system.

Solution: Before we can analyze the work potential, we must fix the state at the end of the process when the container is filled. The first law reduces to

$$0 = \frac{d(mu)_{cv}}{dt} - h_l \dot{m}_{in}$$

for this control volume. Because mass does not leave the control volume, the conservation of mass principle becomes

$$\frac{dm_{cv}}{dt} = \dot{m}_{in}$$

Substituting this into the first law, the result is

$$0 = \frac{d(mu)_{cv}}{dt} - h_l \frac{dm_{cv}}{dt}$$

Rearranging this result and integrating from the beginning condition to the final condition, we get

$$h_l \int_i^f dm_{cv} = \int_i^f d(mu)_{cv}$$

$$h_l(m_{cv,f} - m_{cv,i}) = (mu)_{cv,f} - (mu)_{cv,i}$$

Substituting the mass given by the ideal gas equation of state and using the specific heats to determine the enthalpy and internal energy terms, we gain

$$c_P T_l \left(\frac{P_f}{T_f} - \frac{P_i}{T_i} \right) = c_v (P_f - P_i)$$

When solved for the final temperature of the air in the container, this gives $T_f = 781$ K. The ideal gas equation of state yields $m_i = 0.1189$ kg and $m_f = 4.461$ kg.

For this control volume, the combined laws reduce to

$$\dot{W}_{ptl} = -\frac{d(E - T_0 S)_{cv}}{dt} + \dot{m}_{in}(h - T_0 s)_{in}$$

when we set the entropy-generation term to zero so that we can calculate the work potential. When combined with the conservation of mass equation and integrated over the process, it becomes

$$_i W_{f,ptl} = m_i(u_i - T_0 s_i) - m_f(u_f - T_0 s_f) + (m_f - m_i)(h_l - T_0 s_l)$$

We will select 101 kPa and 25° C for the dead state. By using the specific heats and the integrated Gibbs equation for an ideal gas,

$$s - s_0 = c_P \ln \frac{T}{T_0} - R \ln \frac{P}{P_0}$$

the work potential as measured from the dead state is given by

TIP:

When filled, the air in the container has the potential to produce 402.1 kJ of work by releasing the compressed air through a turbine or other work-producing device to the natural environment.

$$_i W_{f,ptl} = m_i \left[c_v T_i - c_P T_0 \ln \frac{T_i}{T_0} + R T_0 \ln \frac{P_i}{P_0} \right]$$

$$- m_f \left[c_v T_f - c_P T_0 \ln \frac{T_f}{T_0} + R T_0 \ln \frac{P_f}{P_0} \right]$$

$$+ (m_f - m_i) \left[c_P T_l - c_P T_0 \ln \frac{T_l}{T_0} + R T_0 \ln \frac{P_l}{P_0} \right]$$

$$= 0.1189 \left[0.742 \times 293 - 1.029 \times 298 \ln \frac{293}{298} + 0.2870 \times 298 \ln \frac{100}{101} \right]$$

$$- 4.461 \left[0.742 \times 781 - 1.029 \times 298 \ln \frac{781}{298} + 0.2870 \times 298 \ln \frac{10,000}{101} \right]$$

$$+ (4.461 - 0.1189) \left[1.029 \times 573 - 1.029 \times 298 \ln \frac{573}{298} + 0.2870 \times 298 \ln \frac{10,000}{101} \right]$$

$$= 402.1 \text{ kJ}$$

EXAMPLE 6.8

A 0.5 m³ scuba diver's tank is to be filled with atmospheric air (101 kPa, 25° C) by using an adiabatic compressor such as that shown in Figure 6.7. Initially, the air in the tank is at 101 kPa, 25° C. Heat transfer occurs between the air in the tank and the environmental air as the tank is being filled such that the temperature in the tank remains constant at 25° C. Determine the minimum amount of work that must be supplied to the compressor to fill the tank to 1000 kPa.

System: The air contained in the compressor and storage tank. The inlet to the compressor will be denoted by 1.

FIGURE 6.7

Compressor–Storage Tank System of Example 6.8

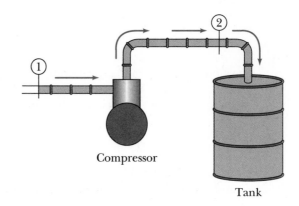

Compressor

Tank

Basic Equations:

$$\frac{dm_{cv}}{dt} = \sum_{in} \dot{m} - \sum_{out} \dot{m}$$

$$\dot{W} = \sum_{j=1}^{k} \dot{Q}_j \left(1 - \frac{T_0}{T_j}\right) - \frac{d(E - T_0 S)_{cv}}{dt} - T_0 \dot{S}_{gen}$$

$$+ \sum_{in}(h - T_0 s + e_k + e_p)\dot{m} - \sum_{out}(h - T_0 s + e_k + e_p)\dot{m}$$

Conditions: The air will be treated as an ideal gas with constant specific heats. There is no entropy generation when the system uses the minimum amount of work.

Solution: When the combined laws are reduced to fit this system, it becomes

$$\dot{W}_{min} = \frac{d(E - T_0 S)_{cv}}{dt} + (h - T_0 s + e_k + e_p)\dot{m}_1$$

because the heat transfer occurs to the environmental air (i.e., $T_j = T_0$), only one stream is going in, and the entropy generation is zero when all irreversibilities have been eliminated so that the system will require the minimum amount of work. When this is multiplied by dt and integrated from the intial time to the final time, the result is

$$W_{min} = m_i(u_i - T_0 s_i) - m_f(u_f - T_0 s_f) + h_1 m_{in}$$

where i refers to the state of the system at the initial time, and f indicates the final time. Similarly, the conservation of mass principle after reduction and integration is

$$m_{in} = m_f - m_i$$

For an ideal gas, $h = u + Pv = u + RT$. Substituting this and the result of the conservation of mass principle into the integrated form of the combined laws produces

$$W_{min} = m_i(u_i - u_1) - m_f(u_f - u_1) + (m_f - m_i)RT_1 + T_0(m_f s_f - m_i s_i)$$

Now $u_i - u_1 = c_v(T_i - T_1) = 0$ and $u_f - u_1 = c_v(T_f - T_1) = 0$. Then

$$W_{min} = (m_f - m_i)RT_1 + T_0(m_f s_f - m_i s_i)$$

Initially, the specific volume of the air in the tank is

$$v_i = \frac{RT_i}{P_i} = \frac{8.314}{28.97} \frac{298}{101} = 0.8467 \text{ m}^3/\text{kg}$$

and the initial mass of air in the tank is

$$m_i = \frac{V}{v_i} = \frac{0.5}{0.8467} = 0.5905 \text{ kg}$$

At the final state,

$$v_f = \frac{RT_f}{P_f} = \frac{8.314}{28.97} \frac{298}{1000} = 0.08552 \text{ m}^3/\text{kg}$$

and the final mass of air in the tank is

$$m_i = \frac{V}{v_f} = \frac{0.5}{0.08552} = 5.846 \text{ kg}$$

The entropy of the air in the tank with respect to the surrounding air (state 1) may be found using the dh form of the Gibbs equation. When this equation is integrated for an ideal gas with constant specific heats from state 1 to the state of the air in the tank, the result is

$$s = c_P \ln \frac{T}{T_0} - R \ln \frac{P}{P_0}$$

At the initial state, the entropy of the air is zero because the pressure and temperature are the same as the environmental air. At the final state,

$$s_f = -R \ln \frac{P_f}{P_0} = -\frac{8.314}{28.97} \ln \frac{1000}{101} = -0.6580 \text{ kJ/kg-K}$$

TIP:
*When irreversibilities
are present, the actual
work required would
be larger than the
minimum work
required by the
amount $T_0 \triangle S_{gen}$. This
product is also
known as the irreversi-
bility of the system.*

Substituting this data into the reduced combined laws gives

$$W_{min} = (m_f - m_i)RT_1 + T_0(m_f s_f - m_i s_i)$$

$$= (5.846 - 0.5905)\frac{8.314}{28.97}298 + 298 \times 5.846 \times (-0.6580)$$

$$= -696.4 \text{ kJ}$$

6.6 Additional Comments

The three most frequently used principles of thermodynamics (i.e., conservation of mass, the first law, and the second law) are presented in their most
general forms in this chapter. These principles form the basic tool set one
draws from to solve a thermodynamic problem. In fact, all thermodynamic
problems use one or more of these principles in one form or another. Students should therefore be very familiar with the results in this chapter and the
techniques for reducing these general statements to fit particular situations.
This is the most important skill that a student of thermodynamics can learn.

In addition to these three principles, one needs property relations for
the working fluid in order to complete a problem solution. We covered the
property relations for the simpler working fluids in Chapter 3. We will cover
more complicated property relations in later discussions.

By combining the first and second laws, we have developed an expression
that may be used to determine the rate at which a transient-flow system produces work or an upper limit on the rate at which a control volume can produce
work. This upper limit determines the change in the work potential of a system
as it undergoes a process and serves as an ideal work production that engineers
can attempt to achieve. As such, this limit can be used to seek optimum systems
and quantify the deviation of a system from the ideal reversible system.

At this point, we have covered the essence of thermodynamics. It only
remains to apply these principles to a variety of applications to sharpen your
skills in using these relations and to illustrate the usefulness of these relations.
The remainder of this book is dedicated to this purpose.

PROBLEMS

Sketch the control volume with clearly labeled inlets and outlets for the following problems. Also, sketch the appropriate $P-V(v)$ and $T-s$ state diagrams for
the following problems, showing all states and processes.

6.1   A spherical hot-air balloon is initially filled with air at 120 kPa, 35° C and has an initial diameter of 3 m. Air enters this balloon at 120 kPa, 35° C with a velocity of 2 m/s through a 1-m diameter opening. How many minutes will it take to inflate this balloon to a 15-m diameter when the pressure and temperature of the air in the balloon remain the same as the air entering the balloon?

6.2 A small positioning control rocket in a satellite is driven by a 2 ft³ container filled with R-12 at −10° F. Upon launch, the container is completely filled with saturated liquid R-12. The rocket is designed for short bursts of 5 s duration. During each burst, the mass flow rate leaving the rocket is 0.05 lb_m/s. How many such bursts can this rocket experience before the quality in the container is 90% or more, presuming that the temperature of the container contents is maintained at −10° F?

6.3 A pneumatic accumulator arranged to maintain a constant pressure as air enters or leaves it is set for 200 psia. Initially, the volume is 0.2 ft³ and the temperature is 80° F. Air is now added to the accumulator until its volume is 1 ft³ and its temperature is 80° F. How much air has been added to the accumulator in lb_m?

6.4 During the inflation and deflation of a safety airbag in an automobile, the gas enters the airbag with a specific volume of 15 ft³/lb_m and at a mass flow rate that varies with time as illustrated in Figure 6.8. The gas leaves this airbag with a specific volume of 13 ft³/lb_m, with a mass flow rate that varies with time as shown in Figure 6.8. Plot the volume of this bag (i.e., airbag size) as a function of time in ft³.

FIGURE 6.8

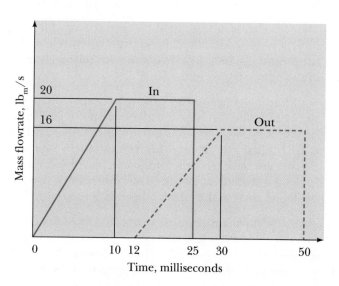

6.5 A rigid container filled with an ideal gas is heated while gas is released so that the temperature of the gas remaining in the container stays constant. This container has a single outlet. Derive an expression for the mass flow rate at the outlet as a function of the rate of pressure change in the container.

6.6   A 2 ft^3 scuba diver's air tank is to be filled with air from a compressed air line at 120 psia, 100° F. Initially, the air in this tank is at 20 psia, 70° F. Presuming that the tank is well insulated, determine the temperature (° F) and mass (lb$_m$) in the tank when it is filled to 120 psia.

6.7   Oxygen tanks are filled by compressing the oxygen gas as shown in Figure 6.7. An initially evacuated 1 m^3 tank is to be filled to 13,000 kPa while the temperature of the oxygen in the tank remains constant at 20° C with this system. An isentropic compressor is used. The oxygen enters this compressor with a constant pressure and temperature of 150 kPa and 20° C. Determine the total work (kJ) required by the compressor and the total heat transfer (kJ) from the oxygen tank.

6.8   An air-conditioning system is to be filled from a rigid container that initially contains 5 kg of liquid R-12 at 25° C. The valve connecting this container to the air-conditioning system is now opened until the mass in the container is 0.25 kg, at which time the valve is closed. During this time, only liquid R-12 flows from the container. Presuming that the process is isothermal while the valve is open, determine the final quality of the R-12 in the container and the total heat transfer (kJ).

6.9   Oxygen is supplied to a medical facility from ten 3 ft^3 compressed oxygen tanks. Initially, these tanks are at 2000 psia, 80° F. The oxygen is removed from these tanks slowly enough that the temperature in the tanks remain at 80° F. After two weeks, the pressure in the tanks is 100 psia. Determine the mass (lb$_m$) of oxygen used and the total heat transfer (Btu) to the tanks.

6.10   The air in an insulated, rigid compressed-air tank whose volume is 0.5 m^3 is initially at 4000 kPa, 20° C. Enough air is now released from the tank to reduce the pressure to 2000 kPa. Following this release, what is the temperature (° C) of the remaining air in the tank?

6.11   Work can be generated by passing the vapor phase of a two-phase substance stored in a tank through a turbine as shown in Figure 6.9. Consider such a system using R-12, which is initially at 80° F, and a 10 ft^3 tank that initially is entirely filled with liquid R-12. The turbine is isentropic, the temperature in the storage tank remains constant as mass is removed from it, and the R-12 leaves the turbine at 10 psia. How much work (Btu) will be produced when the liquid mass in the tank has been half used?

6.12   Two 10 ft^3 adiabatic tanks are connected by a valve. Initially, one tank contains water at 450 psia, 10% quality, while the second contains water at 15 psia, 75% quality. The valve is now opened, allowing the water vapor from the high-pressure tank to move to the low-pressure tank until the pressure in the two becomes equal. Determine the final pressure (in psia) and the mass (in lb$_m$) in each tank.

6.13   Submarines change their depth by adding or removing air from rigid ballast tanks, thereby displacing sea water in the tanks. Consider a submarine that

FIGURE 6.9

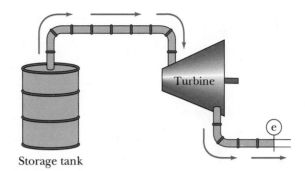

Storage tank

has a 1000 m³ air-ballast tank originally partially filled with 100 m³ of air at 2000 kPa, 15° C. For the submarine to surface, air at 2000 kPa, 20° C is pumped into the ballast tank, until it is entirely filled with air. The tank is filled so quickly that the process is adiabatic and the sea water leaves the tank at 15° C. Determine the final temperature (° C) and mass (kg) of the air in the ballast tank.

6.14   In the preceding problem, presume that air is added to the tank in such a way that the temperature and pressure of the air in the tank remain constant. Determine the final mass (kg) of the air in the ballast tank under this condition. Also determine the total heat transfer (kJ) while the tank is being filled in this manner.

6.15   The weighted piston of the piston-cylinder device in Figure 6.10 maintains the pressure of the piston-cylinder contents at 200 psia. Initially, this system contains no mass. The valve is now opened, and steam from the line flows into the piston cylinder until the volume is 10 ft³. This process is adiabatic, and the steam in the line remains at 300 psia, 450° F. Determine the final temperature (and quality if appropriate) of the steam in the piston cylinder and the total work (Btu) produced as the device is filled.

FIGURE 6.10

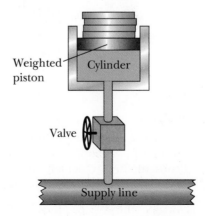

Weighted piston

Cylinder

Valve

Supply line

6.16   Repeat the preceding problem when the line is filled with oxygen at 300 psia, 450° F.

6.17   A spring is added to the piston of Figure 6.10 such that the pressure is 300 kPa when the volume is 0 m³ and 3000 kPa when the volume is 5 m³. Initially, the piston cyclinder volume is 0 m³. Determine the final temperature (and quality if appropriate) when steam in the line (maintained at 1500 kPa, 200° C) is added to the piston cyclinder until the pressure in the device matches that in the line. Also determine the total work (kJ) produced during this adiabatic filling process.

6.18   Repeat the preceding problem when the line is filled with air that is maintained at 2000 kPa, 250° C.

6.19   An engineer has proposed that compressed air be used to "level the load" in an electrical-generation and distribution system. The proposed system is illustrated in Figure 6.11. During those times when electrical-generation capacity exceeds the demand for electrical energy, the excess electrical energy is used to run the compressor and fill the storage tank. When the demand exceeds the generation capacity, compressed air in the tank is passed through the turbine to generate additional electrical energy. Consider this system when the compressor and turbine are isentropic, the tank's temperature stays constant at 70° F, air enters the compressor at 70° F and 1 atm, the tank volume is 1 million cubic feet, and air leaves the turbine at 1 atm. The compressor is activated when the tank pressure is 1 atm, and it remains on until the tank pressure is 10 atm. Calculate the total work (Btu) required to fill the tank and the total heat (Btu) transferred from the air in the tank as it is being filled.

FIGURE 6.11

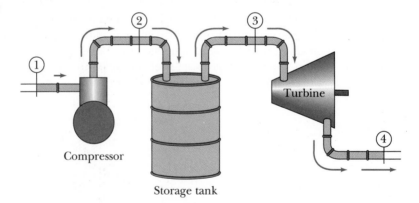

Compressor

Storage tank

Turbine

6.20   The filled compressed-air storage tank of the previous problem is discharged at a later time through the turbine until the pressure in the tank is 1 atm. During this discharge, the temperature of the air in the storage tank remains a constant 70° F. Calculate the total work (Btu) produced by the turbine and the total heat (Btu) added to the air in the tank during this discharge.

6.21   How does the work produced by the turbine during the discharge of the storage tank compare to that used by the compressor during the charging of the storage tank in the previous two problems?

6.22 The air-release flap on a hot-air balloon is used to release hot air from the balloon when appropriate. On one hot-air balloon, the air release opening has an area of 0.5 m², and the filling opening has an area of 1 m². During one flight maneuver, hot air enters the balloon at 100 kPa, 35° C, with a velocity of 2 m/s; the air in the balloon remains at 100 kPa, 35° C; and air leaves the balloon through the air-release flap at velocity 1 m/s. At the start of this maneuver, the volume of the balloon is 75 m³. Determine the volume of the balloon in m³ units and work produced in kJ units by the air inside the balloon as it expands or contracts the balloon skin.

6.23 How does the final specific entropy of the single-phase contents of a rigid, adiabatic container compare to the initial specific entropy when the container is filled through a single opening from a source of working fluid whose properties remain fixed?

6.24 The well-insulated container illustrated in Figure 6.5 is initially evacuated. The supply line contains air that is maintained at 200 psia, 100° F. The valve is opened until the pressure in the container is the same as the pressure in the supply line. Determine the minimum temperature (° F) of the air in the container after it is filled.

6.25 A rigid vessel is filled with a fluid from a source whose properties remain constant. How does the entropy of the surroundings change if the vessel is filled such that the specific entropy of the vessel contents remains constant?

6.26 A rigid vessel is allowed to lose its working fluid through an opening. During this process, the specific entropy of the remaining fluid remains constant. How does the entropy of the environment change as working fluid is lost?

6.27 A rigid, 100-liter steam cooker is arranged with a pressure relief valve set to release vapor and maintain the pressure once the pressure inside the cooker reaches 200 kPa. Initially, this cooker is filled with water at 200 kPa with a quality of 10%. Heat is now added until the quality inside the cooker is 50%. Determine the minimum entropy change (kJ/K) of the thermal-energy reservoir supplying this heat.

6.28 In the preceding problem, the water is stirred at the same time that it is being heated. Determine the minimum entropy change (kJ/K) of the heat-supplying source if 100 kJ of work is done on the water as it is being heated.

6.29 An electric windshield defroster is used to remove 0.25 in. of ice from a windshield. The properties of the ice are $T_{sat} = 32°$ F, $u_{if} h_{if} = 144$ Btu/lb$_m$, and $v = 0.01602$ ft³/lb$_m$. Determine the electrical energy required per square foot of windshield surface area (Btu/ft²) to melt this ice and remove it as liquid water at 32° F, as well as the minimum temperature (° F) at which the defroster may be operated. Assume that no heat is transferred from the defroster or ice to the surroundings.

6.30 Derive an expression for the work potential of the single-phase contents of a rigid adiabatic container when the initially empty container is filled through a single opening from a source of working fluid whose properties remain fixed.

6.31 How much work potential is contained in a rigid vessel filled with 1 kg of liquid R-12, whose temperature remains constant at 25° C, as R-12 vapor is released from the vessel? This vessel exchanges heat with the surrounding atmosphere, which is at 100 kPa, 25° C. The vapor is released until the last of the liquid inside the vessel disappears.

6.32 The 10 ft³ adiabatic container illustrated in Figure 6.5 is initially evacuated. The supply line contains air that is maintained at 200 psia, 100° F. The valve is opened until the pressure in the container is the same as the pressure in the supply line. Determine the work potential (Btu) of the air in this container when it is filled.

6.33 What is the work potential of the air in the filled container of the previous problem if it is filled in such a way that the final pressure and temperature are both the same as in the supply line? The temperature of the surrounding environment is 80° F. Note that the container cannot be adiabatic in this case, and it can exchange heat with the natural environment.

6.34 An initially empty rigid vessel is filled with a fluid from a source whose properties remain constant. Determine the entropy generation if this is done adiabatically and without any work, and the fluid is an ideal gas. Your answer should be in terms of the vessel's volume, the properties of the gas, the dead-state temperature, the initial and final gas pressure and temperatures and the pressure and temperature of the gas-supplying source.

6.35 A rigid 100-liter nitrogen cylinder is provided with a safety relief valve set for 1000 kPa. Initially, this cylinder contains nitrogen at 1000 kPa, 20° C. Heat is now added to the nitrogen from a thermal-energy reservoir at 400° C, and nitrogen is allowed to escape until the mass of nitrogen becomes one-half of its initial mass. Determine the change in the nitrogen's work potential (kJ) as a result of this heating.

6.36 Initially, a rigid, adiabatic, 100-liter nitrogen cylinder is at 1000 kPa, 20° C. A valve is opened that allows nitrogen to escape until one-half of the nitrogen's mass has escaped. How has this changed the work potential (kJ) of the nitrogen in the cylinder?

6.37 The compressed-air storage tank of Figure 6.11 has a volume of 500,000 m³, and it initially contains air at 100 kPa, 20° C. The isentropic compressor proceeds to compress air that enters the compressor at 100 kPa, 20° C until the tank is filled at 600 kPa, 20° C. All heat exchanges are with the surrounding air at 20° C. Calculate the change in the work potential (kJ) of the air stored in the tank. How does this compare to the work required to compress the air as the tank was being filled?

6.38 The air stored in the tank of the previous problem is now released through the isentropic turbine until the tank contents are at 100 kPa, 20° C. The pressure is always 100 kPa at the turbine outlet, and all heat exchanges are with the surrounding air, which is at $20°$ C. How does the total work produced by the turbine compare to the change in the work potential of the air in the storage tank?

PROJECTS

The following projects are more extensive than the problems presented in this chapter. Most are open-ended and require some decisions on the student's part. Many can be expedited with computer-based property tables, computer spread-sheets, and student-written computer programs.

6.1 You are to design a small, directional control rocket to operate in space by providing as many as 100 bursts of 5 seconds each with a mass flow rate of 0.5 lb_m/s at a velocity of 400 ft/s. Storage tanks that will contain up to 3000 psia are available, and the tanks will be located in an environment whose temperature is $40°$ F. Your design criterion is to minimize the volume of the storage tank. Should you use a compressed-air or an R-12 system?

6.2 An air cannon uses compressed air to propel a projectile from rest to a final velocity. Consider an air cannon that is to accelerate a 10-gram projectile to 300 m/s using compressed air, whose temperature cannot exceed $20°$ C. The volume of the storage tank is not to exceed 0.1 m^3. Select the storage volume size and maximum storage pressure that requires the minimum amount of energy to fill the tank.

6.3 Pneumatic nail drivers used in construction require 0.02 ft^3 of air at 100 psia and 1 Btu of energy to drive a single nail. You have been assigned the task of designing a compressed-air storage tank with enough capacity to drive 500 nails. The pressure in this tank cannot exceed 500 psia, and the temperature cannot exceed that normally found at a construction site. What is the maximum pressure to be used in the tank and what is the tank's volume?

6.4 Compare the pumping of water to a higher elevation to isentropic compression of air into a compressed-air tank as means of storing energy for later use.

6.5 To maintain altitude, the temperature of the air inside a hot-air balloon must remain within a $1°$ C band, while the volume cannot vary by more than 1%. At a 300-m altitude, the air in a 1000 m^3 hot-air balloon needs to maintain a $35°$ C average temperature. This balloon loses heat at a rate of 3 kW through the fabric. When the burner is activated, it adds 30 kg/s of $200°$ C, 100 kPa air to the balloon. When the flap that allows air to escape is opened, air leaves the balloon at a rate of 20 kg/s. Design the burner and exhaust-flap control cycles (on time and off time) necessary to maintain the balloon at a 300-m altitude.

Vapor Power and Refrigeration Cycles

Engineers

When the Waters dried an' the earth did appear, . . . The Lord He created the Engineer.

Rudyard Kipling

S ystems that take advantage of the phase changes that occur in certain substances to transform energy and produce mechanical power or a cooling effect are quite common. Practically all electrical power—nuclear, solar, and conventionally fueled electrical-generation stations—is generated by cycles that use water as the working fluid. Almost all air conditioning, cooling, and frozen products are provided by systems that now use chlorofluorocarbons as their working fluids. These systems are widespread and are now the expected equipment in homes, office buildings, and automobiles, as well as in many other applications.

Phase changes are used to transform energy for several reasons. First, a large amount of energy is associated with a phase change. For example, 970.4 Btu (1024 kJ) is required to convert 1 pound-mass (1 kg) of saturated liquid water into a saturated vapor at standard atmospheric pressure. To accomplish the same energy storage with 100 pounds-mass of liquid water would require that we change its temperature by $97.0°$ F ($53.9°$ C). An even larger temperature change would be required if we were to store the same amount of energy in the gaseous phase of water. Second, a unique relationship exists between temperature and pressure during a phase change. During a phase change, an isobaric process is also an isothermal process, which allows us to achieve the isothermal process (a part of all the very best cycles) in practical equipment. Third, the working fluid in these systems is normally not lost from the system, and heat and work are transferred to and from the working fluid in different components of the system. The energy supply and sink reservoirs are then external to the system and can be developed and optimized individually rather than as a part of the system. The combustion of fossil fuels to provide energy, for example, can be done in a way that minimizes its impact on the environment. These advantages are so obvious that even the very earliest practical systems used steam as the working medium. The development of the science of thermodynamics was primarily motivated by these very early steam devices.

Disadvantages also come with the use of phase-changing fluids to transform energy. Most of these are not based on thermodynamics, but on other considerations. Physical and chemical interactions between the fluid and other materials that are used to construct the various system components must always be considered. Using water as the working fluid causes corrosion problems with many of the ferrous-based alloys used to build steel components. Special care must then be taken in selecting the proper component metals and in ensuring that the water chemistry is maintained to avoid excessive and unnecessary corrosion. Many fluids have been developed that optimize the thermodynamics of the systems in which they are used. Unfortunately, many of these are toxic or create other problems that may not be environmentally acceptable. The chlorofluorocarbon dilemma presented in Chapter 3's introduction is but one example of the problems caused by some of these fluids. Thus, the selection of an appropriate working fluid is a trade-off of many considerations.

In this chapter, we consider two basic cycles that use phase-changing working fluids. These are the Rankine cycle and the vapor-compression refrigeration cycle. Once we have looked at the basic cycles, we will consider several modifications to the basic cycle that improve the cycle performance. As in the previous chapters, the first and second laws of thermodynamics are applied to the systems and their components to develop an understanding of how these cycles operate and the factors that affect their performance. The application of exergy analysis is also developed further in this chapter.

7.1   Vapor Cycles

In this chapter, we will examine several cycles that take advantage of phase transformations to convert heat into work or work into a heating or cooling effect. Most of these cycles represent everyday devices such as electrical-generation stations, refrigerators, and air conditioners. Two theoretical cycles—the Carnot and reversed Carnot cycles—are also presented because they give us some insight into what we must do to optimize a cycle's performance. Each cycle is first introduced with irreversible effects such as friction, mixing, and unrestrained expansions neglected. All processes can then be analyzed as reversible. Corrections for irreversible effects are also presented in selected sections.

Vapor cycles consist of components such as pumps, boilers, heat exchangers, liquid–vapor separation units, mixing units, and turbines interconnected to form a system through which the working fluid continuously circulates as the cycle performs its task. Each component in the system becomes a steady-flow device when it is isolated from the cycle. We can analyze the entire cycle by applying first law, second law, and conservation of mass considerations to each steady-flow component.

7.2   The Carnot Cycle

The *Carnot cycle* was first introduced in Section 2.6 in connection with completely reversible devices. The completely reversible cycle and the definition of the thermodynamic temperature were used to prove that the best possible heat engine has a thermodynamic efficiency of

$$\eta_{cr} = 1 - \frac{T_l}{T_h}$$

where T_h is the absolute temperature of the high-temperature thermal-energy reservoir and T_l the absolute temperature of the low-temperature reservoir.

Carnot cycles are made up of two reversible isothermal and two reversible adiabatic processes that form a rectangle on a temperature–entropy state diagram such as that shown in Figure 7.1. Heat is added to the cycle from the high-temperature reservoir during isothermal process 1–2. During this process, the reservoir is warmer than the working fluid by an amount ΔT_h so that heat can be transferred from the energy reservoir to the working fluid, as required by the second law. Work is produced during the adiabatic process, 2–3, as the working fluid is expanded to a lower pressure. Heat is rejected to the low-temperature energy reservoir during isothermal process 3–4. During this process, the reservoir is cooler than the working fluid by an amount ΔT_l. Finally, the working fluid is returned to its original condition through adiabatic compression 4–1. This compression requires some work, although it is less than that produced during the adiabatic expansion. Thus, as one traverses the cycle in a clockwise fashion, the net effect is the production of work while heat transfers from a high-temperature to a low-temperature energy reservoir.

FIGURE 7.1

A Carnot Heat Engine Using a Phase-Changing Working Fluid

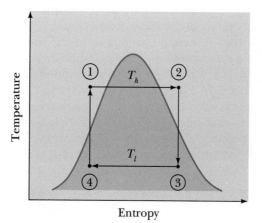

According to the definition of the entropy, the heat added to the cycle by the high-temperature reservoir is

$$q_h = (T_h + \Delta T_h)(s_2 - s_1)$$

Similarly, the heat rejected to the low-temperature reservoir is

$$q_l = (T_l - \Delta T_l)(s_4 - s_3)$$

Now $(s_2 - s_1) = (s_3 - s_4)$, and the Carnot cycle efficiency in the limit as $\Delta T_h \to 0$ and $\Delta T_l \to 0$ is

$$\eta = 1 - \left| \frac{q_l}{q_h} \right| = 1 - \frac{T_l}{T_h}$$

which is the same as that of a completely reversible engine that uses the same thermal-energy reservoirs. The Carnot cycle is therefore the best possible cycle that we can achieve.

To execute the Carnot cycle with a phase-changing working fluid, we would need a boiler that operates isothermally while adding heat to the working fluid, an expansion turbine that operates adiabatically, a condenser (i.e., heat exchanger) that operates isothermally while rejecting heat from the working fluid, and a compression device (i.e., compressor or pump) that operates adiabatically. All these devices must also operate reversibly. Boilers and condensers are constructed to operate at near-constant pressure. If we confine their operation to the saturation region of the working fluid, they will operate at constant temperature as well as constant pressure as the working fluid's phase changes. Operation of these two devices in the saturation region implies that the quality at the end of the expansion process 2–3 is quite low and that the fluid entering the compression device at state 4 is a mixture of liquid and vapor. The presence of liquid droplets in an expansion turbine causes early failure of the moving metal parts by corrosion, erosion, and pitting. This failure is avoided by operating the turbines at 90% or more quality at the outlet. Similar considerations limit the inlet conditions of the compression pump to a pure liquid with no vapor present. Although the Carnot cycle represents the best possible cycle, practical equipment considerations do not allow us to achieve this performance.

The Carnot cycle also can be operated in the reverse (i.e., counterclockwise) sense as illustrated in Figure 7.2. This cycle is known as the *reversed Carnot cycle,* and it produces a refrigeration or heat-pump effect rather than work. Applying the definition of the entropy to process 4–1 gives

$$q_h = (T_h - \Delta T_h)(s_1 - s_4)$$

This results in a heat rejection from the working fluid during the high-temperature isothermal process because $s_4 > s_1$. A similar analysis of process 2–3 yields

$$q_l = (T_l + \Delta T_l)(s_3 - s_2)$$

which is a heat addition to the working fluid. The reversed Carnot cycle then cools the low-temperature energy reservoir while heating the high-temperature reservoir. The net work required to drive the reversed Carnot cycle, according to the first law, as $\Delta T_h \to 0$ and $\Delta T_l \to 0$, is

$$w_{net} = q_l - |q_h| = (T_l - T_h)(s_3 - s_2)$$

where we have taken advantage of the fact that $s_4 - s_1 = s_3 - s_2$.

FIGURE 7.2
Carnot Refrigeration Cycle Using a Phase-Changing Working Fluid

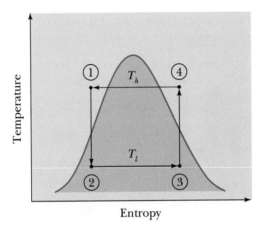

When operated as a refrigerator, the reversed Carnot cycle coefficient of performance (COP) is

$$COP_{ref} = \left| \frac{q_l}{w_{net}} \right| = \frac{T_l}{T_h - T_l}$$

Hence, the maximum refrigerator COP is set by the temperatures of the thermal-energy reservoirs. As a heat pump, the reversed Carnot cycle COP is

$$COP_{hp} = \left| \frac{q_h}{w_{net}} \right| = \frac{T_h}{T_h - T_l}$$

The maximum heat-pump coefficient of performance is also set by the absolute temperatures of the energy reservoirs.

Practical equipment considerations do not allow us to achieve the optimum performance of refrigerators and heat pumps as predicted by the reversed Carnot cycle. Refrigeration heat exchangers (known as *condensers* and *evaporators*) operate at constant pressure, not constant temperature. Compression must occur in the vapor region (or near it) to avoid damaging the compressor while expansion occurs in the liquid region (or near it). These constraints do not allow us to maintain the rectangular shape of the reversed Carnot cycle shown in Figure 7.2.

EXAMPLE 7.1

A Carnot refrigerator using R-12 operates the condenser at 200 psia and the evaporator at 30 psia. Determine this refrigerator's coefficient of performance.

System: The R-12 in a Carnot refrigerator.

Basic Equations:
$$\text{COP}_{ref} = \frac{T_l}{T_h - T_l}$$

Conditions: All internal processes and system–surroundings interactions are reversible. Assume that all processes are executed in the saturation region.

Solution: According to Table B2.2 of Appendix B, the temperature in the evaporator is 11° F, and the temperature in the condenser is 132° F. The coefficient of performance is given by

TIP:

This is the maximum COP$_{ref}$ for this refrigerator. We can only increase it by changing the temperatures (i.e., pressures) in the condenser and evaporator.

$$\text{COP}_{ref} = \frac{T_l}{T_h - T_l}$$
$$= \frac{471}{592 - 471}$$
$$= 3.89$$

7.3 Rankine Cycle

The basic Rankine cycle[1] represents the most fundamental and widely used steam-power cycle. A schematic of this cycle and its equipment, as well as the temperature–entropy state diagram that illustrates the various states and processes, is presented in Figure 7.3. This cycle begins as a compressed liquid enters a boiler (state 1), where it is heated at constant pressure until a superheated vapor is formed (process 1–2). The superheated vapor leaving the boiler is next expanded in an adiabatic turbine to a lower pressure (process 2–3) such that it is nearly a saturated vapor (state 3). Near-saturated vapor leaving the turbine is condensed at constant pressure by cooling the mixture in the condenser until it is a saturated liquid (process 3–4). This saturated liquid is then returned to the boiler pressure (process 4–1) by the pump. The high-pressure, compressed liquid leaving this adiabatic pump at state 1 now reenters the boiler to begin the entire cycle again.

---

[1]Named for William John MacQuorn Rankine (1820–1872), a Scottish engineer and physicist who is best known for his research in molecular physics and thermodynamics. Rankine's *A Manual of the Steam Engine and Other Prime Movers,* written while he served at Glasgow University, is a classic that discusses the cycle that now bears his name. He is also known for his work in soil and fluid mechanics.

FIGURE 7.3
Basic Rankine Cycle

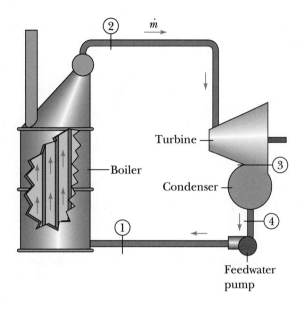

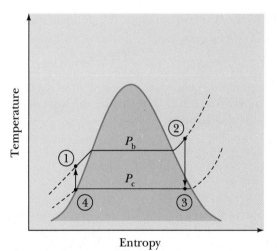

Throughout our discussion of the Rankine cycle, kinetic and potential energies will be neglected in comparison to the enthalpy change. The first law applied to the boiler is then

$$_1q_2 = h_2 - h_1$$

where $_1w_2 = 0$ because this is a single inlet–outlet, steady-flow system operating at constant pressure. The heat added to the Rankine cycle is the increase in the enthalpy of the working fluid as it passes through the boiler in the many tubes that line the boiler walls. Application of the first law to the condenser gives

$$_3q_4 = h_4 - h_3$$

The heat rejected from the Rankine cycle is equal to the decrease in the enthalpy of the working fluid as it passes through the condenser. The thermodynamic efficiency of the basic Rankine cycle is

$$\eta = 1 - \frac{|_3q_4|}{_1q_2} = 1 - \frac{h_3 - h_4}{h_2 - h_1} \qquad (7.1)$$

As the fluid passes through the turbine, it produces the amount of work

$$_2w_3 = h_2 - h_3$$

according to the first law. Similarly, the pump consumes an amount of work

$$_4w_1 = h_4 - h_1$$

During the pumping process, the working fluid is a liquid that can be modeled as an incompressible fluid. In this case, the work required by the pump (see Section 5.7) is also given by

$$_4w_1 = -\int_4^1 v\,dP = -v_4(P_1 - P_4)$$

This latter form for calculating the pump-specific work is often more convenient.

The condenser operates isothermally as well as isobarically as long as state 4 is a saturated liquid and state 3 is a high-quality mixture. But the process experienced by the working fluid in the boiler can never be isothermal while meeting the requirements of the pump and turbine. The minimum temperature of any thermal-energy reservoir serving process 1–2 is T_2. If the isothermal reservoir has a temperature less than this, there is some portion of process 1–2 in which heat is transferred from the cold energy reservoir to the hotter working fluid—which clearly violates the second law. The minimum Carnot efficiency of this cycle is therefore

$$\eta_{cr} = 1 - \frac{T_3}{T_2}$$

which is greater than the actual efficiency of Equation 7.1. This result suggests three ways by which we may improve the efficiency of the Rankine cycle: (1) lower the condenser temperature, (2) increase the temperature at the turbine entrance, and (3) cause process 1–2 to approach an isothermal process. Each method has been used to improve the Rankine cycle's efficiency.

EXAMPLE 7.2

A Rankine cycle that uses water as the working fluid operates the boiler at 6000 kPa, the turbine inlet at 600° C, and the condenser at 50 kPa. The system is sized such that the turbine produces 300 kW of power. Determine the rate at which heat is added in the boiler, the power required to operate the pump, the thermodynamic efficiency, and this cycle's minimum Carnot efficiency.

System: The water passing through the various components of the Rankine cycle.

Basic Equations:
$$_1q_2 - {_1w_2} = h_2 - h_1 + \frac{1}{2}(V_2^2 - V_1^2) + g(z_2 - z_1)$$

$$-{_1w_2} = \int_1^2 v dP + \Delta ke + \Delta pe$$

$$\eta = \frac{w_{net}}{q_{in}}$$

Conditions: All processes will be taken to be steady-state and reversible. Kinetic and potential-energy terms will be neglected.

Solution: The labeling of the various states about the cycle follows the scheme of Figure 7.3. Beginning at the boiler exit, the water-property steam tables give

$$h_2 = 3658.4 \text{ kJ/kg} \quad \text{and} \quad s_2 = s_3 = 7.1677 \text{ kJ/kg-K}$$

Now that we know the entropy at the turbine outlet, the other turbine-outlet properties are

$$x_3 = \frac{s_3 - s_{f,3}}{s_{fg,3}} = \frac{7.1677 - 1.0910}{6.5029} = 0.934$$

and

$$h_3 = h_{f,3} + x_3 h_{fg,3} = 340.5 + 0.934 \times 2305.4 = 2493.7 \text{ kJ/kg}$$

according to the water-property tables. Water enters the pump at state 4 as a saturated liquid. Then,

$$h_4 = h_f = 340.5 \text{ kJ/kg}$$

Finally, we can use the first law and work expression for the pump to fix the enthalpy of the water entering the boiler:

$$h_1 = h_4 + v_4(P_1 - P_4) = 340.5 + 0.00103(6000 - 50) = 346.6 \text{ kJ/kg}$$

The rate at which the working fluid is circulated amongst the cycle components is found by applying the first law to the turbine:

$$\dot{m} = \frac{{}_2\dot{W}_3}{h_2 - h_3} = \frac{300}{3658.4 - 2493.7} = 0.2576 \text{ kg/s}$$

Applying the first law to the boiler gives the rate of heat addition to the cycle:

$${}_1\dot{Q}_2 = \dot{m}(h_2 - h_1) = 0.2576(3658.4 - 346.6)$$
$$= 853.1 \text{ kW}$$

Application of the first law to the pump yields

$${}_4\dot{W}_1 = \dot{m}(h_4 - h_1) = 0.2576(340.5 - 346.6)$$
$$= -1.57 \text{ kW}$$

The thermodynamic efficiency of the Rankine cycle is then

$$\eta = \frac{\dot{W}_{net}}{\dot{Q}_{in}} = \frac{300 - 1.57}{853.1} = 0.350$$

Finally, the minimum Carnot efficiency for the best possible reservoir temperature is

$$\eta_{cr} = 1 - \frac{T_L}{T_H} = 1 - \frac{T_3}{T_2}$$

$$= 1 - \frac{354.3}{873}$$

$$= 0.594$$

TIP:

You should notice that the power required by the pump is a small part of the power produced by the turbine (300 kW versus 1.57 kW). The ratio of the power required to drive the cycle (i.e., pump) to the power produced by the cycle (i.e., turbine) is known as the back–work ratio. The back–work ratio of Rankine cycles is always small.

You should also notice that the actual thermodynamic efficiency is considerably less than the minimum possible Carnot efficiency. This suggests that much can be done to improve the basic Rankine cycle.

7.4 Reheat Rankine Cycle

Increasing the boiler pressure of the Rankine cycle increases the average temperature at which heat is added to the working fluid. The results of the Carnot cycle suggest that this will improve the Rankine cycle's efficiency. The maximum Rankine cycle temperature (i.e., at the boiler exit–turbine inlet) is fixed by the metallurgical properties of the metals used in the boiler and turbine. When we hold the maximum cycle temperature fixed while we increase the boiler pressure, the fraction of liquid in the turbine exhaust also increases and can become unacceptably high. These three conflicting requirements may be met by partially expanding the steam in a turbine to an intermediate pressure and then returning the partially expanded steam back to the boiler for reheating prior to expansion to the condenser pressure in another turbine.

The cycle modification that accomplishes this partial expansion and subsequent reheating is shown in Figure 7.4. The cycle now requires an additional turbine (or turbine section) and an additional heat-transfer section in the boiler to reheat the intermediate pressure steam. At the entrance to the reheater, state 3 in Figure 7.4, the water is near the saturated vapor state for the pressure at which the reheater operates, P_r. This state may be inside the saturated region, at the saturated-vapor line, or in the superheated-vapor region. When the transfer is done in this manner, the boiler pressure has been increased, the maximum metallurigical temperature limit has not been exceeded, and the fraction of liquid in the turbine exhaust is acceptable.

The net heat addition now consists of the heat added to the working fluid on its first pass through the boiler as well as that added during the reheat pass,

$$q_{in} = (h_2 - h_1) + (h_4 - h_3)$$

The heat rejected in the condenser is now

$$q_{out} = h_6 - h_5$$

The thermodynamic efficiency of a reheat Rankine cycle is therefore

$$\eta = 1 - \frac{|q_{out}|}{q_{in}} = 1 - \frac{h_5 - h_6}{(h_2 - h_1) + (h_4 - h_3)}$$

EXAMPLE 7.3

A reheat Rankine cycle that uses water as the working fluid operates a boiler at 6000 kPa, the reheat section at 1000 kPa, each turbine inlet at 400° C, and the condenser at 50 kPa. The system is sized such that the net power produced by the turbines is 300 kW. Determine the rate at which heat is added in the boiler, the power required to operate the pump, and the cycle's thermodynamic efficiency.

System: The water passing through the various components of the Rankine cycle.

Conditions: All processes will be taken to be steady-state and reversible.

FIGURE 7.4
Reheat Rankine Cycle

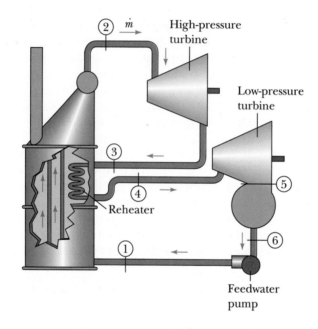

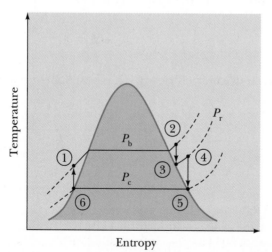

Basic Equations:

$$_1q_2 - {_1}w_2 = h_2 - h_1 + \frac{1}{2}(V_2^2 - V_1^2) + g(z_2 - z_1)$$

$$-{_1}w_2 = \int_1^2 vdP + \Delta ke + \Delta pe$$

$$\eta = \frac{w_{net}}{q_{in}}$$

Solution: The labeling of the various states about the cycle follows the scheme of Figure 7.4. Each state about this cycle is summarized in Table 7.1; the boldface entries represent the information given in the problem statement.

TABLE 7.1
States for Example 7.3

State	P kPa	T °C	x %	h kJ/kg	s kJ/kg-K	v m³/kg
1	**6000**			346.6		$v_1 \approx v_6$
2	**6000**	**400**		3177.2	6.5408	
3	**1000**	179.9	99	2757.5	$s_3 = s_2$	
4	**1000**	**400**		3263.9	7.4651	
5	**50**	81.33	98	2600.2	$s_5 = s_4$	
6	**50**	81.33	0	340.4		0.00103

Application of the first law to the high-pressure turbine gives

$$_2w_3 = h_2 - h_3 = 3177.2 - 2757.5 = 419.7 \text{ kJ/kg}$$

Similarly, the first law when applied to the low-pressure turbine yields

$$_4w_5 = h_4 - h_5 = 3263.9 - 2600.2 = 663.7 \text{ kJ/kg}$$

The mass flow rate through this system is

$$\dot{m} = \frac{\dot{W}}{w_{net,turbs}} = \frac{300}{419.7 + 663.7} = 0.2769 \text{ kg/s}$$

Applying the first law to the boiler yields

$$q_{in} = {_1}q_2 + {_3}q_4 = (h_2 - h_1) + (h_4 - h_3)$$
$$= (3177.2 - 346.6) + (3263.9 - 2757.5) = 3337.0 \text{ kJ/kg}$$

The rate of heat addition in the boiler is then

$$\dot{Q}_{in} = \dot{m}q_{in} = 0.2769 \times 3337.0 = 924.0 \text{ kW}$$

The power required by the pump is

$$\dot{W}_{pump} = \dot{m}(h_6 - h_1) = 0.2769(340.4 - 6.6) = -1.72 \text{ kW}$$

This cycle's thermodynamic efficiency is

$$\eta = \frac{\dot{W}_{net}}{\dot{Q}_{in}} = \frac{300 - 1.72}{924.0} = 0.323$$

7.5 Rankine Cycle with Feedwater Regeneration

Another technique that is used to improve the Rankine cycle's efficiency is preheating the feedwater between the feedwater pump and the boiler entrance. Less of the high-exergy heat produced in the boiler must now be added to the working fluid. However, less steam is now flowing through the turbine, which will cause it to produce less work. The loss of work production is smaller than the reduction in heat addition, and better use is made of the high-exergy heat released by the air–fuel combustion. The net effect is improvement of the thermal efficiency. The basic idea is to use hot steam extracted from the turbine to heat the feedwater in some type of heat exchanger. Two types of heat exchangers—*open-feedwater* and *closed-feedwater heaters*—are used for this purpose.

Hot steam is mixed directly with the feedwater in an open-type feedwater heater. This steam condenses and mixes with the feedwater, causing the feedwater temperature to increase. Ideally, the feedwater is heated enough that it exits the open-feedwater heater as a saturated liquid. This all occurs inside a well-insulated, constant-pressure mixing chamber such as the one schematically illustrated in Figure 7.5. Consequently, the open-feedwater heater is a steady-flow, adiabatic system with two inlets and one outlet arranged so that the pressure is the same at all inlets and the outlet.

FIGURE 7.5

Open Feedwater Heater Schematic

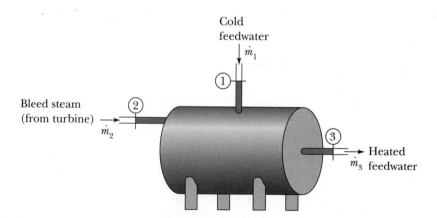

Steam bled from the turbine enters the open-feedwater heater with a mass flow rate of $\dot{m}_2$, while feedwater from a feedwater pump enters the heater with a mass flow rate of $\dot{m}_1$. Feedwater leaves the heater with a mass flow rate of $\dot{m}_3$

at a temperature higher than the feedwater entering at state 1. Applying the conservation of mass principle to this device yields

$$\dot{m}_3 = \dot{m}_1 + \dot{m}_2$$

When the first law is applied to the open-feedwater heater, we obtain

$$\dot{m}_3 h_3 = \dot{m}_1 h_1 + \dot{m}_2 h_2$$

By combining these two expressions, the enthalpy of the heated feedwater at the exit of the feedwater heater is given by

$$h_3 = \frac{\dot{m}_1}{\dot{m}_3} h_1 + \frac{\dot{m}_2}{\dot{m}_3} h_2$$

Ideally, open-feedwater heaters are operated such that the feedwater is heated to a saturated liquid state.

A closed-feedwater heater also warms the feedwater that returns to the boiler with hot steam extracted from the turbine. In this case, the two streams are not allowed to intermix in the heater: We accomplish this by allowing the feedwater stream to pass through a set of tubes while the condensing bleed steam fills the space (i.e., shell) surrounding the tubes (see Figure 7.6). The

FIGURE 7.6
Closed Feedwater
Heater Schematic

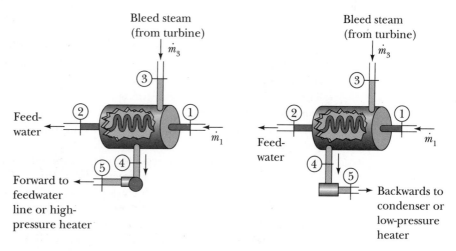

shell is well insulated, so heat is only transferred between the two streams rather than between either stream and the feedwater heater surroundings. Ideally, the feedwater and condensing steam pass through the tubes and shell at constant pressure, with the steam at a pressure different from that of the feedwater. Closed-feedwater heaters are normally operated so that the condensing bleed steam leaves the heater as a saturated liquid. The liquid that results from the condensing steam is returned to the system by routing it either to the turbine condenser or to a lower-pressure feedwater heater after reducing its pressure through an adiabatic throttling device known as a trap. It also may be routed back to the feedwater line or a higher-pressure feedwater heater after its pressure is increased with a pump as in Figure 7.6.

The condensing bleed steam passes through the closed-feedwater heater of Figure 7.6 with a mass flow rate of $\dot{m}_3$, and the feedwater has a mass flow rate of $\dot{m}_1$. Application of the first law to this system (excluding the trap or pump) gives

$$\dot{m}_1(h_2 - h_1) = \dot{m}_3(h_3 - h_4)$$

When a trap is used, $P_5 < P_4$, and the first law applied to the trap tells us that $h_5 = h_4$. When a condensate pump is used, $P_5 > P_4$, and the increase in the condensate's enthalpy across this pump must equal the work applied to the pump. Applying the steady-flow work expression to the pump gives the enthalpy change across the pump,

$$h_5 - h_4 = v_4(P_5 - P_4)$$

According to the second law, heat cannot be transferred from bodies at a low temperature to bodies at a higher temperature of its own accord. Consequently, the temperature of the feedwater (low-temperature fluid) can never exceed the temperature of the condensing bleed steam (high-temperature fluid). In a closed-feedwater heater (see Figure 7.6) this tells us that $T_2 \leq T_4$. The difference between these two temperatures, $T_4 - T_2$, is known as the *terminal temperature difference*. The best possible closed-feedwater heater would permit both streams to exit at the same temperature; that is, the optimum terminal temperature difference for a closed feedwater heater is zero.

A regenerative Rankine cycle with an open-feedwater heater is illustrated in Figure 7.7. Bleed steam is extracted from the turbine at pressure P_3 to heat the feedwater in the open-feedwater heater. The system now requires two pumps, one to raise the pressure from the condenser pressure to the open-feedwater heater pressure, P_3, and the other to raise the pressure from the open-feedwater heater pressure to the boiler pressure. Both pumps must be considered when we calculate the net work required to pump the feedwater from the condenser pressure to the boiler pressure.

Notice that the mass flow rate through the boiler, $\dot{m}_2$, is not the same as the mass flow rate, $\dot{m}_4$, through the latter portion of the turbine, the condenser, and the first feedwater pump. Applying the conservation of mass principle to the turbine, we obtain

$$\dot{m}_4 = \dot{m}_2 - \dot{m}_3$$

FIGURE 7.7

Schematic of a Rankine Cycle with an Open Feedwater Heater

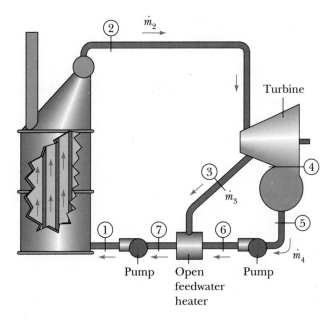

The same principle applied to the open-feedwater heater gives

$$\dot{m}_6 + \dot{m}_3 = \dot{m}_7$$

but

$$\dot{m}_6 = \dot{m}_4 \text{ and } \dot{m}_7 = \dot{m}_1 = \dot{m}_2$$

which reduces the preceding expression to

$$\dot{m}_2 = \dot{m}_3 + \dot{m}_4$$

According to the first law, the rate of heat addition to the cycle in the boiler is

$$\dot{Q}_{in} = \dot{m}_2(h_2 - h_1)$$

and the heat rejected in the condenser is

$$\dot{Q}_{out} = \dot{m}_4(h_5 - h_4)$$

The efficiency of this modified Rankine cycle is therefore

$$\eta = 1 - \left| \frac{\dot{Q}_{out}}{\dot{Q}_{in}} \right| = 1 - \frac{\dot{m}_4(h_4 - h_5)}{\dot{m}_2(h_2 - h_1)}$$

The temperature-entropy state diagram of this system is illustrated in Figure 7.8. Although the areas under reversible processes on this diagram represent heat transfers per unit mass, be careful with this interpretation of areas when the flow rate is different for the various processes. In this case, the product of the area and the process flow rate yields the rate of heat transfer for a reversible process. Rates of heat and work transfer, whether they are all normalized to a common flow rate, such as the boiler flow rate, or not, are more useful for systems that have differing flow rates through the various components.

FIGURE 7.8

T–s State Diagram for a Rankine Cycle with a Single Open-Feedwater Heater

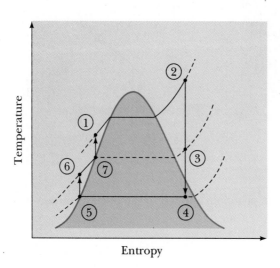

As long as the expansion in the turbine is adiabatic and reversible, all states along the turbine expansion line share the same specific entropy, regardless of the flow rates in the various turbine sections. Hence, the specific entropies at states 3 and 4 are the same as that at state 2 entering the turbine, even though the flow rate from state 2 to 3 is different from the flow rate from state 3 to 4. This fact may be used to fix the states along the turbine expansion line.

EXAMPLE 7.4

A Rankine cycle that uses water as its working fluid operates a boiler at 6000 kPa, the turbine inlet at 600° C, and the condenser at 50 kPa. An open-feedwater heater that operates at 500 kPa is placed in the feedwater return line. The system is sized such that the turbine produces 300 kW of power. Determine the rate at which heat is added in the boiler, the power required to operate the pumps, the thermodynamic efficiency, and the cycle's Carnot efficiency.

System: The water passing through the various components of the Rankine cycle as shown in Figure 7.7.

Basic Equations:

$$\sum_{in} \dot{m} = \sum_{out} \dot{m}$$

$$\dot{Q} - \dot{W} = \sum_{out} \dot{m}(h + \frac{V^2}{2} + gz) - \sum_{in} \dot{m}(h + \frac{V^2}{2} + gz)$$

$$-_1 w_2 = \int_1^2 v dP + \Delta ke + \Delta pe$$

$$\eta = \frac{\dot{W}_{net}}{\dot{Q}_{in}}$$

Conditions: All processes will be taken to be steady-state and reversible.

Solution: The labeling of the various states about the cycle follows the scheme of Figure 7.7. The properties of the various states are given in Table 7.2, where the boldface entries are from the problem statement or from the fact that the fluid at the outlet of a condenser and open-feedwater heater is ideally a saturated liquid. States 3 and 4 are found by taking advantage of the fact that the specific entropy remains constant throughout the turbine expansion.

TABLE 7.2
States for Example 7.4

	P	T	x	h	s	v
State	**kPa**	**°C**	**%**	**kJ/kg**	**kJ/kg-K**	**m^3/kg**
1	**6000**			646.24		
2	**6000**	**600**		3658.4	7.1677	
3	**500**	225.3		2908.8	$s_3 = s_2$	
4	**50**	81.33	93.4	2494.8	$s_4 = s_2$	
5	**50**	81.33	0.0	340.49		0.001030
6	**500**			340.95		
7	**500**	151.9	0.0	640.23		0.001093

States 1 and 6 may be fixed by applying the first law and work expression to the feedwater pumps. For the pump between the condenser and feedwater heater, this gives

$$h_6 = h_5 + v_5(P_6 - P_5)$$
$$= 340.49 + 0.0010300(500 - 50)$$
$$= 340.95 \text{ kJ/kg}$$

Similarly, the enthalpy of the compressed liquid leaving the second feedwater pump is

$$h_1 = h_7 + v_7(P_1 - P_7)$$
$$= 640.23 + 0.001093(6000 - 500)$$
$$= 646.24 \text{ kJ/kg}$$

Applying the conservation of mass principle to the turbine, we obtain

$$\dot{m}_2 = \dot{m}_3 + \dot{m}_4 \qquad (7.2a)$$

Applying the first law to the turbine,

$$\dot{W}_T = \dot{m}_2 h_2 - \dot{m}_3 h_3 - \dot{m}_4 h_4$$

When combined with the conservation of mass principle, this becomes

$$\dot{W}_T = \dot{m}_2(h_2 - h_3) + \dot{m}_4(h_3 - h_4)$$
$$300 = \dot{m}_2(3658.4 - 2908.8) + \dot{m}_4(2908.8 - 2494.8)$$

$$300 = 749.6\dot{m}_2 + 414.0\dot{m}_4 \qquad (7.2b)$$

Applying the first law to the open-feedwater heater yields

$$\dot{m}_2 h_7 = \dot{m}_3 h_3 + \dot{m}_4 h_6$$

$$640.23\dot{m}_2 = 2908.8\dot{m}_3 + 340.95\dot{m}_4 \qquad (7.2c)$$

The simultaneous solution of Equations 7.2a, b, and c gives the three mass flow rates:

$$\dot{m}_2 = 0.2690 \text{ kg/s}$$
$$\dot{m}_3 = 0.0314 \text{ kg/s}$$
$$\dot{m}_4 = 0.2376 \text{ kg/s}$$

Hence, 11.7% of the flow rate through the boiler is bled from the turbine to service the open-feedwater heater.

Now that all the mass flow rates and states are known, we can proceed to calculate the other quantities for the cycle. The rate of heat addition in the boiler when the first law is applied is

$$_1\dot{Q}_2 = \dot{m}_2(h_2 - h_1)$$
$$= 0.2690(3658.4 - 646.24)$$
$$= 810.3 \text{ kW}$$

Using the first law, the power required to operate the first feedwater pump is

$$_5\dot{W}_6 = \dot{m}_4(h_5 - h_6)$$
$$= 0.2376(340.49 - 340.95)$$
$$= -0.1093 \text{ kW}$$

The power required to operate the second feedwater pump is

$$_7\dot{W}_1 = \dot{m}_2(h_7 - h_1)$$
$$= 0.2690(640.23 - 646.24)$$
$$= -1.617 \text{ kW}$$

Applying the definition of the thermodynamic efficiency on a rate basis gives

$$\eta = \frac{\dot{W}_T - {_5}\dot{W}_6 - {_7}\dot{W}_1}{_1\dot{Q}_2}$$
$$= \frac{300 - 0.1093 - 1.617}{810.3}$$
$$= 0.368$$

TIP:

The actual efficiency is 62% of the potential Carnot efficiency, which is an improvement over the basic Rankine cycle of Example 7.2 which used the same boiler and condenser pressures and the same temperature entering the turbine.

The Carnot efficiency for this cycle, using the highest and lowest cycle temperatures, is

$$\eta_{cr} = 1 - \frac{T_l}{T_h}$$
$$= 1 - \frac{354}{873}$$
$$= 0.595$$

Reheating and several feedwater heaters are often used in large steam-power plants. The number and type of feedwater heaters, their bleed pressures, and interconnection are dictated by the cost of the additional units balanced against the economic gains associated with increases in thermodynamic efficiency. Typically, several stages of feedwater heating (rarely more than five) can be justified. An open-feedwater heater operating at a pressure slightly higher than the atmospheric pressure also serves as a deaerator to remove any air that has become entrained in the water. The removal of entrained air is necessary to help prevent corrosion within the various components.

A Rankine system with reheat and two feedwater heaters (one open and one closed) is shown in Figure 7.9. In this case, the saturated condensate from the closed-feedwater heater is trapped back to the open-feedwater heater. Other ways of removing the condensate from the closed-feedwater heater also can be used.

FIGURE 7.9

Schematic of Rankine Cycle with Reheat and Multiple Feedwater Heaters

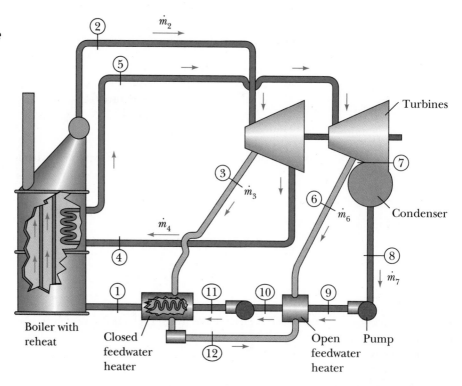

Five different mass flow rates are present in this variation of the Rankine cycle: $\dot{m}_2$, or the flow rate through the first section of the boiler; $\dot{m}_3$, or the bleed steam flow rate servicing the closed-feedwater heater; $\dot{m}_4$, or the flow rate through the reheat section of the boiler; $\dot{m}_6$, or the bleed steam flow rate servicing the open-feedwater heater; and $\dot{m}_7$, or the flow rate through the condenser. The conservation of mass principle and the first law as applied to various cycle components is used to determine these flow rates.

The temperature–entropy state diagram showing the process and states for this variation of the Rankine cycle is presented in Figure 7.10. For this modification of the cycle, we see five pressures in the system: the pressure for boiling and superheating, the pressure for reheating, the pressure in the condenser, the pressure of the open-feedwater heater, and the pressure on the bleed steam side of the closed-feedwater heater. Because the working fluid is reheated, there are two turbine isentropic-expansion processes: that of the first turbine before reheating, and that of the second turbine following reheating. As before, all states along each expansion line share the same specific entropy.

FIGURE 7.10
Temperature-Entropy
State Diagram for
Modified Rankine
Cycle of Figure 7.9

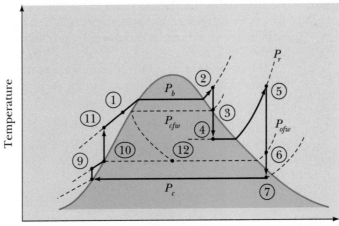

EXAMPLE 7.5

A reheated Rankine cycle with two feedwater heaters arranged as shown in Figure 7.9 uses water as its working fluid. The boiler operates at 900 psia, the reheater at 300 psia, and the condenser at 10 psia. Steam is bled from the first turbine at 350 psia to supply the closed-feedwater heater, and steam is bled from the second turbine at 100 psia to supply the open-feedwater heater. At the entrance of both turbines, the temperature is 600° F. The system is sized for a boiler mass flow rate of 10 lb$_m$/s. Determine the rate at which heat is added in the boiler, the power required to operate the pumps, the net power produced by the cycle, and the cycle's thermodynamic efficiency.

System: The water passing through the various components of the modified Rankine cycle as shown in Figure 7.9.

Basic Equations:

$$\sum_{in} \dot{m} = \sum_{out} \dot{m}$$

$$\dot{Q} - \dot{W} = \sum_{out} \dot{m}(h + \frac{V^2}{2} + gz) - \sum_{in} \dot{m}(h + \frac{V^2}{2} + gz)$$

$$-_1 w_2 = \int_1^2 vdP + \Delta ke + \Delta pe$$

$$\eta = \frac{\dot{W}_{net}}{\dot{Q}_{in}}$$

Conditions: All processes will be taken to be steady-state and reversible. The closed-feedwater heater, open-feedwater heater, and condenser operate ideally with saturated liquid at their outlets. The terminal temperature difference between the closed-feedwater heater outlets will be taken as zero.

Solution: The labeling of the various states about the cycle follows the scheme of Figure 7.9. Properties at the various states of the cycle are presented in Table 7.3; the boldface entries are taken from the problem statement or the above conditions.

The mass flow rates at the locations about the cycle are found by applying the conservation of mass principle to the various components:

For states 4 and 5, $\dot{m} = \dot{m}_2 - \dot{m}_3 = \dot{m}_4$

For states 7 to 9, $\dot{m} = \dot{m}_2 - \dot{m}_3 - \dot{m}_6 = \dot{m}_7$

For states 10 to 2, $\dot{m} = \dot{m}_2$

TABLE 7.3
States for Example 7.5

State	P psia	T °F	x %	h Btu/lb$_m$	s Btu/lb$_m$·°R	v ft^3/lb$_m$
1	**900**	$T_1 = T_{12}$		409.9		
2	**900**	**600**		1260.0	1.4652	
3	**350**		96.4	1176.0	$s_3 = s_2$	
4	**300**		95.0	1163.3	$s_4 = s_2$	
5	**300**	**600**		1314.5	1.6266	
6	**100**			1206.6	$s_6 = s_5$	
7	**10**		89.3	1038.1	$s_7 = s_5$	
8	**10**		**0.0**	161.20		0.01659
9	**100**			161.47		
10	**100**		**0.0**	298.60		0.01774
11	**900**			301.23		
12	**350**	431.82	**0.0**	409.9		

Applying the first law to the closed-feedwater heater gives

$$\dot{m}_3 h_3 + \dot{m}_2 h_{11} = \dot{m}_2 h_1 + \dot{m}_3 h_{12}$$

$$\dot{m}_3 = \frac{h_1 - h_{11}}{h_3 - h_{12}} \dot{m}_2$$

$$= \frac{409.9 - 301.23}{1176.0 - 409.9} 10$$

$$= 1.418 \ \text{lb}_m/\text{s}$$

Similar application of the first law to the open-feedwater heater gives

$$\dot{m}_2 h_{10} = \dot{m}_6 h_6 + (\dot{m}_2 - \dot{m}_3 - \dot{m}_6) h_9 + \dot{m}_3 h_{12}$$

$$\dot{m}_6 = \frac{\dot{m}_2(h_{10} - h_9) + \dot{m}_3(h_9 - h_{12})}{h_6 - h_9}$$

$$= \frac{10(298.60 - 161.47) + 1.418(161.47 - 409.9)}{1206.6 - 161.47}$$

$$= 0.975 \ \text{lb}_m/\text{s}$$

The net heat added in the boiler is found by applying the first law to the boiler:

$$\dot{Q}_{in} = \dot{m}_2(h_2 - h_1) + \dot{m}_4(h_5 - h_4)$$
$$= 10(1260.0 - 409.9) + 8.582(1314.5 - 1163.3)$$
$$= 9799 \ \text{Btu/s}$$

The total power required by the two pumps is

$$\dot{W}_{pumps} = \dot{m}_7(h_8 - h_9) + \dot{m}_{10}(h_{10} - h_{11})$$
$$= 7.607(161.20 - 161.47) + 8.582(298.60 - 301.23)$$
$$= -24.6 \ \text{Btu/s}$$

The net power produced is

$$\dot{W}_{net} = \dot{W}_{turb} + \dot{W}_{pumps}$$
$$= \dot{m}_2 h_2 - \dot{m}_3 h_3 - \dot{m}_4 h_4 + \dot{m}_5 h_5$$
$$\quad - \dot{m}_6 h_6 - \dot{m}_7 h_7 + \dot{W}_{pumps}$$
$$= 10 \times 1260 - 1.418 \times 1176.0 - 8.582 \times 1163.3 + 8.582 \times 1314.5$$
$$\quad -0.975 \times 1206.6 - 7.607 \times 1038.1 - 24.6$$
$$= 3132 \ \text{Btu/s}$$

TIP:

When analyzing a system with multiple feedwater heaters, first determine expressions for the mass flow rates at all locations by applying the conservation of mass principle to the various components. Then proceed by applying the first law to the feedwater heater nearest the boiler and continue to work back toward the condenser.

Finally, the overall thermal efficiency is

$$\eta = \frac{\dot{W}_{net}}{\dot{Q}_{in}}$$

$$= \frac{3132}{9799}$$

$$= 0.320$$

7.6 Optimal Bleed Pressures

The pressure at which steam is extracted from the turbine to service a single feedwater heater can range from the boiler pressure to the condenser pressure. If bleed steam is extracted near the boiler pressure, its temperature is high and the quantity of steam needed to heat the boiler feedwater is small. This reduces the mass flow rate that produces work in the turbine, thereby reducing to some extent the power it produces. Similarly, if we extract bleed steam near the condenser pressure, a larger quantity of steam will be required for feedwater heating. If we extract at this lower pressure, the full mass flow rate will pass through the majority of the turbine expansion, but the mass flow rate passing through the final expansion will be considerably smaller, and the total power produced by the turbine will again be reduced.

This suggests that there is some intermediate bleed pressure at which the power produced by the turbine (and consequently the net thermal efficiency) is a maximum. When there are several stages of feedwater heating, there is an optimal set of bleed pressures that serves each feedwater heater and that maximizes the cycle's net thermal efficiency. The optimum bleed pressure(s) may be found by either trial and error or optimization algorithms such as those of Appendix A.

EXAMPLE 7.6

A steam Rankine cycle uses one closed-feedwater heater with the bleed-steam condensate being trapped back to the condenser for regeneration. At the turbine entrance, the steam is at 600 psia, 600° F. The condenser operates at 10 psia. Determine the optimal pressure for extracting steam from the turbine to service the closed-feedwater heater.

System: The water passing through the various components of the modified Rankine cycle as shown in Figure 7.11.

Basic Equations:

$$\sum_{in} \dot{m} = \sum_{out} \dot{m}$$

$$\dot{Q} - \dot{W} = \sum_{out} \dot{m}(h + \frac{V^2}{2} + gz) - \sum_{in} \dot{m}(h + \frac{V^2}{2} + gz)$$

$$-_1w_2 = \int_1^2 vdP + \Delta ke + \Delta pe$$

$$\eta = \frac{\dot{W}_{net}}{\dot{Q}_{in}}$$

FIGURE 7.11
*Regenerative
Rankine Cycle of
Example 7.6*

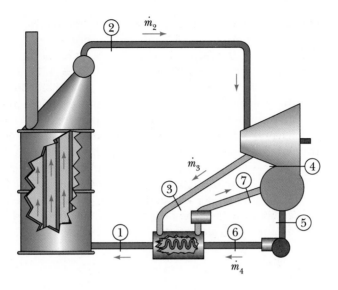

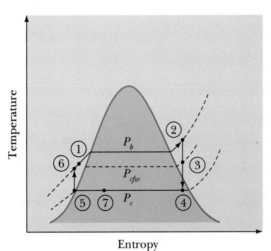

Conditions: All processes will be taken to be steady-flow. The mass flow rate through the boiler will serve as the basis for all calculations.

Solution: The labeling of the various states about the cycle follows the scheme in Figure 7.11. As we change the pressure at which we extract steam from the turbine, only states 1, 3, and 7 will be altered. Because the trap is a throttle device, the enthalpy at state 7 is the same as that at the feedwater heater's bleed-steam outlet. With no terminal temperature difference, $h_7 = h_1$. The properties of states 2, 4, 5, and 6 will not change. The pertinent properties of these fixed states are $h_2 = 1289.5$ Btu/lb$_m$, $s_2 = 1.5320$ Btu/°R, $h_4 = 976.3$ Btu/lb$_m$, $h_5 = 161.2$ Btu/lb$_m$,   and $h_6 = 163.0$ Btu/lb$_m$.

By applying the conservation of mass principle and the first law to the various components as in previous examples, we can determine the net thermal efficiency as we vary the bleed-steam extraction pressure. The mass flow rate through the boiler was set to unity for this calculation. All results are then on the basis of one unit of mass passing through the boiler. Table 7.4 presents the results for various bleed pressures selected by the interval-halving optimization-search technique (see Appendix A). In this table, the flow rate that services the feedwater heater, the heat added in the boiler, and the work produced by the turbine were calculated using the first law:

$$\dot{m}_3 = \frac{h_7 - h_6}{h_3 - h_7}$$
$$q_{in} = h_2 - h_7$$
$$w_{turb} = (h_2 - h_3) + (1 - \dot{m}_3)(h_3 - h_4)$$

The definition of the thermal efficiency,

$$\eta = \frac{w_{turb} + w_{pump}}{q_{in}}$$

was also used.

TABLE 7.4
States for Example 7.6

P_3	h_3	h_7	$\dot{m}_3$	q_{in}	w_{turb}	η
psia	Btu/lb$_m$	Btu/lb$_m$	lb$_m$/lb$_m$-boiler	Btu/lb$_m$-boiler	Btu/lb$_m$-boiler	
600	1289.5	471.7	0.3775	817.8	195.2	0.2365
305	1223.7	395.4	0.2806	894.1	243.6	0.2704
160	1168.8	363.6	0.2491	925.9	265.0	0.2845
85	1119.3	286.6	0.1484	1002.9	292.0	0.2893
48	1077.6	247.6	0.1019	1041.9	302.9	0.2890
66	1100.5	268.7	0.1271	1021.0	297.4	0.2895
75	1109.9	277.6	0.1377	1012.1	294.8	0.2895

Inspection of Table 7.4 reveals that, as the pressure at which steam is extracted from the turbine for feedwater heating decreases, the bleed steam flow rate also decreases. Both the heat that must be supplied by the boiler and the work produced by the turbine increase at differing rates as this pressure is lowered. Maximum thermal efficiency occurs at a bleed steam pressure of 66–75 psia (within round-off errors and iterations conducted). ◼◣

7.7 Rankine Cycle Performance-Reducing Effects

In the previous sections, we neglected the irreversibilities and assumed ideal processes. Actual Rankine cycles deviate from these idealizations to some degree. The most common deviations are pressure drops in the boiler and other piping, the efficiencies of the turbines and pumps, subcooling of the condensates leaving the condenser and open-feedwater heaters, and finite terminal temperature differences at the exits of closed-feedwater heaters. All of these effects degrade any Rankine cycle's overall thermodynamic efficiency.

When a fluid flows through a pipe or tube, the friction acting on the fluid causes its pressure to decrease in the direction of flow. This occurs in the extensive tubing inside the boiler and closed-feedwater heaters and in the piping that connects the various components. Although boiler designers do everything they can to minimize this pressure loss, they cannot completely eliminate it. The pressure at the outlet of the boiler thus is slightly lower than that at the inlet. This is illustrated for the basic Rankine cycle in Figure 7.12, where state 2 represents the actual boiler outlet state and state $2i$ represents the state that would have resulted if the process in the boiler were isobaric. States 2 and $2i$ are at the same temperature. Inspection of the steam tables indicates that the enthalpy of state 2 is greater than that of state $2i$. The amount of heat that must be added to the working fluid in the boiler is then increased as a result of this pressure loss. This additional heating directly reduces the cycle efficiency. Similar pressure loss and reduced cycle efficiency occur in the boiler reheating section.

Irreversibilities in turbines and pumps are reflected in their efficiencies:

$$\eta_{turb} = \frac{w_{act}}{w_{ideal}}$$

$$\eta_{pump} = \frac{w_{ideal}}{w_{act}}$$

FIGURE 7.12
*Effect of Boiler
Pressure Losses on a
Simple Rankine Cycle*

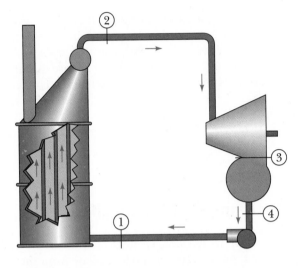

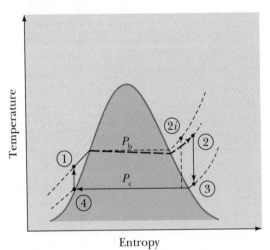

Here the ideal process is isentropic. These are illustrated for the basic Rankine cycle in Figure 7.13, where process 2–3s represents the ideal isentropic turbine process, and process 2–3 is the actual turbine process (state 3 may be located in the vapor region, depending on the turbine efficiency and the turbine inlet state). Notice that states 3s and 3 are both at the condenser pressure. Because the enthalpy of state 3 is greater than that of state 3s, the work produced by the turbine is less than the ideal isentropic work production. The net effect of the irreversible effects in the turbine is to reduce the Rankine cycle's thermodynamic efficiency.

FIGURE 7.13

Effect of Turbine and Pump Efficiencies on a Simple Rankine Cycle

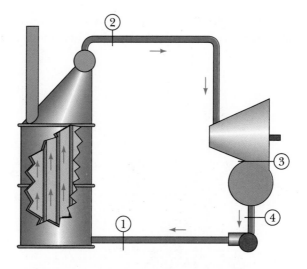

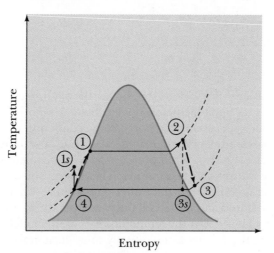

The efficiency of feedwater pumps has a similar effect on the thermodynamic efficiency of a Rankine cycle. This is also illustrated in Figure 7.13 by the ideal feedwater pump process (4–1s) and the actual feedwater pump process (4–1). Irreversibilities in the feedwater pump increase the amount of work required to drive the pump. This in turn reduces the net amount of work produced by the cycle, which results in the Rankine cycle's reduced overall thermodynamic efficiency. Because the work required by any feedwater pump is significantly less than that produced by the turbine, pump irreversibilities are not nearly as critical as turbine irreversibilities.

Subcooling of the feedwater at the exit of condensers and open-feedwater heaters also reduces the Rankine cycle's thermodynamic efficiency. Figure 7.14

FIGURE 7.14

Subcooling of Open Feedwater Heater and Condenser Feedwater

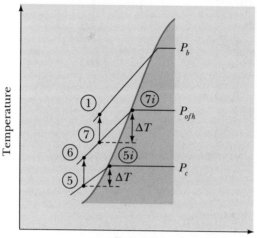

shows this and the various states near the saturated liquid line of a Rankine cycle with an open-feedwater heater. State $5i$ represents the ideal condenser outlet state (i.e., a saturated liquid at the condenser pressure), and state $7i$ the ideal open-feedwater heater outlet state (i.e., a saturated liquid at the open-feedwater pressure). Both outlets may be subcooled to a temperature lower than the saturation temperature. The amount of subcooling is specified by the difference between the outlet temperature when the working fluid is a saturated liquid and the actual outlet temperature, ΔT. The net effect of this subcooling is to lower the temperature T_1 at the boiler inlet as state 1 moves to the left on the P_b line. Consequently, more heat must be added to the working fluid in the boiler, thereby reducing the net thermal efficiency.

In an ideal closed-feedwater heater, the steam bled from the turbine leaves the heater as a saturated liquid, and the feedwater leaves the heater at the same temperature as the condensed bleed steam. Such a heat exchanger would require a very large heat-transfer coefficient or heat-transfer area. This is not the case, and the feedwater leaves a closed-feedwater heater at a temperature lower than that of the condensing bleed steam as shown in Figure 7.15; the terminal temperature difference is illustrated as ΔT.

EXAMPLE 7.7

A reheat Rankine cycle uses one open-feedwater heater for regeneration purposes. The first pass through the boiler operates at 600 psia and heats the water to 600° F. The reheat pass operates at 200 psia and reheats the water to 600° F. The condenser operates at 10 psia, and the open-feedwater heater operates at 25 psia. The high-pressure turbine operates with an efficiency of 90%, and the low-pressure turbine efficiency is 85% (this may be applied between the inlet and bleed-steam extraction point, and between the extraction point and outlet). Both the condenser and open-feedwater heater have a 5° F subcooling at their exits. Determine: (a) the mass of extracted bleed steam per unit mass of water passing through the boiler, (b) the work produced by the two turbines per unit mass of water passing through the boiler, and (c) the cycle's net thermal efficiency.

FIGURE 7.15
Terminal Temperature Difference in a Closed Feedwater Heater

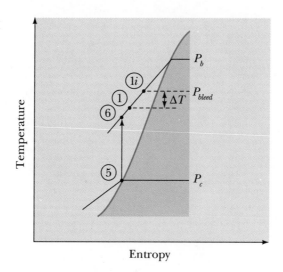

System: The water passing through the various components of the modified Rankine cycle as shown in Figure 7.16.

Basic Equations:

$$\sum_{in} \dot{m} = \sum_{out} \dot{m}$$

$$\dot{Q} - \dot{W} = \sum_{out} \dot{m}(h + \frac{V^2}{2} + gz) - \sum_{in} \dot{m}(h + \frac{V^2}{2} + gz)$$

$$-_1 w_2 = \int_1^2 v\,dP + \Delta ke + \Delta pe$$

$$\eta = \frac{\dot{W}_{net}}{\dot{Q}_{in}}$$

Conditions: All processes will be taken to be steady-state.

Solution: The labeling of the various states about the cycle follows the scheme in Figure
7.16. Properties at the various states about this cycle are presented in Table
7.5; the boldface entries are found in the problem statement.

TABLE 7.5
States for Example 7.7

State	P psia	T °F	x %	h Btu/lb$_m$	s Btu/lb$_m$-°R
1	**600**			205.2	
2	**600**	**600**		1289.5	1.5320
3s	**200**		98.6	1187.1	$s_{3s} = s_2$
3	**200**			1197.3	
4	**200**	**600**		1322.1	1.6767
5s	**25**		97.2	1134.5	$s_{5s} = s_4$
5	**25**			1162.6	1.7155
6s	**10**		95.2	1096.2	$s_{6s} = s_5$
6	**10**		96.2	1106.2	
7	**10**	188.2		156.0	
8	**25**			156.0	
9	**25**	235.1		203.4	

FIGURE 7.16
Reheat and Regenerative Rankine Cycle of Example 7.7

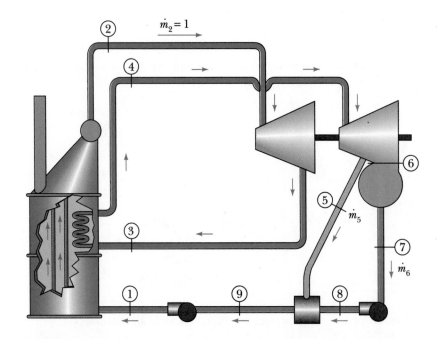

State 3s was determined from the given pressure and the fact that $s_{3s} = s_2$. The enthalpy of the actual state 3 was found from the definition of the turbine efficiency,

$$
\begin{aligned}
h_3 &= h_2 + \eta_{hp}(h_2 - h_{3s}) \\
&= 1289.5 - 0.9(1289.5 - 1187.1) = 1197.3 \text{ Btu/lb}_m
\end{aligned}
$$

State 5s was determined from the given pressure and the fact that $s_{5s} = s_4$. The actual state 5 was then found in the same manner as state 3. Beginning with the actual state 5, states 6s and 6 were found next in the same way as states 5s and 5. The temperature of state 7 was found using the subcooling specification,

$$
\begin{aligned}
T_7 &= T_{sat} - \Delta T \\
&= 193.2 - 5 = 188.2° \text{ F}
\end{aligned}
$$

The enthalpy of state 7 was then found from this temperature. The specific volume at state 7 was found to be $0.016559 \text{ ft}^3/\text{lb}_m$. State 8's enthalpy was found next by combining the first law and the pump work expression:

$$
\begin{aligned}
h_8 &= h_7 + v_7(P_8 - P_7) \\
&= 156.10 + 0.016559\frac{144(25 - 10)}{778} = 156.0 \text{ Btu/lb}_m
\end{aligned}
$$

States 9 and 1 were fixed in the same manner as states 7 and 8.

The mass flow rate through the boiler was set to 1 lb_m/s. All calculations are then on the basis of 1 lb_m/s of flow through the boiler. Applying the conservation of mass principle to the low-pressure turbine gives

$$
\dot{m}_6 = 1 - \dot{m}_5
$$

When applied to the open-feedwater heater, the first law yields

$$
\begin{aligned}
m_5 &= \frac{h_9 - h_8}{h_5 - h_8} \\
&= \frac{203.4 - 156.0}{1162.6 - 156.0} \\
&= 0.0471 \text{ lb}_m \text{ per lb}_m \text{ of boiler flow}
\end{aligned}
$$

The work produced by the two turbines is found by applying the first law to each turbine:

$$w_t = 1(h_2 - h_3) + 1(h_4 - h_5) + (1 - \dot{m}_5)(h_5 - h_6)$$
$$= 1(1289.5 - 1197.3) + 1(1322.1 - 1162.6) + 0.9529(1162.6 - 1106.2)$$
$$= 305.5 \text{ Btu per lb}_m \text{ of boiler flow}$$

Finally, the thermal efficiency is

TIP:

When analyzing complex systems, it is to our advantage to select a common basis for all calculations. In this example, the flow rate through the boiler was selected for this basis.

$$\eta = 1 - \frac{|\dot{Q}_{out}|}{\dot{Q}_{in}}$$

$$= 1 - \frac{(1 - m_5)(h_6 - h_7)}{(h_2 - h_1) + (h_4 - h_3)}$$

$$= 1 - \frac{0.9529(1106.2 - 156.0)}{(1289.5 - 205.2) + (1322.1 - 1197.3)}$$

$$= 25.1\%$$

7.8   Applying Exergy and Work-Potential Accounting to Rankine Cycles

By accounting for the exergy at the various states and the work potential for each process in a Rankine cycle, we can readily spot those processes that do not take full advantage of the work they could produce. We can then focus on those processes that are most ineffective and seek ways to make them more effective. If these processes can be made more effective, the cycle efficiency, fuel economy, and system environmental loading will be improved. Exergy and work-potential accounting, as applied to a Rankine cycle, is illustrated in the following example.

EXAMPLE 7.8

A Rankine cycle for a geothermal-energy source with an open-feedwater heater operates with the steam at 4,000 kPa and 400° C at the turbine inlet. All devices operate ideally except the turbine, which has an efficiency of 94% between the inlet and bleed-steam extraction point and between the bleed-steam extraction point and the outlet. The open-feedwater heater operates at 150 kPa, and the condenser operates at 20 kPa. The energy source is at 400° C and will be assumed to be isothermal; the energy sink is at 20° C and also will be presumed isothermal. Use exergy and work-potential accounting to determine which processes are most ineffective.

System: The water passing through the various components of the Rankine cycle as shown in Figure 7.17.

Basic Equations:

$$\sum_{in} \dot{m} = \sum_{out} \dot{m}$$

$$\dot{Q} - \dot{W} = \sum_{out} \dot{m}\left(h + \frac{V^2}{2} + gz\right) - \sum_{in} \dot{m}\left(h + \frac{V^2}{2} + gz\right)$$

$$-_1 w_2 = \int_1^2 v dP + \Delta ke + \Delta pe$$

$$\psi = (h - h_0) - T_0(s - s_0)$$

Conditions: All processes will be taken to be steady-state and reversible.

FIGURE 7.17
Rankine Cycle for Example 7.8

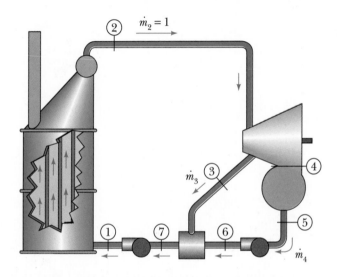

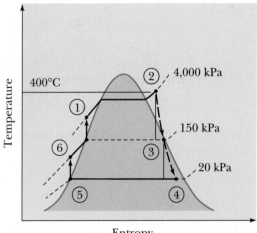

Solution: We will take the mass flow rate through the boiler as the common basis for all calculations. Figure 7.17 shows the numbering scheme for the states around this cycle.

Using the first and second laws, the pertinent properties at all states can be determined as in the previous examples. These are summarized in Table 7.6. The flow exergy, ψ, values in this table are based on a dead state of 101 kPa, 25° C. At this state, water is a liquid, and the dead-state properties may be treated as those of a saturated liquid at 25° C. At state 2, the flow exergy is

$$\psi_2 = (h_2 - h_0) - T_0(s_2 - s_0)$$
$$= (3213.6 - 104.89) - 298(6.7690 - 0.3674)$$
$$= 1201.0 \text{ kJ/kg}$$

according to Section 5.9.

TABLE 7.6
States for Example 7.8

State	P kPa	T °C	x %	h kJ/kg	s kJ/kg-K	ψ kJ/kg
1	4000	250.4		471.1	1.4336	48.49
2	4000	400		3213.5	6.7690	1200.9
3s	150		92.2	2518.5	$s_{3i} = s_2$	
3	150		94.0	2560.2	6.8765	515.6
4s	20		85.3	2262.8	$s_{4i} = s_3$	
4	20		86.0	2280.6	6.9211	222.7
5	20	60.1	0	251.4	0.8320	8.06
6	150			251.5	0.8320	8.16
7	150	111.4	0	467.1	1.4336	44.48

Applying the first law to the open-feedwater heater gives

$$\dot{m}_3 = \frac{h_7 - h_6}{h_3 - h_6} = 0.09339$$

The heat added to the water in the boiler (from the geothermal source) is obtained by applying the first law to the boiler:

$$_1 q_2 = 1(h_2 - h_1) = 2742.4 \text{ kJ/kg-boiler}$$

The heat rejected in the condenser to the sink is found by applying the first law to the condenser:

$$_4 q_5 = \dot{m}_4(h_5 - h_4) = -1839.7 \text{ kJ/kg-boiler}$$

The work produced by the turbine is

$$_1 w_4 = (h_2 - h_3) + (1 - \dot{m}_3)(h_3 - h_4) = 906.8 \text{ kJ/kg-boiler}$$

Similar application of the first law yields the work used by the pumps as $_5w_6 = -0.09$ kJ/kg-boiler and $_7w_1 = -4.00$ kJ/kg-boiler. When these are totaled, the net work produced by this cycle is $w_{net} = 902.7$ kJ/kg-boiler.

Table 7.7 presents the difference in the exergy entering and leaving each device, the work potential, the actual work produced, and each device's lost work potential. Note that the difference between the work potential and the difference between exergies going in and out of each device is simply $q(1 - T_0/T)$ for each thermal-energy reservoir that interacts with the process. The increase in the specific work potential introduced in the boiler is then

$$w_{ptl, boiler} = \psi_1 - \psi_2 + {_1q_2} \left(1 - \frac{T_0}{T_{boiler}} \right)$$

$$= -1155.5 + 2742.4 \left(1 - \frac{298}{673} \right)$$

$$= 372.6 \text{ kJ/kg-boiler}$$

TABLE 7.7
Exergies and Work Potentials for Example 7.8

Device	$\Sigma_{in} \dot{m}\psi - \Sigma_{out} \dot{m}\psi$ kJ/kg-boiler	w_{ptl} kJ/kg-boiler	w_{act} kJ/kg-boiler	$w_{ptl} - w_{act}$ kJ/kg-boiler
Boiler	−1151.6	376.5	0	376.5
Turbine	950.6	950.6	906.8	43.8
Condenser	194.5	255.5	0	255.5
Pump$_{5,6}$	− 0.1	− 0.1	−0.09	−0.01
OFW	13.1	13.1	0	13.1
Pump$_{7,1}$	− 6.1	− 6.1	−4.00	−2.1
Total	**0**	**1589.5**	**902.8**	**686.8**

The total potential work produced by this cycle is 1589.5 kJ/kg-boiler, but the actual work produced was 902.8 kJ/kg-boiler. Hence, 686.8 kJ/kg-boiler of work potential has been lost. The lost work potential of each device is presented in the last column of Table 7.7. This column clearly illustrates that the boiler and condenser are the primary contributors to the lost work potential. As such, these two components represent the best opportunities for improving this cycle's performance. Most of the lost work potential in these two components results from the heat transfer across the temperature difference between the source-energy reservoir and boiler working fluid and between the condenser working fluid and the sink-energy reservoir. The first improvement to seek is reducing these temperature differences as much as possible within the confines of the equipment requirements and the second law. ◼

7.9 Basic Vapor-Compression Refrigeration Cycle

The basic vapor-compression refrigeration system (VCRS) consists of two heat exchangers, a compressor, and a throttle device as shown in Figure 7.18. The high-pressure refrigerant vapor produced by the compressor releases heat as it is condensed to a liquid in the heat exchanger, which is known as the *condenser*. Ideally, the process in the condenser is isobaric. The refrigerant liquid leaving the condenser is next expanded to the lower pressure of the evaporator through the throttle device as shown on the *T–s* state diagram of Figure 7.18. A low-temperature, low-pressure mixture of refrigerant liquid and vapor is produced by this irreversible adiabatic expansion. The liquid portion of this mixture is evaporated in an isobaric heat exchanger known as the *evaporator* as heat is added to the refrigerant mixture from the space or load being cooled. The saturated vapor produced in the evaporator enters the adiabatic compressor, where it is compressed before repeating the cycle.

In refrigeration systems, the vapor phase is being compressed. Compressors, like the feedwater pumps of the previous sections, require single-phase fluids to avoid premature failure from pitting and erosion of their metal parts. To minimize the work consumed by the compressor while it is protected from ingesting liquid refrigerant, the vapor enters the compressor as a saturated vapor. It may also enter the compressor as a superheated vapor, although this will degrade the performance of the cycle to some degree, as we will see in the next section.

You may wonder why a reversible, adiabatic, work-producing expansion process is not being used to expand from the condenser pressure to the evaporator pressure instead of the irreversible throttle device, which does not produce work. In refrigeration systems, we are expanding a liquid rather than a vapor. To avoid premature failure, liquid-expansion devices such as hydraulic turbines cannot have any vapor present during the expansion process. If such a device were used, state 4 would have to be a saturated liquid, and state 3 would be moved further into the compressed liquid region of the *T–s* state diagram. If this were done, the work produced during the expansion would be

$$_3w_4 = v_4(P_3 - P_4)$$

This would be very small compared to the work consumed by the compressor. Thus, the work recovered does not pay for the additional cost of a work-producing expansion device in most cases.

The pressures in the condenser and evaporator are determined by the temperatures we are trying to achieve and the system's working fluid. For example, if we are air-conditioning a building, we would like to cool the air entering the building to approximately 7–10° C (45–50° F) while rejecting heat to the atmospheric air at approximately 35–43° C (95–110° F). Allowing some temperature difference for heat transfer, the evaporator might then operate at 0° C (32° F) and the condenser at 50° C (122° F). Note that this difference

FIGURE 7.18

*Basic
Vapor-Compression
Refrigeration System*

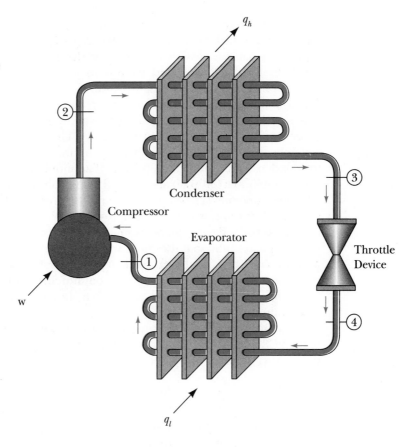

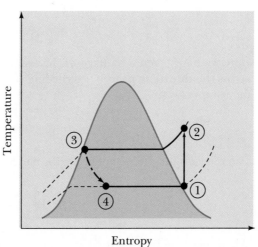

between the temperature of the evaporator and the region being cooled is
necessary to satisfy the Clausius statement of the second law. For the same

reason, we also require a temperature difference between the condenser and the region to which heat is rejected. We also would like the pressures in these units to be as low as possible to hold down the thickness of the tube walls from which these units are built and to minimize the work required to compress the vapor from low pressure to high pressure. Ammonia, chlorofluorocarbons, hydrogenated chlorofluorocarbons (which have been proposed to reduce the depletion of the ozone layer), and water are fluids that meet these as well as other requirements such as reduced toxicity. Thus, if we were to use R-12 in this example, the evaporator would operate at approximately 310 kPa (45 psia) and the condenser at 1200 kPa (180 psia). If ammonia were used, these pressures would be 430 kPa (60 psia) and 2030 kPa (290 psia), respectively. Refrigerant-12 thereby operates at lower pressures and requires a smaller pressure rise (and therefore less work) across the compressor.

The throttle device can be as simple and inexpensive as a long, insulated metal capillary tube that uses fluid friction to reduce pressure. Most systems now use an insulated, thermostatic expansion valve for this purpose. These valves sense the temperature in the evaporator and control the pressure drop between the condenser and evaporator to maintain and control the evaporator temperature.

The first and second laws may be applied to each steady-flow component of the VCRS to determine the amounts of heat transfer, the work required, and the coefficient of performance. Applying the first law to the condenser gives

$$q_h = {}_2q_3 = h_3 - h_2$$

when kinetic and potential energy effects are neglected. Similarly, the heat transfer to the evaporator is given by

$$q_1 = {}_1q_4 = h_4 - h_1 = h_3 - h_1$$

Here we have taken advantage of the fact that the enthalpy does not change across the throttling process. According to the first law, the work required by the compressor is

$$w = {}_1w_2 = h_1 - h_2$$

The refrigerator coefficient of performance is then

$$\text{COP}_r = \frac{h_3 - h_1}{h_2 - h_1}$$

When this cycle is being used as a heat pump, the COP is given by

$$\text{COP}_{hp} = \frac{h_3 - h_2}{h_2 - h_1}$$

EXAMPLE 7.9

A basic vapor-compression heat pump that uses R-12 operates the condensor at 40° C and the evaporator at 1000 kPa. Calculate the specific heat transfer to the evaporator, the specific work consumption, the actual COP, and the minimum Carnot COP for this cycle.

System: The R-12 circulating among the components of the basic vapor-compression refrigeration system in Figure 7.18.

Basic Equations:
$$_1q_2 - _1w_2 = h_2 - h_1 + \frac{1}{2}(V_2^2 - V_1^2) + g(z_2 - z_1)$$

Conditions: All processes shown in the T–s state diagram of Figure 7.18 will be taken to be reversible with the exception of the throttling process. The vapor at the inlet of the compressor is saturated, as is the liquid entering the throttle device.

Solution: The states will be numbered as shown in Figure 7.18. Because states 1 and 3 are a saturated vapor and liquid, respectively, their properties are $h_1 = 347.14$ kJ/kg, $s_1 = 1.5599$ kJ/kg-K, and $h_3 = 240.40$ kJ/kg, according to the R-12 property tables in the appendix. Because $s_2 = s_1$, $h_2 = 373.78$ kJ/kg, and $T_2 = 49.7°$ C by interpolating Table B2.2 of Appendix B. Applying the first law,

$$_4q_1 = h_1 - h_4 = 106.74 \text{ kJ/kg}$$
$$_1w_2 = h_1 - h_2 = -26.64 \text{ kJ/kg}$$
$$_2q_3 = h_3 - h_2 = -133.38 \text{ kJ/kg}$$

The signs indicate the directions of the work and heat transfers. The actual coefficient of performance is

$$COP_{hp} = \left|\frac{_2q_3}{_1w_2}\right| = \frac{133.38}{26.64} = 5.01$$

TIP:

The fact that the actual COP is less than the Carnot COP should come as no surprise because of the irreversibilities in the throttle device and the superheating of the vapor by the compressor.

The Carnot refrigeration cycle with the minimum coefficient of performance and which operates in the same way as this cycle would have thermal-energy reservoirs at 263.2 K and 322.9 K. Its COP would be

$$COP_{hp} = \frac{1}{1 - T_l/T_h}$$
$$= 5.41$$

7.10 Effects that Degrade the Performance of Vapor-Compression Refrigeration Systems

The basic VCRS of the preceding section was operating ideally. Several operational and irreversible effects occur in these systems that degrade their ideal performance.

Actual compressors do not operate isentropically, and we must adjust the work consumed by using the compressor efficiency. The actual work required to drive the compressor is

$$_1 w_2 = \frac{h_{2s} - h_1}{\eta_c} = h_2 - h_1$$

where the right-hand side results from application of the first law to the compressor. State $2s$ is the state for which entropy is the same as that in state 1 and where pressure matches the condenser pressure, as shown in Figure 7.19. Compressor irreversibilities increase the work used by a VCRS and increase the enthalpy of the working fluid as it enters the condenser. This reduces the COP and increases the amount of heat that must be rejected in the condenser.

Working fluid vapor entering the compressor is sometimes slightly superheated to ensure that no liquid enters and damages the compressor. Similarly, the liquid working fluid entering the throttle valve may be slightly subcooled, depending on where the thermostatic valve sensor is located, the temperature of the sink to which the heat is being rejected, and the capacity of the condenser. Both effects are described by a superheat and subcool ΔT, respectively, as illustrated in Figure 7.19.

FIGURE 7.19

Subcooling, Superheating, and Compressor-Efficiency Effects on the VCRS

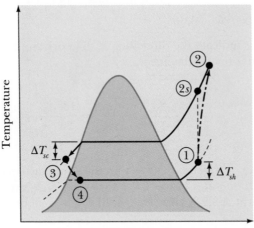

Pressure losses from fluid friction in the connecting lines, condenser, and evaporator also degrade VCRS performance. These are normally minimized by the evaporator and condenser designers. Care must be used in the selection of interconnecting lines, particularly when they are very long. This can become a serious problem when a building's air-conditioning compressor–condenser unit is located outside the building and the evaporator unit is inside the building and far from the exterior unit.

EXAMPLE 7.10

A VCRS that uses ammonia as its working fluid operates a condenser at 250 psia and an evaporator at 30 psia. At the entrance to the compressor the vapor is superheated by 10.6° F; at the entrance to the throttle device, the liquid is subcooled by 10.8° F. The compressor efficiency is 92%. Compare the refrigerator COP of this VCRS to that of the ideal VCRS using the same condenser and evaporator pressures.

System: The system is the R-12 working fluid of the VCRS as illustrated in Figure 7.18.

Basic Equations:
$$\dot{Q} - \dot{W} = \sum_{out} \dot{m}\left(h + \frac{V^2}{2} + gz\right) - \sum_{in} \dot{m}\left(h + \frac{V^2}{2} + gz\right)$$

Conditions: None

Solution: The various states for this cycle are shown in Figure 7.20. From the ammonia tables of Appendix B, $h_1 = 618.33$ Btu/lb$_m$, $s_1 = 1.3507$ Btu/lb$_m$-° R, $h_{2s} = 761.26$ Btu/lb$_m$, and $h_3 = h_4 = 156.7$ Btu/lb$_m$. Applying the first law and the definition of compressor efficiency to the compressor yields

$$h_2 = h_1 + \frac{h_{2s} - h_1}{\eta_c}$$
$$= 773.7 \text{ Btu/lb}_m$$

The heat addition to the evaporator is given by the first law:

$$_4q_1 = h_1 - h_4$$
$$= 416.6 \text{ Btu/lb}_m$$

The work consumed by the compressor is given by

$$_1w_2 = h_1 - h_2$$
$$= -142.9 \text{ Btu/lb}_m$$

The actual COP is then

$$\text{COP}_r = \left|\frac{q_{in}}{w_{in}}\right|$$
$$= \left|\frac{_4q_1}{_1w_2}\right|$$
$$= 2.915$$

FIGURE 7.20
*VCRS State Diagram
for Example 7.10*

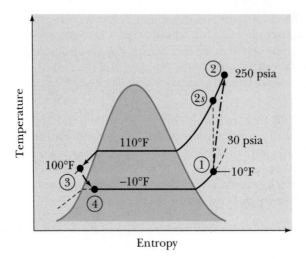

If this cycle had operated ideally with saturated vapor at the entrance to the compressor, saturated liquid at the entrance to the throttle device, and isentropic compression, then $h_3 = h_4 = 169.32$ Btu/lb$_m$, $h_1 = 612.33$ Btu/lb$_m$, and $h_2 = 751.67$ Btu/lb$_m$. The heat added to the evaporator would then be 443.0 Btu/lb$_m$, and the work required by the compressor would be -139.3 Btu/lb$_m$. The coefficient of performance then would be

$$\text{COP}_{ideal} = 3.180$$

The ideal VCRS produces approximately 9% more refrigeration effect for the same compressor work than does the actual VCRS. ◼

EXAMPLE 7.11

A VCRS refrigerator that uses R-12 operates ideally, except for the compressor, whose efficiency is 90%. This system removes heat from a food freezer, which is at $-20°$ C, by operating the evaporator at $-30°$ C. Heat is rejected to the 25° C ambient air by operating the condenser at 30° C. Which VCRS process loses the greatest amount of specific work potential?

System: The system is the R-12 working fluid of the VCRS as illustrated in Figure 7.18.

Basic Equations:

$$\dot{Q} - \dot{W} = \sum_{out} \dot{m}\left(h + \frac{V^2}{2} + gz\right) - \sum_{in} \dot{m}\left(h + \frac{V^2}{2} + gz\right)$$

$$\psi = h - h_0 - T_0(s - s_0)$$

Conditions: The dead state will be taken as a saturated liquid at 25° C.

Solution: We will number the states the same as in Figure 7.21. The various state properties, as determined by the R-12 property tables in Appendix B, are presented
in Table 7.8. The flow exergies were calculated using

$$\psi = h - h_0 - T_0(s - s_0)$$

Applying the first law to the condenser,

$$_2q_3 = h_3 - h_2 = 149.8 \text{ kJ/kg}$$

FIGURE 7.21
*T–s State Diagram for
Example 7.11*

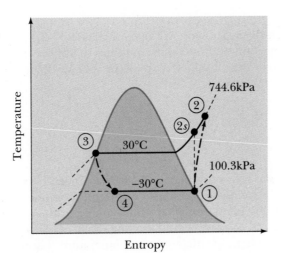

Entropy

TABLE 7.8
*Thermodynamic
States for Example
7.11*

State	h	s	ψ
	kJ/kg	kJ/kg-K	kJ/kg
1	338.1	1.5748	−32.38
2s	374.5	$s_{2s} = s_1$	
2	378.5	1.5872	4.33
3	228.7	1.0988	0.07
4	$h_4 = h_3$	1.1247	− 7.65
0	223.8	1.0826	0

The condenser work potential, which is also the condenser lost work potential
because no work is done by the condenser, is

$$w_{ptl} = \psi_2 - \psi_3 + {}_2q_3 \left(1 - \frac{T_0}{T_s}\right)$$

$$= 4.33 - 0.07 - 148.3 \left(1 - \frac{298}{298}\right)$$

$$= 4.26 \text{ kJ/kg}$$

Similarly, the heat added to the evaporator is 109.4 kJ/kg, and the work potential lost in the evaporator is 5.27 kJ/kg. The work consumed by the compressor is −40.4 kJ/kg, and the work potential lost in the compressor is

$$_1w_{2lost} = \psi_1 - \psi_2 - {_1w_2}$$
$$= -32.38 - 4.33 + 40.4$$
$$= 3.69 \text{ kJ/kg}$$

The throttle device's lost work potential is $\psi_3 - \psi_4 = 7.72$ kJ/kg.

Inspection of these results indicates that the throttle device's irreversibilities make it the most ineffective of the four VCRS processes.

7.11 VCRS Variations

Several variations to the basic VCRS have been developed for specific applications. Systems that serve two or more low temperatures—as, for example, in a freezing plant with an air-conditioning load—require one evaporator for freezing and one for air conditioning. Each evaporator operates at a pressure that is sufficient for the required temperature. Large systems require considerable energy for compression. Systems of this type typically use one or more compression stages to conserve energy. To achieve very low temperatures—as may be required to liquefy gases, for example—several systems using different working fluids may be cascaded. In this arrangement, the evaporator of one system services the condenser of another. Temperatures below the triple point of one working fluid may be achieved in this manner. These are but three of several variations on the basic VCRS.

7.11.1 Multiple Evaporators

A two-evaporator system that services two different loads, each at a different temperature, is shown in Figure 7.22. This system operates at three different pressures: the condenser pressure, the high-temperature evaporator 1 pressure, and the low-temperature evaporator 2 pressure. The high-pressure liquid from the condenser is throttled to the high-temperature evaporator pressure to serve evaporator 1. The vapor leaving evaporator 1 is throttled a second time by a pressure-control valve so that its pressure matches that of evaporator 2. Saturated vapor leaving evaporator 2, after being throttled and evaporated, is mixed with the throttled vapor from evaporator 1 to form the vapor that enters the compressor.

FIGURE 7.22
VCRS with Two Evaporators

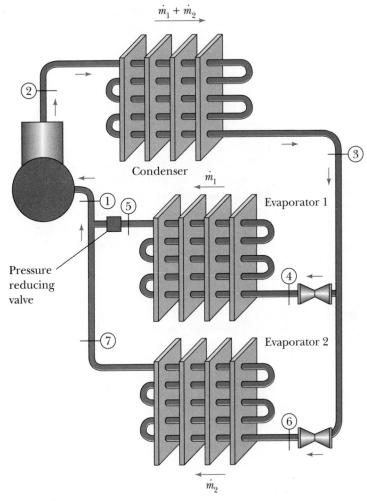

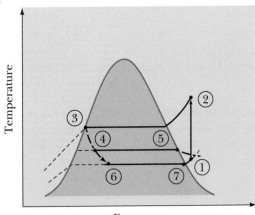

The flow rate is different through the two evaporators. Applying the first law to the point at which the two evaporator streams are mixed to form the stream entering the compressor yields

$$h_1 = \frac{\dot{m}_1 h_5 + \dot{m}_2 h_7}{\dot{m}_1 + \dot{m}_2}$$

The working fluid that enters the compressor is not a saturated vapor. Rather, it is somewhat superheated, as shown in the T–s state diagram of Figure 7.22, and at the same pressure as the fluid in the low-temperature evaporator.

VCRSs with multiple evaporators are analyzed by applying the first and second laws and the conservation of mass principle to various components such as other thermodynamic systems. One needs to account for the difference in the flow rates of the two evaporators when conducting this analysis.

EXAMPLE 7.12

A R-12 VCRS with two evaporators is designed to service a 10 kW cooling load with a 100 kPa evaporator and a 8 kW load with a 0° C evaporator when the condenser is operating at 1000 kPa. Determine the mass flow rate through the condenser, the power required by the compressor, and this cycle's overall COP.

System: The system is the R-12 working fluid of the multievaporator VCRS as illustrated in Figure 7.22.

Basic Equations:

$$\sum_{in} \dot{m} = \sum_{out} \dot{m}$$

$$\dot{Q} - \dot{W} = \sum_{out} \dot{m}(h + \frac{V^2}{2} + gz) - \sum_{in} \dot{m}(h + \frac{V^2}{2} + gz)$$

Conditions: All processes are steady-flow.

Solution: We will number the states as illustrated in Figure 7.22. States 3 through 7 may be determined from the T–s diagram. The pertinent properties at these states are $h_3 = h_4 = h_6 = 240.4$ kJ/kg, $h_5 = 351.5$ kJ/kg,   and $h_7 = 338.1$ kJ/kg. Applying the rate-basis first law to the 0° C evaporator 1 gives

$$\dot{m}_1 = \frac{_4\dot{Q}_5}{h_5 - h_4} = 0.07200 \text{ kg/s}$$

Applying the rate-basis first law to the $-30°$ C evaporator 2 gives

$$\dot{m}_2 = \frac{{}_6\dot{Q}_7}{h_7 - h_6} = 0.1024 \text{ kg/s}$$

The first law applied to the mixing point between states 1, 5, and 7 yields

$$h_1 = \frac{\dot{m}_1 h_5 + \dot{m}_2 h_7}{\dot{m}_1 + \dot{m}_2} = 343.6 \text{ kJ/kg}$$

The entropy at state 1 may now be determined because the pressure and enthalpy of state 1 are known. This entropy is $s_1 = 1.5968$ kJ/kg-K. Based on the entropy at state 1 and the pressure at state 2, the enthalpy at state 2 is 386.2 kJ/kg.

The mass flow rate through the condenser is

$$\dot{m}_c = \dot{m}_1 + \dot{m}_2$$
$$= 0.1744 \text{ kg/s}$$

Applying the first law to the compressor yields

$$\dot{W}_c = \dot{m}_c(h_1 - h_2)$$
$$= -7.42 \text{ kW}$$

The overall COP is then

$$\text{COP}_r = \frac{{}_4\dot{Q}_5 + {}_6\dot{Q}_7}{{}_1\dot{W}_2}$$
$$= 2.43$$

7.11.2 Multiple Compressors

We can reduce the amount of work required to compress the refrigerant vapor by compressing in several pressure stages rather than by using a single pressure stage to compress directly from the evaporator pressure to the condenser pressure. In large refrigeration and heat-pump systems, the energy saved often offsets the cost of the additional equipment used in multiple-compressor systems.

A typical two-stage compressor system is shown in Figure 7.23. A liquid–vapor separation unit is required in addition to the extra compressor. The purpose of this unit is to supply refrigerant vapor to the high-pressure (state 1 to 2) compressor and liquid to the evaporator throttle device (state 5). A mixture of saturated refrigerant liquid and vapor (state 4) enters the separator from the condenser throttle valve. Following the evaporation of the liquid in the evaporator, the resulting vapor is compressed in the low-pressure (state 7 to 8) compressor back to the pressure in the separator chamber. The separation chamber is essentially a rigid container that is well insulated to prevent heat loss to the surroundings.

Multicompressor VCRSs are analyzed by the appropriate application of the conservation of mass principle, the first law, and the second law.

EXAMPLE 7.13

A two-compression–stage VCRS with a liquid–vapor separation unit using R-12 as the refrigerant operates the condenser at 160 psia, the separation unit at 40 psia, and the evaporator $-20°$ F. Calculate this refrigeration unit's overall COP. Compare this COP to that of a single-compression–stage VCRS with the same condenser pressure and evaporator temperature.

System: The R-12 passing through the various components and lines of the VCRS cycle as shown in Figure 7.23.

Basic Equations:

$$\sum_{in} \dot{m} = \sum_{out} \dot{m}$$

$$\dot{Q} - \dot{W} = \sum_{out} \dot{m}\left(h + \frac{V^2}{2} + gz\right) - \sum_{in} \dot{m}\left(h + \frac{V^2}{2} + gz\right)$$

Conditions: All processes will be taken to be steady-state and reversible.

Solution: Using the numbering scheme of Figure 7.23 and Tables B2.1 and B2.2 (Appendix B), the germane properties are $h_1 = 80.38$ Btu/lb$_m$, $s_1 = 0.1678$ Btu/lb$_m$-° R, $h_2 = 91.13$ Btu/lb$_m$, $h_3 = h_4 = 35.05$ Btu/lb$_m$, $h_5 = h_6 = 14.18$ Btu/lb$_m$, $h_7 = 75.47$ Btu/lb$_m$, $s_7 = 0.1718$ Btu/lb$_m$-° R, and $h_8 = 82.46$ Btu/lb$_m$. Selecting the mass flow rate through the evaporator as the basis of all calculations and applying the first law to the separation unit gives

$$\dot{m}_2 = \frac{h_8 - h_6}{h_1 - h_4}$$
$$= 1.503 \text{ lb}_m/\text{lb}_m\text{-evap}$$

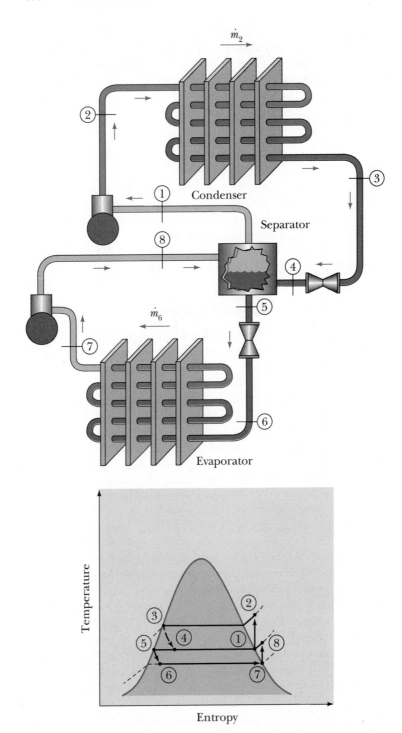

FIGURE 7.23
VCRS with Two Compressors

The cooling effect is $h_7 - h_6 = 61.29$ Btu/lb$_m$-evap. The low-pressure compressor uses $h_7 - h_8 = -6.99$ Btu/lb$_m$-evap of work, and the high-pressure compressor uses $\dot{m}_2(h_1 - h_2) = -16.01$ Btu/lb$_m$-evap. The overall COP of the two-compression–stage refrigeration unit is then

$$\mathrm{COP}_r = \left| \frac{_6q_7}{_1w_2 + {_7}w_8} \right|$$
$$= 2.66$$

With a single compressor, the enthalpy at the compressor outlet becomes 93.49 Btu/lb$_m$, the specific cooling effect is 40.42 Btu/lb$_m$, and the work required by the compressor is -18.02 Btu/lb$_m$. The COP$_r$ for equivalent operation with a single compressor is 2.24. Hence, we get a reduced cooling effect for the same amount of compression work with the single compressor as compared to a two-compression–stage VCRS.

7.11.3 Cascaded Systems

VCRSs may be cascaded as in Figure 7.24. This arrangement is used most commonly to achieve very low temperatures such as those used to liquefy gases. To liquefy a gas, temperatures below the gas's critical point temperature (see Table 3.4) must be reached. Thus, if we wish to liquefy the nitrogen constituent of atmospheric air, we need a refrigeration system that is capable of operating the evaporator at $-210°$ C $(-346°$ F) or lower. The triple-point temperature of the refrigerant used in this evaporator would have to be even lower to avoid the formation of refrigerant solid phase in the system. Without a cascade arrangement, this refrigerant's critical point temperature would have to be greater than the temperature of the thermal-energy reservoir to which heat is rejected to avoid excessive pressures and their associated high compressor operational costs. For example, to serve a condenser at $30°$ C and an evaporator at $-60°$ C with R-12, the condenser operates at 745 kPa and the evaporator at 22 kPa. The compressor must then produce a pressure ratio of approximately 34:1, which is quite large and requires considerable work.

The cascaded arrangement is one method of achieving low temperatures while keeping the compressor operational costs in line. Each VCRS in the cascade uses a different refrigerant specifically selected to work over its individual temperature range. For example, VCRS A of Figure 7.24 might use R-12 to operate its condenser at $30°$ C and its evaporator at $-20°$ C. VCRS B, which may use ammonia as its refrigerant, then can operate its condenser at $-15°$ C and its evaporator at $-50°$ C. Note that the condenser of VCRS B operates at a slightly higher temperature than the evaporator of VCRS A so that the heat released by B will be transferred to A. By cascading the two VCRSs, neither compressor has to compress its vapor over an excessive pressure range. The amount of work required to service low-temperature cooling loads is thereby reduced.

FIGURE 7.24
Two Cascaded VCRSs

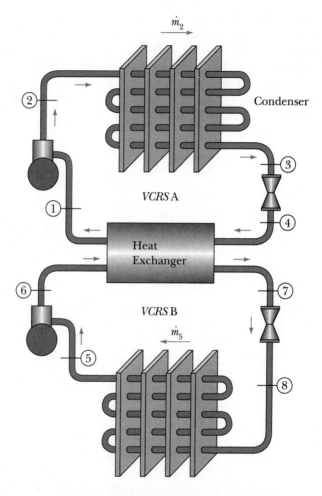

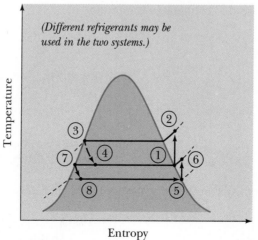

EXAMPLE 7.14

Two VCRSs are cascaded to serve a 33 Btu/s cooling load that requires an evaporator temperature of $-60°$ F while rejecting heat to the ambient air. The system designer elects to use R-12 in the high-temperature VCRS and operate its condenser at 250 psia and its evaporator at $0°$ F. She chooses to use ammonia in the low-temperature VCRS and operate its condenser at 40 psia and its evaporator at $-60°$ F. Calculate the power consumed by and the COP of this cascaded system.

System: The system is the R-12 and ammonia working fluids of the cascade VCRS as shown in Figure 7.24.

Basic Equations:

$$\sum_{in} \dot{m} = \sum_{out} \dot{m}$$

$$\dot{Q} - \dot{W} = \sum_{out} \dot{m}(h + \frac{V^2}{2} + gz) - \sum_{in} \dot{m}(h + \frac{V^2}{2} + gz)$$

Conditions: All processes are steady-flow.

Solution: Following the state-numbering scheme of Figure 7.24, the necessary state data are $h_1 = 77.68$ Btu/lb$_m$, $s_1 = 0.1698$ Btu/lb$_m$-$°$ R, $h_2 = 95.91$ Btu/lb$_m$, $h_3 = h_4 = 44.42$ Btu/lb$_m$, $h_5 = 590.9$ Btu/lb$_m$, $s_5 = 1.4799$ Btu/lb$_m$-$°$ R, $h_6 = 707.0$ Btu/lb$_m$, and $h_7 = h_8 = 56.1$ Btu/lb$_m$.

Applying the rate-basis first law to the ammonia evaporator gives the mass flow rate in the ammonia VCRS:

$$\dot{m}_5 = \frac{_8\dot{Q}_5}{h_5 - h_8}$$
$$= 0.0617 \text{ lb}_m/s$$

The mass flow rate in the R-12 VCRS may be found by applying the first law to the cascaded condenser–evaporator:

$$\dot{m}_2 = \dot{m}_5 \frac{h_6 - h_7}{h_1 - h_4}$$
$$= 1.207 \text{ lb}_m/s$$

The power consumed by the R-12 compressor is

$$_1\dot{W}_2 = \dot{m}_2(h_1 - h_2)$$
$$= -22.00 \text{ Btu/s}$$

The power consumed by the ammonia compressor is

$$_5\dot{W}_6 = \dot{m}_5(h_5 - h_6)$$
$$= -7.16 \text{ Btu/s}$$

Applying our definition, the overall COP is

$$\mathrm{COP}_r = \left| \frac{_8\dot{Q}_5}{_1\dot{W}_2 + \,_5\dot{W}_6} \right|$$
$$= 1.13$$

7.12 Absorption Refrigeration Systems

The cost of operating a VCRS is the work that must be supplied to the compressor(s). In addition to using the work-reducing techniques of the previous section, we also can reduce the work required by compressing the liquid phase rather than the vapor phase, or replace the work form of driving energy with the less valuable heat form, or both. The absorption refrigerator system shown in Figure 7.25 was developed with these objectives in mind.

In an equilibrium situation such as that shown in Figure 7.26, which involves a solution and the vapor of one of its constituents, the concentration of the vapor constituent in the solution depends on the pressure of the vapor and the temperature of the solution (see Chapter 12). In general, this relationship will be like that shown in Figure 7.27 for a typical refrigerant–carrier liquid solution. The figure demonstrates that when the pressure of the refrigerant vapor is held constant, the amount of refrigerant contained in the solution decreases as the temperature of the solution increases. Hence, we can force the refrigerant from the solution to its vapor form by simply heating the solution at constant pressure. The same result also can be accomplished by holding the solution temperature constant while decreasing the vapor pressure. By controlling the vapor pressure and solution temperature, we can thereby cause the refrigerant vapor to enter or leave the liquid solution.

Absorption refrigerators take advantage of the solubility of refrigerants in carrier liquids and the variation of this solubility with temperature, or pressure, or both. Two commonly used solutions are water and lithium bromide (with water serving as the refrigerant) and ammonia and water (with ammonia serving as the refrigerant). The water–lithium bromide absorption system is used for air conditioning where temperatures below the water's triple-point temperature are not required. This avoids freezing the water refrigerant. The variation of the equilibrium concentration of lithium bromide and other water–lithium bromide solution properties are presented in Table B4 in Appendix B. That table not only includes the lithium bromide concentration in the solution and the solution enthalpy with respect to the table's reference state, but also the pressure of the water vapor in equilibrium with the solution. We will use the terms *strong solution* to refer to a solution with a high concentration of refrigerant and *weak solution* to refer to a solution with a low concentration of refrigerant.

The concentration of the weak solution entering the low-pressure absorber of Figure 7.25 is less than the equilibrium concentration of the solution at the temperature in the absorber. Therefore, the weak solution will absorb the refrigerant vapor entering the absorber from the refrigerator evaporator,

FIGURE 7.25
Absorption Refrigeration System

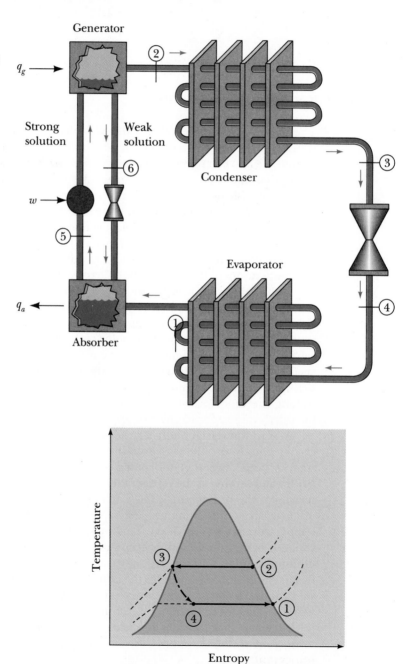

FIGURE 7.26

A Binary Solution in Equilibrium with One of the Constituent's Vapors

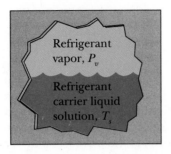

Refrigerant vapor, P_v

Refrigerant carrier liquid solution, T_s

FIGURE 7.27

Equilibrium Concentrations of a Binary Solution

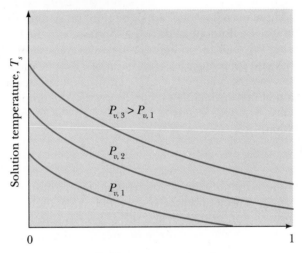

Solution temperature, T_s

$P_{v,3} > P_{v,1}$

$P_{v,2}$

$P_{v,1}$

0   1

Refrigerant concentration

which increases the refrigerant concentration in the solution. This forms a strong solution in the cool absorber. The pressure in the absorber is the same as that in the evaporator, and the absorber needs to be cooled to maintain its temperature. This cool, strong solution is then pumped into the high-pressure generator, which is at the same pressure as the refrigerator condenser. Heat is supplied to the solution in the generator to increase the solution's temperature, decrease the refrigerant's equilibrium concentration, and form a weak solution. This process releases refrigerant vapor, which is routed to the refrigerator's condenser. The resulting high-pressure weak solution is returned to the absorber after passing through a pressure-reducing throttle device to once again absorb refrigerant vapor and repeat the solution cycle.

In addition to absorbing and releasing refrigerant vapor, the absorber and generator also serve to separate the carrier liquid from other refrigeration cycle components. The absorber, solution pump, generator, and solution throttle device replace the compressor of the basic VCRS with a work- and heat-driven refrigerant-compression system. Following separation, refrigerant vapor leaves the generator and enters the condenser as a saturated vapor at the pressure maintained in the generator. The remaining weak solution leaves the generator as an equilibrium solution at the generator conditions. After passing through the refrigerant throttle device and evaporator, refrigerant vapor enters the absorber as a saturated vapor at the pressure maintained in the absorber, where it reenters the solution to form a strong solution. The strong solution then enters the solution pump as an equilibrium solution at the absorber conditions. Some carrier liquid vapor may enter the refrigeration side of this system, depending on the equilibrium properties of the carrier liquid. When a water–lithium bromide solution is used, the vapor space in the absorber and generator are filled with pure water vapor because the lithium bromide cannot be vaporized at the pressures and temperatures occurring in these devices.

Three mass flow rates are present in this system: the flow rate of the strong solution, the flow rate of the weak solution, and the flow rate of the refrigerant passing through the condenser–throttle–evaporator. The processes that the refrigerant undergoes are shown in the T–s state diagram of Figure 7.25. The system may be analyzed by applying the first and second laws and the conservation of mass principle to the various system components. The conservation of mass principle needs to be applied to the individual constituents of the solution.

EXAMPLE 7.15

A water–lithium bromide absorption air conditioner serves a 10 Btu/s cooling load. The evaporator is operated at 40° F, the condenser at 150° F, the absorber solution at 65° F, and the generator solution at 200° F. Determine the rate of heat addition to the generator and the rate of heat rejection from the condenser and absorber.

System: The system is the working fluids of the absorption VCRS shown in Figure 7.25.

Basic Equations:

$$\sum_{in} \dot{m} = \sum_{out} \dot{m}$$

$$\dot{Q} - \dot{W} = \sum_{out} \dot{m}\left(h + \frac{V^2}{2} + gz\right) - \sum_{in} \dot{m}\left(h + \frac{V^2}{2} + gz\right)$$

Conditions: The system is operating in the steady-state.

Solution: The pressure in the generator is the same as that in the condenser. This pressure is the saturation pressure at the condenser temperature, which according to the steam tables is $P_2 = P_6 = P_5 = P_s(150° F) = 3.72$ psia. Similarly, $P_1 = P_4 = P_5 = P_s(40° F) = 0.12$ psia. The enthalpies of the refrigerant (i.e., water) in the refrigerator portion of this system are $h_1 = 1078.9$ Btu/lb$_m$, $h_2 = 1126.1$ Btu/lb$_m$, and $h_3 = h_4 = 118.0$ Btu/lb$_m$.

At state 6, the solution is an equilibrium solution at 3.72 psia and 200° F. According to appendix Table B4, $c_6 = 0.50$ lb_m-LiBr/lb_m-solution and $h_6 = 85.81$ Btu/lb_m. Similarly, at state 5, $P_5 = 0.12$ psia and $T_5 = 65°$ F. Then, $c_5 = 0.45$ lb_m-LiBr/lb_m-solution and $h_5 = 12.88$ Btu/lb_m. Applying the first law to the evaporator gives

$$\dot{m}_1 = \frac{_4\dot{Q}_1}{h_1 - h_4} = 0.01041 \ lb_m/s$$

The rate at which lithium bromide enters the absorber must equal the rate at which it leaves:

$$\dot{m}_5 c_5 = \dot{m}_6 c_6$$

where c is the lithium bromide concentration (mass fraction), $\dot{m}_5$ is the mass flow rate of the strong solution, and $\dot{m}_6$ is the mass flow rate of the weak solution. Applying the conservation of mass principle to the absorber water flows yields

$$\dot{m}_1 + \dot{m}_6(1 - c_6) = \dot{m}_5(1 - c_5)$$

Combining these two conservation of mass statements gives

$$\dot{m}_5 = \frac{\dot{m}_1}{(1 - c_5) - c_5(1 - c_6)/c_6}$$
$$= 0.104 \ lb_m/s$$

and

$$\dot{m}_6 = \frac{c_5}{c_6}\dot{m}_5$$
$$= 0.0936 \ lb_m/s$$

The work required by the pump is

$$\dot{W}_p = \dot{m}_5 v_5(P_6 - P_5)$$
$$= -0.00076 \ Btu/s$$

TIP:

The analysis of this example requires that the conservation of mass principle independently be applied to the two solution constituents (lithium bromide and water). Also note that the work required to pump the strong solution is quite small.

This is small enough that we will neglect the enthalpy change across the pump and treat the enthalpy of the strong solution entering the generator as being the same as that leaving the absorber.

Applying the first law to the various heat-exchange devices gives

$$_2\dot{Q}_3 = \dot{m}_1(h_3 - h_2)$$
$$= -10.49 \ Btu/s$$
$$\dot{Q}_g = \dot{m}_2 h_2 + \dot{m}_6 h_6 - \dot{m}_5 h_5$$
$$= 18.41 \ Btu/s$$

and

$$\dot{Q}_a = \dot{m}_5 h_5 - \dot{m}_6 h_6 - \dot{m}_1 h_1$$
$$= -17.92 \ Btu/s$$

PROBLEMS

Sketch the cycle schematic diagram with clearly labeled states for the following problems. Also sketch the appropriate T–s or P–v state diagrams (or both), showing all states and processes.

7.1 The turbine of a steam Rankine cycle produces 500 kW of power when the boiler is operated at 500 psia and the condenser at 6 psia, and when the temperature at the turbine entrance is 1200° F. Determine the rate of heat addition (Btu/hr) in the boiler, the rate of heat rejection (Btu/hr) in the condenser, the cycle's actual thermodynamic efficiency, and its minimum Carnot efficiency.

7.2 A basic steam Rankine cycle operates the boiler at 500 psia and the condenser at 5 psia. If the quality at the exit of the turbine cannot be less than 85%, what is the maximum thermodynamic efficiency this cycle can have?

7.3 A steam Rankine cycle operates the boiler at 4000 kPa, the condenser at 20 kPa, and the turbine inlet at 700° C. The boiler is sized to provide a steam flow of 50 kg/s. Determine the power (kW) produced by the turbine and consumed by the feedwater pump (kW).

7.4 A Rankine cycle using water as its working fluid operates a boiler at 6000 kPa, the turbine inlet at 600° C, and the condenser at 50 kPa. Compare the thermodynamic efficiency of this cycle when it is operated so that the liquid enters the pump as a saturated liquid against that when the liquid enters the pump 11° C cooler than a saturated liquid at the condenser pressure.

7.5 An engineer has proposed a Rankine cycle that uses R-12 be used to produce work with heat from a low-temperature thermal-energy reservoir. The R-12 boiler operates at 300 psia, the condenser at 60 psia, and the turbine inlet at 180° F. Determine the R-12 mass flow rate in lb_m/s needed for this cycle to produce 1000 hp.

7.6 Ammonia is used as the working fluid in a Rankine cycle that operates a boiler at 2000 kPa and a condenser at 10° C. The mixture at the turbine exit has a quality of 97%. Determine the turbine-inlet temperature (° C), the cycle thermodynamic efficiency, and the cycle's back–work ratio.

7.7 A Rankine cycle uses water as the working fluid, operates its condenser at 50° F, and operates its boiler at 600° F. Calculate the specific work (Btu/lb_m) produced by the turbine, the specific heat (Btu/lb_m) transferred in the boiler, and the cycle's net thermal efficiency when the steam enters the turbine without any superheating.

7.8 A steam Rankine cycle operates a boiler at 2500 psia and a condenser at 5 psia. What is the minimum temperature required at the turbine inlet such that the quality of the steam leaving the turbine is not below 80%? When operated at this temperature, what is this cycle's overall thermal efficiency?

7.9 A steam Rankine cycle operates the condenser at 100 kPa and the boiler at 15,000 kPa. Saturated steam vapor enters the turbine. Determine the specific work (Btu/lb$_m$) produced by the turbine, the specific work (Btu/lb$_m$) used by the feedwater pump, and the net thermal efficiency of this cycle.

7.10 Irreversibilities in the turbine of the previous problem cause the steam quality at the outlet of the turbine to be 80%. Determine the efficiency of this turbine and the net thermal efficiency of this cycle.

7.11 A steam Rankine cycle with reheating operates the inlet of the high-pressure turbine at 600 psia, 600° F; the inlet of the low-pressure turbine at 200 psia, 600° F; and the condenser at 10 psia. The net power produced by this plant is 5000 kW. Determine the rate of heat addition (Btu/hr) and rejection (Btu/hr) and the cycle's net thermal efficiency.

7.12 In the preceding problem, is there any advantage to operating the reheat section of the boiler at 100 psia rather than 200 psia while maintaining the same low-pressure turbine-inlet temperature?

7.13 A reheated steam Rankine cycle operates the high-pressure turbine inlet at 8000 kPa, 440° C; the low-pressure turbine inlet at 500 kPa, 500° C; and the condenser at 10 kPa. Determine the mass flow rate (kg/s) through the boiler needed for this system to produce a net 5000 kW of power.

7.14 A reheated Rankine cycle operates the boiler at 4000 kPa, the reheat section at 500 kPa, and the condenser at 10 kPa. The mixture quality at the exit of both turbines is 90%. Determine the temperature (° C) at the inlet of each turbine and the cycle's thermodynamic efficiency.

7.15 A steam Rankine cycle with reheating operates the boiler at 14,000 kPa, the reheater at 2000 kPa, and the condenser at 100 kPa. The temperature is 440° C at the entrance of the high-pressure and low-pressure turbines of Figure 7.4 (p. 288). This system is designed to produce 2000 kW of power. Determine the mass flow rate (kg/s) through the boiler, the rate of heat transfer (kW) in the boiler, the rate of heat transfer (kW) in the reheater, the power (kW) used by the pumps, and the system's net thermal efficiency.

7.16 A steam Rankine cycle with reheating like that of Figure 7.4 operates the boiler at 2500 psia, the reheater at 60 psia, and the condenser at 5 psia. At the entrance to the high-pressure turbine, the temperature is 1000° F. At the entrance to the low-pressure turbine, the temperature is 600° F. Determine the system's net thermal efficiency.

7.17 How much does the net thermal efficiency of the cycle in the previous problem change when the temperature at the entrance to the low-pressure turbine is increased to 1000° F?

7.18   Turbine bleed steam enters the open-feedwater heater of Figure 7.5 (p. 290) at 20 psia, 250° F, while the cold feedwater enters at 110° F. Determine the ratio of the bleed-steam mass flow rate to the inlet-feedwater mass flow rate required to heat the feedwater to 225° F.

7.19   Ten kilograms per second of cold feedwater enter a 150-kPa open-feedwater heater at 70° C. Bleed steam is available from the turbine at 150 kPa, 160° C. At what rate (kg/s) must bleed steam be supplied to the open-feedwater heater so the feedwater leaves this unit as a saturated liquid?

7.20   The closed-feedwater heater with a pump shown in Figure 7.6 (p. 291) is arranged so that the water at state 5 is mixed with the water at state 2 to form a feedwater that is a saturated liquid at 200 psia. Feedwater enters this heater at a flow rate of 2 lb_m/s and temperature and pressure of 330° F and 200 psia, respectively. Bleed steam is taken from the turbine at 160 psia, 400° F, and enters the condensate pump as a saturated liquid at 160 psia. Determine the mass flow rate (lb_m/s) of bleed steam required to operate this unit.

7.21   The closed-feedwater heater with a pump like that in Figure 7.6 is to heat 7000 kPa feedwater from 260° C to a saturated liquid. The turbine supplies bleed steam at 6000 kPa, 320° C to this unit. This steam is condensed to a saturated liquid before entering the pump. Calculate the amount of bleed steam required to heat 1 kg of feedwater in this unit.

7.22   The closed-feedwater heater with a trap like that shown in Figure 7.6 is to heat feedwater at 400 psia, 375° F to a saturated liquid at 400 psia. The turbine has bleed steam available at 300 psia, 500° F that will be condensed to a saturated liquid before entering the trap. How much bleed steam will be required to heat 1 pound-mass of feedwater in this unit?

7.23   Feedwater at 4000 kPa is heated at a rate of 6 kg/s from 200° C to 245° C in a closed feedwater heater with a trap like that in Figure 7.6. Bleed steam enters this unit at 3000 kPa with a quality of 90% and leaves as a saturated vapor before entering the trap. Calculate the rate (kg/s) at which bleed steam is required.

7.24   The regenerative Rankine cycle with an open-feedwater heater shown in Figure 7.7 (p. 293) uses water as the working fluid. The turbine inlet is operated at 500 psia, 600° F, and the condenser at 5 psia. Steam is supplied to the open-feedwater heater at 40 psia. Determine the work (Btu/lb_m) produced by the turbine, the work (Btu/lb_m) consumed by the pumps, and the heat (Btu/lb_m) rejected in the condenser for this cycle per unit of flow rate through the boiler.

7.25   Determine the change in the thermodynamic efficiency of the regenerative Rankine cycle of the preceding problem when the steam supplied to the open-feedwater heater is at 60 psia rather than 40 psia.

7.26 The regenerative Rankine cycle with a closed-feedwater heater shown in Figure 7.11 (p. 303) uses water as the working fluid, operates the turbine inlet at 3000 kPa and 320° C, and operates the condenser at 20 kPa. Steam is extracted at 1000 kPa to serve the closed-feedwater heater. Calculate the work (kJ/kg) produced by the turbine, the work (kJ/kg) consumed by the pump, and the heat (kJ/kg) added in the boiler for this cycle per unit of boiler flow rate.

7.27 How much does the thermodynamic efficiency of the regenerative Rankine cycle of the preceding problem change when the steam serving the closed-feedwater heater is extracted at 700 kPa rather than 1000 kPa?

7.28 The Rankine cycle with three feedwater heaters shown in Figure 7.28 operates the boiler at 8000 kPa; the condenser at 10 kPa; the reheater at 250 kPa; the high-pressure, closed-feedwater heater at 6000 kPa; the low-pressure, closed-feedwater heater at 3500 kPa; and the open-feedwater heater at 100 kPa. The temperature at the inlet of both turbines is 400° C. Determine the following items for this system per unit of mass flow rate through the boiler.

FIGURE 7.28

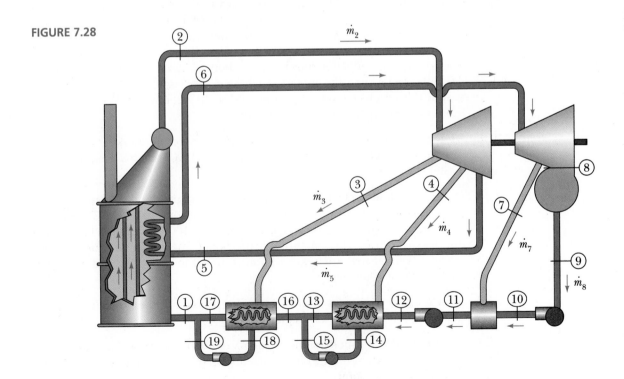

1. The flow (kg/kg-boiler) required to service the high-pressure, closed-feedwater heater.

2. The flow (kg/kg-boiler) required to service the low-pressure, closed-feedwater heater.

3. The flow (kg/kg-boiler) required to service the open-feedwater heater.

4. The flow (kg/kg-boiler) through the condenser.

5. The work (kJ/kg-boiler) produced by the high-pressure turbine.

6. The work (kJ/kg-boiler) produced by the low-pressure turbine.

7. The heat (kJ/kg-boiler) added in the boiler and reheater.

8. The heat (kJ/kg-boiler) rejected in the condenser.

9. The overall thermodynamic efficiency.

7.29   Determine the optimum bleed pressure (psia) for the open-feedwater heater of Problem 7.24.

7.30   Determine the optimum bleed pressure (psia) for the closed-feedwater heater of Problem 7.26.

7.31   Determine the optimum bleed pressures (kPa) for the three feedwater heaters of Problem 7.28.

7.32   A simple Rankine cycle arranged as shown in Figure 7.3 (p. 283) uses water as the working fluid. The boiler operates at 6000 kPa and the condenser at 60 kPa. At the entrance to the turbine, the temperature is 440° C. The efficiency of the turbine is 94%, pressure losses are negligible, and the feedwater leaving the condenser is subcooled by 6° C. The boiler is sized for a mass flow rate of 20 kg/s. Determine the rate at which heat (kW) is added in the boiler, the power (kW) required to operate the pumps, the net power (kW) produced by the cycle, and the thermodynamic efficiency.

7.33   How much would the thermodynamic efficiency of the Rankine cycle of the preceding problem change if there were a 50 kPa pressure drop across the boiler?

7.34   A steam Rankine cycle operates the condenser at 1 psia and the boiler at 2500 psia, while the temperature at the inlet of the turbine is 800° F. The turbine efficiency is 90%, and the cycle is sized to produce 1000 kW of power. Calculate the mass flow rate (lb_m/s) through the boiler, the power produced by the turbine (kW), the rate of heat addition (Btu/hr) in the boiler, and the overall cycle thermal efficiency.

7.35   How much error is caused in the previous problem if the power required by the pumps were completely neglected?

7.36   Determine the thermodynamic efficiency of the regenerative Rankine cycle of Problem 7.26 when the turbine efficiency is 90% before and after the bleed-steam extraction point.

7.37   Determine the thermodynamic efficiency of the regenerative Rankine cycle presented in Problem 7.26 when the turbine efficiency before and after the extraction point is 90% and the condenser condensate is subcooled by 10° F.

7.38 How much additional heat must be supplied to the boiler of Problem 7.26 when the turbine efficiency before and after the extraction point is 90% and there is a 10 psia pressure drop across the boiler?

7.39 Calculate the potential work that is lost in each of the components of the simple Rankine cycle of Problem 7.1 when heat is being rejected to a lake at 40° F and heat is being supplied by a 1500° F energy reservoir.

7.40 Calculate the potential work that is lost in each component of the simple Rankine cycle of Problem 7.3 when heat is rejected to the atmospheric air at 15° C and heat is added from an energy reservoir at 750° C.

7.41 Calculate the potential power that is lost in each compound of the reheat Rankine cycle of Problem 7.11. The sink-energy-reservoir temperature is 50° F and the source-energy-reservoir temperature is 700° F.

7.42 Which component of the reheat Rankine cycle of Problem 7.13 offers the greatest opportunity to regain lost potential power? The sink-reservoir temperature is 10° C, and the source-reservoir temperature is 600° C.

7.43 Which component of the regenerative Rankine cycle of Problem 7.24 loses the greatest amount of potential work? This cycle rejects heat to a river, which has a temperature is 50° F; the source-energy-reservoir temperature is 800° F.

7.44 A basic VCRS refrigerator that uses ammonia as its working fluid operates a condenser at 200 psia and an evaporator at 40° F. Determine this system's COP and the amount of power (kW) required to service a 400 Btu/s cooling load.

7.45 A basic VCRS heat pump uses ammonia as its working fluid. The condenser operates at 250 psia and the evaporator at 60 psia. Determine this system's COP and the rate of heat (Btu/hr) supplied when the compressor consumes 20 kW.

7.46 A 10-kW cooling load is to be served by operating a VCRS with its evaporator at 400 kPa and its condenser at 800 kPa. Calculate the refrigerant flow rate (kg/s) and the compressor power (kW) requirement when R-12 is used.

7.47 A refrigerator uses R-12 as its working fluid, operates its evaporator at −10° F, and operates its condenser at 100 psia. This unit serves a 24,000 Btu/hr cooling load. Determine the rate (lb_m/s) at which the R-12 is to be circulated among the system components. Also calculate the power that this unit will require.

7.48 How much will the working fluid's circulation rate and power requirement of the previous problem change if ammonia is used in place of R-12?

7.49 An air conditioner is to maintain a space at 22° C while operating its condenser at 1000 kPa. Determine the coefficient of performance of this system when it uses R-12 as its working fluid and a temperature difference of 2° C is allowed for the transfer of heat in the evaporator.

7.50   An ammonia heat pump operates its condenser at a pressure of 125 psia while the evaporator temperature is 40° F. What is the COP of this heat pump?

7.51   A R-12 heat pump operates its evaporator at 0° C and its condenser at 1000 kPa. Calculate this system's COP.

7.52   A VCRS refrigerator using R-12 is to operate its condenser at 350 psia and its evaporator at 20° F. If an adiabatic, reversible-expansion device were available and used to expand the condenser liquid to a saturated liquid at the entrance to the evaporator, how much would the COP improve by using this device instead of the throttle device?

7.53   A R-12 VCRS is used to cool a brine solution to −5° C. This solution is pumped to various buildings for air conditioning. Determine the COP of this VCRS when its evaporator is operated at −10° C and its condenser at 600 kPa.

7.54   A basic VCRS heat pump using ammonia as a refrigerant operates its condenser at 1200 kPa and its evaporator at −1.9° C. It operates ideally, except for the compressor, which has an isentropic efficiency of 85%. How much does the compressor efficiency reduce this heat pump's COP as compared to an ideal VCRS?

7.55   What is the effect on the COP of the preceding problem when the vapor entering the compressor is superheated by 2° C and the compressor has no irreversibilities?

7.56   The liquid leaving the condenser of a 100,000 Btu/hr R-12 VCRS heat pump is subcooled by 14° F. The condenser operates at 160 psia and the evaporator at 50 psia. How does this subcooling change the power (kW) required to drive the compressor as compared to an ideal VCRS?

7.57   What is the effect on the compressor power requirement (kW) when the vapor entering the compressor of the previous problem is superheated by 10° F and the condenser operates ideally?

7.58   A VCRS refrigerator using ammonia operates the condenser at 1000 kPa and the evaporator at −10° C. This refrigerator freezes water while rejecting heat to the ambient 22° C air. The compressor has an 85% efficiency. Which process loses the greatest amount of potential work? What would you suggest be done to reduce the lost work potential?

7.59   Rework the preceding problem with a 2.9° C subcooling at the exit of the condenser.

7.60   Rapid freezing of fresh fruit requires air at −30° F. A R-12 VCRS produces this air by operating its evaporator at −37.2° F and its condenser at 180 psia, while rejecting heat to the ambient air at 80° F. If the compressor efficiency is 90% and the vapor entering the compressor is superheated by 10° F, which process loses the greatest amount of work potential?

7.61 Rework the previous problem with 13.5° F of subcooling at the exit of the condenser.

7.62 A two-evaporator VCRS like that in Figure 7.22 (p. 325) uses R-12 as its working fluid and operates one evaporator at −30.1° F, one evaporator at 0° F, and the condenser at 800 kPa. The R-12 is circulated through the compressor at a rate of 0.1 kg/s and the low-temperature evaporator serves a cooling load of 8 kW. Determine the cooling rate (kW) of the high-temperature evaporator and the power (kW) required by the compressor.

7.63 A two-evaporator VCRS uses R-12 as its working fluid and operates one evaporator at −8.2° F, the other at 20° F, and its condenser at 140 psia. The cooling load for the −8.2° F evaporator is 10,000 Btu/hr and that of the 20° F evaporator is 3000 Btu/hr. How much power (kW) is required to operate the compressor of this system?

7.64 The 20° F evaporator of the previous problem is to be replaced with a 50° F evaporator to serve a 48,000 Btu/hr cooling load. How much power is now required to operate the compressor?

7.65 A two-evaporator VCRS like that shown in Figure 7.22 uses ammonia and operates one evaporator at −41.3° F, one evaporator at 30° F, and the condenser at 250 psia. The cooling load of the 30° F evaporator is double that of the −41.3° F evaporator. Determine the specific cooling effect in Btu/lb$_m$ of both evaporators per unit of flow through the compressor, as well as the COP of this system.

7.66 The refrigeration system of the preceding problem cools one reservoir at −30° F and one at 35° F while rejecting heat to a reservoir at 95° F. Which process has the highest rate of lost work potential?

7.67 An ammonia two-compression–stage VCRS with an adiabatic liquid–vapor separator operates the evaporator at −50° C, the condenser at 1200 kPa, and the separation unit at −9.2° C. This system is to serve a 30 kW cooling load. Determine the flow rate (kg/s) through the two compressors, the net power (kW) used by the compressors, and the system's COP.

7.68 An R-12, two-compressor VCRS with an adiabatic liquid–vapor separation unit operates the evaporator at 50° F, the condenser at 250 psia, and the separator at 93.3° F. The compressors use 25 kW of power. Determine the rate of cooling provided by the evaporator and the COP of this cycle.

7.69 An R-12, two-compression stage VCRS with an adiabatic liquid–vapor separator operates the evaporator at −30° F, the condenser at 200 psia, and the separator at 48.6° F. Refrigerant-12 is circulated through the condenser at a rate of 2 lb$_m$/s. Determine the rate of cooling (Btu/hr) and the power (hp) requirement for this system.

7.70   Which process of the preceding problem has the greatest rate of lost work potential when the low-temperature reservoir is at $0°$ F and the high-temperature reservoir is at the same temperature as the refrigerant at the high-pressure compressor outlet?

7.71   Two VCRSs are cascaded as shown in Figure 7.24 (p. 331) to provide cooling at $-40°$ F while operating the high-temperature condenser at 200 psia. The high-temperature VCRS uses water as its working fluid and operates its evaporator at $40°$ F. The low-temperature VCRS uses R-12 as its working fluid and operates its condenser at 60 psia. This system produces a cooling effect of 20 Btu/s. Determine the mass flow rate (lb_m/s) of R-12 and water in their respective systems, as well as this cascaded system's overall COP.

7.72   Analyze the rate of lost work potential for the refrigeration system of the previous problem when the low-temperature reservoir is at $-25°$ F and the high-temperature reservoir is at $90°$ F.

7.73   Calculate the COP of the cascaded system of Example 7.14 when ammonia is used in the high-temperature VCRS and R-12 is used in the low-temperature VCRS. How does this compare to the COP of Example 7.14?

7.74   The refrigeration system of Figure 7.29 is another variation of the basic VCRS that attempts to reduce the compression work. In this system, a heat exchanger is used to superheat the vapor entering the compressor while subcooling the liquid exiting from the condenser. Consider a system of this type that uses R-12 as its refrigerant and operates the evaporator at $-12.5°$ C and the condenser at 800 kPa. Determine the system COP when the heat exchanger provides $12.8°$ C of subcooling at the throttle valve entrance and the refrigerant leaves the evaporator as a saturated liquid.

7.75   How does changing the heat exchanger of the previous problem such that the subcooling is $22.8°$ C change the overall COP?

7.76   A water–lithium bromide absorption refrigerator like that in Figure 7.25 (p. 334) is used to serve a 20 Btu/s cooling load. The evaporator is at $40°$ F, the condenser at $140°$ F, the generator solution at $200°$ F, and the absorber solution at $60°$ F. Determine the refrigerant and strong-solution mass flow rates in lb_m/s. Also determine the rate (Btu/hr) at which the generator must be heated and the absorber cooled.

7.77   A water–lithium bromide absorption refrigerator operates its evaporator at $5°$ C, its condenser at $25°$ C, its generator solution at $50°$ C, and its absorber solution at $20°$ C. Determine the following system parameters per unit of mass passing through the solution pump.

1. The heat (kJ/kg-pump) added to the generator.
2. The heat (kJ/kg-pump) removed from the absorber.
3. The heat (kJ/kg-pump) rejected in the condenser.
4. The cooling effect in kJ/kg-pump.
5. The work (kJ/kg-pump) required to drive the solution pump.

FIGURE 7.29

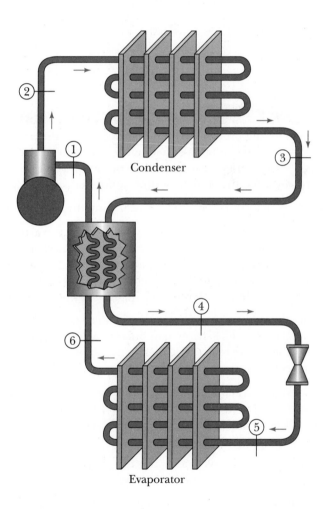

Condenser

Evaporator

7.78 To reduce the generator heat requirement, it has been suggested that a heat exchanger be used to allow heat transfer between the weak and strong solutions. A heat exchanger is mounted between the solution pump and absorber on the strong-solution side, as well as between the generator and solution throttle device on the weak-solution side of a water–lithium bromide absorption refrigeration system rated at 10 Btu/s. This system operates its evaporator at 50° F, its condenser at 140° F, its generator solution at 200° F, and its absorber solution at 70° F. The heat exchanger is sized such that the temperature of the strong solution increases by 100° F as it passes through the heat exchanger. Determine the rate at which heat (Btu/hr) is added to the generator, removed from the absorber, and rejected by the condenser.

7.79 Repeat the previous problem when the evaporator is at 5° C, the condenser is at 35° C, the generator solution is at 50° C, and the solution absorber is at 20° C when serving a 10 kW cooling load with a 10° C rise in strong-solution temperature in the heat exchanger.

PROJECTS

The following projects are more extensive than the problems presented in this chapter. Most are open-ended and require some decisions on the student's part. Many can be expedited with computer-based property tables, computer spread-sheets, and student-written computer programs.

7.1 Is there an optimal pressure for reheating the steam of a Rankine cycle?

7.2 Stack gases exhausting from electrical power plants are at approximately 300° F. Design a basic Rankine cycle that uses water, R-12, or ammonia as the working fluid and that produces the maximum amount of work from this energy source while rejecting heat to the ambient air at 100° F. You are to use a turbine whose efficiency is 92% and whose exit quality cannot be less than 85%.

7.3 A geothermal-energy reservoir is available at, say, 250° C. Design a Rankine cycle that uses water, R-12, or ammonia as the working fluid. Your design criterion is the maximum reasonable thermodynamic efficiency. You may use reheating and regeneration in this design if it helps to meet your design criterion.

7.4 A pressure–enthalpy state diagram (i.e., a Mollier diagram) is frequently used to analyze VCRSs. Using refrigerant property tables, sketch the P–h state diagram and show the saturation lines. Sketch a basic VCRS cycle on this diagram. What is the advantage of this state diagram?

7.5 Develop and discuss techniques that apply the principle of regeneration to improve the performance of VCRSs.

7.6 The heat supplied by a heat pump used to maintain a building's temperature is often supplemented by another source of direct heat. The fraction of the total heat required that is supplied by supplemental heat increases as the temperature of the environmental air (which serves as the low-temperature sink) decreases. Develop a supplemental heat schedule as a function of the environmental air temperature that minimizes the total supplemental and heat-pump energy required to service the building.

7.7   Is there an optimum pressure at which to operate the liquid–vapor separator of a two-compressor–stage VCRS? If so, what is the optimum pressure for the VCRS of Example 7.13 (p. 328)?

7.8   Consider a cascaded VCRS system that uses the same working fluid in both systems. Is there a pressure at which to operate the heat exchanger that will optimize the overall COP?

CHAPTER 8

Gas Power and Refrigeration Cycles

Rewards

The reward of a thing well done

is to have done it.

R. W. Emerson

Systems that use a gaseous working fluid (normally air) are really quite common. They range from units that produce fractional horsepower to units that produce several thousand horsepower. Portable systems are used throughout the transportation industry to move automobiles, trains, and aircraft, while stationary units pump water, provide emergency and peak-load electrical power, power all types of equipment, and serve many other purposes. Smaller-powered, inexpensive portable units are normally based on a piston-cylinder apparatus that operates as a closed system, at least in part. Larger-powered, more expensive units are typically based on steady-flow compressors and turbines that process very large quantities of the working fluid.

The energy source for these devices is the energy released when a fuel is burned with air. Many hydrocarbons are used for this purpose. The physical and chemical processes that occur during the combustion determine what will be left in the combustion products. For example, when a liquid fuel such as gasoline is atomized in a fuel injector or carburetor, some large droplets will be formed. With the combustion process used in a typical automotive engine, the large droplets will not be completely burned and will be, in part, present in the exhaust products. Excessive quantities of these unburned hydrocarbons degrade the natural environment. Consequently, devices such as the air pump and the catalytic converter have been added to automobiles over the past two decades to complete the burning of these droplets.

One means to improve the performance of a spark-ignition engine is to decrease the volume into which the fuel and air are compressed just before combustion is initiated. This can cause preignition and a phenomenon known as *knock*, both of which can mechanically damage an engine. Antiknock compounds have been added to fuels to reduce the occurrence of these phenomena. Unfortunately, these have contained lead, which is released to the atmosphere as one exhaust product generated by the engine. In sufficient quantities, this lead is toxic. This discovery led to the banning of lead-based antiknock compounds. The spark-ignition engine design was then modified to operate at reduced performance to be compatible with the additives that replaced the lead-based compounds.

A chemical reaction that occurs in the combustion of fuels in these engines is the disassociation of nitrogen and oxygen contained in the air and their subsequent recombination to form various nitrous oxides—which also degrade atmospheric quality. These oxides are controlled, in part, by holding down the peak temperatures that occur during the combustion, which curtails the disassociation process. As Carnot pointed out, reducing the temperature of the energy-source reservoir reduces the thermal efficiency of devices that use these sources. Finding a compromise between fuel effectiveness, performance, and environmental acceptability is one of the leading issues facing both the developers of these devices and the fuels of the future. More will be said about the chemistry of combustion in Chapters 11 and 12.

Models of the spark-ignition engine, diesel engine, gas turbine, and refrigeration units that use gases as the working fluid are developed in this chapter. As in previous chapters, these models are formulated from the first and second laws of thermodynamics and the property relationships of the working fluid. Application of these laws to the models will illustrate how such engine design choices as compression ratio, pressure ratio, and others affect the performance and efficiency of a gas power system.

Many power-generation and refrigeration devices use a gas as their working fluid. Examples include internal-combustion engines, gas turbines, and air-cycle refrigerators. These differ from the cycles and systems of Chapter 7 in that the working fluid remains in the gaseous phase throughout the processes executed by the system. Phase transformations are not present.

Most but not all of these systems draw in a gas such as the atmospheric air (sometimes mixed with a fuel), perform a process, add a fuel, combust the fuel to form a new gas consisting of the combustion products, perform a process with the new gas, and finally exhaust the new gas to the environment. Normally, they are not continuous cycles in the sense of the Rankine and vapor-compression cycles of Chapter 7. However, they can be modeled as continuous cycles by making the following presumptions:

1. The working fluid retains the same chemical composition throughout the cycle. Most of the time, we will treat the working fluid as air.

2. The combustion process will be replaced by an equivalent heat addition from a thermal-energy reservoir process.

3. The exhaust and eventual reentry process will be replaced by an equivalent heat rejection to a thermal-energy reservoir process.

We combine these assumptions with the assumption that the working gas can be treated as an ideal gas to form the basis of what is known as *air standard analysis*. Air standard analysis is a common set of assumptions that is used for an initial understanding of gas power and refrigeration systems. Methods of adjusting for combustion and different gases during different cycle processes will be taken up in Chapters 10 and 11.

In this chapter, we also presume that the specific heats of the working gas remain constant throughout the cycle. Means of eliminating this assumption will be presented in Chapter 9. As in the preceding chapters, all processes will be initially treated as reversible processes, and we will introduce means of accounting for irreversibilities at the appropriate point.

8.1 The Ideal Gas Carnot Cycle

When executed with an ideal gas, the Carnot cycle, which consists of two reversible isothermal and two isentropic processes, undergoes the cycle illustrated in Figure 8.1. During the isentropic compression, 1–2, the gas pressure increases and the gas temperature increases from T_l to T_h as the volume of the gas decreases. Work is required to perform this compression. Heat is isothermally added to the ideal gas from an energy reservoir whose temperature is infinitesimally higher than the gas temperature, T_h, as it undergoes process 2–3. The ideal gas is next isentropically expanded in process 3–4 as the pressure and temperature decrease and the volume increases while producing work. Finally, heat is rejected to an energy reservoir whose temperature is infinitesimally lower than the gas temperature, T_l, during the isothermal process 4–1. The work produced during process 3–4 exceeds that consumed during process 1–2, resulting in a net work production that has a magnitude equal to the area enclosed by the cycle processes on a T–s state diagram.

FIGURE 8.1
Ideal Gas Carnot Cycle

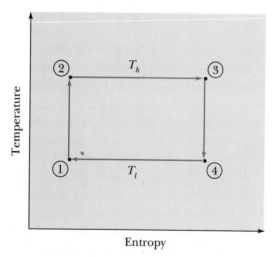

The thermodynamic efficiency of this cycle is

$$\eta = 1 - \left| \frac{q_l}{q_h} \right|$$

$$= 1 - \left| \frac{T_l(s_1 - s_4)}{T_h(s_3 - s_2)} \right|$$

$$= 1 - \frac{T_l}{T_h}$$

where we have applied the definition of the entropy and the fact that $|s_1 - s_4| = |s_3 - s_2|$. As with all completely reversible cycles, the thermodynamic efficiency of the ideal gas Carnot cycle is a function of the energy reservoir temperatures only.

When the ideal gas's specific heats remain constant, the states along the isentropic processes are related by $Pv^k =$ constant and

$$\frac{T_2}{T_1} = \left(\frac{P_2}{P_1}\right)^{(k-1)/k} = \left(\frac{v_1}{v_2}\right)^{k-1}$$

as well as

$$\frac{T_4}{T_3} = \left(\frac{P_4}{P_3}\right)^{(k-1)/k} = \left(\frac{v_3}{v_4}\right)^{k-1}$$

These results reduce the ideal gas Carnot cycle thermal efficiency to

$$\eta = 1 - \left(\frac{P_1}{P_2}\right)^{(k-1)/k} = 1 - \frac{1}{r_p^{(k-1)/k}}$$

or

$$\eta = 1 - \left(\frac{v_2}{v_1}\right)^{k-1} = 1 - \frac{1}{r_v^{k-1}}$$

where r_p is known as the *pressure ratio,* and r_v is known as the *compression ratio.* Consequently, the thermal efficiency is a monotonically increasing function of the pressure increase and volume decrease during the isentropic compression.

EXAMPLE 8.1

An ideal gas Carnot cycle uses argon as the working fluid, rejects 1000 kW of heat to an energy reservoir at –10° C, and increases the pressure by a factor of 10 during the isentropic compression process. Determine the maximum gas temperature, the rate of heat transfer from the high-temperature energy reservoir, and the power produced by this Carnot cycle.

System: Argon undergoing the Carnot cycle as illustrated in Figure 8.1.

Basic Equations:

$$_1q_2 - _1w_2 = u_2 - u_1$$

$$\frac{T_2}{T_1} = \left(\frac{P_2}{P_1}\right)^{(k-1)/k} = \left(\frac{v_1}{v_2}\right)^{k-1}$$

Conditions: The cycle will be modeled using air standard analysis.

Solution: The states will be labeled in the same way as in Figure 8.1. Applying the pressure form of the thermal-efficiency expression for an ideal gas Carnot cycle yields

$$\eta = 1 - \left(\frac{P_1}{P_2}\right)^{(k-1)/k}$$

$$= 1 - \left(\frac{1}{10}\right)^{(1.667-1)/1.667}$$

$$= 0.602$$

The temperature form of the thermal-efficiency expression for completely reversible engines gives

$$T_h = \frac{T_l}{1 - \eta}$$

$$= \frac{263}{1 - 0.602}$$

$$= 661 \text{ K}$$

When the definition of the thermal efficiency is combined with the first law, the result is

$$\dot{Q}_h = \frac{\dot{Q}_l}{1 - \eta}$$

$$= \frac{1000}{1 - 0.602}$$

$$= 2513 \text{ kW}$$

TIP:

This example also could have been solved by using the fact that Pv^k remains constant during the isentropic compression, the definition of the thermal efficiency, and the first law.

Finally, the net power produced is given by the definition of the thermal efficiency:

$$\dot{W} = \eta \dot{Q}_{in}$$

$$= 0.602 \times 2513$$

$$= 1513 \text{ kW}$$

8.2 The Otto Cycle

The Otto cycle[1] is the thermodynamic equivalent of the spark-ignition, internal-combustion engine cycle presented in Section 1.1. It treats the working fluid

---

[1]Named for Nikolaus August Otto (1832–1891), a German engineer who, with his partner E. Langen, developed a major internal-combustion engine that was patented in 1877. This engine was successful in providing power for small shops and industries for which steam power was not then practical.

as if it is purely air with constant specific heats, using the air standard analysis assumptions. Otto cycle processes are illustrated in Figure 8.2. At state 1, the piston cylinder is filled with a fresh charge of air and fuel, and compression is about to begin as the piston moves to top dead center (TDC) while the valves remain closed. This process requires work for the compression. The gas is now compressed adiabatically (process 1–2) into the clearance volume of state 2. At this point, combustion is initiated by the spark device. The combustion occurs so rapidly that the piston does not move far enough to significantly change the volume occupied by the gas during combustion. The combustion process (2–3) can then be modeled as a constant-volume addition of heat, with the gas

FIGURE 8.2
State Diagrams for the Otto Cycle

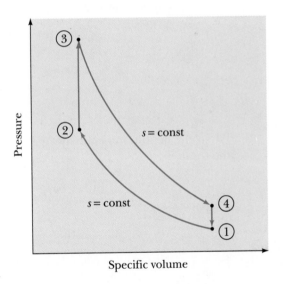

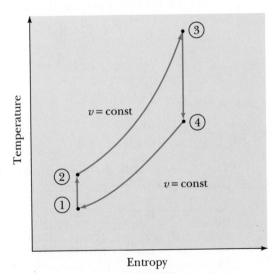

treated as air for an air standard analysis. Following combustion (heat addition), the high-pressure, high-temperature gas is expanded adiabatically (3–4) as the piston moves to bottom dead center (BDC) while producing more work than was required for compression. The exhaust gases are now ejected and a new fuel–air charge drawn into the piston cylinder to repeat the cycle. The gas volume at the point where the exhaust process begins is exactly the same as the volume at the point where the induction of a new charge is completed. We can therefore model the exhaust-intake process as a constant-volume heat-rejection process (4–1) without the gas leaving and reentering the piston cylinder.

The Otto cycle model of the internal-combustion (IC) engine is quite acceptable for early design decisions. With the exception of the exhaust-intake processes, measured pressure–volume state diagrams agree reasonably well with that of Figure 8.2. This model also assists in identifying the primary parameters governing the performance of the IC engine.

Because we are treating all processes as reversible and the system as closed, areas under the processes on the P–v state diagram represent specific work. The area under the expansion process (3–4) is the specific work produced by the expansion. The area under the compression process (1–2) is the specific work consumed for compression. The area enclosed by all of the Otto cycle processes on this diagram is the net specific work produced per cycle of operation.

The compression ratio defined as follows is the primary parameter affecting the Otto cycle performance.

DEFINITION: *Internal-Combustion Engine Compression Ratio* (r_v)

The ratio of the maximum piston-cylinder volume (at BDC) to the minimum volume (at TDC),

$$r_v = \frac{V_1}{V_2}$$

The compression and expansion processes are isentropic, for which Pv^k remains constant when the gas is treated as an ideal gas with constant specific heats. According to the discussion of Section 4.9,

$$\frac{T_2}{T_1} = \left(\frac{v_1}{v_2}\right)^{k-1} \quad \text{and} \quad \frac{T_3}{T_4} = \left(\frac{v_4}{v_3}\right)^{k-1}$$

Applying the closed-system form of the first law to the heat-addition and heat-rejection processes yields

$$_2q_3 = u_3 - u_2 = c_v(T_3 - T_2)$$

and

$$_4q_1 = c_v(T_1 - T_4)$$

The thermodynamic efficiency of the Otto cycle is given by

$$\eta = 1 - \left| \frac{_4q_1}{_2q_3} \right|$$

$$= 1 - \frac{T_4 - T_1}{T_3 - T_2}$$

$$= 1 - \frac{T_1}{T_2}\left[\frac{T_4/T_1 - 1}{T_3/T_2 - 1} \right]$$

Because $v_1 = v_4$ and $v_2 = v_3$, $T_4/T_1 = T_3/T_2$, according to the isentropic process relations, and the thermal efficiency becomes

$$\eta = 1 - \frac{T_1}{T_2}$$

$$= 1 - \frac{1}{r_v^{k-1}}$$

The efficiency of the Otto cycle is a direct function of the compression ratio; that is, we need only increase the compression ratio to increase the efficiency. But two other things happen as we increase the compression ratio: Both the maximum pressure, P_3, and maximum temperature, T_3, also increase as long as we continue to add the same amount of heat during process 2–3. Increasing the maximum pressure implies that we must add material to the piston, cylinder, and the mechanism that moves the piston to withstand the additional loads and stresses. Increasing the maximum temperature increases the formation of undesirable combustion products such as the nitrous oxides, increases the possibility of preignition, and reduces the strength of the metal parts. Consequently, modern IC spark-ignition engines use compression ratios of 6-8:1. Historically, they have been as high as 10-12:1. The addition of knock-reducing but lead-containing compounds to gasoline allowed these high compression ratios. Because lead is not environmentally acceptable, compression ratios have been reduced to accommodate modern antiknock compounds. To reduce the maximum temperature while simultaneously increasing the compression ratio, we could reduce the heat added during process 2–3, and consequently the net work production. This is counterproductive because we have reduced the work output to gain a small benefit in efficiency. Engine compression ratios are then a compromise and so are one of the first parameters selected during the design of an IC engine.

The air standard Otto cycle may be further analyzed by applying the first and second laws and the property relations of ideal gases. Keep in mind that the Otto cycle is executed in a closed system, not a steady-flow system.

EXAMPLE 8.2

An air standard Otto cycle operates with air at the beginning of the compression process at 100 kPa and $20°\,C$, with a volume of $0.0023\ m^3$, an 8:1 compression ratio, and with 1.5 kJ of heat added during the equivalent of the constant-volume combustion process. Determine the temperature and pressure of the air at each stage of the cycle, the cycle thermodynamic efficiency, and the amount of work produced during each cycle.

System: Air contained in a piston-cylinder device undergoing the Otto cycle as illustrated in Figure 8.2.

Basic Equations:

$$_1q_2 - {_1}w_2 = u_2 - u_1$$

$$\eta = \frac{w_{net}}{q_{in}}$$

$$\frac{T_2}{T_1} = \left(\frac{P_2}{P_1}\right)^{(k-1)/k} = \left(\frac{v_1}{v_2}\right)^{k-1}$$

Conditions: The cycle will be modeled using air standard analysis.

Solution: The states will be labeled in the same way as in Figure 8.2. Applying the ideal gas equation of state to the initial state 1 gives the mass of air contained in the piston cylinder as

$$m = \frac{P_1 V_1}{R T_1}$$

$$= \frac{100 \times 0.0023}{0.2870 \times 293}$$

$$= 0.00274\ \text{kg}$$

The temperature at the end of the adiabatic compression may be found by applying the isentropic ideal gas equation across the process:

$$T_2 = T_1 \left(\frac{v_1}{v_2}\right)^{k-1} = T_1 r_v^{k-1}$$

$$= 293 \times 8^{1.4-1}$$

$$= 673\ \text{K}$$

Because Pv^k remains constant during the compression,

$$P_2 = P_1 \left(\frac{v_1}{v_2}\right)^k = P_1 r_v^k$$
$$= 100 \times 8^{1.4}$$
$$= 1838 \text{ kPa}$$

Applying the closed-system first law to the heat-addition process gives

$$T_3 = T_2 + \frac{_2Q_3}{mc_v}$$

$$= 673 + \frac{1.5}{0.00274 \times 0.7165}$$
$$= 1437 \text{ K}$$

The volume at the end of the compression and heat addition strokes is one-eighth that at the beginning of compression. According to the definition of the compression ratio, $V_2 = V_3 = 0.000288 \text{ m}^3$. Now that the temperature and volume at the end of the heat addition are known, the ideal gas equation of state may be used to calculate the pressure:

$$P_3 = \frac{mRT_3}{V_3} = 3853 \text{ kPa}$$

The isentropic ideal gas equations may now be used to calculate the properties of state 4:

$$T_4 = \frac{T_3}{r_v^{k-1}} = 625 \text{ K}$$

and

$$P_4 = \frac{P_3}{r_v^k} = 210 \text{ kPa}$$

The thermodynamic efficiency of this cycle is

$$\eta = 1 - \frac{1}{r_v^{k-1}}$$
$$= 0.565$$

Applying the definition of the thermodynamic efficiency gives

$$w_{net} = \eta q_{in}$$
$$= 0.565 \times 1.5$$
$$= 0.8475 \text{ kJ/cycle}$$

TIP:

Note how the compression ratio enters almost every calculation and is the basic parameter that determines practically every performance aspect of the Otto cycle.

A comparison between the Otto cycle and a cylinder pressure–volume state diagram as measured for a typical spark-ignition IC engine is illustrated in Figure 8.3. The primary differences between these two cycles result from the efficiencies of compression and expansion, heat transfer to the surroundings during the combustion process, and some volume change during the combustion. Figure 8.3 also illustrates the IC engine's exhaust and intake strokes. During the exhaust stroke, the cylinder pressure is elevated some to push the exhaust gases through the exhaust valve. During the intake stroke, the cylinder pressure is depressed somewhat to draw a fresh fuel–air charge through the inlet valves. These two processes do require a small amount of work to accomplish. Recent engine designs employ more than one inlet and exhaust valve to reduce the amount of work required to move the fluids through the valves.

For a closed system undergoing reversible processes, recall that the area enclosed by the cycle on a P–V state diagram is the net work produced per each execution of the cycle

$$W = \oint P\,dV$$

A *mean effective pressure* may be defined such that the net work produced per cycle is equal to this pressure times the difference in the volume at BDC and TDC. The mean effective pressure is defined as follows.

DEFINITION: *Mean Effective Pressure*

$$P_{mean} = \frac{\oint P\,dV}{\Delta V}$$

FIGURE 8.3

Comparison of Otto Cycle to IC Engine's Cycle Indicator Diagram

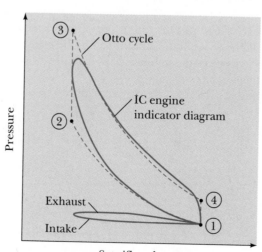

The net work produced per cycle in terms of the mean effective pressure is

$$W = P_{mean} \Delta V$$

Increasing the mean effective pressure for the same displacement volume, ΔV, is one means of increasing the power produced by an engine. Superchargers and turbochargers, which increase the mean effective pressure, are added to engines for this purpose.

EXAMPLE 8.3

An air standard Otto cycle operates with air at the beginning of the compression process at 14.7 psia, 60°F, an 8.5:1 compression ratio, with 100 Btu/lb$_m$ of heat added during the equivalent constant-volume combustion process. The compression process has an isentropic efficiency of 84%, and the expansion process has an isentropic efficiency of 93%. Determine the temperature and pressure of the air at each state of the cycle and the cycle thermodynamic efficiency. Calculate the increase in the power produced by this engine when a supercharger that increases the mean effective pressure by 5 psia is added.

System: Air in a piston-cylinder apparatus undergoing the Otto cycle as illustrated in Figure 8.2.

Basic Equations:

$$_1q_2 - {_1}w_2 = u_2 - u_1$$

$$P_{mean} = \frac{\oint P dV}{\Delta V}$$

$$\frac{T_2}{T_1} = \left(\frac{P_2}{P_1}\right)^{(k-1)/k} = \left(\frac{v_1}{v_2}\right)^{k-1}$$

Conditions: The cycle will be modeled using air standard analysis.

Solution: The T–s state diagram illustrating the various states of this Otto cycle is presented in Figure 8.4. States 2s and 4s are the states the compression and expansion processes would achieve if the compression and expansion processes were isentropic.

Applying the ideal gas isentropic relations to process 1–2s yields

$$T_{2s} = T_1 \left(\frac{v_1}{v_2}\right)^{k-1} = T_1 r_v^{k-1} = 1224° \text{R}$$

and

$$P_{2s} = P_1 \left(\frac{v_1}{v_2}\right)^{k} = P_1 r_v^{k} = 294 \text{ psia}$$

FIGURE 8.4
Temperature-Entropy
State Diagram for
Example 8.3

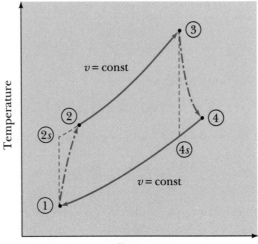

The definition of the compression efficiency, when combined with the first law as applied to the ideal isentropic process, yields

$$T_2 = T_1 + \frac{c_v(T_{2s} - T_1)}{c_v \eta_c}$$

$$= 520 + \frac{1224 - 520}{0.84}$$

$$= 1358° \, R$$

Because the specific volume of state 2 equals that of state 2s, the ideal gas equation of state gives

$$P_2 = P_{2s}\frac{T_2}{T_{2s}} = 326 \text{ psia}$$

The temperature at state 3 may now be determined by applying the first law to process 2–3:

$$T_3 = T_2 + \frac{{}_2 q_3}{c_v} = 1942° \, R$$

Because the specific volume remains constant during process 2–3,

$$P_3 = P_2\frac{T_3}{T_2} = 466 \text{ psia}$$

We can now proceed from state 3 to 4s by applying the isentropic process relations

$$T_{4s} = T_3 \left(\frac{v_3}{v_4}\right)^{k-1} = T_3\frac{1}{r_v^{k-1}} = 825° \, R$$

and

$$P_{4s} = P_3 \left(\frac{v_3}{v_4} \right)^k = P_1 \frac{1}{r_v^k} = 23.2 \text{ psia}$$

The first law and the definition of the isentropic expansion efficiency gives

$$T_4 = T_3 + \eta_e(T_{4s} - T_3) = 902°\text{R}$$

Invoking the ideal gas equation of state and using the fact that $v_{4s} = v_4$ gives

$$P_4 = P_{4s} \frac{T_4}{T_{4s}} = 25.4 \text{ psia}$$

When the first law, as applied to process 2–3 and process 4–1, is combined with the expression for the thermal efficiency, the result is

$$\eta = 1 - \frac{|q_{out}|}{q_{in}}$$
$$= 1 - \frac{c_v(T_4 - T_1)}{c_v(T_3 - T_2)}$$
$$= 0.346$$

The specific work produced during each execution of this cycle is

$$w_{net} = \eta q_{in}$$
$$= 0.346 \times 100$$
$$= 34.6 \text{ Btu/lb}_m = 26{,}920 \text{ ft-lb}_f/\text{lb}_m$$

The volume displaced by the motion of the piston per unit of working fluid mass is

$$\Delta v = v_1 - v_2$$
$$= R \left[\frac{T_1}{P_1} - \frac{T_2}{P_2} \right]$$
$$= \frac{1545}{28.97 \times 144} \left[\frac{520}{14.7} - \frac{1358}{326} \right]$$
$$= 11.56 \text{ ft}^3/\text{lb}_m$$

Without the supercharger, the mean effective pressure would be

$$P_{mean} = \frac{w_{net}}{\Delta v}$$
$$= \frac{26{,}920}{11.56 \times 144}$$
$$= 16.17 \text{ psia}$$

When the mean effective pressure is increased by adding the supercharger, the specific work produced during each cycle will become

$$w_{net} = P_{mean}\Delta v$$
$$= (16.17 + 5) \times 11.56 \times 144$$
$$= 35{,}240 \text{ ft-lb}_f/\text{lb}_m = 45.30 \text{ Btu/lb}_m$$

The addition of the supercharger has increased net work production by 31%.

8.3 The Limited-Pressure Cycle

Inspection of Figure 8.3 suggests that the Otto cycle model of the spark-ignition engine may be improved by introducing a fifth isobaric process that limits the maximum pressure and temperature and more closely replicates the last portion of the heat-addition (combustion) process. This improved model is known as the *limited-pressure cycle*. The limited-pressure cycle is compared to a spark-ignition engine's indicator diagram in Figure 8.5. The additional isobaric process 3–4 may be adjusted to match more closely the pressure–volume measurements of actual engine indicator diagrams.

FIGURE 8.5

IC Engine Indicator Diagram Modeled by a Limited-Pressure Cycle

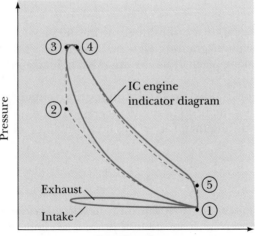

The limited-pressure cycle is also used to model compression-ignition (CI) engines that utilize the increasing gas temperature during the compression process rather than an ignition device to ignite the fuel. In these CI cycles, the fuel is injected into the piston-cylinder device, mixed with air, and ignited at or near the end of the compression. For fuel ignition to occur, the compression must be great enough for the air temperature to ignite the fuel while it is being injected into the air. Consequently, CI engines have higher compression ratios than spark-ignition engines. Their compression ratios are typically in the range of 20:1 or more. As fuel is injected at or near TDC of the compression stroke where the air temperature is the highest, combustion occurs while the piston is stationary near the top and while moving down during the initial stages of the expansion stroke. The net effect is that the combustion process is modeled better by a constant-volume process followed by a constant-pressure process rather than the constant-volume process of the Otto cycle. The extent of the constant-volume and constant-pressure process is determined by when the fuel is injected, the amount of fuel injected, and the amount of time it takes to burn.

The air standard model of the limited-pressure cycle consists of the following five processes:

1. Isentropic compression of the air in the piston cylinder while the inlet and exhaust valves are closed.

2. Constant-volume fuel injection and initial combustion, which is modeled as a reversible constant-volume heat-addition process.

3. Isobaric expansion as the burning of the fuel is completed. This process is modeled as a reversible isobaric heat-addition process.

4. Isentropic expansion of the combustion gases that are modeled as air.

5. The exhausting of spent gases and recharging with fresh air are modeled as a reversible, constant-volume heat-rejection process.

These processes are illustrated in the temperature–entropy and pressure–volume state diagrams of Figure 8.6. Because all of these processes are modeled as taking place while the exhaust and intake valves are closed, the system is closed.

The heat added to the cycle during the constant-volume process is obtained by applying the first law to the process. Presuming that the specific heats are constant during the process, this heat addition is given by

$$_2q_3 = c_v(T_3 - T_2)$$

Application of the first law and the closed-system work expression to the constant-pressure heat-addition process, assuming that the specific heats remain constant, yields

$$_3q_4 = c_P(T_4 - T_3)$$

The net heat addition to the cycle is then

$$q_{in} = c_v(T_3 - T_2) + c_P(T_4 - T_3)$$

FIGURE 8.6
*Limited Pressure
Cycle T–s and
P–v State Diagrams*

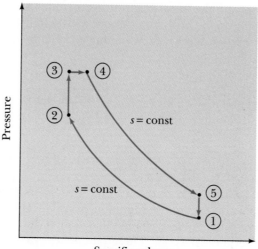

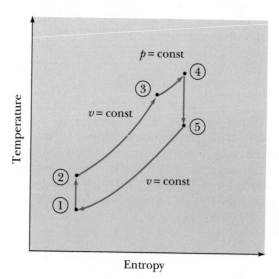

The amount of heat rejected during the constant-volume model of the exhaust-recharge processes is

$$_5 q_1 = -q_{out} = c_v(T_1 - T_5)$$

When these expressions for the heat addition and rejection are substituted into the thermal-efficiency definition, the limited-pressure cycle's thermal efficiency is given by

$$\eta = 1 - \frac{q_{out}}{q_{in}} = 1 - \frac{c_v(T_5 - T_1)}{c_v(T_3 - T_2) + c_P(T_4 - T_3)}$$

By applying the isentropic process relations for ideal gases with constant specific heats to the processes 1–2 and 4–5, as well as the ideal gas equation of state, the temperatures may be eliminated from the thermal-efficiency expression. This yields the result

$$\eta = 1 - \frac{1}{r_v^{k-1}} \left[\frac{\alpha \beta^k - 1}{k\alpha(\beta - 1) + \alpha - 1} \right]$$

where

$$\alpha = \frac{P_3}{P_2} \quad \text{and} \quad \beta = \frac{v_4}{v_3}$$

The first parameter, α, indicates the pressure increase during the constant-volume portion of the heat-addition process. The second parameter, β, is a measure of the volume change that occurs during the isobaric portion of the heat-addition process. These parameters vary with engine designs and the amount of fuel injected during each repetition of the cycle. They can be adjusted to model the engine's indicator diagram.

When $\beta = 1$, no isobaric process exists, and states 3 and 4 become one state. Under these conditions, the limited-pressure cycle's thermal efficiency reduces to

$$\eta = 1 - \frac{1}{r_v^{k-1}}$$

and the limited-pressure cycle becomes identical to the Otto cycle. Similarly, when $\alpha = 1$, there is no constant-volume process, and states 2 and 3 become one and the same state. In this case, all heat is added to the cycle isobarically. This special case of the limited-pressure cycle is known as the *Diesel cycle*, named for its inventor, Rudolf Diesel.[2] Although CI engines are commonly known as Diesel engines, only certain designs under certain operating conditions truly operate on the Diesel cycle. When this is the case, the Diesel cycle's thermal efficiency

$$\eta = 1 - \frac{1}{r_v^{k-1}} \frac{\beta^k - 1}{k(\beta - 1)}$$

is applicable.

The thermal efficiencies of the two extreme limited-pressure cycles (i.e., the Otto and Diesel cycles) using air as the working fluid are compared in Figure 8.7 as a function of compression ratios. From this figure, it is apparent that the Diesel cycle is less efficient than the Otto cycle at the same compression

---

[2]Rudolf Diesel (1858–1913), a German engineer, was inspired by the ideal cycle of Carnot to increase the thermal efficiency of oil-fired engines of the day by compressing the air to a greater extent than normal. His first commercially successful engine was completed in 1897 after four years of development. This engine developed 25 horsepower and was based on patents issued in 1892 and 1893.

ratio. But CI engines operate at higher compression ratios, which increases the mean effective pressure and thermal efficiency. Although CI engines operate at higher compression ratios and higher mean effective pressures, their maximum pressures are not as high as those of spark-ignition engines. This implies that their structure must be somewhat heavier to accommodate the higher mean effective pressures, and they are not subjected to the extreme loads generated by maximum pressures that reduce engine life.

FIGURE 8.7

Thermal Efficiency of Limited-Pressure Cycles

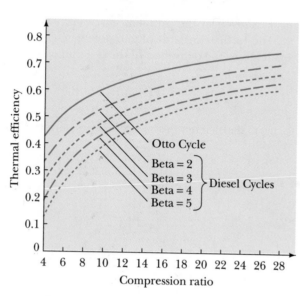

EXAMPLE 8.4

Someone has suggested that for the conditions and gas mixtures in a typical IC engine, $k = 1.3$ is a better value for the specific heat ratio. Consider such an engine using air and operating on the limited-pressure cycle with $r_v = 14$, $\alpha = 1.5$, $\beta = 2$, $P_1 = 100$ kPa, and $v_1 = 0.841$ m³/kg. Determine the maximum pressure and thermal efficiency of this engine. What is the rate of heat addition to this engine when it produces 100 kW of power?

System: Air in a piston-cylinder device undergoing the processes of the limited-pressure cycle as shown in Figure 8.6.

Basic Equations:

$$_1q_2 - {_1w_2} = u_2 - u_1$$

$$\frac{T_2}{T_1} = \left(\frac{P_2}{P_1}\right)^{(k-1)/k} = \left(\frac{v_1}{v_2}\right)^{k-1}$$

Conditions: The cycle will be modeled using the air standard model.

Solution: The specific volume of the air at the end of the compression process is

$$v_2 = \frac{v_1}{r_v} = 0.0601 \text{ m}^3/\text{kg}$$

according to the definition of the compression ratio. At the end of the compression, the pressure is determined by the isentropic ideal gas expression as applied to the compression process,

$$P_2 = P_1 \left(\frac{v_1}{v_2} \right)^k = 3090 \text{ kPa}$$

using k as given in the problem statement. The maximum pressure occurs at the end of the constant-volume process 2–3 of Figure 8.6. This pressure is determined by the definition of the α parameter:

$$P_3 = P_2 \alpha = 4635 \text{ kPa}$$

The thermal efficiency of this limited-pressure cycle is

$$\eta = 1 - \frac{1}{r_v^{k-1}} \left[\frac{\alpha \beta^k - 1}{k \alpha (\beta - 1) + \alpha - 1} \right]$$

$$= 1 - \frac{1}{14^{0.3}} \left[\frac{1.5 \times 2^{1.3} - 1}{1.3 \times 1.5(2 - 1) + 1.5 - 1} \right]$$

$$= 0.502$$

From the definition of thermal efficiency,

$$\dot{Q}_{in} = \frac{\dot{W}}{\eta} = 199.2 \text{ kW}$$

EXAMPLE 8.5

A pure Diesel cycle uses air as its working fluid; begins compression at 15 psia, 70° F; has a compression ratio of 20:1; and has a β of 3. Determine this cycle's thermal efficiency and mean effective pressure.

System: Air in a piston-cylinder device undergoing the processes of the pure Diesel cycle as shown in Figure 8.8.

Basic Equations:

$$_1 q_2 - _1 w_2 = u_2 - u_1$$

$$\frac{T_2}{T_1} = \left(\frac{P_2}{P_1} \right)^{(k-1)/k} = \left(\frac{v_1}{v_2} \right)^{k-1}$$

Conditions: The cycle will be modeled using air standard analysis.

Solution: Using the Diesel cycle limit ($\alpha = 1$) of the limited-pressure cycle gives

$$\eta = 1 - \frac{1}{r_v^{k-1}} \frac{\beta^k - 1}{k(\beta - 1)}$$

$$= 1 - \frac{1}{20^{0.4}} \frac{3^{1.4} - 1}{1.4(3 - 1)}$$

$$= 0.606$$

The full Diesel cycle is shown in Figure 8.8. At state 1, the air's specific volume is

$$v_1 = \frac{RT_1}{P_1} = 13.09 \text{ ft}^3/\text{lb}_m$$

FIGURE 8.8
Diesel Cycle of Example 8.5

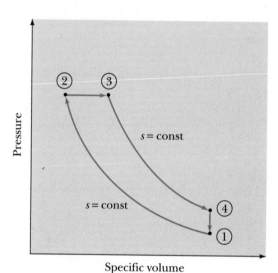

Specific volume

The specific volume at state 2 is

$$v_2 = \frac{v_1}{r_v} = 0.655 \text{ ft}^3/\text{lb}_m$$

and the pressure is then

$$P_2 = P_1 \left(\frac{v_1}{v_2}\right)^k = 994 \text{ psia}$$

At state 3, the specific volume is

$$v_3 = \frac{v_4}{\beta} = \frac{v_1}{\beta} = 4.363 \text{ ft}^3/\text{lb}_m$$

and the pressure at state 4 is

$$P_4 = P_3 \left(\frac{v_3}{v_4}\right)^k = P_2 \left(\frac{v_3}{v_1}\right)^k = P_2 \left(\frac{1}{\beta}\right)^k = 213.5 \text{ psia}$$

The specific work required for the compression is the area under compression process 1–2:

$$_1w_2 = \int_1^2 P\,dv = P_1 v_1^k \int_1^2 \frac{dv}{v^k}$$

$$= \frac{P_1 v_1}{k-1}\left[1 - \left(\frac{v_1}{v_2}\right)^{k-1}\right]$$

$$= -210.3 \text{ Btu/lb}_m$$

The specific work produced during the isobaric process is

$$_2w_3 = \int_2^3 P\,dv = P_3(v_3 - v_2)$$

$$= 682.2 \text{ Btu/lb}_m$$

and the specific work produced during the isentropic expansion is

$$_3w_4 = \frac{P_3 v_3}{k-1}\left[1 - \left(\frac{v_3}{v_4}\right)^{k-1}\right]$$

$$= 713.7 \text{ Btu/lb}_m$$

The net specific work produced by the cycle is then

$$w_{net} = {}_1w_2 + {}_2w_3 + {}_3w_4 = 1185.6 \text{ Btu/lb}_m$$

and the mean effective pressure for the cycle is

$$P_m = \frac{w_{net}}{v_1 - v_2}$$

$$= \frac{509.7(778)}{(13.09 - 0.654)144}$$

$$= 515.1 \text{ psia}$$

8.4 The Stirling Cycle

The Stirling cycle was originally proposed by Robert Stirling[3] as a piston-cylinder device cycle whose thermal efficiency equals that of a completely reversible engine and the Carnot cycle. This was accomplished with a device known as a *regenerator*, which adds or removes heat from the working gas, but does not exchange heat with the engine surroundings. Interest in this cycle over the past 100 years has been constrained by difficulties in constructing devices that can execute the cycle. Recent renewed interest has been the result of the quest for reciprocating engines with improved efficiency and reduced exhaust emissions that also can be manufactured with available technologies and facilities.

A device that conceptually executes the Stirling cycle is illustrated in Figure 8.9. It consists of two pistons that share a working gas through a regeneration unit, made from a porous metal or ceramic through which the gas can pass while it exchanges heat with the metal or ceramic. Its purpose is to internally add or remove heat from the gas at various points in the cycle. When the gas is hot, the regenerator material temperature will increase as the gas is cooled. The hot regenerator material then can be used to reheat the gas at a later point in the cycle when the gas is cool. When the gas and regenerator material are selected as the system, no heat is exchanged between the system and surroundings by the regenerator. It is only temporarily stored in and removed from the regenerator material at appropriate times.

FIGURE 8.9
Stirling Cycle Device

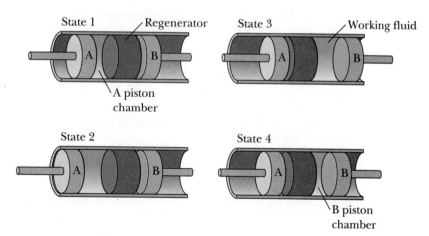

At state 1 of the cycle, the gas is hot and occupies the regenerator and the chamber formed by piston A. Heat is now added externally to the gas in piston A's chamber and the piston is moved in such a way that the gas temperature

---

[3]Robert Stirling (1790–1878), an English parish minister, patented a practical heat engine in 1816 that used air as the working fluid and approached the ideal cycle proposed by Carnot.

remains constant until state 2 is achieved. Pistons A and B are now moved simultaneously in a manner such that the volume of gas does not change as the gas passes through the regenerator. While passing through the regenerator, the gas is cooled as heat is added to the regenerator material. None of this heat leaves the regenerator–gas system, and process 2–3 is an adiabatic constant-volume process. From the gas's perspective, this is a constant-volume process with heat rejection. The gas occupies the regenerator and chamber formed by piston B at state 3. The gas in piston B's chamber is now cooled further externally while piston B is moved to keep the temperature in the chamber constant until state 4 is reached. The two pistons again are simultaneously moved to keep the gas volume constant as the gas passes through the regenerator again. This time, the gas is reheated by the regenerator material while entering piston A's chamber to repeat the cycle.

The cycle executed by this device consists of the following four processes when the regenerator and gas are taken as the system:

1–2	an isothermal heat addition
2–3	an adiabatic, constant-volume process
3–4	an isothermal heat rejection
4–1	an adiabatic, constant-volume process

The processes experienced by the gas are shown in the state diagrams of Figure 8.10.

Using an air standard analysis model for the working fluid only and the Gibbs equation, $Tds = du + Pdv$, of Section 4.9,

$$s_4 - s_1 = c_v \ln \frac{T_4}{T_1} \quad \text{and} \quad s_3 - s_2 = c_v \ln \frac{T_3}{T_2}$$

Because $T_1 = T_2$ and $T_3 = T_4$, this reduces to

$$s_2 - s_1 = s_3 - s_4$$

Applying the definition of entropy to processes 1–2 and 3–4 yields

$$q_{in} = {}_1q_2 = T_1(s_2 - s_1) \quad \text{and} \quad q_{out} = -{}_3q_4 = T_3(s_3 - s_4)$$

The thermal efficiency of the Stirling cycle is then

$$\eta = 1 - \frac{q_{out}}{q_{in}} = 1 - \frac{T_3}{T_1} = 1 - \frac{T_l}{T_h}$$

FIGURE 8.10
Stirling Cycle State Diagrams

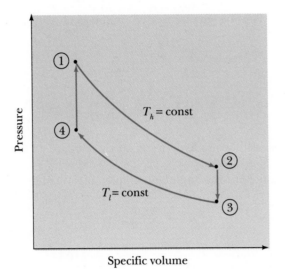

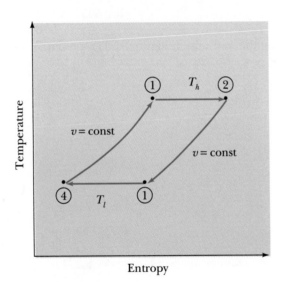

which is the same as that of the Carnot and all other completely reversible cycles. The Stirling cycle therefore achieves the maximum possible thermal efficiency for a given pair of isothermal energy reservoirs.

EXAMPLE 8.6

A Stirling engine contains 1 pound-mass of air, operates between energy reservoirs at 800° F and 50° F, and undergoes a cycle whose maximum and minimum volumes are 15 ft³ and 0.5 ft³, respectively. Determine the thermal efficiency, external heat addition, work produced, and heat transferred between the air and the regenerator of this cycle.

System: Air in a Sterling engine device undergoing the processes as shown in Figure 8.10.

Basic Equations:

$$_1q_2 - _1w_2 = u_2 - u_1$$

$$\frac{T_2}{T_1} = \left(\frac{P_2}{P_1}\right)^{(k-1)/k} = \left(\frac{v_1}{v_2}\right)^{k-1}$$

Conditions: The cycle will be modeled using an air standard analysis model.

Solution: Because the Stirling engine is completely reversible,

$$\eta = 1 - \frac{T_l}{T_h}$$

$$= 1 - \frac{510}{1260}$$

$$= 0.595$$

Following the notation of Figure 8.10 and treating the air as an ideal gas, the work for process 1–2 is

$$_1W_2 = \int_1^2 P\,dV$$

$$= mRT_1 \int_1^2 \frac{dV}{V}$$

$$= mRT_1 \ln \frac{V_2}{V_1}$$

$$= 228{,}600 \text{ Btu}$$

In the same manner, the work for process 3–4 is

$$_3W_4 = \int_3^4 P\,dV$$

$$= mRT_3 \ln \frac{V_4}{V_3}$$

$$= -92{,}500 \text{ Btu}$$

The net work produced by one execution of the cycle is then

$$W_{net} = {_1W_2} + {_3W_4}$$

$$= 136{,}100 \text{ Btu}$$

Applying the first law for a closed system to external heat-addition process 1–2, presuming that the specific heats of the air do not change,

$$_1Q_2 = {_1}W_2 + mc_v(T_2 - T_1)$$
$$= {_1}W_2$$
$$= 228,600 \text{ Btu}$$

The heat transferred to the regenerator from the air occurs during process 4–1. Because there is no work interaction during this constant-volume process, the first law becomes

TIP:

Notice that the air was taken as the system in this example. We can then account for heat transfer between the air and regenerator, although none of this heat is ever transferred to the engine's surroundings.

$$_4Q_1 = m(u_1 - u_4)$$
$$= mc_v(T_1 - T_4)$$
$$= 1 \times 0.1711(800 - 50)$$
$$= 128.3 \text{ Btu}$$

This is also the amount of heat transferred from the regenerator to the air during process 2–3 because no heat is exchanged between the regenerator and system surroundings during the cycle's execution. ◼

In addition to the compression and expansion irreversibilities and the heat transfer during the compression and expansion problems of the preceding reciprocating-engine cycles, the Stirling engine is constrained by our ability to transfer heat quickly between the working fluid and regenerator when there is only a small temperature difference between the two. To accomplish this exchange of heat, the regenerator material would have to be practically massless, or the area of contact between the regenerator material and working gas would have to be extremely large (approaching infinity). Both alternatives are not practical from an engineering viewpoint. In spite of these problems, which limit the Stirling cycle's efficiency, Ford Motor Company, General Motors Corporation, Cummins Engine Company, and Phillips Research Laboratories of the Netherlands have developed working Stirling cycle engines. These engines require additional research and development before they can become competitive with conventional spark-ignition and compression-ignition IC engines.

Stirling cycles clearly illustrate the effect of regeneration on improving thermal efficiency. We will see additional use of regenerators as means of improving the thermal efficiency of cycles in the sections that follow.

8.5 The Brayton Cycle

The Brayton cycle[4] is a model of a simple gas-turbine engine. Gas-turbine engines are steady-flow devices that continuously draw in and compress atmospheric air, mix it with fuel, combust this mixture in a constant-pressure container known as a *combustor* or *combustion chamber*, expand the combustion products through a turbine, and finally exhaust the combustion products back to the atmospheric air at the ambient pressure. A schematic diagram of this equipment is shown in Figure 8.11. The dashed isobaric heat-rejecting component represents the processes of cooling, conversion of exhaust products back to air, and the reintroduction of air into the compressor, all of which occur in the atmosphere.

The temperature–entropy state diagram shown in this figure illustrates the processes that occur in the Brayton cycle. With the air standard analysis assumptions, these processes consist of:

1–2   Isentropic compression of the gas in a steady-flow compressor.

2–3   Reversible, isobaric addition of heat to the gas from an energy reservoir that models the fuel-combustion process.

3–4   Isentropic expansion of the gas in a steady-flow turbine. This turbine produces work, some of which is used to drive the compressor.

4–1   Reversible, isobaric rejection of heat from the gas to an energy reservoir, which models the processes occurring in the atmosphere.

Unlike the cycles presented previously in this chapter, the Brayton cycle components are all steady-flow devices rather than fixed-mass devices. The steady-flow system-analysis techniques presented in Chapter 5 are therefore appropriate.

Applying the steady-flow first law with the air standard analysis assumptions to the turbine yields the specific work produced by the turbine:

$$_3w_4 = c_P(T_3 - T_4)$$

The portion of this work needed to drive the compressor is found by applying the first law to the compressor:

$$_1w_2 = c_P(T_1 - T_2)$$

---

[4]Named for George Bailey Brayton (1830–1892), an American entrepreneur who introduced an engine that operated by injecting compressed air into the combustion chamber after it had passed over a hot grating. Liquid fuel was then injected into this hot air just before it entered the combustion chamber, causing a combustion process of nearly constant pressure.

FIGURE 8.11

*Brayton Cycle
Equipment
Schematic and T–s
State Diagram*

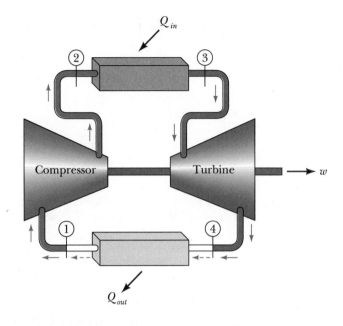

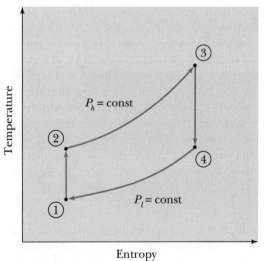

A first law analysis of the high-pressure combustion chamber with the air standard analysis assumptions provides the heat that must be added to drive this system:

$$_2q_3 = c_P(T_3 - T_2)$$

Finally, the heat rejected during the low-pressure isobaric model of the processes occurring in the atmosphere is

$$_4q_1 = c_P(T_1 - T_4)$$

The thermal efficiency of the Brayton cycle is then

$$\eta = 1 - \left| \frac{_4 q_1}{_2 q_3} \right| = 1 - \frac{T_4 - T_1}{T_3 - T_2}$$

This can be rearranged as

$$\eta = 1 - \frac{T_1}{T_2} \frac{T_4/T_1 - 1}{T_3/T_2 - 1}$$

Applying the ideal gas with constant specific-heat, isentropic process expressions to the compression and expansion processes further reduces this result to

$$\eta = 1 - \frac{T_1}{T_2} = 1 - \frac{1}{r_P^{(k-1)/k}}$$

where $r_P = P_2/P_1$ is known as the *pressure ratio*. The thermal efficiency of the Brayton cycle is a monotonic increasing function of the pressure ratio produced by the compressor. Inspection of the Brayton cycle temperature–entropy state diagram reveals that as the pressure ratio is increased, the temperature at the compressor outlet, T_2, also increases. If we maintain a constant heat input to the cycle, the temperature at the outlet of the combustion chamber, T_3, will increase by the same amount. This temperature is limited by the materials used to fabricate the combustion chamber and the components near the hot end of the turbine. The maximum temperature to which these materials can be exposed sets the limit on the thermodynamic performance of gas-turbine engines. Recently, programs to develop ceramic turbine parts have been undertaken to improve the performance of the Brayton cycle.

EXAMPLE 8.7

A simple gas-turbine engine operates with air as the working fluid, a pressure ratio of 18:1, and a maximum temperature of 700° C. The air enters the compressor at 100 kPa, 20° C. Determine the thermal efficiency, specific heat addition, and state of the air exhausted by the turbine when it is modeled by the Brayton cycle.

System: Air in a gas turbine undergoing the processes as shown in Figure 8.11.

Basic Equations:

$$_1 q_2 - _1 w_2 = h_2 - h_1 + \frac{1}{2}(V_2^2 - V_1^2) + g(z_2 - z_1)$$

$$\frac{T_2}{T_1} = \left(\frac{P_2}{P_1} \right)^{(k-1)/k} = \left(\frac{v_1}{v_2} \right)^{k-1}$$

Conditions: The cycle will be modeled using air standard analysis.

Solution: The thermal efficiency is determined by the pressure ratio and working fluid,

$$\eta = 1 - \frac{1}{r_p^{(k-1)/k}}$$

$$= 1 - \frac{1}{18^{0.4/1.4}}$$

$$= 0.562$$

Following the notation of Figure 8.11 and applying the isentropic process relations for an ideal gas with constant specific heats to the compressor, we obtain T_2 :

$$T_2 = T_1 \left(\frac{P_2}{P_1}\right)^{(k-1)/k}$$

$$= 669 \text{ K}$$

The heat added to the combustion chamber per unit of flow through this chamber is found by applying the first law:

$$_2q_3 - {_2w_3} = h_3 - h_2$$

When specialized to the combustion chamber, this becomes

$$_2q_3 = c_P(T_3 - T_2)$$
$$= 1.0035(973 - 669)$$
$$= 305.1 \text{ kJ/kg}$$

Applying the isentropic process relationships to the turbine expansion,

$$T_4 = T_3 \left(\frac{P_4}{P_3}\right)^{(k-1)/k}$$

$$= 973 \left(\frac{1}{18}\right)^{0.4/1.4}$$

$$= 426 \text{ K}$$

and the pressure at the turbine outlet matches the 100 kPa pressure at the compressor inlet.   ◼

The compression and expansion of the gas in this cycle suffer from irreversibilities that can be accounted for by using isentropic efficiencies for both processes. Physically, these irreversibilities result from the shear forces exerted on the gas as it passes through complicated flow passages, the fact that the compression and expansion of the gas is not a completely constrained process, and fluid mixing that occurs as the gas passes between stages in the compressor and turbine. Design improvements in the equipment used to compress and expand the gas are continually improving the overall efficiency of the Brayton cycle. Pressure losses also can occur in both the combustion chamber and the passages connecting the various equipment components. Although these pressure losses are quite small, they do adversely affect the performance of the simple gas turbine.

When we account for the compressor and turbine isentropic efficiencies, the net specific work produced by the Brayton cycle is

$$w_{net} = \eta_t c_P(T_3 - T_{4s}) - \eta_c c_P(T_{2s} - T_1)$$

The cycle's thermal efficiency becomes

$$\eta = \frac{w_{net}}{q_{in}} = \frac{\eta_t(T_3 - T_{4s}) - \eta_c(T_{2s} - T_1)}{T_3 - T_2}$$

where the additional subscript s refers to the state that would have resulted if the compression and expansion processes had been isentropic as illustrated in Figure 8.12.

FIGURE 8.12

Brayton Cycle with an Imperfect Turbine and Compressor

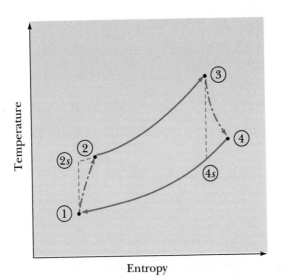

EXAMPLE 8.8

A simple gas-turbine engine operates with air as the working fluid, a pressure ratio of 18:1, and a maximum temperature of 700° C. The air enters the compressor at 100 kPa, 20° C. The compression and expansion isentropic efficiencies are 0.87 and 0.94, respectively. Calculate the specific entropy generation and change in availability for each process in the cycle. The temperature of the energy-source reservoir is the same as the maximum cycle temperature, and the temperature of the energy-sink reservoir is the same as the minimum cycle temperature.

System: Air in a gas turbine undergoing the processes as shown in Figure 8.12.

Basic Equations:

$$_1q_2 - {_1}w_2 = h_2 - h_1 + \frac{1}{2}(V_2^2 - V_1^2) + g(z_2 - z_1)$$

$$\frac{T_2}{T_1} = \left(\frac{P_2}{P_1}\right)^{(k-1)/k} = \left(\frac{v_1}{v_2}\right)^{k-1}$$

$$w_{ptl} = q_j\left(1 - \frac{T_0}{T_j}\right) - \Delta\psi$$

Conditions: The cycle will be modeled using the air standard model.

Solution: The various cycle states will be numbered as in Figure 8.12. Beginning at the compressor inlet and working around the cycle to the turbine outlet,

$$P_2 = P_3 = r_p P_1 = 1800 \text{ kPa}$$

$$T_{2s} = T_1\left(\frac{P_2}{P_1}\right)^{(k-1)/k} = 669 \text{ K}$$

$$T_2 = T_1 + \frac{T_{2s} - T_1}{\eta_c} = 725 \text{ K}$$

$$T_{4s} = T_3\left(\frac{P_3}{P_4}\right)^{(k-1)/k} = 426 \text{ K}$$

$$T_4 = T_3 - \eta_t(T_3 - T_{4s}) = 459 \text{ K}$$

The enthalpy, entropy, and flow exergy of the air, presuming constant specific heats, are

$$h = c_P(T - T_0)$$

$$s = c_P \ln\frac{T}{T_0} - R\ln\frac{P}{P_0}$$

$$\psi = (h - h_0) - T_0(s - s_0)$$

where the second result was obtained from the integrated vdP form of the Gibbs equations. The various states around this cycle are summarized in Table 8.1, which have a dead state of $T_0 = 20°\text{ C}$ and $P_0 = 100$ kPa.

TABLE 8.1
States for Example 8.8

State	P(kPa)	T(K)	h(kJ/kg)	s(kJ/kg-K)	ψ(kJ/kg)
1	100	293	0	0	0
2	1800	725	433.5	0.07967	410.2
3	1800	973	682.4	0.3749	572.6
4	100	459	166.6	0.4504	34.6

The specific entropy generation for a process is

$$_1S_{gen,2} = \frac{1}{T_0}\left[_1q_2\left(1 - \frac{T_0}{T_j}\right) - _1w_2 + (h_1 - T_0s_1) - (h_2 - T_0s_2)\right]$$

according to Section 6.5. In this expression, all quantities are with respect to the system, 1 refers to the inlet, 2 refers to the outlet, and T_j is the temperature of any interacting thermal-energy reservoir. For compression 1–2,

$$_1w_2 = h_1 - h_2 = -433.5 \text{ kJ/kg}$$
$$_1S_{gen,2} = \frac{1}{293}[433.5 + (0 - 433.5) - 293(0 - 0.07967)]$$
$$= 0.07967 \text{ kJ/kg-K}$$
$$\psi_2 - \psi_1 = 410.2 \text{ kJ/kg}$$

For heat-addition process 2–3, presuming that the heat-supplying energy reservoir is at 973 K,

$$_2q_3 = h_3 - h_2 = 248.9 \text{ kJ/kg}$$
$$_2S_{gen,3} = \frac{1}{293}\left[248.9\left(1 - \frac{293}{973}\right) + (433.5 - 682.4) - 293(0.07967 - 0.3749)\right]$$
$$= 0.03942 \text{ kJ/kg-K}$$
$$\psi_3 - \psi_2 = 162.4 \text{ kJ/kg}$$

Across the turbine, 3–4,

$$_3w_4 = h_3 - h_4 = 515.8 \text{ kJ/kg}$$
$$_3S_{gen,4} = \frac{1}{293}[-515.8 + (682.4 - 166.6) - 293(0.3749 - 0.4504)]$$
$$= 0.07550 \text{ kJ/kg-K}$$
$$\psi_4 - \psi_3 = -522.8 \text{ kJ/kg}$$

For the heat-rejection process, 4–1, taking the heat-receiving energy reservoir at 293 K,

$$_4q_1 = h_1 - h_4 = -166.6 \text{ kJ/kg}$$
$$_4S_{gen,1} = \frac{1}{293}[(166.6 - 0) - 293(0.4504 - 0)]$$
$$= 0.1162 \text{ kJ/kg-K}$$
$$\psi_1 - \psi_4 = -34.0 \text{ kJ/kg}$$

The exergy accounting is summarized in Table 8.2. Work potentials were calculated using

$$w_{ptl} = \Delta\psi + q\left(1 - \frac{T_0}{T}\right)$$

The compression could have been done with 410.2 kJ/kg of work, although it actually took 433.5 kJ/kg for this process. Additional work was required to overcome the process irreversibilities as embodied in the compressor efficiency. The net effect of this ineffective use of work was to generate 0.07967 kJ/kg-K of entropy in the compression process. The turbine also does not produce as much work as exergy analysis indicates it can. In the turbine, the impact of the irreversibilities is not quite as extensive as in the compressor; this is seen by comparing the specific entropy generations and lost work potentials of the two components.

TABLE 8.2
Exergy Accounting for Example 8.8

Component	$\Delta\psi$ kJ/kg	w_{ptl} kJ/kg	w_{act} kJ/kg	w_{lost} kJ/kg	s_{gen} kJ/kg-K
Compressor	410.2	−410.2	−433.5	23.3	0.07967
Combustor	162.4	11.5	0	11.5	0.03942
Turbine	−522.8	522.8	515.8	7.1	0.07550
Atmospheric	−34.0	34.0	0	34.0	0.1162
Total		**158.1**	**82.3**	**75.9**	

Adding heat to the constant-pressure process from an isothermal-energy reservoir is ineffective, as shown by the change in the air's exergy during this process. This should have been expected, because the heat transferred from the reservoir to the air must be transferred across a large temperature difference at the beginning of this process. A similar irreversibility occurs in the heat-rejecting process.

8.6 Gas-Turbine Cycles

The thermal efficiency of the basic gas turbine as depicted by the Brayton cycle model may be improved in several ways. For example, a regenerator such as that of the Stirling cycle and anything that we can do to make the isobaric heat-addition and heat-rejection processes approach an isothermal process should improve the gas turbine's thermodynamic efficiency. Several of these improvements, as well as alternate gas-turbine cycles, are discussed in this section.

8.6.1 Regenerators

A regenerating heat exchanger may be added to the basic Brayton cycle to reduce the heat that must be added to the cycle in the combustion process. High-temperature gases at the turbine exhaust are used in this heat exchanger to heat the gases leaving the compressor further before they enter the combustion chamber, as in Figure 8.13. The compressed gas enters the regenerator at temperature T_2 and is heated isobarically to temperature T_3 by the hot turbine exhaust gas. Turbine exhaust gas enters the regenerator at temperature T_5 and is isobarically cooled in the regenerator to temperature T_6. Application

FIGURE 8.13
Brayton Cycle with Regenerative Heating

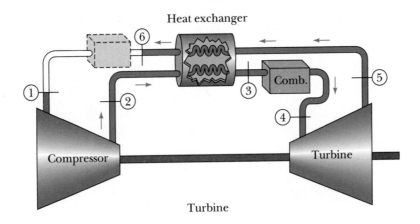

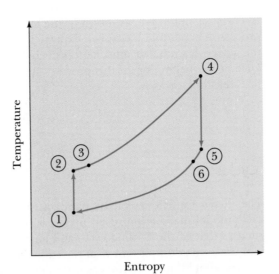

of the first law to the two streams passing through the regenerator, presuming that no heat is exchanged between the streams and the fluid surrounding the regenerator, yields

$$T_3 - T_2 = T_5 - T_6$$

when the specific heat of the turbine exhaust gas matches that of the compressor exit gas.

The two gas streams passing through the regenerator generally flow in the opposite sense. This is known as a *counterflow* arrangement for the regenerative heat exchanger. In this arrangement, the 5–6 stream is the hot stream, which is being cooled by the 2–3 stream. The second law tells us that $T_2 \leq T_6$ and $T_3 \leq T_5$. In a perfect regenerator with an infinite amount of heat transfer surface area, $T_6 = T_2$ and $T_3 = T_5$.

The impact of regeneration on the thermal efficiency of the Brayton cycle may be seen by considering two cycles (one with regeneration and one without) that produce the same amount of work (area enclosed by the cycle on a T–s state diagram). The one without regeneration requires an external heat input that is proportional to $T_4 - T_2$, while the one with regeneration requires an external heat input that is proportional to $T_4 - T_3$. Because the cycle with the regenerator requires a smaller heat input for the same net work production, it has a higher thermodynamic efficiency.

Applying the first law to the combustor reveals the external heat addition as

$$_3q_4 = c_P(T_4 - T_3)$$

Similarly, the heat rejected by the cycle is

$$_6q_1 = c_P(T_1 - T_6)$$

When these are substituted into the heat form of the thermodynamic efficiency expression, the thermodynamic efficiency of the Brayton cycle with regeneration is

$$\eta = 1 - \frac{T_6 - T_1}{T_4 - T_3}$$

EXAMPLE 8.9

A simple gas-turbine engine with regeneration operates with air as the working fluid, a pressure ratio of 6:1, and a maximum temperature of $1100°$ F. The air enters the compressor at 15 psia, $70°$ F. Calculate the specific work production and thermal efficiency of this cycle when the regenerator operates with $T_3 = T_5$.

System: Air in a gas turbine with regeneration undergoing the processes shown in Figure 8.13.

Basic Equations:

$$_1q_2 - {_1w_2} = h_2 - h_1 + \frac{1}{2}(V_2^2 - V_1^2) + g(z_2 - z_1)$$

$$\frac{T_2}{T_1} = \left(\frac{P_2}{P_1}\right)^{(k-1)/k} = \left(\frac{v_1}{v_2}\right)^{k-1}$$

Conditions: The cycle will be modeled using air standard analysis assumptions.

Solution: We will number the states in the same manner as in Figure 8.13. Applying the ideal gas isentropic process relations to the compressor and turbine processes gives

$$T_2 = T_1 \left(\frac{P_2}{P_1} \right)^{(k-1)/k}$$
$$= 884^\circ \, \text{R}$$

$$T_5 = T_3 = T_4 \left(\frac{P_5}{P_1} \right)^{(k-1)/k}$$
$$= 935^\circ \, \text{R}$$

When the first law is applied to the regenerator, and the specific heat at constant pressure used to evaluate enthalpy changes,

$$T_3 - T_2 = T_5 - T_6$$

When solved for T_6, this gives

$$T_6 = T_2 = 884^\circ \, \text{R}$$

Now that all the cycle temperatures are known, the cycle's thermal efficiency is

$$\eta = 1 - \frac{T_6 - T_1}{T_4 - T_3}$$
$$= 0.434$$

By taking advantage of the definition of the thermal efficiency, and the first law as applied to external heat-addition process 3–4,

$$w_{net} = \eta_3 q_4$$
$$= \eta \, c_P (T_4 - T_3)$$
$$= 65.0 \, \text{Btu/lb}_m$$

           ▄

8.6.2   Gas Turbines with Reheating

Reheating the working fluid during the expansion process, such as was done with the Rankine cycle of Chapter 7, will improve the overall performance of gas-turbine engines. This is accomplished by extracting the working fluid after it has been partially expanded in the turbine, mixing and combusting additional fuel, and then completing the turbine expansion. A Brayton cycle modified for one stage of reheating is illustrated in Figure 8.14. In this system, the working fluid exiting turbine A is isobarically reheated to an increased temperature in process 4–5 before it enters turbine B to complete the second stage of the expansion process.

The relationship between T_3 and T_5 is determined by the pressure at which reheating takes place, $P_4 = P_5$, and the amount of heat added during reheating. These temperatures are normally the same as determined by the metallurigical temperature limits that exist at the hot end of the two turbines. Turbine A is also often operated so that it produces just enough work to drive the compressor. All of the work produced by turbine B then may be used for external purposes. When this is the case, notice that the compressor, combustion chamber 2–3, and turbine A form a system whose purpose is to generate a high-pressure, high-temperature gas to supply combustion chamber 4–5 and turbine B. This subsystem is known as a *gas generator*, because it neither produces nor consumes any work and only produces a gas stream that may be used to produce work.

Turbine A produces

$$_3w_4 = h_3 - h_4 = c_P(T_3 - T_4)$$

when the working fluid is an ideal gas with constant specific heats. The extra heat added in the second combustion chamber is given by the first law,

$$_4q_5 = h_5 - h_4 = c_P(T_5 - T_4)$$

and the work produced by turbine B is

$$_5w_6 = c_P(T_5 - T_6)$$

Substituting these results into the definition of the thermal efficiency gives

$$\eta = \frac{(T_3 - T_4) + (T_5 - T_6) - (T_2 - T_1)}{(T_3 - T_2) + (T_5 - T_4)}$$

The maximum amount of work that will be produced by the two turbines will depend on the intermediate pressure at which reheating takes place. In the case where $T_3 = T_5$ and the turbine expansions are reversible polytropic expansions, application of the steady-flow work integral gives

$$w_{net} = RT_3 \frac{n}{n-1} \left\{ \left[1 - \left(\frac{P_4}{P_3}\right)^{n/(n-1)} \right] + \left[1 - \left(\frac{P_6}{P_4}\right)^{n/(n-1)} \right] \right\}$$

where n is the polytropic exponent. Taking the derivative of this expression with regard to P_4 and setting the result to zero reveals that

$$\frac{P_3}{P_4} = \frac{P_4}{P_6}$$

for the maximum work production; that is, the maximum work will be produced when the pressure ratio is the same across each stage of the turbine expansion. This conclusion holds even when there are more than two stages of turbine expansion.

FIGURE 8.14
Brayton Cycle with Reheating

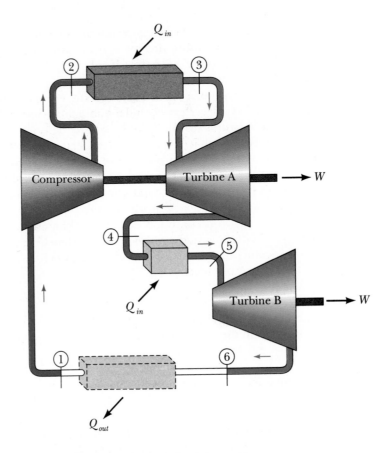

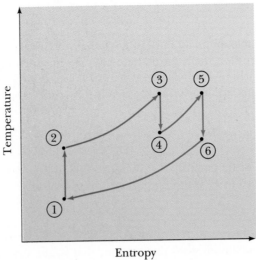

8.6.3   Gas Turbines with Intercooling

The work required to compress a gas from an initial pressure to a final pressure in a steady-flow device can be reduced by compressing in stages, with the gas being cooled between each stage. This is known as *multistage compression with intercooling*. Isentropic compression stages with isobaric intercooling are illustrated in the state diagram of Figure 8.15. Recall that the work for a steady-flow system with a single inlet and outlet is given by

$$_1w_2 = -\int_1^2 v dP$$

when kinetic and potential energy changes are neglected. Thus, the shaded areas of Figure 8.15 are proportional to the amount of work used for compression. The upper P–v state diagram of this figure illustrates the work required without staging and intercooling, and lower the state diagram illustrates the work required with two stages of compression and gas intercooling between stages. The amount of work saved by staging and intercooling is represented by the lightly shaded area of the lower state diagram.

Because compressing in stages with isobaric cooling between stages saves compression work, we might ask whether there is a maximum amount of work that can be saved. For the purpose of addressing this question, we will consider a two-stage polytropic compressor (Pv^n = constant), which first compresses the gas from P_A to P_i and then from P_i to P_B. At the intermediate pressure, P_i, the gas is cooled to the temperature it had at the beginning of the first compression, T_A. Evaluating the work integral for the two compression processes gives the total compression work as

$$W_{comp} = \frac{nRT_1}{n-1} \left\{ \left[1 - \left(\frac{P_i}{P_A}\right)^{n-1/n} \right] + \left[1 - (\frac{P_B}{P_i})^{n-1/n} \right] \right\}$$

Setting the derivative of this expression with respect to P_i to zero produces

$$\frac{P_i}{P_A} = \frac{P_B}{P_i}$$

as the condition for which the work of compression is a minimum; that is, the total compression work is a minimum when the pressure ratio is the same across each stage of compression. In general, when there are j stages of compression and the gas is isobarically cooled to the initial temperature between each stage of compression, the compressor will require the minimum amount of work when

$$r_{p,j} = r_p^{1/j}$$

where $r_{p,j}$ is the pressure ratio of each stage and r_p is the total compression ratio.

FIGURE 8.15
*Effect of Staged
Compression with
Intercooling*

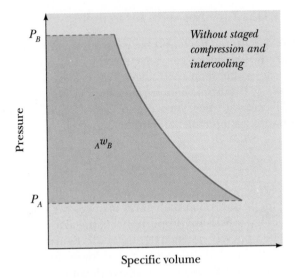

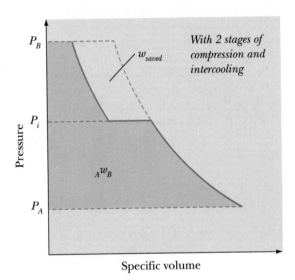

8.6.4   Gas Turbines with Intercooling, Reheating, and Regeneration

Most gas turbines incorporate staged compression with intercooling, staged expansion with reheating, and regeneration to increase the efficiency of the basic Brayton cycle. A two-stage system incorporating these features is shown in Figure 8.16. In this system, the intermediate compressor and turbine pressures are the same, $P_2 = P_3 = P_7 = P_8$; full intercooling is used, $T_1 = T_3$; full reheating is employed, $T_6 = T_8$; and the regenerator operates ideally, $T_5 = T_9$. These are ideal operating conditions, which actual operation tries to achieve.

FIGURE 8.16

*Gas Turbine with
Intercooling,
Reheating, and
Regeneration*

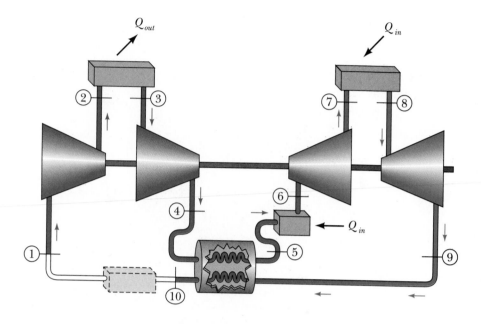

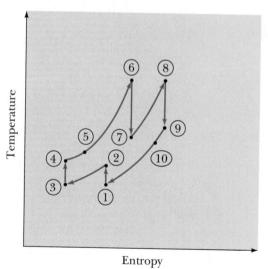

But compressor and turbine irreversibilities, production of heat in the combustion process, and the transfer heat in the intercooler and regenerator limit these conditions.

EXAMPLE 8.10

A three-stage gas-turbine engine with ideal intercooling, reheating, and regeneration operates with air as the working fluid, a total pressure ratio of 8:1 (2:1 per stage), and a maximum temperature of 1100° F. The air enters the compressor at 15 psia, 70° F. Calculate the specific work production and thermal efficiency of this cycle when the regenerator operates perfectly.

System: Air in a gas turbine with intercooling, reheating, and regeneration undergoing the processes as shown in Figure 8.17.

FIGURE 8.17
Three-Stage Gas Turbine for Example 8.10

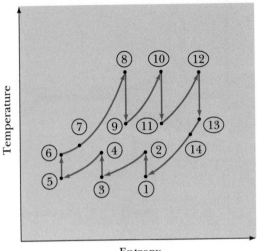

Basic Equations:

$$_1q_2 - {_1}w_2 = h_2 - h_1 + \frac{1}{2}(V_2^2 - V_1^2) + g(z_2 - z_1)$$

$$\frac{T_2}{T_1} = \left(\frac{P_2}{P_1}\right)^{(k-1)/k} = \left(\frac{v_1}{v_2}\right)^{k-1}$$

Conditions: The cycle will be modeled using the air standard analysis model.

Solution: We will number the states in the same manner as in Figure 8.17. We begin by fixing the temperatures of all the states shown in the figure. Because $T_1 = T_3 = T_5$ and each stage of compression has the same pressure ratio, application of the isentropic process equations for an ideal gas with constant specific heats to the compression processes yields

$$\begin{aligned} T_2 = T_4 = T_6 &= T_1 \left(\frac{P_2}{P_1}\right)^{(k-1)/k} \\ &= 530 \times 2^{0.4/1.4} \\ &= 646 \text{ K} \end{aligned}$$

A similar application of the isentropic process relations to the turbine-expansion process yields

$$T_9 = T_{11} = T_{13} = T_8 \left(\frac{P_9}{P_8}\right)^{(k-1)/k} = 1280 \text{ K}$$

Applying the first law to the regenerator, presuming that it operates ideally with $T_7 = T_{13}$, gives $T_6 = T_{14}$. Applying the first law to expansion 8–9 gives

$$
\begin{aligned}
_8w_9 &= h_8 - h_9 \\
&= c_P(T_8 - T_9) \\
&= 67.11 \ \text{Btu/lb}_m
\end{aligned}
$$

Because the temperature drop across the other two stages of turbine expansion is the same as that of 8–9, the total work produced in the turbines is

$$
w_{net,turb} = 3 \times {_8w_9} = 201.3 \ \text{Btu/lb}_m
$$

The work required for compression stage 1–2 is found by applying the first law:

$$
\begin{aligned}
_1w_2 &= h_1 - h_2 \\
&= c_P(T_1 - T_2) \\
&= -27.81 \ \text{Btu/lb}_m
\end{aligned}
$$

Because the temperature change across each stage of compression is the same, the net work required for the total compression is

$$
w_{net,comp} = 3 \times {_1w_2} = -83.42 \ \text{Btu/lb}_m
$$

and the net work produced by this gas turbine is

$$
w_{net} = w_{net,turb} + w_{net,comp} = 117.9 \ \text{Btu/lb}_m
$$

Heat is added in the first combustion chamber (7–8) and the two reheating combustion chambers (9–10 and 11–12). Applying the first law to these three processes and taking advantage of the fact that $T_{10} - T_9 = T_{12} - T_{11}$ gives

$$
\begin{aligned}
q_{net,added} &= c_P[(T_8 - T_7) + 2(T_{10} - T_9)] \\
&= 0.2397[(1560 - 1280) + 2(1560 - 1280)] \\
&= 201.3 \ \text{Btu/lb}_m
\end{aligned}
$$

The thermal efficiency of this gas turbine with intercooling, reheating, and regeneration is then

$$
\eta = \frac{w_{net}}{q_{net,added}} = \frac{117.9}{201.3} = 0.586
$$

EXAMPLE 8.11

A two-stage gas-turbine engine with intercooling, reheating, and regeneration operates with air as the working fluid, a total pressure ratio of 16:1 (4:1 per stage), and a maximum temperature of 600° C. The air enters the compressor at 100 kPa, 10° C. The isentropic efficiency of each compression stage is 90%; for each expansion stage, it is 95%. Calculate the specific work production and thermal efficiency of this cycle when the regenerator operates with a 10° C temperature difference between the two streams.

System: Air in a two-stage gas turbine with intercooling, regeneration, and reheating undergoing the processes as shown in Figure 8.18.

Basic Equations:

$$_1q_2 - {_1}w_2 = h_2 - h_1 + \frac{1}{2}(V_2^2 - V_1^2) + g(z_2 - z_1)$$

$$\frac{T_2}{T_1} = \left(\frac{P_2}{P_1}\right)^{(k-1)/k} = \left(\frac{v_1}{v_2}\right)^{k-1}$$

Conditions: The cycle will be modeled using the air standard analysis model.

Solution: We will number the states in the same manner as in Figure 8.18. We begin by fixing the temperatures of all the states shown in the figure.

The temperature of the ideal state $2s$ (not shown) is given by the isentropic process relationship

$$T_{2s} = T_1 \left(\frac{P_2}{P_1}\right)^{(k-1)/k} = 283(4)^{0.4/1.4} = 421 \text{ K}$$

Applying the first law and the definition of the compressor efficiency to the first stage of compression yields

$$T_2 = T_1 + \frac{T_{2s} - T_1}{\eta_c} = 436 \text{ K}$$

The intercooler cools the air to its original temperature, $T_3 = T_1$. Because air enters both compressors at the same temperature, and the compressors have the same efficiency and compression ratio, the air temperature at the outlet of the compressors is the same, $T_4 = T_2$.

Applying the isentropic process relations across the first turbine stage,

$$T_{7s} = T_6 \left(\frac{P_7}{P_6}\right)^{(k-1)/k} = 873 \left(\frac{1}{4}\right)^{0.4/1.4} = 587 \text{ K}$$

When the first law and the definition of the turbine efficiency are applied to the first turbine stage,

$$T_7 = T_6 - \eta_t(T_6 - T_{7s}) = 601 \text{ K}$$

By the reasoning applied to the second compressor stage, $T_6 = T_8$ and $T_9 = T_7$.

FIGURE 8.18

Two-Stage Gas Turbine for Example 8.11

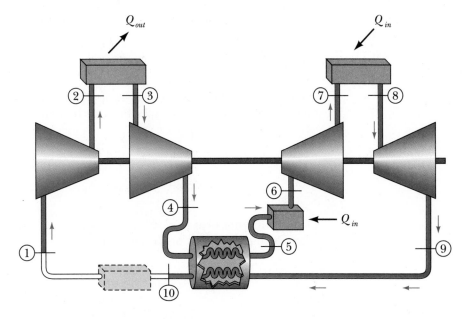

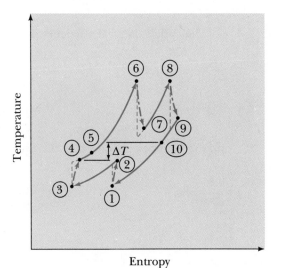

From the problem specification, $T_{10} - T_4 = \Delta T = 10°\,C$. When this specification is combined with the first law as applied to the regenerator,

$$T_5 - T_4 = T_9 - T_{10}$$

The result is

$$T_5 = T_9 - \Delta T = 591 \text{ K}$$

T_{10} is consequently 446 K.

The net heat added to the cycle in the two combustion chambers is

$$q_{in} = {}_5q_6 + {}_7q_8$$

When combined with the first law as applied to these combustion chambers, and presuming that the air's specific heats remain constant, this becomes

$$
\begin{aligned}
q_{in} &= c_P(T_6 - T_5) + c_P(T_8 - T_7) \\
&= 1.0035(873 - 591) + 1.0035(873 - 601) \\
&= 555.9 \text{ kJ/kg}
\end{aligned}
$$

Similarly, the net heat rejected from the cycle in the intercooler and atmospheric exhaust–reentry process model is

$$
\begin{aligned}
q_{out} &= {}_{10}q_1 + {}_2q_3 \\
&= c_P(T_1 - T_{10}) + c_P(T_3 - T_2) \\
&= -317.1 \text{ kJ/kg}
\end{aligned}
$$

The cycle's thermal efficiency is then

$$
\eta = 1 - \left| \frac{q_{out}}{q_{in}} \right|
$$

$$
= 0.430
$$

and the net work produced by this cycle is

$$
\begin{aligned}
w_{net} &= \eta q_{in} \\
&= 238.8 \text{ kJ/kg}
\end{aligned}
$$

8.7 The Ericsson Cycle

Increasing the number of compression–expansion stages in a Brayton cycle with regeneration, reheating, and intercooling improves the thermal efficiency of the cycle as demonstrated in the previous section. What if the number of stages is very large (say, infinity)? What will the thermal efficiency become?

   To answer this question, let us first consider what happens as the number of turbine expansion stages is increased. Two expansion stages are compared to 10 expansion stages in Figure 8.19, the average temperature and overall entropy change are kept the same for both cases. Inspection of this figure reveals that as the number of stages is increased, the series of constant pressure and isentropic processes approaches an isothermal process. In fact, as the number of stages becomes infinite, these processes are replaced by the isothermal process. In the same fashion, increasing the number of compression stages to infinity reduces the compression-intercooling processes to an isothermal process. Hence, the constant-pressure heat-addition process becomes an isothermal heat-addition process, and the constant-pressure heat-rejection process becomes an isothermal heat-rejection process in this limit.

FIGURE 8.19
Comparison of Turbine-Expansion Stages

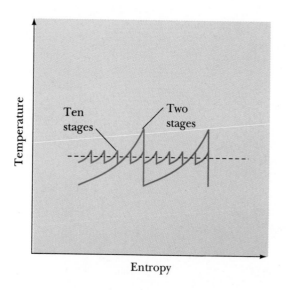

In this limit, both gas streams in the regenerator experience the same temperature change when the gas's specific heats remain constant and the regenerator operates ideally. The heat released by the gas as it passes from the last turbine stage to the first compression stage exactly equals the heat taken on by the gas as it passes from the last compression stage to the first turbine stage. Hence, no heat leaves the system as the gas passes through the regenerator. Heat is only internally transferred from one process to another in the regenerator.

The limiting cycle then consists of the following processes:

1–2   an isothermal process with heat rejection to the surroundings
2–3   an isobaric warming of the gas by internal heat addition
3–4   an isothermal heat addition from an external energy reservoir
4–1   an isobaric cooling of the gas by internal heat transfer

These are illustrated in Figure 8.20. This cycle is known as the *Ericsson cycle*[5].

Based on the definition of the entropy, the thermal efficiency of the Ericsson cycle is

$$\eta = 1 - \frac{|q_l|}{q_h} = 1 - \frac{T_l(s_1 - s_2)}{T_h(s_4 - s_3)}$$

when the states are numbered as in Figure 8.20. Applying the Gibbs equation, $T\,ds = dh - v\,dP$, to process 2–3 yields

$$s_3 - s_2 = c_P \ln \frac{T_3}{T_2} - R \ln \frac{P_3}{P_2}$$

Similarly,

$$s_4 - s_1 = c_P \ln \frac{T_4}{T_1} - R \ln \frac{P_4}{P_1}$$

FIGURE 8.20
Ericsson Cycle

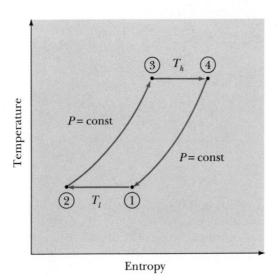

Because $T_1 = T_2$, $T_3 = T_4$, $P_2 = P_3$, and $P_1 = P_4$, these two results are equivalent and can be reduced to

$$s_1 - s_2 = s_4 - s_3$$

---

[5]Named for John Ericsson (1803–1889), a Swedish-American naval engineer who built a "caloric" engine for ship propulsion. Although this engine was technically successful, it was so large that it took up most of the ship's cargo hold. Unfortunately, the prototype ship, the *Ericsson*, sank before conclusive tests could be conducted. Ericsson also designed the *Monitor*, the first iron-clad turreted warship, which was used by the Union navy during the Civil War.

With this result, the thermal efficiency of the Ericsson cycle becomes

$$\eta = 1 - \frac{T_1}{T_h}$$

This is identical to the thermal efficiency of a completely reversible cycle that uses the same T_h and T_l thermal-energy reservoirs. Hence, the Ericsson cycle is a completely reversible cycle with a thermal efficiency as great as possible; it matches the efficiency of the Carnot cycle within the constraints of the thermal-energy reservoir temperatures.

EXAMPLE 8.12   Heat is supplied to an Ericsson cycle at a rate of 1,000,000 Btu/hr from a high-temperature thermal-energy reservoir at 640° F. Air begins this cycle at 14.7 psia, 40° F, and the pressure ratio is 12:1. Determine the power produced by this cycle, the rate of heat rejection from this cycle, and the rate of heat exchange between the two air streams in the regenerator of this cycle.

System: Air in a Ericsson cycle undergoing the processes as shown in Figure 8.20.

Basic Equations:
$$\eta_{cr} = 1 - \frac{T_l}{T_h}$$

$$_1 q_2 = \int_1^2 T \, ds$$

$$_1 q_2 - {_1}w_2 = h_2 - h_1 + \frac{1}{2}(V_2^2 - V_1^2) + g(z_2 - z_1)$$

$$\oint \delta q = \oint \delta w$$

Conditions: The cycle will be modeled using air standard analysis assumptions.

Solution: The thermal efficiency of this cycle is

$$\eta = 1 - \frac{T_l}{T_h} = 1 - \frac{500}{1100} = 0.5455$$

The power produced by this cycle is then

$$\dot{W}_{net} = \eta \dot{Q}_{in} = 0.5455 \times 1,000,000 = 545,500 \text{ Btu/hr} = 214.3 \text{ hp}$$

Applying the cyclic form of the first law to this cycle yields

$$\dot{Q}_{rej} = \dot{Q}_{in} - \dot{W}_{net} = 1,000,000 - 545,500 = 454,500 \text{ Btu/hr}$$

Applying the Gibbs equation to process 3–4 gives

$$s_4 - s_3 = -R\ln\frac{P_4}{P_3}$$

$$= -\frac{53.35}{778}\ln\frac{1}{12}$$

$$= 0.1704 \text{ Btu/lb}_m\text{-}^\circ\text{R}$$

Because process 3–4 is reversible and isothermal, its specific heat transfer is

$$_3q_4 = T_3(s_4 - s_3) = 1100 \times 0.1704 = 187.4 \text{ Btu/lb}_m$$

To accommodate the specified heat-input rate, air must be circulated around the cycle at a mass flow rate of

$$\dot{m} = \frac{\dot{Q}_{in}}{_3q_4} = 5336 \text{ lb}_m/\text{hr}$$

The rate of heat exchange between the two air streams in the regenerator can now be found by applying the steady-flow form of the first law to one of these streams:

$$\dot{Q}_{exc} = \dot{m}(h_3 - h_2)$$
$$= \dot{m}c_P(T_3 - T_2)$$
$$= 5336 \times 0.2397(640 - 40)$$
$$= 767,000 \text{ Btu/hr}$$

8.8 Aircraft-Propulsion Cycles

In a simple Brayton cycle, the work that remains once the compressor work requirements have been met is delivered to the cycle's surroundings to drive some external device. The excess enthalpy remaining after the compressor's needs are met also could be converted into other useful effects such as a thrust force to propel an aircraft. The excess enthalpy is illustrated in Figure 8.21. For ideal gases with constant specific heats, temperature differences represent enthalpy differences. Vertical distances on Figure 8.21 are then proportional to enthalpy differences. Neglecting kinetic-energy and potential-energy changes, vertical distance 1–2 equals vertical distance 3–4, because the work produced by this portion of the turbine expansion (equal to the enthalpy change, according to the first law) equals the work required by the compressor. Vertical distance 4–5 is then proportional to the excess enthalpy difference that may be used outside of the cycle for other purposes.

FIGURE 8.21
Simple Brayton Cycle

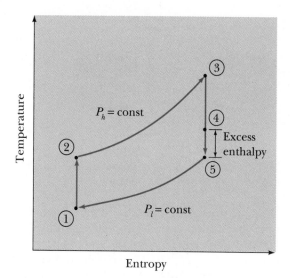

According to Newton's laws of motion, a thrust force can be generated by changing the momentum of a stream. When these laws are applied to a uniform stream with a single inlet and a single outlet, the momentum thrust force is given by

$$F_t = \dot{m}_t (V_e - V_i)$$

where $\dot{m}_t$ is the mass flow rate of the stream, V_e is the exit velocity, and V_i is the inlet velocity. Hence, to generate a momentum thrust force, we need to use the excess enthalpy of the simple Brayton cycle to change the stream's velocity (i.e., kinetic energy). Some of the excess enthalpy also can cause the pressure of the exiting stream to be larger than that of the entering stream. This pressure difference will also contribute to the force that propels the aircraft.

Simple Brayton cycles that consist of a compressor, combustion chamber(s), and a turbine such as those in Figure 8.22 are used to generate the excess enthalpy. More complicated Brayton cycles employing regenerators, reheaters, and intercoolers are normally not used because they require considerable volume and add weight to the aircraft.

Three different devices that convert the excess enthalpy difference into a thrust force are shown in Figure 8.22. The first engine converts the excess enthalpy difference into work, which is then used to turn a propeller. The simple Brayton cycle used to drive the propeller requires a mass flow rate of $\dot{m}_e$, while a mass flow rate of $\dot{m}_t$ passes through the propeller as its velocity is increased somewhat to change the momentum of the stream. The momentum thrust force produced by the turboprop engine is given by

$$F_t = \dot{m}_t (V_e - V_i)$$

FIGURE 8.22

*Aircraft Propulsion
Engines*

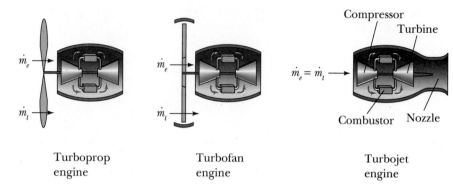

Turboprop
engine

Turbofan
engine

Turbojet
engine

where V_e is the velocity at which the stream leaves the propeller, and V_i is the velocity at which it enters the propeller. Both velocities are measured with regard to an observer attached to the engine. The second engine is essentially the same as the first except that it uses a fan rather than a propeller to generate the momentum thrust force. The fan is more compact than a propeller, and it uses blading similar to that in a compressor to create a larger velocity change while processing a smaller mass flow rate. The proceeding equation for the momentum thrust force also can be used for the turbofan engine. The third device is the pure jet engine, for which the mass flow rate needed to supply the engine is also the mass flow rate generating the thrust force. As this engine uses the smallest mass flow rate for the generation of thrust, it also must produce the largest change in stream velocity to generate an equivalent momentum thrust force. A nozzle is placed after the turbine to accelerate the air in the stream even more. Applying the first law to this nozzle using the state-numbering scheme of Figure 8.21 gives

$$\frac{V_5^2 - V_4^2}{2} = h_4 - h_5 = c_P(T_4 - T_5)$$

while the momentum thrust force is given by

$$F_t = \dot{m}_t(V_5 - V_4)$$

when it is presumed that the velocity of the air stream entering the nozzle is the same as that of the air stream entering the compressor.

Much of the oxygen in the air passing through turbojet engines is not burned in the combustor. Additional fuel may then be injected and burned at the entry to the nozzle, thereby increasing the temperature at the nozzle entry. Expanding this higher-temperature (i.e., enthalpy) gas through the nozzle results in a higher exit velocity, which increases the momentum thrust force. This injection and burning of extra fuel in the nozzle is known as *afterburning*. It is frequently used in high-speed, high-performance aircraft where brief power maneuvers are required and fuel economy is not a major consideration.

The total mass flow rate processed by an aircraft propulsion unit consists of the thrust-generating mass flow rate, $\dot{m}_t$, and the mass flow rate needed to supply the Brayton cycle, $\dot{m}_e$. The ratio of the total mass flow rate to the mass flow rate passing through the Brayton cycle,

$$r_m = \frac{\dot{m}_t + \dot{m}_e}{\dot{m}_e}$$

can be used to characterize aircraft propulsion systems. For a pure jet engine, this ratio is 1:1, because the thrust-generating mass flow rate and engine-service mass flow rate are one and the same. This ratio is very large for turboprop engines because most of the total mass flow rate passes through the propeller, and very little passes through the Brayton cycle. In turbofan engines, this ratio takes on intermediate values (i.e., 2-10:1 is typical).

<table>
<tr><td>

EXAMPLE 8.13

</td><td>

An aircraft with a turbofan propulsion system is flying at a speed of 200 m/s at an altitude where pressure is 60 kPa and temperature is $0°$ C. The compressor has a pressure ratio of 8:1, the maximum Brayton cycle temperature is $400°$ C, the area of the opening servicing the fan and engine is 3 m², and the mass flow rate ratio (r_m) is 4:1. Determine the thrust force generated by this propulsion unit.

</td></tr>
</table>

System: Air in a gas turbine undergoing the processes as shown in Figure 8.21.

Basic Equations:

$$_1q_2 - {}_1w_2 = h_2 - h_1 + \frac{1}{2}(V_2^2 - V_1^2) + g(z_2 - z_1)$$

$$F_t = \dot{m}_t(V_2 - V_1)$$

$$\frac{T_2}{T_1} = \left(\frac{P_2}{P_1}\right)^{(k-1)/k} = \left(\frac{v_1}{v_2}\right)^{k-1}$$

Conditions: The Brayton cycle will be modeled using air standard analysis.

Solution: For the specified inlet conditions, the air's specific volume is

$$v = \frac{RT}{P} = 1.306 \text{ m}^3/\text{kg}$$

Air passes through the opening serving the fan and compressor at the speed of the aircraft. The total mass flow rate is then

$$\dot{m} = \frac{AV}{v} = 459.4 \text{ kg/s}$$

of which

$$\dot{m}_e = \frac{\dot{m}}{r_m} = 114.9 \text{ kg/s}$$

passes through the Brayton cycle and

$$\dot{m}_t = \dot{m} - \dot{m}_e = 344.5 \text{ kg/s}$$

passes through the fan.

Following the state-numbering scheme of Figure 8.21 for the components of the Brayton cycle, the temperature at the end of the compression is

$$T_2 = T_1 \left(\frac{P_2}{P_1} \right)^{(k-1)/k} = 495 \text{ K}$$

Similarly, the temperature at which the air leaves the engine is

$$T_5 = T_3 \left(\frac{P_5}{P_3} \right)^{(k-1)/k} = 371 \text{ K}$$

The temperature drop between states 3 and 4 is the same as that between states 1 and 2. Hence,

$$T_4 = T_3 - (T_2 - T_1) = 451 \text{ K}$$

and the excess enthalpy is

$$h_4 - h_5 = c_P(T_4 - T_5) = 80.28 \text{ kJ/kg}$$

This excess enthalpy is converted into a kinetic-energy change for the fan's thrust-producing stream. Equating the rate of kinetic-energy production to the rate of excess enthalpy production (i.e., applying the first law) gives

$$\dot{m}_t \left(\frac{V_e^2 - V_i^2}{2} \right) = \dot{m}_e(h_4 - h_5)$$

When solved for the velocity at which the air leaves the fan, V_e, this yields

$$V_e = \left(2\frac{\dot{m}_e}{\dot{m}_t}(h_4 - h_5) + V_i^2 \right)^{1/2}$$

$$= \left(2\frac{114.9}{344.5}263.9 \times 1000 + 200^2 \right)^{1/2}$$

$$= 306 \text{ m/s}$$

The momentum thrust force produced by the fan is then

$$F_t = \dot{m}_t(V_e - V_i)$$

$$= \frac{344.5 \times 106.0}{1000}$$

$$= 36.52 \text{ kN}$$

8.9 Combined Cycles

Additional improvements in thermal efficiency and performance may be realized by coupling two or more of the simpler cycles previously presented. There are also unique applications, such as supplying heat for industrial processes and climate control, in which two or more coupled cycles offer significant advantages. Systems of this type are known as *combined cycles*. Several different combined cycles are possible, but we will focus our attention on the combined cycle formed by coupling the Brayton and Rankine cycles.

Most combined cycles are coupled in a cascaded arrangement with the heat being rejected from one cycle serving as the heat supply for another cycle. The hot exhaust gases from a gas turbine can be used in the boiler of a steam-power cycle to produce steam for expansion in a turbine or for distribution to heat loads. The heat rejected by the gas turbine is now the heat supply for the steam cycle.

When the steam or turbine exhaust leaving the boiler is used for heating purposes, and the work produced by the two cycles is used for electrical generation, this is called a *cogeneration system*. Several cogeneration systems have been installed recently around the world. In the United States, such installations have resulted from the economic advantages of these systems and recent legislation that has given them a market for the electricity they produce. Cogeneration systems that produce space heating have the additional advantage that the heating and electrical loads do not occur simultaneously. Electricity is produced during the summer when the demand for electricity to operate air-conditioning systems is high. Comfort heat is produced during the winter when the need is the greatest. Cogeneration systems then have a high utilization when compared to pure heating or pure electrical systems, which are only partially used during certain seasons of the year.

The maximum cycle temperature for a modern steam-power plant is approximately 620° C (1150° F), while the maximum cycle temperature of a modern gas turbine is some 1150° C (2100° F). Advances in the materials used in gas turbines, including ceramics, and techniques for cooling the internal components, including the blades, have led to the high gas-turbine cycle temperatures. Gas turbines then have a greater potential for higher thermal efficiencies. This potential is only partially realized because of the high temperature at which the gases are exhausted to the environment. The average temperature of heat rejection is then quite high. The Rankine steam cycle offers the advantage of rejecting heat at lower temperatures while simultaneously holding the working fluid temperature almost constant.

Quite logically, then, we might combine these two cycles to take advantage of their respective strengths to increase the overall efficiency of the total system. A typical combined gas–steam power cycle is shown in Figure 8.23, which also illustrates the *T–s* state diagram of the two cycles. Inspection of the cycle schematic reveals that the two cycles are cascade-coupled through the heat exchanger where the gas cycle releases heat to produce steam for the simple

FIGURE 8.23
A Basic Combined Gas–Steam Power Cycle

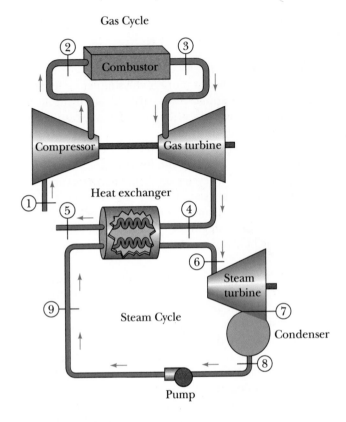

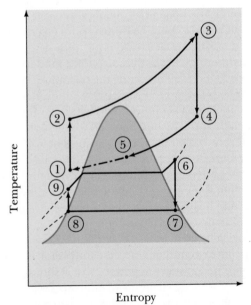

Rankine cycle. The gas cycle is known as a *topping cycle* because it operates at the higher temperatures. The steam cycle serves as the *bottoming cycle* in this case. Feedwater heaters, reheating, and regenerators may be added to this basic cycle to improve cycle performance even more.

Large, 1000-MW-capacity, combined gas–steam power plants typically have thermal efficiencies of 40% to 45%. In comparison, a modern steam-power plant of this capacity would have an efficiency of 35% to 40%. The additional initial cost of the combined cycle equipment can in many cases be recovered from the reduced fuel costs, which are the result of the higher efficiency.

EXAMPLE 8.14

A combined gas–steam power plant is arranged as shown in Figure 8.23. Air enters the gas topping cycle at 14.7 psia, 80° F. The compressor has a pressure ratio of 8:1, and the maximum topping cycle temperature is 2000° F. Turbine exhaust gases leave the heat exchanger at 500° F. The bottoming steam cycle operates the steam side of the heat exchanger at 900 psia, the condenser at 10 psia, and the temperature at the steam turbine inlet at 1100° F. Presuming that all components operate ideally, calculate the following on the basis of a unit of air mass entering the air compressor:

    a. mass of steam circulating through the steam cycle
    b. heat added in the topping cycle combustor
    c. heat rejected in the bottoming cycle condenser
    d. overall thermal efficiency

System: Two systems are necessary. The first is the gas passing through the gas cycle; the second is the water passing through the steam cycle. The state-numbering scheme of Figure 8.23 will be used.

Basic Equations:

$$\sum_{in} \dot{m} = \sum_{out} \dot{m}$$

$$_1q_2 - {_1}w_2 = h_2 - h_1 + \frac{1}{2}(V_2^2 - V_1^2) + g(z_2 - z_1)$$

$$\frac{T_2}{T_1} = \left(\frac{P_2}{P_1}\right)^{(k-1)/k} = \left(\frac{v_1}{v_2}\right)^{k-1}$$

Conditions: All components are operating ideally without any losses. The gas cycle will be modeled with the air standard model.

Solution: We begin by fixing states 1–4 of the gas cycle. Applying the ideal gas isentropic relations to process 1–2 yields

$$T_2 = T_1 \left(\frac{P_2}{P_1}\right)^{(k-1)/k} = 540(8)^{0.4/1.4} = 978° \text{R}$$

According to the pressure ratio specification, the pressure at states 2 and 3 is

$$P_2 = P_3 = 14.7 \times 8 = 117.6 \text{ psia}$$

Applying the ideal gas isentropic relation to process 3–4 while noting that $P_4 = P_5 = P_1 = 14.7$ psia gives

$$T_4 = 2460 \left(\frac{14.7}{117.6} \right)^{0.4/1.4} = 1360° \text{ R}$$

The enthalpies of states 6–9 of the steam cycle are next determined. According to the turbine-inlet specifications, $h_6 = 1565.4$ Btu/lb$_m$. Because process 6–7 is an isentropic process, $x_7 = 0.944$ and $h_7 = 1088.4$ Btu/lb$_m$. State 8 is a saturated liquid with an enthalpy of $h_8 = 161.2$ Btu/lb$_m$. Applying the work integral and the first law to the pump yields

$$h_9 = h_8 + v_8(P_9 - P_8)$$
$$= 161.2 + 0.01659\frac{144}{778}(900 - 10)$$
$$= 163.9 \text{ Btu/lb}_m$$

Setting the mass flow rate of the gas in the gas cycle to unity and applying the first law to the heat exchanger produces

$$\dot{m}_s = \frac{h_4 - h_5}{h_6 - h_9}$$
$$= \frac{c_P(T_4 - T_5)}{h_6 - h_9}$$
$$= \frac{0.2397(1360 - 500)}{1565.4 - 163.9}$$
$$= 0.147 \text{ lb}_m\text{-steam/lb}_m\text{-gas}$$

Applying the first law to the gas-cycle combustor gives the net heat added to the combined cycle as

$$_2q_3 = h_3 - h_2$$
$$= c_P(T_3 - T_2)$$
$$= 0.2397(2460 - 978)$$
$$= 355.2 \text{ Btu/lb}_m\text{-gas}$$

The heat rejected by the Rankine cycle may be found by applying the first law to the condenser:

$$_7q_8 = \dot{m}_s(h_8 - h_7)$$
$$= 0.147(161.2 - 1088.4)$$
$$= -136.3 \text{ Btu/lb}_m\text{-gas}$$

Similarly, the heat rejected by the Brayton topping cycle is

$$_5q_1 = c_p(T_1 - T_5)$$
$$= -100.7 \text{ Btu/lb}_m\text{-gas}$$

TIP:

By setting the mass flow rate of the gas through the gas cycle to unity, all calculations share the same basis of one unit of this flow rate.

The overall combined cycle efficiency is then

$$\eta = 1 - \frac{q_{out}}{q_{in}}$$

$$= 1 - \frac{136.3 + 100.7}{355.2}$$

$$= 0.333$$

8.10 Gas Refrigeration Systems

The use of gases as the working fluid in a refrigeration system is not very common because of the high cost of compressing the gas and low cycle COPs. But certain industrial applications, such as reducing the air temperature before liquefying its various gaseous components, do use gases to create a cooling effect. Gaseous refrigeration systems are also used on aircraft to provide cabin-comfort control as a weight-saving measure. In this section, we will examine two of the more common gas-refrigeration systems. As in previous sections, air standard analysis that assumes ideal gases with constant specific heats will be used unless stated otherwise. Kinetic-energy and potential-energy changes also will be neglected.

8.10.1 The Reversed Brayton Cycle

The first system we will consider is the simple Brayton cycle operated in the reverse sense, as illustrated in Figure 8.24. The system now consists of the turbine, compressor, a high-pressure heat exchanger, and a low-pressure heat exchanger. The air or working gas is first compressed to state 2, is then cooled in the high-pressure heat exchanger to state 3, is next expanded to the lower pressure of state 4, and then is finally reheated in a constant-pressure heat exchanger back to its original temperature and pressure of state 1. The heat transferred to the air in the low-pressure heat exchanger provides the de-sired cooling effect. In some systems, the cold air at the high-pressure heat

FIGURE 8.24
*Reversed Brayton
Cycle*

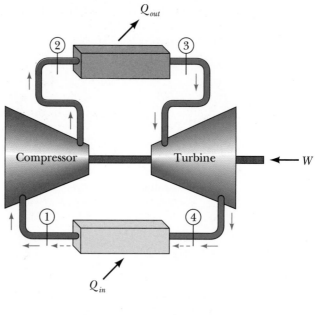

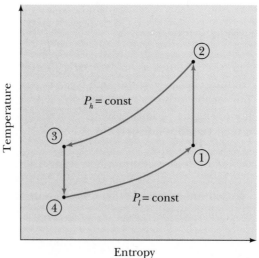

exchanger outlet is delivered directly to the space being cooled after it has
been throttled. In this space, the air temperature increases until the air once
again reenters the compressor. In these systems, the constant-pressure process
models the process occurring in the space being cooled just as it does when a
heat exchanger is used in the system.

Reversed Brayton cycles are analyzed in the same manner as other Brayton cycles by proper application of the first and second laws, definitions, and
properties of the working fluid. The cooling effect is found by applying the

first law to the low-pressure heat exchanger:

$$q_{in} = c_P(T_1 - T_4)$$

Similarly, the heat rejected in the high pressure heat exchanger is given by

$$|q_{out}| = c_P(T_2 - T_3)$$

Because the compression and expansion are isentropic,

$$\frac{T_3}{T_4} = \left(\frac{P_h}{P_l}\right)^{(k-1)/k} = r_P^{(k-1)/k} = \frac{T_2}{T_1}$$

where r_P is the pressure ratio produced by the compressor. From this result, it follows that

$$\frac{T_3}{T_2} = \frac{T_4}{T_1}$$

Applying the definition of the refrigerator coefficient of performance,

$$\text{COP}_r = \frac{Q_{in}}{|Q_{out}| - Q_{in}}$$

$$= \frac{1}{\frac{T_2 - T_3}{T_1 - T_4} - 1}$$

$$= \frac{1}{\frac{T_2}{T_1}\frac{1 - T_3/T_2}{1 - T_4/T_1} - 1}$$

$$= \frac{1}{r_p^{(k-1)/k} - 1}$$

EXAMPLE 8.15

A reversed Brayton cycle cooling unit compresses air at 90 kPa, 20° C to 450 kPa. Following this compression, this air is cooled to 20° C before it enters the turbine for expansion back to 90 kPa. Calculate the refrigerator COP of this cycle and the mass flow rate of air in this system required to service a 20-kW cooling load.

System: Air in a reversed Brayton cycle undergoing the processes as shown in Figure 8.24.

Basic Equations:

$$_1q_2 - _1w_2 = h_2 - h_1 + \frac{1}{2}(V_2^2 - V_1^2) + g(z_2 - z_1)$$

$$\frac{T_2}{T_1} = \left(\frac{P_2}{P_1}\right)^{(k-1)/k} = \left(\frac{v_1}{v_2}\right)^{k-1}$$

Conditions: The cycle will be modeled using the air standard model.

Solution: Because the pressure ratio for this cycle is 5:1,

$$\mathrm{COP}_r = \frac{1}{r_P^{(k-1)/k} - 1}$$

$$= \frac{1}{5^{0.4/1.4} - 1}$$

$$= 1.71$$

Following the state-numbering scheme of Figure 8.24,

$$T_2 = T_1 \left(\frac{P_2}{P_1}\right)^{(k-1)/k}$$

$$= 293 \times 5^{0.4/1.4}$$

$$= 464 \text{ K}$$

and

$$T_4 = T_3 \left(\frac{P_4}{P_3}\right)^{(k-1)/k}$$

$$= 293 \times 0.2^{0.4/1.4}$$

$$= 185 \text{ K}$$

Applying the first law to the cooling process 4–1,

$$_4q_1 = c_P(T_1 - T_4)$$
$$= 1.0035(293 - 185)$$
$$= 108.4 \text{ kJ/kg}$$

Because $_4\dot{Q}_1 = \dot{m}\,_4q_1$,

$$\dot{m} = \frac{_4\dot{Q}_1}{_4q_1}$$
$$= 0.185 \text{ kg/s}$$

8.10.2 Bootstrap Systems

Bootstrap systems add a second stage of compression and heat rejection to the reversed Brayton cycle as shown in Figure 8.25. This reduces the work that must be put into the system just as it did when we added staged compression and intercooling to the simple Brayton cycle. Because this reduces the work required by the system, the coefficient of performance will be improved. Accompanying this improvement in the COP is the cost of the extra compressor. A compromise must be reached between these opposing effects if the bootstrap system is to be optimized.

FIGURE 8.25
*Bootstrap
Gas-Refrigeration
System*

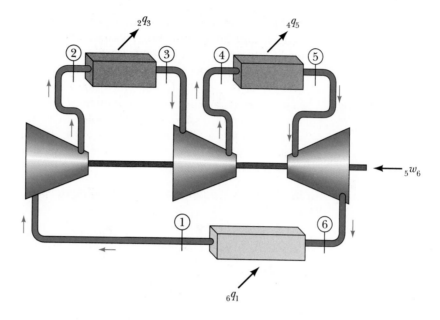

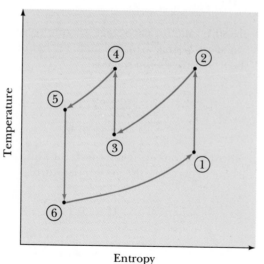

The compression stages of this system operate in the same manner as those of a gas turbine with intercooling as presented in the section "Gas Turbines with Intercooling" (p. 397). For the total compression work to be minimal, the pressure ratio must be the same across each stage of compression when the system operation is ideal.

Also, each stage of compression in an ideal system is an isentropic compression, and the gaseous working fluid is cooled back to its original temperature before undergoing the next stage of compression. In practice, each compression stage is subject to a compression efficiency, the turbine expansion is subject to a turbine-expansion efficiency, there will be some pressure losses in the heat exchangers and interconnecting passages, and the heat exchangers may not perfectly cool or heat the gaseous working fluid. All of these effects can be accounted for by proper application of the first and second laws.

EXAMPLE 8.16

A two-stage bootstrap cooling unit compresses air at 90 kPa, 20° C to 203 kPa. Following this compression, this air is cooled to 20° C before it is further compressed to 459 kPa and then recooled to 20° C. It then enters the turbine for expansion back to 90 kPa. Calculate the refrigerator COP of this cycle and compare it to that of Example 8.15.

System: Air in a bootstrap cooling system undergoing the processes as shown in Figure 8.25.

Basic Equations:

$$_1q_2 - _1w_2 = h_2 - h_1 + \frac{1}{2}(V_2^2 - V_1^2) + g(z_2 - z_1)$$

$$\frac{T_2}{T_1} = \left(\frac{P_2}{P_1}\right)^{(k-1)/k} = \left(\frac{v_1}{v_2}\right)^{k-1}$$

Conditions: The air standard model will be used.

Solution: The pressure ratio for each stage of compression is 2.26:1. Because each compression stage begins with the same air temperature, the temperature at the end of each compression stage is

$$T_2 = T_4 = T_1 \left(\frac{P_2}{P_1}\right)^{(k-1)/k}$$
$$= 370 \text{ K}$$

where the state-numbering scheme is the same as that of Figure 8.25. Following the turbine expansion, the air temperature is

$$T_6 = T_5 \left(\frac{P_6}{P_5}\right)^{(k-1)/k}$$
$$= 293 \times 0.196^{0.4/1.4}$$
$$= 185 \text{ K}$$

Applying the first law to the intercoolers gives the heat rejected from this cycle as

$$q_{out} = c_P(T_5 - T_4) + c_P(T_3 - T_2)$$
$$= 2 \times 1.0035(293 - 370)$$
$$= -154.5 \text{ kJ/kg}$$

The heat added to the cycle is

$$q_{in} = c_P(T_1 - T_6)$$
$$= 108.3 \text{ kJ/kg}$$

and the net work added is

$$w_{in} = q_{in} + q_{out} = -46.2 \text{ kJ/kg}$$

according to the first law. This cycle's coefficient of performance is then

$$COP_r = \left| \frac{q_{in}}{w_{in}} \right|$$
$$= 2.34$$

Notice that this cycle accomplishes the same cooling effect using the same overall pressure ratio and air-inlet conditions as the cycle of Example 8.15. However, the coefficient of performance is 44% higher than that of Example 8.15. The performance advantage of staging the overall compression is quite obvious.

PROBLEMS

Sketch the cycle schematic diagram with clearly labeled states for the following problems. Also sketch the appropriate T–s or P–v state diagrams (or both), showing all states and processes.

8.1 An ideal gas Carnot cycle uses air as the working fluid, accepts heat from an energy reservoir at 1027° C, is repeated 1500 times per minute, and has a compression ratio of 12:1. Determine the maximum temperature (K) of the low-temperature energy reservoir, the cycle's thermal efficiency, and the amount of heat (kJ) that must be supplied each time the cycle is repeated if this device is to produce 500 kW of power.

8.2 The thermal-energy reservoirs of an ideal gas Carnot cycle are at 1240° F and 40° F, and the device executing this cycle rejects 100 Btu of heat each time the cycle is executed. Determine the total heat supplied to and the total work produced by this cycle (both in Btu) each time it is executed.

8.3 An ideal gas Carnot cycle uses helium as the working fluid and rejects heat to a lake, which is at 15° C. Determine the pressure ratio, compression ratio, and minimum temperature (K) of the energy-source reservoir for this cycle to have a thermal efficiency of 50%.

8.4 Repeat the previous problem when the lake is at 60° F and the Carnot cycle's thermal efficiency is to be 60%.

8.5 Can any ideal gas power cycle have a thermal efficiency greater than 55% when using thermal-energy reservoirs at 627° C and 17° C?

8.6 What is the maximum possible thermal efficiency of a gas power cycle when using thermal-energy reservoirs at 940° F and 40° F?

8.7 An Otto cycle has a compression ratio of 12:1, takes in air at 100 kPa, 20° C, and is repeated 1000 times per minute. Determine the thermal efficiency of this cycle and the rate of heat input (kW) if the cycle is to produce 200 kW of power.

8.8 If all of the conditions of the preceding problem remain the same except the compression ratio, which is changed to 10:1, how will the thermal efficiency and rate of heat input change?

8.9 A spark-ignition engine has a compression ratio of 8:1, an isentropic compression efficiency of 85%, and an isentropic expansion efficiency of 95%. At the beginning of the compression, the air in the piston cylinder is at 13 psia and 60° F. The maximum gas temperature is found to be 2300° F by measurement. Determine the thermal efficiency, specific heat addition (Btu/lb_m), and mean effective pressure (psia) of this engine when modeled with the Otto cycle.

8.10 Determine the mean effective pressure (psia) of an Otto cycle that uses air as the working fluid; its state at the beginning of the compression is 14 psia and 60° F, its temperature at the end of the combustion is 1500° F, and its compression ratio is 9:1.

8.11 Determine the rate of heat addition and rejection (Btu/hr) for the Otto cycle of the previous problem when it produces 140 hp and the cycle is repeated 1400 times per minute.

8.12 A typical hydrocarbon fuel produces 42,000 kJ/kg of heat when used in a spark-ignition engine. Determine the compression ratio required for an Otto cycle to use 0.013 grams of fuel to produce 1 kJ of work.

8.13 When we double the compression ratio of an Otto cycle, what happens to the maximum gas temperature and pressure when the state of the air at the beginning of the compression and the amount of heat addition remain the same?

8.14 In a spark-ignition engine, some cooling occurs as the gas is expanded. This may be modeled by using a polytropic process in lieu of the isentropic process. Determine if the polytropic exponent used in this model will be greater than or less than the isentropic exponent.

8.15 A six-cylinder, four-stroke, spark-ignition engine takes in air at 14 psia, 65° F, and is limited to a maximum cycle temperature of 1600° F. Each cylinder has a bore of 3.5 inches, and each piston has a stroke of 3.9 inches. The minimum enclosed volume is 14% of the maximum enclosed volume. How much power (hp) will this engine produce when operated at 2500 rpm?

8.16 A six-cylinder, four-liter, four-stroke, spark-ignition engine takes in air at 90 kPa, 20° C. The minimum enclosed volume is 15% of the maximum enclosed volume. When operated at 2500 rpm, this engine produces 90 hp. Determine the rate of heat addition (Btu/hr) to this engine.

8.17 An Otto cycle has a compression ratio of 7:1. At the beginning of the compression process, $P_1 = 90$ kPa, $T_1 = 27°$ C, and $V_1 = 0.004$ m^3. The maximum cycle temperature is 1127° C. For each repetition of the cycle, calculate the heat rejection (kJ) and the net work production (kJ). Also calculate the thermal efficiency and mean effective pressure (kPa) for this cycle.

8.18 Someone has suggested that the air standard Otto cycle is more accurate if the two isentropic processes are replaced with polytropic processes with $n = 1.3$. Consider such a cycle when the compression ratio is 8:1, $P_1 = 14.3$ psia, $T_1 = 65°$ F, and the maximum cycle temperature is 2000° F. Determine the specific heat (Btu/lb$_m$) added to and rejected from this cycle, as well as the cycle's thermal efficiency.

8.19 How do the results of the previous problem change when isentropic processes are used in place of the polytropic processes?

8.20 Which of the processes of the previous problem loses the greatest amount of work potential? The energy-supply reservoir's temperature is the same as the maximum cycle temperature, and the energy-sink reservoir's temperature is the same as the minimum cycle temperature.

8.21 An air standard limited-pressure cycle has a compression ratio of 14:1, an α of 1.5, and a β of 1.2. Determine the thermal efficiency, amount of heat added (kJ/kg), the maximum gas pressure (kPa), and the maximum gas temperature (° C) when this cycle is operated at 80 kPa and 20° C at the beginning of the compression.

8.22 Determine the amount of heat added (kJ/kg) and the maximum gas pressure (kPa) and temperature (° C) when the state of the air at the beginning of the compression of the preceding problem is 80 kPa and –20° C.

8.23 An air standard limited-pressure cycle has a compression ratio of 20:1, an α of 1.2, and a β of 1.3. Determine the thermal efficiency, amount of heat added (Btu/lb$_m$), and the maximum gas pressure (psia) and temperature ($^\circ$F) when this cycle is operated at 14 psia and 70°F at the beginning of the compression.

8.24 Calculate the specific potential work that is lost each time the cycle of the previous problem is repeated. The cycle's surroundings are at 14.7 psia, 77°F. The energy-source reservoir temperature is the same as the maximum cycle temperature, and the energy-sink reservoir temperature is the same as the minimum cycle temperature.

8.25 If the compression ratio of the previous problem were reduced to 12:1, how would the thermal efficiency, amount of heat added, and the maximum gas pressure and temperature change?

8.26 The α and β parameters determine the amount of heat added to the limited-pressure cycle. Develop an equation for $q_{in}/(c_v T_1 r_v^{(k-1)})$ in terms of k, α, and β.

8.27 An air standard, limited-pressure cycle has a compression ratio of 18:1, an α of 1.1, and a β of 1.1. At the beginning of the compression, $P_1 = 90$ kPa, $T_1 = 18^\circ$C, and $V_1 = 0.003$ m^3. How much power will this cycle produce when it is repeated 4000 times per minute?

8.28 The compression efficiency of the previous problem is 85%, and the expansion efficiency is 90%. How much does the power produced by this cycle change with these process efficiencies?

8.29 Calculate the specific potential work that is lost by each process of the previous problem. Assume standard atmospheric conditions and that the energy-source temperature is the same as the maximum cycle temperature and the energy-sink temperature is the same as the minimum cycle temperature.

8.30 A limited-pressure cycle has a compression ratio of 15:1, an α of 1.1, a β of 1.4, $P_1 = 14.2$ psia, and $T_1 = 75^\circ$F. Calculate the cycle's net specific work (Btu/lb$_m$), specific heat addition (Btu/lb$_m$), and thermal efficiency.

8.31 A pure Diesel cycle has a compression ratio of 20:1 and a β of 1.3. Determine the maximum temperature (K) of the air and the rate of heat addition (kW) to this cycle when it produces 250 kW of power and the state of the air at the beginning of the compression is 90 kPa, 15°C.

8.32 A pure Diesel cycle has a compression ratio of 18:1 and a β of 1.5. Determine the maximum air temperature ($^\circ$R) and the rate of heat addition (Btu/hr) to this cycle when it produces 200 hp of power, the cycle is repeated 1200 times per minute, and the state of the air at the beginning of the compression is 13.8 psia, 65°F.

8.33 Rework the previous problem when the compression efficiency is 90% and the expansion efficiency is 95%.

8.34 Develop an expression for β that expresses it in terms of $q_{in}/(c_p T_1 r_v^{(k-1)})$ for an air standard Diesel cycle.

8.35 A Diesel cycle has a maximum cycle temperature of 2300° F, a β of 1.4, $P_1 = 14.4$ psia, and $T_1 = 50°$ F. This cycle is executed in a four-stroke, eight-cylinder engine that has a cylinder bore of 4 inches and a piston stroke of 4 inches. The minimum volume enclosed in the cylinder is 4.5% of the maximum cylinder volume. Determine the horsepower produced by this engine when it is operated at 1800 rpm.

8.36 A Diesel cycle has a maximum cycle temperature of 2000° C, a β of 1.2, $P_1 = 95$ kPa, and $T_1 = 15°$ C. This cycle is executed in a four-stroke, eight-cylinder engine with a cylinder bore of 10 centimeters and a piston stroke of 12 centimeters. The minimum volume enclosed in the cylinder is 5% of the maximum cylinder volume. Determine the horsepower produced by this engine when it is operated at 1600 rpm.

8.37 Develop an expression for the thermal efficiency of a limited-pressure cycle when operated so that $\alpha = \beta$. What is the thermal efficiency of such an engine when the compression ratio is 20:1 and $\alpha = 2$.

8.38 How can we change α in the preceding problem so that the same thermal efficiency is maintained when the compression ratio is reduced?

8.39 A Stirling cycle operates between thermal-energy reservoirs at 27° C and 527° C. It is filled with 1 kg of air such that the maximum cycle pressure is 2000 kPa and the minimum cycle pressure is 100 kPa. Determine the net work (kJ) produced each time this cycle is executed, and the cycle's thermal efficiency.

8.40 Determine the external rate of heat input (kW) and power (kW) produced by the Stirling cycle of the previous problem when it is repeated 500 times per minute.

8.41 A Stirling cycle uses energy reservoirs at 40° F and 640° F, and hydrogen as the working gas. It is designed so that its minimum volume is .1 ft³, maximum volume is 1ft³, and maximum pressure is 400 psia. Calculate the amount of external heat addition (Btu), external heat rejection (Btu), and heat transfer (Btu) between the working fluid and regenerator for each complete cycle.

8.42 A Stirling cycle filled with air uses a 50° F energy reservoir as a sink. The engine is designed so that the maximum air volume is 0.5 ft³, the minimum air volume is 0.06 ft³, and the minimum pressure is 10 psia. It is to be operated so that the engine produces 2 Btu of net work when 6 Btu of heat are transferred externally to the engine. Determine the temperature (° R) of the source-energy reservoir, the amount of air (lb$_m$) contained in the engine, and the maximum air pressure (psia) during the cycle.

8.43 How would the temperature ($^\circ$R) of the source-energy reservoir and maximum air pressure of the Stirling cycle in the preceding problem change if the engine were to be operated to produce 2.5 Btu of work for the same external-heat input?

8.44 An air standard Stirling cycle operates with a maximum pressure of 600 psia and a minimum pressure of 10 psia. The maximum volume of the air is 10 times the minimum volume. The temperature at states 3 and 4 of Figure 8.10 (p. 381) is 100° F. Calculate the specific heat (Btu/lb$_m$) added to and rejected by this cycle, as well as the net specific work (Btu/lb$_m$) produced by the cycle.

8.45 How much specific heat (Btu/lb$_m$) is stored (and recovered) in the regenerator of the previous problem?

8.46 An air standard Stirling cycle operates with a maximum pressure of 3600 kPa and a minimum pressure of 50 kPa. The maximum volume is 12 times the minimum volume, and the low-temperature sink reservoir is at 20° C. Allowing a 5° C temperature difference between the external reservoirs and the air when appropriate, calculate the specific heat (kJ/kg) added to the cycle and its net specific work (kJ/kg).

8.47 How much specific heat (kJ/kg) is stored (and recovered) in the regenerator of the previous problem?

8.48 Calculate the specific lost work potential (kJ/kg) for each process of Problem 8.46.

8.49 A Brayton cycle operates with minimum and maximum temperatures of 27° C and 727° C. It is filled with 1 kg of air and designed so that the maximum cycle pressure is 2000 kPa and the minimum cycle pressure is 100 kPa. Determine the net work (kJ) produced each time this cycle is executed and the cycle's thermal efficiency.

8.50 Determine the work production and thermal efficiency of the Brayton cycle of the previous problem when the turbine's isentropic efficiency is 90%.

8.51 If the compressor efficiency of Problem 8.49 were 80% in addition to the irreversibilities of the turbine of Problem 8.50, how much will the work production and thermal efficiency of the cycle change?

8.52 Determine the reduction in the work production and thermal efficiency when there is a 50-kPa pressure drop across the combustion chamber in addition to the compressor and turbine irreversibilities of Problem 8.51.

8.53 A simple Brayton cycle uses helium as the working fluid; operates with 12 psia, 60° F at the compressor inlet; has a pressure ratio of 14; and a maximum cycle temperature of 1300° F. How much power (hp) will this cycle produce when the rate at which the helium is circulated about the cycle is 100 pounds-mass per minute?

8.54 If the compression efficiency of the preceding problem were 95%, how much would the power production be reduced?

8.55 The back-work ratio for a gas turbine is defined as the ratio of the work used to power the compressor to the work produced by the turbine. Consider a simple Brayton cycle that uses air as the working fluid, has a pressure ratio of 12:1, has a maximum cycle temperature of 600° C, and operates the compressor inlet at 90 kPa, 15° C. Which will have the greatest impact on the back-work ratio: a compressor efficiency of 90% or a turbine efficiency of 90%?

8.56 Use availability analysis to answer the preceding question.

8.57 A simple Brayton cycle uses argon as its working fluid. At the beginning of the compression, $P_1 = 15$ psia and $T_1 = 80°$ F, the maximum cycle temperature is 1200° F, and the pressure in the combustion chamber is 150 psia. The argon enters the compressor through a 3 ft^2 opening with a velocity of 200 ft/s. Determine the rate of heat addition (Btu/hr) to this engine, the power (hp) produced, and the cycle's thermal efficiency.

8.58 Determine the rate at which entropy is generated by the Brayton cycle of the previous problem. The temperature of the energy-source reservoir is the same as the maximum cycle temperature, and the temperature of the energy-sink reservoir is the same as the minimum cycle temperature.

8.59 An aircraft engine operates as a simple Brayton cycle. Consider such an engine with a pressure ratio of 10:1 when heat is added to the cycle at a rate of 500 kW, air passes through the engine at a rate of 1 kg/s, and the air at the beginning of the compression is at 70 kPa, 0° C. Determine the power (hp) produced by this engine and its thermal efficiency.

8.60 How will the results of the previous problem change if the pressure ratio is increased to 15:1?

8.61 A gas turbine for an automobile is designed with a regenerator as shown in Figure 8.13 (p. 392). Air enters the compressor of this engine at 100 kPa, 20° C. The compressor pressure ratio is 8:1, the maximum cycle temperature is 800° C, and the cold air stream leaves the regenerator 10° C cooler than the hot air stream enters the regenerator. Determine the rates of heat addition (and rejection kW) for this cycle when it produces 150 kW.

8.62 Rework the preceding problem when the compressor efficiency is 87% and the turbine efficiency is 93%.

8.63 Determine the specific lost work potential for each of the processes in Problem 8.62.

8.64 The regenerator of Figure 8.13 is rearranged so that the air streams of states 2 and 5 enter at one end of the regenerator and streams 3 and 6 exit the other end. Consider such a system when air enters the compressor at 14 psia, 70° F; the compressor pressure ratio is 7:1; the maximum cycle temperature is 1240° F;

and the difference between the hot and cold air-stream temperatures is $10°$ F at the end of the regenerator where the cold stream leaves the regenerator. Is the cycle arrangement shown in Figure 8.13 more or less efficient than this arrangement?

8.65 An ideal regenerator ($T_3 = T_6$) is added to a simple Brayton cycle as shown in Figure 8.13. Air enters the compressor of this cycle at 13 psia, $50°$ F; the pressure ratio is 8:1; and the maximum cycle temperature is $1500°$ F. What is the thermal efficiency of this cycle?

8.66 Calculate the thermal efficiency of the previous problem without a regenerator.

8.67 Develop an expression for the thermal efficiency of a Brayton cycle with an ideal regenerator ($T_3 = T_6$ of Figure 8.13). Your final result should only contain the pressure ratio.

8.68 A gas turbine operates with a regenerator and two stages of reheating and intercooling. Air enters this engine at 14 psia, $60°$ F; the pressure ratio for each stage of compression is 3; the air temperature when entering a turbine is $940°$ F; and the regenerator operates perfectly. Determine the mass flow rate (lb_m/s) of the air passing through this engine and the rates of heat addition and rejection (Btu/s) when this engine produces 1000 hp.

8.69 Determine the change in the rate of heat addition to the cycle of the preceding problem when the efficiency of each compressor is 88% and the efficiency of each turbine is 93%.

8.70 Which process of the previous problem loses the greatest amount of work potential? The temperature of the hot-energy reservoir is the same as the maximum cycle temperature and the temperature of the cold-energy reservoir is the same as the minimum cycle temperature.

8.71 Air enters a two-stage gas turbine at 100 kPa, $17°$ C. This system uses a regenerator as well as reheating and intercooling. The pressure ratio across each compressor is 4:1, 300 kJ/kg of heat are added to the air in each combustion chamber, and the regenerator operates perfectly while increasing the temperature of the cold air by $20°$ C. Determine this system's thermal efficiency.

8.72 Rework the previous problem with three stages of ideal compression, expansion, reheating, and intercooling.

8.73 How much would the thermal efficiency of the preceding problem change if the temperature of the cold-air stream leaving the regenerator is $40°$ C lower than the temperature of the hot-air stream entering the regenerator?

8.74 Calculate the lost work potential for each process of Problem 8.73.

8.75 A gas turbine has two stages of ideal compression, expansion, intercooling, and reheating. Air enters the first compressor at 13 psia, $60°$ F; the total pressure ratio (across all compressors) is 12:1; the total rate of heat addition is 500 Btu/s; and the cold air temperatrue is increased by $50°$ F in the regenerator.

Calculate the power (kW) produced by each turbine, power (kW) consumed by each compressor, and the rate of heat rejection (Btu/s).

8.76 Rework the previous problem with each compressor having an efficiency of 85% and each turbine having an efficiency of 90%.

8.77 Compare the thermal efficiency of a two-stage gas turbine with regeneration, reheating, and intercooling to that of a three-stage gas turbine with the same equipment when:

    a. all components operate ideally

    b. air enters the first compressor at 14 psia, 40° F

    c. the total pressure ratio across all stages of compression is 16

    d. the maximum cycle temperature is 1000° F

8.78 An Ericsson cycle operates between thermal-energy reservoirs at 627° C and 7° C while producing 500 kW of power. Determine the rate of heat addition (kW) to this cycle when it is repeated 2000 times per minute.

8.79 If the cycle of the preceding problem is repeated 3000 times per minute while the heat added per cycle remains the same, how much power will the cycle produce?

8.80 The rate of heat addition (i.e., rate of fuel consumption) may be used to control the power produced by an Ericsson or other gas cycle. If the rate of heat addition of Example 8.12 is increased by 10%, what is the impact on the pressure ratio and power produced?

8.81 A turboprop-aircraft propulsion engine operates where the air is at 8 psia, −10° F, on an aircraft flying at a speed of 600 ft/s. The Brayton cycle pressure ratio is 10:1, and the air temperature at the turbine inlet is 940° F. The propeller diameter is 10 ft, and the mass flow rate through the propeller is 20 times that through the compressor. Determine the momentum thrust force (lb_f) generated by this propulsion system.

8.82 How much change would result in the momentum thrust force of the previous problem if the propeller diameter were reduced to 8 ft while the same mass flow rate was maintained through the compressor? *Note:* The mass flow rate ratio will no longer be 20:1.

8.83 A turbofan engine operating on an aircraft flying at 200 m/s at an altitude where the air is at 50 kPa, −20° C, is to produce 50,000 N of momentum thrust. The inlet diameter of this engine is 2.5 m, the compressor pressure ratio is 12:1, and the mass flow rate is 8:1. Determine the air temperature (K) at the turbine inlet needed to produce this thrust.

8.84 A pure jet engine propels an aircraft at 300 m/s through air at 60 kPa, 0° C. The inlet diameter of this engine is 2 m, the compressor pressure ratio is 10, and the temperature at the turbine inlet is 450° C. Determine the velocity (m/s) at the exit of this engine's nozzle and the momentum thrust (N) produced.

8.85 The specific impulse of an aircraft-propulsion system is the momentum thrust force produced per unit of thrust-producing mass flow rate. Consider a pure jet engine that operates where the air is at 10 psia, 30° F, and which propels an aircraft at 1200 ft/s. Determine the specific impulse of this engine when the compressor pressure ratio is 9:1 and the temperature at the turbine inlet is 700° F.

8.86 A combined gas–steam power cycle uses a simple gas turbine for the topping cycle and a simple Rankine cycle for the bottoming cycle. Atmospheric air enters the gas turbine at 101 kPa, 20° C, and the maximum gas cycle temperature is 1100° C. The compressor pressure ratio is 8:1, the compressor efficiency is 85%, and the gas turbine efficiency is 90%. The gas stream leaves the heat exchanger at the saturation temperature of the steam flowing through the heat exchanger. Steam flows through the heat exchanger with a pressure of 6000 kPa, and leaves at 320° C. The steam-cycle condenser operates at 20 kPa, and the steam-turbine efficiency is 90%. Determine the mass flow rate of air through the air compressor required for this system to produce 100 MW of power.

8.87 Determine which components of the system in the previous problem are the most wasteful of work potential.

8.88 One stage of regeneration is added to the gas cycle portion of the combined cycle in Problem 8.86. How much does this change the efficiency of this combined cycle?

8.89 Atmospheric air enters the air compressor of a simple combined gas–steam power system at 14.7 psia, 80° F. The air compressor's compression ratio is 10:1, the gas cycle's maximum temperature is 2100° F, and the air compressor and turbine have an efficiency of 90%. The gas leaves the heat exchanger 50° F hotter than the saturation temperature of the steam in the heat exchanger. The steam pressure in the heat exchanger is 800 psia, and the steam leaves the heat exchanger at 600° F. The steam-condenser pressure is 5 psia and the steam turbine efficiency is 95%. Determine the overall thermal efficiency of this combined cycle.

8.90 It has been suggested that the steam passing through the condenser of the previous problem be routed to buildings during the winter to heat them. When this is done, the pressure in the heating system where the steam is now condensed will have to be increased to 10 psia. How does this change the overall efficiency of the combined cycle?

8.91 During winter, the system of the previous problem must supply 2×10^6 Btu/hr of heat to the buildings. What is the mass flow rate of air through the air compressor and the system's total electrical power production in winter?

8.92 A reversed Brayton cycle cooling system uses air as the working fluid. This air is at 5 psia, −10° F, as it enters the compressor. The compression ratio for this compressor is 4:1, and the temperature at the turbine inlet is 100° F. Determine this cycle's COP.

8.93 Rework the previous problem when the compressor efficiency is 87%, the turbine efficiency is 94%, and the pressure drop across each heat exchanger is 5 psia.

8.94 A reversed Brayton cycle operates with air such that at the beginning of the compression, $P = 100$ kPa and $T = 20°$ C, and at the beginning of the expansion, $P = 500$ kPa and $T = 30°$ C. This system is to provide 15 kW of cooling. Calculate the rate (kg/s) at which air is circulated in this system, as well as the rates (kW) of heat addition and rejection.

8.95 A 20-kW cooling load at 0 ° C is to be serviced by a reversed Brayton cycle using air as the working fluid. The heat released from this cycle is rejected to the surrounding environment at 20° C. At the inlet to the compressor, the air is at 100 kPa, –5° C. Determine the minimum pressure ratio for this system to operate properly.

8.96 A two-stage bootstrap cooling system operates with air at 12 psia, 0° F, entering the first compressor. Each compression stage has a pressure ratio of 4:1, and the two intercoolers can cool the air to 50° F. Calculate the coefficient of performance of this system and the rate (lb_m/s) at which air must be circulated through this system to service a 75,000 Btu/hr cooling load.

8.97 How will the answers of the preceding problem change when the efficiency of each compressor is 85% and the efficiency of the turbine is 95%?

8.98 A three-stage, bootstrap air cooling system operates with 80 kPa, –20° C air entering the first compressor. Each compressor in this system has a pressure ratio of 5:1, and the air temperature at the outlet of all intercoolers is 15° C. Calculate the COP of this system.

PROJECTS

The following projects are more extensive than the problems presented in this chapter. Most are open-ended and require some decisions on the student's part. Many can be expedited with computer-based property tables, computer spreadsheets, and student-written computer programs.

8.1 The amount of fuel introduced into a spark-ignition engine is used in part to control the power produced by the engine. Gasoline produces approximately 22,000 Btu/lb_m when burned with air in a spark-ignition engine. Develop a schedule for gasoline consumption and maximum cycle temperature versus power production for an Otto cycle with a compression ratio of 8:1.

8.2 Exhaust gases from the turbine of a simple Brayton cycle are quite hot and may be used for other thermal purposes. One proposed use is generating saturated steam vapor at 110° C from water at 30° C in a boiler. This steam will be distributed to several buildings on a college campus for heat. A Brayton cycle with a pressure ratio of 6:1 is to be used for this purpose. Plot the power produced, the flow rate of produced steam, and the maximum cycle temperature as functions of the rate at which heat is added to the cycle. The temperature at the turbine inlet is not to exceed 2000° C.

8.3   A gas turbine operates with a regenerator and two stages of reheating and intercooling. This system is designed so that when air enters the compressor at 14 psia, 60° F, the pressure ratio for each stage of compression is 3:1, the air temperature when entering a turbine is 940° F, and the regenerator operates perfectly. At full load, this engine produces 1000 hp. For this engine to service a partial load, the heat addition in both combustion chambers is reduced. Develop an optimal schedule of heat addition to the combustion chambers for partial loads ranging from 500 hp to 1000 hp.

8.4   You have been asked to design a power facility for a lunar-based laboratory. You have selected a simple Brayton cycle that uses argon as the working fluid and has a pressure ratio of 6:1. The heat-rejecting heat exchanger maintains the state at the entrance to the compressor at 50 kPa, -20° C. You have elected to use solar collectors to serve as the heat supply. Tests of these collectors give the temperature increase results shown in Figure 8.26. Develop a plot of the power that will be produced by this system and its thermal efficiency as a function of the argon mass flow rate. Is there a "best" flow rate at which to operate this power plant?

FIGURE 8.26

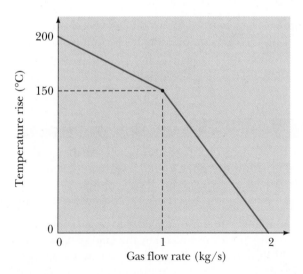

8.5   If the temperature control at the compressor inlet of the previous project is disabled, the temperature at the compressor inlet varies linearly with the gas flow rate. It is −40° C when the flow rate is 0 kg/s and 0° C when the flow rate is 2 kg/s. Develop a schedule of the power produced and the thermal efficiency as the argon flow rate is varied from 0.01 kg/s to 2 kg/s. Is there an optimal flow rate at which to operate this power plant?

8.6 The weight of a diesel engine is directly proportional to the compression ratio ($W = kr_v$) because extra metal must be used to strengthen the engine for the higher pressures. Examine the net specific work produced by a diesel per unit of weight as the pressure ratio is varied and the specific-heat input remains fixed. Do this for several heat inputs and proportionality constants, k. Are there any optimal combinations of k and specific-heat inputs?

Additional Thermodynamic Property Relations and Models

Good Fortune

Some luck lies in not getting what you thought you wanted but getting what you have, which once you have got it you may not be smart enough to see is what you would have wanted had you known.

Garrison Keillor

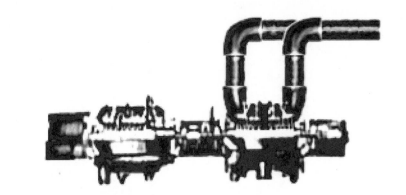

he development of working substances for use in thermodynamic devices has been as important in the evolution of energy conversion as the gains that have been made in the devices themselves. Working fluids and thermodynamic systems are interrelated—like computer hardware and software. Advances in one field leads to advances in the other. This symbiosis leads to the continued improvement of both fields.

An excellent example of this relationship is provided by the history of refrigeration. The vapor compression cycle had been around for a long time, but only in the last 50 years or so has its use become widespread. To be successful, the working fluid used in this cycle must have the appropriate combination of chemical, physical, and thermal characteristics. Early fluids used in the VCRS consisted of whatever was available at the time. These included ammonia, propane, butane, and others. Although these fluids served their purpose, they tended to require large pieces of equipment and special handling. Compressors with high-compression ratios were required, heat exchangers tended to be large, and well-trained individuals aware of the toxic, explosive, and other characteristics of the fluid were needed to operate these systems. Chemists looking for fluids that met all of these requirements of a refrigerant discovered the chlorofluorocarbons. These fluids permitted reduced equipment size, improved coefficients of performance, and better safety, among other considerations. The CFCs were not the optimal solution, but they were a considerable improvement over the other known fluids. As the CFC's became more available, VCRSs became more plentiful and widely used. Today, we again seek a refrigerant that not only meets the requirements of the vapor-compression refrigeration system (VCRS), but also meets the requirements of the environment.

In this chapter, we examine how we can interrelate the various thermodynamic properties. The process begins with the measurement of certain physical properties and how they vary with other properties. Typically, we measure the equation of state, $f(P, T, v) = 0$; the saturation line, $P_{sat} = f(T)$; a specific heat $c_v = f(T)$; and the saturated liquid specific volume, $v_f = f(T)$. This information is then used to build tables of properties in tabular, computer code, or other forms. We will assume that the basic measurements have been made in this chapter and will proceed to illustrate how tables of properties are derived from this information. Alternative equations of state, ideal gas with variable specific heats tables, and the principle of corresponding states are also presented.

9.1 Introduction

Property data normally are first determined by experimental measurements and presented in tabular or graphical form. The most commonly measured properties are pressure, temperature, specific volume, and the specific heats. These measurements are then used to formulate the mathematical relationships between the pressure, temperature, and specific volume that we know as the *equation of state*. The data are also used to develop mathematical relationships between the specific heats and other properties such as the saturation pressure and temperature. The resulting equations of state and specific-heat relationships are sometimes theoretical, such as that of the ideal gas, and sometimes nothing more than statistical curve fits to measured data. However these equations of state and specific-heat relationships are generated, our goal is to use them to generate complete tables of properties that include the first law properties of internal energy and enthalpy and the second law property of entropy. To accomplish this, we need general relationships that can be applied to an equation of state and specific-heat functions to produce other quantities, such as the entropy. These relationships are developed in this section.

General property equations may be developed by taking advantage of the state postulate and the mathematical properties of exact differentials and partial differentials. You may wish to review the mathematical properties of exact and partial differentials as presented in Appendix A before you proceed. We will restrict ourselves to simple compressible substances in this development so that we may use the result of the state postulate, which holds that "any thermodynamic property is at most a function of two other independent thermodynamic properties for a simple compressible substance."

The best possible property relation is a mathematical relation that can be integrated, differentiated, and manipulated as desired for first and second law analysis. Even empirical fits to experimental data are useful for this purpose. Computer models are less useful because numerical integration and differentiation can be inaccurate and are difficult to alter. Tabular and graphical relations are also limited by the accuracy of reading a graph, and tables require interpolation. Still, we must use the data in whatever form they are available.

Many times, new substances are very similar to other known substances. Known property models are then adjusted to the new substance by new empirical constants. Occasionally, a large class of substances can be modeled by a single relationship, as we have done with the ideal gases. Property relations that treat a class of substances are the most useful for obvious reasons. Sometimes, a new substance is treated as belonging to a class of substances such as the ideal gases until experiments can be performed and better data developed.

Remember these three points from this discussion:

1. Property data are continually emerging and improving.
2. Property data obtained from measurements and presented in tabular

form are always the most trustworthy, although not always the most convenient.

3. All mathematical, empirical, and computer forms of property data are always restricted to certain ranges of properties. None of these forms are accurate under all conditions.

9.2 Gibbs and Maxwell Equations

In addition to the internal energy, enthalpy, and entropy, we will need the Helmholtz[1] and Gibbs functions. The specific *Helmholtz function, a,* is defined as

$$a \equiv u - Ts$$

and the specific *Gibbs function, g,* is defined as

$$g \equiv h - Ts$$

These definitions are the natural consequence of the definition of the closed-system exergy, flow exergy, and combined first and second laws. Both of these quantities are properties of a substance because they are both mathematical combinations of other properties. For simple compressible substances, the specific Helmholtz and Gibbs functions only depend on two other independent properties, according to the state postulate.

There are four Gibbs equations. In Section 4.9, we derived the first two of these,

$$T ds = du + P dv$$
$$T ds = dh - v dP$$

by applying the first and second laws to a reversible process. We also have made extensive use of the first two Gibbs equations to interrelate the properties of ideal gases with constant specific heats.

The third Gibbs equation is found by taking the differential of the Helmholtz function:

$$da = du - T ds - s dT$$

---

[1]Hermann Ludwig Ferdinand von Helmholtz (1821–1894) was a German philosopher and scientist who made contributions to the study of the conservation of energy, hydrodynamics, optics, electrodynamics and electricity, meteorological physics, dynamics, and sound. He is regarded as one founder of the concept of energy conservation, an honor he shares with Meyer, Joule, and Lord Kelvin.

This differential is exact because a is a property of the substance. When this is combined with the $Tds = du + Pdv$ Gibbs equation, the differential of the Helmholtz function becomes

$$da = -Pdv - sdT$$

which is known as the third Gibbs equation. In the same manner, the differential of the Gibbs function is

$$dg = dh - Tds - sdT$$

which is an exact differential. Combining this result with the $Tds = dh - vdP$ Gibbs equation produces

$$dg = vdP - sdT$$

or the fourth Gibbs equation.

The complete set of the *Gibbs equations* follows:

$$
\begin{aligned}
du &= Tds - Pdv \\
dh &= Tds + vdP \\
da &= -Pdv - sdT \\
dg &= vdP - sdT
\end{aligned}
\tag{9.1}
$$

Inspection of the right-hand side of the Gibbs equations indicates that the only variables involved are P, v, T, and s, which suggests that they may be used to relate the entropy directly to the equation of state.

All of the quantities on the left-hand side of the Gibbs equations are exact differentials. This being the case, the two coefficients on the right-hand side must satisfy the test for exactness (see Appendix A). This test yields the following results.

$$
\begin{aligned}
\left.\frac{\partial T}{\partial v}\right|_s &= -\left.\frac{\partial P}{\partial s}\right|_v \\
\left.\frac{\partial T}{\partial P}\right|_s &= \left.\frac{\partial v}{\partial s}\right|_P \\
\left.\frac{\partial P}{\partial T}\right|_v &= \left.\frac{\partial s}{\partial v}\right|_T \\
\left.\frac{\partial v}{\partial T}\right|_P &= -\left.\frac{\partial s}{\partial P}\right|_T
\end{aligned}
\tag{9.2}
$$

These are part of a set of equations known as the *Maxwell equations*.[2] Their usefulness lies in the fact that they may be used to relate entropy changes to the
more readily measured pressure, temperature, and specific-volume changes.
For example, consider the third Maxwell equation. We may conduct an experiment in which a fixed quantity of substance is confined in a rigid vessel (i.e.,
v remains constant). As this vessel is heated, the variation of the pressure and
temperature in the vessel may then be measured as the temperature increases.
The results of this experiment would provide $\partial P/\partial T|_v$. The third Maxwell
equation then can be integrated to obtain the variation of the entropy with
specific volume during an isothermal change of state.

EXAMPLE 9.1	Estimate the difference in the specific entropy of water vapor at 300 kPa, 280° C and 700 kPa, 280° C. Verify this estimate by comparing it to results obtained from a property table.

System: An arbitrary quantity of water in vapor form.

Basic Equations:

$$\left.\frac{\partial T}{\partial v}\right|_s = -\left.\frac{\partial P}{\partial s}\right|_v$$

$$\left.\frac{\partial T}{\partial P}\right|_s = \left.\frac{\partial v}{\partial s}\right|_P$$

$$\left.\frac{\partial P}{\partial T}\right|_v = \left.\frac{\partial s}{\partial v}\right|_T$$

$$\left.\frac{\partial v}{\partial T}\right|_P = -\left.\frac{\partial s}{\partial P}\right|_T$$

Conditions: None.

Solution: The entropy can be taken to be a function of temperature and pressure, $s = s(T, P)$, according to the state postulate. The differential of the entropy is then

$$ds = \left.\frac{\partial s}{\partial T}\right|_P dT + \left.\frac{\partial s}{\partial P}\right|_T dP$$

This result as applied to the problem becomes

$$ds = \left.\frac{\partial s}{\partial P}\right|_T dP$$

---

[2]Named for James Clark Maxwell (1831–1879), a British physicist who is best known for his research
in electricity and magnetism. While serving on the faculty of Cambridge University, he wrote a
textbook titled *Theory of Heat* in 1871 that postulated the equations of thermodynamics, which are
named for him. These are not to be confused with *Maxwell's equation* that serves as the foundation
of electromagnetic theory. Maxwell was truly a "renaissance man" of science who studied and
contributed to Newtonian physics as well as to thermodynamics and electromagnetic theory.

because the temperature is the same at the two states, $dT = 0$. The partial derivative of this expression may be replaced by using the fourth Maxwell equation to produce

$$ds = -\left.\frac{\partial v}{\partial T}\right|_P dP$$

Integrating this result over the isothermal process that connects the two states gives

$$s_2 - s_1 = -\int_1^2 \left.\frac{\partial v}{\partial T}\right|_P dP$$

As an estimate, we can evaluate this integral by using an average value of the partial derivative taken at 500 kPa:

$$s_2 - s_1 \simeq -\left.\frac{\partial v}{\partial T}\right|_{P=500} (P_2 - P_1)$$

A property table may be used to estimate the average partial derivative by using a finite-difference expression for the partial derivative at tabulated temperatures on either side of the desired temperature:

$$\left.\frac{\partial v}{\partial T}\right|_{P=500} \simeq \frac{v(500\ \text{kPa}, 320°\ \text{C}) - v(500\ \text{kPa}, 240°\ \text{C})}{320 - 240}$$

$$\simeq \frac{541.6 - 464.6}{80 \times 1000}$$

$$\simeq 9.625(10^{-4})\ \text{m}^3/\text{kg-C}$$

Substituting this into the previous expression gives the estimate of the entropy change:

$$s_2 - s_1 \simeq -9.625(10^{-4})(700 - 300)$$
$$\simeq -0.3850\ \text{kJ/kg-K}$$

According to the property tables,

$$s_2 - s_1 = 7.2233 - 7.6299$$
$$= -0.4066\ \text{kJ/kg-K}$$

TIP:

By using the fourth Maxwell equation, the entropy change could be evaluated directly from the equation of state.

Comparing these two results indicates that the estimate underpredicts the entropy difference by approximately 5.5%.

9.3 General Expressions for *du, dh,* and *ds*

The Gibbs and Maxwell equations will now be used to develop general equations for the change in specific internal energy, enthalpy, and entropy. Our goal is to develop these equations so that only an equation of state and specific-heat relationships are necessary to evaluate these equations.

9.3.1 General Expression for *du*

According to the state postulate, the entropy of a substance can be taken to be a function of the temperature and specific volume, $s = s(T, v)$. The differential of the entropy is then

$$ds = \left.\frac{\partial s}{\partial T}\right|_v dT + \left.\frac{\partial s}{\partial v}\right|_T dv$$

When this is substituted into the first Gibbs equation, $du = Tds - Pdv$, the result is

$$du = T\left.\frac{\partial s}{\partial T}\right|_v dT + \left[T\left.\frac{\partial s}{\partial v}\right|_T - P\right] dv$$

The internal energy can be taken to be a function of the temperature and specific volume, $u = u(T, v)$, according to the state postulate. The differential of the specific internal energy is then

$$du = c_v dT + \left.\frac{\partial u}{\partial v}\right|_T dv$$

where the definition of the specific heat at constant volume, $c_v = \partial u / \partial T|_v$, has been used. Comparing this result to the previous expression for the differential of the specific internal energy reveals that

$$\left.\frac{\partial s}{\partial T}\right|_v = \frac{c_v}{T} \tag{9.3}$$

and

$$\left.\frac{\partial u}{\partial v}\right|_T = T\left.\frac{\partial s}{\partial v}\right|_T - P$$

The third Maxwell equation allows us to replace $\partial s/\partial v|_T$ with $\partial P/\partial T|_v$, so that this last result becomes

$$\left.\frac{\partial u}{\partial v}\right|_T = T\left.\frac{\partial P}{\partial T}\right|_v - P$$

With these results, the differential for the internal energy is as follows:

$$du = c_v dT + \left[T \frac{\partial P}{\partial T}\bigg|_v - P \right] dv \qquad (9.4)$$

This result can be integrated between two states to obtain the difference in the specific internal energy of the two states:

$$u_2 - u_1 = \int_1^2 c_v dT + \int_1^2 \left[T \frac{\partial P}{\partial T}\bigg|_v - P \right] dv$$

These integrals can be performed once we know the equation of state for the substance and how the substance's c_v varies with temperature.

EXAMPLE 9.2

Develop an expression for the change in the internal energy of an ideal gas whose specific heat at constant volume is given by

$$c_v = \sum_{i=0}^{5} A_i (T - T_0)^{i/2}$$

where the A_i's are empirical constants and T_0 is a reference temperature.

System: An arbitrary quantity of a gas.

Basic Equations:
$$du = c_v dT + \left[T \frac{\partial P}{\partial T}\bigg|_v - P \right] dv$$

Conditions: The gas will be treated as an ideal gas.

Solution: Because this substance is an ideal gas, $Pv = RT$. Hence,

$$\frac{\partial P}{\partial T}\bigg|_v = \frac{R}{v}$$

and

$$
\begin{aligned}
T \frac{\partial P}{\partial T}\bigg|_v - P &= T\frac{R}{v} - P \\
&= P - P \\
&= 0
\end{aligned}
$$

and the differential of the internal energy becomes

$$du = c_v dT$$

This proves that the internal energy of an ideal gas depends only on the gas temperature, as Section 3.12 established.

Substituting the given expression for c_v into $du = c_v dT$ and integrating between two states gives

$$u_2 - u_1 = \int_1^2 \sum_{i=0}^{5} A_i (T - T_0)^{i/2} \, dT$$

$$= \sum_{i=0}^{5} \int_1^2 A_i (T - T_0)^{i/2} \, dT$$

$$= \sum_{i=0}^{5} \frac{2A_i}{i+2} [(T_2 - T_0)^{(i+2)/2} - (T_1 - T_0)^{(i+2)/2}] \qquad \blacksquare$$

9.3.2   General Expression for *dh*

In the same manner as that of the previous section, we begin by taking the entropy to be a function of two other properties. In this section, we will select temperature and pressure so that $s = s(T, P)$. The differential of the specific entropy is then

$$ds = \left. \frac{\partial s}{\partial P} \right|_T dP + \left. \frac{\partial s}{\partial T} \right|_P dT$$

Substituting this result for the ds of the second Gibbs equation, $dh = T ds + v dP$, we obtain

$$dh = T \left. \frac{\partial s}{\partial T} \right|_P dT + \left[T \left. \frac{\partial s}{\partial P} \right|_T + v \right] dP$$

By taking the enthalpy to be a function of the pressure and temperature, the differential of the specific enthalpy is

$$dh = c_P dT + \left. \frac{\partial h}{\partial P} \right|_T dP$$

where the definition of the specific heat at constant pressure, $c_P = \partial h / \partial T|_P$, has been used. Comparing these two expressions for dh reveals that

$$\left. \frac{\partial s}{\partial T} \right|_P = \frac{c_P}{T} \qquad\qquad (9.5)$$

and

$$\left. \frac{\partial h}{\partial P} \right|_T = \left[T \left. \frac{\partial s}{\partial P} \right|_T + v \right]$$

The fourth Maxwell equation substituted into this last result yields

$$\left.\frac{\partial h}{\partial P}\right|_T = \left[v - T\left.\frac{\partial v}{\partial P}\right|_P\right]$$

The differential of the enthalpy is then given by the following:

$$dh = c_P dT + \left[v - T\left.\frac{\partial v}{\partial T}\right|_P\right]dP \qquad (9.6)$$

The significance of this result is that it may be used to calculate the enthalpy from the substance's equation of state and constant-pressure specific heat.

EXAMPLE 9.3

Develop an expression for the change in the specific enthalpy of an ideal gas whose specific heat at constant pressure is given by

$$c_P = \sum_{i=0}^{4} A_i(T - T_0)^{i/3}$$

where the A_i's are empirical constants and T_0 is a reference temperature.

System: An arbitrary quantity of a gas.

Basic Equations:
$$dh = c_P dT + \left[v - T\left.\frac{\partial v}{\partial T}\right|_P\right]dP$$

Conditions: The gas will be treated as an ideal gas.

Solution: Because this substance is an ideal gas, $Pv = RT$. Hence,

$$\left.\frac{\partial v}{\partial T}\right|_P = \frac{R}{P}$$

and

$$\left.v - T\frac{\partial v}{\partial T}\right|_P = v - T\frac{R}{P}$$
$$= v - v$$
$$= 0$$

and the differential of the enthalpy becomes

$$dh = c_P dT$$

This proves that the enthalpy of an ideal gas depends only on the gas temperature, as Section 3.12 pointed out. Substituting the given expression for c_P into this result and integrating between two arbitrary states gives

$$h_2 - h_1 = \int_1^2 \sum_{i=0}^4 A_i (T - T_0)^{i/3} dT$$

$$= \sum_{i=0}^4 \int_1^2 A_i (T - T_0)^{i/3} dT$$

$$= \sum_{i=0}^4 \frac{3A_i}{i+3} [(T_2 - T_0)^{(i+3)/3} - (T_1 - T_0)^{(i+3)/3}]$$

9.3.3 General Expression for *ds*

General expressions for the entropy may be readily obtained from the results of the preceding sections. The differential of the specific entropy is

$$ds = \left. \frac{\partial s}{\partial T} \right|_v dT + \left. \frac{\partial s}{\partial v} \right|_T dv$$

when the entropy is taken to be a function of the temperature and specific volume, $s = s(T, v)$. Substituting Equation 9.3 and the third Maxwell equation into this result produces the following:

$$ds = \frac{c_v}{T} dT + \left. \frac{\partial P}{\partial T} \right|_v dv \tag{9.7}$$

This allows us to evaluate the entropy from a c_v as a function of temperature relation and the equation of state.

Alternatively, the entropy could be taken as a function of pressure and temperature, $s = s(T, P)$. The differential of the entropy would then be

$$ds = \left. \frac{\partial s}{\partial T} \right|_P dT + \left. \frac{\partial s}{\partial P} \right|_T dP$$

When the third Maxwell equation and Equation 9.5 are substituted into this result, it becomes the following:

$$ds = \frac{c_P}{T} dT - \left. \frac{\partial v}{\partial T} \right|_P dP \tag{9.8}$$

This permits the evaluation of the entropy in terms of the specific heat at constant pressure and the equation of state.

EXAMPLE 9.4

Develop an expression for the change in the specific entropy of an ideal gas that has a specific heat at constant volume given by

$$c_v = \sum_{i=0}^{3} \frac{A_i}{T^i}$$

where the A_i's are empirical constants.

System: An arbitrary quantity of a gas.

Basic Equations:

$$ds = \frac{c_v}{T} dT + \frac{\partial P}{\partial T}\bigg|_v dv$$

Conditions: The gas will be treated as an ideal gas.

Solution: Because this substance is an ideal gas, $Pv = RT$. Hence,

$$\frac{\partial P}{\partial T}\bigg|_v = \frac{R}{v}$$

and

$$ds = \frac{c_v}{T} dT + \frac{\partial P}{\partial T}\bigg|_v dv$$

$$= \frac{c_v}{T} dT + \frac{R}{v} dv$$

Substituting the given c_v expression into this result and integrating between two general states gives

$$s_2 - s_1 = \int_1^2 \sum_{i=0}^{3} \frac{A_i}{T^{i+1}} dT + R \int_1^2 \frac{dv}{v}$$

$$= -\sum_{i=0}^{3} \frac{A_i}{i}\left(\frac{1}{T_2^i} - \frac{1}{T_1^i}\right) + R \ln \frac{v_2}{v_1}$$

9.4 Specific Heat Relations

In Section 3.12 we demonstrated that the specific heats for ideal gases are interrelated by

$$c_P - c_v = R$$

It is only natural to inquire whether any general relationships exist between the specific heats. Toward this end, we can equate the two ds expressions of Equations 9.7 and 9.8. After multiplying these by the temperature, equating the results, and performing some rearrangement, the result is

$$(c_p - c_v)dT = T\left.\frac{\partial v}{\partial T}\right|_P dP + T\left.\frac{\partial P}{\partial T}\right|_v dv$$

Taking pressure as a function of temperature and specific volume, $P = P(T, v)$, the pressure differential is

$$dP = \left.\frac{\partial P}{\partial T}\right|_v dT + \left.\frac{\partial P}{\partial v}\right|_T dv$$

When substituted into the previous expression, this produces

$$(c_P - c_v)dT = T\left.\frac{\partial v}{\partial T}\right|_P \left.\frac{\partial P}{\partial T}\right|_v dT + T\left[\left.\frac{\partial v}{\partial T}\right|_P \left.\frac{\partial P}{\partial v}\right|_T + \left.\frac{\partial P}{\partial T}\right|_v\right] dv$$

The quantity in the brackets is zero because

$$\left.\frac{\partial v}{\partial T}\right|_P \left.\frac{\partial P}{\partial v}\right|_T = -\left.\frac{\partial P}{\partial T}\right|_v \qquad (9.9)$$

according to the rotational partial derivative relationship

$$\left.\frac{\partial z}{\partial x}\right|_y \left.\frac{\partial x}{\partial y}\right|_z \left.\frac{\partial y}{\partial z}\right|_x = -1$$

of Appendix A. The difference between the specific heats is then given by

$$c_P - c_v = T\left.\frac{\partial v}{\partial T}\right|_P \left.\frac{\partial P}{\partial T}\right|_v$$

Equation 9.9 may be used to further reduce this result to the following:

$$c_P - c_v = -T\left(\left.\frac{\partial v}{\partial T}\right|_P\right)^2 \left.\frac{\partial P}{\partial v}\right|_T \qquad (9.10)$$

The partial derivatives of Equation 9.10 are related to two important and readily measured thermodynamic properties: *volume expansivity* and *isothermal compressibility*. The volume expansivity is normally denoted by β and is defined as follows.

DEFINITION: *Volume Expansivity*

$$\beta \equiv \frac{1}{v} \left. \frac{\partial v}{\partial T} \right|_P$$

This can be readily measured by measuring the change in volume of an isobaric system such as a weighted piston-cylinder device as its temperature is changed by heating or cooling. The isothermal compressibility is typically denoted by κ, and is defined as follows.

DEFINITION: *Isothermal Compressibility*

$$\kappa \equiv -\frac{1}{v} \left. \frac{\partial v}{\partial P} \right|_T$$

This can be measured by measuring the change in volume of an isothermal system as the pressure is changed. In terms of these two new properties, the specific-heat difference becomes as follows:

$$c_P - c_v = \frac{vT\beta^2}{\kappa} \qquad (9.11)$$

This result and some experimental observations provide us with three important conclusions. First, as the temperature approaches absolute zero, the two specific heats approach one another and eventually become equal. This is the case no matter what substance is involved. Second, if the substance is incompressible, $dv = 0$; consequently, β is also zero. The two specific heats are then, in general, equal. Although we pointed this out in Section 3.12, Equation 9.11 proves that the specific heats are equal for any incompressible substance. Third, experimental observations demonstrate that κ is positive for all known substances. Hence,

$$c_P \geq c_v$$

according to these results. The specific heat at constant pressure is always greater than or equal to the specific heat at constant volume regardless of the substance involved or its state.

EXAMPLE 9.5

Develop an expression for the volume expansivity, isothermal compressibility, and specific-heat difference of an ideal gas.

System: An arbitrary quantity of a gas.

Basic Equations:

$$\beta \equiv \frac{1}{v} \frac{\partial v}{\partial T}\bigg|_P$$

$$\kappa \equiv -\frac{1}{v} \frac{\partial v}{\partial P}\bigg|_T$$

$$c_P - c_v \equiv \frac{vT\beta^2}{\kappa}$$

Conditions: The gas will be treated as an ideal gas.

Solution: Because this substance is an ideal gas, $Pv = RT$. Hence,

$$\beta = \frac{1}{v} \frac{\partial v}{\partial T}\bigg|_P$$

$$= \frac{R}{Pv}$$

$$= \frac{1}{T}$$

and

$$\kappa = -\frac{1}{v} \frac{\partial v}{\partial P}\bigg|_T$$

$$= -\frac{1}{v}\left(\frac{-RT}{P^2}\right)$$

$$= \frac{1}{P}$$

Substituting these two results into Equation 9.11 yields

$$c_P - c_v = \frac{Pv}{T}$$

$$= R$$

This is the same result as that of Section 3.12, where the definition of the enthalpy and the ideal gas equation of state were used to derive an expression for the specific-heat difference.

9.5 Other Thermodynamic Properties

The Gibbs and Maxwell equations may be combined in many different ways to interrelate or introduce new properties of substances. Most of these combinations are of either limited or no practical use. But two of these are useful for determining the change in properties across a phase transition and measuring the specific heat at constant pressure. These relations are respectively known as the *Clapeyron equation* and the *Joule–Thomson coefficient;* both are presented and discussed in this section.

9.5.1 The Joule–Thomson Coefficient

The apparatus in Figure 9.1 is used to measure the Joule[3]–Thomson[4] coefficient. In this apparatus, the substance that will have its Joule–Thomson coefficient measured is forced to flow through a porous plug or another pressure-reducing device. The porous plug is well insulated so that no heat is exchanged with the environment. Also, there is no work interaction with the surroundings because the system boundary is rigid and no shafts protrude through the system boundary. As the substance flows through the porous plug, the pressure is reduced such that $P_2 < P_1$. The temperature of the substance also may change as the substance flows through the porous plug. Both the pressure drop and the temperature change that occur across the plug may be measured accurately with simple instruments such as pressure gauges and thermometers.

FIGURE 9.1
Joule–Thomson Coefficient Apparatus

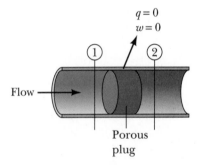

When the steady-flow form of the first law is applied to the system of Figure 9.1, it tells us that the enthalpy at the outlet is the same as the enthalpy at the inlet. In other words, this is a constant-enthalpy throttling process.

The Joule–Thomson coefficient is defined as follows.

---

[3]James Prescott Joule (1818–1899) was an English physicist who was basically self-taught. Through many experiments, he established the equivalence of electrical, mechanical, and chemical energies. The conversion constant, 1 Btu = 778.169 ft-lb$_f$, was named the Joule constant in recognition of this work.

[4]William Thomson Kelvin who was also known as Lord Kelvin.

DEFINITION: *Joule–Thomson Coefficient,* μ

$$\mu \equiv \left.\frac{\partial T}{\partial P}\right|_h$$

This is related to the pressure and temperature changes in the apparatus of Figure 9.1 by

$$\mu = \lim_{P_2 \to P_1} \left.\frac{T_2 - T_1}{P_2 - P_1}\right|_h$$

The Joule–Thomson coefficient can then be measured by extrapolating the temperature change–pressure drop ratio as measured in the Joule–Thomson apparatus with a very small pressure drop (i.e., zero).

Alternatively, we could fix the enthalpy of the process by setting P_1 and T_1 and then holding the enthalpy constant as P_2 is varied. The outlet temperature would then change as the outlet pressure is varied. The results of these measurements would typically appear as shown in Figure 9.2. In this figure, each line is a line of constant enthalpy, and the slope of the tangent drawn at any point along one of these lines is the Joule–Thomson coefficient.

FIGURE 9.2

Temperature–Pressure State Diagram Illustrating the Inversion Line

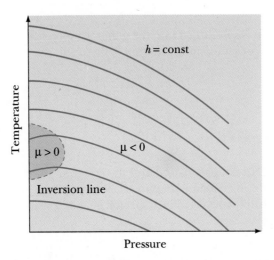

At intermediate values of the enthalpy, the temperature passes through a maximum where the tangent to the constant-enthalpy process line becomes horizontal. At this point, the Joule–Thomson coefficient is zero. The line connecting the points of zero Joule–Thomson coefficient is known as the *inversion line.* This name comes from the fact that the Joule–Thomson coefficient to the left of the inversion line is positive, while it is negative to the right of this line. For states to the left of the inversion line, the temperature at the outlet of the Joule–Thomson device is less than the temperature at the inlet. For states to the right of the inversion line, the temperature at the outlet will be greater than at the inlet. For states along the inversion line, the temperature does not change across the porous plug of the Joule–Thomson throttling device. The inversion line marks the transition in the behavior of the substance in this system from a cooling-like (i.e., lowered temperature) to a heating-like (i.e., increased temperature) effect.

The throttle device used in the vapor-compression refrigeration system (VCRS) of Chapter 7 is one application of the Joule–Thomson effect. When the refrigerant passes through this device, its temperature must decrease as the pressure is decreased in order for the VCRS to operate properly. Hence, refrigerants used in these systems must have a positive Joule–Thomson coefficient for the state at the entrance of the throttle device.

The relationship between the Joule–Thomson coefficient and the specific heat at constant pressure may be found by taking the temperature to be a function of the pressure and enthalpy. The differential of the temperature is then

$$dT = \left.\frac{\partial T}{\partial P}\right|_h dP + \left.\frac{\partial T}{\partial P}\right|_P dh$$

$$= \mu \, dP + \frac{dh}{c_P}$$

When this is substituted into the general expression for the differential of the enthalpy presented in Equation 9.6, the result is:

$$c_P = \frac{1}{\mu}\left[T\left.\frac{\partial v}{\partial T}\right|_P - v\right] \tag{9.12}$$

This directly couples c_P to μ and the equation of state. Once the equation of state is known for a substance, measurements of the Joule–Thomson coefficient can be readily converted into specific heat at constant-pressure data using this result.

EXAMPLE 9.6

Determine the Joule–Thomson coefficient for an ideal gas.

System: An arbitrary quantity of a gas.

Basic Equations:

$$c_P = \frac{1}{\mu}\left[T\left.\frac{\partial v}{\partial T}\right|_P - v\right]$$

Conditions: The gas will be treated as an ideal gas.

Solution: Because the gas is ideal, $Pv = RT$, and

TIP:

This result is expected: We already demonstrated that the enthalpy of an ideal gas is a function of the temperature only. Hence, we cannot change the temperature of an ideal gas in any constant-enthalpy process such as the Joule–Thomson throttling process.

$$\frac{\partial v}{\partial T}\Big|_P = \frac{R}{P}$$

The Joule–Thomson coefficient is then

$$\mu = \frac{1}{c_P}\left[T\frac{\partial v}{\partial T}\Big|_P - v\right]$$
$$= \frac{1}{c_P}\left[\frac{RT}{P} - v\right]$$
$$= 0$$

EXAMPLE 9.7

Estimate the Joule–Thomson coefficient of water vapor at 10,000 kPa, 480° C.

System: An arbitrary quantity of water vapor.

Basic Equations:
$$\mu \equiv \frac{\partial T}{\partial P}\Big|_h$$

Conditions: A finite difference will be used to approximate the partial derivative and allow us to use the property tables of Appendix B.

Solution: In finite difference form,

TIP:

The positive sign indicates that the temperature change is in the same direction as the pressure change during a throttling process. Water vapor at this state might then serve as a refrigerant because its temperature decreases as its pressure is decreased in a throttle device.

$$\mu \approx \frac{\Delta T}{\Delta P}\Big|_h$$

At the given state, $h = 3321.4$ kJ/kg. At 12,000 kPa, the temperature is 490° C when $h = 3321.4$ kJ/kg by interpolating the property tables. The Joule–Thomson coefficient is then approximately

$$\mu \approx \frac{490 - 480}{12,000 - 10,000}$$
$$\approx 0.0050 \text{ K/kPa}$$

9.5.2 Clausius–Clapeyron Equation

During a phase change, the pressure and temperature are not independent of one another, as Section 3.6 demonstrated. This section develops the relationship between the saturation temperature and pressure, as well as the change in extensive properties during a phase conversion. Maxwell's third equation will be useful for this purpose.

During a phase transition, the entropy can be taken as a function of the temperature and specific volume, according to the state postulate. The differential of the entropy is then

$$ds = \left.\frac{\partial s}{\partial T}\right|_v dT + \left.\frac{\partial s}{\partial v}\right|_T dv$$

The temperature remains constant during a phase change and the first term on the right-hand side of this result is zero. According to the third Maxwell equation, $\partial s/\partial v|_T = \partial P/\partial T|_v$, which reduces the preceding result to

$$ds = \left.\frac{\partial P}{\partial T}\right|_v dv$$

Because the pressure only depends on the temperature when the substance is undergoing a phase transition, the partial derivative of this result is independent of v and becomes an ordinary derivative:

$$ds = \left.\frac{dP}{dT}\right|_{sat} dv$$

When this is integrated across the mixture dome from a saturated liquid to a saturated vapor, the result is

$$s_g - s_f = \left.\frac{dP}{dT}\right|_{sat} (v_g - v_f),$$

and therefore

$$\left.\frac{dP}{dT}\right|_{sat} = \frac{s_g - s_f}{v_g - v_f} \qquad (9.13)$$

The first and second laws, as applied to reversible processes in closed systems, produce a second result that may be used to reduce this result further. According to the second law, $ds = \delta q/T$, while the first law as applied to an isobaric state change (as in a phase transformation) in a closed system is $\delta q = dh$. Combining these two laws and integrating the result across the mixture dome gives

$$s_g - s_f = \int_f^g ds = \int_f^g \frac{dh}{T} = \frac{h_g - h_f}{T}$$

When this is combined with Equation 9.13, the rate at which the saturation pressure changes with temperature is given by

$$\frac{dP}{dT}\bigg|_{sat} = \frac{h_{fg}}{Tv_{fg}} \tag{9.14}$$

This is known as the *Clausius–Clapeyron equation.*

   The Clausius–Clapeyron equation is important because it relates the rate of saturation-pressure change with temperature to the enthalpy and specific-volume change as a substance is converted isobarically from a saturated liquid to a saturated vapor. It can then be used to construct phase-state relations and diagrams, such as those of Figures 3.6 and 3.7 (page 89). Alternatively, if the phase-state relation [i.e., $p_{sat}(T)$] is known, the Clausius–Clapeyron equation may be used to determine such latent properties as the latent heat of vaporization, h_{fg}.

   Although we have developed the Clausius–Clapeyron equation for a liquid–vapor phase change, it may be applied to any phase conversion. For example, for a sublimation process where the solid phase is directly converted to the gaseous phase, Clapeyron's equation becomes

$$\frac{dP}{dT}\bigg|_{sat} = \frac{h_g - h_i}{T(v_g - v_i)} = \frac{h_{ig}}{Tv_{ig}}$$

where subscript g refers to the saturated gaseous phase and subscript i refers to the saturated solid phase.

EXAMPLE 9.8

Use the Clausius–Clapeyron equation to estimate the saturation pressure of ammonia at 30° C based on saturation conditions at 26° C. Use an appropriate property table to assess the accuracy of this estimate.

System: An arbitrary quantity of ammonia.

Basic Equations:

$$\frac{dP}{dT}\bigg|_{sat} = \frac{h_{fg}}{Tv_{fg}}$$

Conditions: None.

Solution: According to the saturated ammonia table (Table 3.1) and the Clausius–Clapeyron equation, the rate of change of saturation pressure with temperature at 26° C is

$$\frac{dP}{dT}\bigg|_{sat} = \frac{h_{fg}}{Tv_{fg}}$$
$$= \frac{1160.8 \times 1000}{299.15(124.3 - 1.663)}$$
$$= 31.64 \text{ kPa/K}$$

Integrating this result over the temperature change from $26°$ C to $30°$ C gives

$$P_{30} = P_{26} + 31.64(30 - 26)$$
$$= 1034.5 + 126.6$$
$$= 1161.1 \text{ kPa}$$

The ammonia saturation table indicates that the actual saturation pressure at $30°$ C is 1167.1 kPa; this estimate is in error by approximately 0.5%. ◾

9.6 Generating Tables of Properties

The results of the previous sections may be used to generate tables of properties such as those contained in Appendix B. In this section we will illustrate how this is done by going through the steps involved in generating tables of properties for liquid–vapor saturation states and the vapor phase. The same methods may be applied to other phase-transition saturation states and single-phase states.

The following information will be required to generate saturation and vapor tables of properties for a substance:

1. A mathematical relation between the specific volume (or density) of the saturated liquid as a function of the liquid's temperature (or pressure), $v_f = v_f(T)$.

2. A mathematical relation between the saturation pressure and temperature, $P_{sat} = P_{sat}(T)$.

3. An equation of state for the vapor phase, $P = P(T, v)$, which includes states along the saturated-vapor line.

4. The relation between one of the vapor-phase specific heats (consider c_P) at very low pressures (i.e., as $P \to 0$) as a function of temperature, $c_P^0 = c_P^0(T)$.

All of these are normally obtained from experimental measurements.

The first item is readily measured by measuring the mass, volume, and temperature of a system once the liquid in the system has been heated to a saturated liquid condition. The temperature at which saturation occurs is controlled and varied through the pressure applied to the liquid. Once several v_f, T data points have been measured, statistical or other empirical data-fitting methods can be applied to the data points to obtain $v_f = v_f(T)$. The second item is generated in essentially the same way.

Perhaps the most difficult item to obtain is item 3 because it involves numerous measurements of P, T, and v under a wide range of conditions. These measurements must be made under low- and high-pressure and low- and high-temperature conditons, as well as when the vapor is saturated. Once these data are available, statistical or empirical techniques (or both) that can accommodate two independent variables are used to develop the required mathematical equation of state. The kinetic theory of gases, statistical theory, intermolecular force models, and quantum mechanics are useful for developing appropriate equation-of-state models.

Item 4 is measured in a device such as the Joule–Thomson throttling device with the pressure of the vapor being very low. The vapor pressure is kept extremely low (approaching zero absolute pressure) so that the specific volume will be very large. All gaseous forms become ideal gases when the specific volume is large (approaching infinity) and the distance between the microscopic particles making up the gas become very large. In this way, the gas behaves as an ideal gas, and the specific heats will depend only on the gas temperature, thereby simplifying the amount of data that must be measured and the techniques used to reduce these data to a mathematical form.

Once these relations are available, we can begin to evaluate properties at various states such as states 1, 4, 5, and 6 of Figure 9.3. The first step is to select a reference state at which the internal energy and entropy are set to zero. This is necessary because the equations of Section 9.3 only predict changes in internal energy, enthalpy, and entropy, and we need a reference state against which we can measure these changes. Although this choice is arbitrary, typical practice for saturated liquid–vapor and vapor tables is to select the state where the liquid is saturated at the triple-point pressure for this reference state, which is labeled as state 0 in Figure 9.3. The temperature at this reference state is found with the saturation pressure–temperature relation, $P_{sat} = P_{sat}(T)$, and the specific volume of this reference state is determined with the saturated liquid–specific volume–temperature relation, $v_f = v_f(T)$. According to the definition of the enthalpy, $h_0 = P_0 v_0$.

FIGURE 9.3
Temperature–Entropy State Diagram

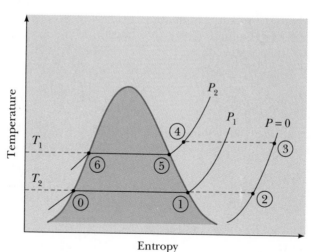

Next we need to work across the saturation dome to the saturated-vapor state shown as state 1 in Figure 9.3. The pressure and temperature of state 1 are the same as those of the reference state. This pressure and temperature is used in the vapor equation of state, $P = P(T, v)$, to calculate the specific volume of state 1. The specific volume change associated with the phase conversion can now be calculated by subtracting v_0 from v_1. The Clausius–Clapeyron saturation-pressure derivative, $dP/dT|_{sat}$, is found by taking the derivative of $P_{sat} = P_{sat}(T)$. All of this information may now be substituted into the Clausius–Clapeyron equation to calculate h_{fg} and $h_1 = h_0 + h_{fg}$. State 1's internal energy is now found from the definition of the enthalpy, $u = h - Pv$. The latent entropy change, s_{fg}, can be calculated with Equation 9.13; s_1 is then fixed by $s_1 = s_0 + s_{fg}$.

We can now work from saturated vapor state 1 to a general vapor state such as state 4 (state 5 is also a general vapor state). The integration path 1–2–3–4 will be used to integrate the relations of Section 9.3. This path is selected to take advantage of the simplifications associated with developing specific-heat relations at very low pressures where the vapor acts as an ideal gas. To demonstrate the procedure, we will consider the difference in the enthalpy of states 1 and 4,

$$h_4 - h_1 = \int_1^2 dh + \int_2^3 dh + \int_3^4 dh$$

where dh is given by Equation 9.6.

Over the 1–2 portion of this integration path, $dT = 0$, and Equation 9.6 gives

$$h_2 - h_1 = \int_1^2 \left[v - T \left. \frac{\partial v}{\partial T} \right|_P \right] dP$$

where the quantity in the brackets may be evaluated with the vapor equation of state. For the zero-pressure portion of the integration path, $dP = 0$, and Equation 9.6 becomes

$$h_3 - h_2 = \int_2^3 c_P^0 dT$$

which is integrated using the measured $c_P^0 = c_P^0(T)$. Finally, the 3–4 portion of the integration path is evaluated in the same way as the 1–2 portion:

$$h_4 - h_3 = \int_3^4 \left[v - T \left. \frac{\partial v}{\partial T} \right|_P \right] dP$$

Combining these results,

$$h_4 - h_1 = \int_1^2 \left[v - T \left. \frac{\partial v}{\partial T} \right|_P \right] dP + \int_2^3 c_{P0} dT + \int_3^4 \left[v - T \left. \frac{\partial v}{\partial T} \right|_P \right] dP$$

The enthalpy of state 4 is then

$$h_4 = h_1 + (h_4 - h_1)$$

The entropy at state 4 can be determined using the same integration method and Equation 9.8. The internal energy of state 4 is found using the definition of the enthalpy, $v = h - Pv$.

Saturated-vapor state 5 is related to state 1 in the same way as state 4 except that the pressure and temperature of state 5 must be related by $P_{sat} = P_{sat}(T)$. The entire saturated-vapor line may be generated in the same way as state 5. To complete the saturated-liquid line, we can work back from state 5 to state 6 by using $v_f = v_f(T)$ and the Clausius–Clapeyron equation.

9.7 Alternate Equations of State

Equations of state are relations between pressure, volume, and temperature. We have made considerable use of the equation of state for ideal gases; in molar form, it is

$$P\bar{v} = \Re T$$

where $\bar{v}$ is the molar specific volume. This is a result of the kinetic theory of gases, which considers gases to be made up of point masses that only interact through elastic collisions with one another. This concept is valid as long as the spacing between the molecules is quite large (in a microscopic sense). As this spacing becomes smaller and smaller when the gas cools to lower temperatures or the pressure increases, intermolecular forces resulting from mutual attraction become more important. The volume of the molecules also becomes a larger fraction of the total volume as the intermolecular distance becomes smaller. As the molecules are crowded into a smaller volume, the ideal gas equation of state becomes inaccurate.

Van der Waals[5] introduced corrections to the ideal gas equation of state for the molecular volume and intermolecular forces in 1873. He suggested that, as the intermolecular spacing decreases, the molar specific volume be reduced to account for the volume of the molecules, $\bar{v} - b$, and that the pressure be corrected for the intermolecular attractive force, $P + a/\bar{v}^2$. The *Van der Waals equation of state* is then:

$$\left(P + \frac{a}{\bar{v}^2}\right)(\bar{v} - b) = \Re T \tag{9.15}$$

[5]Johannes Diderik Van Der Waals (1837–1923) was a Dutch physicist who built up a kinetic theory of the fluid state by combining Laplace's theory of capillarity with the kinetic theory of gases. From this starting point he arrived at an equation of state which gave an explanation of critical phenonema. Eventually, he arrived at an equation that was the same for all substances. Van Der Waals also discoverd the law of binary mixtures.

Here a and b must be in units consistent with P, T, and $\bar{v}$. Values of a and b for selected substances are presented in Table 9.1. Actually, the Van der Waals a and b coefficients depend on the gas and temperature. The coefficients presented in Table 9.1 are for common temperatures found in practice.

When the specific volume is large, as is the case when the pressure is low or the temperature is high, the Van der Waals equation becomes equivalent to the ideal gas equation of state. In this event, $a/\bar{v}^2 \to 0$ and $\bar{v} \gg b$. The Van der Waals equation then reduces to the ideal gas equation of state:

$$P\bar{v} = \Re T$$

TABLE 9.1
Van der Waals Gas Coefficients

Gas	Molecular Weight	a $\dfrac{\text{psia-ft}^6}{\text{lb}_m\text{-mole}^2}$	b $\dfrac{\text{ft}^3}{\text{lb}_m\text{-mole}}$	a $\dfrac{\text{kPa-m}^6}{\text{kg-mole}^2}$	b $\dfrac{\text{m}^3}{\text{kg-mole}}$
Air	28.97	5,032	0.584	135.2	0.0365
n-Butane, C_4H_{10}	58.124	51,580	1.862	1385.8	0.1162
Carbon Dioxide, CO_2	44.01	13,601	0.686	365.8	0.0428
Carbon Monoxide, CO	28.011	5,472	0.632	147.0	0.0395
Ethane, C_2H_6	30.020	20,740	1.041	557.3	0.0650
Ethylene, C_2H_4	28.054	17,020	0.922	457.3	0.0575
Hydrogen, H_2	2.016	923.3	0.426	24.81	0.0266
Methane, CH_4	16.043	8,490	0.684	228.1	0.0427
Nitrogen, N_2	28.013	5,084	0.618	136.6	0.0386
Oxygen, O_2	31.999	5,135	0.508	138.0	0.0317
Propane, C_3H_8	44.097	34,820	1.144	935.5	0.0901
Water, H_2O	18.015	20,580	0.488	552.8	0.0305

All equations of state for gases must display this same behavior.

The Van der Waals equation of state is a satisfactory alternative to the ideal gas equation of state as long as the intermolecular distance does not become too small—that is, when the state of the gas is in the vicinity of the critical point, the specific volume is small, or a phase transformation is about to occur. Do not expect the Van der Waals equation of state to be accurate under these conditions when the a and b coefficients of Table 9.1 are used.

We can take advantage of the characteristics of an isotherm passing through the critical point to develop new values for a and b to use with the Van der Waals equation of state in the region of the critical point. On a P–$\bar{v}$ state diagram, the isothermal line becomes horizontal and is an inflection point when passing through the critical point as illustrated in Figure 9.4. Any equation of state must then satisfy

FIGURE 9.4

The Critical Isotherm

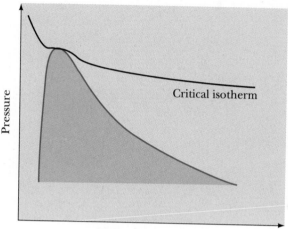

at the critical pressure and specific volume. Solving Equation 9.15 for P, differentiating, and applying these two conditions yields the expressions

$$\frac{\partial P}{\partial \overline{v}}\bigg|_{T=T_c} = 0$$

$$\frac{\partial^2 P}{\partial \overline{v}^2}\bigg|_{T=T_c} = 0$$

$$a = 3 P_c \overline{v}_c^2 = \frac{9}{8} \Re T_c \overline{v}_c = \frac{27}{64} \frac{\Re^2 T_c^2}{P_c}$$

$$b = \frac{\overline{v}_c}{3} = \frac{\Re T_c}{8 P_c}$$

This fixes the a and b coefficients for the Van der Waals equation of state near the critical point. These coefficients are then determined completely by the critical pressure and molar specific volume. Selected critical-state properties are presented in Table 9.2.

TABLE 9.2

Selected Critical State Properties

Gas	T_c		P_c		$\bar{v}_c$	
	K	°R	kPa	psia	m³/kg-mole	ft³/lb$_m$-mole
n-Butane, C_4H_{10}	425.2	765.2	3800	551	0.2547	4.08
Carbon Dioxide, CO_2	304.2	547.5	7390	1071	0.0943	1.51
Carbon Monoxide, CO	133	240	3500	507	0.0930	1.49
Ethane, C_2H_6	305.5	549.8	4880	708	0.1480	2.37
Ethylene, C_2H_4	282.4	508.3	5120	512	0.1242	1.99
Hydrogen, H_2	33.3	59.9	1300	188.1	0.0649	1.04
Methane, CH_4	191.1	343.9	4640	673	0.0993	1.59
Nitrogen, N_2	126.2	227.1	3390	492	0.0899	1.44
Oxygen, O_2	154.78	278.6	5080	736	0.0780	1.25
Propane, C_3H_8	370	665.9	4260	617	0.1998	3.20
Water, H_2O	647.27	1165.3	22,090	3206.2	0.0568	0.91

This method of determining equation-of-state coefficients can be used for any equation of state that contains only two coefficients. Several investigators have suggested two-coefficient equations of state. Bertholet suggested

$$P = \frac{\Re T}{\bar{v} - b} - \frac{a}{T\bar{v}^2} \tag{9.16}$$

This is the same as Van der Waals equation except for the temperature in the denominator of the second term. For low pressures, *Bertholet's equation* is accurate to 1% when the value of a is the same as that of Equation 9.15 and $b = 9\Re T_c/28P_c$ is used. Dieterici suggested the following:

$$P = \frac{\Re T}{\bar{v} - b}e^{-a/\Re T\bar{v}} \tag{9.17}$$

This is quite accurate along the critical isotherm. In regions away from the critical point, however, *Dieterici's equation* can produce very large errors. Using the inflection-point conditions at the critical point, the coefficients of Dieterici's equation are

$$a = 4\frac{\Re^2 T_c^2}{P_c e^2}$$

$$b = \frac{\Re T_c}{P_c e^2}$$

One of the more recent successful two-coefficient equations of state is that of Redlich and Kwong:

$$P = \frac{\Re T}{\overline{v} - b} - \frac{a}{\overline{v}(\overline{v} + b)\, T^{1/2}} \tag{9.18}$$

This is a purely empirical result. In terms of the critical-state properties, the *Redlich–Kwong coefficients* are given by

$$a = 0.42748\frac{\Re^2 T_c^{5/2}}{P_c}$$

$$b = 0.08664\frac{\Re T_c}{P_c}$$

This equation of state is the more accurate of the two-coefficient equations of state presented.

EXAMPLE 9.9

Compare the specific volume of water vapor at 20,000 kPa, 400° C, as predicted by (a) the property tables, (b) the ideal gas equation of state, (c) the Van der Waals equation of state, and (d) the Redlich–Kwong equation of state.

System: Water vapor at the given state.

Conditions: The water vapor is in equilibrium.

Basic Equations:

$$P\overline{v} = \Re T$$

$$\left(P + \frac{a}{\overline{v}^2}\right)(\overline{v} - b) = \Re T$$

$$P = \frac{\Re T}{\overline{v} - b} - \frac{a}{\overline{v}(\overline{v} + b)\, T^{1/2}}$$

Solution: According to the property tables,

$$v = 0.00994 \text{ m}^3/\text{kg}$$

Treating the water vapor as an ideal gas gives

$$v = \frac{\Re T}{MP} = \frac{8.314 \times 673}{18.015 \times 20,000} = 0.01553 \text{ m}^3/\text{kg}$$

This is clearly unacceptable.

The given state is near the critical point. The appropriate coefficients for the Van der Waals equation of state are then

$$a = \frac{27}{64}\frac{\Re^2 T_c^2}{P_c} = \frac{27 \times 8.314^2 \times 647.27^2}{64 \times 22{,}090} = 553.07 \text{ kPa-m}^6/\text{kg-mole}^2$$

$$b = \frac{\Re T_c}{8P_c} = \frac{8.314 \times 647.27}{8 \times 22{,}090} = 0.03045 \text{ m}^3/\text{kg-mole}$$

An iterative solution of the Van der Waals equation of state yields

$$\bar{v} = 0.1858 \text{ m}^3/\text{kg-mole}$$

or

$$v = \frac{\bar{v}}{M} = \frac{0.1858}{18.015} = 0.01031 \text{ m}^3/\text{kg}$$

The coefficients for the Redlich–Kwong equation of state are

$$a = 0.42748\frac{\Re^2 T_c^{5/2}}{P_c} = 14{,}258 \text{ kPa-m}^6\text{-K}^{1/2}/\text{kg-mole}^2$$

$$b = 0.08664\frac{\Re T_c}{P_c} = 0.02111 \text{ m}^3/\text{kg-mole}$$

TIP:

When comparing these answers, observe that the ideal gas equation of state is in significant error while the error in the Van der Waals and Redlich–Kwong equations of state are quite reasonable (less than 3%).

An iterative solution of the Redlich–Kwong equation of state yields

$$\bar{v} = 0.1806 \text{ m}^3/\text{kg-mole}$$

or

$$v = \frac{\bar{v}}{M} = 0.01003 \text{ m}^3/\text{kg}$$

EXAMPLE 9.10

Compare the temperature of water vapor at 100 psia, 4.592 ft^3/lb$_m$ as predicted by (a) the property tables, (b) the ideal gas equation of state, (c) the Van der Waals equation of state, and (d) the Redlich–Kwong equation of state.

System: Water vapor at the given state.

Conditions: The water vapor is in equilibrium.

Basic Equations:

$$P\bar{v} = \Re T$$

$$\left(P + \frac{a}{\bar{v}^2}\right)(\bar{v} - b) = \Re T$$

$$P = \frac{\Re T}{\bar{v} - b} - \frac{a}{\bar{v}(\bar{v} + b)T^{1/2}}$$

Solution: According to the property tables of Appendix B, the temperature of the water vapor is

$$T = 350°\,\text{F}$$

and this state is very close to the saturated liquid–vapor region. Now,

$$\bar{v} = vM = 4.592 \times 18.015 = 82.72 \text{ ft}^3/\text{lb}_m\text{-mole}$$

The ideal gas equation of state predicts a temperature of

$$T = \frac{P\bar{v}}{\Re} = \frac{100 \times 144 \times 82.72}{1545} = 771°\,\text{R} = 311°\,\text{F}$$

This is in significant error when compared to the steam table value.

According to the Van der Waals equation of state using the coefficients of Table 9.1, the temperature of the water vapor is

$$T = \frac{1}{\Re}\left(P + \frac{a}{\bar{v}^2}\right)(\bar{v} - b)$$

$$= \frac{144}{1545}\left(100 + \frac{20{,}580}{82.72^2}\right)(82.72 - 0.488)$$

$$= 790°\,\text{R} = 330°\,\text{F}$$

This is closer to the actual temperature of 350° F.

The appropriate coefficients for the Redlich–Kwong equation of state are

$$a = 0.42748\frac{\Re^2 T_c^{5/2}}{P_c} = 102.45 \times 10^6 \text{ ft}^4\text{-lb}_f\text{-}° \text{R}^{1/2}/(\text{lb}_m\text{-mole})^2$$

TIP:

Note that near saturation conditions, the respective equations of state still contain some error.

$$b = 0.08664\frac{\Re T_c}{P_c} = 0.3379 \text{ ft}^3/\text{lb}_m\text{-mole}$$

Iteratively solving the Redlich–Kwong equation gives

$$T = 336°\,\text{F}$$

This is the most accurate of the equations of state discussed.   ◾

The two-coefficient equations of state are simple, but their ranges of applicability are restrictive, particularly at high densities. Benedict, Webb, and Rubin developed an eight-coefficient equation of state:

$$P = \frac{\Re T}{\overline{v}} + \frac{\Re T B_0 - A_0 - C_0/T^2}{\overline{v}^2}$$
$$+ \frac{\Re T b - a}{\overline{v}^3} + \frac{a\alpha}{\overline{v}^6} + \frac{c}{\overline{v}^3 T^2}\left(\frac{1+\gamma}{\overline{v}^2}\right)e^{-\gamma/\overline{v}^2} \qquad (9.19)$$

where A_0, B_0, C_0, a, b, c, α, and γ are experimentally determined empirical coefficients. Values of these coefficients for selected substances are presented in Table 9.3. This equation was developed originally for use with the hydrocarbons. It has subsequently been adopted for other gases.

TABLE 9.3
Coefficients for the Benedict–Webb–Rubin Equation of State

Constant	N_2	CO	CO_2	CH_4	C_4H_{10}
a, kN-m^7/kg-mole^3	2.54	3.71	13.86	5.00	190.70
A_0, kN-m^4/kg-mole^2	106.73	135.87	277.30	187.90	1022.0
b, m^6/kg-mole^3	0.002328	0.002632	0.00721	0.00338	0.0400
B_0, m^3/kg-mole	0.04074	0.05454	0.04991	0.04260	0.12440
c, kN-m^7-K^4/kg-mole^3	7.379×10^4	1.054×10^5	1.511×10^6	2.578×10^5	3.20×10^7
C_0, kN-m^4-K^4/kg-mole^2	8.164×10^5	8.673×10^5	1.404×10^7	2.286×10^6	1.006×10^8
α, m^9/kg-mole^3	1.272×10^{-4}	1.350×10^{-4}	8.470×10^{-5}	1.244×10^{-4}	1.101×10^{-3}
γ, m^6/kg-mole^2	0.0053	0.0060	0.00539	0.0060	0.0340

EXAMPLE 9.11

A 1m^3 vessel contains one-quarter of a kilogram of methane at 100° C. Compare the pressure in the vessel as predicted by the (a) ideal gas equation of state, (b) the Redlich–Kwong equation of state, and (c) the Benedict–Webb–Rubin equation of state.

System: The methane contained in the vessel.

Conditions: The methane is in equilibrium.

Basic Equations:

$$P\overline{v} = \Re T$$

$$P = \frac{\Re T}{\overline{v} - b} - \frac{a}{\overline{v}(\overline{v} + b)T^{1/2}}$$

$$P = \frac{\Re T}{\overline{v}} + \frac{\Re T B_0 - A_0 - C_0/T^2}{\overline{v}^2}$$

$$+ \frac{\Re T b - a}{\overline{v}^3} + \frac{a\alpha}{\overline{v}^6} + \frac{c}{\overline{v}^3 T^2}\left(\frac{1+\gamma}{\overline{v}^2}\right)e^{-\gamma/\overline{v}^2}$$

Solution: The molar specific volume of the methane is

$$\bar{v} = \frac{V}{mM} = \frac{1}{0.25 \times 16.043} = 0.2493 \text{ m}^3/\text{kg-mole}$$

Solving the ideal gas equation of state for the pressure yields

$$P = \frac{\Re T}{\bar{v}}$$

$$= \frac{8.314 \times 373}{0.2493}$$

$$= 12{,}439 \text{ kPa}$$

The Redlich–Kwong coefficients for methane are

$$a = 0.42748 \frac{\Re^2 T_c^{5/2}}{P_c} = 3214.9 \text{ m}^4\text{-kN-K}^{1/2}/(\text{K-mole})^2$$

$$b = 0.08664 \frac{\Re T_c}{P_c} = 0.02967 \text{ m}^3/\text{kg-mole}$$

The pressure as predicted by the Redlich–Kwong equation of state is

$$P = \frac{\Re T}{\bar{v} - b} - \frac{a}{\bar{v}(\bar{v} + b) T^{1/2}}$$

$$= \frac{8.314 \times 373}{0.2493 - 0.02967} - \frac{3214.9}{0.2493(0.2493 + 0.02967)373^{1/2}}$$

$$= 11{,}726 \text{ kPa}$$

The Benedict–Webb–Rubin equation of state predicts a pressure of

$$P = \frac{\Re T}{\bar{v}} + \frac{\Re T B_0 - A_0 - C_0/T^2}{\bar{v}^2}$$

$$+ \frac{\Re T b - a}{\bar{v}^3} + \frac{a\alpha}{\bar{v}^6} + \frac{c}{\bar{v}^3 T^2}\left(\frac{1 + \gamma}{\bar{v}^2}\right)e^{-\gamma/\bar{v}^2}$$

$$= \frac{8.314 \times 373}{0.2493} + \frac{8.314 \times 373 \times 0.04260 - 187.9 - 2.286 \times 10^6/373^2}{0.2493^2}$$

$$+ \frac{8.314 \times 373 \times 0.00338 - 5.00}{0.2493^3} + \frac{5.00 \times 0.0001244}{0.2493^6}$$

$$+ \frac{2.578 \times 10^5}{0.2493^3 \times 373^2}\left(\frac{1 + 0.0060}{0.2493^2}\right)e^{-0.0060/0.2493^2}$$

$$= 13{,}390 \text{ kPa}$$

Series expansions are also used to generate equations of state. Expansions of the form

$$P\bar{v} = \Re T \left(1 + \frac{B}{\bar{v}} + \frac{C}{\bar{v}^2} + \frac{D}{\bar{v}^3} + \cdots \right) \qquad (9.20)$$

are known as *virial equations of state,* and the coefficients, B, C, D, $\cdots$ are known as the *virial coefficients.* In general, the virial coefficients are functions of the gas temperature. Experimental measurements or theoretical techniques based on statistical mechanics or kinetic theory are used to develop expressions for the virial coefficients. When all of the virial coefficients are zero, the virial equation of state is the same as the ideal gas equation of state. The extra terms in the series expansion then represent the deviation from ideal gas behavior.

Virial equations of state also can be generated by series expansions on other properties such as pressure:

$$P\bar{v} = \Re T(1 + B'P + C'P^2 + D'P^3 + \cdots)$$

These equations of state are quite popular for computer models because of their series form and their ready adjustment to conform with experimental data.

Many other equations of state have been proposed to overcome some of the limitations of the equations of state discussed. By now, it should be apparent that no equation of state is adequate for a substance under all conditions. Always consult the technical literature when selecting the appropriate equation of state needed for a particular substance under a given set of conditions.

9.8 Compressibility Factors

The simplicity of the ideal gas equation of state is extremely attractive, and some theorists have suggested that a factor be introduced in this equation to compensate for the conditions under which it is inaccurate. This factor is known as the *compressibility factor,* and it changes the ideal gas equation of state to

$$Pv = ZRT \qquad (9.21)$$

The compressibility factor, Z, represents the deviation from ideal gas behavior.

Under conditions where a gas behaves as an ideal gas (i.e., low pressure, high temperature), Z approaches unity. When the substance has a low specific volume or the state of the substance is near the saturation region or critical point, Z deviates from unity. It is under these conditions that Z corrects for the real gas behavior. Hence, we can expect Z to be a function of the gas state:

$$Z = Z(P, T)$$

The compressibility factor for nitrogen is illustrated in Figure 9.5 as a function of pressure and temperature. For example, at a pressure of 600 atmospheres and a temperature of 150 K, the compressibility factor of nitrogen is 0.594, according to this figure. If we were to use the ideal gas equation of state to predict the specific volume of nitrogen gas under these conditions, the answer would be in error by approximately 40%.

FIGURE 9.5   *Nitrogen Compressibility Factor*

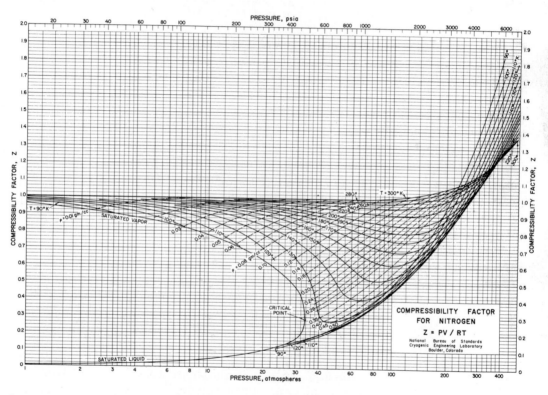

Source: 1960. From the National Institute of Standards and Technology (formerly National Bureau of Standards), Thermodynamics Division, Boulder, Colorado.

Inspection of the compressibility factor of Figure 9.5 reveals the conditions under which nitrogen behaves as an ideal gas. As long as the temperature is greater than approximately 260 K and the pressure is less than some 100 atmospheres, the compressibility factor is approximately unity and nitrogen behaves as an ideal gas. Clearly, the greatest deviations occur near the saturation region and when the pressure is high or the temperature is low.

To gain a better understanding of the compressibility factor, consider a Redlich–Kwong gas. Rearranging the Redlich–Kwong equation of state gives

$$Z = \frac{P\bar{v}}{\Re T} = \frac{\bar{v}}{\bar{v} - b} - \frac{a}{(\bar{v} + b)\Re T^{3/2}}$$

$$= \frac{1}{1 - x} - \left(\frac{a}{b\Re T^{3/2}}\right)\left(\frac{x}{1 - x}\right)$$

where $x = b/\bar{v}$. The coefficient of the second term may be reduced by introducing the definitions of the Redlich–Kwong coefficients,

$$\frac{a}{b\Re T^{3/2}} = \frac{4.934}{T_r^{3/2}}$$

where T_r is known as the *reduced temperature*,

$$T_r = \frac{T}{T_c}$$

Similarly,

$$x = \frac{b}{\bar{v}} = \frac{bP}{Z\Re T} = \frac{0.0876P_r}{ZT_r}$$

where P_r is known as the *reduced pressure*,

$$P_r = \frac{P}{P_c}$$

When these are substituted, the Redlich–Kwong equation of state becomes

$$Z = \frac{ZT_r}{ZT_r - 0.0876P_r} - \frac{4.939}{T_r^{3/2}}\left(\frac{0.0876P_r}{ZT_r + 0.0876P_r}\right)$$

The compressibility factor for a Redlich–Kwong gas then depends only on the reduced pressure and reduced temperature.

The behavior of the Redlich–Kwong gas (as well as others) suggests, in general, the following:

PRINCIPLE:

$$Z = Z(P_r, T_r)$$

This important result is known as the *principle of corresponding states*. It is important because it tells us that the compressibility factor is determined by the reduced pressure and temperature. The compressibility factor is a universal function of P_r and T_r, and substances only differ in their critical pressure and temperature. A plot of the compressibility factor as a function of P_r and T_r is presented in Figure 9.6. Experimentally measured data points for several gases are also presented in the figure. Compare the data points to the average curves: The principle of corresponding states is quite accurate for most gases.

FIGURE 9.6   *Generalized Compressibility Factor*

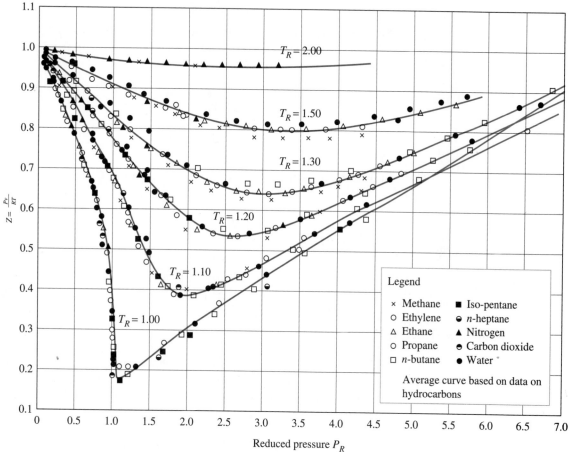

Source: Gour-Jen Su, "Modified Law of Corresponding States," *Ind. Eng. Chem.* (Intern. Ed.), 38:803(1946).

Generalized compressibility charts are presented in Figures C.1 of Appendix C. They are useful when all that is known about a substance is the critical pressure and temperature. Gaseous states can be reasonably predicted using the generalized compressibility charts.

EXAMPLE 9.12

Compare the specific volume of water vapor at 28,000 kPa, 600° C, as predicted by (a) the property tables and (b) the principle of corresponding states.

System: Water vapor at the given state.

Conditions: The water vapor is in equilibrium.

Basic Equations: Property Tables
                              Generalized Compressibility Factor Chart

Solution: According to the property tables,

$$v = 0.01241 \text{ m}^3/\text{kg}$$

The critical pressure and temperature of water are 22,090 kPa and 647.27 K, respectively, according to Table 9.2. The reduced pressure and temperature for this state are

$$P_r = \frac{28,000}{22,090} = 1.27$$

$$T_r = \frac{873}{647.27} = 1.35$$

The compressibility factor from the charts in Appendix C is 0.86, and the specific volume is

$$v = \frac{ZRT}{P}$$

$$= \frac{0.86 \times 8.314 \times 873}{18.016 \times 28000}$$

$$= 0.01238 \text{ m}^3/\text{kg}$$

9.9 Ideal Gas Specific Heats

The previous discussion of ideal gases introduced the constant-pressure specific heat, c_p, and constant-volume specific heat, c_v, which, as long as the gas behaves as an ideal gas, are functions only of temperature. To this point, we have treated the specific heats as if they are constant and have used the typical environmental values of Table 3.5.

The variation of the constant pressure specific heat of several gases under ideal gas conditions is shown in Figure 9.7. Examination of this figure demonstrates that the constant-pressure specific heat of monotonic gases is practically independent of temperature, while the specific heat of diatomic and triatomic gases changes significantly over the temperature range presented.

Portions of the curves in Figure 9.7 may be empirically fit by polynomials in temperature,

$$\bar{c}_p^0 = a + bT + cT^2 + \cdots$$

FIGURE 9.7

*Constant-Pressure
Specific Heats of
Selected Ideal Gases*

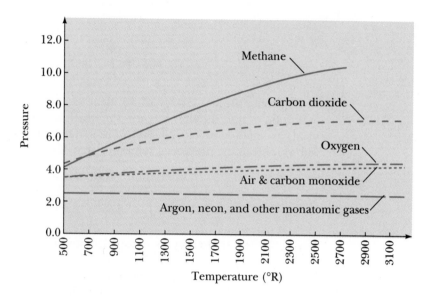

where $\bar{c}_P^0$ is the ideal gas specific heat at constant pressure. The polynomial-fit coefficients for selected gases are presented in Table B.5 of Appendix B.

Other commonly used empirical fits to $\bar{c}_P^0$ include

$$\bar{c}_P^0 = \sum_{i=1}^{n} a_i T^{i-n}$$

$$\bar{c}_P^0 = \sum_{i=1}^{n} a_i T^{i-n} + a_0 e^{\beta/T} \left[\frac{\beta/T}{e^{\beta/T} - 1} \right]$$

$$\bar{c}_P^0 = \sum_{i=1}^{n} a_i [\ln(T/T_r)]^{i-n}$$

where the a_i, a_0, n, β, and T_r are empirically determined constants. Most gaseous substances can be modeled with these functions or their combinations.

9.10   Properties of Ideal Gases with Temperature-Variable Specific Heats

Tables of properties for ideal gases with variable specific heats can be generated from the ideal gas equation of state and an empirically determined specific heat–temperature relationship, such as those of the previous section.

Given the specific heat at constant pressure of an ideal gas as a function of temperature, $c_P^0(T)$, the enthalpy of the ideal gas with respect to a reference state, where $T = T_{ref}$ and $h = 0$, is

$$h(T) = \int_{T_{ref}}^{T} c_P^0(T)\,dT$$

The specific internal energy of an ideal gas may be found by applying the definition of the enthalpy, $h = u + Pv$, which becomes $h = u + RT$ when the ideal gas equation of state is used to eliminate Pv. The specific internal energy with respect to the preceding reference state is then given by

$$u(T) = h(T) - RT + RT_{ref}$$
$$= \int_{T_{ref}}^{T} c_P^0(T)\,dT - R(T - T_{ref})$$

The $\int c_P^0(T)\,dT$ is thereby used to calculate both the specific enthalpy and internal energy.

The Gibbs equation, $T\,ds = dh - v\,dp$, may be used to determine the entropy of an ideal gas with variable specific heats. Substituting the specific heat at constant pressure and the ideal gas equation of state into the Gibbs equation yields

$$ds = c_P^0\frac{dT}{T} - R\frac{dP}{P}$$

This result may be integrated between two states to yield

$$s_2 - s_1 = \int_1^2 \frac{c_P^0}{T}\,dT - R\ln\frac{P_2}{P_1}$$

Inspection of this result reveals that the entropy change is determined by two terms, one depending on the temperature and the other on the pressure. The temperature-dependent portion of this expression can be simplified by introducing the quantity $s^0(T)$, which is the portion of the entropy change attributable to the gas's temperature change. When expressed with respect to a reference state where the temperature is T_{ref}, this portion of the entropy is given by

$$s^0(T) = \int_{T_{ref}}^{T} \frac{c_P^0}{T}\,dT$$

In terms of this quantity,

$$s_2 - s_1 = s^0(T_2) - s^0(T_1) - R\ln\frac{P_2}{P_1}$$

When the specific heats remain constant, the isentropic change of state for an ideal gas is one in which Pv^k remains constant. Because the isentropic state change is so important and represents the processes in so many devices, we will develop special relationships to use with the isentropic state change when the ideal gas's specific heats are not constant. With variable specific heats, the Gibbs equation produces the following result for an isentropic change of state when $ds = 0$:

$$\frac{dh}{T} = R\frac{dP}{P}$$

Introducing the specific heat and integrating this result yields

$$s^0(T_2) - s^0(T_1) = R\ln\frac{P_2}{P_1}$$

This can be written as

$$s^0(T_2) - s^0(T_1) = R\ln\frac{P_2}{P_{ref}} - R\ln\frac{P_1}{P_{ref}}$$

by introducing a reference pressure, P_{ref}. Equating the respective terms on each side of this equation produces the result

$$s^0(T) = R\ln P_r$$

where the relative pressure, $P_r \equiv P/P_{ref}$, has been introduced. Solving this for the relative pressure,

$$P_r(T) = e^{s^0(T)/R}$$

Note that the relative pressure is only a function of temperature like the enthalpy and internal energy.

For an isentropic state change,

$$s^0(T_2) - s^0(T_2) = R\ln\frac{P_2}{P_1}$$

or

$$\frac{P_2}{P_1} = \exp\left(\frac{s^0(T_2) - s^0(T_1)}{R}\right)$$

according to the preceding discussion. The expression for the relative pressure may be restated as

$$\exp\left(\frac{s^0(T_2) - s^0(T_1)}{R}\right) = \frac{\exp(s^0(T_2)/R)}{\exp(s^0(T_1)/R)} = \frac{P_r(T_2)}{P_r(T_1)}$$

Then,

$$\frac{P_2}{P_1} = \frac{P_r(T_2)}{P_r(T_1)} \tag{9.22}$$

for an isentropic change of state involving an ideal gas whose specific heats are variable. A relative specific volume may be defined in a similar manner such that

$$\frac{v_2}{v_1} = \frac{v_r(T_2)}{v_r(T_1)}$$

for an ideal gas with variable specific heats when undergoing an isentropic change of state.

EXAMPLE 9.13

The constant-pressure specific heat of an ideal gas is given by

$$c_P = a_1 T + a_2 T^2 + a_3 T^3$$

where T is the absolute temperature. Develop expressions for $s^0(T)$ and P_r for this gas using a temperature of absolute zero for the reference state.

System: A fixed quantity of an ideal gas.

Basic Equations:

$$s^0(T) = \int_{T_{ref}}^{T} \frac{c_P(T)}{T} dT$$

Conditions: This is an ideal gas whose specific heats vary with temperature.

Solution: $s^0(T)$ is given by

$$s^0(T) = \int_{T_{ref}}^{T} \frac{c_P(T)}{T} dT$$

$$= \int_{0}^{T} a_1 dT + \int_{0}^{T} a_2 T dT + \int_{0}^{T} a_3 T^2 dT$$

$$= a_1 T + \frac{a_2}{2} T^2 + \frac{a_3}{3} T^3$$

TIP:

Results such as these may be used to build tables of properties for these gases. This table only needs temperature as its argument. Different gases are accommodated through the values of R and the a coefficients.

Now,

$$P_r(T) = \exp(s^0(T)/R)$$

$$= \exp\left[\frac{a_1}{R} T + \frac{a_2}{2R} T^2 + \frac{a_3}{3R} T^3 \right]$$

A property table for air as an ideal gas with variable specific heats is included in Appendix B.

EXAMPLE 9.14

Air is expanded from 1000 kPa, 827° C, to 100 kPa in an isentropic turbine. Determine the air temperature at the turbine exhaust and the work produced by the turbine if (a) the air has constant specific heats and (b) the specific heats of the air are variable.

System: The contents of the control volume formed by a steady-flow, isentropic turbine with one inlet and one outlet.

Basic Equations:
$$_1q_2 - _1w_2 = h_2 - h_1 + \frac{1}{2}(V_2^2 - V_1^2) + g(z_2 - z_1)$$
$$Pv = RT$$

Conditions: This is an isentropic, steady-flow system. The problem will be solved treating the air as an ideal gas, with constant specific heats taken at room temperature and specific heats that vary with the air temperature.

Solution: When the specific heats of the air are taken as constant, the isentropic process is one in which Pv^k remains constant. Combining this with the ideal gas equation of state gives

$$T_2 = T_1 \left(\frac{P_2}{P_1}\right)^{k-1/k}$$

$$= 1100 \left(\frac{100}{1000}\right)^{0.4/1.4}$$

$$= 570 \text{ K}$$

Applying the first law to the turbine gives

$$_1w_2 = h_1 - h_2$$
$$= c_P(T_2 - T_1)$$
$$= 1.029(1100 - 570)$$
$$= 545.4 \text{ kJ/kg}$$

Treating the air as an ideal gas with variable specific heats, the relative pressure at the turbine exit is

$$P_{r2} = P_{r1}\frac{P_2}{P_1}$$

$$= 166.2\frac{100}{1000}$$
$$= 16.62$$

Using this result to enter Table B6.1 of Appendix B produces the temperature at the turbine outlet:

$$T_2 = 604 \text{ K}$$

TIP:
*Note that the constant
specific-heat model
underpredicts the work
by less than 1% in this
example while the
temperature is in
considerable error.*

Application of the first law to the turbine using the enthalpies of Table 6.1 yields

$$_1w_2 = h_1 - h_2$$
$$= 1161.11 - 611.47$$
$$= 549.64 \text{ kJ/kg}$$

9.11 Generalized Enthalpy- and Entropy-Departure Charts

In Section 9.3, we developed general relationships for the generation of the specific enthalpy, internal energy, and entropy of a substance. These relations only require an equation of state and specific-heat functions. In this section, we will show how this technique is applied to determining the specific enthalpy and entropy of a general gas using (1) the general compressibility chart as the equation of state and (2) the specific heat at constant pressure of an ideal gas as the required specific-heat relation.

The difference in the enthalpy of two general gaseous states, such as states 1 and 2 of Figure 9.8, may be found in principle by evaluating the general enthalpy differential of Section 9.3.2:

$$dh = c_P dT + \left[v - T \left.\frac{\partial v}{\partial T}\right|_P\right] dP$$

FIGURE 9.8
*T–s Diagram
Illustrating Use of
Enthalpy and
Entropy Departures*

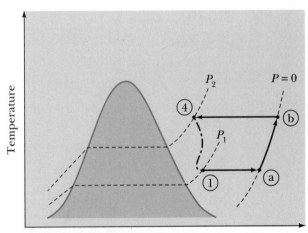

This can be accomplished by selecting some arbitrary path, such as the dashed path of Figure 9.8, along which this differential may be simplified and integrated. The path 1–a–b–2 of Figure 9.8 is particularly convenient for this purpose. The 1–a portion is an isothermal process in which the pressure of the gas is reduced from P_1 to zero. During the a–b portion of this path, the gas pressure is held fixed at zero while the temperature is changed to T_2. Finally, the b–2 portion is an isothermal process in which the gas pressure is increased from zero to P_2. The pressure of zero is only valid in concept and may only be approached in practice. Its utility lies in the fact that all gases become ideal gases at very low pressures. During the a–b process, the general gas behaves as an ideal gas.

The 1–a process isothermally reduces the general gas to an ideal gas. Because $dT = 0$ along this process, the enthalpy difference is given by integrating the general enthalpy differential along this process:

$$h_a - h_1 = \int_{P_1}^{0} \left[v - T \left. \frac{\partial v}{\partial T} \right|_P \right] dP$$

This can be completed once the gas's equation of state is known. The quantity $h_a - h_1$ determined in this manner is known as the *enthalpy departure* because it measures the departure of the enthalpy of the actual gas from the enthalpy of an ideal gas at the same temperature. Careful inspection of this result reveals that the enthalpy departure depends on the gas pressure and temperature—P_1 and T_1, in this case.

Along the a–b portion of the total process that connects states 1 and 2, $dP = 0$. The integral along this portion of the process of the general enthalpy differential is

$$h_b - h_a = \int_{T_1}^{T_2} c_P^0 \, dT$$

where c_P^0 is the ideal gas specific heat at constant pressure. This result is identical to that of the previous section. Ideal gas with variable specific-heat enthalpy equations or tables can be used to evaluate this portion of the enthalpy difference.

The last portion of the process that connects state 1 to state 2 is process b–2, which is isothermal just like 1–a. The enthalpy difference for this process is

$$h_2 - h_b = \int_{0}^{P_2} \left[v - T \left. \frac{\partial v}{\partial T} \right|_P \right] dP$$

which is the negative of the enthalpy departure for state 2.

The enthalpy difference between states 1 and 2 is

$$h_2 - h_1 = (h_2 - h_b) + (h_b - h_a) + (h_a - h_1)$$
$$= [(h_a - h_1) - (h_b - h_2)] + (h_b - h_a)$$

In words, the enthalpy difference is the difference between the enthalpy departures of the two states plus the enthalpy change if the real gas had been an ideal gas.

Enthalpy departure is a function of the gas pressure and temperature, as pointed out earlier, which may be evaluated from the equation of state. The principle of corresponding states suggests that all gases may be reduced to one gas in terms of reduced pressure and temperature. The enthalpy departure of this universal gas is presented in Figure C.4 of Appendix C.

Applying the same approach to the difference of the specific entropy between states 1 and 2 of Figure 9.8 gives

$$s_2 - s_1 = [(s_a - s_1) - (s_b - s_2)] + (s_b^0 - s_a^0)$$

where the entropy departure is given by

$$s_a - s_1 = -\int_0^{P_1} \frac{\partial v}{\partial T}\bigg|_P dP$$

and the ideal gas entropy at zero pressure is given by

$$s^0 = \int_{T_{ref}}^T \frac{c_P^0}{T} dT$$

A chart based on the principle of corresponding states for determining the entropy departure of a universal gas is presented as Figure C.5 in Appendix C. Ideal gas with variable specific-heat entropy tables or relations may be used to evaluate $s_b^0 - s_a^0$.

EXAMPLE 9.15

The third stage of an oxygen compressor adiabatically compresses 0.15 kilograms of oxygen per second from 10,000 kPa, 14° C, to 15,000 kPa, 127° C. Determine the power required by this compressor and the rate at which it generates entropy.

System: The oxygen within a control volume formed by a steady-flow compressor with one inlet and one outlet.

Basic Equations:
$$_1q_2 - _1w_2 = h_2 - h_1 + \frac{1}{2}(V_2^2 - V_1^2) + g(z_2 - z_1)$$

Conditions: The fluid's kinetic and potential energies will be neglected.

Solution: According to Table 9.2, $T_c = 154.8$ K, and $P_c = 5080$ kPa for oxygen. At the compressor inlet, $T_{r,1} = 287/154.8 = 1.85$ and $P_{r,1} = 10,000/5080 = 1.97$. The enthalpy departure at the compressor inlet is then $(\bar{h}^0 - \bar{h})_1 = 0.7\Re T_c = 900$ kJ/kg-mole, according to Figure C.4 of Appendix C. Similarly, the entropy departure at the compressor inlet is $(\bar{s}^0 - \bar{s})_1 = 0.33\Re = 2.74$ kJ/kg-mole-K, according to Figure C.5. At the compressor outlet, $T_{r,1} = 400/154.8 = 2.58$ and $P_{r,1} = 15,000/5080 = 2.95$. Hence, $(\bar{h}^0 - \bar{h})_2 = 0.48\Re T_c = 618$ kJ/kg-mole and $(\bar{s}^0 - \bar{s})_2 = 0.23\Re = 1.92$ kJ/kg-mole-K.

Using the oxygen ideal gas Table B.6.6 of Appendix B, $\bar{h}_2^0 - \bar{h}_1^0 = 11{,}709 - 8356 = 3353$ kJ/kg-mole and $\bar{s}_2^0 - \bar{s}_1^0 = 213.72 - 203.86 = 9.86$ kJ/kg-mole-K.

Application of the first law to the compressor produces

$$_1\dot{W}_2 = \frac{\dot{m}(\bar{h}_1 - \bar{h}_2)}{M}$$

$$= \frac{\dot{m}[(\bar{h} - \bar{h}^0)_1 + (\bar{h}_1^0 - \bar{h}_2^0) + (\bar{h}^0 - \bar{h})_2]}{M}$$

$$= \frac{0.15[-900 - 3353 + 618]}{32}$$

$$= -17.0 \text{ kW}$$

The only entropy being generated in this system is that of the oxygen passing through the compressor because it is adiabatic. The rate at which the oxygen generates entropy is given by

$$\Delta \dot{s} = \frac{\dot{m}(\bar{s}_2 - \bar{s}_1)}{M}$$

$$= \frac{\dot{m}[(\bar{s} - \bar{s}^0)_2 + (\bar{s}_2^0 - \bar{s}_1^0) + (\bar{s}^0 - \bar{s})_1]}{M}$$

$$= \frac{0.15[-1.92 + 9.86 + 2.74]}{32}$$

$$= 0.0501 \text{ kW/kg-K}$$

PROBLEMS

In Problems 9.1 through 9.12, the specific heats may be taken as constant and the equation of state is

$$Pv = RT + aP$$

where R is the ideal gas constant and a is the coefficient determined from experimental measurements.

9.1 Compare the difference in the specific internal energy and volume of air (kJ/kg) at 100 kPa, 20° C, and 600 kPa, 300° C, as predicted by the above equation of state with $a = 1$ N-m/g and by the ideal gas equation of state.

9.2   Compare the difference in the specific enthalpy (kJ/kg) of air at 100 kPa, 20° C, and 600 kPa, 300° C, as predicted by the above equation of state with $a = 0.01 \text{ m}^3/\text{kg}$ and by the ideal gas equation of state.

9.3   Compare the difference in the specific entropy (kJ/kg-K) of air at 100 kPa, 20° C, and 600 kPa, 300° C, as predicted by the above equation of state with $a = 0.01 \text{ m}^3/\text{kg}$ and by the ideal gas equation of state.

9.4   Compare the difference in the specific internal energy (kJ/kg) of air at 100 kPa, 20° C, and 600 kPa, 300° C, as predicted by the above equation of state with $a = 0.1 \text{ m}^3/\text{kg}$ and by the ideal gas equation of state.

9.5   Compare the difference in the specific enthalpy (kJ/kg) of air at 100 kPa, 20° C, and 600 kPa, 300° C, as predicted by the above equation of state with $a = 0.1 \text{ m}^3/\text{kg}$ and by the ideal gas equation of state.

9.6   Compare the difference in the specific entropy (kJ/kg-K) of air at 100 kPa, 20° C, and 600 kPa, 300° C, as predicted by the above equation of state with $a = 0.1 \text{ m}^3/\text{kg}$ and by the ideal gas equation of state.

9.7   Compare the difference in the specific internal energy (kJ/kg) of helium at 100 kPa, 20° C, and 600 kPa, 300° C, as predicted by the above equation of state with $a = 0.01 \text{ m}^3/\text{kg}$ and by the ideal gas equation of state.

9.8   Compare the difference in the specific enthalpy (kJ/kg) of helium at 100 kPa, 20° C, and 600 kPa, 300° C, as predicted by the above equation of state with $a = 0.01 \text{ m}^3/\text{kg}$ and by the ideal gas equation of state.

9.9   Compare the difference in the specific entropy (kJ/kg-K) of helium at 100 kPa, 20° C, and 600 kPa, 300° C, as predicted by the above equation of state with $a = 0.01 \text{ m}^3/\text{kg}$ and by the ideal gas equation of state.

9.10   Compare the difference in the specific internal energy (kJ/kg) of helium at 100 kPa, 20° C, and 600 kPa, 300° C, as predicted by the above equation of state with $a = 0.01 \text{ m}^3/\text{kg}$ and by the ideal gas equation of state.

9.11 Compare the difference in the specific enthalpy (kJ/kg) of helium at 100 kPa, 20° C, and 600 kPa, 300° C, as predicted by the above equation of state with $a = 0.1$ m³/kg and by the ideal gas equation of state.

9.12 Compare the difference in the specific entropy (kJ/kg-K) of helium at 100 kPa, 20° C, and 600 kPa, 300° C, as predicted by the above equation of state with $a = 0.1$ m³/kg and by the ideal gas equation of state.

9.13 Derive an expression for the volume expansivity of a substance whose equation of state is $Pv = RT + aP$.

9.14 Derive an expression for the isothermal compressibility of a substance whose equation of state is $Pv = RT + aP$.

9.15 Derive an expression for the specific-heat difference of a substance whose equation of state is $Pv = RT + aP$.

9.16 Demonstrate that

$$c_P = T \left.\frac{\partial P}{\partial T}\right|_s \left.\frac{\partial v}{\partial T}\right|_P$$

9.17 Prove that

$$c_v = -T \left.\frac{\partial P}{\partial T}\right|_v \left.\frac{\partial v}{\partial T}\right|_s$$

9.18 Show that

$$c_P - c_v = T \left.\frac{\partial P}{\partial T}\right|_v \left.\frac{\partial v}{\partial T}\right|_P$$

9.19 Prove that

$$\left.\frac{\partial P}{\partial T}\right|_s = \frac{k}{k-1} \left.\frac{\partial P}{\partial T}\right|_v$$

9.20 Show how you would evaluate T, v, u, a, and g from the thermodynamic function

$$h = h(s, P)$$

9.21 The Helmholtz function of a substance has the form

$$a = -RT \ln \frac{v}{v_0} - cT_0 \left(1 - \frac{T}{T_0} + \frac{T}{T_0} \ln \frac{T}{T_0}\right)$$

where T_0 and v_0 are the temperature and specific volume at a reference state. Show how to obtain P, h, s, c_v, and c_P from this expression.

9.22   Derive an expression for the volume expansivity of a substance whose equation of state is

$$P = \frac{RT}{v - b} - \frac{a}{v(v + b)T^{1/2}}$$

where a and b are empirical constants.

9.23   Derive an expression for the isothermal compressibility of a substance whose equation of state is

$$P = \frac{RT}{v - b} - \frac{a}{v(v + b)T^{1/2}}$$

where a and b are empirical constants.

9.24   Derive an expression for the specific-heat difference of a substance whose equation of state is

$$P = \frac{RT}{v - b} - \frac{a}{v(v + b)T^{1/2}}$$

where a and b are empirical constants.

9.25   Derive an expression for the volume expansivity of a substance whose equation of state is

$$P = \frac{RT}{v - b} - \frac{a}{v^2 T}$$

where a and b are empirical constants.

9.26   Derive an expression for the isothermal compressibility of a substance whose equation of state is

$$P = \frac{RT}{v - b} - \frac{a}{v^2 T}$$

where a and b are empirical constants.

9.27   Show that

$$\beta = \kappa \left. \frac{\partial P}{\partial T} \right|_v$$

9.28   Demostrate that

$$k = \frac{c_P}{c_v} = -\frac{v\kappa}{\left. \frac{\partial v}{\partial P} \right|_s}$$

9.29 Derive an expression for the specific-heat difference of a substance whose equation of state is

$$P = \frac{RT}{v - b} - \frac{a}{v^2 T}$$

where a and b are empirical constants.

9.30 Demonstrate that the Joule–Thomson coefficient is given by

$$\mu = \frac{T^2}{c_P} \left. \frac{\partial (v/T)}{\partial T} \right|_P$$

9.31 What is the most general equation of state for which the Joule–Thomson coefficient is always zero?

9.32 Estimate the Joule–Thomson coefficient ($°R$/psia) of ammonia at 40 psia, $80°F$.

9.33 Estimate the Joule–Thomson coefficient ($°R$/psia) of ammonia at 40 psia, $20°F$.

9.34 Estimate the Joule–Thomson coefficient (K/kPa) of R-12 at 200 kPa, $100°C$.

9.35 Estimate the Joule–Thomson coefficient (K/kPa) of R-12 at 200 kPa, $0°C$.

9.36 The equation of state of a gas is given by

$$v = \frac{RT}{P} - \frac{a}{T} + b$$

where a and b are constants. Use this equation of state to derive an equation for the Joule–Thomson coefficient inversion line.

9.37 The equation of state of a gas is given by

$$v = \frac{RT}{P} - \frac{bP}{T^2}$$

where b is a constant. Use this equation to derive an expression for the Joule–Thomson coefficient inversion line.

9.38 Two grams of a saturated liquid are converted to a saturated vapor by heat in a weighted piston device arranged to maintain the pressure at 200 kPa. During the phase conversion, the system volume increases by 1000 cm^3, 5 kilojoules of heat are required, and the temperature of the substance stays constant at $80°C$. Estimate the boiling-point temperature (K) of this substance when its pressure is 180 kPa.

9.39 Estimate the saturation pressure (kPa) of the substance in the preceding problem when its temperature is $100°C$.

9.40 Estimate s_{fg} (kJ/kg-K) of the substance in Problem 9.38 when its temperature is 90° C.

9.41 Show that

$$c_{P,g} - c_{P,f} = T \left. \frac{\partial (h_{fg}/T)}{\partial T} \right|_P + v_{fg} \left. \frac{dP}{dT} \right|_{sat}$$

9.42 One-half pound-mass of a saturated vapor is converted to a saturated liquid by being cooled in a weighted piston-cylinder device maintained at 50 psia. During the phase conversion, the system volume decreases by 1.5 ft³, 250 Btu of heat are removed, and the temperature remains fixed at 15° F. Estimate the boiling temperature (° R) of this substance when its pressure is 60 psia.

9.43 Estimate the saturation pressure (psia) of the substance in the preceding problem when its temperature is 10° F.

9.44 Estimate s_{fg} (Btu/lb$_m$-° R) of the substance in Problem 9.42 when its temperature is 20° F.

9.45 A table of properties for methyl chloride lists the saturation pressure as 116.7 psia at 100° F. At 100° F, this table also lists $h_{fg} = 154.85$ Btu/lb$_m$ and $v_{fg} = 0.86332$ ft³/lb$_m$. Estimate the saturation pressures (psia) at 90° F and 110° F.

9.46 A saturation table for R-22 lists the following for $-40°$ C, $P = 105.6$ kPa; $h_{fg} = 233.67$ kJ/kg and $v_{fg} = 5.481$m³/kg. Estimate the saturation pressure (kPa) of R-22 at $-50°$ C and $-30°$ C.

9.47 A saturation table for R-22 lists the following for 0° F: $P = 2.555$ psia; $h_{fg} = 84.38$ Btu/lb$_m$, and $v_{fg} = 1.384$ ft³/lb$_m$. Estimate the saturation pressure (psia) of R-22 at $-50°$ C and $-30°$ C.

9.48 One hundred grams of carbon monoxide are contained in a weighted piston-cylinder device. Initially, the carbon monoxide is at 1000 kPa, 200° C. It is then heated until its temperature is 500° C. Determine the final volume (m³) of the carbon monoxide treating it as (a) an ideal gas, (b) a Redlich–Kwong gas, and (c) a Benedict–Webb–Rubin gas.

9.49 One pound-mass of carbon dioxide is heated in a constant-pressure apparatus. Initially, the carbon dioxide is at 1000 psia, 200° F, and it is heated until its temperature becomes 800° F. Determine the final volume (ft³) of the carbon dioxide treating it as (a) an ideal gas, (b) a Redlich–Kwong gas, and (c) a Benedict–Webb–Rubin gas.

9.50 Methane is heated in a rigid container from 100 kPa, 20° C, to 400° C. Determine the final pressure (kPa) of the methane treating it as (a) an ideal gas, (b) a Redlich–Kwong gas, and (c) a Benedict–Webb–Rubin gas.

9.51 Carbon monoxide is heated in a rigid container from 14.7 psia, 70° F, to 800° F. Determine the final pressure (psia) of the carbon monoxide treating it as (a) an ideal gas, (b) a Redlich–Kwong gas, and (c) a Benedict–Webb–Rubin gas.

9.52 One kilogram of carbon dioxide is compressed from 1000 kPa, 200° C, to 3000 kPa in a piston-cylinder device arranged to execute a polytropic process ($n = 1.2$). Determine the final temperature (° C), treating the carbon dioxide as (a) an ideal gas and (b) a Van der Waals gas.

9.53 One kilogram of carbon dioxide is compressed from 1000 kPa, 200° C, to 3000 kPa in a piston-cylinder device arranged to execute a polytropic process ($n = 1.2$). Determine the final temperature (° C), treating the carbon dioxide as (a) an ideal gas and (b) a Van der Waals gas.

9.54 One-half pound-mass of methane is compressed from 1000 psia, 300° F, to 2000 psia in a piston-cylinder device that executes a polytropic process ($n = 1.6$). Determine the final temperature (° F), treating the methane as (a) an ideal gas and (b) a Redlich–Kwong gas.

9.55 What are the first eight virial coefficients of a Benedict–Webb–Rubin gas? *Hint:* Expand the exponential term in a power-series expansion.

9.56 Can the Redlich–Kwong equation of state be expressed in a virial form?

9.57 Methane is maintained at 4000 kPa, 20° C. Compare the specific volume (m^3/kg) of the methane under this condition as predicted by (a) the ideal gas equation of state, (b) the Redlich–Kwong equation of state, and (c) the Benedict–Webb–Rubin equation of state, as well as with (d) the compressibility factor.

9.58 Nitrogen is maintained at 400 psia, −100° F. Compare the specific volume (ft^3/lb_m) of this nitrogen as predicted by (a) the ideal gas equation of state, (b) the Redlich–Kwong equation of state, and (c) the Benedict–Webb–Rubin equation of state, as well as with (d) the compressibility factor.

9.59 Ethane in a rigid vessel is to be heated from 50 psia, 100° F, until its temperature is 600° F. What is the final pressure (psia) of the ethane as predicted by the compressibility factor?

9.60 Ethylene is heated at constant pressure from 5000 kPa, 20° C, to 200° C. Using the compressibility factor, determine the change in the ethylene's specific volume (m^3/kg) as a result of this heating.

9.61 One kilogram of carbon dioxide is compressed from 1000 kPa, 200° C, to 3000 kPa in a piston-cylinder device arranged to execute a polytropic process ($n = 1.2$). Use the compressibility factor to determine the final temperature in ° C.

9.62   Saturated steam vapor at $600°$ F is heated at constant pressure until its volume has doubled. Determine the final temperature ($°$ F), using the ideal gas equation of state, the compressibility charts, and the steam tables.

9.63   Saturated steam vapor at $400°$ F is heated at constant pressure until its volume has doubled. Determine the final temperature ($°$ C), using the ideal gas equation of state, the compressibility charts, and the steam tables.

9.64   Ammonia at 250 psia, $300°$ F, is expanded in an isothermal process to 125 psia. Determine the final specific volume (ft^3/lb_m), using the ideal gas equation of state, the compressibility factor, and ammonia tables.

9.65   Ammonia at 2000 kPa, $160°$ C, is expanded in an isothermal process to 1000 kPa. Determine the final specific volume (m^3/kg), using the ideal gas equation of state, the compressibility factor, and ammonia tables.

9.66   Methane at 1000 psia, $100°$ F, is heated at constant pressure until its volume has increased by 50%. Determine the final temperature ($°$ F), using the ideal gas equation of state and the compressibility factor. Which of these two results is the more accurate?

9.67   Methane at 10,000 kPa, $100°$ C, is heated at constant pressure until its volume has increased by 50%. Determine the final temperature ($°$ C), using the ideal gas equation of state and the compressibility factor. Which of these two results is the more accurate?

9.68   Develop expressions for h, u, s^0, P_r, and v_r for an ideal gas whose c_P^0 is given by

$$c_P^0 = \sum_{i=0}^{n} a_i T^{i-n} + a_0 e^{\beta/T} \left[\frac{\beta/T}{e^{\beta/T} - 1} \right]$$

where a_i, n, and β are empirical constants.

9.69   Develop expressions for h, u, s^0, P_r, and v_r for an ideal gas whose c_P^0 is given by

$$c_P^0 = \sum_{i=0}^{n} a_i [\ln(T/T_r)]^{i-n}$$

where a_i, n, and T_r are empirical constants.

9.70   Water vapor at 3000 psia, $1500°$ F, is converted to 1000 psia, $1000°$ F. Calculate the change in the specific entropy ($\text{Btu/lb}_m\text{-}°$ R) and enthalpy (Btu/lb_m) of this water vapor, using (a) the departure charts and (b) the property tables.

9.71   Water vapor at 1000 kPa, $600°$ C, is converted to 500 psia, $400°$ C. Calculate the change in the specific entropy (kJ/kg-K) and enthalpy (kJ/kg) of this water vapor, using (a) the departure charts and (b) the property tables.

9.72 Saturated water vapor at 500° F is expanded while its pressure is kept constant until its temperature is 1000° F. Calculate the change in the specific enthalpy (Btu/lb$_m$) and entropy (Btu/lb$_m$-° R), using (a) the departure charts and (b) the property tables.

9.73 Saturated water vapor at 300° C is expanded while its pressure is kept constant until its temperature is 700° C. Calculate the change in the specific enthalpy (kJ/kg) and entropy (kJ/kg-K), using (a) the departure charts and (b) the property tables.

9.74 Methane at 50 psia, 100° F, is compressed in a steady-flow device to 500 psia, 1100° F. Calculate the change in the specific entropy (Btu/lb$_m$-° R) of the methane and the specific work (Btu/lb$_m$) required for this compression, (a) using the departure charts and (b) treating the methane as an ideal gas with temperature-variable specific heats.

9.75 Determine the second law effectiveness of the compression in the previous problem.

9.76 Propane at 500 kPa, 100° C, is compressed in a steady-flow device to 4000 kPa, 500° C. Calculate the change in the specific entropy (kJ/kg-K) of the propane and the specific work (kJ/kg) required for this compression, (a) using the departure charts and (b) treating the propane as an ideal gas with temperature-variable specific heats.

9.77 Determine the second law effectiveness of the compression in the previous problem.

9.78 Methane is to be compressed adiabatically and reversibly from 50 psia, 100° F, to 500 psia. Calculate the specific work (Btu/lb$_m$) required for this compression treating the methane as an ideal gas with temperature-variable specific heats.

9.79 Propane is to be compressed adiabatically and reversibly from 500 kPa, 100° C, to 4000 kPa. Calculate the specific work (kJ/kg) required for this compression treating the propane as an ideal gas with temperature-variable specific heats.

9.80 Oxygen is adiabatically and reversibly expanded in a nozzle from 200 psia, 600° F, to 70 psia. Determine the velocity at which the oxygen leaves the nozzle, assuming that it enters with negligible velocity, treating the oxygen as an ideal gas with temperature-variable specific heats.

9.81 Ethylene at 1000 kPa, $0°$ C, is to be compressed adiabatically and reversibly to 10,000 kPa. Calculate the specific work (kJ/kg) that will be required for this compression.

9.82 Ethylene at 100 psia, $-50°$ F, is to be adiabatically and reversibly compressed to 1000 kPa. Calculate the specific work (Btu/lb$_m$) that will be required for this compression.

9.83 Air is compressed in a steady-flow, isentropic process from 14.7 psia, $80°$ F, to 500 psia. Calculate the specific work (Btu/lb$_m$) required for this compression, treating the air as an ideal gas with (a) constant specific heats and (b) variable specific heats.

9.84 Air is compressed in a steady-flow, isentropic process from 100 kPa, $20°$ C, to 5000 kPa. Calculate the specific work (kJ/kg) required for this compression, treating the air as an ideal gas with (a) constant specific heats and (b) variable specific heats.

9.85 A Brayton cycle operates with air entering the compressor at 13 psia, $20°$ F, a pressure ratio of 12:1, and temperature of $1000°$ F at the turbine entrance. Calculate the net specific work (Btu/lb$_m$) produced by this cycle, treating the air as an ideal gas with (a) constant specific heats and (b) variable specific heats.

9.86 A Brayton cycle operates with air entering the compressor at 70 kPa, $0°$ C, a pressure ratio of 15:1 and a temperature of $600°$ C at the turbine entrance. Calculate the net specific work (kJ/kg) produced by this cycle, treating the air as an ideal gas with (a) constant specific heats and (b) variable specific heats.

9.87 Air is expanded in an isentropic process from 1000 psia, $600°$ F to 10 psia in a closed system. Determine the specific work potential (Btu/lb$_m$) for this process, treating the air as an ideal gas with (a) constant specific heats and (b) variable specific heats.

9.88 Air is expanded in an isentropic process from 1000 kPa, $200°$ C, to 10 kPa in a closed system. Determine the specific work potential (kJ/kg) for this process, treating the air as an ideal gas with (a) constant specific heats and (b) variable specific heats.

9.89 Carbon dioxide is compressed isentropically in a steady-flow device from 15 psia, $80°$ F, to 2500 psia. Calculate the specific work (Btu/lb$_m$) required for this compression and the final temperature ($°$ F), treating the carbon dioxide as an ideal gas with (a) constant specific heats and (b) temperature-variable specific heats.

9.90 Carbon dioxide is compressed isentropically in a steady-flow device from 100 kPa, 30° C, to 10,000 kPa. Calculate the specific work (kJ/Kg) required for this compression and the final temperature (° C), treating the carbon dioxide as an ideal gas with (a) constant specific heats and (b) temperature-variable specific heats.

9.91 An Otto cycle with a compression ratio of 8:1 begins its compression at 13.5 psia, 50° F. The maximum cycle temperature is 1500° F. Determine the thermal efficiency of this cycle, using air standard analysis and treating the air as an ideal gas with temperature-variable specific heats. Also calculate the thermal efficiency of this cycle, treating the air as if the specific heats are constant.

9.92 An Otto cycle with a compression ratio of 7:1 begins its compression at 90 kPa, 15° C. The maximum cycle temperature is 1000° C. Determine the thermal efficiency of this cycle, using air standard analysis and treating the air as an ideal gas with temperature-variable specific heats. Also calculate the thermal efficiency of this cycle, treating the air as if the specific heats are constant.

9.93 A pure Diesel cycle has a compression ratio of 22:1 and begins its compression at 85 kPa, 15° C. The maximum cycle temperature is 1200° C. Calculate the thermal efficiency of this cycle, using air standard analysis and treating the air with temperature-variable specific heats. Also calculate the thermal efficiency if the specific heats are constant.

9.94 A pure Diesel cycle has a compression ratio of 20:1 and begins its compression at 13 psia, 45° F. The maximum cycle temperature is 1800° F. Calculate the thermal efficiency of this cycle, using air standard analysis and treating the air with temperature-variable specific heats. Also calculate the thermal efficiency if the specific heats are constant.

9.95 Carbon monoxide enters an isentropic diffuser at 200 psia, 600° F, with a velocity of 2000 ft/s. It leaves the diffuser at 20 psia. Calculate the exit velocity (ft/s), treating the carbon monoxide as an ideal gas with (a) constant specific heats and (b) temperature-variable specific heats.

9.96 Hydrogen enters an isentropic diffuser at 1000 kPa, 600° C, with a velocity of 1000 m/s. It leaves the diffuser at 1500 kPa. Calculate the exit velocity (m/s), treating the hydrogen as an ideal gas with (a) constant specific heats and (b) temperature-variable specific heats.

PROJECTS

The following projects are more extensive than the problems presented in this chapter. Most are open-ended and require some decisions on the student's part. Many can be expedited with computer-based property tables, computer spreadsheets, and student-written computer programs.

9.1   The empirical equation of state for most CFC refrigerants is

$$P = \frac{RT}{v-b} + \sum_{i=2}^{5} \frac{A_i + B_i T + C_i e^{\kappa T/T_c}}{(v-b)^i} + \frac{A_6 + B_6 + C_6 e^{\kappa T/T_c}}{e^{\alpha v}(1 + ce^{\alpha v})}$$

where A_i, B_i, C_i, b, c, α, and κ are empirical constants, and T_c is the absolute critical temperature. The zero-pressure specific heat at constant volume can be fit empirically by

$$c_v^0 = \sum_{i=1}^{4} G_i T^{i-1} + \frac{G_5}{T^2}$$

where the G_i's are empirical constants. The saturation pressure–temperature relationship is

$$\ln P_{sat} = F_1 + \frac{F_2}{T} + F_3 \ln T + F_4 T + F_5 \frac{\gamma - T}{T} \ln(\gamma - T)$$

where the F's and γ are experimentally determined constants. The saturated-liquid specific volume is related to the temperature by

$$\frac{1}{v_f} = \sum_{i=1}^{5} D_i X^{(i-1)/3} + D_6 X^{1/2} + D_7 X^2$$

where the D's are constants and

$$X = 1 - \frac{T}{T_c}$$

Set up the equations and methods needed to generate saturation and vapor tables of properties using this information.

9.2   Refrigerant-13 may be modeled with the equations of the previous project. The general data for R-13 are $M = 104.47$, $T_c = 302.0$ K, $P_c = 3869.7$ kPa, $v_c = 0.00173073$ m^3/kg, and $T_0 = 150$ K. The constants in Table 9.4 are for T in K, v in m^3/kg, P in Pa, and c_v in J/(kg-K). Write a computer program that may be used to determine the saturation properties of R-13.

TABLE 9.4

Empirical Constants for Refrigerant-13

P–T–v		P_{sat}		c_v^0,	
b =	$2.9956353561 \times 10^{-4}$	F_1 =	6.4251423714×10^1	G_1 =	6.7072536000×10^1
A_2 =	$-8.2852902033 \times 10^1$	F_2 =	$-3.14161026958 \times 10^3$	G_2 =	2.1274805520
A_3 =	$9.8725559035 \times 10^{-2}$	F_3 =	-7.1723439130	G_3 =	$-1.5722103888 \times 10^{-3}$
A_4 =	$-1.0744963678 \times 10^{-4}$	F_4 =	$1.0548780505 \times 10^{-2}$	u_0 =	1.5514715×10^5
A_5 =	$3.4587476371 \times 10^{-8}$	F_5 =	$2.8030109130 \times 10^{-1}$	s_0 =	1.0479552×10^3
A_6 =	$5.0873683031 \times 10^{11}$	γ =	303.33		
B_2 =	$1.1326045344 \times 10^{-1}$				
B_3 =	$-1.7124042321 \times 10^{-4}$				
B_4 =	$2.5233687878 \times 10^{-7}$				
B_5 =	$-8.7021612764 \times 10^{-11}$				
B_6 =	$-9.2279585051 \times 10^8$	v_f			
C_2 =	$-4.8938250202 \times 10^2$				
C_3 =	$9.5943985617 \times 10^{-1}$	$D_1 = 5.77786670 \times 10^2$			
C_4 =	0.0	$D_2 = 8.71328290 \times 10^2$			
C_5 =	$-2.5327694644 \times 10^{-7}$	$D_3 = 0.0$			
C_6 =	0.0	$D_4 = 1.36361900 \times 10^2$			
κ =	4.0	$D_5 = 0.0$			
α =	10011.5625	$D_6 = 4.14557270 \times 10^2$			
R =	79.5900	$D_7 = 1.53601492 \times 10^2$			

9.3 Write a computer program that can be used to determine the properties of superheated R-13 vapor; use the data of the previous project.

9.4 Many hydrocarbons are described by the empirical equation of state:

$$P = \frac{RT}{v} + \left(B_0 RT - A_0 - \frac{C_0}{T^2} + \frac{D_0}{T^3} - \frac{E_0}{T^4} \right) \frac{1}{v^2} + \left(bRT - a - \frac{d}{T} \right) \frac{1}{v^3}$$

$$+ \alpha \left(a + \frac{d}{T} \right) \frac{1}{v^6} + c \frac{1 + \gamma/v^2}{T^2 v^3} e^{\gamma/v^2}$$

The saturation pressure is given by the empirical equation

$$\ln \frac{P}{P_c} = \left(\frac{T_c}{T} - 1 \right) \sum_{i=1}^{8} F_i \left(\frac{T}{T_p} - 1 \right)^{i-1}$$

The saturated-liquid specific volume of these hydrocarbons is given by

$$v_f = \frac{1}{\sum_{i=1}^{5} D_i X^{(i-1)/3} + D_6 X^{0.5} + D_7 X^2}$$

where

$$X = 1 - \frac{T}{T_c}$$

Constant-volume specific heat at very low pressures for these hydrocarbons is represented by

$$c_v^0 = \sum_{i=1}^{6} G_i T^{i-2}$$

Develop the equations and methods to generate saturation and superheated vapor tables for these hydrocarbons.

9.5 Propane can be modeled by the equations of the previous project. The general constants for propane are $M = 44.09$, $T_c = 369.83$ K, $P_c = 4236.2$ kPa, $v_c = 0.00506637$ m^3/kg, and $T_0 = 200$ K. The empirical constants for the equations of the previous project are presented in Table 9.5. The constants in Table 9.5 are for T in K, v in m^3/kg, P in Pa, and c_v in J/(kg-K). Write a computer program that may be used to determine the saturation properties of propane.

TABLE 9.5

Empirical Constants for Propane

$P-T-v$	P_{sat}	v_f	c_v^0
$R = 1.887326 \times 10^2$	$F_1 = -6.2309993$	$D_1 = 1.9738193 \times 10^2$	$G_1 = 2.0582170 \times 10^5$
$B_0 = 1.366892 \times 10^{-3}$	$F_2 = -4.4226860 \times 10^{-1}$	$D_2 = -2.1307184 \times 10^1$	$G_2 = -1.9109547 \times 10^3$
$A_0 = 2.579108 \times 10^2$	$F_3 = -1.8839624$	$D_3 = 3.3522024 \times 10^3$	$G_3 = 1.1622054 \times 10^1$
$C_0 = 3.401044 \times 10^7$	$F_4 = 3.6383362 \times 10^{-1}$	$D_4 = -7.7040243 \times 10^3$	$G_4 = -9.795151 \times 10^{-3}$
$D_0 = 1.076728 \times 10^9$	$F_5 = 1.5177354 \times 10^1$	$D_5 = 7.5224059 \times 10^3$	$G_5 = 4.5167026 \times 10^{-6}$
$E_0 = 3.375879 \times 10^{10}$	$F_6 = 1.121655 \times 10^2$	$D_6 = -2.5663363 \times 10^3$	$G_6 = -8.6345035 \times 10^{-10}$
$b = 1.096523 \times 10^{-5}$	$F_7 = 2.7635840 \times 10^2$		$u_0 = 4.2027216 \times 10^5$
$a = 7.856721 \times 10^{-1}$	$F_8 = 2.3585357 \times 10^2$		$s_0 = 2.1673997 \times 10^3$
$d = 1.639769 \times 10^2$	$T_F = 300.$		
$c = 1.661103 \times 10^5$	$P_c = 4.2359300 \times 10^6$		
$\alpha = 5.728034 \times 10^{-9}$			
$\gamma = 9.157270 \times 10^{-6}$			

9.6 Write a computer program that can be used to determine the properties of superheated propane vapor, using the data of the previous project.

9.7 Design and complete a table for methane as an ideal gas with temperature-variable specific heats.

9.8 Design and complete a table for propane as an ideal gas with temperature-variable specific heats.

9.9 How large can the pressure ratio of a Brayton cycle become before it is no longer accurate to use air standard analysis with constant specific heats?

9.10 How large can the compression ratio of an Otto cycle become before air standard analysis with constant specific heats is no longer valid?

Homogeneous Gaseous Mixtures

Solving Problems

Where principle is put to work, not as a recipe or formula, there will always be style.

Le Corbusier

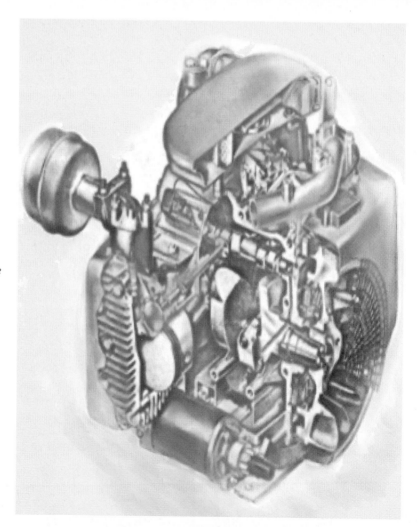

A wide variety of working fluids can be formed simply by blending mixtures of other working fluids. By varying the relative amounts of the constituents used in the mixture, we can alter and tailor the properties of the mixture to fit a specific purpose. Because an infinite number of combinations of constituents and blending ratios are available, it is impossible to generate property tables for any but the most common and widely used mixtures. Instead, we must develop the means of combining the properties of the individual constituents in the mixture in relation to their relative amounts to predict the mixture's effective properties.

Atmospheric air is a common example of a gaseous mixture. It is formed of nitrogen, oxygen, argon, water vapor, and trace amounts of several other gases. To this point, we have treated it as if it were a single gas by using an effective molecular weight of 28.97, by using several other effective properties such as specific heats, and by neglecting the effects of the water vapor. Although this model has served us well, it cannot deal with human comfort when air humidity becomes important, and it cannot effectively handle cooling towers such as those used by some electrical power plants. This model of atmospheric air also is not accurate at very low temperatures such as those occurring in air-liquefaction plants, or very high pressures. We obviously need to develop further the methods we use to generate the effective properties of atmospheric air (as well as other mixtures of gases).

Combustion gases released by automobiles, aircraft, power plants, manufacturing plants, homes, processing plants, and many other sources constitute the second most predominant mixture of gases. Combustion gases consist of mixtures of nitrogen, oxygen, carbon dioxide, carbon monoxide, water vapor, nitrous oxides, sulfur dioxide, and traces of other gases. The exact blend of a specific combustion gas depends primarily on the fuel used and the amount of air used to burn the fuel. The mixing of the fuel and air during combustion, the amount of time allowed for the process, and the temperatures incurred during combustion also affect the resulting mixture. Once these combustion gases are introduced into the atmosphere and mixed with atmospheric air, problems may arise. For example, if the fuel contains any sulfur, the combustion gases will contain a sulfur dioxide constituent. When this constituent blends with the water in atmospheric air, weak sulfuric acid is formed. This acid will precipitate out of the air with rain and dew, causing the phenomenon known as *acid rain*. The severity of acid rain depends on the amount of sulfur dioxide released, the length of time over which it accumulates (as in lakes), and the ability of the atmosphere to diffuse the sulfur dioxide over a very large area. The ultimate cure for this problem is to prevent the release of sulfur dioxide by either removing the sulfur from the fuel or removing the sulfur dioxide from the combustion gases before they are released to the atmosphere.

In this chapter, we will develop methods to determine the effective properties of gaseous mixtures, including mixtures of gases and vapors. Means of characterizing mixtures in terms of partial pressures and partial volumes, as well as molar and mass fractions, are also presented. Applications of the first law and second law to these mixtures are also presented, and several common devices that process these mixtures are discussed.

10.1 Introduction

Often the working fluid that fills a system is a homogeneous mixture of two or more other substances. A common example is atmospheric air, which is a homogeneous mixture of nitrogen, oxygen, water vapor, argon, and several trace constituents. Other examples include the fuel-air mixtures (a portion of which may be in the liquid-phase form) introduced into various devices and the combustion products produced by the burning of these fuel–air mixtures. The first and second laws of thermodynamics presented earlier in this book are applicable to systems that utilize mixtures. Determining the properties to be used is the only factor that will change when the system working fluid is a mixture.

According to the conservation of mass principle, the total mass and number of moles of the mixture equals the sum of the mass and number of moles of each constituent in the mixture. In Chapter 3, we demonstrated how a mixture can be characterized by either its mole fractions, y_i, or its mass fractions, x_i. Either technique characterizes the relative amounts of the various constituents in a mixture on a mass basis. The conservation of mass principle tells us that

$$\sum_{i=1}^{n} x_i = 1$$

and

$$\sum_{i=1}^{n} y_i = 1$$

These two ways of specifying the portions of the mixture are interrelated by the molecular weights of the constituents,

$$M_i = \frac{m_i}{N_i}$$

where M_i is the molecular weight of constituent i, m_i is the constituent's mass, and N_i is the number of moles of the constituent. The apparent molecular weight of a mixture of several constituents is the total mass of the mixture divided by the total number of moles in the mixture,

$$M = \frac{m}{N} = \frac{\sum_{i=1}^{n} m_i}{\sum_{i=1}^{n} N_i}$$

But $m_i = M_i N_i$, which reduces the expression to

$$M = \frac{\sum_{i=1}^{n} N_i M_i}{N} = \sum_{i=1}^{n} y_i M_i$$

As a result, the apparent molecular weight is the mole-fraction weighted sum of the molecular weights of the constituents.

The apparent molecular weight of a mixture may be used to interrelate the mole and mass fractions. Dividing the mass of a constituent in the mixture by the mass of the mixture gives

$$x_i = \frac{m_i}{m} = \frac{N_i M_i}{NM}$$

$$= \frac{M_i}{M} y_i$$

Thus, the mass fraction is the product of the mole fraction and the ratio of the constituent's molecular weight to the mixture's apparent molecular weight.

Intuitively, we might suspect that other extensive properties of a mixture are simply the sum of the constituent's extensive properties like the mass property. For example, when the particles that form the substance behave independently of one another as they do in an ideal gas, then the simple summing approach is appropriate. The additive rule works in this situation because the addition of more, and perhaps foreign, particles has no effect on the previous particles. If the addition of more particles alters the behavior of the particles already present, then the additive rule is not correct. This can occur if the added particles have an effect on the intermolecular attractive or repulsive forces, the internal state of the particles, or other interactions. This is illustrated in Figure 10.1. In this figure, an extensive property, Z, of a mixture of two substances is plotted as the fraction of constituent A in the mixture is varied from 0 to 1 (and the fraction of B from 1 to 0). When the mixture is 100% constituent A, the value of the extensive property would be Z_A, and when the mixture is 100% constituent B, the property would be Z_B. If the particles of the resulting mixture behave independently, then the relationship between mixture property Z and the fraction of constituent A would be linear, as illustrated in Figure 10.1. The more general case where the particles of the two constituents interact with one another also is shown in the figure.

FIGURE 10.1

*Relationship
Between Extensive
Property Z and
Fraction of
Constituent A or B in
a Binary Mixture*

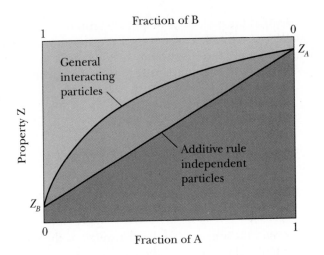

10.2 Amagat Volume Model for Ideal Gas Mixtures

Ideal gases have the characteristic that the distance between the molecules is very large (on a molecular scale) and that the elastic collisions between molecules are independent of the structure of the constituent molecules. In other words, these gases behave as a collection of independent particles that undergo elastic collisions whatever the construction of the molecules forming the gas. The way in which these gases are blended should have no effect on these collisions as long as the specific volume—and consequently the intermolecular distance—of the resulting mixture is sufficiently large so that the resulting mixture also behaves as an ideal gas.

Amagat applied this rationale to a mixture of two or more ideal gases. This model envisions each gas in the mixture as if it occupies its own volume, v_i, at the same pressure, P, and temperature, T, as the mixture itself. The volumes that the constituents occupy if they are alone are known as their *partial volumes*. The homogeneous mixture formed from these constituents has the same P and T as the constituents and fills a V that Amagat suggested is simply the sum of the partial volumes of the constituents,

$$V = \sum_{i=1}^{m} V_i$$

as illustrated in Figure 10.2.

According to the conservation of mass principle, the total number of moles in a mixture must equal the sum of the number of moles of each constituent used to form the mixture:

$$N = \sum_{i=1}^{n} N_i$$

FIGURE 10.2
Amagat Mixture of Ideal Gases

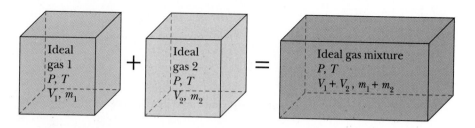

Applying the ideal gas equation of state to the total mixture—and to each constituent as if it existed alone at its partial volume—converts this expression to

$$\frac{PV}{\Re T} = \sum_{i=1}^{n} \frac{PV_i}{\Re T}$$

When the common pressure and temperature terms are eliminated, the expression becomes

$$V = \sum_{i=1}^{n} V_i$$

just as Amagat proposed. Amagat's model thus is valid for mixtures of ideal gases but may not be valid for other types of mixtures. Similarly, the ratio of the partial volume of the ith constituent to the total volume of the mixture is

$$\frac{V_i}{V} = \frac{N_i \Re T}{P} \frac{P}{N \Re T} = y_i$$

This result is the basis of volume analysis as a means of determining the mole fractions of mixtures of ideal gases.

Because the resulting mixture also behaves as an ideal gas, the effective gas constant of the mixture is

$$R_{mix} = \frac{\Re}{M} = \frac{\Re}{\sum_{i=1}^{n} y_i M_i}$$

EXAMPLE 10.1

An Orsat analyzer measures the volume fractions of carbon dioxide, oxygen, and carbon monoxide in a dried mixture of gases that result from burning carbon and hydrogen. An Orsat analysis of the dried exhaust products from an automobile engine yields 12% CO_2, 4% O_2, and 0.5% CO by volume. Presuming that the remainder of this gaseous mixture is nitrogen, determine the mixture's gas constant and the constituents' mass fractions.

System: The dried mixture of gases consisting of carbon dioxide, carbon monoxide, oxygen, and nitrogen.

Basic Equations:
$$y_i = \frac{V_i}{V}$$

Conditions: The constituents and mixture behave as ideal gases.

Solution: These calculations are best organized in a tabular format as in Table 10.1. This permits the use of spreadsheet software to analyze this problem. The effective molecular weight of the mixture is 30.08, and the mass fractions are tabulated in the last column of Table 10.1.

TABLE 10.1
Calculations of Example 10.1

GAS	y_i	M_i	$y_i M_i$	$x_i = y_i M_i / M$
CO_2	0.120	44	5.28	0.176
CO	0.005	28	0.14	0.005
O_2	0.040	32	1.28	0.043
N_2	0.835	28	23.38	0.777
			$M = 30.08$	1.000

10.3 Dalton's Partial-Pressure Model for Ideal Gases

John Dalton[1] suggested that each constituent gas in a mixture of gases fills the entire mixture volume rather than just the partial volume and has the same temperature as the mixture. Each constituent gas will then exist at its own *partial pressure*, P_i, as if the other constituent gases are not present, as illustrated in Figure 10.3. Again, the total number of moles in the mixture must equal the sum of the number of moles of each constituent,

$$\frac{PV}{\Re T} = \sum_{i=1}^{n} \frac{P_i V}{\Re T}$$

When the common terms are canceled, this becomes

$$P = \sum_{i=1}^{n} P_i$$

[1]Dalton (1766–1844) was an English chemist and physicist who is best known for his contributions to modern atomic theory. His hallmark work of 1803, *Absorption of Gases by Water and Other Liquids*, was the first to contain the law of partial pressures that was subsequently named in his honor.

where P_i is each constituent's partial pressure. The pressure exerted by the mixture of ideal gases is therefore the sum of the partial pressures each constituent would exert if it filled the entire mixture volume at the same temperature as the mixture.

FIGURE 10.3

Dalton's Partial-Pressure Model of Ideal Gas Mixtures

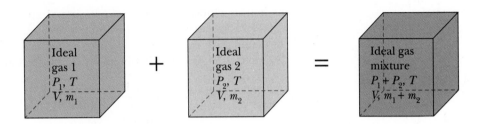

Applying the ideal gas equation of state to one constituent and the total mixture yields

$$\frac{N_i}{N} = \frac{P_i v/\Re T}{PV/\Re T} = \frac{P_i}{P} = y_i$$

Hence, Dalton's model is valid for mixtures of ideal gases but may not be for other mixtures. The ratio of the constituent's partial pressure to the total pressure exerted by the mixture is also the mole fraction of that constituent. Specification or measurement of partial pressures is another means of determining mole fractions.

Both the Amagat and Dalton models of mixtures of ideal gases are used to determine the constituents' mole fractions. Either model may be used to determine other properties of ideal gases, such as internal energy or entropy. Normally, the most convenient model is used, but the Dalton partial-pressure model tends to be the more widely applied.

EXAMPLE 10.2

An initially evacuated 20-liter scuba-diving tank is first filled with air until its pressure is 1700 kPa and its temperature is 20° C. Helium is next added to this tank until the pressure is 1900 kPa and the temperature is 20° C. Determine (a) the total mass of the gases in this tank after the helium is added and (b) the mass fraction of the helium in the resulting mixture.

System: The mixture of gases consisting of oxygen, nitrogen, and helium in the tank after it is filled.

Basic Equations:
$$y_i = \frac{P_i}{P}$$

Conditions: The constituents and mixture behave as ideal gases.

Solution: The mass contained in the tank when it is first filled with air only is

$$m_{air} = \frac{PV}{RT} = \frac{1700(20/1000)}{293(8.314/28.97)}$$
$$= 0.4043 \text{ kg}$$

Once the helium is added, the partial pressure of the helium is $P_{He} = 1900 - 1700 = 200$ kPa, because the volume and temperature of the air–helium mixture is the same as those of the air before the helium is added. Applying the ideal gas equation of state to the helium constituent alone gives

$$m_{He} = \frac{P_{He}V}{RT} = \frac{200(20/1000)}{293(8.314/4)}$$
$$= 0.00657 \text{ kg}$$

The total mass of the final mixture is then

$$m = m_{air} + m_{He} = 0.4109 \text{ kg}$$

and the mass fraction of the helium in the final mixture is

$$x_{He} = \frac{0.00657}{0.4109} = 0.0160$$

10.4 Other Properties of Ideal Gas Mixtures

The total extensive property of an ideal gas mixture is simply the sum of the values of the extensive property of each constituent in the mixture, as demonstrated by the Amagat mixture model:

$$Z = \sum_{i=1}^{n} Z_i \qquad (10.1)$$

The total volume of a mixture of ideal gases is then the sum of the partial volumes occupied by each constituent when it exists independently of other constituents at the same pressure and temperature as the total mixture:

$$V = \sum_{i=1}^{n} V_i(T, P)$$

Applying this relationship to the total internal energy of a mixture of ideal gases gives

$$U(T, V) = \sum_{i=1}^{n} U_i(T, V_i) = \sum_{i=1}^{n} m_i u_i(T, v_i) = \sum_{i=1}^{n} N_i \bar{u}_i(T, \bar{v}_i)$$

where $u_i(T, v_i)$ is the mass specific internal energy of a constituent at the mixture temperature and constituent partial volume, and $\bar{u}_i(T, \bar{v}_i)$ is the molar specific internal energy. But recall that the internal energy of an ideal gas depends only on the temperature of the gas, as discussed in Section 9.3. The specific internal energy of a mixture of ideal gases is therefore

$$u(T) = \frac{U(T)}{m} = \sum_{i=1}^{n} \frac{m_i}{m} u_i(T) = \sum_{i=1}^{n} x_i u_i(T)$$

where the constituent internal energies may be found in an ideal gas table for each constituent at the mixture temperature. When ideal gas tables such as those of Appendix B are used to determine the mixture's specific internal energy, each constituent table must use the same reference temperature for its construction.

Temperature-variable, constant-volume specific heats for each constituent also may be used to determine the mixture-specific internal energy. Substituting

$$u_i(T) = \int_{T_{ref}}^{T} c_{v,i}(T)dT$$

into the preceding expression for the mixture's specific internal energy yields

$$u(T) = \sum_{i=1}^{n} x_i \int_{T_{ref}}^{T} c_{v,i}(T)dT$$

where $u(T)$ of the mixture is with respect to the same T_{ref} as each constituent.

If the constant-volume specific heat of the constituents remains constant over the temperature range of concern or average c_v's are used, then the preceding integrations simplify to

$$u(T) = \sum_{i=1}^{n} x_i c_{v,i}(T - T_{ref})$$

Introducing an effective constant-volume specific heat for the mixture, such that the specific internal energy of the mixture with regard to T_{ref} is

$$u(T) = c_v(T - T_{ref})$$

reduces the previous result to the following:

$$c_v = \sum_{i=1}^{n} x_i c_{v,i} \qquad (10.2)$$

The mass-based effective-mixture specific heat at constant volume of the mixture is the mass-fraction weighted sum of the constituents' constant-volume specific heats. Similarly, the molar-based effective specific heat at constant volume of a mixture is given by the following:

$$\bar{c}_v = \sum_{i=1}^{n} y_i \bar{c}_{v,i} \qquad (10.3)$$

Applying the additive rule to the total enthalpy of a mixture of several gas constituents produces

$$H(T, P) = \sum_{i=1}^{n} H_i(T, P_i) = \sum_{i=1}^{n} m_i h_i(T, P_i) = \sum_{i=1}^{n} N_i \bar{h}_i(T, P_i)$$

where $h_i(T, P_i)$ is the mass specific enthalpy of a constituent at the mixture temperature and constituent partial pressure, and $\bar{h}_i(T, P_i)$ is the molar specific enthalpy. The specific enthalpy of the mixture is then

$$h(T, P) = \frac{H(T, P)}{m} = \sum_{i=1}^{n} x_i h_i(T, P_i)$$

When the constituents behave as ideal gases, the enthalpies depend only on temperature. This result then reduces to

$$h(T) = \sum_{i=1}^{n} x_i h_i(T)$$

where ideal gas tables such as those of the appendix may be used to determine the constituent enthalpies.

If temperature-variable specific heats at constant pressure are used, then this result becomes

$$h(T) = \sum_{i=1}^{n} x_i \int_{T_{ref}}^{T} c_{P,i}(T) dT$$

When the temperature range is such that the specific heats at constant pressure remain constant or average c_P's are employed, then an effective specific heat at constant pressure given by the following may be used:

$$c_P = \sum_{i=1}^{n} x_i c_{P,i} \qquad (10.4)$$

The effective molar constant-pressure specific heat is then the following:

$$\bar{c}_P = \sum_{i=1}^{n} y_i \bar{c}_{P,i} \tag{10.5}$$

The total entropy of a mixture according to the additive rule is

$$S = \sum_{i=1}^{n} S_i(T, P_i) = \sum_{i=1}^{n} m_i s_i(T, P_i) = \sum_{i=1}^{n} N_i \bar{s}_i(T, P_i)$$

where $s_i(T, P_i)$ is the mass specific entropy of a constituent at the mixture temperature and constituent partial pressure, and $\bar{s}_i(T, P_i)$ is the molar specific entropy. All entropies are measured with respect to an appropriate reference state. The specific entropy of the mixture is then

$$s = \frac{S}{m} = \sum_{i=1}^{n} x_i s_i(T, P_i)$$

Using the Gibbs equation, $T ds = dh - v dP$, and the ideal gas equation of state, the entropy of one of the constituents with respect to an appropriate reference state is given by

$$s_i(T, P_i) = \int_{T_{ref}}^{T} \frac{dh_i}{T} - R_i \ln \frac{P_i}{P_{ref}}$$

which reduces the expression for the mixture's specific entropy to

$$s = \sum_{i=1}^{n} x_i \left(\int_{T_{ref}}^{T} \frac{dh_i}{T} - R_i \ln \frac{P_i}{P_{ref}} \right)$$

When the $\int dh_i / T$ for each constituent is evaluated with ideal gas tables or empirical expressions for the specific heats, this becomes the following:

$$s = \sum_{i=1}^{n} x_i \left(s_i^0(T) - R_i \ln \frac{P_i}{P_{ref}} \right) \tag{10.6}$$

In the event that c_P remains constant or average specific heats are used, this relation becomes the following:

$$s = \sum_{i=1}^{n} x_i \left(c_{P,i} \ln \frac{T}{T_{ref}} - R_i \ln \frac{P_i}{P_{ref}} \right) \tag{10.7}$$

EXAMPLE 10.3

One pound-mass of nitrogen is mixed with one pound-mass of hydrogen and one pound-mass of carbon dioxide. This mixture is placed in a rigid vessel at 15 psia, 80° F. The mixture is now heated to 440° F. Treating the mixture as an ideal gas with temperature-variable specific heats, determine the mixture's final pressure, the final partial pressures of the mixture's constituents, and the amount of heat added to the mixture.

System: A three pound-mass mixture of nitrogen, hydrogen, and carbon dioxide.

Basic Equations:

$$Pv = RT$$
$$_1Q_2 - {_1W_2} = U_2 - U_1$$
$$y_i = \frac{P_i}{P}$$

Conditions: The constituents and mixture will be modeled as ideal gases with temperature-variable specific heats.

Solution: We begin by determining the apparent molecular weight of the mixture and the constituent mole fractions. These calculations are summarized in Table 10.2.

TABLE 10.2
Calculations of Example 10.3

GAS	x_i	M_i	x_i/M_i	y_i	y_iM_i
N_2	0.3333	28.0134	0.01190	0.06440	1.804
H_2	0.3333	2.0159	0.1653	0.8946	1.803
CO_2	0.3333	43.9999	0.007575	0.04100	1.804
			0.18478		$M = 5.411$

Using Appendix Table B.6, the mixture's initial internal energy is

$$U_1 = \sum_{i=1}^{n} m_i \frac{\bar{u}_i(T_1)}{M_i}$$

$$= 1\frac{2678.3}{28.0134} + 1\frac{2590.0}{2.0159} + 1\frac{2983.4}{43.9999}$$

$$= 1448.6 \text{ Btu}$$

The final mixture pressure will be

$$P_2 = \frac{mRT_2}{V}$$

$$= P_1 \frac{T_2}{T_1}$$

$$= 15\frac{900}{540}$$

$$= 25 \text{ psia}$$

Each constituent's partial pressure is then

$$P_i = y_i P$$
$$P_{N_2} = 0.06440 \times 25 = 1.610 \text{ psia}$$
$$P_{H_2} = 0.8946 \times 25 = 22.37 \text{ psia}$$

and

$$P_{CO_2} = 0.04099 \times 25 = 1.025 \text{ psia}$$

At the final state, the mixture's internal energy is

$$U_2 = \sum_{i=1}^{n} m_i \frac{\overline{u}_i(T_2)}{M_i}$$

$$= 1\frac{4481.4}{28.0134} + 1\frac{4381.8}{2.0159} + 1\frac{5810.2}{43.9999}$$
$$= 2465.6 \text{ Btu}$$

No compression or expansion work is done on or by the gas mixture as it is heated because the container is rigid. Hence, the first law becomes

$$_1Q_2 = U_2 - U_1$$
$$= 2465.6 - 1448.6$$
$$= 1017.0 \text{ Btu}$$

EXAMPLE 10.4

Oxygen and hydrogen are mixed in a steady-flow system to form a mixture at 5000 kPa, 227° C, with an oxygen mole fraction of 0.67. This mixture then passes through an adiabatic turbine until its pressure is 100 kPa. Determine the maximum specific work that can be produced by this turbine and the minimum turbine outlet temperature when the mixture is modeled as an ideal gas with constant specific heats.

System: Oxygen–hydrogen mixture passing through a steady-flow, adiabatic turbine.

Basic Equations:

$$Pv = RT$$
$$_1q_2 - {_1w_2} = h_2 - h_1 + \Delta e_{ke} + \Delta e_{pe}$$

Conditions: The constituents and mixture will be treated as ideal gases with constant specific heats.

Solution: This problem can be solved as with any ideal gas once the apparent properties of the mixture are determined. The calculation of the apparent properties are summarized in Table 10.3.

TABLE 10.3

Calculations of Example 10.4

GAS	y_i	M_i	y_iM_i	x_i	R	x_iR	c_P	x_ic_P	c_v	x_ic_v
O_2	0.67	32.00	21.44	0.97	0.2598	0.2520	0.9190	0.8914	0.6592	0.6394
H_2	0.33	2.02	0.67	0.03	4.1158	0.1235	14.321	0.4296	10.197	0.3059
			$M = 22.11$			$R = 0.3755$		$c_P = 1.321$		$c_v = 0.9453$

The apparent specific heat ratio of this mixture is then

$$k = \frac{c_P}{c_v} = \frac{1.321}{0.9453} = 1.397$$

The isentropic process between the initial state and the final pressure as illustrated in Figure 10.4 produces the greatest amount of work and exhausts the mixture at its lowest possible temperature. Using the isentropic process relations for ideal gases with constant specific heats, the minimum outlet temperature is

$$T_2 = T_1 \left(\frac{P_2}{P_1} \right)^{(k-1)/k}$$

$$= 500 \left(\frac{100}{5000} \right)^{0.397/1.397}$$

$$= 164 \text{ K}$$

FIGURE 10.4

Isentropic Process of Example 10.4

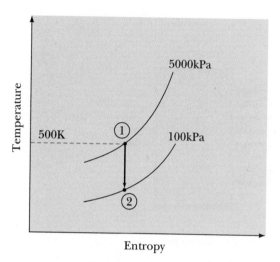

Applying the first law between the initial and final states, recalling that the process is adiabatic, and neglecting the kinetic- and potential-energy changes gives

$$
\begin{aligned}
_1w_2 &= h_1 - h_2 \\
&= c_P(T_1 - T_2) \\
&= 1.321(500 - 164) \\
&= 443.9 \text{ kJ}
\end{aligned}
$$

10.5 Mixing Ideal Gases and Irreversibility

We have stated several times that mixing is an irreversible process that degrades the ability of a system to produce work. This conclusion is apparent when the mixing is done by stirring or mixing devices that rely on fluid friction to blend the constituents. But it is also true when fluid friction is not required to blend the constituents in a system such as the steady-flow mixing chamber illustrated in Figure 10.5. In this chamber, n streams of ideal gases at P, T, are adiabatically mixed to form a mixture stream whose pressure and temperature are also P and T, respectively. Adiabatic mixing is selected here for study because it does not introduce entropy changes associated with heat transfer. As long as the process is adiabatic and all the streams to be mixed are at the same temperature, the resulting mixture stream is also at the same temperature, according to the first law. Application of the second law and the concepts of exergy will give us insight into how perfect this mixing is.

As constituent i passes through the mixing chamber, its entropy change is

$$
\Delta \dot{S}_i = \dot{N}_i[\bar{s}_{i,m} - \bar{s}_i] = \dot{N}_i\left[\bar{s}_i{}^0(T_m) - \bar{s}_i{}^0(T) - \Re \ln \frac{P_i}{P_{ref}} + \Re \ln \frac{P}{P_{ref}}\right]
$$

FIGURE 10.5
*A Steady-Flow
Mixing Chamber*

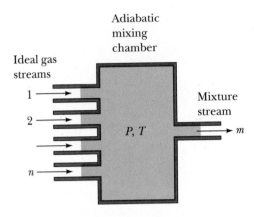

according to the Gibbs equation. The s^0 terms cancel because $T_m = T_0$. The properties of logarithms further reduce this expression to

$$\Delta \dot{S}_i = -\dot{N}_i \Re \ln \frac{P_i}{P}$$

Now, according to Dalton's model, $P_i/P = y_i$, and the preceding equation reduces to

$$\Delta \dot{S}_i = -\dot{N}_i \Re \ln y_i$$

Additional simplification of this expression can be accomplished by putting it on a per-unit-of-mixture-mass basis. This is done by dividing both sides by the molar mass flow rate of the mixture, $\dot{N}_m$. When this is done, the result is

$$\Delta \bar{s}_i = -\Re y_i \ln y_i$$

The total entropy change when several streams of ideal gases are mixed in an adiabatic, constant-pressure process is the sum of the change experienced by each gas constituent:

$$\Delta \bar{s} = -\sum_{i=1}^{n} \Re y_i \ln y_i$$

The universal gas constant, $\Re$, is always a positive number, and the mole fractions are fractions between 0 and 1. Inspecting the preceding equation demonstrates that the net entropy change resulting from this mixing process is then always a positive number (i.e., an entropy increase)—as it must be according to the increase in entropy principle. This number can never be zero as long as two or more ideal gases are being mixed. Because the total entropy change is always positive and never zero, *the mixing process is always an irreversible process.*

A measure of the irreversibility associated with a constant-pressure, adiabatic mixing of two or more ideal gases may be obtained by calculating the flow exergy in the resulting mixture stream in comparison to that in the constituent streams. The total change of the flow exergy across the mixing chamber represents the work potential that is associated with the process occurring in the mixing chamber. For the conditions of this discussion, each constituent has its pressure changed from the total mixture pressure to its partial pressure while the temperature remains constant in the mixture chamber. The rate at which the flow exergy of the ith constituent changes in the mixing chamber is then

$$\dot{\Psi} = -\dot{N}_i T_0 (\bar{s}_{i,m} - \bar{s}_i)$$

according to the presentation in Section 5.9. Substituting the previous expression for the change in the constituent's entropy and placing the final result on a per unit of mixture mass basis reduces this result to

$$\Delta \overline{\psi}_i = \Re_i T_0 y_i^2 \ln y_i$$

and the total change in the flow exergy of all the constituent streams is

$$\Delta \overline{\psi}_i = \sum_{i=1}^{n} \Re_i T_0 y_i^2 \ln y_i$$

As expected, an inspection of this result shows that the flow exergy of the mixture is less than the sum of that in the constituent streams. Consequently, some work potential has been lost in the mixing process because of the irreversibilities occurring in the process.

EXAMPLE 10.5

A 5 kg/s, 200 kPa, 100° C oxygen stream is adiabatically mixed with a 10 kg/s, 200 kPa, 20° C helium stream to form a mixture stream at 200 kPa. Using constant specific heats, determine the temperature of the mixture stream and the work potential lost or gained.

System: A steady-flow chamber in which oxygen and helium are mixed adiabatically at constant pressure.

Basic Equations:
$$Pv = RT$$

$$\dot{Q} - \dot{W} = \sum_{out} \dot{m}\left(h + \frac{V^2}{2} + gz\right) - \sum_{in} \dot{m}\left(h + \frac{V^2}{2} + gz\right)$$

$$\dot{W}_{ptl} = \sum_{in} \dot{m}\psi - \sum_{out} \dot{m}\psi + \sum \dot{Q}_j \left(1 - \frac{T_0}{T_j}\right)$$

Conditions: The constituents and mixture will be treated as ideal gases with constant specific heats.

Solution: The conservation of mass principle gives the mixture stream mass flow rate as 15 kg/s. The mass fraction of oxygen in the mixture is then 0.3333, and the mass fraction of the helium in the mixture is 0.6667. The effective molecular weight and specific heat at constant pressure of the mixture are calculated in Table 10.4.

TABLE 10.4
Calculations of Example 10.5

GAS	x_i	M_i	x_i/M_i	y_i	$y_i M_i$	c_P	$x_i c_P$
O_2	0.3333	32.00	0.01042	0.05887	1.884	0.9190	0.3063
He	0.6667	4.0026	0.1666	0.9412	3.767	5.192	3.462
			0.1770		$M = 5.651$		$c_P = 3.768$

As applied to the mixing chamber when the kinetic and potential energies are neglected, the first law is

$$\dot{m}_m h_m = \dot{m}_{O_2} h_{O_2} + \dot{m}_{He} h_{He}$$

because the chamber is adiabatic and no shaft work is being done in the chamber. When we measure the enthalpy with respect to 0 K, use constant specific heats, and place our equation on the basis of the mass flow rate of the mixture stream, this becomes

$$c_{P,m} T_m = x_{O_2} c_{P,O_2} T_{O_2} + x_{He} c_{P,He} T_{He}$$

The temperature of the mixture stream is then

$$T_m = \frac{x_{O_2} c_{P,O_2} T_{O_2} + x_{He} c_{P,He} T_{He}}{c_{P,m}}$$

$$= \frac{0.3333 \times 0.9190 \times 373 + 0.6667 \times 5.192 \times 293}{3.768}$$

$$= 300 \text{ K}$$

We will use 101 kPa, 25° C for the dead state as discussed in Section 4.13. Invoking the integrated Gibbs equation and presuming constant specific heats, the difference between the specific entropy of the oxygen at the inlet and at the dead state is

$$(s - s_0)_{O_2} = c_P \ln \frac{T}{T_0} - R \ln \frac{P}{P_0}$$

$$= 0.9190 \ln \frac{373}{298} - \frac{8.314}{32.00} \ln \frac{200}{101}$$

$$= 0.02880 \text{ kJ/kg-K}$$

Similarly, the specific-entropy difference for the other two streams are

$$(s - s_0)_{He} = -1.5070 \text{ kJ/kg-K}$$

and

$$(s - s_0)_m = -0.9799 \text{ kJ/kg-K}$$

According to Section 5.9, the work potential associated with this mixing is given by

$$\dot{W}_{ptl} = \dot{m}_{O_2}[(h - h_0) - T_0(s - s_0)]_{O_2} + \dot{m}_{He}[(h - h_0) - T_0(s - s_0)]_{He}$$
$$- \dot{m}_m[(h - h_0) - T_0(s - s_0)]_m$$

When placed on a basis of one unit of mixture mass, this becomes

$$w_{ptl} = x_{O_2}[(h - h_0) - T_0(s - s_0)]_{O_2} + x_{He}[(h - h_0) - T_0(s - s_0)]_{He}$$
$$- [(h - h_0) - T_0(s - s_0)]_m$$

Presuming constant specific heats, this result becomes

$$w_{ptl} = x_{O_2}[c_P(T - T_0) - T_0(s - s_0)]_{O_2}$$
$$+ x_{He}[c_P(T - T_0) - T_0(s - s_0)]_{He}$$
$$- [c_P(T - T_0) - T_0(s - s_0)]_m$$
$$= 0.3333[0.9190(373 - 298) - 298(0.02880)]$$
$$+ 0.6667[5.192(293 - 298) - 298(-1.5070)]$$
$$- [3.768(300 - 298) - 298(-0.9799)]$$
$$= 2.664 \text{ kJ/kg}$$

10.6 Gas and Vapor Mixtures

In some cases, one or more gas constituents in a mixture can change to the liquid phase in the temperature range of interest. The gaseous phase of these constituents are known as *vapors*. The vapors in these mixtures exist at their partial pressures, which will be very low when the mixture pressure is small or the vapor's mole fraction is small. When the vapor's partial pressure is small, the vapor, like all gases, will behave as an ideal gas as demonstrated in Section 9.8. Under these conditions, gas–vapor mixtures can be modeled as ideal gas mixtures with proper consideration given to the fact that the liquid phase of the vapor constituent can condense out of the mixture and also evaporate into the mixture. We will develop the analysis of these mixtures by considering ordinary atmospheric air. This mixture will be modeled as a mixture of dry air (i.e., oxygen, nitrogen, argon, and trace gasses) and water vapor, with each being treated as an ideal gas. We will call this dry air–water vapor mixture *humid air* even when the water vapor content is very low.

The aggregate properties of humid air are the total mixture pressure and the mixture's dry-bulb temperature. The total mixture pressure is the pressure as measured with an ordinary pressure-measuring device such as a barometer or pressure gauge. As with all mixtures of ideal gases, the total mixture pressure is the sum of the partial pressures of the constituents according to Dalton's model. The dry-bulb temperature of the mixture is that measured with an ordinary thermometer, such as a liquid-in-glass thermometer, whose sensing element is not dampened by any evaporating liquid. In the case of humid air, the thermometer's sensing element is not wetted with liquid water as might occur, for example, when dew is forming.

The amount of water vapor in a water vapor–dry air mixture is specified by the humidity ratio, ω, which is defined as follows:

DEFINITION: *Humidity Ratio*

The humidity ratio is the ratio of the mass of water vapor in humid air to the mass of dry air in the mixture:

$$\omega = \frac{m_v}{m_{da}}$$

This is not the same as the mass fraction of the previous sections: The denominator of the humidity ratio is the mass of only one of the mixture's constituents (i.e., the dry air) rather than the mass of all of the mixture's constituents. Using Dalton's partial pressure model and invoking the ideal gas equation of state for both the gas and vapor constituents, the humidity ratio can be expressed as

$$\omega = \frac{M_v P_v V/\Re T}{M_{da} P_{da} V/\Re T} = \frac{M_v P_v}{M_{da} P_{da}} = 0.622 \frac{P_v}{P - P_v}$$

where P is the pressure of the total mixture and P_v is the partial pressure of the water vapor.

Gas–vapor mixtures are said to be saturated when conditions are such that the vapor's liquid phase can exist in equilibrium with the vapor in the mixture; that is, there is no net conversion of liquid to vapor by evaporation or conversion of vapor to liquid by condensation. This condition would occur when the vapor's partial pressure is the saturation pressure corresponding to the temperature of the mixture. This may be illustrated by considering the process undergone by the water vapor in humid air as liquid water is evaporated into the mixture while the mixture dry-bulb temperature and total pressure remain constant. This process is sketched on the water vapor temperature–entropy state diagram of Figure 10.6. Initially, the water vapor in the mixture exists at the vapor partial pressure, $P_{v,1}$, and the mixture dry-bulb temperature, T_1. Because this state is not on the saturated vapor line, the mixture is not saturated and additional water vapor can be added to the mixture. As water vapor is added to the mixture by evaporation while the mixture dry-bulb temperature remains constant at T_1, the partial pressure of the water vapor increases as the water vapor mole fraction increases. We can continue to add water vapor until the partial pressure of the water vapor is P_s, which is on the saturated vapor line. At this point, the mixture is saturated in the sense that no more water vapor can be added.

FIGURE 10.6
State Diagram Illustrating the Saturation of Atmospheric Air

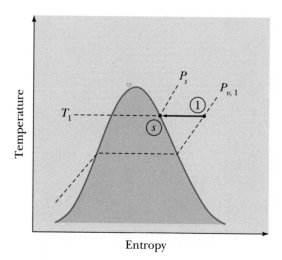

The *relative humidity*, ϕ, is defined as the ratio of the partial pressure of the water vapor in the mixture, P_v, to the partial pressure of the water vapor when the mixture is saturated at the mixture dry-bulb temperature, P_s:

DEFINITION: *Relative Humidity, ϕ*

$$\phi = \frac{P_v}{P_s}$$

Relative humidity measures the ability of a humid air mixture to absorb additional vapor. When relative humidity is 100%, the mixture is saturated and no further vapor can be added. When relative humidity is 0%, no vapor is present in the mixture. Do not confuse relative humidity with the humidity ratio. Relative humidity is a measure of the potential of humid air to take on additional vapor, while the humidity ratio is the mass of vapor in the mixture as a fraction of the mass of the other gases in the mixture.

We also can change the state of a gas–vapor mixture by cooling it while keeping the mixture's pressure constant. The process experienced by the water-vapor constituent when the humid air is cooled in this manner is illustrated in Figure 10.7. As the temperature of the mixture is reduced, the state of the vapor moves down the constant-pressure line until the state of the water-vapor constituent reaches the saturated-vapor line. At this point the mixture becomes saturated and cannot be cooled any more while retaining the same amount of vapor. We can only change this partial pressure by altering the mole fraction of the water-vapor constituent through the addition or removal of water from the humid air while it remains saturated.

FIGURE 10.7

State Diagram Illustrating the Cooling of Water Vapor in Atmospheric Air

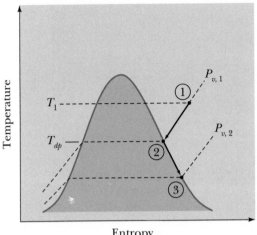

The temperature at which the mixture becomes saturated during constant-pressure cooling is known as the *dew-point temperature, T_{dp}*. It is so named because it is the process experienced by atmospheric air as it cools and begins to form dew. The dew-point temperature is shown in Figure 10.7.

EXAMPLE 10.6	Humid atmospheric air at 14.7 psia has a temperature of 70° F and relative humidity of 57%. Determine the air's humidity ratio and dew-point temperature.

System: A mixture of dry air and water vapor.

Conditions: The air and water vapor will be treated as ideal gases.

Solution: According to the water property tables, the saturation pressure for the water vapor at the mixture temperature of 70° F is 0.3632 psia. Applying the definition of the relative humidity, the water vapor's partial pressure is

$$P_v = \phi P_s$$
$$= 0.57 \times 0.3632$$
$$= 0.2070 \text{ psia}$$

Applying the ideal gas equation of state to the water vapor and air constituents gives the humidity ratio as

$$\omega = \frac{m_v}{m_{da}} = 0.622 \frac{P_v}{P - P_v}$$

$$= 0.622 \frac{0.2070}{14.7 - 0.2070}$$
$$= 0.008884 \text{ lb}_m/\text{lb}_m\text{-da}$$

The saturation temperature for the water vapor partial pressure of 0.207 psia is approximately 54° F. The dew-point temperature is then

$$T_{dp} \sim 54° \text{ F}$$

Once the mixture is saturated and the dew-point temperature reached, further cooling at constant mixture pressure will cause the vapor's liquid phase to be formed as condensate or dew. The partial pressure of the vapor in the gaseous mixture is now reduced as the vapor's mole fraction is decreased by the amount of vapor that has been converted to the liquid phase. The state of the vapor in the remaining mixture now moves down along the saturated vapor line as illustrated by the states between 2 and 3 in Figure 10.7. At all states between 2 and 3, the vapor constituent of the mixture remains saturated at a relative humidity of 100%, and we can only move between these states by adding or removing liquid water from the humid air.

The temperature of a humid air mixture of dry air and water vapor whose relative humidity is not 100% (i.e., unsaturated) also can be reduced by evaporating liquid water into the mixture adiabatically. A device that accomplishes this process is shown in Figure 10.8. As humid air passes through this adiabatic device, liquid water is evaporated into this air until it becomes saturated. During this process, the amount of water evaporated per unit of air passing through the duct is given by $\omega_3 - \omega_1$, according to the conservation of mass principle when it is applied to the water constituent. To evaporate this water, an amount of heat, $(\omega_3 - \omega_1)h_{fg}$, must be added to the water. Because the duct is adiabatic, this heat can only come from the dry air–water vapor mixture by decreasing the enthalpy (i.e., temperature) of the mixture. The saturated humid air leaving the duct is then at a temperature lower than that at which it entered the duct. The temperature at which the saturated humid air leaves this duct is known as the *adiabatic saturation temperature*. This temperature depends on the humidity ratio at which the humid air enters the duct, it thus serves as a means for measuring the humidity ratio of a humid air mixture.

FIGURE 10.8

A Device that Adiabatically Saturates Humid Air

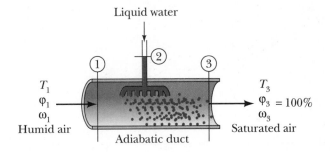

In principle, the adiabatic saturation temperature serves to measure the humidity ratio of humid air. In practice, it is seldom used because the duct must be very long (in theory, infinitely long) for the mass and heat transfer processes to saturate the mixture. It is more practical to measure a temperature known as the *wet-bulb temperature*. The simplest device for measuring the wet-bulb temperatrue, T_{wb}, consists of a thermometer whose sensing element is covered with a porous wick (a cloth wick is normally used). Liquid water is added to this wick until it is completely wet (hence, the name *wet-bulb* temperature). Humid air whose humidity ratio is to be determined is now blown over the wet wick. Liquid water is evaporated from the wick into this air. As this water is evaporated, an amount of heat given by the product of the evaporation rate and enthalpy of vaporization is removed from the thermometer's sensing element. This heat loss lowers the temperature of the sensing element. Because the air is now warmer then the sensing element, it transfers heat to the element, thereby attempting to raise its temperature. After a short period, the sensing element's temperature reaches a steady-state where the rate of the heat gain from the warm air equals the rate of heat loss because of evaporation. The thermometer shows the wet-bulb temperature once this steady-state is reached.

10.7 Psychrometric Chart

The dry-bulb temperature–humidity ratio state diagram for an air–water vapor mixture is known as a *psychrometric chart*. A full psychrometric chart for standard atmospheric pressure is contained in Appendix C; a skeleton outline of this psychrometric chart is shown in Figure 10.9. Lines of constant relative humidity, wet-bulb temperature, specific volume, and specific enthalpy are illustrated on the skeleton chart. All specific quantities such as the specific volume and enthalpy are on a per-unit-air-mass basis rather than a per-unit-of-mixture-mass basis. The merits of this convention will become apparent in the next section.

FIGURE 10.9
Skeleton Psychrometric Chart

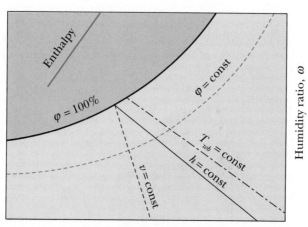

Dry–bulb temperature

Because mixtures are not pure (in the sense that we can vary the mixing proportions) and simple compressible substances, more than two independent properties are needed to fix the state of these substances. In addition to the two independent properties of the state postulate, we need $N - 1$ additional independent properties, when N is the number of constituents in the mixture. One additional independent property is then required for the binary dry air–water vapor mixture. The total mixture pressure normally serves this purpose for psychrometry. The psychrometric charts of Appendix C are only valid when the total mixture pressure is standard atmospheric pressure. Other psychrometric charts or equations of this and the previous section are required when the total pressure is not standard atmospheric pressure.

The enthalpy of humid air is given by

$$H = m_{da} h_{da}(T, P_{da}) + m_v h_v(T, P_v)$$

where the specific enthalpies of the constituents are evaluated at the mixture dry-bulb temperature and their respective partial pressures. Treating the constituents as ideal gases, taking h_v to be the same as h_g of the steam tables, and putting the final result on the basis of one unit of dry-air mass reduces this to

$$h = h_{da}(T) + \omega h_g(T)$$

Constant specific heats are normally used to evaluate $h_{da}(T)$ in this expression. Psychrometric charts and tables typically use the same reference state for the air constituent as is used in the water property tables (i.e., $32°\,F$ or $0°\,C$).

Similarly, the volume of a mixture of dry air and water vapor is given by

$$V = m_{da} v_{da}(T, P_{da}) + m_v v_v(T, P_v)$$

When placed on a unit-of-dry-air-mass basis, this becomes

$$v = v_{da}(T, P_{da}) + \omega v_v(T, P_v)$$

The ideal gas equation of state is used to evaluate the specific volumes of both constituents in this result.

Means of obtaining the quantities of a humid air mixture from a psychrometric chart are illustrated in Figure 10.10 for arbitrary state 1. The humidity ratio and dry-bulb temperatures are read directly from the chart axis. Relative humidity and specific volume are read from the scales internal to the chart. The mixture's specific enthalpy is read from the external enthalpy scale by using the enthalpy guidelines. Wet-bulb temperatures are read by following the constant wet-bulb temperature guidelines to the saturation line (i.e., $\phi = 100\%$) and then vertically down to the dry-bulb temperature scale. The dew-point temperature is read by following a horizontal guideline to the saturation axis and then moving vertically down to the dry-bulb temperature scale. Some psychrometric charts have a dew-point and wet-bulb temperature scale inscribed on the saturation-line axis.

FIGURE 10.10
Reading Properties on a Psychrometric Chart

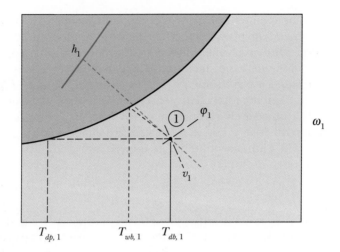

EXAMPLE 10.7

Humid atmospheric air at 101 kPa has a dry-bulb temperature of 30° C and a dew-point temperature of 21° C. Determine this air's humidity ratio, specific enthalpy, and relative humidity.

System: A mixture of dry air and water vapor.

Conditions: The psychrometric chart will be used to obtain the requested properties.

Solution: Using the given temperatures to locate the state of this mixture on the psychrometric chart allows us to read the following values from the chart:

$$\omega = 0.015 \text{ kg/kg-da}$$
$$h = 70 \text{ kJ/kg-da}$$

and

$$\phi = 58\%$$

10.8 Analysis of Humid-Air Processes

Several processes are used to alter the condition of humid air. Many of them are used to alter the atmospheric air to conditions that are more comfortable for humans. Humans are comfortable over a limited range of temperature and relative humidity. High relative humidity tends to accentuate discomfort because it retards the perspiration–evaporation cooling process. Similarly, extremely low relative humidity tends to lead to health problems. Methods of controlling relative humidity and humid air temperature are commonly used to provide comfortable environments for humans.

Dry air is also useful for drying many commercial products such as paint, foodstuffs, and lumber. Dry air also can be used to create a cooling effect through evaporative cooling, which is used in such applications as power plant cooling and air conditioning in semiarid and desert climates. Air of high relative humidity is useful for easing certain health conditions, controlling static electricity, and other purposes.

Humid-air processes can be analyzed using the conservation of mass principle, first law, and second law, just like any other system. The only difference is the basis of the analysis and the determination of the mixture quantities. One unit of dry air mass rather than a unit of mixture mass will be used as the basis of all calculations. Mixture quantities will be determined using the methods presented in the preceding sections.

The majority of the humid-air processes occur in a system that is an adaptation of the one shown in Figure 10.11. This system processes the humid air in an open duct or other device in which the total mixture pressure remains constant. This pressure is usually the same as that of the surrounding atmosphere. Humid air flows steadily through this system while its water content and temperature are changed. Liquid water streams that add or remove water from the humid air, while perhaps changing their temperatures, also flow steadily through the system. Heat also may be added to or removed from the system, and the system may produce or require shaft work.

FIGURE 10.11
General Humid-Air Processing System

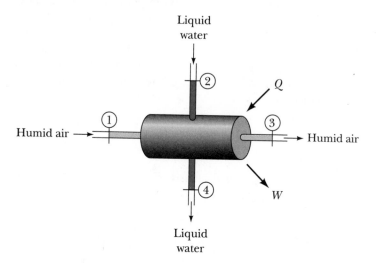

When the conservation of mass principle is applied to the dry-air constituent of the humid-air mixture flowing through this system, the result is

$$\dot{m}_{da,1} = \dot{m}_{da,3}$$

In words, the mass flow rate of the dry-air constituent only is conserved between the inlet, 1, and the outlet, 3. The net mass flow rate between this inlet and outlet will not be the same because the outlet may contain more or less water vapor than the inlet. This change in the net mass flow rate is determined by the amount of water evaporated into or condensed out of the humid air stream inside the system. Applying the conservation of mass principle to the water constituent of this system yields

$$\dot{m}_{v,1} + \dot{m}_{f,2} = \dot{m}_{v,3} + \dot{m}_{f,4}$$

or

$$\dot{m}_{v,3} - \dot{m}_{v,1} = \dot{m}_{f,2} - \dot{m}_{f,4}$$

Subscript v refers to the water vapor in the humid air, and subscript f refers to the saturated-liquid water stream. Dividing this result by the mass flow rate of the dry air reduces it to

$$\omega_3 - \omega_1 = \hat{m}_2 - \hat{m}_4$$

where $\hat{m}$ is the amount of liquid water per unit of dry air mass. This important result tells us that the change in the humidity ratio of the mixture is dictated by the net flow rate of liquid water per unit of dry air passing through the system. If the net specific water flow rate is positive, $\hat{m}_2 > \hat{m}_4$, then the air leaving the system is more humid (contains more water) than when it entered the system. If it is negative, $\hat{m}_2 < \hat{m}_4$, then the humid air is drier when it leaves the system.

Applying the rate form of the first law, while taking advantage of the fact that the specific enthalpy of the humid air is on the basis of the mass of the air constituent, gives

$$\dot{Q} - \dot{W} = \dot{m}_{da}(h_3 - h_1) + \dot{m}_{f,4}h_4 - \dot{m}_{f,2}h_2$$

When placed on the basis of the dry-air mass flow rate passing through the system, this becomes

$$q - w = (h_3 - h_1) + \hat{m}_4 h_4 - \hat{m}_2 h_2$$

where q and w are per unit of dry air mass.

10.8.1 Evaporative Cooling

By adding heat to the general system, it may be used to cool its surroundings while the entering liquid water is simultaneously evaporated into the humid-air stream. The humidity ratio of the humid-air stream will then be increased by the amount of water that is evaporated. In the steady state, no liquid water leaves the system ($\dot{m}_4 = 0$), because the liquid that enters at section 2 is completely evaporated into the humid air to provide the cooling effect. The conservation of mass principle then becomes

$$\omega_3 - \omega_1 = \hat{m}_2$$

When there is no shaft work or exiting liquid stream, the first law becomes

$$q = (h_3 - h_1) - \hat{m}_2 h_2$$

When combined with the conservation of mass principle, this gives

$$q = (h_3 - h_1) - (\omega_3 - \omega_1) h_2$$

The humid air's enthalpy is given by $h(T) = h_{da}(T) + \omega h_g(T)$. When this is substituted into the reduced first law and the air constituent is treated with constant specific heats, the result is

$$q = c_{P,da}(T_3 - T_1) + \omega_3 h_g(T_3) - \omega_1 h_g(T_1) - (\omega_3 - \omega_1) h_2$$

EXAMPLE 10.8

Humid atmospheric air at 101 kPa with a dry-bulb temperature of 30° C and a relative humidity of 40% enters an evaporative cooler. The water being evaporated enters the cooler at 26° C. This air leaves the cooler at 26° C with a relative humidity of 90%. Determine the amount of evaporation and the amount of heat added to the humidified air.

System: A mixture of dry air and water vapor as it passes through an evaporative cooler.

Basic Equations:
$$q - w = (h_3 - h_1) + \hat{m}_4 h_4 - \hat{m}_2 h_2$$
$$\omega_3 - \omega_1 = \hat{m}_2 - \hat{m}_4$$

Conditions: The psychrometric chart will be used to obtain the requested quantities.

Solution: According to the psychrometric chart,

$$\omega_3 = 0.0192 \text{ kg/kg-da}$$

and

$$\omega_1 = 0.0105 \text{ kg/kg-da}$$

Adapting the conservation of mass principle to this evaporative cooler gives

TIP:

*In this example, 16.0 kJ
of heat per each
kilogram of air
constituent passing
through the system
must be added to the
system. This heat
would be removed
from the surroundings,
thereby producing
a cooling effect.*

$$\begin{aligned} \hat{m}_2 &= \omega_3 - \omega_1 \\ &= 0.0192 - 0.0105 \\ &= 0.0087 \text{ kg/kg-da} \end{aligned}$$

According to the psychrometric chart, $h_1 = 57.5$ kJ/kg-da and $h_3 = 74.4$ kJ/kg-da. The enthalpy of the water entering the humidifier is 109.07 kJ/kg, according to the water property tables. The combined conservation of mass and first law reduces to

$$\begin{aligned} q &= (h_3 - h_1) - \hat{m}_2 h_f(T_2) \\ &= (74.4 - 57.5) - 0.0087 \times 109.07 \\ &= 16.0 \text{ kJ/kg-da} \end{aligned}$$

10.8.2   Adiabatic Humidification Process

Adding water vapor to humid air to make it more humid is normally done in an adiabatic system with no shaft work and no exiting liquid stream. The entering liquid water temperature is usually the same or greater than the temperature of the entering humid air. Steam is sometimes used in place of the entering liquid stream because it is already in the vapor form and will more readily mix with the humid air.

These conditions reduce the conservation of mass principle as applied to the system of Figure 10.11 to

$$\omega_3 - \omega_1 = \hat{m}_2$$

and the first law to

$$(h_3 - h_1) = \hat{m}_2 h_2$$

Combining these results gives

$$(h_3 - h_1) = (\omega_3 - \omega_1) h_2$$

The addition of water vapor to the humid air then increases the humidity ratio and the enthalpy of the mixture. Simultaneously, the relative humidity, wet-bulb temperature, and dew-point temperature also increase. Substituting the expression for the humid-air enthalpy converts the combined first law and conservation of mass principle to

$$c_{P,da}(T_3 - T_1) + \omega_3 h_g(T_3) - \omega_1 h_g(T_1) = (\omega_3 - \omega_1) h_f(T_2)$$

This provides a higher degree of accuracy than the psychrometric chart when additional accuracy is required. This equation typically involves an iterative solution (see Appendix A) to determine simultaneously the humid-air outlet temperature and humidity ratio. When $(\omega_3 - \omega_1) h_2$ is small in comparison to

$h_3 - h_1$, as is the case when liquid water is used to humidify the air stream, then the humidification process is one of near-constant mixture enthalpy. Inspection of a psychrometric chart demonstrates that the humidification process of near-constant enthalpy is also approximately a constant wet-bulb temperature process, as shown in Figure 10.12. The wet-bulb temperature is then a good start for an iterative solution when humidification is accomplished with liquid water.

FIGURE 10.12
Approximate Adiabatic Humidification Process

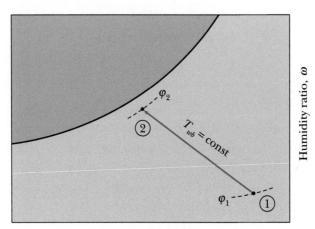

EXAMPLE 10.9

Humid atmospheric air at standard atmospheric pressure with a dry-bulb temperature of 80° F and a relative humidity of 30% enters an adiabatic humidifier. Water to be evaporated enters the cooler at 60° F. Humidified air leaves with a relative humidity of 90%. Determine the amount of water evaporated and the temperature of the humidified air.

System: A mixture of dry air and water vapor as it passes through an adiabatic humidifier.

Basic Equations:
$$q - w = (h_3 - h_1) + \hat{m}_4 h_4 - \hat{m}_2 h_2$$
$$\omega_3 - \omega_1 = \hat{m}_2 - \hat{m}_4$$

Conditions: The mixture constituents will be treated as ideal gases using constant specific heats for the air constituent.

Solution: Beginning with the psychrometric chart, the entering humid-air humidity ratio and wet-bulb temperature are

$$\omega_1 = 0.0066 \ \text{lb}_m/\text{lb}_m\text{-da}$$
$$T_{wb,1} = 60° \ \text{F}$$

The enthalpy of the water entering the system to be evaporated is

$$h_f(T_2) = 28.08 \ \text{Btu/lb}_m$$

while the enthalpy of the water vapor in the humid air entering the humidifier is

$$h_g(T_1) = 1096.4 \text{ Btu/lb}_m$$

according to the water property tables. Using the inlet wet-bulb temperature as the first estimate of the wet-bulb temperature at the humid air outlet,

$$\omega_3 = 0.0106 \text{ lb}_m/\text{lb}_m\text{-da}$$
$$T_{db,3} = 62°\text{ F}$$

according to the psychrometric chart. Then,

$$h_g(T_3) = 1088.6 \text{ Btu/lb}_m$$

by the water property tables. The combined first law and conservation of mass principle is

$$c_{P,da}(T_3 - T_1) + \omega_3 h_g(T_3) - \omega_1 h_g(T_1) = (\omega_3 - \omega_1) h_f(T_2)$$

When the data from the first estimate of the humid-air outlet conditions are substituted into this result,

$$0.2397(62 - 80) + 0.0106 \times 1088.6 - 0.0066 \times 1096.4$$
$$= (0.0106 - 0.0066)28.08$$
$$-0.011 \neq 0.11$$

and the left-hand and right-hand sides are not equal.

As a second estimate for the outlet of the humid air, we will use $T_{db,3} = 64°$ F. Then

$$\omega_3 = 0.0115 \text{ lb}_m/\text{lb}_m\text{-da}$$
$$h_g(T_3) = 1089.4 \text{ Btu/lb}_m$$

When these are substituted into the combined first law, the final result is

$$0.2397(64 - 80) + 0.0115 \times 1089.4 - 0.0066 \times 1096.4$$
$$= (0.0115 - 0.0066)28.08$$
$$1.45 \neq 0.14$$

and again the left-hand and right-hand sides are not equal, although the sign of the left-hand side has changed. Hence, we can conclude that the outlet dry-bulb temperature is between 62° F and 64° F and actually closer to 62° F. To refine this answer further, we will require a more detailed water property table and psychrometric chart. We will accept

$$T_{db,3} = 62°\text{ F}$$

as sufficiently accurate. Then,

$$\hat{m}_2 = \omega_3 - \omega_1$$
$$= 0.0106 - 0.0066$$
$$= 0.004 \text{ lb}_m/\text{lb}_m\text{-da}$$

TIP:

If this system had been operated such that the outlet air had been saturated, $\phi_2 = 100\%$, then the temperature at the outlet would have been the adiabatic saturation temperature discussed in Section 10.6. This example demonstrates that the wet-bulb and adiabatic saturation temperatures are almost equal.

10.8.3   Cooling Towers

Cooling towers are used to reject heat from power and refrigetation cycles to atmospheric air in locations where the relative humidity is low, such as in semiarid and desert areas. A small portion of the water from the cycle coolant system is evaporated adiabatically into the atmospheric air to reduce the temperature of the coolant water before it reenters the cycle. This process is similar to an adiabatic humidification process except that only a small portion of the inlet liquid water stream is evaporated. The remainder of the liquid water stream leaves the system at a reduced temperature.

Schematically, this system is the same as the general system of Figure 10.11, and the conservation of mass principle as applied to the water constituents in this system reduces to

$$\omega_3 - \omega_1 = \hat{m}_2 - \hat{m}_4$$

When combined with this result, the first law becomes

$$(h_3 - h_1) = \hat{m}_2 h_2 - \hat{m}_4 h_4$$

where it is presumed that no shaft work is added to the system and the process is adiabatic. Combining these two results produces

$$(h_3 - h_1) = \hat{m}_2(h_2 - h_4) - (\omega_3 - \omega_1)h_4$$

EXAMPLE 10.10

A coal-fired electrical-generation station rejects heat at a rate of 1×10^{10} Btu/hr using cooling water that leaves the plant at $90°$ F and reenters at $60°$ F. This water is cooled by dry desert air, which enters a cooling tower at $56°$ F, 20% relative humidity, and leaves at $80°$ F, 90% relative humidity. Determine the volume flow rate of the air entering the cooling tower and the rate of evaporation.

System: The volume enclosed by a cooling tower through which humid air and liquid water pass.

Basic Equations:
$$q - w = (h_3 - h_1) + \hat{m}_4 h_4 - \hat{m}_2 h_2$$
$$\omega_3 - \omega_1 = \hat{m}_2 - \hat{m}_4$$

Conditions: The water property tables and definitions of various humid air properties will be used to obtain the required properties to illustrate this alternative to psychrometric charts. Both the air and water components will be treated as ideal gases, and c_P of the air will be treated as a constant.

Solution: At the humid air inlet, $P_s(T_1) = P_v(T_1) = 0.2219\,\text{psia}$ and $P_{v,1} = \phi_1 P_s(T_1) = 0.2 \times 0.2219 = 0.04438\,\text{psia}$. The humidity ratio at this inlet is

$$\omega_1 = \frac{0.622 P_v(T_1)}{P - P_v(T_1)}$$

$$= \frac{0.622 \times 0.04438}{14.7 - 0.04438}$$

$$= 0.00188\ \text{lb}_m/\text{lb}_m\text{-da}$$

and the humid-air enthalpy at this inlet is

$$h_1 = c_{P,da}(T_1 - 32) + \omega_1 h_g(T_1)$$
$$= 0.2397(56 - 32) + 0.00188 \times 1085.9$$
$$= 7.79 \text{ Btu/lb}_m\text{-da}$$

Similarly, at the humid-air outlet, $P_s(T_3) = 0.5073$ psia, $P_{v,3} = 0.4566$ psia,

$$\omega_3 = \frac{0.622 \times 0.4566}{14.7 - 0.4566}$$
$$= 0.01994 \text{ lb}_m\text{-da}$$

and

$$h_3 = 0.2397(90 - 32) + 0.01994 \times 1096.4$$
$$= 35.85 \text{ Btu/lb}_m\text{-da}$$

Combining the first law and conservation of mass principle gives

$$\hat{m}_2 = \frac{h_3 - h_1 + (\omega_3 - \omega_1)h_4}{h_2 - h_4}$$
$$= \frac{35.85 - 7.79 + (0.01994 - 0.00188)28.08}{58.07 - 28.08}$$
$$= 0.9526 \text{ lb}_m/\text{lb}_m\text{-da}$$

According to the conservation of the water constituent mass equation,

$$\hat{m}_4 = \hat{m}_2 - (w_3 - w_1)$$
$$= 0.9524 - (0.01994 - 0.00188)$$
$$= 0.9343 \text{ lb}_m/\text{lb}_m\text{-da}$$

The first law, when applied to the liquid water stream passing through the tower is

$$q = \hat{m}_4 h_f(T_4) - \hat{m}_2 h_f(T_2)$$
$$= 0.9343 \times 28.08 - 0.9524 \times 58.07$$
$$= -29.07 \text{ Btu/lb}_m\text{-da}$$

The mass flow rate of dry air passing through the cooling tower is then

$$\dot{m}_{da} = \frac{\dot{Q}}{q}$$
$$= \frac{-1 \times 10^{10}}{-29.07}$$
$$= 3.44 \times 10^8 \text{lb}_m\text{-da/hr}$$

The specific volume of the air as it enters the cooling tower is

$$v_1 = \frac{RT_1}{P_{da,1}}$$

$$= \frac{1545 \times 516}{28.97(14.7 - 0.04438)144}$$
$$= 13.04 \text{ ft}^3/\text{lb}_m$$

and the volume flow rate at the humid-air inlet is

$$\dot{V}_1 = \dot{m}_{da}v_1$$
$$= 4.49 \times 10^9 \text{ ft}^3/\text{hr}$$

The rate at which water is being evaporated is given by $\dot{m}_2 - \dot{m}_4$; according to the conservation of mass principle, this is

$$\dot{m}_{evap} = \dot{m}_{da}(\omega_3 - \omega_1)$$
$$= 3.44 \times 10^8(0.01994 - 0.00188)$$
$$= 6.21 \times 10^6 \text{ lb}_m/\text{hr}$$

10.8.4   Dehumidification

Water vapor can be removed from excessively humid air in a variety of different ways. Some solid and liquid substances such as silica gel and water–lithium chloride solutions absorb water vapor from humid air under the proper conditions. These substances are known as *desiccants*. The water-vapor content of humid air also can be reduced by cooling the air to temperatures below the dew-point temperature. This cooling first saturates the mixture and then causes the water vapor to condense in the form of liquid water or dew. The cooling method of lowering the water-vapor content in humid air will be discussed in this chapter.

The process of cooling humid air to a temperature below its dew point while maintaining the mixture pressure constant is shown on the psychrometric chart in Figure 10.13. The initial humid but unsaturated air at state 1 of Figure 10.13 is first cooled to the dew-point temperature, state d, where it becomes saturated. During this portion of the process, the humidity ratio remains constant because no water is being added to or removed from the humid air. Simultaneously, the relative humidity increases while the temperature decreases. Further reduction of the temperature below the dew point can only be accomplished as water is removed from the mixture while the humid air remains saturated at a relative humidity of 100%. When state 3 is reached, the humidity ratio of the humid air has been decreased to ω_3, and the mass of the water condensed out of the mixture per unit of dry air mass is $\omega_1 - \omega_3$.

FIGURE 10.13
*Constant-Pressure
Cooling of Humid Air*

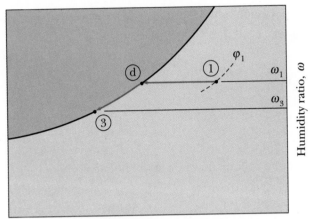

Dry–bulb temperature

EXAMPLE 10.11

Ten thousand cubic meters of atmospheric air at $25°$ C, 70% relative humidity, are cooled to $15°$ C. The liquid water generated during this cooling leaves the system at an average temperature of $17°$ C. Determine the amount of water removed from this air and the amount of cooling required.

System: Ten thousand cubic meters of humid air whose pressure remains constant at standard atmospheric pressure.

Basic Equations:

$$q - w = (h_3 - h_1) + \hat{m}_4 h_4 - \hat{m}_2 h_2$$
$$\omega_3 - \omega_1 = \hat{m}_2 - \hat{m}_4$$

Conditions: The steam tables and definitions of various humid-air quantities will again be used to obtain the required properties to illustrate this alternative to a psychrometric chart. Both the air and water components will be treated as ideal gases, and c_P of the air will be treated as a constant.

Solution: Following the numbering scheme of Figure 10.11, the humid-air conditions at the humid-air entrance are $P_s(T_1) = 3.169$ kPa and $P_{v,1} = \phi P_s(T_1) = 0.7 \times 3.169 = 2.218$ kPa, $T_{dp,1} = T_s(P_{v,1}) \simeq 19°$ C,

$$\omega_1 = \frac{0.622 P_v(T_1)}{P - P_v(T_1)}$$

$$= \frac{0.622 \times 2.218}{101 - 2.218}$$

$$= 0.0140 \text{ kg/kg-da}$$

and

$$h_1 = c_{P,da}(T_1 - 0) + \omega_1 h_g(T_1)$$
$$= 1.0035(25 - 0) + 0.0140 \times 2547.2$$
$$= 60.75 \text{ kJ/kg-da}$$

Similarly, the humid-air conditions at the humid-air exit are $\phi = 100\%$, $P_s(T_3) = 1.705$ psia $= P_{v,3}$,

$$\omega_3 = \frac{0.622 \times 1.705}{101 - 1.705}$$
$$= 0.01068 \text{ kg/kg-da}$$

and

$$h_3 = 1.0035(15 - 0) + 0.01068 \times 2528.9$$
$$= 42.06 \text{ kJ/kg-da}$$

The average enthalpy of the water extracted from the humid air is $h_4 = h_f(T_4) = 71.38$ kJ/kg.

The initial specific volume of the dry air is

$$v_{da,1} = \frac{RT}{P_{da}}$$

$$= \frac{8.314 \times 298}{28.97(101 - 2.218)}$$
$$= 0.866 \text{ m}^3/\text{kg}$$

and the mass of the air in this system is

$$m_{da} = \frac{V}{v_{da,1}}$$
$$= 11.55 \times 10^3 \text{ kg}$$

The total water removed from this air is

$$m_{removed} = m_{da}(\omega_1 - \omega_3)$$
$$= 11.55 \times 10^3(0.0140 - 0.01068)$$
$$= 38.4 \text{ kg}$$

Applying the first law, the total heat that must be removed to cool and dehumidify this humid air is

$$Q = m_{da}[(h_3 - h_1) + \hat{m}_4 h_4]$$
$$= m_{da}[(h_3 - h_1) + (\omega_1 - \omega_4)h_4]$$
$$= 11.55 \times 10^3[(42.06 - 60.75) + (0.014 - 0.01068)71.38]$$
$$= -213.1 \times 10^3 \text{ kJ}$$

10.8.5   Air Conditioning

The term *air conditioning* refers to the practice of maintaining certain air conditions in spaces such as buildings, hospital operating rooms, and electronic clean rooms for the comfort of the occupants or the satisfactory operation of equipment in the space. Maintaining these conditions is done with equipment that modifies the normal atmospheric air by raising or lowering its temperature, adding or removing water vapor, removing dust, or meeting any other special requirements. In this section, we will look specifically at the lowering of the temperature and relative humidity of the atmospheric air to provide a comfortable environment.

Lowering the temperature and relative humidity of humid air is done in two steps. First the air is cooled to a temperature below the dew point as discussed in the section on dehumidification. The water content is adjusted in this step to the desired final humidity ratio. Following this dehumidification, the air is usually quite cold and is saturated at a relative humidity of 100%. This cold, "wet" air is quite uncomfortable for most purposes and must be warmed to a comfortable level in a second process. The second step either heats the dehumidified air or mixes it with warm humid air while maintaining the total mixture pressure constant.

Constant-mixture-pressure dehumidification and reheating processes are shown in Figure 10.14. The dehumidification process in which the air is cooled is the same as process 1–d–3 of the previous section (see Figure 10.13). The reheating process requires that we add heat to the air to increase the air's temperature. While the air is being heated, no water vapor is added or removed from the air. Hence, the air's humidity ratio remains constant as the relative humidity is decreased in the constant-humidity ratio heating process i–3 of Figure 10.14.

FIGURE 10.14
Constant-Mixture-Pressure Dehumidification and Reheating

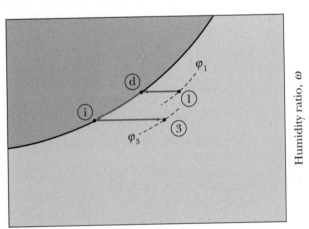

Dry–bulb temperature

Reducing the first law to the conditions of this reheating process when no shaft work is applied to the system yields

$$_iq_3 = h_3 - h_i = c_{P,da}(T_3 - T_i) + \omega_3 h_g(T_3) - \omega_i h_g(T_2)$$

This can be used to determine the amount of reheating required.

EXAMPLE 10.12

A manufacturing "clean room" requires 100,000 ft^3/hr of air at standard atmospheric pressure that has been conditioned from 90° F, 90% relative humidity, to 75° F, 60% relative humidity. The condensate generated leaves the air-conditioning unit at an average temperature of 70° F. Using a psychrometric chart, determine the rate of cooling needed to dehumidify the air and the rate of heating required to bring the dehumidified air to the required conditions.

System: The humid air as it passes through the dehumidification and reheating equipment.

Basic Equations:

$$q - w = (h_3 - h_1) + \hat{m}_4 h_4 - \hat{m}_2 h_2$$
$$\omega_3 - \omega_1 = \hat{m}_2 - \hat{m}_4$$

Conditions: The state numbering scheme of Figures 10.14 and 10.11 will be followed. Also, $\hat{m}_2 = 0$, because no water is being added to the humid air.

Solution: Using the psychrometric chart,

$$\omega_1 = 0.0279 \text{ lb}_m/\text{lb}_m\text{-da} \quad \omega_i = \omega_3 \quad\quad \omega_3 = 0.0112 \text{ lb}_m/\text{lb}_m\text{-da}$$
$$h_1 = 52.3 \text{ Btu/lb}_m\text{-da} \quad h_i = 26.7 \text{ Btu/lb}_m\text{-da} \quad h_3 = 30.2 \text{ Btu/lb}_m\text{-da}$$

The partial pressure of the water vapor in the humid air as it enters the air-conditioning equipment is $p_{v,1} = \phi p_s(T_1) = 0.9 \times 0.6988 = 0.629$ psia. The average enthalpy of the condensate is $h_4 = h_f(T_4) = 38.09 \text{ Btu/lb}_m$, according to the water property tables. The specific volume of the air as it enters the unit is $v_1 = RT_1/(P_\infty - P_{v,1}) = 14.49 \text{ ft}^3/\text{lb}_m$. The mass flow rate of air through this system is

$$\dot{m}_{da} = \frac{\dot{V}_{da}}{v_1}$$

$$= \frac{100,000}{14.49}$$

$$= 6900 \text{ lb}_m/\text{hr}$$

TIP:

The net effect is that the air must be cooled. Some of this cooling goes into condensing the water vapor. This cooling is called latent heat removal. *The remainder of the cooling reduces the temperature of the humid air. This portion of the cooling is known as* sensible heat removal.

Applying the first law to the 1–d–i process of Figure 10.14 and combining it with the conservation of mass principle yields

$$_1\dot{Q}_2 = \dot{m}_{da}[(h_i - h_1) + (\omega_i - \omega_1)h_4]$$
$$= 6910[(26.7 - 52.3) + (0.0112 - 0.0279)38.09]$$
$$= -181.200 \text{ Btu/hr}$$

A similar application of the first law to the i–3 heating process produces

$$_2\dot{Q}_3 = \dot{m}_{da}(h_3 - h_i)$$
$$= 6900(30.2 - 26.7)$$
$$= 24,150 \text{ Btu/hr}$$

10.9 Mixtures of Non-Ideal Gases

The *compressibility factor* and the *principle of corresponding states* were introduced in Chapter 9 as a means of correcting ideal gas results to account for actual gas behavior. These techniques were developed for single gases such as oxygen, methane, and propane. An infinite number of new gases can be formed by mixing two or more of these single gases in various proportions, so it is not possible to develop a compressibility chart, enthalpy-departure chart, or entropy-departure chart for every possible mixture of gases that we may incur. Instead, we need some type of process such as that developed in the preceding sections for mixtures of ideal gases to determine the properties of these mixtures from the properties of their constituent gases. Unfortunately, there is no simple way of doing this, and we must balance simplicity with accuracy to obtain the level of results required for the problem at hand. In this section, we will examine three of the simpler methods that are accurate under certain, but not all, conditions. We will leave the more accurate and more complicated methods for the several speciality treatises on this subject.

The first simple method is known as *Dalton's rule of additive pressures*. Under this rule, we presume that each constituent occupies the full mixture volume at the same temperature as the mixture. Each constituent will then exert its own partial pressure as predicted by a compressibility chart or a real gas equation of state such as the Redlich–Kwong equation. Dalton's rule states that the total pressure of the gas mixture is the sum of these partial pressures:

$$P = \sum_{i=1}^{n} P_i$$

For example, if we use the compressibility factor, the partial pressure of the ith constituent is given by

$$P_i = Z_{i,T,v} \frac{N_i \Re T}{V}$$

where $Z_{i,T,v}$ is the compressibility factor of constituent i obtained by using the temperature of the mixture and the specific volume of the constituent when it fills the entire volume of the mixture. When a general compressibility-factor chart based on the principle of corresponding states is used, we would enter this chart at the reduced temperature, T_r, and psuedo reduced specific volume, v_r. The total pressure of the mixture is then given by

$$P = \frac{\Re T}{V} \sum_{i=1}^{n} N_i Z_{i,T,v}$$

An effective compressibility factor for the mixture may be defined such that

$$P = Z_{eff} \frac{N \Re T}{V}$$

When compared to the preceding result, this yields

$$Z_{eff} = \sum_{i=1}^{n} y_i Z_{i,T,v}$$

Dalton's rule then predicts that the effective compressibility factor for a mixture of non-ideal gases is the mole fraction weighted sum of the compressibility factors of each constituent as if it filled the complete volume of the mixture and had the same temperature.

The second simple model for mixtures of non-ideal gases is known as *Amagat's rule of additive volumes*. Under this rule, each constituent is envisioned as having the same pressure and temperature as the total mixture. Each constituent would then occupy only a fraction of the total mixture volume as determined by a compressibility factor or real gas equation of state. Under the additive-volume rule, the total volume of the mixture is the sum of the partial volumes occupied by the constituents

$$V = \sum_{i=1}^{n} V_i$$

If the compressibility-factor–modified ideal gas equation of state is used, then the partial volume of constituent i is given by

$$V_i = Z_{i,T,P} \frac{N_i \Re T}{P}$$

where $Z_{i,T,P}$ is the compressibility factor of this constituent as evaluated using the mixture pressure and temperature. Under the additive-volume rule, the effective compressibility factor for the mixture can be shown to be

$$Z_{eff} = \sum_{i=1}^{n} y_i Z_{i,T,P}$$

Both of these approaches have limitations. Dalton's rule of additive pressures essentially neglects intermolecular forces and effectively treats each constituent as an ideal gas that is unaware of the presence of the other constituent gases. One should expect that the accuracy of this rule improves as the properties of the mixture approach those of an ideal gas. The additive pressure rule is most accurate when the pressure of the mixture is low and the specific volume of the mixture is high. Amagat's rule of additive volumes accounts for intermolecular forces because it is based on the total pressure of the mixture. Consequently, it produces poorer results when the pressure of the mixture is low and the specific volume of the mixture is high. In general, experience indicates that the additive-volume rule gives better results than the additive-pressure rule except when the mixture is approaching ideal gas behavior.

Another simple rule that is accurate to approximately 10% is known as *Kay's rule,* which defines a pseudocritical temperature and pressure for the mixture as

$$T_c' = \sum_{i=1}^{n} y_i T_{c,i}$$

and

$$P_c' = \sum_{i=1}^{n} y_i P_{c,i}$$

These are used with the principle of corresponding states to determine the mixture's compressibility factor. This rule is satisfactory for initial estimates and adequate in situations where 10% accuracy is acceptable. If higher accuracy is desired, then the literature should be consulted for more accurate methods.

EXAMPLE 10.13

The pressure of dry atmospheric air at 180 K, 0.00356 m³/kg, is experimentally measured as 10,100 kPa. Neglecting the trace components of the air, compare this measured pressure to the pressure as (a) predicted by the ideal gas equation of state, (b) Dalton's rule of additive pressures and the Redlich–Kwong equation of state, (c) Amagat's law of additive volumes and the Redlich–Kwong equation of state, and (d) Kay's pseudocritical pressure and temperature.

System: One kilogram-mole of air.

Conditions: The air will be modeled as a mixture of oxygen and nitrogen only.

Basic Equations:
$$P\bar{v} = Z\Re T$$

Solution: According to Example 3.1, $M = 28.84$, $y_{N_2} = 0.79$, and $y_{O_2} = 0.21$ when the trace elements are neglected in standard atmospheric air. Based on the molecular weight, the specified specific volume is 0.1027 m³/kg-mole.

(a) Treating the air as an ideal gas with $Z = 1$,

$$P = \frac{\Re T}{\bar{v}} = \frac{8.314 \times 180}{0.1027} = 14{,}570 \text{ kPa}$$

which is 44% greater than the experimentally measured value.

(b) If the nitrogen constituent were to fill the entire volume, its molar specific volume would be 0.1027/0.79 = 0.130 m³/kg-mole. Similarly, if the oxygen constituent were to fill the entire volume, its molar specific volume would be 0.489 m³/kg-mole. The coefficients of the nitrogen Redlich–Kwong equation of state are

$$a = 0.4278 \frac{\Re^2 T_c^{5/2}}{P_c} = 0.4278 \frac{8.314^2 \times 126^{5/2}}{3400} = 1550 \text{ kPa-m}^6/(\text{kg-mole})^2$$

and

$$b = 0.08664 \frac{\Re T_c}{P_c} = 0.08664 \frac{8.314 \times 126}{3400} = 0.0267 \text{ m}^3/\text{kg-mole}$$

using the critical pressures and temperatures of Table 9.2. These coefficients for the oxygen constituent are

$$a = 0.4278 \frac{8.314^2 \times 154^{5/2}}{5000} = 1740 \text{ kPa-m}^6/(\text{kg-mole})^2$$

and

$$b = 0.08664 \frac{8.314 \times 154}{5000} = 0.0222 \text{ m}^3/\text{kg-mole}$$

Dalton's additive-pressure rule gives the mixture pressure as

$$P = P_{N_2} + P_{O_2}$$

$$= \left[\frac{\Re T}{\overline{v} - b} - \frac{a}{\overline{v}(\overline{v} + b) T^{1/2}} \right]_{N_2} + \left[\frac{\Re T}{\overline{v} - b} - \frac{a}{\overline{v}(\overline{v} + b) T^{1/2}} \right]_{O_2}$$

$$= \frac{8.314 \times 180}{0.130 - 0.0267} - \frac{1550}{0.130(0.130 + 0.0267)180^{1/2}}$$

$$+ \frac{8.314 \times 180}{0.489 - 0.0222} - \frac{1740}{0.489(0.489 + 0.0222)180^{1/2}}$$

$$= 11{,}503 \text{ kPa}$$

This is 14% higher than the measured pressure.

(c) Amagat's additive-volume rule states that the mixture volume is the sum of the partial volumes of the constituents when they are at the mixture pressure and temperature. Hence,

$$\overline{v} = \sum_{i=1}^{n} y_i v_i$$

$$0.1027 = 0.79 \overline{v}_{N_2} + 0.21 \overline{v}_{O_2} \tag{a}$$

Applying the Redlich–Kwong equation of state to each constituent gives

$$P_{N_2} = \frac{1497}{\bar{v}_{N_2} - 0.0267} - \frac{115.5}{\bar{v}_{N_2}(\bar{v}_{N_2} + 0.0267)} \tag{b}$$

and

$$P_{O_2} = \frac{1497}{\bar{v}_{O_2} - 0.0222} - \frac{129.7}{\bar{v}_{O_2}(\bar{v}_{O_2} + 0.0222)} \tag{c}$$

When the additive-volume rule is satisfied, this should yield $P = P_{O_2} = P_{N_2}$.

An iterative solution (see Appendix A) is required. As a first iteration, take $\bar{v}_{N_2} = 0.10$ m^3/kg-mole, which gives $P_{N_2} = 11,310$ kPa by Equation b above. Solving Equation a for $\bar{v}_{O_2}$ with this $\bar{v}_{N_2}$ gives $\bar{v}_{O_2} = 0.113$ m^3/kg-mole; when this is substituted into Equation c, we get $P_{O_2} = 8000$ kPa, which does not agree with P_{N_2}. As a second iteration, take $\bar{v}_{N_2} = 0.11$ m^3/kg-mole. Then $P_{N_2} = 10,290$ kPa, $\bar{v}_{O_2} = 0.0754$ m^3/kg-mole, and $P_{O_2} = 10,520$ kPa, which does not equal P_{N_2}. Next, take $\bar{v}_{N_2} = 0.109$ m^3/kg-mole, which yields $P_{N_2} = 10,380$ kPa, $\bar{v}_{O_2} = 0.0790$ m^3/kg-mole, and $P_{O_2} = 10,100$ kPa. The mixture pressure as predicted by the additive volume rule is then approximately 10,200 kPa, which is within 1% of the measured pressure.

(d) Using the data of Table 9.2, the pseudocritical pressure of this mixture is

$$P'_c = \sum_{i=1}^{n} y_i P_{c,i} = 0.79 \times 3400 + 0.21 \times 5000 = 3736 \text{ kPa}$$

and the pseudocritical temperature is

$$T'_c = 0.79 \times 126 + 0.21 \times 154 = 132 \text{ K}$$

The mixture's reduced specific volume is then

$$v'_r = \frac{\bar{v}P'_c}{\Re T'_c} = \frac{0.1027 \times 3736}{8.314 \times 132} = 0.350$$

and its reduced temperature is

$$T_r = \frac{T}{T'_c} = \frac{180}{132} = 1.36$$

Using the figures of Appendix C, the compressibility factor of this mixture is $Z = 0.70$. Kay's method predicts that the pressure will be

$$P = \frac{Z\Re T}{P} = \frac{0.70 \times 8.314 \times 180}{0.1027} = 10,200 \text{ kPa}$$

This is 1% higher than the measured value.

PROBLEMS

Sketch the appropriate P–V(v) and T–s state diagrams or psychrometric charts for the following problems, showing all states and processes.

10.1 A mixture of gases consists of 100 grams of oxygen, 1 kilogram of carbon dioxide, and 500 grams of helium. This mixture is maintained at 100 kPa, 27° C. Determine the apparent molecular weight of this mixture, the volume it occupies, the partial volume of the oxygen, and the partial pressure of the helium.

10.2 The mass fractions of a mixture of gases are 15% nitrogen, 5% helium, 60% methane, and 20% ethane. Determine the mole fractions of each constituent, the mixture's apparent molecular weight, the partial pressure of each constituent when the mixture pressure is 200 psia, and the apparent specific heats of the mixture when the mixture temperature is 300° F.

10.3 The mixture of the previous problem is compressed from 20 psia, 100° F, in an isentropic process to 200 psia. Determine the final mixture temperature (° F) and the specific work (Btu/lb$_m$) required.

10.4 A mixture of gases consists of 30% hydrogen, 40% helium, and 30% nitrogen by volume. Calculate the mass fractions and apparent molecular weight of this mixture.

10.5 The mixture of the previous problem is expanded isentropically from 5000 kPa, 600° C, to 200 kPa. Calculate the specific work (kJ/kg) produced by this expansion.

10.6 The volumetric analysis of a mixture of gases is 30% oxygen, 40% nitrogen, 10% carbon dioxide, and 20% methane. Calculate the apparent specific heats (Btu/lb$_m$-° R) and molecular weight of this mixture of gases.

10.7 The gaseous mixture of the previous problem is heated from 20° C to 200° C while flowing through a tube in which the pressure is maintained at 150 kPa. Determine the specific heat transfer (kJ/kg) to the mixture.

10.8 A mixture of hydrocarbon gases is composed of 60% methane, 25% propane, and 15% butane by weight. Determine the volume occupied by 100 lb$_m$ of this mixture when its pressure is 500 psia and its temperature is 100° F.

10.9 The gaseous mixture of the previous problem is compressed from 100 kPa, 20° C, to 1000 kPa in a reversible, isothermal, steady-flow compressor. Calculate the specific work (kJ/kg) and heat transfer (kJ/kg) for this compression.

10.10 The mixture of Problem 10.1 is heated from 50° F to 500° F while its pressure is maintained constant at 50 psia. Determine the change in the volume of the mixture (ft^3) and the total heat (Btu) transferred to the mixture.

10.11   The gaseous mixture of Problem 10.2 is enclosed in a 10 m³ rigid, well-insulated vessel at 200 kPa, 20° C. A paddle wheel in the vessel is turned until 100 kJ of work have been done on the mixture. Calculate the mixture's final pressure (kPa) and temperature (° C).

10.12   A mixture of gases is assembled by first filling an evacuated 5 ft³ tank with neon until the pressure is 5 psia. Oxygen is added next until the pressure increases to 15 psia. Finally, nitrogen is added until the pressure increases to 20 psia. During each step of the tank's filling, the contents are maintained at 140° F. Determine the mass of each constituent in the resulting mixture, the apparent molecular weight of the mixture, and the fraction of the tank volume occupied by nitrogen.

10.13   A portion of the gas in the tank of the previous problem is placed in a spring-loaded piston-cylinder device. The piston diameter and spring are selected for this device such that the volume is 0.1 m³ when the pressure is 200 kPa and 1.0 m³ when the pressure is 10000 kPa. Initially, the gas is added to this device until the pressure is 200 kPa and the temperature is 10° C. The device is now heated until the pressure is 500 kPa. Calculate the total work (kJ) and heat transfer (kJ) for this process.

10.14   The piston-cylinder device of the previous problem is filled with a mixture whose mass is 70% nitrogen and 30% carbon dioxide. Initially, this mixture is at 400 kPa and 30° C. The gas is heated until the volume has doubled. Calculate the total work (kJ) and heat transfer (kJ) for this process.

10.15   Calculate the total work (kJ) and heat transfer (kJ) required to triple the initial pressure of the mixture of the previous problem as it is heated in the spring-loaded piston-cylinder device.

10.16   A 5 ft³ scuba diver's tank is filled with a mixture of oxygen, nitrogen, and helium. The mass fractions of these constituents are 60% N_2, 30% O_2, and 10% He. Determine the mass of the mixture in the tank when the pressure and temperature are 300 psia and 70° F.

10.17   What is the partial volume of the oxygen, nitrogen, and helium in the previous problem?

10.18   An engineer has proposed mixing extra oxygen with normal air in internal-combustion engines to control some of the exhaust products. If an additional 5% (by volume) of oxygen is mixed with standard atmospheric air, how will this change the mixture's molecular weight?

10.19   A 30% (by mass) ethane and 70% methane mixture is to be blended in a 100-m³ tank at 130 kPa and 25° C. If the tank is initially evacuated, to what pressure should ethane be added before methane is added?

10.20   The dry stack gas of an electrical-generation station boiler has the following Orsat analysis: 15% CO_2, 15% O_2, and 1% CO. This gas passes through a

10 ft^2 metering duct at a velocity of 20 ft/s at standard atmospheric pressure and 200° F. Determine this gas mixture's mass flow rate.

10.21 A mixture of air and methane is formed in the inlet manifold of a natural gas–fueled internal-combustion engine. The mole fraction of the methane is 15%. This engine is operated at 3000 rpm and has a 5-liter displacement. Determine the mass flow rate of this mixture in the manifold where the pressure and temperature are 80 kPa and 20° C.

10.22 Natural gas (95% methane and 5% ethane by volume) flows through a 36-inch-diameter pipeline with a velocity of 10 ft/s. The pressure in the pipeline is 100 psia, and the temperature is 60° F. Calculate the mass (lb_m/hr) and volumetric (ft^3/hr) flow rates in this pipe.

10.23 Separation units often use membranes, absorbers, and other devices to reduce the mole fraction of selected constituents in gaseous mixtures. Consider a mixture of hydrocarbons that consists of 60% (by volume) methane, 20% ethane, and 10% propane. After passing through a separator, the mole fraction of the propane is reduced to 1%. The mixture pressure before and after the separation is 100 kPa. Determine the change in the partial pressures of all the constituents in the mixture.

10.24 Atmospheric contaminants are often measured in parts per million (by volume). What would the partial pressure of R-12 be in atmospheric air at 14.7 psia, 70° F, to form a 100-ppm contaminant?

10.25 During the expansion process of the Otto cycle, the gas is a mixture whose volumetric composition is 30% nitrogen, 10% oxygen, 35% water, and 25% carbon dioxide. Calculate the thermal efficiency of this cycle when the air at the beginning of the compression is at 90 kPa, 15° C, the compression ratio is 8:1, and the maximum cycle temperature is 1100° C. Model the heat-addition and heat-rejection processes using gas properties that are the average of the air and expansion gas properties.

10.26 How does the thermal efficiency of the previous problem compare to that predicted by air standard analysis?

10.27 During the expansion process of the Otto cycle, the gas is a mixture whose volumetric composition is 25% nitrogen, 7% oxygen, 28% water, and 40% carbon dioxide. Calculate the thermal efficiency of this cycle when the air at the beginning of the compression is at 12 psia, 55° F, the compression ratio is 7:1 and the maximum cycle temperature is 1600° F. Model the heat-addition and heat-rejection processes using gas properties that are the average of the air and expansion gas properties.

10.28 How does the thermal efficiency of the previous problem compare to that predicted by air standard analysis?

10.29 The gas passing through the turbine of a simple Brayton cycle has the volumetric composition 20% nitrogen, 5% oxygen, 40% carbon dioxide, and 35%

water. Calculate the thermal efficiency of this cycle when the air enters the compressor at 10 psia, 40° F, the pressure ratio is 6:1, and the temperature at the turbine inlet is 1400° F. Use the average of the air and turbine gas properties to model the heat-addition and heat-rejection processes.

10.30 How does the thermal efficiency of the previous problem compare to that predicted by air standard analysis?

10.31 The gas passing through the turbine of a simple Brayton cycle has the volumetric composition 30% nitrogen, 10% oxygen, 40% carbon dioxide, and 20% water. Calculate the thermal efficiency of this cycle when the air enters the compressor at 100 kPa, 25° C, the pressure ratio is 8:1, and the temperature at the turbine inlet is 1000° C. Use the average of the air and turbine gas properties to model the heat-addition and heat-rejection processes.

10.32 How does the thermal efficiency of the previous problem compare to that predicted by air standard analysis?

10.33 The mixture of gases in Problem 10.1 is expanded from 1000 kPa, 327° C, to 100 kPa in an adiabatic, steady-flow turbine of 90% efficiency. Calculate the second law effectiveness of this expansion and the specific potential work (kJ/kg) thus lost.

10.34 The mixture in Problem 10.2 is expanded in an adiabatic, steady-flow turbine of 85% efficiency. The mixture enters this turbine at 400 psia, 500° F, and leaves at 20 psia. Calculate the second law effectiveness of this expansion and the specific lost potential work (Btu/lb$_m$).

10.35 The gaseous mixture in Problem 10.1 is compressed to 2500 psia, 70° F. Determine the mass (lb$_m$) of this gas contained in a 10 ft^3 tank (a) treating it as an ideal gas mixture, (b) using a compressibility factor based on Dalton's law of additive pressures, (c) using a compressibility factor based on the law of additive volumes, and (d) using Kay's pseudocritical pressure and temperature.

10.36 The gaseous mixture in Problem 10.1 is compressed to 10,000 psia, 20° C. Determine the mass (kg) of this gas contained in a 1 m^3 tank (a) treating it as an ideal gas mixture, (b) using a compressibility factor based on Dalton's law of additive pressures, (c) using a compressibility factor based on the law of additive volumes, and (d) using Kay's pseudocritical pressure and temperature.

10.37 The gaseous mixture in Problem 10.6 flows through a 1-inch-diameter pipe at 1500 psia, 70° F, with a velocity of 10 ft/s. Determine the volumetric (ft^3/hr) and mass (lb$_m$/hr) flow rates of this mixture (a) treating it as an ideal gas mixture, (b) using a compressibility factor based on Dalton's law of additive pressures, (c) using a compressibility factor based on the law of additive volumes, and (d) using Kay's pseudocritical pressure and temperature.

10.38 The gaseous mixture in Problem 10.6 flows through a 2-centimeter-diameter pipe at 8000 kPa, 15° C, with a velocity of 3 m/s. Determine the volumetric (m^3/hr) and mass (kg/hr) flow rates of this mixture (a) treating it as an ideal gas mixture, (b) using a compressibility factor based on Dalton's law of additive pressures, (c) using a compressibility factor based on the law of additive volumes, and (d) using Kay's pseudocritical pressure and temperature.

10.39 A gaseous mixture consists of 75% methane and 25% ethane by mass. One million cubic feet of this mixture is trapped in a geological formation as natural gas at 300° F, 2000 psia. Determine the mass (lb_m) of this gas (a) treating it as an ideal gas mixture, (b) using a compressibility factor based on Dalton's law of additive pressures, (c) using a compressibility factor based on the law of additive volumes, and (d) using Kay's pseudocritical pressure and temperature.

10.40 The natural gas of the previous problem is pumped 6000 ft. to the surface. At the surface, the gas pressure is 20 psia and its temperature is 200° F. Using Kay's rule and the enthalpy-departure method, calculate the work required to pump this gas.

10.41 In an air-liquefaction plant, it is proposed that the pressure and temperature of air that is initially at 1500 psia, 40° F be adiabatically reduced to 15 psia, −200° F. Using Kay's rule and the departure charts, determine whether this is possible. If so, then how much specific work (Btu/lb_m) will this process produce?

10.42 In a liquid-oxygen plant, it is proposed that the pressure and temperature of air that is initially at 9000 kPa, 10° C, be adiabatically reduced to 50 kPa, −130° C. Using Kay's rule and the departure charts, determine whether this is possible. If so, then how much specific work (kJ/kg) will this process produce?

10.43 Atmospheric air at a pressure of 1 atmosphere and a dry-bulb temperature of 90° F has a relative humidity of 70%. Using a psychrometric chart, determine (a) the wet-bulb temperature, (b) the humidity ratio, (c) the specific enthalpy, (d) the dew-point temperature, and (e) the water-vapor pressure.

10.44 Atmospheric air at a pressure of 1 atmosphere and dry-bulb temperature of 30° C has a relative humidity of 80%. Using a psychrometric chart, determine (a) the wet-bulb temperature, (b) the humidity ratio, (c) the specific enthalpy, (d) the dew-point temperature, and (e) the water-vapor pressure.

10.45 Atmospheric air at a pressure of 1 atmosphere and dry-bulb temperature of 90° F has a wet-bulb temperature of 85° F. Using a psychrometric chart, determine (a) the relative humidity, (b) the humidity ratio, (c) the specific enthalpy, (d) the dew-point temperature, and (e) the water-vapor pressure.

10.46 Determine the adiabatic saturation temperature (° F) of the humid air in the previous problem.

10.47 Atmospheric air at a pressure of 1 atmosphere and dry-bulb temperature of 32° C has a wet-bulb temperature of 78° C. Using a psychrometric chart, determine (a) the relative humidity, (b) the humidity ratio, (c) the specific enthalpy, (d) the dew-point temperature, and (e) the water-vapor pressure.

10.48 Determine the adiabatic saturation temperature ($^\circ$ C) of the humid air in the previous problem.

10.49 Atmospheric air at a pressure of 1 atmosphere and dry-bulb temperature of 90° F has a dew-point temperature of 75° F. Using a psychrometric chart, determine (a) the relative humidity, (b) the humidity ratio, (c) the specific enthalpy, (d) the wet-bulb temperature, and (e) the water-vapor pressure.

10.50 Determine the adiabatic saturation temperature ($^\circ$ F) of the humid air in the previous problem.

10.51 Atmospheric air at a pressure of 1 atmosphere and dry-bulb temperature of 28° C has a dew-point temperature of 20° F. Using a psychrometric chart, determine (a) the relative humidity, (b) the humidity ratio, (c) the specific enthalpy, (d) the wet-bulb temperature, and (e) the water-vapor pressure.

10.52 Determine the adiabatic saturation temperature ($^\circ$ C) of the humid air in the previous problem.

10.53 Humid air with a temperature of 100° F and relative humidity of 80% is cooled at constant pressure to the dew-point temperature. Determine the specific cooling (Btu/lb$_m$-da) required for this process.

10.54 Humid air with a temperature of 30° C and relative humidity of 60% is cooled at constant pressure to the dew-point temperature. Determine the specific cooling (kJ/kg-da) required for this process.

10.55 Humid atmospheric air at 90° F, 90% relative humidity, is cooled to 50° F while the mixture pressure remains constant. Calculate the amount of water (lb$_m$/lb$_m$-da) removed from the air and the cooling requirement (Btu/lb$_m$-da) when the liquid water leaves the system at 60° F.

10.56 Humid atmospheric air at 30° C, 80% relative humidity, is cooled to 20° C while the mixture pressure remains constant. Calculate the amount of water (kg/kg-da) removed from the air and the cooling requirement (kJ/kg-da) when the liquid water leaves the system at 22° C.

10.57 Ten thousand cubic feet per hour of atmospheric air at 85° F with a dew point temperature of 70° F are to be cooled to 60° F. Determine the rate (lb$_m$/hr) at which condensate leaves this system and the cooling rate (Btu/hr) when the condensate leaves the system at 65° F.

10.58 Ten thousand cubic meters per hour of atmospheric air at 28° C with a dew-point temperature of 25° C are to be cooled to 18° C. Determine the rate (kg/hr) at which condensate leaves this system and the cooling rate (kW) when the condensate leaves the system at 20° C.

10.59 Humid air at 100 psia and a humidity ratio of 0.025 lb$_m$/lb$_m$-da is expanded to 15 psia in an isentropic nozzle. How much of the initial water vapor has been converted to liquid water at the nozzle outlet?

10.60 Humid air at 100 kPa, 20° C, and 90% relative humidity is compressed in a steady-flow, isentropic compressor to 800 kPa. What is the relative humidity of the air at the compressor outlet?

10.61 During a summer day in Phoenix, Arizona, the air temperature is 110° F and relative humidity is 15%. Water at 70° F is evaporated into this air to produce air at 75° F, 80% relative humidity. How much water (lb_m/lb_m-da) is required and how much cooling (Btu/lb_m-da) has been produced?

10.62 If the system of the previous problem is operated as an adiabatic system and the air produced by this system has a relative humidity of 70%, what is the temperature (° F) of the air produced?

10.63 During a summer day in El Paso, Texas, the air temperature is 40° C and relative humidity is 20%. Water at 20° C is evaporated to produce air at 25° C, 80% relative humidity. How much water (kg/kg-da) is required, and how much cooling (kJ/kg-da) has been produced?

10.64 If the system of the previous problem is operated as an adiabatic system and the air produced by this system has a relative humidity of 80%, what is the temperature (° C) of the air produced?

10.65 Desert dwellers often wrap their heads with a water-soaked porous cloth. On a desert where the temperature is 120° F and relative humidity is 10%, what is the temperature (° F) of this cloth?

10.66 In the previous problem, what is the temperature in ° C of the cloth on a desert when the temperature is 45° C and relative humidity is 15%?

10.67 Water at 30° C is to be cooled in a cooling tower which it enters at a rate of 5 kg/s. Humid air enters this tower at 15° C with a relative humidity of 25% and leaves at 18° C with a relative humidity of 95%. Determine the mass flow rate (kg/s) of dry air through this tower and the water's outlet temperature (° C).

10.68 How much specific work potential (kJ/kg-da) is lost in the cooling tower of the previous problem?

10.69 Water at 100° F is to be cooled in a cooling tower which it enters at a rate of 10,000 lb_m/hr. Humid air enters this tower at 60° F, 20% relative humidity, with a dry air flow rate of 500 lb_m/hr and leaves at 75° F. Determine the relative humidity at which the air leaves the tower and the water's exit temperature (° F).

10.70 Water enters a cooling tower at 95° F and at a rate of 3 lb_m/s. Humid air enters this tower at 65° F with a relative humidity of 30% and leaves at 75° F with a relative humidity of 80%. Determine the mass flow rate (lb_m/s) of dry air through this tower and the outlet temperature (° F) of the water.

10.71 How much specific work potential (kJ/kg-da) is lost in the cooling tower of the previous problem?

10.72 Water at $32°$ C is to be cooled in a cooling tower which it enters at a rate of 4 kg/s. Humid air enters this tower at $15°$ C, 20% relative humidity, with a dry-air flow rate of 0.12 kg/s and leaves at $20°$ C. Determine the relative humidity at which the air leaves the tower and the water's exit temperature ($°$ C).

10.73 On a summer day in New Orleans, Louisiana, the temperature is $35°$ C, and relative humidity is 95%. This air is to be conditioned to $24°$ C, 60% relative humidity. Determine the amount of cooling (kJ) required and the water removed (kg) per 1000 m^3 of dry air processed at the entrance to this system.

10.74 How far will the temperature ($°$ C) of the humid air of the previous problem have to be reduced to produce the desired dehumidification?

10.75 A simple vapor-compression system using R-12 as the working fluid is used to provide the cooling required in the previous two problems. It operates its evaporator at $5°$ C and its condenser saturation temperature at $40°$ C. The condenser rejects its heat to the New Orleans summer air. Calculate the potential work (kJ) lost in the total system per 1000 m^3 of dry air processed.

10.76 A summer day temperature in Atlanta, Georgia, is typically $90°$ F with a relative humidity of 85%. Determine the cooling required to condition 1000 ft^3 of this air to $75°$ F, 50% relative humidity. What is the minimum humid air temperature ($°$ F) required to meet this cooling requirement?

10.77 A simple vapor-compression refrigeration system using R-12 provides the cooling of the previous problem. This system operates its condenser such that the saturation temperature is $15°$ F higher than the summer air to which it rejects heat and its evaporator is $10°$ F cooler than the lowest humid air temperature in the system. What is the system's COP?

10.78 Humid air at 150 kPa, $40°$ C, 70% relative humidity, is cooled at constant pressure in a pipe to its dew-point temperature. Calculate the specific-heat transfer (kJ/kg-da) required for this process.

10.79 Humid air at 40 psia, $50°$ F, 90% relative humidity, is heated in a pipe at constant pressure to $120°$ F. Calculate the relative humidity at the pipe outlet and the specific amount of heat (Btu/lb_m-da) required.

10.80 Saturated humid air at 200 kPa, $15°$ C, is heated to $30°$ C as it flows through a 4-centimeter-diameter pipe with a velocity of 20 m/s and at constant pressure. Calculate the relative humidity at the pipe outlet and the rate of heat transfer (kW) to the air.

10.81 Calculate the rate at which the exergy of the humid air of the previous problem is increased.

10.82 Saturated humid air at 70 psia, 200° F, is cooled to 100° F as it flows through a 3-inch-diameter pipe with a velocity of 50 ft/s and at constant pressure. Calculate the rate (lb_m/s) at which liquid water is formed inside this pipe and the rate at which the pipe is cooled (Btu/hr).

10.83 Two humid air streams are adiabatically mixed at 1 atmosphere pressure to form a third stream. The first stream has a temperature of 100° F, a relative humidity of 90%, and a volumetric flow rate of 3 ft^3/s, while the second stream has a temperature of 50° F, a relative humidity of 30%, and a volumetric flow rate of 1 ft^3/s. Calculate the third stream's temperature (° F) and relative humidity.

10.84 Calculate the rate of increase for the total entropy of the previous problem.

10.85 Two humid air streams are adiabatically mixed at 1 atmosphere pressure to form a third stream. The first stream has a temperature of 40° C, a relative humidity of 40%, and a volumetric flow rate of 3 liter/s; the second stream has a temperature of 15° C, a relative humidity of 80%, and a volumetric flow rate of 1 liter/s. Calculate the third stream's temperature (° C) and relative humidity.

10.86 Calculate the rate at which potential work (kW) is lost in the previous problem.

10.87 Rather than reheating as discussed in Section 10.8.5, dehumidified humid air is to be mixed with warmer air at 1 atmosphere pressure. Saturated humid air at 50° F is to be mixed with atmospheric air at 90° F, 80% relative humidity, to form air at 70° F. Determine the proportions at which these two streams are to be mixed and the relative humidity of the resulting air.

10.88 Rather than reheating dehumidified humid air, it is to be mixed with warm air at 1 atmosphere pressure. Saturated humid air at 10° C is to be mixed with atmospheric air at 35° C, 80% relative humidity, to form air of 70% relative humidity. Determine the proportions at which these two streams are to be mixed and the temperature (° C) of the resulting air.

10.89 A typical winter day in Fairbanks, Alaska, has a temperature of 0° F and a relative humidty of 60%. What is the relative humidity inside a home where this air has been heated to 70° F?

10.90 The relative humidity of the air in the home of the previous problem is to be restored to 60% by evaporating 60° F water into the air. How much heat (Btu) is required to do this in a home of 16,000 ft^3 volume?

10.91 A typical winter day in Moscow has a temperature of −10° C and a relative humidity of 40%. What is the relative humidity inside a dacha that has air that has been heated to 18° C?

10.92 The relative humidity inside the dacha of the previous problem is to be brought to 60% by evaporating water at 20° C. How much heat (kJ) is required for this purpose per cubic meter of air in the dacha?

PROJECTS

The following projects are more extensive than the problems presented in this chapter. Most are open-ended and require some decisions on the student's part. Many can be expedited with computer-based property tables, computer spreadsheets, and student-written computer programs.

10.1   Figure 10.1 indicated that the simple additive rule may not be appropriate for the volume of binary mixtures of gases. Prove this for a pair of gases of your choice at several different temperatures and pressures using Kay's rule and the principle of corresponding states.

10.2   The operation of a cooling tower is governed by the principles of fluid mechanics, heat transfer, and mass transfer, as well as thermodynamics. The laws of thermodynamics do place bounds on the conditions under which satisfactory operation may be expected, while the other sciences determine equipment sizes and other factors. Use the second law as expressed by the increase in entropy or other appropriate principle and the first law to place bounds on the humid air at its inlet in comparison to the conditions at the liquid-water inlet. Do the same for the humid-air outlet conditions as compared to the liquid-water outlet conditions.

10.3   Someone suggested that the water–lithium bromide solutions used in the absorption refrigeration system of Chapter 7 be used to dehumidify humid air. This would be done by bringing a weak solution into contact with humid air. This could be accomplished by spraying the weak solution into the air, for example. As the water in the air is absorbed by this solution, the humidity ratio of the air is decreased. Develop the thermodynamic criterion for the conditions at the weak-solution inlet, humid-air inlet, strong-solution outlet, and humid-air outlet needed for this system to work. Also determine the need for heat addition or removal.

10.4   A hurricane is a large heat engine driven by the exchange of water with humid air. Evaporation of ocean water occurs as the air approaches the eye of the storm, and condensation occurs as rain near the eye of the storm. Develop a plot of the wind speed near the eye of the storm as a function of the amount of water released from the air as rain. On this plot, indicate the minimum air temperature and relative humidity necessary to sustain each wind speed. *Hint:* As an upper bound, all of the energy released by the condensing water would be converted into kinetic energy.

10.5   The daily change in the temperature of the atmosphere tends to be smaller in locations where the relative humidity is high. Demonstrate why this occurs by calculating the change in the temperature of a fixed quantity of air when a fixed quantity of heat is removed from the air. Plot this temperature change as a function of the initial relative humidity and be sure that the air temperature reaches or exceeds the dew-point temperature. Do the same when a fixed amount of heat is added to the air.

10.6 A pressurized mixture of nitrogen and argon is supplied to a directional control nozzle on a space satellite. Plot the gas velocity at the nozzle exit as a function of the argon mass fraction with fixed pressure and temperature at the entrance and pressure at the exit. The force produced by this nozzle is proportional to the product of the mass flow rate and velocity at the exit. Is there an optimal argon mass fraction that produces the greatest force?

Thermochemistry and Combustion

A Thought

*Knowledge and timber shouldn't
be much used till they are
seasoned.*

              Oliver Wendell Holmes

B y applying the laws of thermodynamics to chemical reactions, we can develop an understanding of the energy produced or consumed by the reactions and the constraints placed on them. Those reactions that produce energy are of particular interest because they can be used as energy-source reservoirs for the several systems presented in the previous chapters.

Combustion is the most widely used chemical reaction for the production of energy. Table 1.1 shows that the United States used approximately 81 quadrillion Btu of energy (approximately 38 million barrels of oil) daily in 1990. Eighty-eight percent of this was provided by fossil fuels: coal, crude oil, and natural gas. These fuels must, of course, be burned in some device to release their energy. In addition to the energy released by their combustion, we must deal with the combustion products that are released into the atmosphere. These products are primarily carbon dioxide and water vapor, which are innocuous provided that there is enough plant life available to convert the carbon dioxide back to oxygen. It is the trace substances—including carbon monoxide, nitrous oxides, sulfur dioxide, unburned fuel, and ash residues, among many others—that create the environmental quality problems that have been discussed in the previous chapter opening essays. An understanding of the combustion process and its chemistry thus is important from the perspectives of both energy production and environmental quality.

In view of the large amount of energy consumed and the problems being created by using fossil fuels, we should not be surprised to find considerable activity dedicated to the development of alternative fuels. These fuels range from garbage and biofuels to the alcohols and hydrogen. Both garbage and biofuels are quite bulky and contain low levels of energy in relationship to their volume and weight. One informal study of corn stalks as fuel for electrical-power generation in Nebraska found that a 300-megawatt plant would require all the corn stalks within a 500-mile radius. Garbage has similar characteristics, and the combustible components must be separated from other materials. This separation is sometimes very costly, if not impossible for some items such as multilayer plastic films. All of the alternatives generate a different set of trace substances in their combustion products. For example, the burning of alcohols can generate formaldehyde and other aldehydes. If garbage is not properly separated, then heavy metals, sulfur compounds, aldehydes, and many other gases will be present in the combustion products. Only hydrogen, which burns completely and forms water vapor, offers a completely clean alternative. Unfortunately, hydrogen is produced by decomposing water, which may require as much, if not more, energy than can be produced by burning the hydrogen.

Hence, the alternative fuels need to be considered carefully and developed further before they will become major entities in the energy picture.

In this chapter, we will focus on the application of the first law to chemical reactions and introduce the implications of the second law for these same reactions. Our primary emphasis will be on the major chemical reactions that occur in combustion. The techniques that we learn by studying the combustion reaction can be applied to any other chemical reaction. The topics presented here are thus general and have wide application.

11.1 Chemical and Combustion Reactions

Thermochemistry is the study of the energy production or requirements associated with chemical reactions. Those reactions that produce energy are known as *exothermic reactions,* while those that require energy are known as *endothermic reactions.* Our interest lies in the exothermic reactions because they produce heat that may be used as the energy supply for a thermodynamic power or refrigeration system. Although many exothermic reactions produce heat, the overwhelming choices by several orders of magnitude are two simple reactions: the rapid oxidations of carbon and hydrogen, which are known as *combustion.* We will limit this chapter's study to combustion reactions and the production of fuels. However, the principles developed here may be applied to any chemical reaction.

All of the fuels of the world, with the exception of nuclear fuels, are formed of carbon and hydrogen, along with small quantities of other elements mixed in various proportions and taking different forms. These fuels are the legacy of prehistorical biological production and, to a lesser degree, the current biological production of organic compounds. As time and geological and other events have acted on these biological remains, fuels including coal, crude oil, and natural gases have been formed and are now available for our use.

Coal is a solid fuel whose principle constituent is carbon. The compositions of several coals are presented in Table 11.1. The term *ash* in this table refers to the noncombustible solid matter in the coal. Inspection of this table reveals that all coals also contain a small amount of hydrogen that also can be burned. All of the coals listed in the table also contain traces of sulfur. This is not untypical of a coal and presents a special problem because this sulfur will also burn and form gaseous sulfur dioxide; combined with water in the atmosphere, this compound forms weak sulfuric acid, which is a principle constituent of acid rain. Low-sulfur coals thus are highly desirable.

TABLE 11.1
Composition of Typical Dry U.S. Coals by Mass Fraction (in %)

COAL	CARBON	HYDROGEN	OXYGEN	NITROGEN	SULFUR	ASH
Pennsylvania Anthracite	84.36	1.89	4.40	0.63	0.89	7.83
Virginia Semianthracite	69.27	3.91	7.50	0.66	0.63	18.03
Colorado Semibituminous	79.61	4.66	4.76	1.83	0.52	8.62
Illinois Bituminous	67.40	5.31	15.11	1.44	2.36	8.38
Utah Sub-bituminous	61.40	5.79	25.31	1.09	1.41	5.00
Texas Lignite	39.25	6.93	41.11	0.72	0.79	11.20

Liquid fuels are primarily derived from crude oil, which is a mixture of many different hydrocarbons. The approximate composition of crude oil (by mass fraction) at the wellhead is as follows:

83–87% Carbon   0–7% Oxygen and Nitrogen
11–16% Hydrogen   0–4% Sulfur

The many hydrocarbons in crude oil are separated during the refining process. Hydrocarbons that result from the refining process also are blends of several hydrocarbons; these can be represented by C_xH_y, where x is the average carbon content and y is the average hydrogen content. Octane, C_8H_{18}, is the average hydrocarbon in gasoline, and dodecane, $C_{12}H_{26}$, is the average hydrocarbon in diesel fuel. Other liquid fuels can be represented in a similar manner. Crude-oil derivatives often contain traces of sulfur. As the exhaust emission requirements for motor vehicles change, the amount of sulfur permitted in motor vehicle fuels is being reduced to avoid the formation of sulfur compounds in atmospheric air. These compounds are one of the more irritating constituents of *smog*.

Natural gases extracted from ground wells are mixtures whose primary constituents are methane, CH_4, and ethane, C_2H_6. Their compositions as produced at the well vary considerably, as Table 11.2 shows, and contain small amounts of other gases. Natural gases produced at the well are refined so that the gas delivered to the user is almost pure methane. This refining also removes any sulfur that may be contained in the produced gas. As a result, sulfur compounds are not present in the exhausts from the combustion of processed natural gas. And this fuel is gaseous, which makes it easy to thoroughly mix it with oxygen so that its combustion will be nearly complete with little or no undesirable residuals in its exhaust gases. The ease of transporting this fuel in pipelines, its clean combustion characteristics, and absence of sulfur make natural gas expensive in comparison to other fuels and ideal for the space-heating market.

TABLE 11.2
Composition of Typical U.S. Natural Gases at the Well by Mole Fraction (in %)

SOURCE	CO$_2$	N$_2$	CH$_4$	C$_2$H$_6$
Kentucky	3.8	2.9	23.6	69.7
California	6.5	—	77.5	16.0
Oklahoma	2.4	1.8	64.1	31.7
Texas	—	4.6	74.5	20.9
Utah	3.6	5.6	90.8	—
Louisiana	0.9	1.5	97.6	—

The composition of fuels may be specified on a mass basis (i.e., mass fraction) or mole basis (i.e., mole fraction). One basis may be converted to the other as presented in Section 3.1. The mole basis will be used throughout the remainder of this chapter because it is consistent with the methods used to analyze chemical-reaction balances. Apparent molecular weights of the fuels also may be calculated using the methods of Section 3.1.

EXAMPLE 11.1

Determine the apparent molecular weight of the Illinois bituminous coal of Table 11.1, neglecting the ash constituent.

System: The Illinois coal described in Table 11.1.

Basic Equations:

$$M = \frac{m}{N}$$

Conditions: None.

Solution: First, the mass fractions of Table 11.1 are adjusted to a basis that does not include the ash constituent. Then the mass fractions are converted to mole fractions using the constituent molecular weights. Finally, the apparent molecular weight is calculated by summing the products of the constituent mole fraction and molecular weight. These calculations are summarized in Table 11.3.

TABLE 11.3
Calculations of Example 11.1

	x_i	M_i	x_i/M_i	$y_i = x_i/(NM_i)$	y_iM_i
Carbon	0.7356	12	0.06130	0.6309	7.571
Hydrogen	0.0580	2	0.02900	0.2984	0.597
Oxygen	0.1649	32	0.00515	0.0530	1.696
Nitrogen	0.0157	28	0.00056	0.0058	0.162
Sulfur	0.0258	32	0.00081	0.0083	0.266
	1.0000		$N = 0.09717$		$M = 10.292$

11.1.1 Stoichiometric Reactions

Stoichiometric reactions are those in which the exact amount of reactants are brought together and sufficient time is allowed for the reaction to be completed. Stoichiometric reactions thus are perfect reactions that may or may not occur in practice. The primary stoichiometric combustion reactions are

$$C + O_2 \rightarrow CO_2$$
$$4H + O_2 \rightarrow 2H_2O$$
$$S + \dot{O}_2 \rightarrow SO_2,$$

The last reaction is not a desirable one and occurs when a fuel contains traces of sulfur. The perfect combustion of carbon would produce only carbon dioxide. Carbon monoxide, CO, is an intermediate product in this reaction. Given sufficient time and oxygen, carbon monoxide and carbon will completely burn to carbon dioxide. Water is the result of the perfect combustion of hydrogen. The combustion of hydrogen is a very rapid process. Consequently, it is extremely rare when this reaction is *not* completed. Sulfur dioxide will be formed when sulfur is burned with oxygen.

Applying the conservation of mass principle to each elemental species of a chemical reaction states that *the number of moles of an elemental species in the reactants must equal the number of moles of the same elemental species in the products formed by the reaction regardless of how these elemental species are combined in compounds.* This principle is applied to each individual elemental species in the reactants to determine the amounts of products that will be formed. This does not tell us which products will be formed. We must rely on other principles for this determination.

The process of balancing the number of moles of an individual elemental species in the reactants with those in the products is known as a *species balance.* For example, the stoichiometric combustion of a hydrocarbon consisting of x moles of carbon and y moles of hydrogen reacting with the most common form of oxygen (i.e., O_2) is

$$C_xH_y + pO_2 \rightarrow qH_2O + rCO_2$$

where the p, q, and r coefficients are to be determined. Once determined, these coefficients dictate the amount of oxygen *required for* the reaction and the amount of carbon dioxide and water *formed by* the reaction. The carbon species balance gives

$$r = x$$

the hydrogen species balance gives

$$q = \frac{y}{2}$$

and the oxygen species balance gives

$$2p = q + 2r$$

The simultaneous solution of these species balances yields the amount of oxygen required:

$$p = \frac{y}{4} + x$$

The completely balanced combustion reaction of this hydrocarbon is then

$$C_xH_y + \left(\frac{y}{4} + x\right)O_2 \rightarrow \frac{y}{2}H_2O + xCO_2$$

Thus, the stoichiometric combustion of one mole of methane, $x = 1$ and $y = 4$, requires two moles of oxygen and produces two moles of water and one mole of carbon dioxide.

11.1.2   Practical Combustion

The stoichiometric reactions of the preceding discussion are the reactions that would occur in well-controlled laboratory conditions. When combustion occurs in thermodynamic equipment, four factors will alter these laboratory-type reactions:

1. Imperfect fuels containing sulfur, ash, and other contaminants
2. Using air for the combustion oxygen supply
3. Using excess air to control the reactions and flame temperatures
4. Incomplete combustion because of inadequate time for burning or inadequate oxygen supply or mixing.

When present, any or all of these factors must be properly accounted for in our reaction balances.

To illustrate the impact of an imperfect fuel on the reaction balance, we will consider a fuel mixture of methane, sulfur, and carbon dioxide. This mixture has been analyzed such that the mole fraction of the methane is known to be x and the mole fraction of the sulfur is known to be y. The mole fraction of the carbon dioxide is then $1 - x - y$. One mole of this imperfect fuel has the composition $xCH_4 + yS + (1 - x - y)CO_2$. We will momentarily presume that this combustion is going to have pure oxygen available for the oxidizer. As carbon, hydrogen, and sulfur are available in the fuel, we can expect carbon dioxide, water, and sulfur dioxide in the combustion products. The general reaction balance is

$$[xCH_4 + yS + (1 - x - y)CO_2] + pO_2 \rightarrow qCO_2 + rH_2O + sSO_2$$

where the p, q, r, and s coefficients can be determined from the carbon, hydrogen, sulfur, and oxygen species balances:

$$\begin{aligned}
\text{Carbon:} & \quad x + (1 - x - y) = q \\
\text{Hydrogen:} & \quad 4x = 2r \\
\text{Sulfur:} & \quad y = s \\
\text{Oxygen:} & \quad 2(1 - x - y) + 2p = 2q + r + 2s
\end{aligned}$$

The solution of these species balances reduces the reaction balance to

$$[x\text{CH}_4 + y\text{S} + (1 - x - y)\text{CO}_2] + (2x + y)\text{O}_2 \rightarrow (1 - y)\text{CO}_2 + 2x\text{H}_2\text{O} + y\text{SO}_2$$

Note that the mass basis of this expression is one mole of fuel. The coefficient on each term in this reaction balance represents the molar mass of each compound per mole of fuel in the reaction. These coefficients are also the number of moles of each constituent when one mole of fuel is burned.

Instead of pure oxygen, atmospheric air is used in practical combustion equipment because it is so much more readily available. The trace gases and water vapor present in the atmosphere constitute approximately 1% of the air constituents. These may be neglected for the purpose of combustion energy analysis. We will model atmospheric air as if its only constituents are nitrogen and oxygen; with only these two constituents, each mole of O_2 is mixed with 3.76 moles of N_2. Atmospheric air can then be chemically represented by $O_2 + 3.76N_2$. The nitrogen portion of this mixture is inert as long as the temperatures incurred during the combustion are not too high. When this is the case, the same number of moles of N_2 must appear on the products side of the reaction balance as on the reactants side. For example, the reaction balance on the combustion of dodecane (i.e., diesel fuel) using atmospheric air is

$$\text{C}_{12}\text{H}_{26} + \frac{37}{2}(\text{O}_2 + 3.76\text{N}_2) \rightarrow 12\text{CO}_2 + 13\text{H}_2\text{O} + \frac{3.76 \times 37}{2}\text{N}_2$$

when correctly balanced using the carbon, hydrogen, oxygen, and nitrogen species balances. Note that the coefficient on the air term, 37/2, is the same as when only oxygen is used for the combustion.

If the temperature of the combusting gases becomes great enough, the inert nitrogen will disassociate into nitrogen ions that will readily combine with varying numbers of oxygen ions to form nitrous oxides. If these gases are cooled too quickly as in an internal-combustion engine, they will appear in the combustion gases. Several different nitrous oxides normally appear in exhaust gases. These constituents can be represented as a single NO_x, with x representing the average oxygen content of all the various nitrous oxides in the products. We will neglect the formation of the nitrous oxides in this chapter, although these are major contributors to degraded air quality.

Excess air (i.e., oxygen) is always supplied with the fuel to ensure that the maximum amount of fuel will burn. This excess air also may be used to control the maximum flame temperature. The amount of excess air to be supplied with the fuel depends on several factors, including the fuel, the equipment used to prepare the fuel, and the combustion equipment.

When excess air is used, not all of the supplied oxygen will be combined with the carbon, hydrogen, and (possibly) sulfur of the fuel. The extra oxygen will appear in the products as diatomic oxygen. For example, when octane is burned with an amount x of excess air, the general combustion balance is

$$C_8H_{18} + p(1 + x)(O_2 + 3.76N_2) \rightarrow qCO_2 + rH_2O + pxO_2 + sN_2$$

where p is the coefficient that would result if the octane were burned with a stoichiometric amount of oxygen. When all the species balances have been satisfied, this becomes

$$C_8H_{18} + \frac{25}{2}(1 + x)(O_2 + 3.76N_2) \rightarrow 8CO_2 + 9H_2O + \frac{25}{2}xO_2$$
$$+ \frac{25(1 + x)3.76}{2}N_2$$

The amount of air supplied in relation to the amount of fuel burned can also be specified by the *air–fuel ratio* (A–F) or its reciprocal, the *fuel–air ratio* (F–A). The air–fuel ratio is normally defined as the mass of air supplied per mass of fuel burned. For example, the air–fuel ratio for the combustion of the octane fuel presented in the previous paragraph is given by

$$A\text{--}F = \frac{25}{2}\frac{M_{air}}{M_f}(1 + x) = \frac{25}{2}\frac{28.97}{114}(1 + x) = 3.177(1 + x)$$

The stoichiometric air–fuel ratio is then 3.177, and will be higher when excess air is used for combustion. Occasionally, but not often, molar masses are used in these ratios.

Incomplete combustion can have two effects: unburned fuel and carbon monoxide in the combustion products. The sources of the unburned fuel are an inadequate supply of oxygen, inadequate mixing of the oxygen with the fuel, and insufficient time for the combustion to be completed. The presence of carbon monoxide in the products of combustion results from the lack of time for complete combustion. In industrial boilers, adequate time, oxygen supply, and the mixing of fuel and air normally are not a problem. But in internal-combustion engines, all of these factors of incomplete combustion are present.

EXAMPLE 11.2

Methane is rapidly burned with 80% excess air. Insufficient time is available for the combustion to be completed, and 5% of the carbon (by volume) in the methane forms carbon monoxide. Determine the apparent molecular weight of the products of combustion and the air–fuel ratio.

System: One mole of methane when burned with 80% excess air.

Basic Equations:
$$M = \frac{m}{N}$$
Elemental species balances

Conditions: None.

Solution: As a result of the incomplete combustion of the carbon, the mixture $0.95CO_2 + 0.05CO$ will be present in the products of combustion. The amount of oxygen supplied with the fuel will be more than the stoichiometric requirement because 80% excess air is provided. The general reaction balance is

$$CH_4 + 1.8p(O_2 + 3.76N_2) \rightarrow q(0.95CO_2 + 0.05CO) + rH_2O + 0.8pO_2 + sN_2$$

where p is the stoichiometric oxygen requirement. From the carbon balance, $q = 1$, while the hydrogen balance gives $r = 2$. The oxygen balance is

$$3.6p = (2 \times 0.95 + 0.05)q + r + 1.6p$$
$$= 1.95 + 2 + 1.6p$$
$$p = 1.975$$

and the nitrogen balance yields

$$s = 1.8 \times 3.76p$$
$$= 8.743$$

The completed reaction balance is then

$$CH_4 + 3.555(O_2 + 3.76N_2) \rightarrow 1(0.95CO_2 + 0.05CO) + 2H_2O + 1.58O_2 + 8.743N_2$$

where the coefficient on each constituent represents the number of moles of that constituent per mole of fuel burned. The calculation of the apparent molecular weight of the combustion products is presented in Table 11.4.

TABLE 11.4

Calculations for Example 11.2

	N_i	$y_i = N_i/N$	M_i	$y_i M_i$
CO_2	0.95	0.0713	44	3.137
CO	0.05	0.0038	28	0.106
H_2O	2.00	0.1502	18	2.704
O_2	1.58	0.1186	32	3.795
N_2	8.74	0.6561	28	18.371
	13.32			$M = 28.113$

For each mole of methane burned, 16.92 moles of air must be supplied. The air–fuel ratio is then

$$A\text{–}F = \frac{16.92}{1}\frac{28.97}{16.01} = 30.62\frac{\text{kg-air}}{\text{kg-fuel}} \quad \text{or} \quad \frac{\text{lb}_m\text{-air}}{\text{lb}_m\text{-fuel}}$$

11.2 Basis for First and Second Law Analysis

When the working fluid is a simple, pure substance such as air or water, we need not be too concerned about the basis of first and second law properties such as enthalpy and entropy. The same substance will be present at the beginning and ending of a process or at the inlets and outlets of the system, and its properties are all measured with regard to the same reference level. Because the first and second laws deal with changes, the datum from which properties are measured will add and subtract from the equations correctly. However, when the substances at the end of a process are different from those at the beginning or the substances at the outlets differ from those at the inlets as when a chemical reaction occurs, then a common property datum is necessary for the various substances if we are to account for the energy transfers and entropy generations properly.

We were exposed to the concept of a common basis for properties when we studied mixtures of ideal gases and vapors. The enthalpies of the psychrometric chart and first law applications of the air and water vapor were all measured with respect to a saturated liquid at $T = 32° F = 0° C$. This basis was selected because it is the datum for the water property tables of Appendix B, Table B1.1 to Table B1.4.

Enthalpies on a mole basis at an arbitrary state 1 are given by

$$\bar{h}_1 = \bar{h}^0 + \int_0^1 d\bar{h}$$

where 0 is the reference state and $\bar{h}^0$ is the molar enthalpy at the reference state. The most common reference state is that of *the most stable form of the elemental substance (O_2, N_2, H_2, etc.) at a pressure of one atmosphere and a temperature of 77° F or 25° C*, which is known as the *standard state*. The enthalpy of the most stable form of an elemental substance is assigned a value of zero at this state. The enthalpy of an elemental substance at an arbitrary state 1 is then given by

$$\bar{h}_1 = \int_0^1 d\bar{h} = \Delta\bar{h}^0$$

where $\Delta\bar{h}^0$ is the molar enthalpy measured with respect to the standard state.

When compounds such as H_2O or CH_4 are formed from elemental substances such as C, H_2, and O_2 and are brought to the standard state, their enthalpy will not be zero because some energy was produced or consumed during the reaction when the compound was formed. This enthalpy is known as the *enthalpy of formation*. The enthalpy of formation may be best appreciated by considering the steady-flow reactor of Figure 11.1. The most stable forms of the elements from which the compound is formed enter this reactor at the standard temperature and pressure. The compound that is formed by the chemical reaction in the reactor also leaves at the standard temperature and pressure. For the inlets and outlet to all be at the same temperature and pressure, heat must be added or removed from the reactor at some rate. This rate of heat transfer divided by the formed compound's molar mass flow rate is the compound's enthalpy of formation. The sign convention for the enthalpy of formation is the same as that of the heat transfer; that is, exothermic reactions, which release heat to the surroundings, such as those that occur during combustion, will have negative enthalpies of reaction, and endothermic reactions will have positive enthalpies of formation.

FIGURE 11.1

Steady-Flow Reactor for Determining Heats of Formation

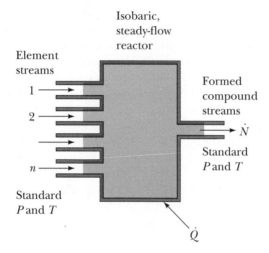

In the absence of a shaft-work interaction with the reactor of Figure 11.1, and when all streams are at the reference pressure and temperature, the first law as applied to this reactor is

$$\dot{Q} = \dot{N}_c \overline{h}_c^0 - \sum_{i=1}^{n} \dot{N}_i \overline{h}_i^0$$

This reduces to

$$\overline{q} = \overline{h}_f^0 = \overline{h}_c^0 - \sum_{i=1}^{n} z_i \overline{h}_i^0$$

where $\overline{q}$ is the heat transfer per mole of formed compound, $\overline{h}_f^0$ is the compound's enthalpy of formation, and the z_i's are the number of moles of each element forming the compound per mole of formed compound. By definition, the reference enthalpies of the elements of which the compound is formed are zero. Thus,

$$\overline{h}_c^0 = \overline{h}_f^0$$

and the enthalpy of the compound at the standard state, $\overline{h}_c^0$, is the enthalpy of formation. The enthalpies of formation of several compounds are presented in appendix Table B.7. The enthalpy of a substance at an arbitrary state is therefore

$$\bar{h} = \bar{h}_f^0 + \Delta \bar{h}^0$$

Entropy is a measure of disorder; that is, the more disordered something is, the greater its entropy. This is best demonstrated by comparing the entropy of a saturated vapor to that of a saturated liquid. The entropy of the vapor is higher than that of the liquid. Because the liquid and vapor are at the same temperature and pressure, this difference can result only from the fact that we are less certain about the position of the molecules, their velocities, and other factors in the gas than we are for the liquid.

Inspection of any table of properties reveals that as the temperature of a substance is reduced while pressure is held constant, for instance, then entropy decreases even when a phase change is not present. The quantum theory predicts that as the temperature is reduced, the number of quantum energy states available to the atom become fewer and fewer. Thus, as the temperature of the atom is reduced, we become more certain about the condition of the atom.

These observations have led to the *third law of thermodynamics,* which was stated by Nernst[1] as follows:

PRINCIPLE:

The entropy of a pure substance in thermodynamic equilibrium approaches zero as the temperature approaches absolute zero.

A temperature of absolute zero is the common reference state for the entropy of all pure substances. Entropies that use a temperature of absolute zero for the reference state in accordance with the third law of thermodynamics are known as *absolute entropies.*

The absolute entropy of a substance at the standard state is known as the *standard entropy.* Standard entropies of several substances that appear in combustion reactions are tabulated in appendix Table B.7. The standard entropy is somewhat more convenient because the reference state is the same as that

---

[1]Walther Hermann Nernst (1864–1941), a German chemist, won the Nobel Prize in 1920 for his thermochemical work. While serving at the University of Berlin in 1906, he announced his heat theorem, which later became the third law of thermodynamics. This theorem demonstrated that the maximum work obtained from a process can be calculated from the heat evolved at temperatures near absolute zero.

for the enthalpies. In terms of the standard entropy, the entropy of a substance at an arbitrary state 1 is given by

$$\bar{s}_1 = \bar{s}^0 + \int_0^1 ds = \bar{s}^0 + \Delta\bar{s}^0$$

where the last term may be evaluated using the means developed in Chapters 3 and 9 or with the Gibbs relations.

11.3 First Law Analysis of Chemical Reactions

A typical steady-flow system in which a chemical reaction occurs is shown in Figure 11.2. One or more feedstock streams that contain the reacting elements enter the reactor. Although these streams may be at different temperatures and pressures, they also can enter at the same temperature and pressure. Once in the reactor, heat may be added or removed, the feedstocks may be exposed to catalysts, or other means may be used to effect the desired reaction. The reaction products then leave the reactor at a common pressure and temperature. This system may be analyzed by using the multi-inlet and outlet rate form of the first law:

$$\dot{Q} - \dot{W} = \sum_{prod} \dot{N}(\bar{h} + \bar{e}_k + \bar{e}_p) - \sum_{reac} \dot{N}(\bar{h} + \bar{e}_k + \bar{e}_p)$$

FIGURE 11.2
*A General
Steady-Flow Reactor*

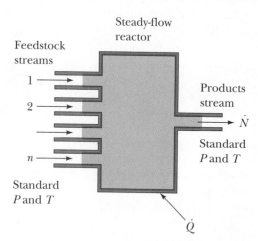

The basis of this equation can be converted to a basis of the molar mass of one of the feedstock streams. For example, the combustion heat generated per mole of fuel burned is often more convenient than the rate at which heat is generated. Dividing the rate form of the first law with the molar mass flow rate of the basis constituent yields the result

$$\bar{q} - \bar{w} = \sum_{prod} z_i(\bar{h} + \bar{e}_k + \bar{e}_p)_i - \sum_{reac} z_i(\bar{h} + \bar{e}_k + \bar{e}_p)_i$$

where $\bar{q}$ is the heat transfer per mole of substance in the basis stream and $\bar{w}$ is the shaft work per mole of the substance in the basis stream. The z coefficients are the molar mass of substance in each stream per unit of molar mass of substance in the basis stream. If the reaction balance is done on the basis of one mole of mass in the basis stream, the z coefficients are the same as those of the reaction balance.

Typically, the kinetic and potential energies of the streams are small when compared to the enthalpies. The first law then becomes

$$\bar{q} - \bar{w} = \sum_{prod} z_i \bar{h} - \sum_{reac} z_i \bar{h}_i$$

According to the previous subsection, the enthalpy of a substance at an arbitrary state is given by

$$\bar{h}(P,T) = \bar{h}_f^0 + \int_0^{P,T} dh = \bar{h}_f^0 + \Delta \bar{h}^0(P,T)$$

where $\Delta \bar{h}^0(P,T)$ is the difference between the enthalpy of the substance at the arbitrary state and the standard state. This enthalpy difference may be calculated using any of the techniques developed for pure substances to this point. When this is substituted into the first law, the result can be rearranged as follows:

$$\bar{q} - \bar{w} = \sum_{prod} z_i \Delta \bar{h}_i^0(P,T) + \sum_{prod} z_i \bar{h}_{f,i}^0 - \sum_{react} z_i \bar{h}_{f,i}^0 - \sum_{react} z_i \Delta \bar{h}_i^0(P,T) \qquad (11.1)$$

In words, this equation states that

$$\begin{bmatrix} \text{The heat added} \\ \text{per unit of} \\ \text{basis substance} \\ \text{mass} \end{bmatrix} - \begin{bmatrix} \text{The work produced} \\ \text{per unit of} \\ \text{basis substance} \\ \text{mass} \end{bmatrix} = \begin{bmatrix} \text{The enthalpy difference} \\ \text{generated as the} \\ \text{products are converted} \\ \text{to the standard state} \end{bmatrix}$$

$$+ \begin{bmatrix} \text{The difference between} \\ \text{the heats of formation} \\ \text{of the products} \\ \text{and reactants} \end{bmatrix} - \begin{bmatrix} \text{The enthalpy difference} \\ \text{generated as the} \\ \text{reactants are converted} \\ \text{to the standard state} \end{bmatrix}$$

The application of the first law then may be viewed as a three-step process:

1. Take the reacting substances to the standard state.
2. Account for the reaction by converting the reactants at the standard state to products at the standard state using the heats of formation.
3. Take the product substances to the state of the products.

Of course, we must weigh the contribution of each substance (both products and reactants) by the coefficients of the appropriate reaction balance.

EXAMPLE 11.3

A natural gas fuel consists of 90% methane and 10% ethane by volume. This fuel is burned with 50% excess air. Both the air and fuel are introduced to the combustion chamber at 1 atmosphere pressure and 25° C. The combustion is complete and the products of combustion are exhausted from the combustion chamber at 1 atmosphere pressure and 100° C. Determine the heat released by the combustion per mole of fuel burned.

System: A general reaction chamber such as that shown in Figure 11.2.

Basic Equations:

$$\bar{q} - \bar{w} = \sum_{prod} z_i \Delta \bar{h}_i^0(P,T) + \sum_{prod} z_i \bar{h}_{f,i}^0 - \sum_{react} z_i \bar{h}_{f,i}^0 - \sum_{prod} z_i \Delta \bar{h}_i^0(P,T)$$

Elemental species balances

Conditions: The basis of both the reaction balance and the first law will be one mole of fuel, $0.9CH_4 + 0.1C_2H_6$. Because the temperature changes are not very large, enthalpy differences from the standard state will be calculated using constant specific heats at 25° C (see appendix Table B.5).

Solution: On the basis of one mole of fuel, the reaction balance may be written as

$$(0.9CH_4 + 0.1C_2H_6) + 1.5p(O_2 + 3.76N_2) \rightarrow qCO_2 + rH_2O + 0.5pO_2 + 1.5 \times 3.76pN_2$$

The carbon species balance gives $q = 1.1$, and the hydrogen species balance gives $r = 2.1$. The oxygen species balance is

$$3p = 2q + r + p$$

When solved for p, this gives $p = 2.15$. The final reaction balance is then

$$(0.9CH_4 + 0.1C_2H_6) + 3.225O_2 + 12.126N_2 \rightarrow 1.1CO_2 + 2.1H_2O + 1.075O_2 + 12.126N_2$$

Employing the coefficients of the reaction balance, the first law expands to

$$\bar{q} = (1.1\bar{h}_{CO_2} + 2.1\bar{h}_{H_2O} + 1.075\bar{h}_{O_2} + 12.126\bar{h}_{N_2})_{prod}$$
$$- (0.9\bar{h}_{CH_4} + 0.1\bar{h}_{C_2H_6} + 3.225\bar{h}_{O_2} + 12.126\bar{h}_{N_2})_{react}$$

Because all the reactants are at the standard state, their enthalpies are the enthalpies of formation, which are zero for the elemental reactants O_2 and N_2. The enthalpy of the products is the sum of the enthalpy of formation and the difference between the enthalpies at the given state and at the standard state. Because these enthalpies are to be determined using constant specific heats, this enthalpy difference is given by $\bar{c}_P(T - T_0)$. The first law then becomes

TIP:

The heat transfer is the difference between the sum of the enthalpies weighted by the reaction-balance coefficients for the products minus those for the reactants.

$$\begin{aligned}
\bar{q} &= 1.1[\bar{h}^0_{f,CO_2} + \bar{c}_{P,CO_2}(T - T_0)] + 2.1[\bar{h}^0_{f,H_2O} + \bar{c}_{P,H_2O}(T - T_0)] \\
&\quad + 1.075\bar{c}_{P,O_2}(T - T_0) + 12.126\bar{c}_{P,N_2}(T - T_0) - 0.9\bar{h}^0_{f,CH_4} - 0.1\bar{h}^0_{f,C_2H_6} \\
&= 1.1(-393{,}520 + 37.712 \times 75) + 2.1(-241{,}820 + 33.655 \times 75) \\
&\quad + 1.075 \times 29.409 \times 75 + 12.126 \times 29.073 \times 75 \\
&\quad - [0.9(-74{,}850) + 0.1(-84{,}680)] \\
&= -827{,}700 \text{ kJ/kg-mole of fuel}
\end{aligned}$$

11.3.1 Heating Values

The maximum amount of heat that a fuel can produce as a result of combustion is one measure of the fuel's effectiveness. Inspection of the verbal expression of the first law as applied to a chemical reaction reveals that this would occur when the reactants enter and the products leave at the standard state. All of the difference between the heats of formation of the products and reactants (weighted by the reaction-balance coefficients) is then released as heat. None of this released energy is used to change the enthalpies of products or reactants. Only the complete products of combustion—CO_2, H_2, and maybe SO_2—would be present in the products if the maximum amount of heat is released. If this were not the case, then the heats of formation of incomplete combustion products weighted by their respective reaction-balance coefficients enter the first law in a manner that reduces the net heat produced. As a minimum, the stoichiometric amount of oxygen must be included in the reactants for the combustion to be complete. A stoichiometric amount of air or excess air also can be used because this only introduces nitrogen or excess oxygen, both of which are elements. Because the heat of formation of all elements is zero, this neither adds nor detracts from the total heat produced.

The net heat produced by a fuel when burned in this manner is known as a *heating value*. Actually, two heating values exist: the *higher heating value* (HHV) and the *lower heating value* (LHV). The difference between these two values is the phase of the water in the products. When the water in the products is presumed to be all vapor, then the heating value will be the lower one. When the water in the products is taken to be all liquid, the heating value will be the higher one. At the standard state, carbon dioxide in the products will always be gaseous. But the water formed by the combustion will either be all vapor or vapor and liquid at the standard state. The latter case will occur when the quantity of water in the products is great enough so that the product mixture is saturated as discussed in Chapter 10. Excess water then would appear in the products as a liquid.

The higher heating value is given by

$$\text{HHV} = \frac{z_{CO_2}\overline{h}^0_{f,CO_2} + z_{H_2O}\overline{h}^0_{f,H_2O(l)} - \overline{h}^0_{f,fuel}}{M_{fuel}}$$

and the lower heating value is given by

$$\text{LHV} = \frac{z_{CO_2}\overline{h}^0_{f,CO_2} + z_{H_2O}\overline{h}^0_{f,H_2O(g)} - \overline{h}^0_{f,fuel}}{M_{fuel}}$$

where the molecular weight of the fuel, M_{fuel}, has been introduced to put the heating values on a per-unit-of-fuel-mass basis. The difference between the two heating values is $z_{H_2O}h_{fg}/M_{fuel}$, which is the energy released when all of the water vapor in the products is condensed.

EXAMPLE 11.4

Determine the HHV and LHV of propane in USCS units. Also determine the amount of liquid and vapor water in the products when the fuel is completely burned with oxygen only.

System: A general reaction chamber such as that shown in Figure 11.2.

Basic Equations:
$$\overline{q} - \overline{w} = \sum_{prod} z_i \Delta\overline{h}^0_i(P, T) + \sum_{prod} z_i\overline{h}^0_{f,i} - \sum_{react} z_i\overline{h}^0_{f,i} - \sum_{react} z_i \Delta\overline{h}^0_i(P, T)$$

Elemental species balances

Conditions: Because we are determining the heating values, all inlets and outlets will be at one atmosphere pressure and a temperature of 77° F. The combustion will be complete with no carbon monoxide or other incomplete combustion substances in the products.

Solution: When completely reacted with oxygen only, the reaction balance is

$$C_3H_8 + 5O_2 \rightarrow 3CO_2 + 4H_2O$$

When the conditions under which the LHV is determined are applied, the first law becomes

$$\text{LHV} = \frac{z_{CO_2}\overline{h}^0_{f,CO_2} + z_{H_2O}\overline{h}^0_{f,H_2O(g)} - \overline{h}^0_{f,C_3H_8}}{M_{C_3H_8}}$$

$$= \frac{3(-169,300) + 4(-104,040) - (-44,680)}{44.10}$$

$$= -19,940 \text{ Btu/lb}_m\text{-propane}$$

A similar application of the first law to determine the HHV yields

$$\text{HHV} = \frac{z_{CO_2}\overline{h}^0_{f,CO_2} + z_{H_2O}\overline{h}^0_{f,H_2O(l)} - \overline{h}^0_{f,C_3H_8}}{M_{C_3H_8}}$$

$$= \frac{3(-169,300) + 4(-122,970) - (-44,680)}{44.10}$$

$$= -21,660 \text{ Btu/lb}_m\text{-propane}$$

The mole fraction of all the water in the products is

$$y_{H_2O} = \frac{4}{4+3} = 0.5714$$

When the carbon dioxide–water vapor gaseous mixture is saturated, the partial pressure of the water vapor would be the saturation pressure at the standard temperature of 77° F. This pressure is 0.460 psia, according to the steam tables. The maximum mole fraction the water vapor can have in the gaseous products is then

$$y_{H_2O,g,max} = \frac{P_s}{P_{max}} = \frac{0.460}{14.696} = 0.0313$$

The mole fraction of the liquid water in the products then will be

$$y_{H_2O,f} = 0.5714 - 0.0313 = 0.5401$$

The fraction of the water that appears in the products as liquid is

$$f_f = \frac{0.5401}{0.5714} = 0.945$$

and the fraction that appears as a vapor is

$$f_g = 1 - f_f = 1 - 0.945 = 0.055$$

TIP:

The relative amounts of water appearing as liquid and vapor can be indicated in the reaction balance by writing

$C_3H_8 + 5O_2 \rightarrow 3CO_2$
$\quad + 4[0.945H_2O(f)$
$\quad + 0.055H_2O(g)]$

11.3.2   Adiabatic Flame Temperature

Sometimes the maximum temperature that can be produced by combustion is a more important engineering consideration than the amount of heat that can be produced by combustion. This is the case in certain chemical and industrial processes such as steel manufacturing and flame cutting. The temperature of the products produced by a reaction will be as large as possible when the general steady-flow reactor of Figure 11.2 is well insulated such that the reaction process is adiabatic. Presuming that the reactants are supplied to the reactor at the standard state and that there is no shaft work interaction, the first law reduces to

$$\sum_{prod} z_i \Delta \bar{h}_i^0 (P, T) = \sum_{reac} z_i \bar{h}_{f,i}^0 - \sum_{prod} z_i \bar{h}_{f,i}^0$$

In words, the difference between the actual enthalpy of the products and their enthalpy at the standard state equals the net enthalpy generated by the reaction.

The temperature of the products (i.e., adiabatic flame temperature) may be calculated from this result once a model for $\Delta \bar{h}_i^0(P, T)$ has been selected. When the products behave as ideal gases with constant specific heats, then $\bar{h}_i^0(P, T) = \bar{c}_{P,i}(T - T_0)$. Substituting this model for the enthalpies into the first law, the first law and rearranged result gives

$$T_{af} - T_0 = \frac{\sum_{reac} z_i \bar{h}_{f,i}^0 - \sum_{prod} z_i \bar{h}_{f,i}^0}{\sum_{prod} z_i \bar{c}_{P,i}}$$

where T_{af} is the *adiabatic flame temperature*. Note that the adiabatic flame temperature may be determined explicitly when a constant specific heat model is used. Also note that as the number of substances in the products such as incomplete combustion products and excess air residuals are increased, the adiabatic flame temperature will be decreased. Using excess air is one means of controlling flame temperatures.

Adiabatic flame temperatures are usually high enough that the use of the constant specific heat model does not yield reliable results. Improved enthalpy models such as the ideal gas tables of appendix Table B.6 then should be used. An explicit solution for the adiabatic flame temperature is not possible when tables such as ideal gas tables are employed. An iterative solution for the adiabatic flame temperature will then be necessary to satisfy the first law.

Adiabatic flame temperatures are normally calculated presuming complete combustion and using air rather than pure oxygen as the oxygen supply. When the air supplied to the combustion provides the stoichiometric amount of oxygen, the resulting adiabatic flame temperature is known as the *theoretical adiabatic flame temperature*. This is the maximum possible flame temperature: No excess air or incomplete combustion products are present in the products.

EXAMPLE 11.5

Determine the theoretical adiabatic flame temperature of methane in SI units.

System: A general reaction chamber such as that shown in Figure 11.2.

Basic Equations:

$$\bar{q} - \bar{w} = \sum_{prod} z_i \Delta \bar{h}_i^0(P, T) + \sum_{prod} z_i \bar{h}_{f,i}^0 - \sum_{react} z_i \bar{h}_{f,i}^0 - \sum_{react} z_i \Delta \bar{h}_i^0(P, T)$$

Elemental species balances

Conditions: To determine the adiabatic flame temperature, no heat will be transferred with the reactor chamber. The pressures at the inlet and outlet will be presumed to be low enough that all products and reactants behave as ideal gases. Specific heats of the product constituents will be allowed to vary so that appendix Table B.6 may be used. The combustion will be complete and with no carbon monoxide or other incomplete combustion substances in the products.

Solution: The completed reaction balance on the basis of one mole of fuel is

$$CH_4 + 2(O_2 + 3.76N_2) \rightarrow 1CO_2 + 2H_2O + 7.52N_2$$

When the reactants enter the reaction chamber at the standard state and there is no shaft work or heat interaction, the first law reduces to

$$\sum_{prod} z_i \Delta \bar{h}_i^0 = \sum_{reac} z_i \bar{h}_{f,i}^0 - \sum_{prod} z_i \bar{h}_{f,i}^0$$
$$= (-74,850) - 1(-393,520) - 2(-241,820)$$
$$= 802,310 \text{ kJ/kg-mole}$$

Our task is to find the temperature of the products such that this equation is satisfied. We begin by assuming that this temperature is 2580 K. The left-hand side of this equation, according to appendix Table B.6, is

$$\sum_{prod} z_i \Delta \bar{h}_i^0 = (136,200 - 9,400) + 2(113,300 - 9,900) + 7.52(85,900 - 8,700)$$
$$= 914,000 \text{ kJ/kg-mole}$$

which is larger than the right-hand side of this equation. For the second iteration, we will use 2000 K for the temperature of the products. Then

$$\sum_{prod} z_i \Delta \bar{h}_i^0 = 659,000 \text{ kJ/kg-mole}$$

which is too low. Interpolating between these two results suggests that our next iteration for the temperature of the product should be 2330 K. This iteration yields

$$\sum_{prod} z_i \Delta \bar{h}_i^0 = 803,200 \text{ kJ/kg-mole}$$

This is within 0.2% of the value of the right-hand side of the first law. Therefore, the adiabatic flame temperature, with acceptable accuracy, is

$$T = 2330 \text{ K}$$

11.3.3   Combustion Bomb Calorimeters

Experimental measurements of heats of formation and heating values can be obtained from a device known as a *bomb calorimeter* (see Figure 11.3). This device consists of a heavy metal container in which the fuel can be mixed with pure oxygen and then burned. Structurally, this container must be strong enough to withstand the pressures generated by the combustion while maintaining a constant volume. A water bath completely surrounds this container. Both the water bath and the container are placed in a well-insulated vessel so that any heat generated by the combustion increases the internal energy of the container metal and the water bath and does not escape to the surrounding environment.

FIGURE 11.3
Typical Combustion Bomb Calorimeter

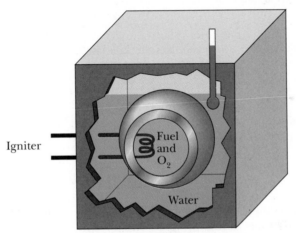

Insulated vessel

Before actual measurements are taken, the mass of all the parts of the system (i.e., fuel, container, and water) is carefully determined. The fuel is then placed in the container, the container is closed, and pressurized oxygen is added to the container. An amount of oxygen in excess of the stoichiometric requirement is always placed in the container to ensure that the fuel's combustion will be complete. The container is next placed in the water bath, and an insulated cover is put into place to prevent heat loss to the surroundings. Sufficient time is now allowed for the water, fuel and oxygen container, and container contents to come to a single temperature (a small amount of stirring of the water may be necessary to mix the water thoroughly). This temperature is then measured. The fuel–oxygen mixture is then ignited using a small electrical igniter inside the container. Again, sufficient time is allowed for the water, container, and container contents to come to a new single temperature (some slight stirring of the water again may be necessary). The final temperature is now measured. Applying the first law to the container metal and the water gives

$$Q = [(mc)_{cont} + (mc)_{water}](T_2 - T_1)$$

This system is configured such that T_2 is within a few degrees of T_1, and both are close to the standard state temperature. The heat released by the combustion of the fuel may be calculated from the two measured temperatures and the masses and specific heats of the container and water using this result.

When the first law is applied to the contents of the container, it becomes

$$Q = \sum_{prod} N_i \bar{u}_i - \sum_{reac} N_i \bar{u}_i$$

According to the definition of the enthalpy,

$$\bar{h} = \bar{h}_f^0 + \Delta \bar{h}^0 = \bar{u} + P\bar{v}$$

On rearrangement, this becomes

$$\bar{u} = \bar{h}_f^0 + \Delta \bar{h}^0 - P\bar{v}$$

Substituting this result into the first law and grouping like terms gives

$$Q = \left[\sum_{prod} N_i \bar{h}_{f,i}^0 - \sum_{reac} N_i \bar{h}_{f,i}^0 \right] + \left[\sum_{prod} N_i \Delta \bar{h}_i^0 - \sum_{reac} N_i \Delta \bar{h}_i^0 \right] - \left[\sum_{prod} N_i P \bar{v}_i - \sum_{reac} N_i P \bar{v}_i \right]$$

This result can be placed on a mole-of-fuel basis by dividing every term by the N_i of the fuel. On this basis,

$$\bar{q} = \left[\sum_{prod} z_i \bar{h}_{f,i} - \sum_{reac} z_i \bar{h}_{f,i}^0 \right] + \left[\sum_{prod} z_i \Delta \bar{h}_i^0 - \sum_{reac} z_i \Delta \bar{h}_i^0 \right] - \left[P_{prod} \sum_{prod} z_i \bar{v}_i - P_{reac} \sum_{reac} z_i \bar{v}_i \right]$$

Neglecting the small volume occupied by any liquid fuel in the reactants and liquid water in the products and applying the Amagat volume model to the reactant and product gases, $\sum_{prod} z_i \bar{v}_i = \sum_{reac} z_i \bar{v}_i = \bar{V}$, where $\bar{V}$ is the volume of the container divided by the number of moles of fuel placed in the container. Then

$$\bar{q} = \left[\sum_{prod} z_i \bar{h}_{f,i}^0 - \sum_{reac} z_i \bar{h}_{f,i}^0 \right] + \left[\sum_{prod} z_i \Delta \bar{h}_i^0 - \sum_{reac} z_i \Delta \bar{h}_i^0 \right] - [P_{prod} - P_{reac}] \bar{V}$$

The heat released in the container per unit of fuel mass is therefore the appropriately weighted difference in the heats of formation plus the enthalpy deviations from the standard state minus the pressure–volume product difference.

If the experiment is conducted so that the state of the products and reactants is close to the standard state, then the enthalpy deviations from the standard state will be small compared with the enthalpy of formation terms. The second group of terms then can be justifiably neglected. The experiment can be controlled so that the $(P_{prod} - P_{reac})\bar{V}$ is also small and thus neglected. When this is the case,

$$\bar{q} = \left[\sum_{prod} z_i \bar{h}_{f,i}^0 - \sum_{reac} z_i \bar{h}_{f,i}^0 \right]$$

Because the combustion was done with excess pure oxygen and ample time was allowed for the completion of the reactions, the products will consist of carbon dioxide, water, and unconsumed oxygen. The enthalpy of formation of these products is well known. The enthalpy of formation of the unknown hydrocarbon fuel is then ideally given by

$$\overline{h}_{f,fuel} = (z h_f)_{CO_2} + (z h_f)_{H_2O} - \overline{q}$$

where $\overline{q}$ is determined from the temperature rise of the water bath and the mass of the fuel placed in the container. In practice, small corrections are applied to this result to adjust for the assumptions embedded in this equation.

EXAMPLE 11.6

A combustion bomb calorimeter uses 15 lb_m of water, and the container is made of 2 lb_m of steel. A quantity (0.005 pound-mass) of liquid octane (i.e., gasoline) is placed in the container, which is then closed and charged with excess oxygen. After the charged container has been placed in the water bath and a new equilibrium temperature has been established, the water bath temperature is measured as 75.0° F. Following ignition, the new water equilibrium temperature is measured as 81.8° F. Use these data to determine octane's heat of formation.

System: A combustion bomb calorimeter such as that in Figure 11.3.

Basic Equations:
$$\overline{q} - \overline{w} = \sum_{prod} z_i \Delta \overline{h}_i^0 (P, T) + \sum_{prod} z_i \overline{h}_{i,f}^0 - \sum_{react} z_i \overline{h}_{f,i}^0 - \sum_{react} z_i \Delta \overline{h}_i^0 (P, T)$$

Elemental species balances

Conditions: Because the temperatures of the reactants and products are very close to the standard state temperature, the enthalpy departures from the standard state will be neglected. The excess oxygen in both the reactants and products will not contribute to the net energy balances and will be neglected. At the final temperature, most of the water in the products will be in liquid form. We presume that all of the water in the products is liquid.

Solution: The stoichiometric reaction balance is

$$C_8H_{18} + \frac{25}{2}O_2 \rightarrow 9H_2O + 8CO_2$$

The temperature rise of the water and container may be used to calculate the heat released by the reaction:

$$\begin{aligned}
Q &= [(mc)_{cont} + (mc)_{water}](T_2 - T_1) \\
&= [2 \times 0.107 + 15 \times 1](81.8 - 75.0) \\
&= 103.5 \text{ Btu}
\end{aligned}$$

Reducing the first law to the conditions of the problem and rearranging gives

$$\bar{h}_{f,fuel} = \sum_{prod} z_i h_{f,i}^0 - \bar{q}$$

$$= \sum_{prod} z_i h_{f,i} - \frac{QM}{m}$$

$$= 9(-123,000) + 8(-169,300) - \frac{-103.5 \times 114.2}{0.005}$$

$$= -97,500 \text{ Btu/lb}_m\text{-mole}$$

11.4 Second Law Analysis of Chemical Reactions

To this point, we have not determined the conditions under which a reaction will occur and, if it does occur, how far it will proceed. Instead, we have used our experience to answer these questions. Here we find that the second law of thermodynamics also may be used to find our answers.

The *increase in entropy principle* corollary of the second law,

$$\Delta S_{sys} + \Delta S_{surr} \geq 0$$

is the more convenient form for application to chemical reactions. When this is applied to the general steady-flow reactor of Figure 11.2 on a rate basis, the result is

$$\sum_{prod} \dot{N}_i \bar{s}_i - \sum_{reac} \dot{N}_i \bar{s}_i + \dot{S}_{surr} \geq 0$$

The entropy of the surroundings will be altered by any heat transfer between the system and surroundings. When heat is transferred to the system at rate $\dot{Q}$ while the surroundings are at a constant temperature T_∞, the rate of entropy increase of the surroundings is $-\dot{Q}/T_\infty$ when $\dot{Q}$ is taken with respect to the system. Then

$$\sum_{prod} \dot{N}_i \bar{s}_i - \sum_{reac} \dot{N}_i \bar{s}_i - \frac{\dot{Q}}{T_\infty} \geq 0$$

The entropy of one of the constituents of the reactants or products at an arbitrary state is

$$\bar{s} = \bar{s}^0 + \Delta \bar{s}^0$$

where $\bar{s}^0$ is the constituent's absolute entropy at the standard state, and $\Delta \bar{s}^0$ is the difference between the constituent's entropy at the arbitrary and standard states. The increase in entropy principle thus can be written as follows:

$$\sum_{prod} \dot{N}_i(\bar{s}^0 + \Delta\bar{s}^0) - \sum_{reac} \dot{N}_i(\bar{s}^0 + \Delta\bar{s}^0) - \frac{\dot{Q}}{T_\infty} \geq 0 \qquad (11.2)$$

The basis of this equation may be changed to the molar mass of one of the constituents by dividing by its $\dot{N}$. The resulting increase in entropy principle is

$$\sum_{prod} z_i(\bar{s}^0 + \Delta\bar{s}^0) - \sum_{reac} z_i(\bar{s}^0 + \Delta\bar{s}^0) - \frac{\bar{q}}{T_\infty} \geq 0$$

where the z_i's are obtained from the appropriate reaction balance.

Because each constituent behaves as a pure substance in the product and reactant mixtures, $\Delta\bar{s}^0$ may be evaluated by using the techniques presented for pure substances. For example, if a constituent can be modeled as an ideal gas with variable specific heats, then

$$\Delta\bar{s}^0 = \int_0^1 \frac{\bar{c}_p dT}{T} - \Re\ln\frac{P_1}{P_0} = \bar{s}^0 - \Re\ln\frac{P_1}{P_0}$$

where $\int_0^1 \bar{c}_p dT/T$ is the $\bar{s}_0$ of Table B.6 in the appendix.

EXAMPLE 11.7

We propose to burn gaseous methyl alcohol with a stoichiometric amount of air as a means of boiling water at 300° C. The reactant mixture enters this process at the standard state, and the products leave at 1 atmosphere pressure and 400° C. Is this proposal valid?

System: A steady-flow, constant-pressure reactor burning methyl alcohol with air while heating the 300° C surroundings.

Basic Equations:

$$\sum_{prod} z_i(\bar{s}^0 + \Delta\bar{s}^0) - \sum_{reac} z_i(\bar{s}^0 + \Delta\bar{s}^0) - \frac{\bar{q}}{T_\infty} \geq 0$$

$$\bar{q} - \bar{w} = \sum_{prod} z_i \Delta\bar{h}_i^0(P,T) + \sum_{prod} z_i\bar{h}_{f,i}^0 - \sum_{reac} z_i\bar{h}_{f,i}^0 - \sum_{reac} z_i\Delta\bar{h}_i^0(P,T)$$

Elemental species balances

TABLE 11.5
Calculations of Example 11.7

		$\bar{h}$ kJ/kg-mole	P_i atmospheres	$\bar{s}_i$ kJ/kg-mole-K
Products	CO_2	−371,690	0.116	266.63
	H_2O	−228,630	0.231	226.38
	N_2	11,590	0.653	218.61
Reactants	O_2	0	0.184	219.11
	N_2	0	0.693	156.25
	CH_3OH	−200,890	0.123	257.12

Conditions: All of the constituents will be modeled as ideal gases with temperature-variable specific heats. The water in the products will be in vapor form.

Solution: The complete reaction balance on the basis of 1 kilogram-mole of methyl alcohol is

$$CH_3OH + \frac{3}{2}(O_2 + 3.76N_2) \rightarrow CO_2 + 2H_2O + \frac{3}{2}3.76N_2$$

The enthalpy of any constituent is given by

$$\bar{h} = \bar{h}_f^0 + \Delta\bar{h}^0$$
$$= \bar{h}_f^0 + [\bar{h}(T) - \bar{h}(T_0)]$$

When applied to the CO_2 in the products using appendix Tables B.6 and B.7, this gives

$$\bar{h}_{CO_2} = (-393,520) + [25,790 - 3960]$$
$$= -371,690 \text{ kJ/kg-mole}$$

The enthalpies of the remainder of the constituents are presented in Table 11.5.
The partial pressure of the methyl alcohol in the reactant mixture is

$$P_{CH_3OH} = \frac{1}{1 + 1.5 + 1.5 \times 3.76} = 0.123 \text{ atmosphere}$$

Partial pressures of all the constituents also are listed in Table 11.5.

In the products, the entropy of the nitrogen using appendix Tables B.6 and B.7 is

$$\bar{s} = \bar{s}^0 + \Delta\bar{s}^0$$

$$= \bar{s}^0 + [\bar{s}_0(T) - \bar{s}_0(T_0)] - \Re\ln\frac{P}{P_0}$$

$$= 191.50 + [215.51 - 191.45] - 8.314\ln 0.653$$
$$= 218.61 \ \text{kJ/kg-mole-K}$$

The entropy of all constituents also are listed in Table 11.5.

Applying the first law without any shaft work, the heat released by the combustion is

$$\bar{q} = \sum_{prod} z_i\bar{h}_i - \sum_{reac} z_i\bar{h}_i$$

$$= 1(-371,690) + 2(-228,630) + 5.64(11,590) - 1(-200,890)$$
$$= -562,690 \ \text{kJ/kg-mole-fuel}$$

The net entropy change is

$$\Delta\bar{s} = \sum_{prod} z_i\bar{s}_i - \sum_{reac} z_i\bar{s}_i - \frac{\bar{q}}{T_\infty}$$

$$= 1(266.63) + 2(226.38) + 5.64(218.61)$$

$$-1(257.12) - 1.5(219.11) - 5.64(156.25) - \frac{-562,690}{573}$$

$$= 1467.3 \ \text{kJ/kg-mole-fuel-K}$$

This process is then *possible* because the total entropy change of everything involved in the combustion and heat transfer to the boiling water is positive.

In the preceding discussion and example, we assumed that the reaction will be completed. But in many circumstances, a reaction may only be partially completed. For example, consider the high-temperature reaction of carbon monoxide with water vapor. The stoichiometric balance for this reaction is

$$CO + H_2O \rightarrow CO_2 + H_2$$

if the reaction is completed. But under many circumstances, this reaction will not reach completion without violating the second law. When this is the case, the incomplete reaction balance is

$$CO + H_2O \rightarrow f[CO_2 + H_2] + (1 - f)[CO + H_2O]$$

where f represents the fraction of the carbon monoxide that is converted to carbon dioxide and serves to measure the degree of reaction completion. The value of f depends on the conditions under which the reaction is executed. As the reaction progresses from $f = 0$ (no completion) to $f = 1$ (fully completed), the entropy of each stage of this progression must be larger than that of the previous stage to satisfy the increase in entropy principle. Hence, the final equilibrium value of f when the system is adiabatic is the value that maximizes

$$\Delta \bar{s} = \sum_{prod} z_i(\bar{s}^0 + \Delta \bar{s}^0) - \sum_{reac} z_i(\bar{s}^0 + \Delta \bar{s}^0) - \frac{\bar{q}}{T_\infty}$$

according to the second law.

EXAMPLE 11.8

A synthetic gaseous fuel may be generated from coal by using a *reforming process* in which very hot carbon is brought into contact with steam to produce a fuel whose major combustables are carbon monoxide and hydrogen. Consider such a process where the carbon (coal) and steam enter the reactor at 1 atmosphere pressure and 1600 K. The products consist of hydrogen and carbon monoxide, which leave the reactor at 1 atmosphere pressure. Determine the degree to which the reaction is completed and the temperature of the products when this reaction is adiabatic.

System: A general steady-flow reaction chamber such as that shown in Figure 11.2.

Basic Equations:
$$\bar{q} - \bar{w} = \sum_{prod} z_i \Delta \bar{h}_i^0(P,T) + \sum_{prod} z_i \bar{h}_{f,i}^0 - \sum_{react} z_i \bar{h}_{f,i}^0 - \sum_{react} z_i \Delta \bar{h}_i^0(P,T)$$

$$\sum_{prod} z_i(\bar{s}^0 + \Delta \bar{s}^0) - \sum_{reac} z_i(\bar{s}^0 + \Delta \bar{s}^0) - \frac{1\bar{q}_2}{T_\infty} \geq 0$$

Elemental species balances

Conditions: The solid carbon reactant will be treated with a constant specific heat of 0.448 kJ/kg-K (5.376 kJ/kg-mole-K) and the water reactant as an ideal gas with temperature-variable specific heats. The products will be treated as ideal gases, also with temperature-variable specific heats.

Solution: The stoichiometric reaction is

$$C + H_2O \rightarrow CO + H_2$$

The incomplete reaction balance in terms of the fraction of the carbon that has been converted into carbon monoxide in the products is

$$C + H_2O \rightarrow f(CO + H_2) + (1 - f)(C + H_2O)$$

Because the reaction is adiabatic and no shaft work is involved, the first law reduces to

$$\bar{c}_{P,C}(T_{reac} - T_0) + (\bar{h}_f^0 + \Delta\bar{h}^0)_{H_2O} = f[(\bar{h}_f^0 + \Delta\bar{h}^0)_{CO} + (\Delta\bar{h}^0)_{H_2O}]$$
$$+ (1 - f)[\bar{c}_{P,C}(T_{prod} - T_0) + (\bar{h}_f^0 + \Delta\bar{h}^0)_{H_2O}]$$

When the enthalpies of formation, specific heats, standard state enthalpies, and standard state temperature are substituted into this expression, it reduces to

$$181{,}900 = f[\bar{h}_{CO} + \bar{h}_{H_2} - 128{,}900] + (1 - f)[5.376 T_{prod} + \bar{h}_{H_2O} - 240{,}200]$$

This will require an iterative solution for the temperature of the products because some of the products are being modeled with temperature-variable specific heats.

Without any heat transfer, the total increase in the entropy of this system is

$$\Delta\bar{s} = \sum_{prod} z_i \left(\bar{s}_i^0 + \Delta\bar{s}_i^0 - \Re \ln \frac{P_i}{P_0} \right) - \sum_{reac} z_i \left(\bar{s}_i^0 + \Delta\bar{s}_i^0 - \Re \ln \frac{P_i}{P_0} \right)$$

where P_i is the Dalton partial pressure of each constituent. Expanding this expression for the conditions of this reaction yields

$$\Delta\bar{s} = f[\bar{s}_{0,CO} + \bar{s}_{0,H_2} - 8.314 \ln P_{CO} - 8.314 \ln P_{H_2} + 328.13]$$

$$+ (1 - f)5.376 \ln \frac{T}{298} + (1 - f)[\bar{s}_{0,H_2O} - 8.314 \ln P_{H_2O} + 194.44] - 457.06$$

The answer is obtained by the following process. First, an f value is selected. The temperature of the products is found next by iteratively solving the first law as in Example 11.5. Once the value of the temperature of the products is known, $\Delta\bar{s}$ may be calculated for the selected f. This process is repeated for several f's until the f at which $\Delta\bar{s}$ is a maximum is found. This f is the one at which the products will emerge from the reactor in equilibrium. The results of these calculations are presented in Table 11.6.

TABLE 11.6

Calculations of Example 11.8

f	T_{prod} K	P_{CO} Atmospheres	P_{H_2} Atmospheres	P_{H_2O} Atmospheres	$\Delta \bar{s}$ kJ/kg-mole-K
0.0	1600	0.0	0.0	1.0	0.00
0.10	1083	0.09	0.09	0.82	9.13
0.20	818	0.17	0.17	0.66	22.98
0.25	686	0.20	0.20	0.60	27.91
0.30	557	0.23	0.23	0.54	32.81
0.35	424	0.26	0.26	0.48	31.35

These results are plotted in Figure 11.4, which illustrates that the equilibrium value of f is slightly more than 0.30. The temperature of the products will then be approximately 560–570 K, and the final reaction balance is

$$C + H_2O \rightarrow 0.3(CO + H_2) + 0.7(C + H_2O)$$

FIGURE 11.4

Entropy Change for Reaction in Example 11.8

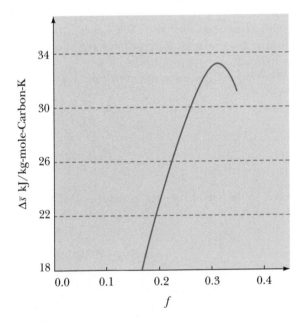

When applying the increase in entropy principle to chemical reactions, we must keep in mind that many reactions occur in stages, with the products of one stage serving as the reactants of the next. This is the case even when the reactions occur at the explosive rates associated with combustion. Reactions of this type will stop at that point where the total entropy of everything can no longer increase, according to the increase in entropy principle. For example, consider the general two-stage reaction

$$X + Y \rightarrow XY$$
$$XY + Y \rightarrow XY_2$$

which is typical of many reactions, including the combustion of carbon (X=C and Y=O). At any point during this reaction after all of X has completed the first-stage reaction, the reaction balance is

$$X + 2Y \rightarrow (1 - f)XY + fXY_2 + (1 - f)Y$$

where f represents the fraction of reactant X that has completed the second stage.

The increase in entropy principle when applied to this reaction on the basis of the mass of reactant X becomes

$$(1 - f)(\bar{s}^0 + \Delta \bar{s}^0)_{XY} + f(\bar{s}^0 + \Delta \bar{s}^0)_{XY_2} + (1 - f)(\bar{s}_0 + \Delta \bar{s}^0)_Y$$

$$-(\bar{s}^0 + \Delta \bar{s}^0)_X - 2(\bar{s}^0 + \Delta \bar{s}^0)_Y - \frac{\bar{q}}{T_\infty} \geq 0$$

The value of the left-hand side of this expression will depend on the extent of the completion of the second-stage reaction, f. This value as a function of f is shown in Figure 11.5. Inspection reveals that during the early stages of this reaction when the fraction of XY_2 in the products is low, the entropy increases. However, at a certain point during the reaction, any further increase in the fraction of XY_2 will cause the entropy to begin to decrease. This reaction must stop at the equilibrium fraction, f_{equil} of XY_2 to satisfy the second law. The final products will then contain the XY and Y constituents in the products, as well as the desired XY_2 constituent.

FIGURE 11.5

Entropy Change for a General Two-Stage Reaction

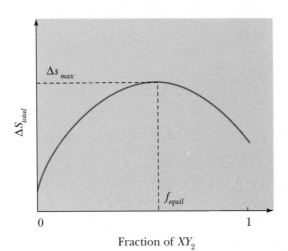

Fraction of XY_2

Figure 11.5 also indicates that if the reaction could have been completed ($f = 1$), then the final entropy would still exceed the initial entropy. We may have then concluded that the complete reaction is possible. This conclusion is erroneous, and we must consider each step of the reaction before we can conclude the extent to which the reaction will be completed.

11.5 Application of Exergy Analysis to Chemical Reactions

Extending the steady-flow exergy analysis of Section 5.9 to the general reaction chamber of Figure 11.2, the potential power that could be produced by the chemical reaction in the chamber is given by

$$\dot{W}_{ptl} = \sum_{prod} \dot{N}_i \bar{\phi}_i - \sum_{reac} \dot{N}_i \bar{\phi}_i + \sum \dot{Q}_i \left[1 - \frac{T_i}{T_0} \right]$$

where each transfer $\dot{Q}_i$ occurs at temperature T_i, $\bar{\phi}$ is the molar stream exergy of each constituent, and the system's dead state is the same as its standard state. On the basis of 1 mole of a basis reactant, this becomes

$$\bar{w}_{ptl} = \sum_{reac} z_i \bar{\phi}_i - \sum_{prod} z_i \bar{\phi}_i + \sum \bar{q}_i \left[1 - \frac{T_i}{T_0} \right]$$

where $\bar{w}_{ptl}$ and $\bar{q}_i$ are the specific potential work and heat transfer on the basis of the selected reactant. As in previous chapters, the potential work represents the maximum shaft work the system is capable of producing while maintaining the inlet and outlet states. The *second law effectiveness* is the actual shaft work produced by the system divided by the potential work. The lost work potential (i.e., irreversibility) is the difference between the work potential and actual shaft-work production.

When the definition of the stream exergy is substituted into the work-potential expression, it becomes

$$\bar{w}_{ptl} = \sum_{reac} z_i \left[(\bar{h} - \bar{h}^0) - T_0(\bar{s} - \bar{s}^0) \right]_i$$

$$- \sum_{prod} z_i \left[(\bar{h} - \bar{h}^0) - T_0(\bar{s} - \bar{s}^0) \right]_i + \sum \bar{q}_i \left[1 - \frac{T_i}{T_0} \right]$$

Now $\bar{h} - \bar{h}^0 = \bar{h}_f^0 + \Delta\bar{h}^0$ and $\bar{s} - \bar{s}^0 = \bar{s} + \Delta\bar{s}^0$. Hence, the work potential can be written as

$$\bar{w}_{ptl} = \sum_{prod} z_i \left[(\bar{h}_f^0 - T_0\bar{s}^0) + (\Delta\bar{h}^0 - T_0\Delta\bar{s}^0) \right]_i$$

$$- \sum_{react} z_i \left[(\bar{h}_f^0 - T_0\bar{s}^0) + (\Delta\bar{h}_i - T_0\Delta\bar{s}^0) \right]_i + \sum \bar{q}_i \left[1 - \frac{T_i}{T_0} \right]$$

The group $\bar{h}_f^0 - T_0\bar{s}^0$ in the first term is known as the *Gibbs free energy of formation*, g_f^0, and is tabulated for selected substances in Table B.7. In the second term, the group $\Delta\bar{h}^0 - T_0\Delta\bar{s}^0$ is the Gibbs free energy of each constituent measured with respect to the standard state. In terms of these Gibbs free energies, the work potential is given by

$$\bar{w}_{ptl} = \sum_{reac} z_i \left[\bar{g}_f^0 + \Delta \bar{g}^0 \right]_i - \sum_{prod} z_i \left[\bar{g}_f^0 + \Delta \bar{g}^0 \right]_i + \sum \bar{q}_i \left[1 - \frac{T_i}{T_0} \right]$$

When the reactor is controlled such that the reactants and products are at the standard state, and the reaction process is isothermal, then the work potential is given by

$$\bar{w}_{ptl} = \sum_{reac} z_i \bar{g}_{f,i}^0 - \sum_{prod} z_i \bar{g}_{f,i}^0$$

because $T_i = T_0$ and $\Delta \bar{g}^0 = 0$. For example, the work potential for the stoi-chiometric combustion of methane under these conditions is 818,000 kJ/kg-mole-methane (341,600 Btu/lb$_m$-mole-methane). The widespread use of com-bustion as a means of producing work is readily apparent in view of the large amount of potential work generated by combustion, as illustrated by this and the next example.

EXAMPLE 11.9

A stationary gas turbine operating on the simple Brayton cycle uses methane as the fuel. The fuel enters the combustor at 77° F. This fuel is mixed with 70% excess air in the combustor, where it is ignited and burned adiabatically. This unit operates with air at 77° F, 14.7 psia, entering the compressor inlet and a pressure ratio of 8:1 across the compressor. Determine the fraction of the potential work produced in the combustor that is converted into net work in this system.

System: The steady-flow, simple Brayton cycle shown in Figure 11.6.

FIGURE 11.6
Gas Turbine of Example 11.9

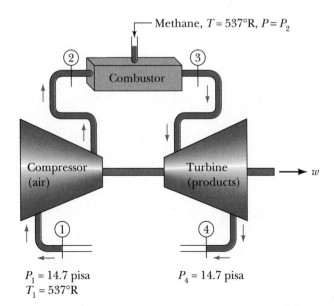

Methane, $T = 537°R$, $P = P_2$

Combustor

Compressor (air)

Turbine (products)

w

$P_1 = 14.7$ pisa
$T_1 = 537°R$

$P_4 = 14.7$ pisa

Basic Equations:

$$\bar{q} - \bar{w} = \sum_{prod} z_i \Delta \bar{h}_i^0(P,\ T) + \sum_{prod} z_i \bar{h}_{f,i}^0 - \sum_{react} z_i \bar{h}_{f,i}^0 - \sum_{react} z_i \Delta \bar{h}_i^0(P,T)$$

$$\sum_{prod} z_i(\bar{s}^0 + \Delta \bar{s}^0) - \sum_{react} z_i(\bar{s}^0 + \Delta \bar{s}^0) - \frac{\bar{q}}{T_\infty} \geq 0$$

Elemental species balances

Conditions: The reactants and products will be treated as ideal gases with variable specific heats. The compression of the air and expansion of the combustion products will be considered isentropic. An adiabatic combustion occurs in the combustor. An average specific heat of 14.3 Btu/lb$_m$-mole-° R will be used to model the gaseous methane fuel.

Solution: At the compressor outlet, the pressure is $P_2 = P_1 P_r = 117.6$ psia. The entropy of the air at the compressor outlet is the same as at the inlet because the compressor is isentropic. Then

$$s_{0,2} - R\ln\frac{P_2}{P_0} = s_{0,1} - R\ln\frac{P_1}{P_0}$$

or
$$s_{0,2} = s_{0,1} - R\ln\frac{P_1}{P_2}$$

$$= 0.59929 - \frac{1545}{28.97 \times 778}\ln\frac{1}{8}$$
$$= 0.74184 \text{ Btu/lb}_m\text{-° R}$$

According to Table B.6.1, $T_2 = 966°$ R.
The reaction balance is

$$CH_4 + 1.7 \times 2(O_2 + 3.76N_2) \rightarrow CO_2 + 2H_2O$$
$$+ 0.7 \times 2O_2 + 1.7 \times 2 \times 3.76N_2$$

TABLE 11.7
Calculations for Example 11.9

		z_i	$\bar{g}_f^0$ Btu/lb$_m$-mole	$\Delta\bar{h}^0$ Btu/lb$_m$-mole	$T_0[\bar{s}_0(T) - \bar{s}_0(T_0)]$ Btu/lb$_m$-mole	P_i psia	$\Re T_0 \ln\frac{P_i}{P_0}$ Btu/lb$_m$-mole
77° F	CH$_4$	1	−21,860	0	0	117.6	2218
React	O$_2$	3.4	0	3,107	2,276	24.7	553
966° R	N$_2$	12.78	0	3,006	2,208	92.9	1966
Prod	CO$_2$	1	−169,680	34,674	11,522	6.8	−822
3277° R	H$_2$O	2	−98,350	27,390	9,128	13.7	−75
	O$_2$	1.4	0	22,540	7,703	9.6	−454
	N$_2$	12.8	0	21,366	7,321	87.5	1902

Applying the first law to the adiabatic combustor yields

$$\sum_{prod} z_i(\bar{h}_f^0 + \Delta\bar{h}^0)_i = \sum_{reac} z_i(\bar{h}_f^0 + \Delta\bar{h}^0)_i$$

which must be solved iteratively using the technique of Example 11.5 and data from appendix Tables B.6 and B.7. This solution gives the temperature of the products as 3277° R. Because the combustion process is adiabatic, the work potential generated in the combustor is

$$\bar{w}_{ptl} = \sum_{reac} z_i(\bar{g}_f^0 + \Delta\bar{h}^0 - T_0\Delta\bar{s}^0)_i - \sum_{prod} z_i(\bar{g}_f^0 + \Delta\bar{h}^0 - T_0\Delta\bar{s}^0)_i$$

where

$$T_0\Delta\bar{s}^0 = T_0[\bar{s}_0(T) - \bar{s}_0(T_0)] - \Re T_0 \ln\frac{P_i}{P_0}$$

Appropriate summing of the data in Table 11.7 gives

$$\bar{w}_{ptl} = 181,690 \text{ Btu/lb}_m\text{-mole-methane}$$

Only the products of combustion are expanded in the turbine. Because this expansion is isentropic,

$$\bar{s}_4 = \bar{s}_3$$

where the entropies must be measured with respect to a common datum. In terms of the quantities in appendix Table B.6, this result may be written as

$$\sum z_i\left(\bar{s}_0(T_4) - \bar{s}_0(T_0) - \Re \ln\frac{P_{i,4}}{P_0}\right) = \sum z_i\left(\bar{s}_0(T_3) - \bar{s}_0(T_0) - \Re \ln\frac{P_{i,3}}{P_0}\right)$$

The pressure and temperature are now known at the turbine inlet, so the right-hand side of this expression is known. Similarly, the pressure term of the left-hand side of this equation is known because the pressure at the outlet of the turbine is the same as at the compressor inlet. We must then find the set of $\bar{s}^0(T_4)$ that satisfies this equation. This can be done by iterating on the temperature of the products at state 4. This solution gives $T_4 = 2080°$ R.

Applying the first law to the turbine yields

$$\begin{aligned}
_3\bar{w}_4 &= \sum z_i(\bar{h}_{0,3} - \bar{h}_{0,4}) \\
&= (38,703 - 22,704) + 2(31,651 - 18,261) \\
&\quad +1.4(26,275 - 15,847) + 12.78(25,096 - 15,169) \\
&= 184,917 \text{ Btu/lb}_m\text{-mole-methane}
\end{aligned}$$

Applying the first law to the compressor gives

$$\begin{aligned}
_1\bar{w}_2 &= M(h_1 - h_2) \\
&= 28.97(128.4 - 232.7) \\
&= -3022 \text{ Btu/lb}_m\text{-mole-air}
\end{aligned}$$

One mole of fuel is added for each 16.18 moles of air passing through the compressor, according to the reaction balance. Then

$$_1\bar{w}_2 = 16.18(-3022) = -48{,}896 \text{ Btu/lb}_m\text{-mole-methane}$$

The net work produced by this system is

$$\bar{w}_{net} = \bar{w}_{turb} + \bar{w}_{comp} = 184{,}917 - 48{,}896 = 136{,}021 \text{ Btu/lb}_m\text{-mole-methane}$$

which is

$$F = \frac{136{,}021}{181{,}690} = 0.749$$

of the work potential generated by the combustion.

PROBLEMS

In the following problems, combustion occurs at atmospheric pressure and is complete; use temperature-variable specific heats unless stated otherwise.

11.1 Propane fuel is burned with a stoichiometric amount of oxygen only. Determine the mass fractions of carbon dioxide and water in the products. Also calculate the mass of water in the products per unit of fuel mass burned ($\text{lb}_m\text{-H}_2\text{O}/\text{lb}_m\text{-C}_3\text{H}_8$).

11.2 n-Butane is burned with a stoichiometric amount of oxygen only. Determine the mole fraction of carbon dioxide and water in the products. Also calculate the molar mass of carbon dioxide in the products per molar mass of fuel burned ($\text{kg-mole-CO}_2/\text{kg-mole-C}_4\text{H}_{10}$).

11.3 n-Octane is burned with a stoichiometric amount of oxygen. Calculate the mass fraction of each product and the mass of water in the products per unit mass of fuel burned ($\text{lb}_m\text{-H}_2\text{O}/\text{lb}_m\text{-C}_8\text{H}_{18}$).

11.4 Acetylene is burned with 10% excess oxygen in a cutting torch. Determine the mass fraction of each product. Calculate the mass of oxygen used per unit mass of acetylene burned ($\text{lb}_m\text{-O}_2/\text{lb}_m\text{-C}_2\text{H}_2$).

11.5 Propane fuel is burned with a stoichiometric amount of air. Determine the mass fraction of each product. Also calculate the mass of water in the products ($\text{lb}_m\text{-H}_2\text{O}/\text{lb}_m\text{-C}_3\text{H}_8$) and mass of air required ($\text{lb}_m\text{-air}/\text{lb}_m\text{-C}_3\text{H}_8$) per unit of fuel mass burned.

11.6 The Colorado coal of Table 11.1 is burned with a stoichiometric amount of air. Neglecting the ash constituent, calculate the mass fraction of the products and the mass of air required per unit mass of coal burned (kg-air/kg-fuel).

11.7 Determine the fuel–air ratio when the Colorado coal of Table 11.1 is burned with 50% excess air.

11.8   n-Butane fuel is burned with a stoichiometric amount of air. Determine the mass fractions of each product. Also calculate the mass of carbon dioxide in the products ($kg\text{-}CO_2/kg\text{-}C_3H_8$) and mass of air required ($kg\text{-}air/kg\text{-}C_4H_{10}$) per unit of fuel mass burned.

11.9   The Pennsylvania coal of Table 11.1 is burned with a stoichiometric amount of air. Calculate the mole fractions of the products, the apparent molecular weight of the product gas, and the mass of air required per unit mass of coal burned ($lb_m\text{-}air/lb_m\text{-}fuel$), neglecting the ash constituent.

11.10 Calculate the air–fuel ratio of the previous problem.

11.11 n-Octane is burned with a stoichiometric amount of air. Calculate the mass fraction of each product and the mass of water in the products per unit mass of fuel burned ($lb_m\text{-}H_2O/lb_m\text{-}C_8H_{18}$). Also calculate the mass fraction of each reactant.

11.12 Methyl alcohol is burned with a stoichiometric amount of air. Calculate the mole fraction of each product and the apparent molecular weight of the product gas. Also calculate the mass of water in the products per unit mass of fuel burned ($lb_m\text{-}H_2O/lb_m\text{-}CH_3OH$).

11.13 Utah coal (Table 11.1) is burned with a stoichiometric amount of air, but the combustion is incomplete: 5% (by volume) of the carbon in the fuel forms carbon monoxide. Calculate the mass fraction of the products, the apparent molecular weight of the products, and the mass of air required per unit mass of fuel burned ($lb_m\text{-}air/lb_m\text{-}fuel$).

11.14 Ethyl alcohol is burned with a stoichiometric amount of air. The combustion is incomplete, with 10% (by volume) of the carbon in the fuel forming carbon monoxide and 5% of the hydrogen forming OH. Calculate the apparent molecular weight of the products.

11.15 Propane fuel is burned with 30% excess air. Determine the mole fraction of each product. Also calculate the mass of water in the products ($kg\text{-}H_2O/kg\text{-}C_3H_8$) and the air–fuel ratio.

11.16 n-Butane fuel is burned with 100% excess air. Determine the mole fraction of each product. Also calculate the mass of carbon dioxide in the products ($lb_m\text{-}CO_2/lb_m\text{-}C_4H_{10}$) and the air–fuel ratio.

11.17 n-Octane is burned with 50% excess air. Calculate the mass fractions of each product and the mass of water in the products per unit mass of fuel burned ($kg\text{-}H_2O/kg\text{-}C_8H_{18}$). Also calculate the mass fraction of each reactant.

11.18 The Illinois coal of Table 11.1 is burned with 40% excess air. Calculate the mass of air required per unit mass of coal burned ($lb_m\text{-}air/lb_m\text{-}fuel$) and the apparent molecular weight of the product gas, neglecting the ash constituent.

11.19   Ethyl alcohol is burned with 70% excess air. Calculate the mole fraction of both reactants and products. Also calculate the mass of water and oxygen contained in the products per unit mass of fuel burned (lb_m-H_2O/lb_m-C_2H_5OH).

11.20   Calculate the air–fuel ratio of the previous problem.

11.21   Methyl alcohol is burned with 50% excess air. The combustion is incomplete, with 10% (by volume) of the carbon in the fuel forming carbon monoxide. Calculate the mole fraction of carbon monoxide in the products and the apparent molecular weight of the products.

11.22   n-Octane is burned with 100% excess air, 15% (by volume) of the carbon in the fuel forms carbon monoxide. Calculate the mole fractions of the products and the dew-point temperature ($^\circ$ C) of the water vapor in the products when the products are at 1 atmosphere pressure.

11.23   The Utah coal of Table 11.1 is burned with 20% excess air in an atmospheric pressure boiler. Calculate the mass of water in the products per unit mass of coal burned (lb_m/lb_m-fuel), and the dew-point temperature ($^\circ$ F) of the water vapor in the products.

11.24   n-Octane is burned with 60% excess air in an automobile engine. Assuming complete combustion and that the exhaust system pressure is 1 atmosphere, determine the minimum temperature ($^\circ$ F) of the combustion products before liquid water will begin to form in the exhaust system.

11.25   The Utah coal of Table 11.1 is burned with 25% excess air in an industrial boiler. Assuming complete combustion and that the pressure in the boiler smokestack is 1 atmosphere, calculate the minimum temperature ($^\circ$ F) of the combustion products before liquid water begins to form in the smokestack.

11.26   Methane fuel is burned with 50% excess air in a space-heating furnace. The pressure in the chimney is 1 atmosphere. Presuming complete combustion, determine the temperature ($^\circ$ C) of the combustion products at which liquid water will begin to form in the chimney.

11.27   Propane fuel is burned with an air–fuel ratio of 18:1 in an atmospheric-pressure heating furnace. Determine the heat released per kilogram of fuel burned when the temperature of the products is such that liquid water just begins to form in the products.

11.28   The Colorado coal of Table 11.1 is burned with an industrial boiler with 10% excess air. The temperature and pressure in the smokestack are 120° F and 1 atmosphere, respectively. Calculate the fraction of the water in the combustion products that is liquid and the fraction that is vapor.

11.29   n-Butane is burned with a stoichiometric amount of air in a cook stove. The products of combustion are at 1 atmosphere pressure and 40° C. What fraction of the water in these products is liquid?

11.30 Propane fuel is burned with a stoichiometric amount of air in a water heater. The products of combustion are at 1 atmosphere pressure and 120° F. What fraction of the water vapor in the products is vapor?

11.31 Compare the adiabatic flame temperature (° C) of propane fuel burned with a stoichiometric amount of air and burned with 50% excess air. The reactants are at standard pressure and temperature.

11.32 What is the adiabatic flame temperature (° F) of methane when burned with 30% excess air?

11.33 Estimate the flame temperature (° C) of an acetylene cutting torch that uses a stoichiometric amount of pure oxygen.

11.34 The Pennsylvania coal of Table 11.1 is burned in an individual boiler. This combustion is incomplete: 3% (by volume) of the carbon in the products forms carbon monoxide. What is the impact of the incomplete combustion on the adiabatic flame temperature as compared to when the combustion is complete?

11.35 n-Octane is burned with 100% excess air in a constant-pressure burner. The air and fuel enter this burner at standard conditions, and the combustion products leave at 500° F. Calculate the specific heat generated (Btu/lb$_m$-fuel) by this combustion.

11.36 The Texas coal of Table 11.1 is burned with 40% excess air in a constant-pressure power-plant boiler. The coal and air enter this boiler at standard conditions, and the combustion products in the smokestack are at 250° F. Calculate the specific heat generated (Btu/lb$_m$-fuel) in this boiler.

11.37 Propane fuel is burned in a space heater with 50% excess air. The fuel and air enter this heater at 1 atmosphere pressure and 18° C, and the combustion products leave at 1 atmosphere pressure and 100° C. Calculate the molar specific heat generated (kJ/kg-mole-fuel) in this heater.

11.38 n-Octane is burned in the constant-pressure, adiabatic combustor of an aircraft engine with 40% excess air. The air enters this combustor at 600 kPa, 300° C, and the fuel is injected into the combustor at 25° C. Estimate the temperature (° C) at which the combustion products leave the combustor.

11.39 A railroad has experimented with burning powdered coal in a gas turbine combustor. The air was introduced to the combustor at 200 psia, 250° F, while the powdered coal was injected at 77° F. The combustion was adiabatic and at constant pressure. Based on the Colorado coal of Table 11.1, what is the estimated temperature of the combustion products?

11.40 The combustion products of the previous problem are expanded in an isentropic turbine to 20 psia. Calculate the specific work (Btu/lb$_m$-fuel) produced by this turbine.

11.41 Ethane is burned at atmospheric pressure using air as the oxidizer. Only enough air is supplied to oxidize the hydrogen. Determine the heat released or absorbed (kJ/kg-mole-fuel) when the products and reactants are at 25° C, and the water appears in the products as water vapor.

11.42 What is the minimum temperature of the combustion products of the previous problem that will ensure that the water in the products will be in vapor form?

11.43 Propane fuel at 25° C is mixed with 50% excess air in an atmospheric pressure combustor. The air enters at 500° C, and the products leave at 1500° C. The volumetric flow rate of the air is 1 ft^3/s. Determine the rate (kW) at which heat is released by this combustor.

11.44 Calculate the higher and lower heating values (HHV and LHV) of the Illinois coal of Table 11.1.

11.45 Calculate the HHV and LHV of n-octane fuel.

11.46 Calculate the HHV and LHV of methane fuel.

11.47 Calculate the HHV and LHV of the Utah coal of Table 11.1.

11.48 Calculate the HHV and LHV of propane fuel.

11.49 Develop an expression for the HHV of a gaseous alkane C_nH_{2n+2} in terms of n.

11.50 Methane undergoes complete combustion with a stoichiometric amount of air in a constant-volume rigid container. Initially, the air and methane are at 14.7 psia, 77° F. The combustion products are at 800° F. How much heat has been released (Btu/lb$_m$-mole-fuel)?

11.51 The effective mass specific-heat product of a constant-volume bomb calorimeter is 100 kJ/K. Wheat straw is being considered as an alternative fuel and so is tested in the calorimeter. Ten grams of this straw are placed in the calorimeter. After charging the calorimeter with oxygen and burning the straw, we find that the temperature of the calorimeter has increased by 1.8° C. Determine the straw's heating value. How does this compare to the HHV of propane fuel?

11.52 Ethyl alcohol is burned with 200% excess air in an adiabatic, constant-volume container. Initially, the air and ethyl alcohol are at 100 kPa, 25° C. Determine the final pressure and temperature of the combustion products.

11.53 Methane is burned with 300% excess air in an adiabatic, constant-volume container. Initially, the air and methane are at 14.7 psia, 77° F. Determine the final pressure and temperature of the combustion products.

11.54 A steam boiler heats liquid water at 200° C to superheated steam at 4000 kPa, 400° C. Methane fuel is burned at atmospheric pressure with 50% excess air.

The fuel and air enter the boiler at 25° C, and the combustion products leave at 220° C. Calculate the following:

    a. The amount of steam generated per unit of fuel mass burned (kg-steam/kg-fuel)

    b. The change in the exergy of the combustion streams (kJ/kg-fuel)

    c. The change in exergy of the steam stream (kJ/kg-fuel)

    d. The lost work potential (kJ/kg-fuel)

11.55 Repeat the previous problem using the Utah coal of Table 11.1.

11.56 n-Octane is burned in an automobile engine with 200% excess air. Air enters this engine at 1 atmosphere pressure, 77° F. Liquid fuel at 77° F is mixed with this air before combustion. The exhaust products leave the exhaust system at 1 atmosphere pressure and 160° F. What is the maximum amount of work (Btu/lb$_m$-fuel) that can be produced by this engine?

11.57 The automobile engine of the previous problem is to be converted to natural gas (i.e., methane). Presuming that all factors remain the same, what is the maximum work that can be produced by the modified engine (Btu/lb$_m$-fuel)?

11.58 The automobile engine of Problem 11.56 is converted to use methyl alcohol as fuel. Presuming that all factors remain the same, what is the maximum work that can be produced by the modified engine (Btu/lb$_m$-fuel)?

11.59 n-Octane is burned in the constant-pressure combustor of an aircraft engine with 30% excess air. The air enters this combustor at 600 kPa, 300° C, liquid fuel is injected into the combustor at 30° C, and the products of combustion leave at 600 kPa, 1200° C. How much has the exergy been increased by this combustion?

11.60 n-Octane is burned with a stoichiometric amount of air. Determine the maximum work (kJ/kg-mole-fuel) that can be produced by this combustion when the air, fuel, and products are at 25° C.

11.61 How much is the maximum work of the previous problem changed when 100% excess air is used for the combusiton?

11.62 If the n-octane of Problem 11.60 is burned with 100% excess air and 10% of the carbon forms carbon monoxide, what is this fuel's maximum work-producing capability?

11.63 Methane is burned with a stoichiometric amount of air. Determine the maximum work (Btu/lb$_m$-fuel) that can be produced by this combustion when the air, fuel, and combustion products are at 77° F?

11.64 How much is the maximum work of the previous problem changed when 100% excess air is used for the combustion?

11.65   If the methane of Problem 11.63 is burned with 100% excess air and 10% of the carbon forms carbon monoxide, what is this fuel's maximum work-producing capability?

PROJECTS

The following projects are more extensive than those presented in this chapter. Most are open-ended and require some decisions on the student's part. Many can be expedited with computer-based property tables, computer spreadsheets, and student-written computer programs.

11.1   The Virginia coal of Table 11.1 is burned in an electric power plant that produces 300 megawatts of power. This plant uses a steam Brayton cycle with reheat and a single open-feedwater heater. The boiler operates at 600 psia and delivers steam at 825° F to the turbine throttle. The reheater also provides 825° F steam to the low-pressure turbine. The condenser operates at 1.5 in-Hg absolute. Steam is extracted at 180° F for the feedwater heater. The minimum flame temperature must exceed the minimum water temperature in the boiler, and the maximum flame temperature must exceed the maximum water temperature. The air and coal are introduced into the boiler at the dead-state conditons. Exhaust gases leave the boiler at 1 atmosphere pressure and a temperature greater than the dew-point temperature of these gases. A sufficient supply of cooling water at 70° F is avaiable for the condenser. Use exergy accounting to determine the best excess amount of air to use and the cooling water flow rate.

11.2   An electrical utility is changing from the Pennsylvania coal of Table 11.1 to the Illinois coal as fuel for its boilers. With the Pennsylvania coal, the boilers used 15% excess air. Develop a schedule for the new coal showing the heat released, the smokestack dew-point temperature, adiabatic flame temperature, and carbon-dioxide production for various amount of excess air. Use this schedule to determine how to operate with the new coal as closely as possible to the conditions of the old coal. Is there anything else that will have to be changed to use the new coal?

11.3   A steam boiler is designed to convert saturated liquid water at 3500 kPa to steam at 3450 kPa, 400° C. This boiler burns natural gas (i.e., methane) at 1 atmosphere pressure and 25° C. Combustion air is also supplied at 1 atmosphere pressure and 25° C. The flow of combustion air is adjusted to maintain the temperature of the combustion products at 300° C. On a common basis of 1 kilogram of steam produced, calculate the change in the exergy of the combustion and water streams for excess air amounts ranging from 0 to 200%. Also calculate the exergy lost in this boiler as the excess air amount is varied.

11.4   An economizer (a heat exchanger that preheats the combustion air by cooling the combustion products) is added to the boiler of the preceding project. With this economizer, the temperature of the combustion products is reduced to 200° C. All other factors remain the same. Air at 1 atmosphere pressure and 25° C passes through the economizer, where it is heated before entering the boiler. Compare the loss of exergy of the economizer–boiler combination to that of the boiler only in the previous project as the amount of excess air is varied.

Introduction to Equilibrium

Mission of Technology

Concern for man himself and his fate must always form the chief interest of all technical endeavors, . . . in order that the creations of our mind shall be a blessing and not a curse to mankind. Never forget this in the midst of your diagrams and equations.

Albert Einstein

Given sufficient time, a system will always come to equilibrium. At that point, it will no longer change its condition of its own accord and will always return to this condition when perturbed. The only way that we can change this condition is by the manner in which the system is constrained and controlled. If we change the pressure of the system through some mechanism in the surroundings, like adding more weight to a weighted piston, the system will seek a new state of equilibrium that will remain stable as long as the system is not altered by some event that occurs in the surroundings. Equilibrium conditions depend on how the system is constrained and controlled. For example, the products produced by a chemical reaction may be different depending on the pressure and temperature at which the reaction occurs. The system constraints are the difference between a combustion process that only forms carbon dioxide, water vapor, and nitrogen versus one that produces carbon monoxide, nitrous oxides, and other environmentally unacceptable substances. These constraints also determine the yield from chemical reactions, separation of a blend of two or more liquids by selective evaporation, and other engineering processes.

Equilibrium can only exist when a system is static and not changing its state. When a process is being executed, the forces which drive the system must be unbalanced for the process to occur in the direction desired. It is only when the speed at which the process occurs is slower than the speed at which a system can reestablish equilibrium that a system can approach equilibrium. The concept of the *quasi-equilibrium* process was introduced in Chapter 1 in recognition that equilibrium can only be approached, but not achieved, in practical work-producing and other processes.

As a system progresses toward equilibrium, the entropy of the system and any surroundings that interact with the system will be changing. To satisfy the second law as embodied in the increase in entropy principle, the system cannot progress beyond the point at which the total entropy begins to decrease. Conditions for equilibrium then can be determined through application of the second law. General conditions for equilibrium are developed in this manner in this chapter. The utility of the Gibbs function and chemical potential in determining equilibrium conditions are also presented.

The general conditions for equilibrium are applied to two specific systems in this chapter. The first system is isobaric and filled with two or more substances that behave as ideal solutions. Each substance in this system can have as many as two phases. When the system conditions are such that there is no natural tendency for one phase to convert to the other, the system is said to be in *phase equilibrium*. Conditions for phase equilibrium are developed in this chapter. The second system is a chemical reaction that occurs as the temperature and pressure are held constant. Once this reaction comes to the condition that the proportions of the constituents in the products does not change, the system is said to be in *chemical equilibrium*. This chapter develops the means of determining the reaction products' equilibrium composition.

12.1 Introduction

In Chapter 1, we stated that *for equilibrium to exist, the contents of a system cannot change their condition of their own accord.* A system can also be said to be *in equilibrium if when its state is disturbed and the system is released from this disturbed state, it will return to its original, undisturbed state.* This second view of equilibrium is illustrated for the marble-in-a-bowl and pendulum systems in Figure 12.1. These two systems demonstrate three types of equilibrium—stable equilibrium, metastable equilibrium, and unstable equilibrium. Stable equilibrium exists when the marble is located at the bottom of the bowl and when the pendulum center of mass is directly below the pivot point. Figure 12.1 shows that if the position of the marble in the bowl or pendulum is changed by a finite amount and then released, both will eventually return to their original equilibrium positions.

FIGURE 12.1
Simple Systems Demonstrating the Three Forms of Equilibrium

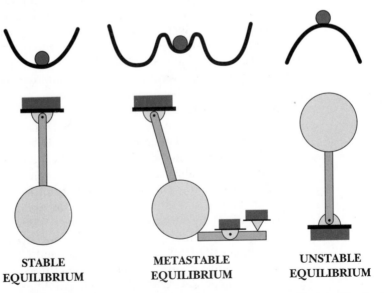

STABLE
EQUILIBRIUM

METASTABLE
EQUILIBRIUM

UNSTABLE
EQUILIBRIUM

Metastable equilibrium is illustrated by the marble in a bowl with a cusp or the pendulum supported by a rotating lever. The marble can be moved a small, but finite, amount within the cusp and when released will return to its original position in the cusp. If the marble is displaced too far in the cusp such that it goes beyond the lip of the cusp, it will seek a new position in the bowl when released and will not return to its original position. The same can be said for the pendulum supported by the lever. If the lever is not rotated too far, the whole system will return to its original state when the lever is released. Once the lever is rotated to the point that the pendulum can slip past the lever, the system will never return of its own accord to the original equilibrium state.

A marble sitting directly on top of a bowl or an inverted pendulum with its center of mass directly above the pivot point are in an unstable equilibrium condition. As long as these two systems are not disturbed, they will retain their original states. But, these states cannot be maintained if even the smallest disturbance occurs. Even an infinitesimal movement of the marble or pendulum from the original state will cause the marble to roll off the bowl or the pendulum to seek the position shown at the far left of Figure 12.1.

We can define a stable equilibrium state as one that *when released from a large finite disturbance will always return to its original state.* A metastable equilibrium state can be defined as one that *will return to its original state when the disturbance is small and will not return to the original state if the disturbance is too large.* When a system is in an unstable equilibrium state, *even the smallest disturbance will cause the system to leave its original state and assume a new state.* These considerations apply to thermodynamic systems as well as mechanical and other systems. Throughout the remainder of this chapter, we will focus on stable equilibrium and leave the other two forms of equilibrium for the specialty treatises on these subjects.

This definition of stable equilibrium is quite satisfactory when we consider mechanical equilibrium where all the forces acting on a system are in balance. As long as these forces are in balance, no force, including the pressure and stresses, will change of its own accord, and, according to Newton's law of motion, the velocity property of the system will not change. The only way that we can change the system's state is by externally altering one or more of the forces acting on its boundary. We then can change the pressure of a fluid in a weighted piston-cylinder device by adjusting the piston's weight. Once this adjustment has been made, the system will come to a new state of mechanical equilibrium.

Several times we have said that thermal equilibrium exists when the temperature of the system will not change of its own accord. The basis of this statement is intuitive and drawn from our everyday experience. However, this is a direct consequence of the second law of thermodynamics as can be demonstrated by considering what happens when a hot piece of metal is dropped into an insulated cup of a cooler liquid. The metal has a mass, m_m, a specific heat, c_m, and an initial temperature, $T_{1,m}$. Similarly, the liquid has a mass, m_l; a specific heat, c_l; and an initial temperature, $T_{1,l}$. Physically, this process is one in which the temperature of the metal will decrease to $T_{2,m}$ and the liquid temperature will increase to $T_{2,l}$ by an exchange of heat between the metal and liquid but not with the surroundings. Applying the first law to the metal and liquid, and equating the results gives

$$Q = m_m c_m (T_{1,m} - T_{2,m}) = m_l c_l (T_{2,l} - T_{1,l})$$

Treating the metal and liquid as incompressible substances, and applying the appropriate Gibbs equation gives the change of the metal's entropy as

$$S_{2,m} - S_{1,m} = m_m c_m \ln \frac{T_{2,m}}{T_{1,m}}$$

and the change of the liquid's entropy as

$$S_{2,l} - S_{1,l} = m_l c_l \ln \frac{T_{2,l}}{T_{1,l}}$$

The increase in entropy principle form of the second law as applied to this system is

$$m_m c_m \ln \frac{T_{2,m}}{T_{1,m}} + m_l c_l \ln \frac{T_{2,l}}{T_{1,l}} \geq 0$$

Using the first law to eliminate the masses and specific heats reduces this to

$$\frac{T_{2,l} - T_{1,l}}{T_{1,m} - T_{2,m}} \ln \frac{T_{2,m}}{T_{1,m}} + \ln \frac{T_{2,l}}{T_{1,l}} \geq 0$$

This can be satisfied only as long as $T_{2,m}$ is decreasing, $T_{2,l}$ is increasing, and $T_{2,m} \geq T_{2,l}$. Heat will continue to be transferred until $T_{2,m} = T_{2,l}$. At this point, the process can go no farther of its own accord without decreasing the entropy of everything. This, of course, is impossible. The system has then reached a new state of thermal equilibrium.

Using the second law to examine equilibrium has been adequate for the most part to this point. It began to become cumbersome, as demonstrated by Example 11.8, when determining the equilibrium composition of the products of a chemical reaction. A general statement or test for equilibrium is then needed to treat equilibrium problems. This statement should encompass all forms of equilibrium. Appropriate application of the first and second laws to isolated systems that do not interact with their surroundings will be used to develop this general statement of equilibrium.

12.2 General Equilibrium Considerations

A closed system consisting of a simple, compressible substance that can change its state while undergoing phase transformations and chemical reactions will be considered to develop a criterion for equilibrium. The increase in entropy principle as applied to this system is

$$\Delta S + \Delta S_\infty \geq 0$$

where S is the entropy of the system, and S_∞ is the entropy of the surroundings. As this system executes an infinitesimal, reversible process along the path to an equilibrium state, it exchanges an amount of heat

$$\delta Q = \delta W + dU$$

with the surroundings according to the first law. The entropy change of the surroundings is then

$$dS_\infty = -\frac{\delta Q}{T_\infty} = -\frac{\delta W + dU}{T_\infty}$$

where T_∞ is the uniform temperature of the surroundings. When this result is substituted into the increase of entropy principle, it becomes

$$T_\infty dS - \delta W - dU \geq 0$$

The compression and expansion work produced by the system in this infinitesimal, reversible process changes the volume of the isobaric surroundings. This work is given by

$$\delta W = P_\infty dV$$

where P_∞ is the pressure of the surroundings. Substituting this into our previous expression and rearranging produces

$$dU + P_\infty dV - T_\infty dS \leq 0$$

This important result states that when the quantity $dU + P_\infty dV - T_\infty dS$ is positive, no process can occur. The process then can only occur as long as this quantity is negative and cannot proceed any farther once the quantity becomes zero. The system then is again in equilibrium because it cannot change its state any further without violating the second law. Once the system is in mechanical and thermal equilibrium with its surroundings, $P_\infty = P$, and $T_\infty = T$, this result reduces to

$$dU + PdV - TdS \leq 0 \tag{12.1}$$

This is a general expression for the direction of naturally occurring processes involving simple, compressible substances.

Combining this important result with the defining expressions for the enthalpy ($H = U + PV$), Gibbs function ($G = H - TS$), and the Helmholtz function ($A = U - TS$) yields a complete set of expressions for determining the direction of processes in a simple, compressible substance. For example, the differential of the Helmholtz function is

$$dA = dU - TdS - SdT$$

When combined with the previous result this gives an alternate statement of the second law,

$$dA + PdV + SdT \leq 0$$

The complete set is:

$$
\begin{aligned}
(a) \quad & dU + PdV - TdS \leq 0 \\
(b) \quad & dH - VdP - TdS \leq 0 \\
(c) \quad & dA + PdV + SdT \leq 0 \\
(d) \quad & dG - VdP + SdT \leq 0
\end{aligned}
\qquad (12.2)
$$

This set is general and may be used in the above forms. They may be reduced even more to be of direct utility for phase- and chemical-equilibrium considerations. This is accomplished by applying these results to systems subject to additional constraints.

For example, consider a closed system that is constrained in such a manner that neither its volume nor its internal energy can be changed as it undergoes a process. A simple compressible substance contained in an insulated, rigid vessel with no means for a work interaction or heat transfer is constrained in this way. Phase transformations, chemical reactions, and other events may be occurring inside this vessel, but the volume and internal energy are not altered. When Equation 12.2a is reduced to fit this situation, it becomes

$$(dS)_{U,V} \geq 0$$

The entropy change of a simple compressible substance can increase only until it reaches a maximum extremum when the internal energy and volume are held constant.

Applying this approach to all of Equations 12.2 gives

PRINCIPLE: General Conditions for Equilibrium

$$
\begin{aligned}
(a) \quad & (dS)_{U,V} \geq 0 \\
(b) \quad & (dH)_{S,P} \leq 0 \\
(c) \quad & (dA)_{T,V} \leq 0 \\
(d) \quad & (dG)_{T,P} \leq 0
\end{aligned}
$$

Each of these relations applies to a closed system and may be used under the proper circumstances. Each indicate the direction in which a property must change to satisfy the second law when the system is constrained by holding two other properties constant. These results also state that when the rate of change of some property becomes zero (i.e., an extremum), the system will be in equilibrium. For example, any process that occurs in a closed system such that the pressure and total entropy cannot change can only progress in such a manner that the total enthalpy will decrease according to Condition b above. Equilibrium will be established once it is no longer possible for the total enthalpy to decrease any more and a minimum extremum has been reached.

The last of these conditions is perhaps the most widely used because its constraints represent the conditions under which phase transformations and many chemical reactions occur. This last result states that the change in the Gibbs function of a simple, compressible substance whose temperature and pressure are held fixed can only be negative or zero. Once the state of this substance is such that the Gibbs function has become a minimum extremum, the substance will be in equilibrium. This is demonstrated in Figure 12.2 for an

FIGURE 12.2
Use of Gibbs Function to Determine Equilibrium Conditions

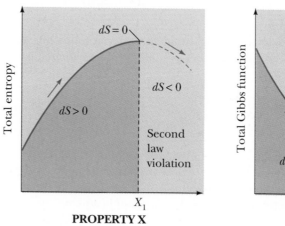 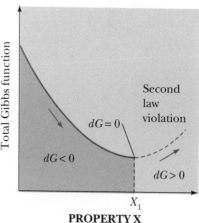

isolated system whose pressure and temperature cannot change. The X property in this figure measures the direction of the process and the degree of the process completion. In a chemical reaction, X could be the fraction of reactants that have been converted to products. In a phase-transformation situation, X could be the fraction of the vapor phase that has been converted to liquid. The left-hand figure illustrates what happens to the total entropy of the isolated system. It also shows that once the total system entropy has reached a maximum at $X = X_1$, the system has come to a new equilibrium state—as it must, according to the increase in entropy principle. The right-hand figure illustrates the total Gibbs function of the system. In this case, the total Gibbs function will decrease until it can decrease no more when $X = X_1$. A new equilibrium state is then reached by the system.

12.3 Chemical Potential and Equilibrium

The $(dG)_{T,P} = 0$ expression can be used to determine equilibrium conditions for systems that contain many constituents. The total Gibbs function for a multiconstituent system is given by

$$G = G(P, T, N_1, N_2, \ldots, N_n)$$

because it depends on the two intensive properties, P and T, according to the state postulate, as well as the relative proportions of the constituents in the system as specified by the number of moles or each constituent, N_i. The size of the system may be adjusted without changing the relative proportions of the constituents by multiplying the number of moles of each constituent by a common factor, α. The total Gibbs function, like all other extensive properties, will then change by the same factor, α:

$$\alpha G = G(P, T, \alpha N_1, \alpha N_2, \ldots, \alpha N_n)$$

Differentiating this result with respect to α while keeping the pressure, temperature, and number of moles of the various constituents constant gives

$$G = \left(\frac{\partial G}{\partial (\alpha N_1)} \right)_{P,T,N_i} N_1 + \left(\frac{\partial G}{\partial (\alpha N_2)} \right)_{P,T,N_i} N_2 + \cdots + \left(\frac{\partial G}{\partial (\alpha N_n)} \right)_{P,T,N_i} N_n$$

where the N_i in the subscript notation indicates that all mole fractions except the current one are held constant while performing the differentiation. This result must hold for all system sizes, including the system for which $\alpha = 1$. For the system whose $\alpha = 1$, this result becomes

$$G = \left(\frac{\partial G}{\partial N_1} \right)_{P,T,N_i} N_1 + \left(\frac{\partial G}{\partial N_2} \right)_{P,T,N_i} N_2 + \cdots + \left(\frac{\partial G}{\partial N_n} \right)_{P,T,N_i} N_n$$

The *chemical potential, $\overline{\mu}_i$*, is defined as follows:

DEFINITION: *Chemical Potential*

$$\overline{\mu}_i = \left(\frac{\partial G}{\partial N_i} \right)_{P,T,N_i}$$

This reduces the previous result to

$$G = \bar{\mu}_1 N_1 + \bar{\mu}_2 N_2 + \cdots + \bar{\mu}_n N_n$$

The chemical potential is an intensive property (just as a specific property is an intensive property) and thus a function of two independent properties (e.g., P, T) when the substance is a simple compressible substance according to the state postulate. At constant pressure and temperature, the differential of the preceding expression is as follows:

$$(dG)_{P,T} = \bar{\mu}_1 dN_1 + \bar{\mu}_2 dN_2 + \cdots + \bar{\mu}_n dN_n \qquad (12.3)$$

When a system contains several constituents, this result expresses the condition for equilibrium in terms of the chemical potential and the change in the amount of each constituent in the system. It will prove to be useful when examining systems of two or more phases where one phase may be converting to another, or in systems in which a chemical reaction is converting one constituent into another.

12.4 Phase Equilibrium

An isolated system consisting of one or more nonreacting constituents such as that shown in Figure 12.3 can contain two or more phases even when the system is constrained to a constant pressure and temperature. In this situation, it is possible that one phase may transform to another (e.g., liquid evaporating to the vapor phase). Once the state of the system has reached the point where there is no net natural transformation between the phases, *phase equilibrium* exists. Such equilibrium exists when the state of the system is such that the Gibbs function is a minimum, where

$$(dG)_{T,P} = 0$$

FIGURE 12.3
A Closed System with Equilibrium Between the Phases

Phase 1

Phase 2

When the system consists of n nonreacting constituents each of which can have m phases, then the total Gibbs function for the system is

$$G = \sum_{i=1}^{n} \sum_{j=1}^{m} N_{i,j} \bar{g}_{i,j}(P,T)$$

where $N_{i,j}$ is the number of moles of substance i in the jth phase, and $\bar{g}_{i,j}(P,T)$ is the value of the specific molar Gibbs function of the ith substance in the jth phase. The differential of the total molar Gibbs function is

$$(dG)_{P,T} = \sum_{j=1}^{m} \sum_{i=1}^{n} \bar{g}_{i,j}(P,T) dN_{i,j}$$

because the molar specific Gibbs function of each constituent phase does not depend on the number of constituent moles in that phase. The criterion for equilibrium in a constant pressure and temperature system that can undergo phase transformations is then

$$(dG)_{P,\,T} = \sum_{j=1}^{m} \sum_{i=1}^{n} \bar{g}_{i,j}(P,T) dN_{i,j} = 0$$

In this situation, the *chemical potential* is the molar specific Gibbs function of each constituent's phase.

The constituents are not allowed to chemically react. Then, the number of moles of each constituent cannot change regardless of their distribution among the various phases. Hence, the total number of moles of constituent i is

$$N_i = \sum_{j=1}^{m} N_{i,j}$$

Taking the differential of this expression,

$$dN_i = 0 = \sum_{j=1}^{m} dN_{i,j}$$

which can be used to help determine the conditions of equilibrium.

12.4.1   Single-Constituent Systems

When the system consists of one constituent that can have two phases (say, liquid and vapor), then the total Gibbs function for the system is given by

$$G = \bar{g}_1 N_1 + \bar{g}_2 N_2$$

whose differential is

$$dG = \bar{g}_1 \, dN_1 + \bar{g}_2 \, dN_2$$

because the $\bar{g}$ of each phase does not change as mass is added or removed from each phase. The total number of moles of the constituent is given by

$$N_1 + N_2 = N$$

the differential of which is

$$dN_1 + dN_2 = 0$$

Eliminating dN_1 from the Gibbs function differential and setting the result to zero gives the condition for phase equilibrium as follows:

$$dG = \bar{g}_1(-dN_2) + \bar{g}_2 \, dN_2 = 0 \tag{12.4}$$

or

$$\bar{g}_1 = \bar{g}_2 \quad \text{(molar basis)}$$
$$\text{or} \tag{12.5}$$
$$g_1 = g_2 \quad \text{(mass basis)}$$

Phase equilibrium will exist between the two phases of the constituent when the specific Gibbs functions (either on a molar or mass basis) of the phases are equal.

The second law requires that

$$(dG)_{P,T} \leq 0$$

which when combined with Equation 12.3, gives

$$(\bar{g}_2 - \bar{g}_1) \, dN_2 \leq 0$$

This finding illustrates the direction in which phase transformations will occur. For example, if $\bar{g}_2 > \bar{g}_1$, then dN_2 must be negative, which tells us that a transformation is occurring in which phase 2 is being converted into phase 1. This transformation will continue until $\bar{g}_2 = \bar{g}_1$ and phase equilibrium exists. The specific Gibbs function (i.e., chemical potential in this case) is then a measure of the potential for the transfer of mass between phases just as temperature measures the potential for heat transfer.

EXAMPLE 12.1   A two-phase mixture of water is maintained at 300° F. Compare the molar-specific Gibbs function of the liquid and vapor phases.

System: A two-phase mixture of water.

Basic Equations:
$$\bar{g} = M(h - Ts)$$

Conditions: None

Solution: Using the water property tables of Appendix B, the molar specific Gibbs function of the liquid phase is

TIP:

The molar-specific Gibbs functions of the phases must be equal when the phases are in equilibrium—as they are in this example. The equality of the specific Gibbs function of the phases is often used in computer and other property tables to relate the properties of saturated phases.

$$
\begin{aligned}
\bar{g}_f &= M(h_f - Ts_f) \\
&= 18.02(269.7 - 759.67 \times 0.4372) \\
&= -1125 \text{ Btu/lb}_m\text{-mole}
\end{aligned}
$$

Similarly, the molar-specific Gibbs function for the vapor phase is

$$
\begin{aligned}
\bar{g}_g &= M(h_g - Ts_g) \\
&= 18.02(1180.2 - 759.67 \times 1.6356) \\
&= -1123 \text{ Btu/lb}_m\text{-mole}
\end{aligned}
$$

This result is equal to that of the liquid phase within the accuracy of the property tables and round-off errors.

Expanding the specific Gibbs function in terms of the enthalpy, temperature, and entropy reduces Equation 12.5 to the form

$$h_1 - Ts_1 = h_2 - Ts_2$$

or

$$s_2 - s_1 = \frac{h_2 - h_1}{T}$$

When this is combined with Equation 9.13, the result is

$$\frac{dP}{dT}\bigg|_{\substack{phase \\ change}} = \frac{h_2 - h_1}{T(v_2 - v_1)}$$

This is the Clausius–Clapeyron equation of Section 9.5.2.

When more than two phases are present (for example, at the triple point) the total Gibbs function of the system is

$$G = \bar{g}_1 N_1 + \bar{g}_2 N_2 + \bar{g}_3 N_3$$

while the total number of moles is

$$N = N_1 + N_2 + N_3$$

Conditions for phase equilibrium may now be found by finding the extrema of the total Gibbs function, subject to the one constraint that the total number of moles does not change. Applying the method of Lagrange multipliers (see Appendix A) gives the single Lagrange multiplier, λ, as

$$\lambda = \bar{g}_1 = \bar{g}_2 = \bar{g}_3$$

In words, the specific Gibbs function must be the same for all phases for phase equilibrium to exist.

12.4.2   Multiconstituent Systems

When the system consists of n constituents each of which can have m phases, the total Gibbs function for the system is

$$G = \sum_{i=1}^{n} \sum_{j=1}^{m} N_{i,j} \bar{g}_{i,j}(P,T)$$

The total number of moles of each constituent is given by

$$N_i = \sum_{j=1}^{m} N_{i,j}$$

Applying the method of Lagrange multipliers of Appendix A to determine the extrema of the total Gibbs function subject to the constraints on the n total number of moles of each constituent yields the Lagrange multipliers

$$\lambda_1 = \bar{g}_{1,1} = \bar{g}_{1,2} = \cdots = \bar{g}_{1,m}$$
$$\lambda_2 = \bar{g}_{2,1} = \bar{g}_{2,2} = \cdots = \bar{g}_{2,m}$$
$$\vdots = \vdots$$
$$\lambda_n = \bar{g}_{n,1} = \bar{g}_{n,2} = \cdots = \bar{g}_{n,m}$$

Hence, the specific Gibbs function of each constituent must be the same in all phases for phase equilibrium to exist. For example, when a system consists of two constituents and two phases (say, liquid and vapor), this result states that the specific Gibbs function of the liquid and vapor phases of constituent 1 must be equal. The specific Gibbs function of the liquid and vapor phases of constituent 2 also must be equal to one another but not necessarily to that of constituent 1. Phase equilibrium will exist when these conditions are met.

   The compositions (i.e., mole or mass fraction) of the various phases in a multiconstituent system, in general, are not the same. This is illustrated for the oxygen–nitrogen (i.e., air) system in Figure 12.4. The experimentally de-termined upper line of this figure represents the mole fraction of the vapor phase while the lower line represents the mole fraction of the liquid phase. Thus, at 86 K, the vapor phase is approximately 50% nitrogen and 50% oxy-gen by mole fraction, and the liquid phase is approximately 80% oxygen and 20% nitrogen. The composition of the liquid phase is different from that of the vapor phase, but note that

$$y_{N_2} + y_{O_2} = 1 \quad \text{(liquid phase)}$$
$$y_{N_2} + y_{O_2} = 1 \quad \text{(vapor phase)}$$

FIGURE 12.4

Phase Equilibrium Diagram for Oxygen–Nitrogen System

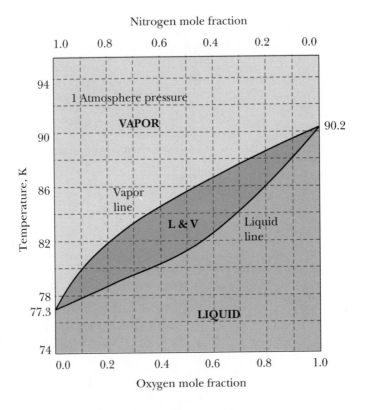

The utility of the phase equilibrium diagram can be demonstrated by con-sidering the events that occur as a mixture of gases is cooled at constant pres-sure. This is illustrated in the skeleton phase equilibrium diagram in Figure 12.5 for a mixture of oxygen and nitrogen. The process begins with a mixture of the two gases at 90 K and 1 atmosphere pressure with a 20% oxygen mole fraction (approximate composition of atmospheric air). Initially, the cooling decreases the temperature of the two gases until the temperature equals that at point A (approximately 82 K). At this temperature, liquid will begin to con-dense out of the mixture of two gases.

FIGURE 12.5

Skeleton Oxygen–Nitrogen Phase Equilibrium Diagram

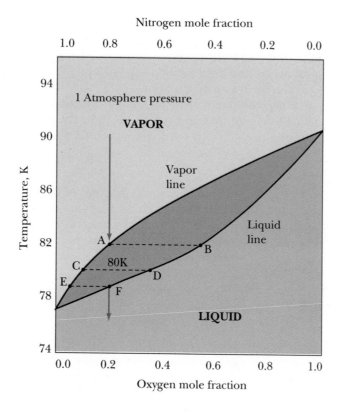

The very first liquid formed will have the concentrations of point B (about 55% oxygen). As we continue to reduce the temperature, more liquid will be formed. The concentrations in the liquid phase (and gaseous phase) will simultaneously change because the oxygen and nitrogen condense out of the gaseous mixture at different rates. At 80 K, the concentrations of the gaseous mixture are those of point C, while the concentrations of the liquid phase are those of point D. Notice that as the condensation occurs, the liquid mixture is richer in oxygen than the gaseous mixture. This process might then be used to partially separate the oxygen from the nitrogen.

Continued cooling will eventually convert all of the gases into liquid form. In this example, this occurs at point F where the temperature is just above 78 K. The concentrations of the last portion of the gaseous mixture to be condensed are those of point E. Once all of the gases have been condensed, the concentrations in the liquid mixture become the same as those of the original mixture of oxygen and nitrogen, and further cooling decreases the temperature of the liquid mixture.

Phase equilibrium conditions such as those in Figure 12.4 normally are determined experimentally. It is possible to calculate the equilibrium concentrations of a two-phase, two-constituent system when the phases behave as ideal solutions. An *ideal solution* is defined as a solution whose total volume is equal to the sum of the partial volumes of the constituents forming the solution. An ideal solution then can be modeled in a manner similar to the Amagat mixture model of Section 10.2. Consequently, a mixture of two ideal gases is an ideal solution. Many mixtures of two nonreacting liquids (e.g., water and ethyl alcohol) are also ideal solutions. Solid–liquid solutions (e.g., sugar and water) normally are not ideal solutions.

When the vapor phase of a multiconstituent system behaves as an ideal gas, the relationship between the constituent partial pressure and its mole fraction is

$$P_i = y_{g,i} P_{total}$$

where $y_{g,i}$ is the vapor-phase mole fraction of the constituent and P_{total} is the pressure of the vapor phase. This model for the vapor phase is also Dalton's partial-pressure model for an ideal gas mixture as presented in Section 10.3. The vapor-phase partial pressure of a constituent is related to the constituent's saturation pressure by the following:

$$P_i = y_{f,i} P_{sat,i}(T) \qquad\qquad (12.6)$$

where $y_{f,i}$ is the mole fraction of the constituent in the liquid phase, and $P_{sat,i}(T)$ is the constituent's saturation pressure at the temperature of the mixture. This important result is known as *Raoult's Law*, which can be shown to be valid when both the liquid and vapor phases of the system are ideal solutions.

EXAMPLE 12.2

At 115 K, the saturation pressure of nitrogen is 1940 kPa, and the saturation pressure of methane is 110 kPa. These are blended to form a mixture of liquid and vapor whose temperature is 115 K such that the mole fraction of the nitrogen in the liquid phase is 0.3. Determine the pressure of this mixture and the mole fraction of the nitrogen in the vapor phase.

System: A two-phase, two-constituent mixture of nitrogen and methane.

Basic Equations:
$$P_i = y_{f,i} P_{sat,i}(T)$$
$$P_i = y_{g,i} P_{total}$$

Conditions: The liquid and vapor phases will be treated as ideal solutions.

Solution: The mole fraction of the methane in the liquid phase is

$$y_{f,CH_4} = 1 - y_{f,N_2} = 0.7$$

According to Raoult's law, the partial pressures of the constituents in the vapor phase are:

$$P_{N_2} = y_{g,N_2} P_{sat,N_2}(T) = 0.3 \times 1940 = 582 \text{ kPa}$$
$$P_{CH_4} = 0.7 \times 110 = 77 \text{ kPa}$$

Because the vapor phase is an ideal solution (i.e., ideal gas),

$$P_{total} = P_{N_2} + P_{CH_4} = 659 \text{ kPa}$$

The mole fraction of the nitrogen in the vapor phase will then be

$$y_{N_2} = \frac{P_{N_2}}{P_{total}} = \frac{582}{659} = 0.883$$

Applying the method of Lagrange multipliers also yields the following result:

PRINCIPLE: Gibbs phase rule

$$n_v = n - m + 2$$

where n_v is the number of independent variables for a multiconstituent system with several phases.

The number of independent variables for a single-constituent system with two phases is then 1. This can be either the pressure or the temperature, but not both. A phase-equilibrium condition then can only be changed by changing this one variable. This was presented without proof in Chapter 3 when the saturation tables for water were first discussed. A liquid–vapor saturation condition could be altered only by changing the pressure or temperature, according to these tables. The complete state of a mixture of two phases is determined by the value of the single independent variable and a specification of the relative proportions of the two phases. The quality was introduced in Chapter 3 as a means of specifying the relative proportions of the liquid and vapor phases.

When the Gibbs phase rule is applied to a single-constituent system with three phases (such as occurs at a triple point), the number of independent variables is zero. This makes triple points unique: They cannot be altered by changing some property such as pressure or temperature. It is this uniqueness that makes them ideally suited to the purposes of defining temperature scales.

The number of independent variables required for a two-phase, two-constituent system such as the nitrogen–oxygen system of Figure 12.4 is two, according to the Gibbs phase rule. Dalton's partial-pressure model for the vapor phase and Raoult's law demonstrate that the total pressure and temperature of the system control the equilibrium conditions when the two phases behave as ideal solutions. When the mixture pressure is fixed as in Figure 12.4, the compositions of the two phases can be changed only by changing the mixture temperature, also as shown in Figure 12.4.

EXAMPLE 12.3

A two-phase mixture of water and ammonia is at $100°$ F and 160 psia in the generator of an absorption refrigerator system. Determine the mole fraction of the ammonia in the liquid and vapor phases.

System: A two-phase, two-constituent mixture of water and ammonia.

Basic Equations:
$$P_i = y_{f,i} P_{sat,i}(T)$$
$$P_i = y_{g,i} P_{total}$$

Conditions: The liquid and vapor phases will be treated as ideal solutions, and phase equilibrium is presumed to exist.

Solution: The total pressure of the vapor phase is the sum of the partial pressures of the two constituents. Then

$$P_{total} = y_{f,NH_3} P_{sat,NH_3}(T) + (1 - y_{f,NH_3}) P_{sat,H_2O}(T)$$

When solved for y_{f,NH_3} this gives

$$\begin{aligned} y_{f,NH_3} &= \frac{P_{total} - P_{sat,H_2O}(T)}{P_{sat,NH_3}(T) - P_{sat,H_2O}(T)} \\ &= \frac{160 - 0.9503}{212 - 0.9503} \\ &= 0.754 \end{aligned}$$

The partial pressure of the ammonia in the vapor phase is

$$P_{g,NH_3} = y_{f,NH_3} P_{sat,NH_3}(T) = 0.754 \times 212 = 159.9 \text{ psia}$$

and the ammonia mole fraction in the vapor phase is

$$y_{g,NH_3} = \frac{P_{g,NH_3}}{P_{total}} = \frac{159.9}{160} = 0.999$$

TIP:

According to the Gibbs phase rule, the number of independent variables for this mixture is

$$n_V = 2 - 2 + 2 = 2$$

which, in this example, are the mixture pressure and temperature.

This vapor is then almost pure ammonia, but it does contain a small amount of water that may present a freezing problem in the refrigeration side of this system. ▬

12.5 Chemical Equilibrium

The chemical reactions considered in Chapter 11 were one-directional in that the reactants only produced products. These reactions were mostly fully completed (with the exception of those in Section 11.4), with all of the reactants being converted to products. General chemical reactions can be two-directional, with some of the products recombining to form reactants as in a disassociation–recombination reaction. Consequently, the reaction may never be fully completed. However, when this does occur, an equilibrium will be achieved where the amount of reacting reactants equals the amount of recombining products. We will consider this general chemical reaction in this chapter for the case of two reacting compounds, A and B, and two product compounds, C and D. This general, two-directional chemical reaction may be written as

$$\nu_A A + \nu_B B \rightleftharpoons \nu_C C + \nu_D D$$

where the ν's are stoichiometric coefficients for the reaction (see Section 11.1). The results that we will obtain for this reaction can be readily extended to more than two reactant and product compounds. Each compound in this reaction is made of chemical elements such as carbon, hydrogen, and oxygen. When n of these elements are in the two reacting compounds, n species balances must always be satisfied, even when the reaction reaches equilibrium.

This isobaric and isothermal reaction will be in equilibrium when

$$(dG)_{P,T} = \overline{\mu}_A dN_A + \overline{\mu}_B dN_B + \overline{\mu}_C dN_C + \overline{\mu}_D dN_D = 0$$

according to Equation 12.3. At any point in the progress of the reaction as it comes to equilibrium, the system consists of N_A moles of compound A, N_B moles of compound B, N_C moles of compound C, and N_D moles of compound D as illustrated in Figure 12.6. The number of moles of each compound is always proportional to the reaction stoichiometric coefficients because the n species balances must always be satisfied. Then

$$N_A = \varepsilon \nu_A$$
$$N_B = \varepsilon \nu_B$$
$$N_C = \varepsilon \nu_C$$
$$N_D = \varepsilon \nu_D$$

where ε is the proportionality constant that may be changed to control the size of the system by starting with more or less reactants. As the reaction progresses, the change in the number of moles of the various compounds is

$$dN_A = -\nu_A d\varepsilon$$
$$dN_B = -\nu_B d\varepsilon$$
$$dN_C = \nu_C d\varepsilon$$
$$dN_D = \nu_D d\varepsilon$$

because the stoichiometric coefficients are fixed. During the progression of the reaction, the number of moles of reactant compounds, A and B, decreases while the number of moles of product compounds, C and D, increases as indicated by the signs on the right-hand sides of these equations. The change in the total Gibbs function of the products and reactants caused by this change from reactants to products is then

$$(dG)_{P,T} = (\nu_C \overline{\mu}_C + \nu_D \overline{\mu}_D - \nu_A \overline{\mu}_A - \nu_B \overline{\mu}_B)d\varepsilon = 0$$

when the reaction proceeds at constant pressure and temperature.

Once the reaction has progressed to the point that equilibrium is established, the condition for equilibrium, $(dG)_{P,T} = 0$, gives the important result

$$\nu_C \overline{\mu}_C + \nu_D \overline{\mu}_D - \nu_A \overline{\mu}_A - \nu_B \overline{\mu}_B = 0$$

or

$$\nu_C \overline{\mu}_C + \nu_D \overline{\mu}_D = \nu_A \overline{\mu}_A + \nu_B \overline{\mu}_B$$

FIGURE 12.6
General Chemical Reaction System

N_A, N_B Moles of reactants

N_C, N_D Moles of products

Extending this result to an arbitrary number of reactant and product compounds gives the following as the chemical equilibrium criterion for isothermal and isobaric chemical reactions:

$$\sum_{reacts} \nu_i \overline{\mu}_i = \sum_{prods} \nu_i \overline{\mu}_i \quad \text{(constant } P \text{ and } T) \qquad (12.7)$$

The total Gibbs function for a reaction involving m reactant and product constituents is

$$G = N_1 \overline{g}_1 + N_2 \overline{g}_2 + \cdots + N_m \overline{g}_m$$

Constituent compound j's chemical potential is then

$$\overline{\mu}_j = \left(\frac{\partial G}{\partial N_j} \right)_{P,T,N_i} = \overline{g}_j$$

This result allows us to write the criterion for chemical equilibrium as follows:

$$\sum_{reacts} \nu_i \bar{g}_i = \sum_{prods} \nu_i \bar{g}_i \quad \text{(constant } P \text{ and } T) \tag{12.8}$$

12.6 Chemical Equilibrium for Ideal Gas Reactions

The results of the previous sections apply to all types of chemical reactions. In this section, we will apply these results to reacting systems whose reacting compounds, reactant and product, may be treated as ideal gases. This is the case when a gaseous fuel such as methane is burned with air, or when a gas is heated to a high enough temperature to disassociate. The mixture formed by these compounds consequently also behaves as an ideal gas. Dalton's partial-pressure model and the methods of Chapter 10 can be used to model the mole fractions and properties of the reacting compounds, as well as the mixture of these compounds, when this is the case.

The specific molar Gibbs function of one of these compounds is given by

$$\bar{g} = \bar{h} - T\bar{s}$$

It is convenient to express the specific molar Gibbs function of each constituent in terms of $\bar{g}^*$, which is taken at *1 atmosphere pressure and the temperature of the mixture;* and $\Delta\bar{g}^*$, which corrects $\bar{g}^*$ from 1 atmosphere pressure to the partial pressure of the constituent in the mixture of constituents. In terms of these quantities,

$$\bar{g} = \bar{g}^*(T) + \Delta\bar{g}^*(T, P_i)$$
$$= [\bar{h}^*(T) - T\bar{s}^*(T)] + [\Delta\bar{h}^*(T, P_i) - T\Delta\bar{s}^*(T, P_i)]$$

Using the standard state enthalpies and entropies of Chapter 11, the first term of the above result can be written as

$$\bar{g}^*(T) = \bar{g}_f^0 + \Delta\bar{g}(T)$$
$$= \bar{g}_f^0 + [\Delta\bar{h}(T) - T\Delta\bar{s}_0(T)]$$

where the Gibbs function of formation, $\bar{g}_f^0$, is presented for many substances in Table B.7 of Appendix B. The quantities $\Delta\bar{h}(T)$ and $\Delta\bar{s}_0(T)$ are the difference between these quantities at the mixture temperature and the standard temperature for reaction calculations (77° F or 25° C). Table B.6 of Appendix B can be used to evaluate these quantities for ideal gases with variable specific heats.

Because each constituent behaves as an ideal gas, $\Delta \bar{h}^*(T, P_i) = 0$. Then

$$\Delta \bar{g}^*(T, P_i) = -T\Delta \bar{s}^*(T, P_i)$$

$$= -\int_{ref}^{i} Td\bar{s}$$

The Gibbs equation, $Td\bar{s} = d\bar{h} - \bar{v}dP$, may be used to integrate this expression from the reference 1 atmosphere pressure to the constituent partial pressure while keeping the mixture temperature constant, or

$$\Delta \bar{g}^*(T, P_i) = \int_{P_{ref}}^{P_i} \bar{v}dP$$

$$= \Re T \int_{P_{ref}}^{P_i} \frac{dP}{P}$$

$$= \Re T \ln \frac{P_i}{P_{ref}}$$

where P_i is the partial pressure of the constituent in the reacting mixture. The mole fraction of one of the compounds in the mixture of gaseous reactant and product compounds is $y_i = N_i / \Sigma N_i$. In terms of the mole fraction, the constituent partial pressure when all of the reacting compounds behave as ideal gases is

$$P_i = y_i P$$

where P is the fixed total pressure of the mixture. Hence,

$$\bar{g}_i = \bar{g}_i^* + \Re T \ln \frac{y_i P}{P_{ref}}$$

when all of the reacting constituents behave as ideal gases.

When the expression for the molar-specific Gibbs function for compounds that behave as ideal gases is substituted into the condition for chemical equilibrium for two reactants and two products, it becomes

$$\nu A \left(\bar{g}_A^* + \Re T \ln \frac{y_A P}{P_{ref}} \right) + \nu B \left(\bar{g}_B^* + \Re T \ln \frac{y_B P}{P_{ref}} \right)$$

$$= \nu_C \left(\bar{g}_C^* + \Re T \ln \frac{y_C P}{P_{ref}} \right) + \nu_D \left(\bar{g}_D^* + \Re T \ln \frac{y_D P}{P_{ref}} \right)$$

This can be rearranged as

$$\ln \left[\frac{y_C^{\nu_C} y_D^{\nu_D}}{y_A^{\nu_A} y_B^{\nu_B}} \left(\frac{P}{P_{ref}} \right)^{\nu_C + \nu_D - \nu_A - \nu_B} \right] = -\frac{\Delta G^*}{\Re T}$$

where, in general,

$$\Delta G^* = \sum_{prods} v_i \overline{g}_i^* - \sum_{reacts} v_i \overline{g}_i^*$$

We may define the equilibrium constant, K_P, as follows:

DEFINITION: *Equilibrium Constant*

$$\ln K_P \equiv -\frac{\Delta G^*}{\Re T}$$

When the preceding expression for ΔG^* is substituted into the right-hand side of this definition, it becomes

$$\ln K_P = -\frac{\sum_{prods} v_i \overline{g}_i^* - \sum_{reacts} v_i \overline{g}_i^*}{\Re T}$$

In terms of the equilibrium constant, the condition for chemical equilibrium becomes

$$\frac{y_C^{v_C} y_D^{v_D}}{y_A^{v_A} y_B^{v_B}} \left(\frac{P}{P_{ref}}\right)^{v_C + v_D - v_A - v_B} = K_P$$

In general, this result may be written as follows:

$$\frac{\prod_{prods} y_i^{v_i}}{\prod_{reacts} y_i^{v_i}} \left(\frac{P}{P_{ref}}\right)^{\sum_{prods} v_i - \sum_{reacts} v_i} = K_P \qquad (12.9)$$

This is the condition for chemical equilibrium when the reactant and product compounds behave as ideal gases. When coupled with the elemental species balances, this result may be solved for the mole fractions of the various compounds when the reaction has come to equilibrium.

EXAMPLE 12.4

Water vapor will disassociate into hydrogen and oxygen gases at elevated temperatures. Determine the equilibrium constant and the mole fraction of hydrogen gas present when this reaction occurs at (a) 101 kPa, 25° C, and (b) 101 kPa, 2000 K.

System: A reacting system of water vapor, hydrogen, and oxygen gas.

Basic Equations:

$$\ln K_P = -\frac{\Delta G^*}{\Re T}$$

$$\frac{\prod_{prods} y_i^{\nu_i}}{\prod_{reacts} y_i^{\nu_i}} \left(\frac{P}{P_{ref}}\right)^{\sum_{prods} \nu_i - \sum_{reacts} \nu_i} = K_P$$

Conditions: The water vapor, hydrogen, and oxygen will be treated as ideal gases.

Solution: The stoichiometric reaction for this system is

$$H_2O \rightleftharpoons H_2 + \frac{1}{2}O_2$$

At 298 K, $\overline{g}^0 = \overline{g}_f^0$, and the equilibrium constant will be

$$\ln K_P = -\frac{\nu_{H_2}\overline{g}_{f,H_2}^0 + \nu_{O_2}\overline{g}_{f,O_2}^0 - \nu_{H_2O}\overline{g}_{f,H_2O}^0}{\Re T}$$

$$= -\frac{0 + 0.5 \times 0 - (-228,590)}{8.314 \times 473}$$

$$= -92.2$$

using the $\overline{g}_f^0$ values of Table B.7. In the same manner, at 2000 K

$$\begin{aligned}
\overline{g}_{H_2}^0 &= \overline{g}_f^0 + \Delta\overline{g} = \overline{g}_f^0 + \Delta(\overline{h} - T\overline{s}) \\
&= 0 + (61,419 - 2000 \times 188.313) - (8463 - 298 \times 130.554) \\
&= -284,760 \text{ kJ/kg-mole}
\end{aligned}$$

$$\begin{aligned}
\overline{g}_{O_2}^0 &= 0 + (67,846 - 2000 \times 268.586) - (8679 - 298 \times 204.971) \\
&= -416,920 \text{ kJ/kg-mole}
\end{aligned}$$

$$\begin{aligned}
\overline{g}_{H_2O}^0 &= -228,590 + (82,693 - 2000 \times 264,562) - (9899 - 298 \times 188.694) \\
&= -628,870 \text{ kJ/kg-mole}
\end{aligned}$$

where appendix Tables B.6 and B.7 have been used to obtain the required properties. Then, the equilibrium constant is

$$\ln K_P = -\frac{-284,760 - 0.5 \times 416,920 + 628,870}{8.314 \times 2000}$$

$$= -8.15$$

We will let α be the fraction of the water vapor that has been disassociated into hydrogen and oxygen once the reaction comes to equilibrium (α is also known as the *degree of reaction completion*). According to the stoichiometric reaction statement, the number of moles of each constituent in the resulting equilibrium mixture is then

$$
\begin{aligned}
\text{Water:}\quad &1 - \alpha \\
\text{Hydrogen:}\quad &\alpha \\
\text{Oxygen:}\quad &\frac{\alpha}{2}
\end{aligned}
$$

$$
\textbf{Total for mixture:}\quad \frac{2 + \alpha}{2}
$$

The mole fractions of each constituent in the equilibrium mixture are then:

$$
\text{Water:}\quad \frac{2(1 - \alpha)}{2 + \alpha}
$$

$$
\text{Hydrogen:}\quad \frac{2\alpha}{2 + \alpha}
$$

$$
\text{Oxygen:}\quad \frac{\alpha}{2 + \alpha}
$$

Substituting these results into the mole fraction expression for the equilibrium constant while noting that $P = P_{ref} = 101$ kPa gives

TIP:

Observe that as the equilibrium constant approaches zero and becomes positive (i.e., increasing temperature), the reaction is more nearly complete. Also note that, by using , the degree of reaction completion, we have automatically satisfied all of the individual elemental species balances.

$$
\frac{y_{H_2} y_{O_2}^{0.5}}{y_{H_2O}} = K_P
$$

$$
\frac{2\left[\dfrac{\alpha}{2 + \alpha}\right]\left[\dfrac{\alpha}{2 + \alpha}\right]^{0.5}}{2\left[\dfrac{1 - \alpha}{2 + \alpha}\right]} = K_P
$$

$$
\frac{\alpha^3}{(1 - \alpha)^2 (2 + \alpha)} = K_P
$$

An iterative solution of this equation, noting that $0 \le \alpha \le 1$, according to the actual reaction equation, gives $\alpha \approx 0.0$ when $T = 25°$ C, and $\alpha = 0.0798$ when $T = 2000$ K. Hence, practically no hydrogen is present when the temperature is 25° C, and only 8% (by mole fraction) of the reacting mixture is hydrogen when the temperature is 2000 K.   ◼

   Equilibrium constants may be calculated as in Example 12.4. Appendix table B.8 lists the results of these equilibrium constant calculations for several common disassociation reactions at several temperatures. This table presumes the total pressure of the reactant–product mixture to be 1 atmosphere. A USCS unit system table of equilibrium constants is not included because the equilibrium constant lacks dimensions.

EXAMPLE 12.5

Methane gas is heated at 5 atmospheres pressure and 1000 K. This heating causes the methane to disassociate into carbon and hydrogen gases. Determine the equilibrium mole fractions of each gas, including the remainder of the methane, in the resulting mixture.

System: A reacting system of methane, carbon, and hydrogen gas.

Basic Equations:

$$\frac{\prod_{prods} y_i^{\nu i}}{\prod_{reacts} y_i^{\nu i}} \left(\frac{P}{P_{ref}}\right)^{\sum_{prods} \nu_i - \sum_{reacts} \nu_i} = K_P$$

Conditions: All of the constituents in the product mixture will be treated as ideal gases.

Solution: The stoichiometric methane disassociation reaction is

$$C + 2H_2 \rightleftharpoons CH_4$$

At a pressure of 1 atmosphere and 1000 K, the equilibrium constant for this reaction is $\ln K_P = 2.328$, according to Table B.8.

If we let α equal the fraction of the methane gas that has disassociated into carbon and hydrogen, then the number of moles of each constituent in the final mixture is

Methane:	$1 - \alpha$
Carbon:	α
Hydrogen:	2α
Mixture total:	$1 + 2\alpha$

The mole fraction of each constituent in the final mixture is then

$$\text{Methane:} \quad y_{CH_4} = \frac{1 - \alpha}{1 + 2\alpha}$$

$$\text{Carbon:} \quad y_C = \frac{\alpha}{1 + 2\alpha}$$

$$\text{Hydrogen:} \quad y_{H_2} = \frac{2\alpha}{1 + 2\alpha}$$

Substituting these data into the expression for the equilibrium constant gives

$$\frac{y_{CH_4}^1}{y_C^1 y_{H_2}^2}\left(\frac{P}{P_{ref}}\right)^{2+1-1} = K_P$$

$$\frac{\dfrac{1-\alpha}{1+2\alpha}}{\left(\dfrac{\alpha}{1+2\alpha}\right)\left(\dfrac{2\alpha}{1+2\alpha}\right)^2}\left(\frac{5}{1}\right)^2 = e^{2.328}$$

$$\frac{1-\alpha}{\alpha}\left(\frac{1+2\alpha}{2\alpha}\right)^2 25 = 10.257$$

When solved by iteration, this gives $\alpha = 0.8590$. Substituting this result into the previous expressions for the mixture's constituent mole fractions gives

$$\text{Methane:}\quad y_{CH_4} = \frac{1-0.8590}{1+2\times 0.8590} = 0.0519$$

$$\text{Carbon:}\quad y_C = \frac{0.8590}{1+2\times 0.8590} = 0.3160$$

$$\text{Hydrogen:}\quad y_{H_2} = \frac{2\times 0.8590}{1+2\times 0.8590} = 0.6321$$

12.7 Equilibrium Constants of Ideal Gases

The equilibrium constant can be expressed in three different ways,

$$\ln K_P = -\frac{\Delta G^*}{\Re T}$$

$$K_P = \frac{\prod_{prods} y_i^{\nu_i}}{\prod_{reacts} y_i^{\nu_i}}\left(\frac{P}{P_{ref}}\right)^{\sum_{prods}\nu_i - \sum_{reacts}\nu_i}$$

$$K_P = \frac{\prod_{prods} N_i^{\nu_i}}{\prod_{reacts} N_i^{\nu_i}}\left(\frac{P}{N_{total}}\right)^{\sum_{prods}\nu_i - \sum_{reacts}\nu_i}$$

The last of these is obtained from the definition of the mole fractions, $y_i = N_i/N_{total}$. Each form is useful. The first permits us to calculate the equilibrium constant using the thermodynamic properties of the constituents, and the stoichiometric reaction coefficients. This was demonstrated in Example 12.4.

Tables of equilibrium constants, such as appendix Table B.8, can be developed from other thermodynamic data using the first of these expressions. The last two forms for the equilibrium constant permit us to determine the equilibrium composition of the constituents by either mole fraction or number of moles. Elemental species balances also must be satisfied. These are automatically satisfied by using the *degree of reaction completion* as illustrated in Examples 12.4 and 12.5.

Six considerations must be kept in mind when using equilibrium constants and elemental species balances to determine the equilibrium composition of the resulting mixture of ideal gases.

First, we have defined the equilibrium constant as

$$\frac{\prod_{prods} y_i^{\nu_i}}{\prod_{reacts} y_i^{\nu_i}} = K_P$$

but it can also have been defined as

$$\frac{\prod_{reacts} y_i^{\nu_i}}{\prod_{prods} y_i^{\nu_i}} = K_P$$

This latter expression is often termed the *equilibrium constant for the reverse reaction* because this equilibrium constant is obtained by interchanging the reactants and products. Hence, if K_P is the equilibrium constant for the reaction

$$\nu_A A + \nu_B B \rightleftharpoons \nu_C C + \nu_D D$$

then $1/K_P$ is the equilibrium constant of the reaction

$$\nu_C C + \nu_D D \rightleftharpoons \nu_A A + \nu_B B$$

In fact, several authors, reference tables, and handbooks use this reverse-reaction convention for K_P. Either convention is satisfactory as long as the appropriate form of the reaction equation and equilibrium condition are used.

Second, equilibrium constants are calculated for specific reactions. For example, the stoichiometric disassociation reaction for carbon dioxide may be written as either

$$CO_2 \rightleftharpoons CO + \frac{1}{2}O_2$$

or

$$2CO_2 \rightleftharpoons 2CO + O_2$$

Both are chemically correct and satisfy the elemental species balances, but the equilibrium constant of the second expression is the square of that of the first expression because the coefficients on all terms have been doubled. The exponents in the expression for the equilibrium constant will also be doubled when the second expression is used. In general, if a stoichiometric reaction can be written as

$$\nu_A A + \nu_B B \rightleftharpoons \nu_C C + \nu_D D$$

it also can be written as

$$z(\nu_A A + \nu_B B) \rightleftharpoons z(\nu_C C + \nu_D D)$$

where z is an arbitrary multiplier. If ΔG^* is the difference between the molar-specific Gibbs function of the products and reactants for the basic reaction, then $z\Delta G^*$ is the same difference for the second reaction (multiplied by z). According to the definition of the equilibrium constant,

$$\ln K_P^+ = -\frac{z\Delta G^*}{\Re T} = z \ln K_P$$

or

$$K_P^+ = K_P^z$$

where K_P^+ is the equilibrium constant of the multiplied reaction. As long as we are careful to use the correct equilibrium constant and matching stoichiometric coefficients in the equilibrium expression, then the correct equilibrium composition will be obtained using either expression. When using tabulated equilibrium constants, be sure to use the form of the reaction equation that was used to calculate the tabulated constants.

Third, equilibrium constants, K_P, are normally calculated for a product mixture pressure of 1 atmosphere. When using tabular results, be sure to check the pressure basis for the equilibrium constants because not all data sources use 1 atmosphere as the basis for their calculations.

Fourth, if any inert gases are present in the reaction, they have no effect on the equilibrium constant. For example, when an inert gas, say E, is introduced into the two-reactant, two-product reaction of the previous sections, the stoichiometric reaction balance is

$$\nu_A A + \nu_B B + \nu_E E \rightleftharpoons \nu_C C + \nu_D D + \nu_E^* E$$

Elemental species balances for the elements forming compound E reveal that $\nu_E = \nu_E^*$. Then

$$\Delta G^* = \sum_{prods} \nu_i \overline{g}_i^* + \nu_E \overline{g}_E^* - \sum_{reacts} \nu_i \overline{g}_i^* - \nu_E^* \overline{g}_E^*$$

where the terms associated with the inert E compound have been separated from the summations. This expression clearly demonstrates that the inert E compound cancels out of the calculation of ΔG^*, and consequently is eliminated from the equilibrium constant. Inert gases will alter the equilibrium composition of the final products as demonstrated by considering the mole-fraction expression for the equilibrium constant as applied to the above reaction,

$$\frac{y_C^{\nu_C} y_D^{\nu_D} y_E^{\nu_E}}{y_A^{\nu_A} y_B^{\nu_B} y_E^{\nu_E}} \left(\frac{P}{P_{ref}} \right)^{\nu_C + \nu_D + \nu_E - \nu_A - \nu_B - \nu_E} = K_P$$

Although it appears that the effect of compound E cancels out of this expression, the presence of this inert gas does alter the mole fractions of compounds A, B, C, and D in the final mixture. The presence of inert compounds in the reaction will then alter the equilibrium composition of the final mixture, but not the equilibrium constant.

Fifth, equilibrium constants of ideal gases are functions of temperature only because

$$\ln K_P = \frac{\Delta G^*(T)}{\Re T}$$

Pressure does not affect the equilibrium constants of ideal gases. The equilibrium constant of the following three reactions are then the same:

$$CO_2 \rightleftharpoons C + \frac{1}{2} O_2 \quad \text{at 1 atmosphere}$$

$$CO_2 \rightleftharpoons C + \frac{1}{2} O_2 \quad \text{at 10 atmospheres}$$

$$CO_2 + 2N_2 \rightleftharpoons C + \frac{1}{2} O_2 + 2N_2 \quad \text{at 5 atmospheres}$$

The last reaction includes inert nitrogen, which does not change the equilibrium constant. Although the pressure of the products does not alter the equilibrium constant, it will change the equilibrium composition through the pressure term of the equilibrium-constant expressions.

Sixth, as the equilibrium constant approaches zero, becomes positive, and then becomes larger (i.e., increasing temperature), the reaction becomes more complete, as demonstrated in Example 12.4. Inspection of the mole-fraction form of the equilibrium-constant expressions reveals that as the equilibrium constant becomes larger, the mole fraction of the products is increasing, while the mole fraction of the reactants is decreasing. The fraction of reactants in the resulting mixture will then decrease to the point where the reactants are effectively absent from the mixture. Once $\ln K_P > 7$, the mole fractions of the reactants in the mixture are so small that the reaction is effectively complete. Similarly, when $\ln K_P < -7$, the mole fractions of the products in the mixture are so small that there is effectively no reaction.

12.8 Simultaneous Reactions

Often the reactions are more complicated than the simple reactions of tabular data such as Table B.8. For example, the disassociation of water can involve the hydroxyl molecule, as well as hydrogen and oxygen, at certain temperatures. The stoichiometric statement of this reaction is

$$3H_2O \rightleftharpoons 2OH + \frac{1}{2}O_2 + 2H_2$$

This reaction may be modeled by two simultaneously occurring simpler reactions

$$OH \rightleftharpoons \frac{1}{2}O_2 + \frac{1}{2}H_2$$

$$H_2O \rightleftharpoons \frac{1}{2}O_2 + H_2$$

These have readily available equilibrium constants.

If we multiply the preceding OH reaction by 2, interchange the products and reactants, and multiply the H_2O reaction by 3, they become

$$O_2 + H_2 \rightleftharpoons 2OH$$

$$3H_2O \rightleftharpoons \frac{3}{2}O_2 + 3H_2$$

Summing these two reactions gives

$$3H_2O + O_2 + H_2 \rightleftharpoons 2OH + \frac{3}{2}O_2 + 3H_2$$

When the common O_2 and H_2 terms in the products and reactants are canceled, this result reduces to

$$3H_2O \rightleftharpoons 2OH + \frac{1}{2}O_2 + 2H_2$$

This is the original disassociation of H_2O into OH, H_2, and O_2. Practically all chemical reactions can be generated from simpler simultaneous reactions in this manner.

To illustrate the method of determining the equilibrium constant of a complex reaction that consists of two simultaneous reactions, we will consider these two simultaneous reactions:

$$\text{Reaction 1} \quad \nu_A A \rightleftharpoons \nu_{B,1} B + \nu_C C$$
$$\text{Reaction 2} \quad \nu_D D \rightleftharpoons \nu_{B,2} B + \nu_E E$$

The equilibrium constant for reaction 1 is $K_{P,1}$, and that for reaction 2 is $K_{P,2}$. A more complex reaction can be generated from these two simple reactions by multiplying reaction 1 by z_1 and reaction 2 by z_2 and then summing the results to produce

$$z_1 \nu_A A + z_2 \nu_D D \rightleftharpoons z_1 \nu_{B,1} B + z_2 \nu_{B,2} B + z_1 \nu_C C + z_2 \nu_E E$$

The ΔG^* for the reaction formed from this combination of the two simpler simultaneous reactions is then

$$\Delta G^* = z_1 \Delta G_1^* + z_2 \Delta G_2^*$$

Substituting this expression into the definition of the ideal gas equilibrium constant gives

$$\ln K_P = -z_1 \frac{\Delta G_1^*}{\Re T} - z_2 \frac{\Delta G_2^*}{\Re T}$$

$$= \ln K_{P,1}^{z_1} + \ln K_{P,2}^{z_2}$$

or

$$K_P = K_{P,1}^{z_1} K_{P,2}^{z_2}$$

In words, the equilibrium constant of the simultaneous reaction is the product of the equilibrium constants of the two simpler reactions when they are summed. Each equilibrium constant of the simpler reactions must be raised to an appropriate power to adjust for any multipliers that may be required to form the proper combination. The reciprocal of the equilibrium constant will be necessary if a reverse reaction is used to generate the complete reaction. When more than two simple simultaneous reactions are needed to form the complete reaction, this method can be applied repeatedly.

EXAMPLE 12.6

Derive an expression for the equilibrium constant of the reaction

$$CO_2 \rightleftharpoons CO + \frac{1}{2} O_2$$

in terms of the equilibrium constants presented in Table B.8.

System: A reacting system of carbon dioxide, carbon monoxide, and oxygen gases.

Basic Equations:
$$\frac{\prod_{prods} y_i^{\nu_i}}{\prod_{reacts} y_i^{\nu_i}} \left(\frac{P}{P_{ref}}\right)^{\sum_{prods} \nu_i - \sum_{reacts} \nu_i} = K_P$$

Conditions: All constituents will be treated as ideal gases.

Solution: The desired reactions can be formed from the two simpler simultaneous reactions

$$CO_2 \rightleftharpoons C + O_2 \qquad K_P = K_{P,1}$$

$$C + \frac{1}{2}O_2 \rightleftharpoons CO \qquad K_P = \frac{1}{K_{P,2}}$$

where $K_{P,1}$ and $K_{P,2}$ are presented in Table B.8. Note that the second reaction is the reverse of the reaction in Table B.8. Adding these two equations gives

$$CO_2 + C + \frac{1}{2}O_2 \rightleftharpoons C + O_2 + CO$$

When the common terms are canceled, this becomes

$$CO_2 \rightleftharpoons CO + \frac{1}{2}O_2$$

The equilibrium constant for the overall reaction is the product of the equilibrium constants of the two simpler constants. In terms of the equilibrium constants presented in Table B.8, the overall equilibrium constant is

$$K_P = \frac{K_{P,1}}{K_{P,2}}$$

    In general, the overall equilibrium constant and equilibrium composition for a reaction formed from simultaneous simpler reactions, including inert gases, can be determined using the following steps.

1. Write the complete reaction, including inert gases, and use elemental species balances to balance this reaction statement. This result will be used to determine the mole fractions of the final equilibrium mixture.

2. Select the simpler simultaneous reactions that will be combined to form the complete reaction without the inert gases. Also obtain the equilibrium constants for each simpler reaction. If any of these reactions are the reverse of those in the data set used, such as Table B.8, be sure to use the reciprocal of the equilibrium constant presented in the data set.

3. Select two reactions to be combined to form all or part of the complete reaction. Multiply each by appropriate factors, and then add the results to form an intermediate reaction or the complete reaction. The equilibrium constant for this reaction is the product of the equilibrium constants of the simpler reactions, each raised to the power of the multiplier used.

4. If there are more than two simple simultaneous reactions, then proceed to combine the intermediate reaction generated in step 3 with another of the simpler reactions in the same manner as in step 3.

5. Once all of the simpler reactions have been combined by repeating step 4, the resulting reaction should be the same as the complete reaction of step 1 without the inert gases. The equilibrium constant generated by repeating step 4 is the overall equilibrium constant for this resulting reaction.

6. Use the degree of reaction completion to obtain the mole fractions of the constituents of the complete reaction of step 1.

7. Apply the mole fraction form of the expressions for the equilibrium constant using the reaction generated in step 4 for the products, reactants, and stoichiometric coefficients, the mole fractions of step 6, and the overall equilibrium constant of step 5.

8. Proceed to solve the equation generated in step 7 for the degree of reaction completion.

This process is demonstrated further in the following examples.

EXAMPLE 12.7 A mixture of 1 kg-mole of oxygen and 3.76 kg-moles of nitrogen (i.e., atmospheric air) is heated at 1 atmosphere to 3000 K. Determine the number of moles of the various constituents that will be present in the resulting mixture.

System: A reacting system of ideal gases.

Basic Equations:

$$\frac{\prod_{prods} y_i^{\nu_i}}{\prod_{reacts} y_i^{\nu_i}} \left(\frac{P}{P_{ref}}\right)^{\sum_{prods} \nu_i - \sum_{reacts} \nu_i} = K_P$$

Conditions: All disassociation and recombination reactions will be considered, although some may not be relevant.

Solution: The first step is to determine which reactions might be involved. In this case, the molecular oxygen and nitrogen can disassociate into atomic oxygen and nitrogen. These atomic species can then recombine to form N_xO_y compounds. All of the potential reactions of Table B.8 and their equilibrium constants for 3000 K are:

$$O_2 \rightleftharpoons 2O \qquad K_P = e^{-4.370} = 0.01265$$
$$N_2 \rightleftharpoons 2N \qquad K_P = e^{-22.372} = 1.928(10^{-10})$$

$$\frac{1}{2}O_2 + \frac{1}{2}N_2 \rightleftharpoons NO \qquad K_P = e^{-2.102} = 0.1222$$

$$O_2 + \frac{1}{2}N_2 \rightleftharpoons NO_2 \qquad K_P = e^{-8.897} = 1.368(10^{-4})$$

$$\frac{1}{2}O_2 + N_2 \rightleftharpoons N_2O \qquad K_P = e^{-11.868} = 7.011(10^{-6})$$

Inspection of these equilibrium constants reveals that all of the reactions are unimportant ($\ln K_P < -7.0$), except for the disassociation of the oxygen and recombination to form NO. Including these two reactions, the actual reaction is

$$\frac{1}{2}N_2 + O_2 + 3.76N_2 \rightleftharpoons NO + O + 3.26N_2$$

where a portion of the N_2 is an inert gas that has no effect on the equilibrium constant but does alter the mole fraction of the final mixture.

Multiplying the O_2 disassociation reaction by $1/2$ and adding the result to the NO recombination reaction gives

$$\text{Reaction 1} \qquad \frac{1}{2}O_2 \rightleftharpoons O$$

$$\text{Reaction 2} \qquad \frac{1}{2}O_2 + \frac{1}{2}N_2 \rightleftharpoons NO$$

$$\textbf{Combined reaction} \qquad O_2 + \frac{1}{2}N_2 \rightleftharpoons O + NO$$

This is the same as the actual reactions without the inert portion of the N_2. The equilibrium constant for this combined reaction is

$$K_P = K_{P,1}^{1/2} K_{P,2} = (0.01265)^{1/2} 0.1222 = 0.01375$$

Letting α represent the degree of reaction completion, the number of moles of each constituent in the actual resulting mixture is

NO: α

O: α

N_2: $3.26\alpha + 3.76(1-\alpha) = 3.76 - \dfrac{\alpha}{2} = \dfrac{7.52 - \alpha}{2}$

O_2: $1 - \alpha$

Total: $4.76 + \dfrac{\alpha}{2} = \dfrac{9.52 + \alpha}{2}$

The respective mole fractions in the final mixture are then

$$\text{NO:}\quad y_{NO} = \frac{2\alpha}{9.52 + \alpha}$$

$$\text{O:}\quad y_O = \frac{2\alpha}{9.52 + \alpha}$$

$$\text{N}_2\text{:}\quad y_{N_2} = \frac{7.52 - \alpha}{9.52 + \alpha}$$

$$\text{O}_2\text{:}\quad y_{O_2} = \frac{2(1 - \alpha)}{9.52 + \alpha}$$

Using the combined reaction equation (without the inert portion of the N_2) and the above mole fractions, the expression for the overall equilibrium constant becomes

$$\frac{\dfrac{2\alpha}{9.52 + \alpha}\dfrac{2\alpha}{9.52 + \alpha}}{\dfrac{2(1 - \alpha)}{9.52 + \alpha}\left(\dfrac{7.52 - \alpha}{9.52 + \alpha}\right)^{1/2}} = K_P$$

$$\frac{2\alpha^2}{(1 - \alpha)(7.52 - \alpha)^{1/2}} = 0.01375$$

Solving this for α gives $\alpha = 0.6193$. The constituent mole fractions in the final mixture are then

$$\text{NO:}\quad y_{NO} = \frac{2 \times 0.6193}{9.52 + 0.6193} = 0.122$$

$$\text{O:}\quad y_O = \frac{2 \times 0.6193}{9.52 + 0.6193} = 0.122$$

$$\text{N}_2\text{:}\quad y_{N_2} = \frac{7.52 - 0.6193}{9.52 + 0.6193} = 0.775$$

$$\text{O}_2\text{:}\quad y_{O_2} = \frac{2(1 - 0.6193)}{9.52 + 0.6193} = 0.075$$

TIP:

NO is one of the NO_x air pollutants formed by internal-combustion engines. This example demonstrates the large quantities of NO that can be formed at high temperatures. Inspection of Table B.8 reveals that as long as the air temperature does not exceed approximately 1300 K, then NO formation will be practically nonexistent.

EXAMPLE 12.8

One kg-mole of carbon is burned at 1 atmosphere pressure with one kg-mole of O_2 to form CO, CO_2, and O products at 4000 K. Determine the equilibrium mole fractions of the resulting mixture of products and reactants.

System: A reacting system of carbon dioxide, carbon monoxide, carbon, and oxygen gases.

Basic Equations:

$$\frac{\prod_{prods} y_i^{v_i}}{\prod_{reacts} y_i^{v_i}} \left(\frac{P}{P_{ref}}\right)^{\sum_{prods} v_i - \sum_{reacts} v_i} = K_P$$

Conditions: All the constituents in the resulting mixture will be treated as ideal gases.

Solution: This reaction can be modeled by the system shown in Figure 12.7. In this model, some of the energy released by the combustion reaction

$$C + O_2 \rightarrow CO_2$$

increases the temperature of the CO_2 and O_2 such that they disassociate into CO and O. The overall reaction may be written as

$$C + O_2 \rightleftharpoons \frac{1}{2}CO + \frac{1}{2}CO_2 + \frac{1}{2}O$$

This reaction can be formed from the three simpler reactions listed below with their equilibrium constants at 4000 K.

Reaction 1 $C + O_2 \rightleftharpoons CO_2$ $K_{P,1} = \dfrac{1}{e^{-11.856}} = 140{,}900$

Reaction 2 $O_2 \rightleftharpoons 2O$ $K_{P,2} = e^{0.783} = 2.188$

Reaction 3 $C + \dfrac{1}{2}O_2 \rightleftharpoons CO$ $K_{P,3} = \dfrac{1}{e^{-13.449}} = 693{,}100$

FIGURE 12.7
Reactions for Example 12.8

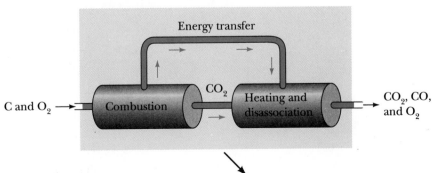

Net reaction

Energy transfer

C and O_2 → Combustion → CO_2 → Heating and disassociation → CO_2, CO, and O_2

$-Q$

When reaction 1 is multiplied by 1/2, reaction 2 is multiplied by 1/4, and the two results added, the new reaction is

$$\frac{1}{2}C + \frac{3}{4}O_2 \rightleftharpoons \frac{1}{2}CO_2 + \frac{1}{2}O \qquad K_P = K_{P,1}^{1/2} K_{P,2}^{1/4}$$

When reaction 3 is multiplied by 1/2 and the result added to the above intermediate reaction, we obtain the final reaction,

$$C + O_2 \rightleftharpoons \frac{1}{2}CO + \frac{1}{2}CO_2 + \frac{1}{2}O \qquad K_P = K_{P,1}^{1/2} K_{P,2}^{1/4} K_{P,3}^{1/2}$$

The overall equilibrium constant is then

$$\begin{aligned}
K_P &= K_{P,1}^{1/2} K_{P,2}^{1/4} K_{P,3}^{1/2} \\
&= 140{,}900^{1/2} 2.188^{1/4} 693{,}100^{1/2} \\
&= 380{,}000
\end{aligned}$$

Using the degree of reaction completion, α, the number of moles of each constituent in the resulting mixture will be

$$CO: \quad \frac{\alpha}{2}$$

$$CO_2: \quad \frac{\alpha}{2}$$

$$O: \quad \frac{\alpha}{2}$$

$$O_2: \quad 1 - \alpha$$

$$C: \quad 1 - \alpha$$

Mixture total: $\dfrac{4 - \alpha}{2}$

The representative mole fractions in the resulting mixture are

$$CO: \quad \frac{\alpha}{4 - \alpha}$$

$$CO_2: \quad \frac{\alpha}{4 - \alpha}$$

$$O: \quad \frac{\alpha}{4 - \alpha}$$

$$O_2: \quad \frac{2(1 - \alpha)}{4 - \alpha}$$

$$C: \quad \frac{2(1 - \alpha)}{4 - \alpha}$$

Substituting these results into the expression for the equilibrium constant gives

$$
\frac{\left(\dfrac{\alpha}{4-\alpha}\right)^{1/2}\left(\dfrac{\alpha}{4-\alpha}\right)^{1/2}\left(\dfrac{\alpha}{4-\alpha}\right)^{1/2}}{\dfrac{2(1-\alpha)}{4-\alpha}\dfrac{2(1-\alpha)}{4-\alpha}} = K_P
$$

$$
\frac{\alpha^{3/2}(4-\alpha)^{1/2}}{4(1-\alpha)^2} = 380{,}000
$$

When this is solved for α, the result is

$$
\alpha = 0.9989
$$

and the reaction is practically completed. The resulting mole fractions are

$$
\text{CO:} \quad \frac{0.9989}{4-0.9989} = 0.3328
$$

$$
\text{CO}_2\text{:} \quad \frac{0.9989}{4-0.9989} = 0.3328
$$

$$
\text{O:} \quad \frac{0.9989}{4-0.9989} = 0.3328
$$

$$
\text{O}_2\text{:} \quad \frac{2(1-0.9989)}{4-0.9989} = 0.00073
$$

$$
\text{C:} \quad \frac{2(1-0.9989)}{4-0.9989} = 0.00073
$$

12.9 Effect of Temperature on Ideal Gas Equilibrium Constants

Section 12.7 showed that the equilibrium constant of an ideal gas is purely a function of temperature. Section 12.6 defined the equilibrium constant as

$$
\ln K_P = -\frac{\Delta G^*(T)}{\Re T}
$$

According to the definition of the Gibbs function, $\Delta G^*(T) = \Delta H^*(T) - T\Delta S^*(T)$, which permits us to write the equilibrium constant as

$$
\ln K_P = -\frac{\Delta H^*(T)}{\Re T} + \frac{\Delta S^*(T)}{\Re}
$$

Differentiating this result with regard to T gives

$$\frac{d(\ln K_P)}{dT} = \frac{\Delta H^*(T)}{\Re T^2} - \frac{d(\Delta H^*(T))}{\Re T dT} + \frac{d(\Delta S^*(T))}{\Re dT}$$

The Gibbs equation, $Tds = dh - vdP$, becomes $Tds = dh$ for a constant-pressure process such as that of the chemical reactions under consideration. Because ΔS^* and ΔH^* are made up of sums of stoichiometric coefficients and molar-specific enthalpies and entropies, $Tds = dh$ becomes $Td(\Delta s^*) = d(\Delta H^*)$. The last two terms of the above equation result then cancel and it becomes as follows:

$$\frac{d(\ln K_P)}{dT} = \frac{\Delta H^*(T)}{\Re T^2} = \frac{\overline{h}_r}{\Re T^2} \qquad (12.10)$$

Here $\overline{h}_r$ is the *heat of reaction* as given by the difference between the reaction-coefficient weighted molar-specific enthalpies of the products and reactants. This important result is known as the *van't Hoff Equation* whose utility lies in relating the heat of reaction to the equilibrium constant.

To integrate the van't Hoff equation, we need to know how the heat of reaction varies with temperature. For small temperature changes, the heat of reaction can be taken as constant. In this event, the van't Hoff equation integrates to

$$\ln \frac{K_{P,1}}{K_{P,2}} = \frac{\overline{h}_r}{\Re}\left(\frac{1}{T_1} - \frac{1}{T_2}\right)$$

This result demonstrates two things. First, it allows us to determine heats of reaction from equilibrium constants that are sometimes easier to obtain. Second, it shows that for exothermic reactions such as combustion where $\overline{h}_r < 0$ the reaction will be less complete as the temperature is increased.

EXAMPLE 12.9   Estimate the heat of reaction when hydrogen gas is burned with a stoichiometric amount of oxygen gas at 2500 K and 1 atmosphere using (a) the first law and (b) the van't Hoff equation.

System: A reacting system of hydrogen and oxygen gases.

Basic Equations:

$$\bar{q} - \bar{w} = \sum_{prods} z_i[\bar{h}^0_{f,i} + \Delta \bar{h}^0_i(P,T)] - \sum_{reacts} z_i[\bar{h}^0_{f,i} + \Delta \bar{h}^0_i(P,T)]$$

$$\ln \frac{K_{P,1}}{K_{P,2}} = \frac{\bar{h}_r}{\Re}\left(\frac{1}{T_1} - \frac{1}{T_2}\right)$$

Conditions: All constituents will be treated as ideal gases.

Solution: The appropriate chemical reaction for this problem is

$$H_2 + \frac{1}{2}O_2 \rightarrow H_2O$$

The heat of reaction is the heat released as this reaction occurs in a steady-flow, isobaric combustion reactor. Applying the first law to this system when there is no shaft work gives

$$\bar{q} = \bar{h}_r = \sum_{prods} z_i[\bar{h}^0_{f,i} + \Delta \bar{h}^0_i(P,T)] - \sum_{reacts} z_i[\bar{h}^0_{f,i} + \Delta \bar{h}^0_i(P,T)]$$

$$= 1[-241,820 + (109,001 - 9899)] - 1[0 + (78,966 - 8,463)]$$

$$- \frac{1}{2}[0 + (86,970 - 8,679)]$$

$$= -252,370 \text{ kJ/kg-mole-}H_2$$

TIP:

These results are in good agreement in view of the fact that the temperature difference selected for the van't Hoff equation was 200 K. Better accuracy would be obtained if this temperature differerence were reduced.

The heats of formation were obtained from Table B.7, and the enthalpies as measured with regard to the reference state were obtained from Table B.6.

Applying the van't Hoff equation to 2400 K and 2600 K using Table B.8 for the equilibrium constants gives

$$\ln \frac{0.003607}{0.009523} = \frac{\bar{h}_r}{8.314}\left(\frac{1}{2400} - \frac{1}{2600}\right)$$

which gives

$$\bar{h}_r = -251,800 \text{ kJ/kg-mole-}H_2$$

PROBLEMS

12.1   Calculate the value of the Gibbs function (in Btu/lb_m) for saturated steam at $300°\,F$ as a saturated liquid, saturated vapor, and a mixture of liquid and vapor with a quality of 60%. Demonstrate that phase equilibrium exists.

12.2   Calculate the value of the Gibbs function (in kJ/kg) for saturated R-12 at $0°\,C$ as a saturated liquid, saturated vapor, and a mixture of liquid and vapor with a quality of 30%. Demonstrate that phase equilibrium exists.

12.3   A liquid-vapor mixture of R-12 is at $-10°\,C$ with a quality of 40%. Determine the value of the specific Gibbs function (in kJ/kg) when the two phases are in equilibrium.

12.4   A liquid-vapor mixture of ammonia is at 40 psia with a quality of 70%. Determine the value of the specific Gibbs function (in kJ/kg) when the two phases are in equilibrium.

12.5   An oxygen–nitrogen mixture consists of 30 kg of oxygen and 40 kg of nitrogen. This mixture is cooled to 84 K at 1 atmosphere pressure. Determine the mass of the oxygen in the liquid and gaseous phases.

12.6   What is the total mass of the liquid phase of Problem 12.5?

12.7   At what temperature will the liquid phase of an oxygen–nitrogen mixture at 100 kPa have a nitrogen mass fraction of 60%?

12.8   At what temperature will the gaseous phase of an oxygen–nitrogen mixture at 100 kPa have a nitrogen-mole fraction of 30%? What is the mass fraction of the oxygen in the liquid phase at this temperature?

12.9   An ammonia–water mixture is at $10°\,C$. Determine the pressure of the ammonia vapor when the mole fraction of the ammonia in the liquid is (a) 20% and (b) 80%.

12.10   An ammonia–water absorption refrigeration unit operates its evaporator at $0°\,C$ and its condenser at $46°\,C$. The vapor mixture in the generator and absorber is to have an ammonia-mole fraction of 96%. Assuming ideal behavior, determine the operating pressure in the (a) generator and (b) absorber. Also determine the mole fraction of the ammonia in the (c) strong-liquid mixture being pumped from the absorber and the (d) weak-liquid solution being drained from the generator.

12.11   Rework Problem 12.10 when the temperature in the evaporator is increased to $6°\,C$ and the temperature in the condenser is reduced to $40°\,C$.

12.12   Air at $70°$ F, 100 psia is blown through a porous media that is saturated with liquid water also at $70°$ F. Determine the maximum partial pressure of the water evaporated into the air as it emerges from the porous media.

12.13   Foam products such as shaving cream are made from liquid mixtures whose ingredients are primarily water and a refrigerant such as R-12. Consider a liquid mixture of water and R-12 with a water mass fraction of 90% that is at $20°$ C. What is the mole fraction of the water and R-12 vapor in the gas that fills the bubbles that form the foam?

12.14   One lb_m-mole of R-12 is mixed with 1 lb_m-mole of water in a closed container that is maintained at 14.7 psia, $77°$ F. Determine the mole fraction of the R-12 in (a) the liquid phase and (b) the vapor phase.

12.15   One kg-mole of ammonia is mixed with 2 kg-mole of water in a closed container that is maintained at 100 kPa, $25°$ C. Determine the mole fraction of the ammonia in (a) the liquid phase and (b) the vapor phase.

12.16   A mixture of ideal gases consists of the following gases by mole fraction: 10% CO_2, 60% H_2O, and 30% CO. Determine the chemical potential of the CO in this mixture when the mixture pressure is 10 atmospheres and its temperature is 800 K.

12.17   A mixture of ideal gases is made up of 30% N_2, 30% O_2, and 40% H_2O by mole fraction. Determine the chemical potential of the N_2 when the mixture pressure is 5 atmospheres and its temperature is $1000°$ R.

12.18   A mixture of ideal gases is blended in a rigid vessel that is initially evacuated and is maintained at a constant temperature of $77°$ F. First, nitrogen is added until the pressure is 20 psia, next carbon dioxide is added until the pressure is 35 psia, and finally NO is added until the pressure is 50 psia. Determine the chemical potential of the N_2 in this mixture.

12.19   The equilibrium constant for the $CO \rightleftharpoons C + \frac{1}{2}O_2$ reaction at 100 kPa, 1600 K is K_P. Use this information to determine the equilibrium constant for the following reactions.

$$(a) \quad C + \frac{1}{2}O_2 \rightleftharpoons CO \qquad 100 \text{ kPa, 1600 K}$$

$$(b) \quad C + \frac{1}{2}O_2 \rightleftharpoons CO \qquad 500 \text{ kPa, 1600 K}$$

$$(c) \quad 2CO \rightleftharpoons 2C + O_2 \qquad 100 \text{ kPa, 1600 K}$$

$$(d) \quad 2CO \rightleftharpoons 2C + O_2 \qquad 500 \text{ kPa, 1600 K}$$

12.20 The equilibrium constant for the $H_2O \rightleftharpoons H_2 + \frac{1}{2}O_2$ reaction at 15 psia, 2000° R is K_P. Use this information to determine the equilibrium constant for the following reactions.

$$\text{(a)} \quad H_2 + \frac{1}{2}O_2 \rightleftharpoons H_2O \qquad \text{15 psia, 2000° R}$$

$$\text{(b)} \quad H_2 + \frac{1}{2}O_2 \rightleftharpoons H_2O \qquad \text{100 psia, 1600° R}$$

$$\text{(c)} \quad 3H_2O \rightleftharpoons 3H_2 + \frac{3}{2}O_2 \qquad \text{15 psia, 1600° R}$$

$$\text{(d)} \quad 3H_2O \rightleftharpoons 3H_2 + \frac{3}{2}O_2 \qquad \text{200 kPa, 2000 K}$$

12.21 Use the Gibbs function to determine the equilibrium constant of the $H_2O \rightleftharpoons H_2 + \frac{1}{2}O_2$ reaction at (a) 600 K and (b) 2000 K. How do these compare to the equilibrium constants of Table B.8?

12.22 Use the Gibbs function to determine the equilibrium constant of the $H_2O \rightleftharpoons H_2 + \frac{1}{2}O_2$ reaction at (a) 1440° R and (b) 3960° R. How do these compare to the equilibrium constants of Table B.8?

12.23 An inventor claims that she can produce hydrogen gas by the reversible reaction

$$2H_2O \rightleftharpoons 2H_2 + O_2$$

Determine the mole fractions of the hydrogen and oxygen produced when this reaction occurs at 4000 K and 10 kPa.

12.24 Will the amount of hydrogen gas produced in Problem 12.23 be increased when the reaction occurs at 100 kPa rather than 10 kPa?

12.25 How will the amount of hydrogen gas produced in Problem 12.23 be changed if inert nitrogen is mixed with the water vapor such that the original mole fraction of the nitrogen is 20%?

12.26 One means of producing carbon dioxide is through the reaction

$$C + O_2 \rightleftharpoons CO_2$$

Determine the yield of carbon dioxide (by mole fraction) when this is done in a reactor maintained at 14.7 psia, 6840° R.

12.27 If the pressure of the reactor of Problem 12.26 is increased to 100 psia, how much will the carbon-dioxide yield change?

12.28 How much will the carbon-dioxide yield of Problem 12.26 change if atmospheric air provides the oxygen required for the reaction?

12.29 At what temperature will nitrogen be 20% disassociated at (a) 10 kPa and (b) 1000 kPa?

12.30 At what temperature will oxygen be 15% disassociated at (a) 3 psia and (b) 100 psia?

12.31 Determine the composition of the products of the disassociation reaction

$$CO_2 \rightleftharpoons CO + O$$

when the products are at 1 atmosphere pressure and 2500 K.

12.32 Three moles of nitrogen are added to the 1 mole of CO_2 in Problem 12.31. Determine the equilibrium composition of the products at the same temperature and pressure with the additional nitrogen.

12.33 Show that as long as the extent of the reaction, α, for the disassociation reaction

$$X_2 \rightleftharpoons 2X$$

is smaller than one, α is given by

$$\alpha = \frac{\sqrt{K_p}}{2}$$

12.34 A gaseous mixture of 50% (by mole fraction) methane and 50% nitrogen is heated to $1800°$ R as its pressure is maintained at 14.7 psia. Determine the equilibrium composition (by mole fraction) of the resulting mixture.

12.35 A gaseous mixture of 30% methane (by mole fraction) and 70% carbon dioxide is heated at 1 atmosphere pressure to 1200 K. What is the equilibrium composition (by mole fraction) of the resulting mixture?

12.36 The reaction

$$N_2 + O_2 \rightleftharpoons 2NO$$

occurs in internal-combustion engines. Determine the equilibrium mole fraction of NO when the pressure is 101 kPa and the temperature is 1000 K.

12.37 One pound-mass mole of oxygen is heated from 1 atmosphere pressure, $537°$ R to 10 atmospheres pressure, $3960°$ R. Calculate the total amount of heat (in Btu) required if (a) disassociation is neglected, and (b) disassociation is included.

12.38 The oxygen of the previous problem is replaced with air. Compare the total heat required (Btu) to heat this air in the same manner (a) neglecting disassociation and (b) including disassociation.

12.39 Determine the equilibrium constant for the reaction

$$CH_4 + 2O_2 \rightleftharpoons CO_2 + 2H_2O$$

when the reaction occurs at 100 kPa, 2000 K.

12.40 What is the equilibrium-mole fraction of the water vapor in Problem 12.39?

12.41 Determine the equilibrium constant for the reaction

$$CH_4 + 2O_2 \rightleftharpoons CO_2 + 2H_2O$$

when the reaction occurs at 100 psia, $5400°$ R.

12.42 Determine the equilibrium partial pressure of the carbon dioxide in Problem 12.41.

12.43 Ten kilogram-moles of methane gas are heated from 1 atmosphere pressure, 298 K to 1 atmosphere pressure, 1000 K. Calculate the total amount of heat required (in kJ) when (a) disassociation is neglected, and (b) when disassociation is included.

12.44 Solid carbon at $77°$ F is burned with a stoichiometric amount of air that is at 1 atmosphere pressure and $77°$ F. Determine the number of moles of CO_2 formed per mole of carbon when only CO_2, CO, O_2, and N_2 are present in the products and the products are at 1 atmosphere pressure, $1400°$ F.

12.45 Determine the amount of heat released per lb_m of carbon (Btu/lb_m-carbon) by the combustion of Problem 12.44.

12.46 Methane gas is burned with 30% excess air. This fuel enters a steady-flow combustor at 101 kPa, $25°$ C mixed with the air. The products of combustion leave this reactor at 101 kPa, 1600 K. Determine the equilibrium composition of the products of combustion and the amount of heat (in kJ/kg-mole-methane) released by this combustion when the products contain some NO.

12.47 Propane gas is burned with 20% excess air. The air–fuel mixture enters a steady-flow combustor at 14.7 psia, $77°$ F. The products of combustion leave the reactor at 14.7 psia, $3600°$ R. What is the equilibrium composition (by mass fraction) of the products when the products contain some NO?

12.48 How much heat (in kJ/kg-fuel) is released by the combustion of Problem 12.47?

12.49 Propane gas is burned at 1 atmosphere pressure with a stoichiometric amount of oxygen supplied by atmospheric air. What is the equilibrium composition (by mole fraction) of the resulting products of combustion if the temperature is 1600 K and the products contain some NO?

12.50 How much heat (in kJ/kg-fuel) is released by the combustion of Problem 12.49?

12.51 What is the final temperature (in K) of the products of combustion of Problem 12.49 if the combustion is done without any heat transfer?

12.52 Gaseous octane is burned with 40% excess air in an automobile engine. During combustion, the pressure is 600 psia and the temperature reaches $3600°$ R. Determine the equilibrium composition of the products of combustion when the products contain some NO.

12.53 The pressure and temperature of the products of combustion of Problem 12.52 are reduced to 14.7 psia, 600° F during the expansion stroke. Determine the equilibrium composition of the products of combustion as they enter the exhaust system and are released to the atmosphere.

PROJECTS

The following projects are more extensive than the problems presented in this chapter. Most are open-ended and require some decisions on the student's part. Many can be expedited with computer-based property tables, computer spreadsheets, and student-written computer programs.

12.1 An engineer suggested that high-temperature disassociation of water be used to produce a hydrogen fuel. A reactor–separator has been designed that can accommodate temperatures as high as 4000 K and pressures as much as 5 atmospheres. Water enters this reactor–separator at 25° C. The separator separates the various constituents in the mixture into individual streams whose temperature and pressure match those of the reactor–separator. These streams are then cooled to 25° C and stored in atmospheric pressure tanks with the exception of any remaining water, which is returned to the reactor to repeat the process again. Hydrogen gas from these tanks is later burned with a stoichiometric amount of air to provide heat for an electrical-power plant. The parameter that characterizes this system is the ratio of the heat released by burning the hydrogen to the amount of heat used to generate the hydrogen gas. Select the operating pressure and temperature for the reactor–separator that maximizes this ratio. Can this ratio ever be bigger than unity?

12.2 A gas turbine (Brayton cycle) at a natural-gas pipeline pumping station uses natural gas (methane) as its fuel. Air is drawn into the turbine at 101 kPa, 25° C and the compression ratio of the turbine is 8. The natural gas fuel is injected into the combustor such that the excess air is 40%. Determine the net specific work produced by this engine and the engine's overall thermal efficiency.

12.3 One means of producing liquid oxygen from atmospheric air is to take advantage of the phase-equilibrium properties of oxygen–nitrogen mixtures. This system is illustrated in Figure 12.8. In this cascaded-reactors system, dry-atmospheric air is cooled in the first reactor until liquid is formed. According to the phase-equilibrium properties, this liquid will be richer in oxygen than in the vapor phase. The vapor in the first reactor is discarded while the oxygen-enriched liquid leaves the first reactor and is heated in a heat exchanger until it is again a vapor. The vapor mixture enters the second reactor where it is again cooled until a liquid that is further enriched in oxygen is formed. The vapor from the second reactor is routed back to the first reactor while the liquid is routed to another heat exchanger and another reactor to repeat the process once again. The liquid formed in the third reactor is very rich in oxygen. If all three reactors are operated at 1 atmosphere pressure, select the three temperatures that produce the greatest amount of 99% pure oxygen.

FIGURE 12.8

Dry atmospheric air

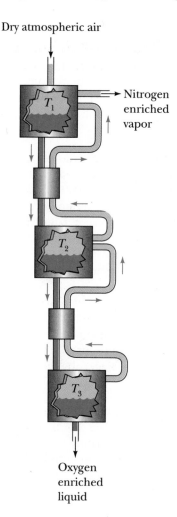

Nitrogen enriched vapor

Oxygen enriched liquid

12.4   To protect the atmosphere, it has been suggested that hydrogen be used as a fuel in aircraft that fly at high elevations. This would avoid the formation of carbon dioxide and other carbon-based combustion products. The combustion can of a Brayton cycle operates at about 400 kPa at these altitudes. Assume that new materials that allow for maximum temperature of 2600 K are available, and the atmospheric composition at these altitudes is 21% oxygen and 79% nitrogen by volume. The presence of NO_x in the exhaust gases is critical at these altitudes and cannot exceed 0.1% by volume. Excess air supply is used to control the maximum temperature of the combustion process. Determine the quantity of excess air to be used so that neither the maximum temperature nor the maximum allowable NO_x specification is exceeded. What is the NO_x mole fraction if the maximum temperature specification governs? If the NO_x specification governs, what is the temperature of the combustion gases?

Mathematics Review

A.1 Mathematics Review

The purpose of this appendix is to briefly review selected topics from mathematics that are used in thermodynamics. Not all needed topics are included, and this review may be inappropriate for some students.

A.1.1 Logarithms and Exponents

Logarithms and exponents are used extensively throughout thermodynamics analyses. The accepted symbol for "$\log_{10}$" is **log**, and the accepted symbol for the natural logarithm, "$\log_e$," is **ln**. Useful logarithmic and exponent identities include the following:

$$\log_a xy = \log_a x + \log_a y \qquad a^x a^y = a^{x+y}$$

$$\log_a \frac{x}{y} = \log_a x - \log_a y \qquad \frac{a^x}{a^y} = a^{x-y}$$

$$\log_a x^y = y \log_a x \qquad (a^x)^y = a^{xy}$$

A.1.2 Integration

Integration is the process for determining areas under curves as shown in Figure A.1. For example, if we plot the component of force in the direction of motion versus the system displacement, then the area under the resulting curve (i.e., integral) is the work. Integration is a very useful tool in thermodynamics.

Useful integrals include:

$$\int_{x_1}^{x_2} x^n\,dx = \frac{x_2^{n+1} - x_1^{n+1}}{n+1}$$

$$= \frac{x_2^{n+1}}{n+1}\left[1 - \left(\frac{x_1}{x_2}\right)^{n+1}\right]$$

$$= \frac{x_1^{n+1}}{n+1}\left[\left(\frac{x_2}{x_1}\right)^{n+1} - 1\right] \qquad n \neq -1$$

$$\int_{x_1}^{x_2} \frac{dx}{x} = \ln x_2 - \ln x_1$$

$$= \ln \frac{x_2}{x_1} = -\ln \frac{x_1}{x_2}$$

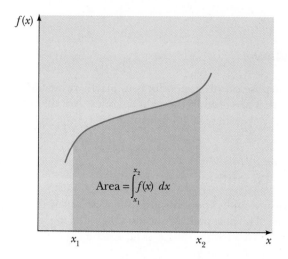

$$\text{Area} = \int_{x_1}^{x_2} f(x) \, dx$$

A.1.3   Functions of Several Variables

In thermodynamics, we often deal with quantities that depend on two or more variables. The state postulate tells us that any property of a simple compressible substance is a function of two independent properties. These relationships are given by $f(x, y, z) = 0$, which can be rearranged as $x = f_1(y, z)$ or $y = f_2(x, z)$ or $z = f_3(x, y)$. The preferred form depends on the problem being addressed.

Figure A.2 illustrates a functional relationship between a dependent variable, z, and two independent variables, x and y. This functional relationship generates a three-dimensional surface. If we fix one of the independent variables—say, y is fixed at y_1—then we slice this three-dimensional figure with a plane that passes through y_1. Two such slices are shown in Figure A.2.

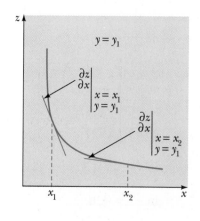

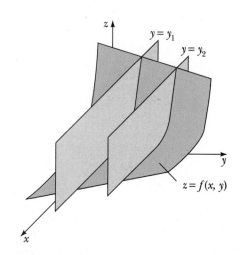

A *partial derivative* is the rate of change of the dependent variable with respect to one of the independent variables when all other independent variables remain constant; that is,

$$\left.\frac{\partial z}{\partial y}\right|_x \equiv \lim_{\Delta y \to 0} \frac{z(x, y + \Delta y) - z(x, y)}{\Delta y}$$

Partial derivatives are obtained by ordinary differentiation rules while treating all independent variables, except the one the derivative is being taken with respect to, as constants.

Physically, the partial derivative represents the slope of the function as observed in one of the slicing planes as shown in Figure A.2. The shown slopes are obtained by

$$\left.\frac{\partial z}{\partial x}\right|_{\substack{x = x_1 \\ y = y_1}} \quad \text{and} \quad \left.\frac{\partial z}{\partial x}\right|_{\substack{x = x_2 \\ y = y_1}}$$

for the y_1 plane at the points $x = x_1$ and $x = x_2$, respectively. These partial derivatives yield new functions with z as the dependent variable and x and y as independent variables. The subscript notation is redundant with only three variables, but it is very convenient with more than three, as is frequently the case in thermodynamics.

One also can take partial derivatives of partial derivative functions. These second derivatives are denoted as

$$\frac{\partial}{\partial x}\left[\left.\frac{\partial z}{\partial x}\right|_y\right]_y = \frac{\partial^2 z}{\partial x^2}$$

$$\frac{\partial}{\partial y}\left[\left.\frac{\partial z}{\partial x}\right|_y\right]_x = \frac{\partial^2 z}{\partial y \partial x}$$

The result is independent of the order in which the differentiation is done if the function is single-valued and continuous, as is the usual case in thermodynamics. When this is the case,

$$\frac{\partial^2 z}{\partial x \partial y} = \frac{\partial^2 z}{\partial y \partial x}$$

Other useful partial derivative relations are:

$$\left.\frac{\partial z}{\partial x}\right|_y \left.\frac{\partial x}{\partial y}\right|_z \left.\frac{\partial y}{\partial z}\right|_x = -1$$

and

$$\left.\frac{\partial y}{\partial x}\right|_z = \frac{1}{\left.\dfrac{\partial x}{\partial y}\right|_z}$$

A.1.4 Differentials

The differential of a function of several variables is:

$$dz = \left.\frac{\partial z}{\partial x}\right|_y dx + \left.\frac{\partial z}{\partial y}\right|_x dy + \cdots$$

where one term must be included for each independent variable. This expression gives the change in the dependent variable (z) when the independent variables $(x, y, \ldots)$ are changed by an infinitesimal amount (more precisely, it is the change in z in the limit as the change in $x, y, \ldots$ approaches 0).

Differentials are either *exact* or *inexact*. Consider a general relationship between a dependent variable, z, and two independent variables, x and y:

$$z = z(x, y)$$

The differential of z is then

$$dz = \left.\frac{\partial z}{\partial x}\right|_y dx + \left.\frac{\partial z}{\partial y}\right|_x dy$$
$$= M\,dx + N\,dy$$

which is an exact differential if

$$\frac{\partial M}{\partial y} = \frac{\partial N}{\partial x}$$

and $z(x, y)$ is single-valued and continuous. We will denote exact differentials as dz. If this condition is not met, then the differential is an inexact differential, which we will denote as δz. Functions whose differentials are exact are known as *point functions,* and functions whose differentials are inexact are known as *path functions.*

The change in the dependent variable for a finite change in the independent variables—say, as x changes from x_1 to x_2 and y changes from y_1 to y_2—is given by

$$\Delta z = \int_1^2 dz$$

If the differential of z is exact, then this result is independent of the integration path used to get from 1 to 2. In this case,

$$\Delta z = \int_1^2 dz$$
$$= \int_1^2 \left.\frac{\partial z}{\partial x}\right|_y dx + \int_1^2 \left.\frac{\partial z}{\partial y}\right|_x dy$$
$$= z_2 - z_1$$

When the differential of z is inexact, then this result depends on the integration path used to go from point 1 to point 2,

$$\int_1^2 \delta z = {}_1 z_2$$

where ${}_1 z_2$ stands for the amount of z realized as we move from the point (x_1, y_1) to the point (x_2, y_2) along a given connecting path.

A.1.5   Single-Variable Linear Interpolation

The divisions used in tables of properties and other tables do not fit all situations, and it often becomes necessary to interpolate between entries in these tables. When the desired quantity is a function of only one independent variable, such as the temperature of Table B.6, interpolation is particularly easy. To illustrate how this is done, let us determine the enthalpy of an ideal gas at a temperature T that is between temperature entries T_1 and $T_2 (T_2 > T > T_1)$. The correct table shows that the enthalpy for temperature T_1 is h_1, and that for temperature T_2 is h_2. Linear interpolation gives the desired enthalpy at the intermediate temperature T as

$$h(T) = \frac{T - T_1}{T_2 - T_1}(h_2 - h_1) + h_1$$

A.1.6   Two-Variable Linear Interpolation

Most thermodynamic quantities are functions of two independent variables. For example, the enthalpy of superheated steam is a function of the steam pressure and steam temperature. When this is the case, the tables are set up in terms of the two independent variables, and interpolation is required when the given pair of independent variables does not match the divisions of the table being used. Consider the problem of determining the enthalpy of steam when the specified pressure is P, which lies between two entries in the tables, $P_1 < P < P_2$, and the specified temperature also lies between two table entries, $T_1 < T < T_2$. We will need four enthalpies from the tables. These are $h_{1,1} = h(P_1, T_1)$, $h_{1,2} = h(P_1, T_2)$, $h_{2,1} = h(P_2, T_1)$ and $h_{2,2} = h(P_2, T_2)$.

A linear interpolation can now be performed on the temperature while the pressure is held constant at pressure P_1,

$$h_1 = \frac{T - T_1}{T_2 - T_1}(h_{1,2} - h_{1,1}) + h_{1,1}$$

and at pressure P_2,

$$h_2 = \frac{T - T_1}{T_2 - T_1}(h_{2,2} - h_{2,1}) + h_{2,1}$$

These results are next used to perform a linear interpolation on the pressure,

$$h(P, T) = \frac{P - P_1}{P_2 - P_1}(h_2 - h_1) + h_1$$

which yields the desired enthalpy at the specified pressure and temperature.

A.1.7   Iterative Solutions

Frequently, we cannot reduce a solution to a closed mathematical expression because we need to obtain certain data in the expression from a table or other source. In such a case, an iterative process must be used to solve the expression. There are many methods to conduct an iterative solution when programming the solution on a computer. Two of the simpler methods, trial-and-error and interval-halving, are particularly convenient for manual iterative solutions.

When only one variable in the final expression is to be determined—say, the temperature, T—then the final expression can always be written in the general form

$$f(T) = 0$$

When this function, $f(T)$, is plotted as shown in Figure A.3, then the solution is the value of T at which the function crosses the T axis. We must find the value of T at which this crossing occurs.

FIGURE A.3

A General Function of One Independent Variable

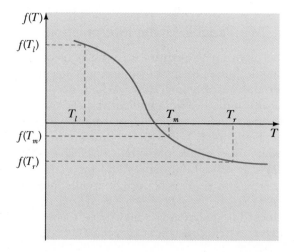

With the trial-and-error method, we simply substitute values of T and data from tables and other sources taken at the selected value of T until we find an $f(T)$ as close to zero as necessary. The selection of trial T's is not random, and the findings from each selection of T gives us considerable insight into how to select the next T value. For example, if we select the value T_l for the first trial, we will obtain the value $f(T_l)$ when the function is evaluated. This result clearly is not zero, and we would select the value T_r for the second trial. When the second value is substituted into the function, the result is $f(T_r)$, which is again not zero. However, the sign of the function has changed from positive for T_l to negative for T_r. This tells us that the value of T where the function crosses the axis is somewhere between T_l to T_r. This fact would be used to select T_m for the third trial. This process can now be continued until we obtain a value of T with acceptable accuracy.

The interval-halving technique is simply a formally structured adaptation of the trial-and-error method. To begin this technique, we must find one value of T that is to the left of the answer—say T_l—and one that is to the right—say, T_r. This is normally done by arbitrary selection. For the next trial, we use a value of T that is halfway between T_l and T_r:

$$T_m = \frac{T_l + T_r}{2}$$

When the function is evaluated for this trial value of T_m, the sign of the result will either be the same as that of $f(T_l)$ or $f(T_r)$. If the sign is the same as for $f(T_l)$, then the left half of the interval $T_l - T_m$ does not contain the solution and may be discarded. This half of the interval is discarded by making the new T_l equal to T_m. The right half of the interval can be eliminated in the same manner when the sign of $f(T_m)$ is the same as that of $f(T_r)$. This process can now be repeated until the value of T_m is within the accuracy required.

Both methods presume only one value of T that satisfies

$$f(T) = 0$$

This is the most commonly occurring case when solving thermodynamic problems.

A.1.8 Method of Lagrange Multipliers

Finding the extrema (i.e., maximum, minimum, or inflection point) of a function of several variables when subject to constraints is a commonly incurred problem. Mathematically, this problem can be stated as one of finding values for the set of n variables, $x_1, x_2, \ldots, x_n$, such that the function

$$f(x_1, x_2, \ldots, x_n) = 0$$

when subject to the m constraints

$$\phi_j(x_1, x_2, \ldots, x_n) = 0 \qquad j = 1, 2, \ldots, m$$

such that f is at an extrema. The number of constraints must be less than the number of independent variables, $m < n$. At an extremum point,

$$df = \frac{\partial f}{\partial x_1} dx_1 + \frac{\partial f}{\partial x_2} dx_2 + \cdots + \frac{\partial f}{\partial x_n} dx_n$$

$$= \sum_{i=1}^{n} \frac{\partial f}{\partial x_i} dx_i = 0$$

A brute force technique for solving this problem is to solve the set of constraint equations for m variables in terms of the remaining $n - m$ variables. When these are substituted into the function, a new function f_1 is formed such that

$$f_1(x_1, x_2, \ldots, x_{n-m}) = 0$$

where the variables $x_1, x_2, \cdots, x_{n-m}$ are now independent of one another and are no longer interrelated to one another by the constraint equations. Since these variables are independent of one another, the new function's differential is zero when

$$\frac{\partial f}{\partial x_i} = 0 \qquad i = 1, 2, \ldots, n - m$$

The solution of the $n - m$ equations generated in this manner provides the value of the x variables for the point at which f has an extremum.

Lagrange developed a simpler and more useful way of solving this problem. He suggested that a linear combination of the differentials of the function, f, and constraint functions, ϕ_j, be used. The differential of the constraint functions are

$$d\phi_j = \sum_{i=1}^{n} \frac{\partial \phi_j}{\partial x_i} dx_i = 0$$

When each of these differentials are multiplied by an undetermined coefficient (known as the Lagrange multiplier), λ_j, and added to the differential of the function, the result is

$$\sum_{i=1}^{n} \frac{\partial f}{\partial x_i} dx_i + \sum_{j=1}^{m} \left[\lambda_j \sum_{i=1}^{n} \frac{\partial \phi_j}{\partial x_i} dx_i \right] = 0$$

Expanding this result gives

$$
\left(\frac{\partial f}{\partial x_1} + \lambda_1 \frac{\partial \phi_1}{\partial x_1} + \lambda_2 \frac{\partial \phi_2}{\partial x_1} + \cdots + \lambda_m \frac{\partial \phi_m}{\partial x_1} \right) dx_1
$$

$$
+ \left(\frac{\partial f}{\partial x_2} + \lambda_1 \frac{\partial \phi_1}{\partial x_2} + \lambda_2 \frac{\partial \phi_2}{\partial x_2} + \cdots + \lambda_m \frac{\partial \phi_m}{\partial x_2} \right) dx_2
$$

$$
+ \cdots
$$

$$
+ \left(\frac{\partial f}{\partial x_n} + \lambda_1 \frac{\partial \phi_1}{\partial x_n} + \lambda_2 \frac{\partial \phi_2}{\partial x_n} + \cdots + \lambda_m \frac{\partial \phi_m}{\partial x_n} \right) dx_n = 0
$$

This result demonstrates that if the Lagrange multipliers are chosen such that

$$
\frac{\partial f}{\partial x_i} + \lambda_1 \frac{\partial \phi_1}{\partial x_i} + \lambda_2 \frac{\partial \phi_2}{\partial x_i} + \cdots + \lambda_m \frac{\partial \phi_m}{\partial x_i} = 0 \qquad i = 1, 2, \ldots, n
$$

together with the constraint equations

$$
\phi_j(x_1, x_2, \ldots, x_n) = 0
$$

a set of $n + m$ variables $x_1, x_2, \ldots, x_n, \lambda_1, \lambda_2, \ldots, \lambda_m$ can be determined to satisfy the necessary conditon for the extrema of the function $f(x_1, x_2, \ldots, x_n)$.

PROBLEMS

A.1 Evaluate $\ln 6 + \ln 0.5$.

A.2 Evaluate $\log 6 - \log 0.5$.

A.3 Evaluate $\ln 6^2 - \ln 0.5^2$.

A.4 Evaluate $5^2 5^5$.

A.5 Evaluate 5^{2^5}.

A.6 Evaluate $5^3 7^3$.

A.7 Evaluate

$$
\int_1^4 x^2 \, dx
$$

A.8 Evaluate

$$
\int_{0.5}^{1.3} \frac{dx}{x}
$$

A.9 Evaluate

$$\int_1^2 xe^x\,dx$$

A.10 Evaluate

$$\int_1^2 y\,dx$$

along the path for which $yx^n (n \neq 1)$ remains constant. This path begins at $x - 1$ and $y = 2$, and it ends at $x = 2$.

A.11 Evaluate

$$\int_1^2 y\,dx$$

along the path for which yx remains constant. This path begins at $x = 1$ and $y = 1$, and it ends at $x = 4$.

A.12 Evaluate

$$\int_1^2 y\,dx$$

along the path for which $x + y$ remains constant. This path begins at $x = 1$ and $y = 1$, and it ends at $x = 4$.

A.13 Consider the relationship $z = (xy)^2 - (xy)^3$. Evaluate

$$\left.\frac{\partial z}{\partial x}\right|_y \quad\text{and}\quad \left.\frac{\partial z}{\partial y}\right|_x$$

when $x = 1$ and $y = 3$.

A.14 Repeat the previous problem for $x = 3$ and $y = 1$.

A.15 What is the differential, dz, of the function

$$z = 3xy^{0.5} - x^2 y^{0.75}$$

Is this an exact or inexact differential?

A.16 Find the specific internal energy of air at 1272 K by interpolating the values found in appendix Table B.6.1.

A.17 Find the specific entropy of superheated ammonia vapor at 135 psia, 130° F, by interpolating the values found in appendix Table B.3.2.

A.18 Find the specific volume of saturated liquid water when the pressure is 180 kPa by interpolating the values found in appendix Table B.1.2.

A.19 Determine the enthalpy of compressed liquid water at 3200 psia, 150° F, by interpolating the values found in appendix Table B.1.4.

A.20 Solve the following equation for the temperature, T:

$$2\sqrt{T} + \ln T = 1$$

A.21 Solve the following equation for the pressure, P:

$$\sqrt{P} + 2e^P = 4$$

A.22 Find the values of x, y, z, and the Lagrange multiplier for the extrema of the function, $x^2 + 3y = z$, when $x = y^2$.

A.23 Find the values of x, y, z, and the Lagrange multipliers for the extrema of the function, $x^2 + \sin y = 2z$, when $z = 3x$ and $\sin y = 1$.

Property Tables
(SI Units)

TABLE B1.1 (SI UNITS)
Properties of Saturated Water
Primary Argument: Temperature
v in cubic meters per kilogram
u and h in kiloJoules per kilogram
s in kiloJoules per kilogram and Kelvin

Temp. °C T	Pres. kPa P	Specific Volume		Internal Energy		Enthalpy			Entropy	
		Sat. Liq. v_f	Sat. Vap. v_g	Sat. Liq. u_f	Sat. Vap. u_g	Sat. Liq. h_f	h_{fg}	Sat. Vap. h_g	Sat. Liq. s_f	Sat. Vap. s_g
0	0.611	0.001000	206.3	−0.03	2375.3	−0.02	2501.4	2501.3	−0.0001	9.1565
4	0.813	0.001000	157.2	16.77	2380.9	16.78	2491.9	2508.7	0.0610	9.0514
5	0.872	0.001000	147.1	20.97	2382.3	20.98	2489.6	2510.6	0.0761	9.0257
6	0.935	0.001000	137.7	25.19	2383.6	25.20	2487.2	2512.4	0.0912	9.0003
8	1.072	0.001000	120.9	33.59	2386.4	33.60	2482.5	2516.1	0.1212	8.9501
10	1.228	0.001000	106.4	42.00	2389.2	42.01	2477.7	2519.8	0.1510	8.9008
11	1.312	0.001000	99.86	46.20	2390.5	46.20	2475.4	2521.6	0.1658	8.8765
12	1.402	0.001001	93.78	50.41	2391.9	50.41	2473.0	2523.4	0.1806	8.8524
13	1.497	0.001001	88.12	54.60	2393.3	54.60	2470.7	2525.3	0.1953	8.8285
14	1.598	0.001001	82.85	58.79	2394.7	58.80	2468.3	2527.1	0.2099	8.8048
15	1.705	0.001001	77.93	62.99	2396.1	62.99	2465.9	2528.9	0.2245	8.7814
16	1.818	0.001001	73.33	67.18	2397.4	67.19	2463.6	2530.8	0.2390	8.7582
17	1.938	0.001001	69.04	71.38	2398.8	71.38	2461.2	2532.6	0.2535	8.7351
18	2.064	0.001001	65.04	75.57	2400.2	75.58	2458.8	2534.4	0.2679	8.7123
19	2.198	0.001002	61.29	79.76	2401.6	79.77	2456.5	2536.2	0.2823	8.6897
20	2.339	0.001002	57.79	83.95	2402.9	83.96	2454.1	2538.1	0.2966	8.6672
21	2.487	0.001002	54.51	88.14	2404.3	88.14	2451.8	2539.9	0.3109	8.6450
22	2.645	0.001002	51.45	92.32	2405.7	92.33	2449.4	2541.7	0.3251	8.6229
23	2.810	0.001002	48.57	96.51	2407.0	96.52	2447.0	2543.5	0.3393	8.6011
24	2.985	0.001003	45.88	100.70	2408.4	100.70	2444.7	2545.4	0.3534	8.5794
25	3.169	0.001003	43.36	104.88	2409.8	104.89	2442.3	2547.2	0.3674	8.5580
26	3.363	0.001003	40.99	109.06	2411.1	109.07	2439.9	2549.0	0.3814	8.5367
27	3.567	0.001004	38.77	113.25	2412.5	113.25	2437.6	2550.8	0.3954	8.5156
28	3.782	0.001004	36.69	117.42	2413.9	117.43	2435.2	2552.6	0.4093	8.4946
29	4.008	0.001004	34.73	121.60	2415.2	121.61	2432.8	2554.5	0.4231	8.4739
30	4.246	0.001004	32.89	125.78	2416.6	125.79	2430.5	2556.3	0.4369	8.4533
31	4.496	0.001005	31.17	129.96	2418.0	129.97	2428.1	2558.1	0.4507	8.4329
32	4.759	0.001005	29.54	134.14	2419.3	134.15	2425.7	2559.9	0.4644	8.4127
33	5.034	0.001005	28.01	138.32	2420.7	138.33	2423.4	2561.7	0.4781	8.3927
34	5.324	0.001006	26.57	142.50	2422.0	142.50	2421.0	2563.5	0.4917	8.3728
35	5.628	0.001006	25.22	146.67	2423.4	146.68	2418.6	2565.3	0.5053	8.3531
36	5.947	0.001006	23.94	150.85	2424.7	150.86	2416.2	2567.1	0.5188	8.3336
38	6.632	0.001007	21.60	159.20	2427.4	159.21	2411.5	2570.7	0.5458	8.2950
40	7.384	0.001008	19.52	167.56	2430.1	167.57	2406.7	2574.3	0.5725	8.2570
45	9.593	0.001010	15.26	188.44	2436.8	188.45	2394.8	2583.2	0.6387	8.1648

continued

TABLE B1.1 (SI UNITS) *continued*
Properties of Saturated Water
Primary Argument: Temperature
v in cubic meters per kilogram
u and h in kiloJoules per kilogram
s in kiloJoules per kilogram and Kelvin

Temp. °C T	Pres. kPa P	Specific Volume		Internal Energy		Enthalpy			Entropy	
		Sat. Liq. v_f	Sat. Vap. v_g	Sat. Liq. u_f	Sat. Vap. u_g	Sat. Liq. h_f	h_{fg}	Sat. Vap. h_g	Sat. Liq. s_f	Sat. Vap. s_g
50	12.35	0.001012	12.03	209.32	2443.5	209.33	2382.7	2592.1	0.7038	8.0763
55	15.76	0.001015	9.57	230.21	2450.1	230.23	2370.7	2600.9	0.7679	7.9913
60	19.94	0.001017	7.671	251.11	2456.6	251.13	2358.5	2609.6	0.8312	7.9096
65	25.03	0.001020	6.197	272.02	2463.1	272.06	2346.2	2618.3	0.8935	7.8310
70	31.19	0.001023	5.042	292.95	2469.6	292.98	2333.8	2626.8	0.9549	7.7553
75	38.58	0.001026	4.131	313.90	2475.9	313.93	2321.4	2635.3	1.0155	7.6824
80	47.39	0.001029	3.407	334.86	2482.2	334.91	2308.8	2643.7	1.0753	7.6122
85	57.83	0.001033	2.828	355.84	2488.4	355.90	2296.0	2651.9	1.1343	7.5445
90	70.14	0.001036	2.361	376.85	2494.5	376.92	2283.2	2660.1	1.1925	7.4791
95	84.55	0.001040	1.982	397.88	2500.6	397.96	2270.2	2668.1	1.2500	7.4159
100	101.4	0.001044	1.673	418.94	2506.5	419.04	2257.0	2676.1	1.3069	7.3549
110	143.3	0.001052	1.210	461.14	2518.1	461.30	2230.2	2691.5	1.4185	7.2387
120	198.5	0.001060	0.8919	503.50	2529.3	503.71	2202.6	2706.3	1.5276	7.1296
130	270.1	0.001070	0.6685	546.02	2539.9	546.31	2174.2	2720.5	1.6344	7.0269
140	361.3	0.001080	0.5089	588.74	2550.0	589.13	2144.7	2733.9	1.7391	6.9299
150	475.8	0.001091	0.3928	631.68	2559.5	632.20	2114.3	2746.5	1.8418	6.8379
160	617.8	0.01102	0.3071	674.87	2568.4	675.55	2082.6	2758.1	1.9427	6.7502
170	791.7	0.001114	0.2428	718.33	2576.5	719.21	2049.5	2768.7	2.0419	6.6663
180	1002	0.001127	0.1941	762.09	2583.7	763.22	2015.0	2778.2	2.1396	6.5857
190	1254	0.001141	0.1565	806.19	2590.0	807.62	1978.8	2786.4	2.2359	6.5079
200	1554	0.001157	0.1274	850.65	2595.3	852.45	1940.7	2793.2	2.3309	6.4323
210	1906	0.001173	0.1044	895.53	2599.5	897.76	1900.7	2798.5	2.4248	6.3585
220	2318	0.001190	0.08619	940.87	2602.4	943.62	1858.5	2802.1	2.5178	6.2861
230	2795	0.001209	0.07158	986.74	2603.9	990.12	1813.8	2804.0	2.6099	6.2146
240	3344	0.001229	0.05976	1033.2	2604.0	1037.3	1766.5	2803.8	2.7015	6.1437
250	3973	0.001251	0.05030	1080.4	2602.4	1085.4	1716.2	2801.5	2.7927	6.0730
260	4688	0.001276	0.04221	1128.4	2599.0	1134.4	1662.5	2796.9	2.8838	6.0019
270	5499	0.001302	0.03564	1177.4	2593.7	1184.5	1605.2	2789.7	2.9751	5.9301
280	6412	0.001332	0.03017	1227.5	2586.1	1236.0	1543.6	2779.6	3.0668	5.8571
290	7436	0.001366	0.02557	1278.9	2576.0	1289.1	1477.1	2766.2	3.1594	5.7821
300	8581	0.001404	0.02167	1332.0	2563.0	1344.0	1404.9	2749.0	3.2534	5.7045
320	11274	0.001499	0.01549	1444.6	2525.5	1461.5	1238.6	2700.1	3.4480	5.5362
340	14586	0.001638	0.01080	1570.3	2464.6	1594.2	1027.9	2622.0	3.6594	5.3357
360	18651	0.001893	0.006945	1725.2	2351.5	1760.5	720.5	2481.0	3.9147	5.0526
374.14	22090	0.003155	0.003155	2029.6	2029.6	2099.3	0.0	2099.3	4.4298	4.4298

Abridged from J. H. Keenan, F. G. Keyes, P. G. Hill, and J. G. Moore, "Steam Tables," Wiley, New York, 1979. Reprinted by permission of John Wiley & Sons, Inc.

TABLE B1.2 (SI UNITS)

Properties of Saturated Water
Primary Argument: Pressure
v in cubic meters per kilogram
u and h in kiloJoules per kilogram
s in kiloJoules per kilogram and Kelvin

		Specific Volume		Internal Energy		Enthalpy			Entropy	
Pres. kPa P	Temp. °C T	Sat. Liq. v_f	Sat. Vap. v_g	Sat. Liq. u_f	Sat. Vap. u_g	Sat. Liq. h_f	h_{fg}	Sat. Vap. h_g	Sat. Liq. s_f	Sat. Vap. s_g
4	28.96	0.001004	34.80	121.45	2415.2	121.46	2432.9	2554.4	0.4226	8.4746
6	36.16	0.001006	23.74	151.53	2425.0	151.53	2415.9	2567.4	0.5210	8.3304
8	41.51	0.001008	18.10	173.87	2432.2	173.88	2403.1	2577.0	0.5926	8.2287
10	45.81	0.001010	14.67	191.82	2437.9	191.83	2392.8	2584.7	0.6493	8.1502
20	60.06	0.001017	7.649	251.38	2456.7	251.40	2358.3	2609.7	0.8320	7.9085
30	69.10	0.001022	5.229	289.20	2468.4	289.23	2336.1	2625.3	0.9439	7.7686
40	75.87	0.001027	3.993	317.53	2477.0	317.58	2319.2	2636.8	1.0259	7.6700
50	81.33	0.001030	3.240	340.44	2483.9	340.49	2305.4	2645.9	1.0910	7.5939
60	85.94	0.001033	2.732	359.79	2489.6	359.86	2293.6	2653.5	1.1453	7.5320
70	89.95	0.001036	2.365	376.63	2494.5	376.70	2283.3	2660.0	1.1919	7.4797
80	93.50	0.001039	2.087	391.58	2498.8	391.66	2274.1	2665.8	1.2329	7.4346
90	96.71	0.001041	1.869	405.06	2502.6	405.15	2265.7	2670.9	1.2695	7.3949
100	99.63	0.001043	1.694	417.36	2506.1	417.46	2258.0	2675.5	1.3026	7.3594
150	111.4	0.001053	1.159	466.94	2519.7	467.11	2226.5	2693.6	1.4336	7.2233
200	120.2	0.001061	0.8857	504.49	2529.5	504.70	2201.9	2706.7	1.5301	7.1271
250	127.4	0.001067	0.7187	535.10	2537.2	535.37	2181.5	2716.9	1.6072	7.0527
300	133.6	0.001073	0.6058	561.15	2543.6	561.47	2163.8	2725.3	1.6718	6.9919
350	138.9	0.001079	0.5243	583.95	2548.9	584.33	2148.1	2732.4	1.7275	6.9405
400	143.6	0.001084	0.4625	604.31	2553.6	604.74	2133.8	2738.6	1.7766	6.8959
450	147.9	0.001088	0.4140	622.77	2557.6	623.25	2120.7	2743.9	1.8207	6.8565
500	151.9	0.001093	0.3749	639.68	2561.2	640.23	2108.5	2748.7	1.8607	6.8213
600	158.9	0.001101	0.3157	669.90	2567.4	670.56	2086.3	2756.8	1.9312	6.7600
700	165.0	0.001108	0.2729	696.44	2572.5	697.22	2066.3	2763.5	1.9922	6.7080
800	170.4	0.001115	0.2404	720.22	2576.8	721.11	2048.0	2769.1	2.0462	6.6628
900	175.4	0.001121	0.2150	741.83	2580.5	742.83	2031.1	2773.9	2.0946	6.6226
1000	179.9	0.001127	0.1944	761.68	2583.6	762.81	2015.3	2778.1	2.1387	6.5865
1500	198.3	0.001154	0.1318	843.16	2594.5	844.89	1947.3	2792.2	2.3150	6.4448
2000	212.4	0.001177	0.0996	906.44	2600.3	908.79	1890.7	2799.5	2.4474	6.3409
2500	224.0	0.001197	0.07998	959.11	2603.1	962.11	1841.0	2803.1	2.5547	6.2575
3000	233.9	0.001217	0.06668	1004.8	2604.1	1008.4	1795.7	2804.2	2.6457	6.1869
3500	242.6	0.001235	0.05707	1045.4	2603.7	1049.8	1753.7	2803.4	2.7253	6.1253
4000	250.4	0.001252	0.04978	1082.3	2602.3	1087.3	1714.1	2801.4	2.7964	6.0701
4500	257.5	0.001269	0.04406	1116.2	2600.1	1121.9	1676.4	2798.3	2.8610	6.0199
5000	264.0	0.001286	0.03944	1147.8	2597.1	1154.2	1640.1	2794.3	2.9202	5.9734
6000	275.6	0.001319	0.03244	1205.4	2589.7	1213.4	1571.0	2784.3	3.0267	5.8892

continued

TABLE B1.2 (SI UNITS) *continued*
Properties of Saturated Water
Primary Argument: Pressure
v in cubic meters per kilogram
u and h in kiloJoules per kilogram
s in kiloJoules per kilogram and Kelvin

Pres. kPa P	Temp. °C T	Specific Volume		Internal Energy		Enthalpy			Entropy	
		Sat. Liq. v_f	Sat. Vap. v_g	Sat. Liq. u_f	Sat. Vap. u_g	Sat. Liq. h_f	h_{fg}	Sat. Vap. h_g	Sat. Liq. s_f	Sat. Vap. s_g
7000	285.9	0.001351	0.02737	1257.6	2580.5	1267.0	1505.1	2772.1	3.1211	5.8133
8000	295.1	0.001384	0.02352	1305.6	2569.8	1316.6	1441.3	2758.0	3.2068	5.7432
9000	303.4	0.001418	0.02048	1350.5	2557.8	1363.3	1378.9	2742.1	3.2858	5.6772
10000	311.1	0.001452	0.01803	1393.0	2544.4	1407.6	1317.1	2724.7	3.3596	5.6141
11000	318.2	0.001489	0.01599	1433.7	2529.8	1450.1	1255.5	2705.6	3.4295	5.5527
12000	324.8	0.001527	0.01426	1473.0	2513.7	1491.3	1193.6	2684.9	3.4962	5.4924
13000	330.9	0.001567	0.01278	1511.1	2496.1	1531.5	1130.7	2662.2	3.5606	5.4323
14000	336.8	0.001611	0.01149	1548.6	2476.8	1571.1	1066.5	2637.6	3.6232	5.3717
15000	342.2	0.001658	0.01034	1585.6	2455.5	1610.5	1000.0	2610.5	3.6848	5.3098
16000	347.4	0.001711	0.009306	1622.7	2431.7	1650.1	930.6	2580.6	3.7461	5.2455
17000	352.4	0.001770	0.008364	1660.2	2405.0	1690.3	856.9	2547.2	3.8079	5.1777
18000	357.1	0.001840	0.007489	1698.9	2374.3	1732.0	777.1	2509.1	3.8715	5.1044
19000	361.5	0.001924	0.006657	1739.9	2338.1	1776.5	688.0	2464.5	3.9388	5.0228
20000	365.8	0.002036	0.005834	1785.6	2293.0	1826.3	583.4	2409.7	4.0139	4.9269
22090	374.1	0.003155	0.003155	2029.6	2029.6	2099.3	0.0	2099.3	4.4298	4.4298

Abridged from J. H. Keenan, F. G. Keyes, P. G. Hill, and J. G. Moore, "Steam Tables," Wiley, New York, 1979. Reprinted by permission of John Wiley & Sons, Inc.

TABLE B1.3 (SI UNITS)

Properties of Water Vapor
v in cubic meters per kilogram
u and h in kiloJoules per kilogram
s in kiloJoules per kilogram and Kelvin

Temp °C	Pressure = 6 kPa			
	v	u	h	s
36.16	23.739	2425.0	2567.4	8.3304
80	27.132	2487.3	2650.1	8.5804
120	30.219	2544.7	2726.0	8.7840
160	33.302	2602.7	2802.5	8.9693
200	36.383	2661.4	2879.7	9.1398
240	39.462	2721.0	2957.8	9.2982
280	42.540	2781.5	3036.8	9.4464
320	45.618	2843.0	3116.7	9.5859
360	48.696	2905.5	3197.7	9.7180
400	51.774	2969.0	3279.6	9.8435
440	54.851	3033.5	3362.6	9.9633
500	59.467	3132.3	3489.1	10.134

Temp °C	Pressure = 35 kPa			
	v	u	h	s
72.69	4.526	2473.0	2631.4	7.7158
80	4.625	2483.7	2645.6	7.7564
120	5.163	2542.4	2723.1	7.9644
160	5.696	2601.2	2800.6	8.1519
200	6.228	2660.4	2878.4	8.3237
240	6.758	2720.3	2956.8	8.4828
280	7.287	2780.9	3036.0	8.6314
320	7.815	2842.5	3116.1	8.7712
360	8.344	2905.1	3197.1	8.9034
400	8.872	2968.6	3279.2	9.0291
440	9.400	3033.2	3362.2	9.1490
500	10.192	3132.1	3488.8	9.3194

Temp °C	Pressure = 70 kPa			
	v	u	h	s
89.95	2.365	2494.5	2660.0	7.4797
100	2.434	2509.7	2680.0	7.5341
120	2.571	2539.7	2719.6	7.6375
160	2.841	2599.4	2798.2	7.8279
200	3.108	2659.1	2876.7	8.0012
240	3.374	2719.3	2955.5	8.1611
280	3.640	2780.2	3035.0	8.3102
320	3.905	2842.0	3115.3	8.4504
360	4.170	2904.6	3196.5	8.5828
400	4.434	2968.2	3278.6	8.7086
440	4.698	3032.9	3361.8	8.8286
500	5.095	3131.8	3488.5	8.9991

Temp °C	Pressure = 100 kPa			
	v	u	h	s
99.63	1.694	2506.1	2675.5	7.3594
100	1.696	2506.7	2676.2	7.3614
120	1.793	2537.3	2716.6	7.4668
160	1.984	2597.8	2796.2	7.6597
200	2.172	2658.1	2875.3	7.8343
240	2.359	2718.5	2954.5	7.9949
280	2.546	2779.6	3034.2	8.1445
320	2.732	2841.5	3114.6	8.2849
360	2.917	2904.2	3195.9	8.4175
400	3.103	2967.9	3278.2	8.5435
440	3.288	3032.6	3361.4	8.6636
500	3.565	3131.6	3488.1	8.8342

continued

TABLE B1.3 (SI UNITS) *continued*
Properties of Water Vapor
v in cubic meters per kilogram
u and h in kiloJoules per kilogram
s in kiloJoules per kilogram and Kelvin

Temp °C	Pressure = 150 kPa			
	v	u	h	s
111.37	1.159	2519.7	2693.6	7.2233
120	1.188	2533.3	2711.4	7.2693
160	1.317	2595.2	2792.8	7.4665
200	1.444	2656.2	2872.9	7.6433
240	1.570	2717.2	2952.7	7.8052
280	1.695	2778.6	3032.8	7.9555
320	1.819	2840.6	3113.5	8.0964
360	1.943	2903.5	3195.0	8.2293
400	2.067	2967.3	3277.4	8.3555
440	2.191	3032.1	3360.7	8.4757
500	2.376	3131.2	3487.6	8.6466
600	2.685	3301.7	3704.3	8.9101

Temp °C	Pressure = 300 kPa			
	v	u	h	s
133.55	0.606	2543.6	2725.3	6.9919
160	0.651	2587.1	2782.3	7.1276
200	0.716	2650.7	2865.6	7.3115
240	0.781	2713.1	2947.3	7.4774
280	0.844	2775.4	3028.6	7.6299
320	0.907	2838.1	3110.1	7.7722
360	0.969	2901.4	3192.2	7.9061
400	1.032	2965.6	3275.0	8.0330
440	1.094	3030.6	3358.7	8.1538
500	1.187	3130.0	3486.0	8.3251
600	1.341	3300.8	3703.2	8.5892

Temp °C	Pressure = 500 kPa			
	v	u	h	s
151.86	0.3749	2561.2	2748.7	6.8213
180	0.4045	2609.7	2812.0	6.9656
200	0.4249	2642.9	2855.4	7.0592
240	0.4646	2707.6	2939.9	7.2307
280	0.5034	2771.2	3022.9	7.3865
320	0.5416	2834.7	3105.6	7.5308
360	0.5796	2898.7	3188.4	7.6660
400	0.6173	2963.2	3271.9	7.7938
440	0.6548	3028.6	3356.0	7.9152
500	0.7109	3128.4	3483.9	8.0873
600	0.8041	3299.6	3701.7	8.3522
700	0.8969	3477.5	3925.9	8.5952

Temp °C	Pressure = 700 kPa			
	v	u	h	s
165	0.2729	2572.5	2763.5	6.7080
180	0.2847	2599.8	2799.1	6.7880
200	0.2999	2634.8	2844.8	6.8865
240	0.3292	2701.8	2932.2	7.0641
280	0.3574	2766.9	3017.1	7.2233
320	0.3852	2831.3	3100.9	7.3697
360	0.4126	2895.8	3184.7	7.5063
400	0.4397	2960.9	3268.7	7.6350
440	0.4667	3026.6	3353.3	7.7571
500	0.5070	3126.8	3481.7	7.9299
600	0.5738	3298.5	3700.2	8.1956
700	0.6403	3476.6	3924.8	8.4391

continued

TABLE B1.3 (SI UNITS) *continued*
Properties of Water Vapor
v in cubic meters per kilogram
u and h in kiloJoules per kilogram
s in kiloJoules per kilogram and Kelvin

Temp °C	Pressure = 1000 kPa			
	v	u	h	s
179.91	0.1944	2583.6	2778.1	6.5865
200	0.2060	2621.9	2827.9	6.6940
240	0.2275	2692.9	2920.4	6.8817
280	0.2480	2760.2	3008.2	7.0465
320	0.2678	2826.1	3093.9	7.1962
360	0.2873	2891.6	3178.9	7.3349
400	0.3066	2957.3	3263.9	7.4651
440	0.3257	3023.6	3349.3	7.5883
500	0.3541	3124.4	3478.5	7.7622
540	0.3729	3192.6	3565.6	7.8720
600	0.4011	3296.8	3697.9	8.0290
640	0.4198	3367.4	3787.2	8.1290

Temp °C	Pressure = 1500 kPa			
	v	u	h	s
198.32	0.1318	2594.5	2792.2	6.4448
200	0.1325	2598.1	2796.8	6.4546
240	0.1483	2676.9	2899.3	6.6628
280	0.1627	2748.6	2992.7	6.8381
320	0.1765	2817.1	3081.9	6.9938
360	0.1899	2884.4	3169.2	7.1363
400	0.2030	2951.3	3255.8	7.2690
440	0.2160	3018.5	3342.5	7.3940
500	0.2352	3120.3	3473.1	7.5698
540	0.2478	3189.1	3560.9	7.6805
600	0.2668	3293.9	3694.0	7.8385
640	0.2793	3364.8	3783.8	7.9391

Temp °C	Pressure = 2000 kPa			
	v	u	h	s
212.42	0.0996	2600.3	2799.5	6.3409
240	0.1085	2659.6	2876.5	6.4952
280	0.1200	2736.4	2976.4	6.6828
320	0.1308	2807.9	3069.5	6.8452
360	0.1411	2877.0	3159.3	6.9917
400	0.1512	2945.2	3247.6	7.1271
440	0.1611	3013.4	3335.5	7.2540
500	0.1757	3116.2	3467.6	7.4317
540	0.1853	3185.6	3556.1	7.5434
600	0.1996	3290.9	3690.1	7.7024
640	0.2091	3362.2	3780.4	7.8035
700	0.2232	3470.9	3917.4	7.9487

Temp °C	Pressure = 3000 kPa			
	v	u	h	s
233.90	0.0667	2604.1	2804.2	6.1869
240	0.0682	2619.7	2824.3	6.2265
280	0.0771	2709.9	2941.3	6.4462
320	0.0850	2788.4	3043.4	6.6245
360	0.0923	2861.7	3138.7	6.7801
400	0.0994	2932.8	3230.9	6.9212
440	0.1062	3002.9	3321.5	7.0520
500	0.1162	3108.0	3456.5	7.2338
540	0.1227	3178.4	3546.6	7.3474
600	0.1324	3285.0	3682.3	7.5085
640	0.1388	3357.0	3773.5	7.6106
700	0.1484	3466.5	3911.7	7.7571

continued

TABLE B1.3 (SI UNITS) *continued*
Properties of Water Vapor
v in cubic meters per kilogram
u and h in kiloJoules per kilogram
s in kiloJoules per kilogram and Kelvin

Temp °C	Pressure = 4000 kPa			
	v	u	h	s
250.40	0.04978	2602.3	2801.4	6.0701
280	0.05546	2680.0	2901.8	6.2568
320	0.06199	2767.4	3015.4	6.4553
360	0.06788	2845.7	3117.2	6.6215
400	0.07341	2919.9	3213.6	6.7690
440	0.07872	2992.2	3307.1	6.9041
500	0.08643	3099.5	3445.3	7.0901
540	0.09145	3171.1	3536.9	7.2056
600	0.09885	3279.1	3674.4	7.3688
640	0.1037	3351.8	3766.6	7.4720
700	0.1110	3462.1	3905.9	7.6198
740	0.1157	3536.6	3999.6	7.7141

Temp °C	Pressure = 6000 kPa			
	v	u	h	s
275.64	0.03244	2589.7	2784.3	5.8892
280	0.03317	2605.2	2804.2	5.9252
320	0.03876	2720.0	2952.6	6.1846
360	0.04331	2811.2	3071.1	6.3782
400	0.04739	2892.9	3177.2	6.5408
440	0.05122	2970.0	3277.3	6.6853
500	0.06565	3082.2	3422.2	6.8803
540	0.06015	3156.1	3517.0	6.9999
600	0.06525	3266.9	3658.4	7.1677
640	0.06859	3341.0	3752.6	7.2731
700	0.07352	3453.1	3894.1	7.4234
740	0.07677	3528.3	3989.2	7.5190

Temp °C	Pressure = 8000 kPa			
	v	u	h	s
295.06	0.02352	2569.8	2758.0	5.7432
320	0.02682	2662.7	2877.2	5.9489
360	0.03089	2772.7	3019.8	6.1819
400	0.03432	2863.8	3138.3	6.3634
440	0.03742	2946.7	3246.1	6.5190
480	0.04034	3025.7	3348.4	6.6586
520	0.04313	3102.7	3447.7	6.7871
560	0.04582	3178.7	3545.3	6.9072
600	0.04845	3254.4	3642.0	7.0206
640	0.05102	3330.1	3738.3	7.1283
700	0.05481	3443.9	3882.4	7.2812
740	0.05729	3520.4	3978.7	7.3782

Temp °C	Pressure = 10000 kPa			
	v	u	h	s
311.06	0.01803	2544.4	2724.7	5.6141
320	0.01925	2588.8	2781.3	5.7103
360	0.02331	2729.1	2962.1	6.0060
400	0.02641	2832.4	3096.5	6.2120
440	0.02911	2922.1	3213.2	6.3805
480	0.03160	3005.4	3321.4	6.5282
520	0.03394	3085.6	3425.1	6.6622
560	0.03619	3164.1	3526.0	6.7864
600	0.03837	3241.7	3625.3	6.9029
640	0.04048	3318.9	3723.7	7.0131
700	0.04358	3434.7	3870.5	7.1687
740	0.04560	3512.1	3968.1	7.2670

continued

TABLE B1.3 (SI UNITS) *continued*
Properties of Water Vapor
v in cubic meters per kilogram
u and h in kiloJoules per kilogram
s in kiloJoules per kilogram and Kelvin

Temp °C	Pressure = 12000 kPa			
	v	u	h	s
324.75	0.01426	2513.7	2684.9	5.4924
360	0.01811	2678.4	2895.7	5.8361
400	0.02108	2798.3	3051.3	6.0747
440	0.02355	2896.1	3178.7	6.2586
480	0.02576	2984.4	3293.5	6.4154
520	0.02781	3068.0	3401.8	6.5555
560	0.02977	3149.0	3506.2	6.6840
600	0.03164	3228.7	3608.3	6.8037
640	0.03345	3307.5	3709.0	6.9164
700	0.03610	3425.2	3858.4	7.0749
740	0.03781	3503.7	3957.4	7.1746

Temp °C	Pressure = 14000 kPa			
	v	u	h	s
336.75	0.01149	2476.8	2637.6	5.3717
360	0.01422	2617.4	2816.5	5.6602
400	0.01722	2760.9	3001.9	5.9448
440	0.01954	2868.6	3142.2	6.1474
480	0.02157	2962.5	3264.5	6.3143
520	0.02343	3049.8	3377.8	6.4610
560	0.02517	3133.6	3486.0	6.5941
600	0.02683	3215.4	3591.1	6.7172
640	0.02843	3296.0	3694.1	6.8326
700	0.03075	3415.7	3846.2	6.9939
740	0.03225	3495.2	3946.7	7.0952

Temp °C	Pressure = 16000 kPa			
	v	u	h	s
347.44	0.00931	2431.7	2580.6	5.2455
360	0.01105	2539.0	2715.8	5.4614
400	0.01426	2719.4	2947.6	5.8175
440	0.01652	2839.4	3103.7	6.0429
480	0.01842	2939.7	3234.4	6.2215
520	0.02013	3031.1	3353.3	6.3752
560	0.02172	3117.8	3465.4	6.5132
600	0.02323	3201.8	3573.5	6.6399
640	0.02467	3284.2	3678.9	6.7580
700	0.02674	3406.0	3833.9	6.9224
740	0.02808	3486.7	3935.9	7.0251

Temp °C	Pressure = 18000 kPa			
	v	u	h	s
357.06	0.00749	2374.3	2509.1	5.1044
360	0.00809	2418.9	2564.5	5.1922
400	0.01190	2672.8	2887.0	5.6887
440	0.01414	2808.2	3062.8	5.9428
480	0.01596	2915.9	3203.2	6.1345
520	0.01757	3011.8	3378.0	6.2960
560	0.01904	3101.7	3444.4	6.4392
600	0.02042	3188.0	3555.6	6.5696
640	0.02174	3272.3	3663.6	6.6905
700	0.02362	3396.3	3821.5	6.8580
740	0.02483	3478.0	3925.0	6.9623

continued

TABLE B1.3 (SI UNITS) *continued*
Properties of Water Vapor
v in cubic meters per kilogram
u and h in kiloJoules per kilogram
s in kiloJoules per kilogram and Kelvin

Temp °C	Pressure = 20000 kPa			
	v	u	h	s
365.81	0.00583	2293.0	2409.7	4.9269
400	0.00994	2619.3	2818.1	5.5540
440	0.01222	2774.9	3019.4	5.8450
480	0.01399	2891.2	3170.8	6.0518
520	0.01551	2992.0	3302.2	6.2218
560	0.01689	3085.2	3423.0	6.3705
600	0.01818	3174.0	3537.6	6.5048
640	0.01940	3260.2	3648.1	6.6286
700	0.02113	3386.4	3809.0	6.7993
740	0.02224	3469.3	3914.1	6.9052
800	0.02385	3592.7	4069.7	7.0544

Temp °C	Pressure = 24000 kPa			
	v	u	h	s
400	0.00673	2477.8	2639.4	5.2393
440	0.00929	2700.6	2923.4	5.6506
480	0.01100	2838.3	3102.3	5.8950
520	0.01241	2950.5	3248.5	6.0842
560	0.01366	3051.1	3379.0	6.2448
600	0.01481	3145.2	3500.7	6.3875
640	0.01588	3235.5	3616.7	6.5174
700	0.01739	3366.4	3783.8	6.6947
740	0.01835	3451.7	3892.1	6.8038
800	0.01974	3578.0	4051.6	6.9567

Temp °C	Pressure = 28000 kPa			
	v	u	h	s
400	0.00383	2223.5	2330.7	4.7494
440	0.00712	2613.2	2812.6	5.4494
480	0.00885	2780.8	3028.5	5.7446
520	0.01020	2906.8	3192.3	5.9566
560	0.01136	3015.7	3333.7	6.1307
600	0.01241	3115.6	3463.0	6.2823
640	0.01338	3210.3	3584.8	6.4187
700	0.01473	3346.1	3758.4	6.6029
740	0.01558	3433.9	3870.0	6.7153
800	0.01680	3563.1	4033.4	6.8720
900	0.01873	3774.3	4298.8	7.1084

Temp °C	Pressure = 32000 kPa			
	v	u	h	s
400	0.00236	1980.4	2055.9	4.3239
440	0.00544	2509.0	2683.0	5.2327
480	0.00722	2718.1	2949.2	5.5968
520	0.00853	2860.7	3133.7	5.8357
560	0.00963	2979.0	3287.2	6.0246
600	0.01061	3085.3	3424.6	6.1858
640	0.01150	3184.5	3552.5	6.3290
700	0.01273	3325.4	3732.8	6.5203
740	0.01350	3415.9	3847.8	6.6361
800	0.01460	3548.0	4015.1	6.7966
900	0.01633	3762.7	4285.1	7.0372

Abridged from J. H. Keenan, F. G. Keyes, P. G. Hill, and J. G. Moore, "Steam Tables," Wiley, New York, 1979. Reprinted by permission of John Wiley & Sons, Inc.

TABLE B1.4 (SI UNITS)

Properties of Compressed Liquid Water
v in cubic meters per kilogram
u and h in kiloJoules per kilogram
s in kiloJoules per kilogram and Kelvin

Temp	Pressure = 2500 kPa				Temp	Pressure = 5000 kPa			
°C	v	u	h	s	°C	v	u	h	s
20	0.001001	83.80	86.30	0.2961	20	0.0009995	83.65	88.65	0.2956
40	0.001007	167.25	169.77	0.5715	40	0.001006	166.95	171.97	0.5705
80	0.001028	334.29	336.86	1.0737	80	0.001027	333.72	338.85	1.0720
100	0.001042	418.24	420.85	1.3050	100	0.001041	417.52	422.72	1.3030
140	0.001078	587.82	590.52	1.7369	140	0.001077	586.76	592.15	1.7343
180	0.001126	761.16	763.97	2.1375	180	0.001124	759.63	765.25	2.1341
200	0.001156	849.9	852.8	2.3294	200	0.001153	848.1	853.9	2.3255
220	0.001190	940.7	943.7	2.5174	220	0.001187	938.4	944.4	2.5128
223.99	0.001197	959.1	962.1	2.5546	263.99	0.001286	1147.8	1154.2	2.9202

Temp	Pressure = 7500 kPa				Temp	Pressure = 10,000 kPa			
°C	v	u	h	s	°C	v	u	h	s
20	0.0009984	83.50	90.99	0.2950	20	0.0009972	83.36	93.33	0.2945
40	0.001005	166.64	174.18	0.5696	40	0.001003	166.35	176.38	0.5686
80	0.001026	333.15	340.84	1.0704	80	0.001025	332.59	342.83	1.0688
100	0.001040	416.81	424.62	1.3011	100	0.001039	416.12	426.50	1.2992
140	0.001075	585.72	593.78	1.7317	140	0.001074	584.68	595.42	1.7292
180	0.001122	758.13	766.55	2.1308	180	0.001120	756.65	767.84	2.1275
220	0.001184	936.2	945.1	2.5083	220	0.001181	934.1	945.9	2.5039
260	0.001270	1124.4	1134.0	2.8763	260	0.001265	1121.1	1133.7	2.8699
290.59	0.001368	1282.0	1292.2	3.1649	311.06	0.001452	1393.0	1407.6	3.3596

continued

TABLE B1.4 (SI UNITS) *continued*
Properties of Compressed Liquid Water
v in cubic meters per kilogram
u and h in kiloJoules per kilogram
s in kiloJoules per kilogram and Kelvin

Temp °C	Pressure = 15,000 kPa			
	v	u	h	s
20	0.0009950	83.06	97.99	0.2934
40	0.001001	165.76	180.78	0.5666
80	0.001022	331.48	346.81	1.0656
100	0.001036	414.74	430.28	1.2955
140	0.001071	582.66	598.72	1.7242
180	0.001116	753.76	770.50	2.1210
220	0.001175	929.9	947.5	2.4953
260	0.001255	1114.6	1133.4	2.8576
300	0.001377	1316.6	1337.3	3.2260
342.24	0.001658	1585.6	1610.5	3.6848

Temp °C	Pressure = 20,000 kPa			
	v	u	h	s
20	0.0099280	82.77	102.62	0.2923
40	0.0099920	165.17	185.16	0.5646
80	0.001020	330.40	350.80	1.0624
100	0.001034	413.39	434.06	1.2917
140	0.001068	580.69	602.04	1.7193
180	0.001112	750.95	773.20	2.1147
220	0.001169	925.9	949.3	2.4870
260	0.001246	1108.6	1133.5	2.8459
300	0.001360	1306.1	1333.3	3.2071
365.81	0.002036	1785.6	1826.3	4.0139

Temp °C	Pressure = 25,000 kPa			
	v	u	h	s
20	0.0099070	82.47	107.24	0.2911
40	0.0099710	164.60	189.52	0.5626
100	0.001031	412.08	437.85	1.2881
200	0.001134	834.5	862.8	2.2961
300	0.001344	1296.6	1330.2	3.1900

Temp °C	Pressure = 30,000 kPa			
	v	u	h	s
20	0.0098860	82.17	111.84	0.2899
40	0.0099510	164.04	193.89	0.5607
100	0.001029	410.78	441.66	1.2844
200	0.001130	831.4	865.3	2.2893
300	0.001330	1287.9	1327.8	3.1741

Abridged from J. H. Keenan, F. G. Keyes, P. G. Hill, and J. G. Moore, "Steam Tables," Wiley, New York, 1979. Reprinted by permission of John Wiley & Sons, Inc.

TABLE B2.1 (SI UNITS)

Properties of Saturated Refrigerant-12
Primary Argument: Temperature
v in cubic meters per kilogram
u and h in kiloJoules per kilogram
s in kiloJoules per kilogram and Kelvin

Temp. °C T	Pres. kPa P	Specific Volume		Internal Energy		Enthalpy			Entropy	
		Sat. Liq. v_f	Sat. Vap. v_g	Sat Liq. u_f	Sat. Vap. u_g	Sat. Liq. h_f	h_{fg}	Sat. Vap. u_g	Sat. Liq. s_f	Sat. Vap. s_g
−100	1.2	0.0005991	10.115	112.09	294.05	112.09	193.85	305.94	0.6009	1.7204
−90	2.8	0.0006078	4.4211	120.64	297.85	120.64	189.76	310.40	0.6489	1.6849
−80	6.2	0.0006169	2.1408	129.20	301.76	129.20	185.76	314.96	0.6944	1.6561
−70	12.3	0.0006266	1.1285	137.79	305.75	137.79	181.79	319.58	0.7377	1.6326
−60	22.6	0.0006369	0.63847	146.43	324.22	146.45	177.78	324.23	0.7793	1.6134
−50	39.1	0.0006478	0.38340	155.13	313.89	155.15	173.74	328.89	0.8192	1.5978
−40	64.1	0.0006594	0.24208	163.93	333.51	163.93	169.60	333.53	0.8577	1.5851
−30	100.3	0.0006720	0.15948	172.80	338.12	172.80	165.34	338.14	0.8948	1.5748
−20	150.8	0.0006854	0.10891	181.75	326.25	181.75	160.93	342.68	0.9309	1.5666
−10	219.0	0.0007000	0.076692	190.81	330.35	190.81	156.33	347.14	0.9659	1.5599
0	308.4	0.0007159	0.055426	199.77	351.48	200.00	151.51	351.50	1.0000	1.5547
10	423.1	0.0007332	0.040934	209.36	338.41	209.36	146.37	335.73	1.0334	1.5504
20	567.0	0.0007524	0.030794	218.89	342.34	218.89	140.92	359.81	1.0663	1.5470
30	744.6	0.0007738	0.023519	228.65	346.18	228.65	135.05	363.70	1.0988	1.5443
40	960.3	0.0007980	0.018181	238.71	349.90	238.71	128.65	367.35	1.1311	1.5419
50	1218.9	0.0008257	0.014176	248.18	370.69	249.18	121.53	370.71	1.1636	1.5397
60	1525.4	0.0008581	0.011117	260.14	357.74	260.14	113.55	373.69	1.1965	1.5374
70	1885.2	0.0008971	0.008729	270.11	359.70	271.81	104.35	376.15	1.2304	1.5344
80	2303.9	0.0009460	0.006825	284.42	377.86	284.42	93.46	377.88	1.2658	1.5304
90	2787.7	0.0010117	0.005261	295.63	378.45	298.45	80.01	378.46	1.3040	1.5243
100	3343.2	0.0011129	0.003906	314.98	376.85	314.98	61.87	376.86	1.3476	1.5134
111.6	4010.0	0.0017920	0.0017920	340.00	340.00	347.37	0.00	347.37	1.4267	1.4267

Source: J. R. Howell, and R. O. Backius, "Fundamentals of Engineering Thermodynamics," McGraw-Hill, New York, 1987. Generated from Program REFRIG, Wiley Professional Software, Wiley, New York, 1985. Reproduced with permission of McGraw-Hill.

TABLE B2.2 (SI UNITS)
Properties of Refrigerant-12 Vapor
Saturated Liquid and Vapor States are in Italics
v in cubic meters per kilogram
u and h in kiloJoules per kilogram
s in kiloJoules per kilogram and Kelvin

Temp °C	Pressure = 20 kPa $T_{sat} = -62.1°$ C			
	v	u	h	s
Liq	*0.0006347*	*144.62*	*144.64*	*0.7708*
Vap	*0.7154*	*308.95*	*323.25*	*1.6171*
−50	0.7585	314.36	329.53	1.6460
−25	0.8466	326.09	343.02	1.7032
0	0.9339	338.52	357.20	1.7576
25	1.0209	351.60	372.02	1.8096
50	1.1075	365.29	387.44	1.8592
75	1.1940	379.53	403.41	1.9068
100	1.2804	394.28	419.89	1.9525
125	1.3666	409.49	436.83	1.9964
150	1.4528	425.13	454.18	2.0387
175	1.5390	441.14	471.92	2.0794
200	1.6251	457.51	490.01	2.1187

Temp °C	Pressure = 40 kPa $T_{sat} = -49.6°$ C			
	v	u	h	s
Liq	*0.0006483*	*155.50*	*155.53*	*0.8209*
Vap	*0.3755*	*314.07*	*329.09*	*1.5972*
−25	0.4200	325.74	342.53	1.6542
0	0.4645	338.26	356.83	1.7091
25	0.5084	351.40	371.74	1.7612
50	0.5521	365.13	387.22	1.8111
75	0.5956	379.40	403.23	1.8588
100	0.6390	394.17	419.73	1.9046
125	0.6823	409.40	436.69	1.9485
150	0.7255	425.05	454.07	1.9909
175	0.7687	441.07	471.82	2.0316
200	0.8119	457.44	489.92	2.0709

Temp °C	Pressure = 60 kPa $T_{sat} = -41.4°$ C			
	v	u	h	s
Liq	*0.0006578*	*162.66*	*162.69*	*0.8523*
Vap	*0.2575*	*317.43*	*332.88*	*1.5867*
−25	0.2777	325.38	342.04	1.6249
0	0.3079	337.99	356.46	1.6802
25	0.3376	351.19	371.45	1.7327
50	0.3670	364.97	386.99	1.7827
75	0.3962	379.27	403.04	1.8306
100	0.4252	394.06	419.58	1.8764
125	0.4542	409.31	436.56	1.9205
150	0.4831	424.96	453.95	1.9628
175	0.5120	441.00	471.72	2.0036
200	0.5408	457.38	489.83	2.0429

Temp °C	Pressure = 80 kPa $T_{sat} = -35.2°$ C			
	v	u	h	s
Liq	*0.0006654*	*168.14*	*168.19*	*0.8757*
Vap	*0.1970*	*320.00*	*335.76*	*1.5799*
−25	0.2066	325.01	341.54	1.6037
0	0.2296	337.72	356.09	1.6595
25	0.2522	350.99	371.16	1.7122
50	0.2744	364.81	386.76	1.7625
75	0.2965	379.14	402.86	1.8104
100	0.3183	393.95	419.42	1.8564
125	0.3402	409.21	436.43	1.9005
150	0.3619	424.89	453.84	1.9429
175	0.3836	440.93	471.62	1.9837
200	0.4053	457.31	489.74	2.0230

continued

TABLE B2.2 (SI UNITS) *continued*
Properties of Refrigerant-12 Vapor
Saturated Liquid and Vapor States are in Italics
v in cubic meters per kilogram
u and h in kiloJoules per kilogram
s in kiloJoules per kilogram and Kelvin

Temp °C	Pressure = 100 kPa $T_{sat} = -30.1°C$			
	v	u	h	s
Liq	*0.0006719*	*172.66*	*172.73*	*0.8945*
Vap	*0.1600*	*322.10*	*338.10*	*1.5749*
−25	0.1639	324.65	341.04	1.5868
0	0.1826	337.45	355.71	1.6432
25	0.2009	350.78	370.87	1.6962
50	0.2188	364.64	386.53	1.7466
75	0.2366	379.01	402.67	1.7947
100	0.2542	393.84	419.27	1.8408
125	0.2717	409.12	436.29	1.8849
150	0.2892	424.80	453.72	1.9273
175	0.3066	440.85	471.51	1.9682
200	0.3240	457.25	489.64	2.0075

Temp °C	Pressure = 200 kPa $T_{sat} = -12.5°C$			
	v	u	h	s
Liq	*0.0006962*	*188.38*	*188.52*	*0.9572*
Vap	*0.0835*	*329.32*	*346.03*	*1.5615*
0	0.0886	336.04	353.76	1.5904
25	0.0983	349.72	369.38	1.6451
50	0.1077	363.81	385.35	1.6965
75	0.1169	378.34	401.71	1.7452
100	0.1259	393.29	418.47	1.7917
125	0.1348	408.65	435.62	1.8362
150	0.1437	424.39	453.13	1.8788
175	0.1525	440.49	471.00	1.9198
200	0.1613	456.92	489.18	1.9593

Temp °C	Pressure = 400 kPa $T_{sat} = 8.2°C$			
	v	u	h	s
Liq	*0.0007299*	*207.34*	*207.63*	*1.0274*
Vap	*0.0432*	*337.68*	*354.97*	*1.5511*
25	0.0469	347.46	366.22	1.5900
50	0.0521	362.08	382.91	1.6436
75	0.0570	376.96	399.75	1.6938
100	0.0617	392.15	416.85	1.7412
125	0.0664	407.69	434.24	1.7863
150	0.0710	423.56	451.94	1.8294
175	0.0755	439.76	469.95	1.8707
200	0.0800	456.26	488.25	1.9104

Temp °C	Pressure = 600 kPa $T_{sat} = 22.0°C$			
	v	u	h	s
Liq	*0.0007565*	*220.38*	*220.84*	*1.0729*
Vap	*0.0291*	*343.13*	*360.61*	*1.5465*
25	0.0296	344.97	362.74	1.5536
50	0.0335	360.23	380.30	1.6101
75	0.0370	375.52	397.70	1.6619
100	0.0403	390.98	415.17	1.7103
125	0.0436	406.71	432.84	1.7561
150	0.0467	422.72	450.74	1.7997
175	0.0498	439.01	468.89	1.8413
200	0.0529	455.59	487.31	1.8813

continued

TABLE B2.2 (SI UNITS) *continued*
Properties of Refrigerant-12 Vapor
Saturated Liquid and Vapor States are in Italics
v in cubic meters per kilogram
u and h in kiloJoules per kilogram
s in kiloJoules per kilogram and Kelvin

Temp °C	Pressure = 800 kPa T_{sat} = 32.8° C				Temp °C	Pressure = 1000 kPa T_{sat} = 41.7° C			
	v	u	h	s		v	u	h	s
Liq	*0.0007802*	*230.76*	*231.39*	*1.1077*	*Liq*	*0.0008023*	*239.60*	*240.40*	*1.1365*
Vap	*0.0219*	*347.22*	*364.73*	*1.5436*	*Vap*	*0.0174*	*350.49*	*367.93*	*1.5416*
50	0.0241	358.25	377.50	1.5842	50	0.0184	356.09	374.45	1.5620
75	0.0269	374.00	395.55	1.6379	75	0.0209	372.41	393.28	1.6180
100	0.0296	398.77	413.45	1.6874	100	0.0231	388.52	411.65	1.6688
125	0.0321	405.70	431.40	1.7339	125	0.0253	404.67	429.92	1.7162
150	0.0346	421.86	449.51	1.7780	150	0.0273	420.98	448.26	1.7607
175	0.0370	438.26	467.82	1.8200	175	0.0292	437.50	466.74	1.8031
200	0.0393	454.92	486.36	1.8602	200	0.0312	454.24	485.40	1.8436

Temp °C	Pressure = 2000 kPa T_{sat} = 72.9° C				Temp °C	Pressure = 4000 kPa T_{sat} = 110.3° C			
	v	u	h	s		v	u	h	s
Liq	*0.0009100*	*374.93*	*376.75*	*1.5335*	*Liq*	*0.001384*	*334.75*	*340.29*	*1.4126*
Vap	*0.00814*	*360.47*	*376.75*	*1.5335*	*Vap*	*0.002408*	*357.31*	*366.94*	*1.4821*
75	0.00831	362.23	378.86	1.5395					
100	0.01003	381.30	401.37	1.6017	110.3	0.00024	365.98	366.94	1.4821
125	0.01142	399.02	421.86	1.6547	125	0.00406	382.49	398.71	1.5633
150	0.01265	416.31	441.61	1.7027	150	0.00517	404.82	425.51	1.6281
175	0.01379	433.50	461.08	1.7472	175	0.00600	424.38	448.38	1.6803
200	0.01488	450.73	480.48	1.7893	200	0.00671	443.05	469.89	1.7267

Source: J. R. Howell, and R. O. Backius, "Fundamentals of Engineering Thermodynamics," McGraw-Hill, New York, 1987. Generated from Program REFRIG, Wiley Professional Software, Wiley, New York, 1985. Reproduced with permission of McGraw-Hill.

TABLE B3.1 (SI UNITS) ·
Properties of Saturated Ammonia
Primary Argument: Temperature
v in cubic meters per kilogram
h in kiloJoules per kilogram
s in kiloJoules per kilogram and Kelvin

Temp. °C T	Pres. kPa P	Specific Volume		Enthalpy			Entropy	
		Sat. Liq. v_f	Sat. Vap. v_g	Sat Liq. h_f	h_{fg}	Sat. Vap. h_g	Sat. Liq. s_f	Sat. Vap. s_g
−50	40.7	0.001424	2.6371	−33.6	1425.8	1392.3	0.0566	6.4461
−46	51.4	0.001434	2.1206	−13.4	1412.3	1398.9	0.1461	6.3635
−42	64.3	0.001444	1.7207	6.1	1399.2	1405.3	0.2312	6.2845
−38	79.7	0.001454	1.4075	25.4	1386.2	1411.6	0.3140	6.2089
−34	97.9	0.001465	1.1604	44.2	1373.6	1417.7	0.3928	6.1365
−30	119.5	0.001476	0.9637	62.6	1361.1	1423.7	0.4692	6.0670
−26	144.6	0.001487	0.8058	80.9	1348.5	1429.5	0.5437	6.0001
−22	173.9	0.001498	0.6780	99.1	1335.9	1435.0	0.6167	5.9358
−18	207.7	0.001509	0.5739	117.3	1323.1	1440.4	0.6884	5.8739
−14	246.5	0.001521	0.4885	135.6	1310.0	1445.6	0.7590	5.8141
−10	290.9	0.001534	0.4180	153.8	1296.7	1450.5	0.8288	5.7564
−6	341.3	0.001546	0.3595	172.2	1283.1	1455.3	0.8978	5.7005
−2	398.4	0.001559	0.3106	190.7	1269.1	1459.8	0.9661	5.6465
2	462.7	0.001573	0.2695	209.3	1254.8	1464.0	1.0337	5.5941
6	534.8	0.001586	0.2348	228.0	1240.1	1468.1	1.1008	5.5432
10	615.3	0.001601	0.2054	246.8	1225.0	1471.8	1.1672	5.4937
14	704.9	0.001615	0.1803	265.7	1209.6	1475.3	1.2331	5.4455
18	804.1	0.001631	0.1588	284.7	1193.7	1478.5	1.2984	5.3984
22	913.8	0.001646	0.1403	303.9	1177.5	1481.3	1.3630	5.3524
26	1034.5	0.001663	0.1243	323.1	1160.8	1483.9	1.4271	5.3074
30	1167.1	0.001680	0.1105	342.4	1143.7	1486.1	1.4906	5.2632
34	1312.2	0.001698	0.0984	361.9	1126.1	1488.0	1.5536	5.2198
38	1470.5	0.001716	0.0879	381.5	1108.0	1489.4	1.6160	5.1769
42	1642.9	0.001735	0.0787	401.2	1089.3	1490.5	1.6780	5.1346
46	1830.2	0.001755	0.0706	421.1	1070.1	1491.2	1.7396	5.0926
50	2033.1	0.001776	0.0634	441.1	1050.3	1491.3	1.8009	5.0509

Source: J. R. Howell, and R. O. Backius, "Fundamentals of Engineering Thermodynamics," McGraw-Hill, New York, 1987. Generated from Program REFRIG, Wiley Professional Software, Wiley, New York, 1985. Reproduced with permission of McGraw-Hill.

TABLE B3.2 (SI UNITS)

Properties of Ammonia Vapor
Saturated Liquid and Vapor States are in Italics
v in cubic meters per kilogram
h in kiloJoules per kilogram
s in kiloJoules per kilogram and Kelvin

Temp °C	Pressure = 50 kPa T_sat = −46.5° C			Temp °C	Pressure = 100 kPa T_sat = −33.6° C		
	v	h	s		v	h	s
Liq	*0.001433*	*−15.8*	*0.1356*	*Liq*	*0.001466*	*46.1*	*0.4008*
Vap	*2.1759*	*1398.1*	*6.3732*	*Vap*	*1.1381*	*1418.4*	*6.1292*
−20	2.4464	1454.3	6.6077				
−10	2.5471	1475.5	6.6898	−10	1.2622	1470.5	6.3369
0	2.6474	1496.6	6.7687	0	1.3137	1492.4	6.4184
10	2.7472	1517.8	6.8449	10	1.3647	1514.1	6.4966
20	2.8466	1539.1	6.9187	20	1.4153	1535.8	6.5719
30	2.9458	1560.4	6.9902	30	1.4657	1557.5	6.6447
40	3.0447	1581.8	7.0596	40	1.5158	1579.3	6.7152
50	3.1435	1603.3	7.1273	50	1.5658	1601.0	6.7837
60	3.2417	1624.9	7.1931	60	1.6153	1622.9	6.8502
80	3.4389	1668.6	7.3205	80	1.7148	1666.9	6.9786
100	3.6355	1712.9	7.4425	100	1.8137	1711.5	7.1013

Temp °C	Pressure = 150 kPa T_sat = −25.3° C			Temp °C	Pressure = 200 kPa T_sat = −18.9° C		
	v	h	s		v	h	s
Liq	*0.001489*	*84.3*	*0.5573*	*Liq*	*0.001507*	*113.4*	*0.6731*
Vap	*0.7819*	*1430.5*	*5.9882*	*Vap*	*0.5946*	*1439.3*	*5.8870*
−20	0.7977	1442.6	6.0355				
−10	0.8338	1465.5	6.1241	−10	0.6192	1460.3	5.9683
0	0.8690	1488.0	6.2082	0	0.6466	1483.6	6.0553
10	0.9038	1510.4	6.2885	10	0.6733	1506.5	6.1377
20	0.9382	1532.6	6.3655	20	0.6995	1529.2	6.2164
30	0.9723	1554.6	6.4396	30	0.7255	1551.7	6.2919
40	1.0062	1576.7	6.5111	40	0.7513	1574.1	6.3645
50	1.0398	1598.7	6.5804	50	0.7768	1596.4	6.4347
60	1.0734	1620.8	6.6477	60	0.8023	1618.7	6.5027
80	1.1401	1665.2	6.7771	80	0.8527	1663.5	6.6330
100	1.2065	1710.0	6.9005	100	0.9028	1708.5	6.7572

continued

TABLE B3.2 (SI UNITS) *continued*
Properties of Ammonia Vapor
Saturated Liquid and Vapor States are in Italics
v in cubic meters per kilogram
h in kiloJoules per kilogram
s in kiloJoules per kilogram and Kelvin

Temp °C	Pressure = 300 kPa $T_{sat} = -9.2°C$			Temp °C	Pressure = 400 kPa $T_{sat} = -1.9°C$		
	v	h	s		v	h	s
Liq	*0.001536*	*157.3*	*0.8420*	*Liq*	*0.001560*	*191.2*	*0.9679*
Vap	*0.4061*	*1451.5*	*5.7456*	*Vap*	*0.3094*	*1459.9*	*5.6451*
0	0.4238	1474.4	5.8312	0	0.3123	1464.9	5.6634
10	0.4425	1498.7	5.9183	10	0.3270	1490.5	5.7556
20	0.4608	1522.4	6.0008	20	0.3413	1515.4	5.8418
30	0.4787	1545.7	6.0790	30	0.3552	1539.6	5.9233
40	0.4964	1568.8	6.1539	40	0.3688	1563.4	6.0005
50	0.5138	1591.7	6.2259	50	0.3823	1586.9	6.0744
60	0.5311	1614.5	6.2953	60	0.3954	1610.2	6.1452
80	0.5653	1660.0	6.4279	80	0.4216	1656.4	6.2801
100	0.5992	1705.6	6.5534	100	0.4473	1702.6	6.4072

Temp °C	Pressure = 500 kPa $T_{sat} = 4.1°C$			Temp °C	Pressure = 600 kPa $T_{sat} = 9.3°C$		
	v	h	s		v	h	s
Liq	*0.001580*	*219.2*	*1.0694*	*Liq*	*0.001598*	*243.3*	*1.1552*
Vap	*0.2503*	*1466.2*	*5.5668*	*Vap*	*0.2104*	*1471.1*	*5.5026*
20	0.2695	1508.1	5.7138	20	0.2215	1500.6	5.6049
40	0.2923	1557.9	5.8783	40	0.2412	1552.3	5.7756
60	0.3141	1605.9	6.0267	60	0.2598	1601.4	5.9277
80	0.3354	1652.9	6.1637	80	0.2778	1649.3	6.0672
100	0.3562	1699.6	6.2923	100	0.2955	1696.5	6.1974
120	0.3768	1746.3	6.4144	120	0.3129	1743.7	6.3206
140	0.3973	1793.4	6.5313	140	0.3300	1791.1	6.4382
160	0.4176	1841.0	6.6437	160	0.3470	1839.0	6.5513

continued

TABLE B3.2 (SI UNITS) *continued*
Properties of Ammonia Vapor
Saturated Liquid and Vapor States are in Italics
v in cubic meters per kilogram
h in kiloJoules per kilogram
s in kiloJoules per kilogram and Kelvin

Temp °C	Pressure = 800 kPa $T_{sat} = 17.8°$ C			Temp °C	Pressure = 1000 kPa $T_{sat} = 24.9°$ C		
	v	h	s		v	h	s
Liq	*0.001630*	*284.0*	*1.2958*	*Liq*	*0.001658*	*317.8*	*1.4094*
Vap	*0.1596*	*1478.3*	*5.4003*	*Vap*	*0.1285*	*1483.2*	*5.3198*
20	0.1614	1484.7	5.4222				
40	0.1772	1540.6	5.6065	40	0.1387	1528.3	5.4672
60	0.1919	1592.3	5.7668	60	0.1511	1582.9	5.6365
80	0.2059	1641.9	5.9113	80	0.1627	1634.3	5.7864
100	0.2195	1690.4	6.0449	100	0.1739	1684.1	5.9236
120	0.2328	1738.5	6.1704	120	0.1848	1733.1	6.0515
140	0.2459	1786.5	6.2897	140	0.1955	1781.9	6.1725
160	0.2589	1834.9	6.4040	160	0.2060	1830.8	6.2881

Temp °C	Pressure = 1200 kPa $T_{sat} = 30.9°$ C			Temp °C	Pressure = 1400 kPa $T_{sat} = 36.3°$ C		
	v	h	s		v	h	s
Liq	*0.001684*	*347.0*	*1.5054*	*Liq*	*0.001708*	*373.0*	*1.5889*
Vap	*0.1075*	*1486.6*	*5.2530*	*Vap*	*0.0923*	*1488.9*	*5.1955*
40	0.1129	1515.2	5.3458	40	0.0943	1501.4	5.2357
60	0.1238	1573.1	5.5251	60	0.1042	1563.0	5.4265
80	0.1339	1626.6	5.6810	80	0.1132	1618.6	5.5888
100	0.1435	1677.7	5.8219	100	0.1217	1671.2	5.7337
120	0.1528	1727.7	5.9523	120	0.1299	1722.2	5.8667
140	0.1618	1777.2	6.0751	140	0.1378	1772.4	5.9914
160	0.1707	1826.7	6.1921	160	0.1455	1822.5	6.1098
180	0.1795	1876.5	6.3044	180	0.1532	1872.8	6.2232

continued

TABLE B3.2 (SI UNITS) *continued*
Properties of Ammonia Vapor
Saturated Liquid and Vapor States are in Italics
v in cubic meters per kilogram
h in kiloJoules per kilogram
s in kiloJoules per kilogram and Kelvin

Temp °C	Pressure = 1600 kPa			Temp °C	Pressure = 2000 kPa		
	v	h	s		v	h	s
60	0.0895	1552.3	5.3366	60	0.0687	1529.6	5.1742
80	0.0977	1610.4	5.5060	80	0.0760	1593.3	5.3600
100	0.1054	1664.6	5.6552	100	0.0825	1650.9	5.5187
120	0.1127	1716.6	5.7911	120	0.0886	1705.2	5.6606
140	0.1198	1767.7	5.9177	140	0.0945	1757.9	5.7913
160	0.1266	1818.3	6.0375	160	0.1002	1809.9	5.9141
180	0.1334	1869.1	6.1520	180	0.1057	1861.6	6.0308

Source: J. R. Howell, and R. O. Backius, "Fundamentals of Engineering Thermodynamics," McGraw-Hill, New York, 1987. Generated from Program REFRIG, Wiley Professional Software, Wiley, New York, 1985. Reproduced with permission of McGraw-Hill.

TABLE B4 (SI UNITS)
Equilibrium Properties of Water-Lithium Bromide Solution
Water Vapor Saturation Temperature (T_r) in °C
Water Vapor Saturation Pressure (P_r) in kPa
Solution Enthalpy h_s in kJ/kg
Solution Specific Volume (v_s) in m³/kg

Solution Temp		Concentration kg-LiBr/kg-solution						
°C		0.40	0.45	0.50	0.55	0.60	0.65	0.70
	v_s	0.0007164	0.0006870	0.0006599	0.0006308	0.0006042	0.0005764	0.0005511
15	T_r	42.35	34.28	23.24	10.02			
	P_r	8.36	5.41	2.85	1.23			
	h_s	21.66	22.32	27.38	42.05			
20	T_r	51.05	42.55	31.29	17.92	3.57		
	P_r	13.00	8.45	4.57	2.06	0.79		
	h_s	33.78	33.77	38.29	52.38	77.19		
25	T_r	59.75	50.83	39.34	25.82	11.28		
	P_r	19.70	12.86	7.13	3.33	1.34		
	h_s	45.92	45.24	49.22	62.70	86.84		
30	T_r	68.45	59.10	47.39	33.72	18.99	4.75	
	P_r	29.13	19.11	10.84	5.24	2.20	0.86	
	h_s	58.08	56.73	60.15	73.03	96.48	127.99	
35	T_r	77.15	67.37	55.44	41.61	26.70	12.14	
	P_r	42.13	27.78	16.08	8.05	3.51	1.42	
	h_s	70.27	68.23	71.09	83.37	106.13	136.95	
40	T_r	85.85	75.65	63.49	49.51	34.41	19.54	6.75
	P_r	59.72	39.58	23.37	12.05	5.45	2.28	0.98
	h_s	82.48	79.76	82.04	93.71	115.78	145.90	178.36
45	T_r	94.55	83.92	71.54	57.41	42.11	26.94	13.66
	P_r	83.08	55.36	33.29	17.66	8.26	3.56	1.56
	h_s	94.72	91.31	93.01	104.05	125.43	154.86	186.81
50	T_r	103.25	92.19	79.59	65.31	49.82	34.33	20.57
	P_r	113.62	76.10	46.56	25.35	12.24	5.43	2.42
	h_s	106.97	102.87	103.98	114.40	135.08	163.81	195.25
55	T_r	111.95	100.47	87.64	73.21	57.53	41.73	27.48
	P_r	152.93	102.98	64.02	35.73	17.76	8.09	3.67
	h_s	119.25	114.45	114.97	124.76	144.73	172.77	203.70
60	T_r	120.65	108.74	95.70	81.10	65.24	49.12	34.39
	P_r	202.83	137.30	86.66	49.49	25.28	11.82	5.45
	h_s	131.56	126.06	125.96	135.11	154.39	181.73	212.14
65	T_r	129.35	117.01	103.75	89.00	72.95	56.52	41.30
	P_r	265.35	180.56	115.61	67.45	35.34	16.93	7.91
	h_s	143.88	137.68	136.97	145.47	164.04	190.69	220.58

continued

TABLE B4 (SI UNITS) *continued*
Equilibrium Properties of Water-Lithium Bromide Solution
Water Vapor Saturation Temperature (T_r) in °C
Water Vapor Saturation Pressure (P_r) in kPa
Solution Enthalpy h_s in kJ/kg
Solution Specific Volume (v_s) in m³/kg

Solution Temp		Concentration kg-LiBr/kg-solution						
°C	v_s	0.40	0.45	0.50	0.55	0.60	0.65	0.70
70	T_r	138.05	125.29	111.80	96.90	80.66	63.92	48.21
	P_r	342.77	234.45	152.16	90.57	48.60	23.82	11.29
	h_s	156.23	149.32	147.99	155.84	173.70	199.65	229.02
80	T_r	155.45	141.83	127.90	112.70	96.07	78.71	62.04
	P_r	552.39	381.72	253.97	156.78	87.87	44.91	21.88
	h_s	181.00	172.65	170.06	176.58	193.02	217.57	245.89
90	T_r	172.84	158.38	144.00	128.49	111.49	93.50	75.86
	P_r	854.14	596.14	405.59	258.59	150.61	79.92	39.94
	h_s	205.86	196.06	192.17	197.35	212.34	235.50	262.76
100	T_r	190.24	174.93	160.10	144.29	126.91	108.29	89.68
	P_r	1273.68	897.57	623.10	408.85	246.41	135.23	69.22
	h_s	230.81	219.55	214.33	218.12	231.67	253.43	279.62
110	T_r	207.64	191.47	176.20	160.08	142.32	123.09	103.51
	P_r	1839.57	1308.47	925.10	622.83	387.01	218.98	114.65
	h_s	255.85	243.12	236.53	238.92	251.00	271.36	296.47
120	T_r	225.04	208.02	192.31	175.88	157.74	137.88	117.33
	P_r	2582.74	1853.63	1332.50	918.08	586.37	341.11	182.42
	h_s	280.98	266.75	258.77	259.73	270.34	289.30	313.32
130	T_r	242.44	224.56	208.41	191.68	173.16	152.67	131.15
	P_r	3536.00	2559.73	1868.25	1314.34	860.52	513.40	280.11
	h_s	280.98	266.75	258.77	259.73	270.34	289.30	313.32
140	T_r	242.44	224.56	208.41	191.68	173.16	152.67	131.15
	P_r	3536.00	2559.73	1868.25	1314.34	860.52	513.40	280.11
	h_s	280.98	266.75	258.77	259.73	270.34	289.30	313.32
150	T_r	277.24	257.66	240.61	223.27	203.99	182.25	158.80
	P_r	6209.65	4568.67	3424.82	2497.96	1707.09	1064.66	602.60
	h_s	356.93	338.13	325.76	322.27	328.39	343.13	363.82

Source: "ASHRAE Handbook Fundamentals," Atlanta, GA, 1989. Reprinted with permission.

TABLE B5 (SI UNITS)

Temperature-Dependent Molar Constant Pressure Specific Heats

$$\bar{c}_P^0 = a + bT + cT^2 + dT^3$$

T in K

$\bar{c}_P^0$ in kJ/kg-mole-K

Substance		a	$b \times 10^2$	$c \times 10^5$	$d \times 10^9$	Temp. Range, K	$\bar{c}_P^0$ 25° C
Acetylene	C_2H_2	21.8	9.2143	−6.527	18.21	273-1500	43.944
Air		28.11	0.1967	0.4802	−1.966	273-1800	29.071
Ammonia	NH_3	27.568	2.5630	0.99072	−6.6909	273-1500	35.908
Benzene	C_6H_6	−36.22	48.475	−31.57	77.62	273-1500	82.254
Carbon Dioxide	CO_2	22.26	5.981	−3.501	7.469	273-1800	37.172
Carbon Monoxide	CO	28.16	0.1675	0.5372	−2.222	273-1800	29.077
Ethane	C_2H_6	6.900	17.27	−6.406	7.285	273-1500	52.869
Ethanol	C_2H_6O	19.9	20.96	−10.38	20.05	273-1500	73.674
Ethylene	C_2H_4	3.95	15.64	−8.344	17.67	273-1500	43.615
Hydrogen	H_2	29.11	−0.1916	0.4003	−0.8704	273-1800	28.871
Hydrogen Chloride	HCl	30.33	−0.7620	1.327	−4.338	273-1500	29.123
i-Butane	C_4H_{10}	−7.913	41.60	−23.01	49.91	273-1500	96.942
Methane	CH_4	19.89	5.024	1.269	−11.01	273-1500	35.697
Methanol	CH_4O	19.0	9.152	−1.22	−8.039	273-1000	44.977
n-Butane	C_4H_{10}	3.96	37.15	−18.34	35.00	273-1500	99.307
n-Hexane	C_6H_{14}	6.938	55.22	−28.65	57.69	273-1500	147.578
n-Pentane	C_5H_{12}	6.774	45.43	−22.46	42.29	273-1500	123.329
Nitric Oxide	NO	29.34	−0.09395	0.9747	−4.187	273-1500	29.815
Nitrogen	N_2	28.90	−0.1571	0.8081	−2.873	273-1800	29.073
Nitrogen Dioxide	NO_2	22.9	5.715	−3.52	7.87	273-1500	37.013
Nitrous Oxide	N_2O	24.11	5.8632	−3.562	10.58	273-1500	38.699
Oxygen	O_2	25.48	1.520	−0.7155	1.312	273-1800	29.409
Propane	C_3H_8	−4.04	30.48	−15.72	31.74	273-1500	73.670
Propylene	C_3H_6	3.15	23.83	-12.18	24.62	273-1500	63.999
Sulfur	S_2	27.21	2.218	−1.628	3.986	273-1800	32.479
Sulfur Dioxide	SO_2	25.78	5.795	−3.812	8.612	273-1800	39.892
Sulfur Trioxide	SO_3	16.40	14.58	−11.20	32.42	273-1300	50.760
Water Vapor	H_2O	32.24	0.1923	1.055	−3.595	273-1800	33.655
Monotonic Gases		$\overline{c_V^0} = \frac{3}{2}\Re$		$\overline{c_P^0} = \frac{5}{2}\Re$		All Temps.	

Source: B.G., Kyle, "Chemical and Process Thermodynamics," 2/e, 1992, p. 545. Adapted by permission of Prentice-Hall, Englewood Cliffs, NJ.

TABLE B6.1 (SI UNITS)
Properties of Air as an Ideal Gas with Variable Specific Heats
Temperature in Kelvins
Specific enthalpy in kiloJoules per kilogram
Specific internal energy in kiloJoules per kilogram
s_0 *in kiloJoules per kilogram and Kelvin*

T	h	u	s_0	P_r	v_r	T	h	u	s_0	P_r	v_r
200	200.13	142.72	5.2950	0.33468	171.52	1025	1074.73	780.52	6.9955	125.20	2.3499
215	215.16	153.45	5.3674	0.43083	143.24	1040	1091.93	793.41	7.0121	132.68	2.2499
230	23.20	164.18	5.4350	0.54524	121.08	1055	1109.16	806.34	7.0286	140.51	2.1552
245	245.24	174.92	5.4984	0.67987	103.44	1070	1126.44	819.32	7.0448	148.70	2.0654
260	260.28	185.65	5.5580	0.8368	89.188	1085	1143.76	832.33	7.0609	157.26	1.9803
275	275.33	196.40	5.6143	1.0180	77.539	1100	1161.11	845.38	7.0768	166.21	1.8996
290	290.39	207.15	5.6676	1.2258	67.909	1115	1178.50	858.46	7.0925	175.56	1.8230
305	305.45	217.91	5.7182	1.4623	59.867	1130	1195.93	871.59	7.1080	185.32	1.7502
320	320.53	228.68	5.7665	1.7301	53.091	1145	1213.40	884.75	7.1234	195.50	1.6811
335	335.62	239.47	5.8126	2.0314	47.334	1160	1230.90	897.94	7.1386	206.12	1.6154
350	350.73	250.27	5.8567	2.3689	42.407	1175	1248.43	911.17	7.1536	217.19	1.5528
365	365.86	261.09	5.8990	2.7453	38.162	1190	1266.00	924.43	7.1684	228.73	1.4933
380	381.01	271.94	5.9397	3.1633	34.481	1205	1283.60	937.73	7.1831	240.75	1.4367
395	396.18	282.81	5.9788	3.6257	31.270	1220	1301.24	951.06	7.1977	253.26	1.3827
410	411.38	293.70	6.0166	4.1356	28.456	1235	1318.90	964.42	7.2121	266.29	1.3312
425	426.62	304.63	6.0531	4.6962	25.976	1250	1336.60	977.81	7.2263	279.83	1.2822
440	441.88	315.59	6.0884	5.3106	23.781	1265	1354.33	991.24	7.2404	293.92	1.2353
455	457.18	326.58	6.1226	5.9824	21.831	1280	1372.09	1004.69	7.2544	308.57	1.1907
470	472.51	337.61	6.1557	6.7150	20.090	1295	1389.88	1018.17	7.2682	323.78	1.1480
485	487.89	348.68	6.1879	7.5123	18.531	1310	1407.70	1031.69	7.2819	339.59	1.1073
500	503.30	359.79	6.2193	8.378	17.130	1325	1425.54	1045.23	7.2954	356.00	1.0683
515	518.76	370.94	6.2497	9.316	15.868	1340	1443.42	1058.79	7.3088	373.03	1.0311
530	534.27	382.14	6.2794	10.331	14.726	1355	1461.32	1072.39	7.3221	390.70	0.99546
545	549.82	393.39	6.3083	11.426	13.690	1370	1479.25	1086.01	7.3353	409.03	0.96138
560	565.42	404.68	6.3366	12.608	12.749	1385	1497.20	1099.66	7.3483	428.03	0.92876
575	581.07	416.02	6.3641	13.879	11.892	1400	1515.18	1113.34	7.3612	447.73	0.89752
590	596.77	427.42	6.3911	15.245	11.108	1415	1533.19	1127.04	7.3740	468.14	0.86759
605	612.52	438.86	6.4175	16.712	10.391	1430	1551.22	1140.77	7.3867	489.27	0.83891
620	628.32	450.36	6.4433	18.284	9.733	1445	1569.28	1154.52	7.3993	511.16	0.81141
635	644.18	461.91	6.4685	19.967	9.129	1460	1587.36	1168.29	7.4117	533.81	0.78504
650	660.09	473.52	6.4933	21.766	8.5718	1475	1605.46	1182.09	7.4240	557.26	0.75974
665	676.05	485.17	6.5176	23.687	8.0583	1490	1623.59	1195.91	7.4363	581.51	0.73546
680	692.07	496.89	6.5414	25.736	7.5839	1510	1647.79	1214.37	7.4524	615.14	0.70459
695	708.14	508.65	6.5648	27.920	7.1448	1525	1665.97	1228.25	7.4644	641.35	0.68250
710	724.27	520.47	6.5877	30.245	6.7380	1540	1684.17	1242.14	7.4763	668.45	0.66128

continued

TABLE B6.1 (SI UNITS) *continued*
Properties of Air as an Ideal Gas with Variable Specific Heats
Temperature in Kelvins
Specific enthalpy in kiloJoules per kilogram
Specific internal energy in kiloJoules per kilogram
s_0 in kiloJoules per kilogram and Kelvin

T	h	u	s_0	P_r	v_r	T	h	u	s_0	P_r	v_r
725	740.45	532.35	6.6103	32.717	6.3605	1555	1702.39	1256.06	7.4880	696.44	0.64088
740	756.68	544.28	6.6324	35.344	6.0097	1570	1720.64	1270.00	7.4997	725.36	0.62127
755	772.97	556.26	6.6542	38.131	5.6832	1585	1738.90	1283.96	7.5113	755.21	0.60241
770	789.31	568.30	6.6757	41.088	5.3791	1600	1757.19	1297.94	7.5228	786.04	0.58426
785	805.71	580.39	6.6967	44.220	5.0955	1615	1775.49	1311.94	7.5342	817.85	0.56680
800	822.15	592.53	6.7175	47.535	4.8306	1630	1793.82	1325.95	7.5455	850.67	0.54999
815	838.65	604.72	6.7379	51.043	4.5831	1645	1812.16	1339.99	7.5567	884.53	0.53381
830	855.20	616.97	6.7581	54.749	4.3514	1660	1830.52	1354.05	7.5678	919.44	0.51822
845	871.80	629.26	6.7779	58.664	4.1344	1675	1848.90	1368.13	7.5788	955.44	0.50320
860	888.45	641.61	6.7974	62.795	3.9310	1690	1867.30	1382.22	7.5897	992.55	0.48873
875	905.15	654.00	6.8167	67.152	3.7401	1705	1885.72	1396.34	7.6006	1030.8	0.47477
890	921.90	666.45	6.8356	71.742	3.5608	1720	1904.16	1410.47	7.6113	1070.2	0.46132
905	938.70	678.94	6.8544	76.576	3.3922	1735	1922.61	1424.61	7.6220	1110.8	0.44834
920	955.55	691.48	6.8728	81.663	3.2336	1750	1941.09	1438.78	7.6326	1152.6	0.43582
935	972.44	704.07	6.8910	87.013	3.0843	1765	1959.57	1452.96	7.6432	1195.6	0.42374
950	989.38	716.70	6.9090	92.63	2.9436	1780	1978.08	1467.16	7.6536	1239.9	0.41207
965	1006.36	729.38	6.9267	98.54	2.8109	1795	1996.60	1481.38	7.6640	1285.4	0.40081
980	1023.39	742.10	6.9443	104.74	2.6857	1810	2015.14	1495.61	7.6742	1332.3	0.38994
995	1040.46	754.86	6.9615	111.24	2.5674	1825	2033.69	1509.86	7.6844	1380.6	0.37943
1010	1057.57	767.67	6.9786	118.06	2.4556	1840	2052.26	1524.13	7.6946	1430.2	0.36928

Source: Abridged from Keenan, J. H., Chao, J., and Kaye, J., "Gas Tables," Wiley, New York, 1980. Reprinted by permission of John Wiley & Wiley, Inc.

TABLE B6.2 (SI UNITS)
Properties of Carbon Dioxide as an Ideal Gas with Variable Specific Heats
Temperature in Kelvins
$\bar{h}$ in kiloJoules per kilogram-mole
$\bar{u}$ in kiloJoules per kilogram-mole
$\bar{s}_0$ in kiloJoules per kilogram-mole and Kelvin
M = 44.0098

T	$\bar{h}$	$\bar{u}$	$\bar{s}_0$	T	$\bar{h}$	$\bar{u}$	$\bar{s}_0$
200	5952.3	4289.4	199.859	1400	65261.8	53621.7	288.086
220	6609.1	4779.9	202.989	1420	66420.1	54613.7	288.909
240	7285.9	5290.4	205.932	1440	67578.4	55605.7	289.715
260	7982.5	5820.7	208.719	1460	68741.3	56602.2	290.514
280	8698.6	6370.5	211.373	1480	69904.1	57598.8	291.312
298	9359.7	6882.0	214.657				
300	9433.7	6939.4	213.908	1500	71071.5	58599.9	292.093
320	10187.1	7526.5	216.339	1520	72241.1	59603.2	292.867
340	10957.9	8131.0	218.675	1540	73410.8	60606.6	293.631
360	11745.8	8752.6	220.925	1560	74582.7	61612.2	294.380
380	12549.3	9389.8	223.099	1580	75759.2	62622.4	295.128
400	13368.5	10042.8	225.199	1600	76935.6	63632.6	295.868
420	14202.3	10710.2	227.233	1620	78114.4	64645.0	296.600
440	15050.1	11391.7	229.205	1640	79295.4	65659.7	297.323
460	15911.1	12086.5	231.119	1660	80478.6	66676.7	298.038
480	16785.3	12794.3	232.979	1680	81661.9	67693.7	298.753
500	17671.7	13514.5	234.788	1700	82847.5	68713.0	299.452
520	18570.2	14246.7	236.550	1720	84035.3	69734.5	300.150
540	19479.8	14990.0	238.266	1740	85223.1	70756.0	300.832
560	20400.5	15744.4	239.940	1760	86413.2	71779.8	301.514
580	21331.9	16509.5	241.574	1780	87605.5	72805.9	302.187
600	22273.3	17284.6	243.170	1800	88797.9	73832.0	302.852
620	23224.7	18069.7	244.729	1820	89992.5	74860.3	303.517
640	24185.4	18864.2	246.255	1840	91189.4	75890.9	304.166
660	25155.2	19667.7	247.747	1860	92386.3	76921.6	304.814
680	26133.8	20480.0	249.207	1880	93585.5	77954.4	305.463
700	27120.9	21300.8	250.639	1900	94784.7	78987.3	306.095
720	28116.3	22129.9	252.040	1920	95986.1	80022.5	306.727
740	29119.5	22966.8	253.414	1940	97189.8	81059.9	307.350
760	30130.1	23811.2	254.762	1960	98393.6	82097.3	307.965
780	31148.3	24663.1	256.084	1980	99599.6	83137.0	308.572
800	32173.5	25522.0	257.382	2000	100805.5	84176.7	309.179
820	33205.5	26387.7	258.656	2020	102013.8	85218.7	309.778
840	34244.1	27260.0	259.908	2040	103222.1	86260.7	310.377
860	35289.1	28138.7	261.137	2060	104430.3	87302.6	310.967

continued

TABLE B6.2 (SI UNITS) *continued*

Properties of Carbon Dioxide as an Ideal Gas with Variable Specific Heats
Temperature in Kelvins
$\bar{h}$ *in kiloJoules per kilogram-mole*
$\bar{u}$ *in kiloJoules per kilogram-mole*
$\bar{s}_0$ *in kiloJoules per kilogram-mole and Kelvin*
M = 44.0098

T	$\bar{h}$	$\bar{u}$	$\bar{s}_0$	T	$\bar{h}$	$\bar{u}$	$\bar{s}_0$
880	36340.4	29023.7	262.345	2080	105640.8	88346.9	311.549
900	37397.4	29914.5	263.533	2100	106853.6	89393.4	312.131
920	38460.1	30810.8	264.701	2120	108066.4	90439.9	312.705
940	39528.4	31712.9	265.850	2140	109281.5	91488.7	313.278
960	40602.0	32620.2	266.980	2160	110496.6	92537.5	313.844
980	41680.8	33532.7	268.092	2180	111711.7	93586.3	314.401
1000	42764.6	34450.2	269.186	2200	112929.0	94637.3	314.958
1020	43853.9	35373.2	270.260	2220	114146.3	95688.4	315.507
1040	44946.3	36299.3	271.324	2240	115363.7	96739.4	316.055
1060	46043.3	37230.0	272.372	2260	116583.3	97792.8	316.596
1080	47144.8	38165.2	273.403	2280	117802.9	98846.1	317.136
1100	48250.8	39105.0	274.417	2300	119024.8	99901.7	317.668
1120	49361.4	40049.3	275.415	2320	120246.7	100957.3	318.201
1140	50474.3	40996.9	276.396	2340	121468.6	102012.9	318.724
1160	51591.7	41947.0	277.360	2360	122692.7	103070.7	319.240
1180	52715.9	42905.0	278.325	2380	123916.9	104128.6	319.755
1200	53840.2	43862.9	279.272	2400	125143.3	105188.7	320.271
1220	54968.9	44825.4	280.212	2420	126369.7	106248.9	320.778
1240	56100.0	45790.1	281.135	2440	127596.2	107309.0	321.285
1260	57235.6	46759.4	282.041	2460	128824.9	108371.4	321.792
1280	58375.7	47733.2	282.939	2480	130053.6	109433.8	322.291
1300	59515.8	48707.1	283.820	2500	131282.3	110496.3	322.782
1320	60660.5	49685.5	284.693	2520	132511.0	111558.7	323.272
1340	61805.1	50663.8	285.566	2540	133741.9	112623.4	323.755
1360	62954.3	51646.8	286.415	2560	134972.9	113688.0	324.237
1380	64108.1	52634.2	287.254	2580	136203.9	114752.7	324.719

Abridged from Keenan, J. H., Chao, J., and Kaye, J., "Gas Tables," Wiley, New York, 1980. Reprinted by permission of John Wiley & Sons, Inc.

TABLE B6.3 (SI UNITS)
Properties of Carbon Monoxide as an Ideal Gas with Variable Specific Heats
Temperature in Kelvins
$\bar{h}$ in kiloJoules per kilogram-mole
$\bar{u}$ in kiloJoules per kilogram-mole
$\bar{s}_0$ in kiloJoules per kilogram-mole and Kelvin
M = 28.0104

T	$\bar{h}$	$\bar{u}$	$\bar{s}_0$	T	$\bar{h}$	$\bar{u}$	$\bar{s}_0$
200	5813.2	4150.3	185.891	1400	44007.6	32367.5	245.862
220	6395.3	4566.2	188.665	1420	44706.4	32900.0	246.358
240	6977.6	4982.1	191.198	1440	45406.5	33433.7	246.848
260	7559.9	5398.1	193.529	1460	46107.7	33968.7	247.331
280	8142.3	5814.3	195.687	1480	46810.1	34504.8	247.809
298	**8666.7**	**6189.1**	**197.499**				
300	8725.0	6230.7	197.697	1500	47513.7	35042.1	248.281
320	9308.1	6647.5	199.579	1520	48218.4	35580.5	248.748
340	9891.6	7064.7	201.347	1540	48924.1	36119.9	249.209
360	10475.9	7482.7	203.017	1560	49630.9	36660.5	249.665
380	11061.0	7901.5	204.599	1580	50338.8	37202.0	250.116
400	11647.1	8321.4	206.102	1600	51047.6	37744.6	250.562
420	12234.6	8742.6	207.535	1620	51757.4	38288.1	251.003
440	12823.6	9165.3	208.905	1640	52468.2	38832.5	251.439
460	13414.3	9589.7	210.218	1660	53179.8	39377.9	251.870
480	14007.0	10016.0	211.479	1680	63892.4	39924.2	252.297
500	14601.7	10444.5	212.693	1700	54605.8	40471.3	252.719
520	15198.7	10875.2	213.864	1720	55320.1	41019.3	253.137
540	15798.1	11308.3	214.995	1740	56035.2	41568.2	253.550
560	16400.1	11744.0	216.090	1760	56751.2	42117.8	253.959
580	17004.7	12182.4	217.150	1780	57467.9	42668.2	254.364
600	17612.1	12623.5	218.180	1800	58185.4	43219.4	254.765
620	18222.3	13067.4	219.180	1820	58903.6	43771.4	255.162
640	18835.4	13514.2	220.154	1840	59622.6	44324.1	255.555
660	19451.4	13963.9	221.101	1860	60342.2	44877.5	255.944
680	20070.4	14416.6	222.025	1880	61062.6	45431.5	256.329
700	20692.3	14872.2	222.927	1900	61783.7	45986.3	256.710
720	21317.2	15330.8	223.807	1920	62505.4	46541.7	257.088
740	21945.0	15792.3	224.667	1940	63227.8	47097.8	257.463
760	22575.7	16256.8	225.508	1960	63950.8	47654.5	257.833
780	23209.4	16724.2	226.331	1980	64674.4	48211.9	258.201
800	23845.9	17194.4	227.137	2000	65398.6	48769.8	258.565
820	24485.3	17667.4	227.926	2020	66123.4	49328.3	258.925
840	25127.4	18143.3	228.700	2040	66848.8	49887.4	259.283
860	25772.2	18621.8	229.458	2060	67574.8	50447.1	259.637
880	26419.8	19103.1	230.203	2080	68301.3	51007.3	259.988

continued

TABLE B6.3 (SI UNITS) *continued*

Properties of Carbon Monoxide as an Ideal Gas with Variable Specific Heats

Temperature in Kelvins
$\bar{h}$ *in kiloJoules per kilogram-mole*
$\bar{u}$ *in kiloJoules per kilogram-mole*
$\bar{s}_0$ *in kiloJoules per kilogram-mole and Kelvin*
M = 28.0104

T	$\bar{h}$	$\bar{u}$	$\bar{s}_0$	T	$\bar{h}$	$\bar{u}$	$\bar{s}_0$
900	27069.9	19586.9	230.933	2100	69028.3	51568.1	260.336
920	27722.6	20073.4	231.650	2120	69755.9	52129.4	260.680
940	28377.8	20562.3	232.355	2140	70484.0	52691.2	261.022
960	29035.5	21053.7	233.047	2160	71212.6	53253.5	261.361
980	29695.6	21547.4	233.728	2180	71941.7	53816.3	261.697
1000	30357.9	22043.5	234.397	2200	72671.3	54379.6	262.030
1020	31022.6	22541.9	235.055	2220	73401.4	54943.4	262.361
1040	31689.4	23042.4	235.702	2240	74131.9	55507.7	262.688
1060	32358.4	23545.2	236.340	2260	74862.9	56072.4	263.013
1080	33029.5	24049.9	236.967	2280	75594.3	56637.5	263.335
1100	33702.6	24556.8	237.584	2300	76326.2	57203.1	263.655
1120	34377.7	25065.6	238.193	2320	77058.5	57769.1	263.972
1140	35054.7	25576.3	238.792	2340	77791.2	58335.5	264.286
1160	35733.6	26088.9	239.382	2360	78524.4	58902.4	264.598
1180	36414.3	26603.3	239.964	2380	79527.9	59469.7	264.908
1200	37096.7	27119.4	240.537	2400	79991.9	60037.3	265.215
1220	37780.9	27637.3	241.103	2420	80726.2	60605.4	265.520
1240	38466.7	28156.8	241.660	2440	81460.9	61173.8	265.822
1260	39154.1	28678.0	242.210	2460	82196.0	61742.6	266.122
1280	39843.1	29200.7	242.753	2480	82931.5	62311.8	266.420
1300	40533.7	29724.9	243.288	2500	83667.3	62881.3	266.715
1320	41225.7	30250.7	243.816	2520	84403.5	63451.2	267.009
1340	41919.1	30777.8	244.338	2540	85140.0	64021.4	267.300
1360	42613.9	31306.3	244.852	2560	85876.9	64592.0	267.589
1380	43310.1	31836.2	245.361	2580	86614.1	65163.0	267.876

Abridged from Keenan, J. H., Chao, J., and Kaye, J., "Gas Tables," Wiley, New York, 1980.
Reprinted by permission of John Wiley & Sons, Inc.

TABLE B6.4 (SI UNITS)
Properties of Hydrogen as an Ideal Gas with Variable Specific Heats
Temperature in Kelvins
$\bar{h}$ in kiloJoules per kilogram-mole
$\bar{u}$ in kiloJoules per kilogram-mole
$\bar{s}_0$ in kiloJoules per kilogram-mole and Kelvin
M = 2.0158

T	$\bar{h}$	$\bar{u}$	$\bar{s}_0$	T	$\bar{h}$	$\bar{u}$	$\bar{s}_0$
200	5697.2	4034.3	119.328	1400	41548.6	29908.5	176.531
220	6247.7	4418.6	121.947	1420	42186.8	30380.4	176.980
240	6807.4	4811.9	124.383	1440	42827.3	30854.6	177.421
260	7373.8	5212.0	126.645	1460	43467.8	31328.7	177.870
280	7945.2	5617.2	128.773	1480	44112.8	31807.5	178.311
298	**8463.1**	**5985.4**	**130.558**				
300	8520.7	6026.4	130.752	1500	44757.8	32286.2	178.743
320	9099.0	6438.3	132.623	1520	45405.1	32767.2	179.175
340	9679.0	6852.1	134.377	1540	46054.6	33250.4	179.599
360	10260.7	7267.5	136.040	1560	46704.2	33733.7	180.015
380	10843.2	7683.7	137.620	1580	47356.0	34219.2	180.431
400	11426.7	8100.9	139.117	1600	48010.1	34707.1	180.838
420	12010.6	8518.5	140.538	1620	48666.5	35197.1	181.246
440	12594.7	8936.4	141.894	1640	49322.8	35687.2	181.645
460	13179.3	9354.7	143.199	1660	49981.5	36179.6	182.044
480	13764.2	9773.2	144.438	1680	50642.4	36674.2	182.443
500	14349.2	10192.0	145.635	1700	51303.3	37168.8	182.834
520	14934.5	10611.0	146.791	1720	51966.5	37665.7	183.224
540	15520.0	11030.2	147.888	1740	52629.6	38162.6	183.607
560	16105.9	11449.9	148.953	1760	53297.4	38664.0	183.989
580	16691.9	11869.6	149.975	1780	53965.1	39165.5	184.364
600	17278.3	12289.7	150.973	1800	54635.1	39669.2	184.738
620	17865.0	12710.0	151.937	1820	55307.3	40175.1	185.112
640	18451.8	13130.6	152.869	1840	55981.9	40683.4	185.478
660	19039.4	13551.9	153.767	1860	56656.4	41191.6	185.844
680	19627.4	13973.6	154.648	1880	57331.0	41699.9	186.201
700	20216.1	14396.0	155.504	1900	58007.8	42210.4	186.559
720	20805.2	14818.8	156.327	1920	58686.8	42723.2	186.916
740	21395.0	15242.4	157.142	1940	59368.2	43238.2	187.274
760	21985.5	15666.6	157.924	1960	60049.5	43753.3	187.623
780	22576.9	16091.7	158.697	1980	60733.2	44270.6	187.972
800	23168.1	16516.6	159.445	2000	61419.0	44790.2	188.313
820	23760.9	16943.1	160.177	2020	62104.9	45309.8	188.654
840	24355.9	17371.8	160.892	2040	62793.1	45831.7	188.995
860	24951.0	17800.6	161.590	2060	63483.5	46355.9	189.327

continued

TABLE B6.4 (SI UNITS) *continued*

Properties of Hydrogen as an Ideal Gas with Variable Specific Heats
Temperature in Kelvins
$\bar{h}$ *in kiloJoules per kilogram-mole*
$\bar{u}$ *in kiloJoules per kilogram-mole*
$\bar{s}_0$ *in kiloJoules per kilogram-mole and Kelvin*
M = 2.0158

T	$\bar{h}$	$\bar{u}$	$\bar{s}_0$	T	$\bar{h}$	$\bar{u}$	$\bar{s}_0$
880	25546.0	18229.4	162.272	2080	64171.7	46877.7	189.660
900	26143.3	18660.4	162.946	2100	64864.4	47404.1	189.992
920	26742.9	19093.7	163.603	2120	65557.1	47930.6	190.317
940	27342.5	19527.0	164.251	2140	66254.3	48461.5	190.641
960	27942.1	19960.3	164.883	2160	66949.3	48990.2	190.974
980	28544.0	20395.9	165.498	2180	67646.6	49521.2	191.298
1000	29148.1	20833.7	166.105	2200	68343.8	50052.1	191.614
1020	29752.2	21271.5	166.704	2220	69043.3	50585.4	191.930
1040	30358.6	21711.6	167.294	2240	69745.1	51120.9	192.237
1060	30967.3	22154.0	167.876	2260	70449.2	51658.6	192.553
1080	31576.0	22596.4	168.441	2280	71151.0	52194.1	192.861
1100	32186.9	23041.1	169.007	2300	71855.0	52731.9	193.169
1120	32800.1	23488.0	169.556	2320	72561.4	53271.9	193.476
1140	33413.3	23934.9	170.104	2340	73267.7	53812.0	193.784
1160	34028.8	24384.1	170.636	2360	73976.3	54354.3	194.083
1180	34646.6	24835.6	171.169	2380	74684.9	54896.6	194.382
1200	35264.3	25287.1	171.684	2400	75395.8	55441.2	194.682
1220	35884.4	25740.8	172.191	2420	76108.9	55988.1	194.973
1240	36506.7	26196.8	172.698	2440	76822.1	56534.9	195.264
1260	37131.2	26655.1	173.197	2460	77535.2	57081.8	195.555
1280	37758.1	27115.6	173.688	2480	78250.6	57630.9	195.846
1300	38384.9	27576.2	174.178	2500	78966.0	58180.0	196.137
1320	39014.0	28039.0	174.661	2520	79683.7	58731.4	196.428
1340	39645.4	28504.1	175.143	2540	80401.4	59282.8	196.710
1360	40276.8	28969.2	175.609	2560	81119.1	59834.2	196.993
1380	40912.7	29438.8	176.074	2580	81841.3	60390.2	197.268

Abridged from Keenan, J. H., Chao, J., and Kaye, J., "Gas Tables," Wiley, New York, 1980. Reprinted by permission of John Wiley & Sons, Inc.

TABLE B6.5 (SI UNITS)
Properties of Nitrogen as an Ideal Gas with Variable Specific Heats
Temperature in Kelvins
$\bar{h}$ in kiloJoules per kilogram-mole
$\bar{u}$ in kiloJoules per kilogram-mole
$\bar{s}_0$ in kiloJoules per kilogram-mole and Kelvin
M = 28.0134

T	$\bar{h}$	$\bar{u}$	$\bar{s}_0$	T	$\bar{h}$	$\bar{u}$	$\bar{s}_0$
200	5812.8	4149.9	179.842	1400	43605.3	31965.1	239.343
220	6394.9	4565.8	182.616	1420	44296.3	32489.8	239.833
240	6977.1	4981.7	185.149	1440	44988.6	33015.8	240.317
260	7559.3	5397.6	187.479	1460	45682.2	33543.1	240.795
280	8141.6	5813.6	189.637	1480	46377.1	34071.8	241.268
298	**8665.8**	**6188.1**	**191.448**				
300	8724.1	6229.7	191.646	1500	47073.2	34601.6	241.735
320	9306.7	6646.1	193.526	1520	47770.6	35132.7	242.197
340	9889.6	7062.7	195.293	1540	48469.1	35664.9	242.653
360	10472.9	7479.7	196.960	1560	49168.8	36198.3	243.105
380	11056.8	7897.3	198.538	1580	49869.6	36732.8	243.551
400	11641.3	8315.6	200.037	1600	50571.4	37268.4	243.993
420	12226.8	8734.7	201.466	1620	51274.4	37805.1	244.429
440	12813.3	9155.0	202.830	1640	51978.4	38342.7	244.861
460	13401.0	9576.4	204.136	1660	52683.4	38881.4	245.288
480	13990.2	9999.3	205.390	1680	53389.3	39421.1	245.711
500	14581.0	10423.8	206.596	1700	54096.3	39961.8	246.129
520	15173.5	10850.0	207.758	1720	54804.1	40503.4	246.543
540	15767.9	11278.1	208.879	1740	55512.9	41045.9	246.953
560	16364.4	11708.4	209.964	1760	56222.6	41589.3	247.359
580	16963.1	12140.8	211.014	1780	56933.1	42133.5	247.760
600	17564.1	12575.5	212.033	1800	57644.5	42678.6	248.158
620	18167.6	13012.6	213.022	1820	58356.8	43224.5	248.551
640	18773.5	13452.3	213.984	1840	59069.8	43771.3	248.941
660	19381.9	13894.4	214.920	1860	59783.6	44318.8	249.326
680	19993.0	14339.2	215.833	1880	60498.2	44867.1	249.709
700	20606.8	14786.7	216.722	1900	61213.5	45416.1	250.087
720	21223.2	15236.9	217.590	1920	61929.6	45965.9	250.462
740	21842.4	15689.7	218.439	1940	62646.4	46516.4	250.833
760	22464.3	16145.3	219.268	1960	63363.9	47067.6	251.201
780	23088.9	16603.6	220.079	1980	64082.0	47619.5	251.566
800	23716.2	17064.7	220.873	2000	64800.9	48172.1	251.927
820	24346.2	17528.4	221.651	2020	65520.4	48725.3	252.285
840	24978.9	17994.8	222.413	2040	66240.5	49279.1	252.640
860	25614.2	18463.8	223.161	2060	66961.3	49833.6	252.991

continued

Property Tables

TABLE B6.5 (SI UNITS) *continued*
Properties of Nitrogen as an Ideal Gas with Variable Specific Heats
Temperature in Kelvins
$\bar{h}$ *in kiloJoules per kilogram-mole*
$\bar{u}$ *in kiloJoules per kilogram-mole*
$\bar{s}_0$ *in kiloJoules per kilogram-mole and Kelvin*
M = 28.0134

T	$\bar{h}$	$\bar{u}$	$\bar{s}_0$	T	$\bar{h}$	$\bar{u}$	$\bar{s}_0$
880	26252.1	18935.5	223.894	2080	67682.7	50388.7	253.340
900	26892.6	19409.7	224.614	2100	68404.6	50944.4	253.685
920	27535.7	19886.5	225.320	2120	69127.2	51500.7	254.028
940	28181.3	20365.7	226.015	2140	69850.3	52057.5	254.367
960	28829.2	20847.4	226.697	2160	70574.0	52614.9	254.704
980	29479.6	21331.5	227.367	2180	71298.3	53172.9	255.038
1000	30132.4	21818.0	228.027	2200	72023.1	53731.4	255.369
1020	30787.5	22306.8	228.675	2220	72748.4	54290.4	255.697
1040	31444.8	22797.8	229.313	2240	73474.2	54850.0	256.022
1060	32104.3	23291.0	229.941	2260	74200.6	55410.0	256.345
1080	32765.9	23786.4	230.560	2280	74927.4	55970.6	256.665
1100	33429.7	24283.9	231.169	2300	75654.7	56531.6	256.983
1120	34095.5	24783.4	231.769	2320	76382.5	57093.1	257.298
1140	34763.3	25284.9	232.360	2340	77110.8	57655.1	257.611
1160	35433.1	25788.4	232.942	2360	77839.5	58217.5	257.921
1180	36104.8	26293.8	233.516	2380	78568.7	58780.4	258.228
1200	36778.3	26801.0	234.082	2400	79298.3	59343.7	258.534
1220	37453.6	27310.0	234.640	2420	80028.4	59907.5	258.837
1240	38130.7	27820.8	235.191	2440	80758.8	60471.7	259.137
1260	38809.5	28333.3	235.734	2460	81489.7	61036.3	259.436
1280	39489.9	28847.5	236.270	2480	82221.0	61601.3	259.732
1300	40172.0	29363.3	236.798	2500	82952.7	62166.7	260.025
1320	40855.7	29880.6	237.320	2520	83684.8	62732.5	260.317
1340	41540.9	30399.6	237.835	2540	84417.3	63298.7	260.607
1360	42227.6	30920.0	238.344	2560	85150.1	63865.3	260.894
1380	42915.7	31441.8	238.846	2580	85883.4	64432.2	261.179

Abridged from Keenan, J. H., Chao, J., and Kaye, J., "Gas Tables," Wiley, New York, 1980. Reprinted by permission of John Wiley & Sons, Inc.

TABLE B6.6 (SI UNITS)

Properties of Oxygen as an Ideal Gas with Variable Specific Heats
Temperature in Kelvins
$\bar{h}$ *in kiloJoules per kilogram-mole*
$\bar{u}$ *in kiloJoules per kilogram-mole*
$\bar{s}_0$ *in kiloJoules per kilogram-mole and Kelvin*
M = 31.9988

T	$\bar{h}$	$\bar{u}$	$\bar{s}_0$	T	$\bar{h}$	$\bar{u}$	$\bar{s}_0$
200	5815.7	4152.8	193.330	1400	45639.6	33999.4	255.400
220	6398.4	4569.2	196.107	1420	46365.6	34559.1	255.915
240	6981.6	4986.1	198.644	1440	47092.6	35119.9	256.423
260	7565.6	5403.9	200.981	1460	47820.8	35681.7	256.926
280	8150.8	5822.8	203.150	1480	48549.9	36244.6	257.422
298	**8678.9**	**6201.2**	**204.975**				
300	8737.6	6243.3	205.174	1500	49280.1	36808.5	257.912
320	9326.4	6665.8	207.074	1520	50011.3	37373.5	258.396
340	9917.5	7090.6	208.866	1540	50743.6	37939.4	258.874
360	10511.4	7518.3	210.563	1560	51476.8	38506.3	259.348
380	11108.4	7949.0	212.177	1580	52211.0	39074.3	259.815
400	11708.8	8383.0	213.717	1600	52946.2	39643.2	260.278
420	12312.8	8820.7	215.190	1620	53682.4	40213.1	260.735
440	12920.5	9262.2	216.604	1640	54419.5	40783.9	261.187
460	13532.2	9707.6	217.963	1660	55157.6	41355.7	261.634
480	14147.9	10157.0	219.273	1680	55896.7	41928.5	262.077
500	14767.7	10610.5	220.538	1700	56636.7	42502.2	262.515
520	15391.6	11068.1	221.762	1720	57377.6	43076.8	262.948
540	16019.5	11529.7	222.947	1740	58119.5	43652.4	263.377
560	16651.5	11995.4	224.096	1760	58862.3	44228.9	263.801
580	17287.4	12465.1	225.212	1780	59606.0	44806.3	264.222
600	17927.3	12938.7	226.296	1800	60350.6	45384.7	264.638
620	18571.0	13416.1	227.352	1820	61096.2	45963.9	265.050
640	19218.4	13897.2	228.379	1840	61842.6	46544.1	265.457
660	19869.5	14382.0	229.381	1860	62590.0	47125.2	265.861
680	20524.0	14870.2	230.358	1880	63338.3	47707.2	266.262
700	21182.0	15361.9	231.312	1900	64087.5	48290.1	266.658
720	21843.2	15856.8	232.243	1920	64837.5	48873.9	267.051
740	22507.6	16355.0	233.153	1940	65588.5	49458.6	267.440
760	23175.1	16856.1	234.043	1960	66340.4	50044.2	267.825
780	23845.5	17360.3	234.914	1980	67093.2	50630.7	268.208
800	24518.8	17867.3	235.766	2000	67846.8	51218.0	268.586
820	25194.8	18377.0	236.601	2020	68601.4	51806.3	268.962
840	25873.4	18889.3	237.418	2040	69356.8	52395.4	269.334
860	26554.6	19404.2	238.220	2060	70113.1	52985.5	269.703

continued

TABLE B6.6 (SI UNITS) *continued*

Properties of Oxygen as an Ideal Gas with Variable Specific Heats
Temperature in Kelvins
$\bar{h}$ in kiloJoules per kilogram-mole
$\bar{u}$ in kiloJoules per kilogram-mole
$\bar{s}_0$ in kiloJoules per kilogram-mole and Kelvin
M = 31.9988

T	$\bar{h}$	$\bar{u}$	$\bar{s}_0$	T	$\bar{h}$	$\bar{u}$	$\bar{s}_0$
880	27238.2	19921.5	239.006	2080	70870.3	53576.4	270.069
900	27924.2	20441.2	239.776	2100	71628.4	54168.1	270.431
920	28612.4	20963.1	240.533	2120	72387.3	54760.8	270.791
940	29302.8	21487.2	241.275	2140	73147.2	55354.3	271.148
960	29995.2	22013.4	242.004	2160	73907.9	55948.7	271.501
980	30689.7	22541.6	242.720	2180	74669.4	56544.0	271.852
1000	31386.2	23071.8	243.423	2200	75431.8	57140.2	272.201
1020	32084.5	23603.8	244.115	2220	76195.1	57737.2	272.546
1040	32784.6	24137.6	244.795	2240	76959.3	58335.0	272.889
1060	33486.4	24673.2	245.463	2260	77724.3	58933.7	273.229
1080	34190.0	25210.4	246.121	2280	78490.2	59533.3	273.566
1100	34895.2	25749.3	246.768	2300	79256.9	60133.7	273.901
1120	35601.9	26289.8	247.404	2320	80024.4	60735.0	274.233
1140	36310.2	26831.8	248.031	2340	80792.8	61337.1	274.563
1160	37020.0	27375.3	248.648	2360	81562.1	61940.1	274.890
1180	37731.2	27920.2	249.256	2380	82332.2	62543.9	275.215
1200	38443.8	28466.5	249.855	2400	83103.1	63148.5	275.538
1220	39157.8	29014.2	250.445	2420	83874.9	63754.0	275.858
1240	39873.1	29563.2	251.027	2440	84647.5	64360.3	276.176
1260	40589.6	30113.5	251.600	2460	85420.9	64967.4	276.492
1280	41307.5	30665.1	252.165	2480	86195.1	65575.4	276.805
1300	42026.6	31217.9	252.723	2500	86970.2	66184.2	277.116
1320	42746.9	31771.9	253.272	2520	87746.1	66793.8	277.425
1340	43468.3	32327.0	253.815	2540	88522.7	67404.2	277.732
1360	44191.0	32883.4	254.350	2560	89300.2	68015.4	278.037
1380	44914.7	33440.9	254.879	2580	90078.5	68627.4	278.340

Abridged from Keenan, J. H., Chao, J., and Kaye, J., "Gas Tables," Wiley, New York, 1980. Reprinted by permission of John Wiley & Sons, Inc.

TABLE B6.7 (SI UNITS)
Properties of Water Vapor as an Ideal Gas with Variable Specific Heats
Temperature in Kelvins
$\bar{h}$ *in kiloJoules per kilogram-mole*
$\bar{u}$ *in kiloJoules per kilogram-mole*
$\bar{s}_0$ *in kiloJoules per kilogram-mole and Kelvin*
M = 18.0152

T	$\bar{h}$	$\bar{u}$	$\bar{s}_0$	T	$\bar{h}$	$\bar{u}$	$\bar{s}_0$
200	6621.8	4958.9	175.369	1400	53396.6	41756.5	247.290
220	7289.3	5460.1	178.549	1420	54319.8	42513.4	247.945
240	7956.8	5961.3	181.454	1440	55247.2	43274.4	248.593
260	8625.4	6463.7	184.129	1460	56179.0	44040.0	249.236
280	9295.2	6967.2	186.611	1480	57114.7	44809.4	249.872
298	**9899.0**	**7421.3**	**188.697**				
300	9966.1	7471.8	188.925	1500	58054.5	45582.9	250.503
320	10639.0	7978.4	191.097	1520	58998.2	46360.3	251.128
340	11314.0	8487.1	193.143	1540	59946.0	47141.8	251.748
360	11991.5	8998.3	195.079	1560	60897.6	47927.1	252.362
380	12672.0	9512.5	196.918	1580	61853.1	48716.3	252.970
400	13355.6	10029.8	198.672	1600	62811.9	49508.9	253.573
420	14042.6	10550.5	200.347	1620	63774.9	50305.6	254.171
440	14733.3	11074.9	201.954	1640	64741.3	51105.7	254.764
460	15427.8	11603.2	203.498	1660	65711.5	51909.6	255.352
480	16126.2	12135.2	204.983	1680	66685.0	52716.8	255.935
500	16828.4	12671.2	206.417	1700	67662.0	53527.5	256.513
520	17535.2	13211.7	207.803	1720	68642.2	54341.5	257.087
540	18246.1	13756.3	209.144	1740	69626.1	55159.1	257.655
560	18961.3	14305.2	210.445	1760	70613.2	55979.8	258.219
580	19681.0	14858.6	211.708	1780	71603.4	56803.8	258.779
600	20405.3	15416.6	212.935	1800	72596.8	57630.9	259.334
620	21134.1	15979.1	214.130	1820	73593.2	58460.9	259.884
640	21867.4	16546.2	215.294	1840	74592.9	59294.4	260.430
660	22605.6	17118.1	216.430	1860	75595.4	60130.6	260.972
680	23348.2	17694.4	217.538	1880	76601.1	60970.0	261.510
700	24095.7	18257.6	218.622	1900	77609.5	61812.1	262.044
720	24848.1	18861.7	219.681	1920	78620.8	62657.2	262.573
740	25605.3	19452.7	220.719	1940	79634.9	63505.0	263.099
760	26367.3	20048.3	221.735	1960	80651.9	64355.7	263.620
780	27134.3	20649.0	222.731	1980	81671.5	65208.9	264.138
800	27906.2	21254.7	223.708	2000	82693.7	66064.9	264.652
820	28683.2	21865.4	224.667	2020	83718.4	66923.4	265.161
840	29465.2	22481.1	225.609	2040	84746.2	67784.8	265.668
860	30252.1	23101.7	226.535	2060	85776.1	68648.5	266.170

continued

TABLE B6.7 (SI UNITS) *continued*
Properties of Water Vapor as an Ideal Gas with Variable Specific Heats
Temperature in Kelvins
$\bar{h}$ in kiloJoules per kilogram-mole
$\bar{u}$ in kiloJoules per kilogram-mole
$\bar{s}_0$ in kiloJoules per kilogram-mole and Kelvin
M = 18.0152

T	$\bar{h}$	$\bar{u}$	$\bar{s}_0$	T	$\bar{h}$	$\bar{u}$	$\bar{s}_0$
880	31044.1	23727.4	227.446	2080	86808.6	69514.6	266.669
900	31841.2	24358.3	228.342	2100	87843.6	70383.3	267.164
920	32643.6	24994.4	229.223	2120	88881.0	71254.5	267.656
940	33451.0	25635.5	230.092	2140	89920.8	72127.9	268.144
960	34263.7	26281.8	230.947	2160	90962.8	73003.7	268.629
980	35081.3	26933.2	231.790	2180	92007.3	73881.9	269.110
1000	35904.1	27589.7	232.621	2200	93053.8	74762.1	269.588
1020	36732.0	28251.3	233.441	2220	94102.7	75644.7	270.062
1040	37565.0	28918.0	234.249	2240	95153.8	76529.5	270.534
1060	38403.1	29589.8	235.048	2260	96206.9	77416.3	271.002
1080	39246.1	30266.6	235.836	2280	97262.1	78305.2	271.467
1100	40094.4	30948.6	236.614	2300	98319.5	79196.4	271.928
1120	40947.5	31635.3	237.383	2320	99378.8	80089.4	272.387
1140	41805.7	32327.3	238.142	2340	100440.3	80984.6	272.842
1160	42669.0	33024.3	238.893	2360	101503.9	81881.9	273.295
1180	43537.0	33726.1	239.634	2380	102569.3	82781.0	273.745
1200	44410.1	34432.8	240.368	2400	103636.5	83682.0	274.191
1220	45287.7	35144.1	241.093	2420	104705.8	84585.0	274.635
1240	46170.2	35860.4	241.811	2440	105777.1	85490.0	275.076
1260	47057.6	36581.4	242.521	2460	106850.0	86396.6	275.514
1280	47949.5	37307.0	243.223	2480	107924.7	87305.0	275.949
1300	48846.1	38037.4	243.918	2500	109001.3	88215.2	276.381
1320	49747.3	38772.3	244.606	2520	110079.6	89127.3	276.811
1340	50652.8	39511.5	245.287	2540	111159.5	90041.0	277.238
1360	51563.1	40255.5	245.961	2560	112241.5	90956.7	277.662
1380	52477.7	41003.8	246.629	2580	113325.1	91873.9	278.083

Abridged from Keenan, J. H., Chao, J., and Kaye, J., "Gas Tables," Wiley, New York, 1980. Reprinted by permission of John Wiley & Sons, Inc.

TABLE B7 (SI UNITS)

Enthalpy and Gibbs Free Energy of Formation, and Standard Absolute Entropy of Selected Substances

$\overline{h}_f^0$ in kiloJoules per kilogram-mole

$\overline{g}_f^0$ in kiloJoules per kilogram-mole

$\overline{s}^0$ in kiloJoules per kilogram-mole and Kelvin

Substance	Formula	M	$\overline{h}_f^0$	$\overline{g}_f^0$	$\overline{s}^0$
Carbon	$C(s)$	12.0112	0	0	5.74
Hydrogen	$H_2(g)$	2.0159	0	0	130.57
Nitrogen	$N_2(g)$	28.0134	0	0	191.50
Oxygen	$O_2(g)$	31.9988	0	0	205.04
Carbon Monoxide	$CO(g)$	28.0105	−110,530	−137,150	197.56
Carbon Dioxide	$CO_2(g)$	44.0100	−393,520	−394,380	213.67
Water	$H_2O(g)$	18.0153	−241,820	−228,590	188.72
Water	$H_2O(f)$	18.0153	−285,830	−237,180	69.95
Hydrogen Peroxide	$H_2O_2(g)$	34.0147	−136,310	−105,600	232.63
Ammonia	$NH_3(g)$	17.0306	−46,190	−16,590	192.33
Oxygen	$O(g)$	15.9994	249,170	231,770	160.95
Hydrogen	$H(g)$	1.0080	218,000	203,290	114.61
Nitrogen	$N(g)$	14.0067	472,680	455,510	153.19
Hydroxyl	$OH(g)$	17.0074	39,040	34,280	183.75
Methane	$CH_4(g)$	16.0112	−74,850	−50,790	186.16
Acetylene (Ethyne)	$C_2H_2(g)$	26.0382	226,730	209,170	200.85
Ethylene (Ethene)	$C_2H_4(g)$	28.0542	52,280	68,120	219.83
Ethane	$C_2H_6(g)$	30.0701	−84,680	−32,890	229.49
Propylene (Propene)	$C_3H_6(g)$	42.0813	20,410	62,720	266.94
Propane	$C_3H_8(g)$	44.0972	−103,850	−23,490	269.91
n-Butane	$C_4H_{10}(g)$	58.1243	−126,150	−15,710	310.03
n-Pentane	$C_5H_{12}(g)$	72.1514	−146,440	−8,200	348.40
n-Octane	$C_8H_{18}(g)$	114.2327	−208,450	17,320	463.67
n-Octane	$C_8H_{18}(f)$	114.2327	−249,910	6,610	360.79
Benzene	$C_6H_6(g)$	78.1147	82,930	129,660	269.20
Methyl Alcohol	$CH_3OH(g)$	32.0424	−200,890	−162,140	239.70
Methyl Alcohol	$CH_3OH(f)$	32.0424	−238,810	−166,290	126.80
Ethyl Alcohol	$C_2H_5OH(g)$	44.0536	−235,310	−168,570	282.59
Ethyl Alcohol	$C_2H_5OH(f)$	44.0536	−277,690	−174,890	160.70

Abridged from Wark, K., Jr., "Thermodynamics," McGraw-Hill, New York, 1988 as taken from JANAF, "Thermochemical Tables," Dow Chemical Company, Midland, MI, 1971; "Selected Values of Chemical Thermodynamic Properties," NBS Technical Note 270-3, 1968; and "API Research Project 44," Carnegie Press, 1953. Reproduced with permission of McGraw-Hill.

TABLE B8 (SI UNITS)
Logarithms to Base e of the Equilibrium Constant K_p
The Equilibrium Constant is Defined as $K_p \equiv \prod_{\text{prods}} y_i^{v_i} / \prod_{\text{reacts}} y_i^{v_i}$

(1) $H_2 \leftrightarrow 2N$

(2) $O_2 \leftrightarrow 2O$

(3) $N_2 \leftrightarrow 2N$

(4) $H_2O \leftrightarrow H_2 + \frac{1}{2}O_2$

(5) $\frac{1}{2}H_2 + \frac{1}{2}O_2 \leftrightarrow OH$

(6) $CO_2 \leftrightarrow C + O_2$

(7) $CO \leftrightarrow C + \frac{1}{2}O_2$

(8) $\frac{1}{2}O_2 + \frac{1}{2}N_2 \leftrightarrow NO$

(9) $O_2 + \frac{1}{2}N_2 \leftrightarrow NO_2$

(10) $\frac{1}{2}O_2 + N_2 \leftrightarrow N_2O$

(11) $C + 2H_2 \leftrightarrow CH_4$

Temp °K	°R	1	2	3	4	5	6	7	8	9	10	11
600	1080	−75.230	−85.536	−175.369	−42.904	−5.913	−79.220	−32.968	−16.394	−14.071	−25.421	−4.607
800	1440	−53.139	−60.332	−127.766	−30.599	−3.970	−59.476	−27.433	−12.072	−12.473	−21.306	−0.336
1000	1800	−39.816	−45.163	−99.140	−23.169	−2.814	−47.617	−24.083	−9.353	−11.513	−18.814	2.328
1200	2160	−30.887	−35.018	−80.024	−18.188	−2.049	−39.703	−21.826	−7.541	−10.871	−17.136	4.147
1400	2520	−24.476	−27.755	−66.342	−14.615	−1.510	−34.044	−20.196	−6.245	−10.405	−15.920	5.462
1600	2880	−19.646	−22.298	−56.068	−11.927	−1.110	−29.795	−18.959	−5.273	−10.055	−14.999	6.454
1800	3240	−15.879	−18.043	−48.064	−9.832	−0.799	−26.489	−17.985	−4.518	−9.781	−14.271	7.228
2000	3600	−12.848	−14.635	−41.658	−8.151	−0.553	−23.839	−17.198	−3.912	−9.560	−13.684	7.847
2200	3960	−10.366	−11.840	−36.404	−6.774	−0.352	−21.670	−16.544	−3.417	−9.381	−13.198	8.351
2400	4320	−8.289	−9.510	−32.024	−5.625	−0.189	−19.860	−15.994	−3.005	−9.229	−12.789	8.771
2600	4680	−6.526	−7.534	−28.317	−4.654	−0.048	−18.329	−15.522	−2.657	−9.102	−12.436	9.123
2800	5040	−5.015	−5.839	−25.130	−3.818	0.069	−17.011	−15.112	−2.360	−8.992	−12.135	9.424
3000	5400	−3.698	−4.370	−22.372	−3.092	0.170	−15.869	−14.753	−2.102	−8.897	−11.868	9.685
3200	5760	−2.547	−3.085	−19.950	−2.457	0.258	−14.870	−14.435	−1.877	−8.814	−11.633	9.910
3400	6120	−1.529	−1.948	−17.813	−1.897	0.334	−13.986	−14.149	−1.679	−8.743	−11.425	10.111
3600	6480	−0.622	−0.939	−15.911	−1.398	0.401	−13.198	−13.894	−1.504	−8.681	−11.239	10.286
3800	6840	0.193	−0.032	−14.212	−0.951	0.461	−12.492	−13.661	−1.347	−8.625	−11.069	10.445
4000	7200	0.926	0.783	−12.673	−0.548	0.513	−11.856	−13.449	−1.207	−8.575	−10.917	10.587

Source: Values obtained from Strehlow, Roger A., "Combustion Fundamentals," McGraw-Hill, New York, 1984. Based on thermodynamic data given in JANAF, "Thermochemical Tables," The Dow Chemical Company, Midland, MI 1971. Reproduced with permission of Roger Strehlow.

Property Tables (USCS Units)

TABLE B1.1 (USCS UNITS)
Properties of Saturated Water
Primary Argument: Temperature
v in cubic feet per pound-mass
u and h in British thermal units per pound-mass
s in British thermal units per pound-mass and degree Rankine

Temp. °F T	Pres. psia P	Specific Volume		Internal Energy		Enthalpy			Entropy	
		Sat. Liq. v_f	Sat. Vap. v_g	Sat Liq. u_f	Sat. Vap. u_g	Sat. Liq. h_f	h_{fg}	Sat. Vap. h_g	Sat. Liq. s_f	Sat. Vap. s_g
32	0.0886	0.01602	3305	−0.01	1021.2	−0.01	1075.4	1075.4	−0.00003	2.1870
35	0.0999	0.01602	2948	2.99	1022.2	3.00	1073.7	1076.7	0.00607	2.1764
40	0.1217	0.01602	2445	8.02	1023.9	8.02	1070.9	1078.9	0.01617	2.1592
45	0.1475	0.01602	2037	13.04	1025.5	13.04	1068.1	1081.1	0.02618	2.1423
50	0.1780	0.01602	1704	18.06	1027.2	18.06	1065.2	1083.3	0.03607	2.1259
52	0.1917	0.01603	1589	20.06	1027.8	20.07	1064.1	1084.2	0.04000	2.1195
54	0.2064	0.01603	1482	22.07	1028.5	22.07	1063.0	1085.1	0.04391	2.1131
56	0.2219	0.01603	1383	24.08	1029.1	24.08	1061.9	1085.9	0.04781	2.1068
58	0.2386	0.01603	1292	26.08	1029.8	26.08	1060.7	1086.8	0.05169	2.1005
60	0.2563	0.01604	1207	28.08	1030.4	28.08	1059.6	1087.7	0.05555	2.0943
62	0.2751	0.01604	1129	30.09	1031.1	30.09	1058.5	1088.6	0.05940	2.0882
64	0.2952	0.01604	1056	32.09	1031.8	32.09	1057.3	1089.4	0.06323	2.0821
66	0.3165	0.01604	988.4	34.09	1032.4	34.09	1056.2	1090.3	0.06704	2.0761
68	0.3391	0.01605	925.8	36.09	1033.1	36.09	1055.1	1091.2	0.07084	2.0701
70	0.3632	0.01605	867.7	38.09	1033.7	38.09	1054.0	1092.0	0.07463	2.0642
72	0.3887	0.01606	813.7	40.09	1034.4	40.09	1052.8	1092.9	0.07839	2.0584
74	0.4158	0.01606	763.5	42.09	1035.0	42.09	1051.7	1093.8	0.08215	2.0526
76	0.4446	0.01606	716.8	44.09	1035.7	44.09	1050.6	1094.7	0.08589	2.0469
78	0.4750	0.01607	673.3	46.09	1036.3	46.09	1049.4	1095.5	0.08961	2.0412
80	0.5073	0.01607	632.8	48.08	1037.0	48.09	1048.3	1096.4	0.09332	2.0356
82	0.5414	0.01608	595.0	50.08	1037.6	50.08	1047.2	1097.3	0.09701	2.0300
84	0.5776	0.01608	559.8	52.08	1038.3	52.08	1046.0	1098.1	0.1007	2.0245
86	0.6158	0.01609	527.0	54.08	1038.9	54.08	1044.9	1099.0	0.1044	2.0190
88	0.6562	0.01609	496.3	56.07	1039.6	56.07	1043.8	1099.9	0.1080	2.0136
90	0.6988	0.01610	467.7	58.07	1040.2	58.07	1042.7	1100.7	0.1117	2.0083
92	0.7439	0.01611	440.9	60.06	1040.9	60.06	1041.5	1101.6	0.1153	2.0030
94	0.7914	0.01611	415.9	62.06	1041.5	62.06	1040.4	1102.4	0.1189	1.9977
96	0.8416	0.01612	392.4	64.05	1042.2	64.06	1039.2	1103.3	0.1225	1.9925
98	0.8945	0.01612	370.5	66.05	1042.8	66.05	1038.1	1104.2	0.1261	1.9874
100	0.9503	0.01613	350.0	68.04	1043.5	68.05	1037.0	1105.0	0.1296	1.9822

continued

TABLE B1.1 (USCS UNITS) *continued*

Properties of Saturated Water
Primary Argument: Temperature
v in cubic feet per pound-mass
u and h in British thermal units per pound-mass
s in British thermal units per pound-mass and degree Rankine

Temp. °F	Pres. psia	Specific Volume		Internal Energy		Enthalpy			Entropy	
		Sat. Liq.	Sat. Vap.	Sat Liq.	Sat. Vap.	Sat. Liq.		Sat. Vap.	Sat. Liq.	Sat. Vap.
T	P	v_f	v_g	u_f	u_g	h_f	h_{fg}	h_g	s_f	s_g
110	1.276	0.01617	265.1	78.02	1046.7	78.02	*1031.3	1109.3	0.1473	1.9574
120	1.695	0.01621	203.0	87.99	1049.9	88.00	1025.5	1113.5	0.1647	1.9336
130	2.225	0.01625	157.2	97.97	1053.0	97.98	1019.8	1117.8	0.1817	1.9109
140	2.892	0.01629	122.9	107.95	1056.2	107.96	1014.0	1121.9	0.1985	1.8892
150	3.722	0.01634	97.0	117.95	1059.3	117.96	1008.1	1126.1	0.2150	1.8684
160	4.745	0.01640	77.2	127.94	1062.3	127.96	1002.2	1130.1	0.2313	1.8484
170	5.996	0.01645	62.0	137.95	1065.4	137.97	996.2	1134.2	0.2473	1.8293
180	7.515	0.01651	50.2	147.97	1068.3	147.99	990.2	1138.2	0.2631	1.8109
190	9.343	0.01657	41.0	158.00	1071.3	158.03	984.1	1142.1	0.2787	1.7932
200	11.529	0.01663	33.6	168.04	1074.2	168.07	977.9	1145.9	0.2940	1.7762
210	14.13	0.01670	27.82	178.1	1077.0	178.1	971.6	1149.7	0.3091	1.7599
212	14.70	0.01672	26.80	180.1	1077.6	180.2	970.3	1150.5	0.3121	1.7567
220	17.19	0.01677	23.15	188.2	1079.8	188.2	965.3	1153.5	0.3241	1.7441
230	20.78	0.01685	19.39	198.3	1082.6	198.3	958.8	1157.1	0.3388	1.7289
240	24.97	0.01692	16.33	208.4	1085.3	208.4	952.3	1160.7	0.3534	1.7143
250	29.82	0.01700	13.83	218.5	1087.9	218.6	945.6	1164.2	0.3677	1.7001
260	35.42	0.01708	11.77	228.6	1090.5	228.8	938.8	1167.6	0.3819	1.6864
270	41.85	0.01717	10.07	238.8	1093.0	239.0	932.0	1170.9	0.3960	1.6731
280	49.18	0.01726	8.65	249.0	1095.4	249.2	924.9	1174.1	0.4099	1.6602
290	57.53	0.01735	7.47	259.3	1097.7	259.4	917.8	1177.2	0.4236	1.6477
300	66.98	0.01745	6.472	269.5	1100.0	269.7	910.4	1180.2	0.4372	1.6356
310	77.64	0.01755	5.632	279.8	1102.1	280.1	903.0	1183.0	0.4507	1.6238
320	89.60	0.01765	4.919	290.1	1104.2	290.4	895.3	1185.8	0.4640	1.6123
330	103.00	0.01776	4.312	300.5	1106.2	300.8	887.5	1188.4	0.4772	1.6010
340	117.93	0.01787	3.792	310.9	1108.0	311.3	879.5	1190.8	0.4903	1.5901
350	134.53	0.01799	3.346	321.4	1109.8	321.8	871.3	1193.1	0.5033	1.5793
360	152.92	0.01811	2.961	331.8	1111.4	332.4	862.9	1195.2	0.5162	1.5688
370	173.23	0.01823	2.628	342.4	1112.9	343.0	854.2	1197.2	0.5289	1.5585
380	195.60	0.01836	2.339	353.0	1114.3	353.6	845.4	1199.0	0.5416	1.5483
390	220.20	0.01850	2.087	363.6	1115.6	364.3	836.2	1200.6	0.5542	1.5383
400	247.1	0.01864	1.866	374.3	1116.6	375.1	826.8	1202.0	0.5667	1.5284
410	276.5	0.01878	1.673	385.0	1117.6	386.0	817.2	1203.1	0.5792	1.5187
420	308.5	0.01894	1.502	395.8	1118.3	396.9	807.2	1204.1	0.5915	1.5091
430	343.3	0.01909	1.352	406.7	1118.9	407.9	796.9	1204.8	0.6038	1.4995
440	381.2	0.01926	1.219	417.6	1119.3	419.0	786.3	1205.3	0.6161	1.4900

continued

TABLE B1.1 (USCS UNITS) *continued*
Properties of Saturated Water
Primary Argument: Temperature
v in cubic feet per pound-mass
u and h in British thermal units per pound-mass
s in British thermal units per pound-mass and degree Rankine

Temp. °F T	Pres. psia P	Specific Volume		Internal Energy		Enthalpy			Entropy	
		Sat. Liq. v_f	Sat. Vap. v_g	Sat Liq. u_f	Sat. Vap. u_g	Sat. Liq. h_f	h_{fg}	Sat. Vap. h_g	Sat. Liq. s_f	Sat. Vap. s_g
450	422.1	0.01943	1.1011	428.6	1119.5	430.2	775.4	1205.6	0.6282	1.4806
460	466.3	0.01961	0.9961	439.7	1119.6	441.4	764.1	1205.5	0.6404	1.4712
470	514.1	0.01980	0.9025	450.9	1119.4	452.8	752.4	1205.2	0.6525	1.4618
480	565.5	0.02000	0.8187	462.2	1118.9	464.3	740.3	1204.6	0.6646	1.4524
490	620.7	0.02021	0.7436	473.6	1118.3	475.9	727.8	1203.7	0.6767	1.4430
500	680.0	0.02043	0.6761	485.1	1117.4	487.7	714.8	1202.5	0.6888	1.4335
520	811.4	0.02091	0.5605	508.5	1114.8	511.7	687.3	1198.9	0.7130	1.4145
540	961.5	0.02145	0.4658	532.6	1111.0	536.4	657.5	1193.8	0.7374	1.3950
560	1131.8	0.02207	0.3877	557.4	1105.8	562.0	625.0	1187.0	0.7620	1.3749
580	1324.3	0.02278	0.3225	583.1	1098.9	588.6	589.3	1178.0	0.7872	1.3540
600	1541.0	0.02363	0.2677	609.9	1090.0	616.7	549.7	1166.4	0.8130	1.3317
620	1784.4	0.02465	0.2209	638.3	1078.5	646.4	505.0	1151.4	0.8398	1.3075
640	2057.1	0.02593	0.1805	668.7	1063.2	678.6	453.4	1131.9	0.8681	1.2803
660	2362.0	0.02767	0.1446	702.3	1042.3	714.4	391.1	1105.5	0.8990	1.2483
680	2705.0	0.03032	0.1113	741.7	1011.0	756.9	309.8	1066.7	0.9350	1.2068
700	3090	0.03666	0.0744	801.7	947.7	822.7	167.5	990.2	0.9902	1.1346
705.4	3204	0.05053	0.0503	872.6	872.6	902.5	0.0	902.5	1.0580	1.0580

Abridged from J. H. Keenan, F. G. Keyes, P. G. Hill, and J. G. Moore, "Steam Tables," Wiley, New York, 1979. Reprinted by permission of John Wiley & Sons, Inc.

TABLE B1.2 (USCS UNITS)
Properties of Saturated Water
Primary Argument: Pressure
v in cubic feet per pound-mass
u and h in British thermal units per pound-mass
s in British thermal units per pound-mass and degree Rankine

Pres. psia P	Temp. °F T	Specific Volume		Internal Energy		Enthalpy			Entropy	
		Sat. Liq. v_f	Sat. Vap. v_g	Sat. Liq. u_f	Sat. Vap. u_g	Sat. Liq. h_f	h_{fg}	Sat. Vap. h_g	Sat. Liq. s_f	Sat. Vap. s_g
0.40	72.84	0.01606	792.0	40.94	1034.7	40.94	1052.3	1093.3	0.0800	2.0559
0.60	85.19	0.01609	540.0	53.26	1038.7	53.27	1045.4	1098.6	0.1029	2.0213
0.80	94.35	0.01611	411.7	62.41	1041.7	62.41	1040.2	1102.6	0.1195	1.9968
1.0	101.70	0.01614	333.6	69.74	1044.0	69.74	1036.0	1105.8	0.1327	1.9779
1.2	107.88	0.01616	280.9	75.90	1046.0	75.90	1032.5	1108.4	0.1436	1.9626
1.5	115.65	0.01619	227.70	83.65	1048.5	83.65	1028.0	1111.7	0.1571	1.9438
2.0	126.04	0.01623	173.75	94.02	1051.8	94.02	1022.1	1116.1	0.1750	1.9198
3.0	141.43	0.01630	118.72	109.38	1056.6	109.39	1013.1	1122.5	0.2009	1.8861
4.0	152.93	0.01636	90.64	120.88	1060.2	120.89	1006.4	1127.3	0.2198	1.8624
5.0	162.21	0.01641	73.53	130.15	1063.0	130.17	1000.9	1131.0	0.2349	1.8441
6.0	170.03	0.01645	61.98	137.98	1065.4	138.00	996.2	1134.2	0.2474	1.8292
7.0	176.82	0.01649	53.65	144.78	1067.4	144.80	992.1	1136.9	0.2581	1.8167
8.0	182.84	0.01653	47.35	150.81	1069.2	150.84	988.4	1139.3	0.2675	1.8058
9.0	188.26	0.01656	42.41	156.25	1070.8	156.27	985.1	1141.4	0.2760	1.7963
10.0	193.19	0.01659	38.42	161.20	1072.2	161.23	982.1	1143.3	0.2836	1.7877
14.696	211.99	0.01672	26.80	180.10	1077.6	180.15	970.4	1150.5	0.3121	1.7567
15	213.03	0.01672	26.29	181.14	1077.9	181.19	969.7	1150.9	0.3137	1.7551
20	227.96	0.01683	20.09	196.19	1082.0	196.26	960.1	1156.4	0.3358	1.7320
25	240.08	0.01692	16.31	208.44	1085.3	208.52	952.2	1160.7	0.3535	1.7142
30	250.34	0.01700	13.75	218.84	1088.0	218.93	945.4	1164.3	0.3682	1.6996
35	259.30	0.01708	11.90	227.93	1090.3	228.04	939.3	1167.4	0.3809	1.6873
40	267.26	0.01715	10.50	236.03	1092.3	236.16	933.8	1170.0	0.3921	1.6767
45	274.46	0.01721	9.40	243.37	1094.0	243.51	928.8	1172.3	0.4022	1.6673
50	281.03	0.01727	8.52	250.08	1095.6	250.24	924.2	1174.4	0.4113	1.6589
55	287.10	0.01733	7.79	256.28	1097.0	256.46	919.9	1176.3	0.4196	1.6513
60	292.73	0.01738	7.177	262.1	1098.3	262.25	915.8	1178.0	0.4273	1.6444
65	298.00	0.01743	6.657	267.5	1099.5	267.67	911.9	1179.6	0.4345	1.6380
70	302.96	0.01748	6.209	272.6	1100.6	272.79	908.3	1181.0	0.4412	1.6321
75	307.63	0.01752	5.818	277.4	1101.6	277.61	904.8	1182.4	0.4475	1.6265
80	312.07	0.01757	5.474	282.0	1102.6	282.21	901.4	1183.6	0.4534	1.6214
85	316.29	0.01761	5.170	286.3	1103.5	286.58	898.2	1184.8	0.4591	1.6165
90	320.31	0.01766	4.898	290.5	1104.3	290.76	895.1	1185.9	0.4644	1.6119
95	324.16	0.01770	4.654	294.5	1105.0	294.76	892.1	1186.9	0.4695	1.6076
100	327.86	0.01774	4.434	298.3	1105.8	298.61	889.2	1187.8	0.4744	1.6034
110	334.82	0.01781	4.051	305.5	1107.1	305.88	883.7	1189.6	0.4836	1.5957

continued

Property Tables

TABLE B1.2 (USCS UNITS) *continued*
Properties of Saturated Water
Primary Argument: Pressure
v in cubic feet per pound-mass
u and h in British thermal units per pound-mass
s in British thermal units per pound-mass and degree Rankine

Pres. psia P	Temp. °F T	Specific Volume Sat. Liq. v_f	Sat. Vap. v_g	Internal Energy Sat Liq. u_f	Sat. Vap. u_g	Enthalpy Sat. Liq. h_f	h_{fg}	Sat. Vap. h_g	Entropy Sat. Liq. s_f	Sat. Vap. s_g
120	341.30	0.01789	3.730	312.3	1108.3	312.67	878.5	1191.1	0.4920	1.5886
130	347.37	0.01796	3.457	318.6	1109.4	319.04	873.5	1192.5	0.4999	1.5821
140	353.08	0.01802	3.221	324.6	1110.3	325.05	868.7	1193.8	0.5073	1.5761
150	358.48	0.01809	3.016	330.2	1111.2	330.75	864.2	1194.9	0.5142	1.5704
160	363.60	0.01815	2.836	335.6	1112.0	336.16	859.8	1196.0	0.5208	1.5651
170	368.47	0.01821	2.676	340.8	1112.7	341.3	855.6	1196.9	0.5270	1.5600
180	373.13	0.01827	2.533	345.7	1113.4	346.3	851.5	1197.8	0.5329	1.5553
190	377.59	0.01833	2.405	350.4	1114.0	351.0	847.5	1198.6	0.5386	1.5507
200	381.86	0.01839	2.289	354.9	1114.6	355.6	843.7	1199.3	0.5440	1.5464
250	401.04	0.01865	1.845	375.4	1116.7	376.2	825.8	1202.1	0.5680	1.5274
300	417.43	0.01890	1.544	393.0	1118.2	394.1	809.8	1203.9	0.5883	1.5115
350	431.82	0.01912	1.327	408.7	1119.0	409.9	795.0	1204.9	0.6060	1.4978
400	444.70	0.01934	1.162	422.8	1119.5	424.2	781.2	1205.5	0.6218	1.4856
450	456.39	0.01955	1.033	435.7	1119.6	437.4	768.2	1205.6	0.6360	1.4746
500	467.13	0.01975	0.928	447.7	1119.4	449.5	755.8	1205.3	0.6490	1.4645
550	477.07	0.01994	0.842	458.9	1119.1	460.9	743.9	1204.8	0.6611	1.4551
600	486.33	0.02013	0.770	469.4	1118.6	471.7	732.4	1204.1	0.6723	1.4464
700	503.23	0.02051	0.656	488.9	1117.0	491.5	710.5	1202.0	0.6927	1.4305
800	518.36	0.02087	0.569	506.6	1115.0	509.7	689.6	1199.3	0.7110	1.4160
900	532.12	0.02123	0.501	523.0	1112.6	526.6	669.5	1196.0	0.7277	1.4027
1000	544.75	0.02159	0.446	538.4	1109.9	542.4	650.0	1192.4	0.7432	1.3903
1100	556.45	0.02195	0.401	552.9	1106.8	557.4	631.0	1188.3	0.7576	1.3786
1200	567.37	0.02232	0.362	566.7	1103.5	571.7	612.3	1183.9	0.7712	1.3673
1300	577.60	0.02269	0.330	579.9	1099.8	585.4	593.8	1179.2	0.7841	1.3565
1400	587.25	0.02307	0.302	592.7	1096.0	598.6	575.5	1174.1	0.7964	1.3461
1500	596.39	0.02346	0.277	605.0	1091.8	611.5	557.2	1168.7	0.8082	1.3359
1600	605.06	0.02386	0.255	616.9	1087.4	624.0	538.9	1162.9	0.8196	1.3258
1700	613.32	0.02428	0.236	628.6	1082.7	636.2	520.6	1156.9	0.8307	1.3159
1800	621.21	0.02472	0.218	640.0	1077.7	648.3	502.1	1150.4	0.8414	1.3060
1900	628.76	0.02517	0.203	651.3	1072.3	660.1	483.4	1143.5	0.8519	1.2961
2000	636.00	0.02565	0.188	662.4	1066.6	671.9	464.4	1136.3	0.8623	1.2861
2250	652.90	0.02698	0.157	689.9	1050.6	701.1	414.8	1115.9	0.8876	1.2604
2500	668.31	0.02860	0.131	717.7	1031.0	730.9	360.5	1091.4	0.9131	1.2327
2750	682.46	0.03077	0.107	747.3	1005.9	763.0	297.4	1060.4	0.9401	1.2005
3000	695.52	0.03431	0.084	783.4	968.8	802.5	213.0	1015.5	0.9732	1.1575
3203.6	705.44	0.05053	0.05053	872.6	872.6	902.5		902.5	1.0580	1.0580

Abridged from J. H. Keenan, F. G. Keyes, P. G. Hill, and J. G. Moore, "Steam Tables," Wiley, New York, 1979.

TABLE B1.3 (USCS UNITS)

Properties of Water Vapor

v in cubic feet per pound-mass
u and h in British thermal units per pound-mass
s in British thermal units per pound-mass and degree Rankine

Temp °F	Pressure = 1 psia				Temp °F	Pressure = 5 psia			
	v	u	h	s		v	u	h	s
101.7	333.6	1044.0	1105.8	1.9779					
150	362.6	1060.4	1127.5	2.0151	162.2	73.53	1063.0	1131.0	1.8441
200	392.5	1077.5	1150.1	2.0508	200	78.15	1076.3	1148.6	1.8715
250	422.4	1094.7	1172.8	2.0839	250	84.21	1093.8	1171.7	1.9052
300	452.3	1112.0	1195.7	2.1150	300	90.24	1111.3	1194.8	1.9367
400	511.9	1147.0	1241.8	2.1720	400	102.24	1146.6	1241.2	1.9941
500	571.5	1182.8	1288.5	2.2235	500	114.20	1182.5	1288.2	2.0458
600	631.1	1219.3	1336.1	2.2706	600	126.15	1219.1	1335.8	2.0930
700	690.7	1256.7	1384.5	2.3142	700	138.08	1256.5	1384.3	2.1367
800	750.5	1294.9	1433.7	2.3550	800	150.01	1294.7	1433.5	2.1775
900	809.9	1333.9	1483.8	2.3932	900	161.94	1333.8	1483.7	2.2158
1000	869.5	1373.9	1534.8	2.4294	1000	173.86	1373.9	1534.7	2.2520

Temp °F	Pressure = 10 psia				Temp °F	Pressure = 14.7 psia			
	v	u	h	s		v	u	h	s
193.2	38.42	1072.2	1143.3	1.7877					
200	38.85	1074.7	1146.6	1.7927	212.0	26.80	1077.6	1150.5	1.7567
250	41.93	1092.6	1170.2	1.8272	250	28.42	1091.5	1168.8	1.7832
300	44.99	1110.4	1193.7	1.8592	300	30.52	1109.6	1192.6	1.8157
400	51.03	1146.1	1240.5	1.9171	400	34.67	1145.6	1239.9	1.8741
500	57.04	1182.2	1287.7	1.9690	500	38.77	1181.8	1287.3	1.9263
600	63.03	1218.9	1335.5	2.0164	600	42.86	1218.6	1335.2	1.9737
700	69.01	1256.3	1384.0	2.0601	700	46.93	1256.1	1383.8	2.0175
800	74.98	1294.6	1433.3	2.1009	800	51.00	1294.4	1433.1	2.0584
900	80.95	1333.7	1483.5	2.1393	900	55.07	1333.6	1483.4	2.0967
1000	86.91	1373.8	1534.6	2.1755	1000	59.13	1373.7	1534.5	2.1330
1100	92.88	1414.7	1586.6	2.2099	1100	63.19	1414.6	1586.4	2.1674

continued

TABLE B1.3 (USCS UNITS) *continued*

Properties of Water Vapor

v in cubic feet per pound-mass
u and h in British thermal units per pound-mass
s in British thermal units per pound-mass and degree Rankine

Temp °F	Pressure = 20 psia				Temp °F	Pressure = 40 psia			
	v	u	h	s		v	u	h	s
228.0	20.09	1082.0	1156.4	1.7320					
250	20.79	1090.3	1167.2	1.7475	267.3	10.50	1092.3	1170.0	1.6767
300	22.36	1108.7	1191.5	1.7805	300	11.04	1105.1	1186.8	1.6993
350	23.90	1126.9	1215.4	1.8110	350	11.84	1124.2	1211.8	1.7312
400	25.43	1145.1	1239.2	1.8395	400	12.62	1143.0	1236.4	1.7606
500	28.46	1181.5	1286.8	1.8919	500	14.16	1180.1	1284.9	1.8140
600	31.47	1218.4	1334.8	1.9395	600	57.04	1182.2	1287.7	1.9690
700	34.47	1255.9	1383.5	1.9834	700	17.20	1255.1	1382.4	1.9063
800	37.46	1294.3	1432.9	2.0243	800	18.70	1293.7	1432.1	1.9474
900	40.45	1333.5	1483.2	2.0627	900	20.20	1333.0	1482.5	1.9859
1000	43.44	1373.5	1534.3	2.0989	1000	21.70	1373.1	1533.8	2.0223
1100	46.42	1414.5	1586.3	2.1334	1100	23.20	1414.2	1585.9	2.0568

Temp °F	Pressure = 60 psia				Temp °F	Pressure = 80 psia			
	v	u	h	s		v	u	h	s
292.7	7.18	1098.3	1178.0	1.6444					
300	7.26	1101.3	1181.9	1.6496	312.1	5.47	1102.6	1183.6	1.6214
350	7.82	121.4	1208.2	1.6830	350	5.80	1118.5	1204.3	1.6476
400	8.35	1140.8	1233.5	1.7134	400	6.22	1138.5	1230.6	1.6790
500	9.40	1178.6	1283.0	1.7678	500	7.02	1177.2	1281.1	1.7346
600	10.43	1216.3	1332.1	1.8165	600	7.79	1215.3	1330.7	1.7838
700	11.44	1254.4	1381.4	1.8609	700	8.56	1253.6	1380.3	1.8285
800	12.45	1293.0	1431.2	1.9022	800	9.32	1292.4	1430.4	1.8700
900	13.45	1332.5	1481.8	1.9408	900	10.08	1332.0	1481.2	1.9087
1000	14.45	1372.7	1533.2	1.9773	1000	10.83	1372.3	1532.6	1.9453
1100	15.45	1413.8	1585.4	2.0119	1100	11.58	1413.5	1584.9	1.9799
1200	16.45	1455.8	1638.5	2.0448	1200	12.33	1455.5	1638.1	2.0130

continued

TABLE B1.3 (USCS UNITS) *continued*

Properties of Water Vapor
v in cubic feet per pound-mass
u and h in British thermal units per pound-mass
s in British thermal units per pound-mass and degree Rankine

Temp °F	Pressure = 100 psia				Temp °F	Pressure = 120 psia			
	v	u	h	s		v	u	h	s
327.9	4.434	1105.8	1187.8	1.6034	341.3	3.730	1108.3	1191.1	1.5886
350	4.592	1115.4	1200.4	1.6191	350	3.783	1112.2	1196.2	1.5950
400	4.934	1136.2	1227.5	1.6517	400	4.079	1133.8	1224.4	1.6288
450	5.265	1156.2	1253.2	1.6812	450	4.360	1154.3	1251.2	1.6590
500	5.587	1175.7	1279.1	1.7085	500	4.633	1174.2	1277.1	1.6868
600	6.216	1214.2	1329.3	1.7582	600	5.164	1213.2	1327.8	1.7371
700	6.834	1252.8	1379.2	1.8033	700	5.682	1252.0	1378.2	1.7825
800	7.445	1291.8	1429.6	1.8449	800	6.195	1291.2	1428.7	1.8243
900	8.053	1331.5	1480.5	1.8838	900	6.703	1330.9	1479.8	1.8633
1000	8.657	1371.9	1532.1	1.9204	1000	7.208	1371.5	1531.5	1.9000
1100	9.260	1413.1	1584.5	1.9551	1100	7.711	1412.8	1584.0	1.9348
1200	9.861	1455.2	1637.7	1.9882	1200	8.213	1454.9	1637.3	1.9679

Temp °F	Pressure = 140 psia				Temp °F	Pressure = 160 psia			
	v	u	h	s		v	u	h	s
353.1	3.221	1110.3	1193.8	1.5761	363.6	2.836	1112.0	1196.0	1.5651
400	3.466	1131.4	1221.2	1.6088	400	3.007	1128.8	1217.8	1.5911
450	3.713	1152.4	1248.6	1.6399	450	3.228	1150.5	1246.1	1.6230
500	3.952	1172.7	1275.1	1.6682	500	3.440	1171.2	1273.0	1.6518
550	4.184	1192.6	1300.9	1.6945	550	3.646	1191.3	1299.2	1.6784
600	4.412	1212.1	1326.4	1.7191	600	3.848	1211.1	1325.0	1.7034
700	4.860	1251.2	1377.1	1.7648	700	4.243	1250.4	1376.0	1.7494
800	5.301	1290.5	1427.9	1.8068	800	4.631	1289.9	1427.0	1.7916
900	5.739	1330.4	1479.1	1.8459	900	5.015	1329.9	1478.4	1.8308
1000	6.173	1371.0	1531.0	1.8827	1000	5.397	1370.6	1530.4	1.8677
1100	6.605	1412.4	1583.6	1.9176	1100	5.776	1412.1	1583.1	1.9026
1200	7.036	1454.6	1636.9	1.9507	1200	6.154	1454.3	1636.5	1.9358

continued

TABLE B1.3 (USCS UNITS) *continued*
Properties of Water Vapor
v in cubic feet per pound-mass
u and h in British thermal units per pound-mass
s in British thermal units per pound-mass and degree Rankine

Temp °F	Pressure = 180 psia				Temp °F	Pressure = 200 psia			
	v	u	h	s		v	u	h	s
373.1	2.533	1113.4	1197.8	1.5553	381.9	2.289	1114.6	1199.3	1.5464
400	2.648	1126.2	1214.4	1.5749	400	2.361	1123.5	1210.8	1.5600
450	2.850	1148.5	1243.4	1.6078	450	2.548	1146.4	1240.7	1.5938
500	3.042	1169.6	1270.9	1.6372	500	2.724	1168.0	1268.8	1.6239
550	3.228	1190.0	1297.5	1.6642	550	2.893	1188.7	1295.7	1.6512
600	3.409	1210.0	1323.5	1.6893	600	3.058	1208.9	1322.1	1.6767
700	3.763	1249.6	1374.9	1.7357	700	3.379	1248.8	1373.8	1.7234
800	4.110	1289.3	1426.2	1.7781	800	3.693	1288.6	1425.3	1.7660
900	4.453	1329.4	1477.7	1.8175	900	4.003	1328.9	1477.1	1.8055
1000	4.793	1370.2	1529.8	1.8545	1000	4.310	1369.8	1529.3	1.8425
1100	5.131	1411.7	1582.6	1.8894	1100	4.615	1411.4	1582.2	1.8776
1200	5.467	1454.0	1636.1	1.9227	1200	4.918	1453.7	1635.7	1.9109

Temp °F	Pressure = 250 psia				Temp °F	Pressure = 300 psia			
	v	u	h	s		v	u	h	s
401.0	1.845	1116.7	1202.1	1.5274	417.4	1.544	1118.2	1203.9	1.5115
450	2.002	1141.1	1233.7	1.5632	450	1.636	1135.4	1226.2	1.5365
500	2.150	1163.8	1263.3	1.5948	500	1.766	1159.5	1257.5	1.5701
550	2.290	1185.3	1291.3	1.6233	550	1.888	1181.9	1286.7	1.5997
600	2.426	1206.1	1318.3	1.6494	600	2.004	1203.2	1314.5	1.6266
700	2.688	1246.7	1371.1	1.6970	700	2.227	1244.6	1368.3	1.6751
800	2.943	1287.0	1423.2	1.7401	800	2.442	1285.4	1421.0	1.7187
900	3.193	1327.6	1475.3	1.7799	900	2.653	1326.3	1473.6	1.7589
1000	3.440	1368.7	1527.9	1.8172	1000	2.860	1367.7	1526.5	1.7964
1100	3.685	1410.5	1581.0	1.8524	1100	3.066	1409.6	1579.8	1.8317
1200	3.929	1453.0	1634.8	1.8858	1200	3.270	1452.2	1633.8	1.8653
1300	4.172	1496.3	1689.3	1.9177	1300	3.473	1495.6	1688.4	1.8973

continued

TABLE B1.3 (USCS UNITS) *continued*

Properties of Water Vapor
v in cubic feet per pound-mass
u and h in British thermal units per pound-mass
s in British thermal units per pound-mass and degree Rankine

Temp °F	Pressure = 350 psia				Temp °F	Pressure = 400 psia			
	v	u	h	s		v	u	h	s
431.8	1.327	1119.0	1204.9	1.4978	444.7	1.162	1119.5	1205.5	1.4856
450	1.373	1129.2	1218.2	1.5125	450	1.175	1122.6	1209.6	1.4901
500	1.491	1154.9	1251.5	1.5482	500	1.284	1150.1	1245.2	1.5282
550	1.600	1178.3	1281.9	1.5790	550	1.383	1174.6	1277.0	1.5605
600	1.703	1200.3	1310.6	1.6068	600	1.476	1197.3	1306.6	1.5892
700	1.898	1242.5	1365.4	1.6562	700	1.650	1240.4	1362.5	1.6397
800	2.085	1283.8	1418.8	1.7004	800	1.816	1282.1	1416.6	1.6844
900	2.267	1325.0	1471.8	1.7409	900	1.978	1323.7	1470.1	1.7252
1000	2.446	1366.6	1525.0	1.7787	1000	2.136	1365.5	1523.6	1.7632
1100	2.624	1408.7	1578.6	1.8142	1100	2.292	1407.8	1577.4	1.7989
1200	2.799	1451.5	1632.8	1.8478	1200	2.446	1450.7	1631.8	1.8327
1300	2.974	1495.0	1687.6	1.8799	1300	2.599	1494.3	1686.8	1.8648

Temp °F	Pressure = 450 psia				Temp °F	Pressure = 500 psia			
	v	u	h	s		v	u	h	s
456.4	1.033	1119.6	1205.6	1.4746	467.1	0.928	1119.4	1205.3	1.4645
500	1.123	1145.1	1238.5	1.5097	500	0.992	1139.7	1231.5	1.4923
550	1.215	1170.7	1271.9	1.5436	550	1.079	1166.7	1266.6	1.5279
600	1.300	1194.3	1302.5	1.5732	600	1.158	1191.1	1298.3	1.5585
700	1.458	1238.2	1359.6	1.6248	700	1.304	1236.0	1356.7	1.6112
800	1.608	1280.5	1414.4	1.6701	800	1.441	1278.8	1412.1	1.6571
900	1.752	1322.4	1468.3	1.7113	900	1.572	1321.0	1466.5	1.6987
1000	1.894	1364.4	1522.2	1.7495	1000	1.701	1363.3	1520.7	1.7371
1100	2.034	1406.9	1576.3	1.7853	1100	1.827	1406.0	1575.1	1.7731
1200	2.172	1450.0	1630.8	1.8192	1200	1.952	1449.2	1629.8	1.8072
1300	2.308	1493.7	1685.9	1.8515	1300	2.075	1493.1	1685.1	1.8395
1400	2.444	1538.1	1741.7	1.8823	1400	2.198	1537.6	1741.0	1.8704

continued

TABLE B1.3 (USCS UNITS) *continued*
Properties of Water Vapor
v in cubic feet per pound-mass
u and h in British thermal units per pound-mass
s in British thermal units per pound-mass and degree Rankine

Temp °F	Pressure = 600 psia				Temp °F	Pressure = 700 psia			
	v	u	h	s		v	u	h	s
486.3	0.770	1118.6	1204.1	1.4464					
500	0.795	1128.0	1216.2	1.4592	503.2	0.656	1117.0	1202.0	1.4305
550	0.875	1158.2	1255.4	1.4990	550	0.728	1149.0	1243.2	1.4723
600	0.946	1184.5	1289.5	1.5320	600	0.793	1177.5	1280.2	1.5081
700	1.073	1231.5	1350.6	1.5872	700	0.907	1226.9	1344.4	1.5661
800	1.190	1275.4	1407.6	1.6343	800	1.011	1272.0	1402.9	1.6145
900	1.302	1318.4	1462.9	1.6766	900	1.109	1315.6	1459.3	1.6576
1000	1.411	1361.2	1517.8	1.7155	1000	1.204	1358.9	1514.9	1.6970
1100	1.517	1404.2	1572.7	1.7519	1100	1.296	1402.4	1570.2	1.7337
1200	1.622	1447.7	1627.8	1.7861	1200	1.387	1446.2	1625.8	1.7682
1300	1.726	1491.7	1683.4	1.8186	1300	1.476	1490.4	1681.7	1.8009
1400	1.829	1536.5	1739.5	1.8497	1400	1.565	1535.3	1738.1	1.8321

Temp °F	Pressure = 800 psia				Temp °F	Pressure = 900 psia			
	v	u	h	s		v	u	h	s
518.4	0.569	1115.0	1199.3	1.4160	532.1	0.501	1112.6	1196.0	1.4027
550	0.615	1138.8	1229.9	1.4469	550	0.527	1127.5	1215.2	1.4219
600	0.678	1170.1	1270.4	1.4861	600	0.587	1162.2	1260.0	1.4652
650	0.732	1197.2	1305.6	1.5186	650	0.639	1191.1	1297.5	1.4999
700	0.783	1222.1	1338.0	1.5471	700	0.686	1217.1	1331.4	1.5297
800	0.876	1268.5	1398.2	1.5969	800	0.772	1264.9	1393.4	1.5810
900	0.964	1312.9	1455.6	1.6408	900	0.851	1310.1	1451.9	1.6257
1000	1.048	1356.7	1511.9	1.6807	1000	0.927	1354.5	1508.9	1.6662
1100	1.130	1400.5	1567.8	1.7178	1100	1.001	1398.7	1565.4	1.7036
1200	1.210	1444.6	1623.8	1.7526	1200	1.073	1443.0	1621.7	1.7386
1300	1.289	1489.1	1680.0	1.7854	1300	1.144	1487.8	1678.3	1.7717
1400	1.367	1534.2	1736.6	1.8167	1400	1.214	1533.0	1735.1	1.8031

continued

TABLE B1.3 (USCS UNITS) *continued*

Properties of Water Vapor
v in cubic feet per pound-mass
u and h in British thermal units per pound-mass
s in British thermal units per pound-mass and degree Rankine

Temp °F	Pressure = 1000 psia				Temp °F	Pressure = 1200 psia			
	v	u	h	s		v	u	h	s
544.8	0.446	1109.9	1192.4	1.3903	567.4	0.362	1103.5	1183.9	1.3673
600	0.514	1153.7	1248.8	1.4450	600	0.402	1134.4	1223.6	1.4054
650	0.564	1184.7	1289.1	1.4822	650	0.450	1170.9	1270.8	1.4490
700	0.608	1212.0	1324.6	1.5135	700	0.491	1201.3	1310.2	1.4837
800	0.688	1261.2	1388.5	1.5664	800	0.562	1253.7	1378.4	1.5402
900	0.761	1307.3	1448.1	1.6120	900	0.626	1301.5	1440.4	1.5876
1000	0.831	1352.2	1505.9	1.6530	1000	0.685	1347.5	1499.7	1.6297
1100	0.898	1369.8	1562.9	1.6908	1100	0.743	1393.0	1557.9	1.6682
1200	0.963	1441.5	1619.7	1.7261	1200	0.798	1438.3	1615.5	1.7040
1300	1.027	1486.5	1676.5	1.7593	1300	0.853	1483.8	1673.1	1.7377
1400	1.091	1531.9	1733.7	1.7909	1400	0.906	1529.6	1730.7	1.7696
1600	1.215	1624.4	1849.3	1.8499	1600	1.011	1622.6	1847.1	1.8290

Temp °F	Pressure = 1400 psia				Temp °F	Pressure = 1600 psia			
	v	u	h	s		v	u	h	s
587.3	0.302	1096.0	1174.1	1.3461					
600	0.318	1110.9	1193.1	1.3642	605.1	0.255	1087.4	1162.9	1.3258
650	0.367	1155.5	1250.5	1.4171	650	0.303	1137.8	1227.4	1.3852
700	0.406	1189.6	1294.8	1.4562	700	0.342	1177.0	1278.1	1.4299
800	0.471	1245.8	1367.9	1.5168	800	0.403	1237.7	1357.0	1.4953
900	0.529	1295.6	1432.5	1.5661	900	0.456	1289.5	1424.4	1.5468
1000	0.582	1342.8	1493.5	1.6094	1000	0.504	1338.0	1487.1	1.5913
1100	0.632	1389.1	1552.8	1.6487	1100	0.549	1385.2	1547.7	1.6315
1200	0.681	1435.1	1611.4	1.6851	1200	0.592	1431.8	1607.1	1.6684
1300	0.728	1481.1	1669.6	1.7192	1300	0.634	1478.3	1666.1	1.7029
1400	0.774	1527.2	1727.8	1.7513	1400	0.675	1524.9	1724.8	1.7354
1600	0.865	1620.8	1844.8	1.8111	1600	0.755	1619.0	1842.6	1.7955

continued

TABLE B1.3 (USCS UNITS) *continued*

Properties of Water Vapor
v in cubic feet per pound-mass
u and h in British thermal units per pound-mass
s in British thermal units per pound-mass and degree Rankine

Temp °F	Pressure = 1800 psia				Temp °F	Pressure = 2000 psia			
	v	u	h	s		v	u	h	s
621.2	0.218	1077.7	1150.4	1.3060	636.0	0.188	1066.6	1136.3	1.2861
650	0.251	1117.0	1200.4	1.3517	650	0.205	1091.1	1167.2	1.3141
700	0.291	1163.1	1259.9	1.4042	700	0.249	1147.7	1239.8	1.3782
750	0.322	1198.6	1305.9	1.4430	750	0.280	1187.3	1291.1	1.4216
800	0.350	1229.1	1345.7	1.4753	800	0.307	1220.1	133.8	1.4562
900	0.399	1283.2	1416.1	1.5291	900	0.353	1276.8	1407.6	1.5126
1000	0.443	1333.1	1480.7	1.5749	1000	0.395	1328.1	1474.1	1.5598
1100	0.484	1381.2	1542.5	1.6159	1100	0.433	1377.2	1537.2	1.6017
1200	0.524	1428.5	1602.9	1.6534	1200	0.469	1425.2	1598.6	1.6398
1300	0.561	1475.5	1662.5	1.6883	1300	0.503	1472.7	1659.0	1.6751
1400	0.598	1522.5	1721.8	1.7211	1400	0.537	1520.2	1718.8	1.7082
1600	0.670	1617.2	1840.4	1.7817	1600	0.602	1615.4	1838.2	1.7692

Temp °F	Pressure = 2500 psia				Temp °F	Pressure = 3000 psia			
	v	u	h	s		v	u	h	s
668.3	0.1306	1031.0	1091.4	1.2327	695.5	0.0840	968.8	1015.5	1.1575
700	0.1684	1098.7	1176.6	1.3073	700	0.0977	1003.9	1058.1	1.1944
750	0.2030	1155.2	1249.1	1.3686	750	0.1483	1114.7	1197.1	1.3122
800	0.2291	1195.7	1301.7	1.4112	800	0.1757	1167.6	1265.2	1.3675
900	0.2712	1259.9	1385.4	1.4752	900	0.2160	1241.8	1361.7	1.4414
1000	0.3069	1315.2	1457.2	1.5262	1000	0.2485	1301.7	1439.6	1.4967
1100	0.3393	1366.8	1523.8	1.5704	1100	0.2772	1356.2	1510.1	1.5434
1200	0.3696	1416.7	1587.7	1.6101	1200	0.3036	1408.0	1576.6	1.5848
1300	0.3984	1465.7	1650.0	1.6465	1300	0.3285	1458.5	1640.9	1.6224
1400	0.4261	1514.2	1711.3	1.6804	1400	0.3524	1508.1	1703.7	1.6571
1500	0.4531	1562.5	1772.1	1.7123	1500	0.3754	1557.3	1765.7	1.6896
1600	0.4795	1610.8	1832.6	1.7424	1600	0.3978	1606.3	1827.1	1.7201

continued

TABLE B1.3 (USCS UNITS) *continued*

Properties of Water Vapor
v in cubic feet per pound-mass
u and h in British thermal units per pound-mass
s in British thermal units per pound-mass and degree Rankine

Temp °F	Pressure = 3500 psia				Temp °F	Pressure = 4000 psia			
	v	u	h	s		v	u	h	s
650	0.0249	663.5	679.7	0.8630	650	0.0245	657.7	675.8	0.8574
700	0.0306	759.5	779.3	0.9506	700	0.0287	742.1	763.4	0.9345
750	0.1046	1058.4	1126.1	1.2440	750	0.0633	960.7	1007.5	1.1395
800	0.1363	1134.7	1223.0	1.3226	800	0.1052	1095.0	1172.9	1.2740
900	0.1763	1222.4	1336.5	1.4096	900	0.1462	1201.5	1309.7	1.3789
1000	0.2066	1287.6	1421.4	1.4699	1000	0.1752	1272.9	1402.6	1.4449
1100	0.2328	1345.2	1496.0	1.5193	1100	0.1995	1333.9	1481.6	1.4973
1200	0.2566	1399.2	1565.3	1.5624	1200	0.2213	1390.1	1553.9	1.5423
1300	0.2787	1451.1	1631.7	1.6012	1300	0.2414	1443.7	1622.4	1.5823
1400	0.2997	1501.9	1696.1	1.6368	1400	0.2603	1495.7	1688.4	1.6188
1500	0.3199	1552.0	1759.2	1.6699	1500	0.2784	1546.7	1752.8	1.6526
1600	0.3395	1601.7	1821.6	1.0701	1600	0.2959	1597.1	1816.1	1.6841

Temp °F	Pressure = 4400 psia				Temp °F	Pressure = 4800 psia			
	v	u	h	s		v	u	h	s
650	0.0242	653.6	673.3	0.8535	650	0.0239	649.8	671.0	0.8499
700	0.0278	732.7	755.3	0.9257	700	0.0271	725.1	749.1	0.9187
750	0.0415	870.8	904.6	1.0513	750	0.0352	832.6	863.9	1.0154
800	0.0844	1056.5	1125.3	1.2306	800	0.0667	1011.2	1070.5	1.1827
900	0.1270	1183.7	1287.1	1.3548	900	0.1109	1164.8	1263.4	1.3310
1000	0.1552	1260.8	1387.2	1.4260	1000	0.1385	1248.3	1371.4	1.4078
1100	0.1784	1324.7	1469.9	1.4809	1100	0.1608	1315.3	1458.1	1.4653
1200	0.1989	1382.8	1544.7	1.5274	1200	0.1802	1375.4	1535.4	1.5133
1300	0.2176	1437.7	1614.9	1.5685	1300	0.1979	1431.7	1607.4	1.5555
1400	0.2352	1490.7	1682.3	1.6057	1400	0.2143	1485.7	1676.1	1.5934
1500	0.2520	1542.5	1747.6	1.6399	1500	0.2300	1538.2	1742.5	1.6282
1600	0.2681	1593.4	1811.7	1.6718	1600	0.2450	1589.8	1807.4	1.6605

Abridged from J. H. Keenan, F. G. Keyes, P. G. Hill, and J. G. Moore, "Steam Tables," Wiley, New York, 1979. Reprinted by permission of John Wiley & Sons, Inc.

TABLE B1.4 (USCS UNITS)
Properties of Compressed Liquid Water
v in cubic feet per pound-mass
u and h in British thermal units per pound-mass
s in British thermal units per pound-mass and degree Rankine

Temp °F	Pressure = 500 psia			
	v	u	h	s
32	0.015994	0.00	1.49	0.00000
50	0.015998	18.02	19.50	0.03599
100	0.016106	67.87	69.36	0.12932
150	0.016318	117.66	119.17	0.21457
200	0.016608	167.65	169.19	0.29341
300	0.017416	268.92	270.53	0.43641
400	0.018608	373.68	375.40	0.56604
467.1	0.019748	447.70	449.53	0.64904

Temp °F	Pressure = 1000 psia			
	v	u	h	s
32	0.015967	0.03	2.99	0.00005
50	0.015972	17.99	20.94	0.03592
100	0.016082	67.70	70.68	0.12901
150	0.016293	117.38	120.40	0.21410
200	0.016580	167.26	170.32	0.29281
300	0.017379	268.24	271.46	0.43552
400	0.018550	372.55	375.98	0.56472
544.8	0.021591	538.39	542.38	0.74320

Temp °F	Pressure = 1500 psia			
	v	u	h	s
32	0.015939	0.05	4.47	0.00007
50	0.015946	17.95	22.38	0.03584
100	0.016058	67.53	71.99	0.12870
150	0.016268	117.10	121.62	0.21364
200	0.016554	166.87	171.46	0.29221
300	0.017343	267.58	272.39	0.43463
400	0.018493	371.45	376.59	0.56343
500	0.02024	481.8	487.4	0.6853
596.4	0.02346	605.0	611.5	0.8082

Temp °F	Pressure = 2000 psia			
	v	u	h	s
32	0.015912	0.06	5.95	0.00008
50	0.015920	17.91	23.81	0.03575
100	0.016034	67.37	73.30	0.12839
150	0.016244	116.83	122.84	0.21318
200	0.016527	166.49	172.60	0.29162
300	0.017308	266.93	273.33	0.43376
400	0.018439	370.38	377.21	0.56216
500	0.02014	479.8	487.3	0.6832
636.0	0.02565	662.4	671.9	0.8623

Temp °F	Pressure = 2500 psia			
	v	u	h	s
32	0.015885	0.08	7.43	0.00009
50	0.015895	17.88	25.23	0.03566
100	0.016010	67.20	74.61	0.12808
150	0.016220	116.56	124.07	0.21272
200	0.016501	166.11	173.75	0.29104
300	0.017274	266.29	274.28	0.43290
400	0.018386	369.34	377.84	0.56092
668.3	0.02861	717.7	730.9	0.9131

Temp °F	Pressure = 3000 psia			
	v	u	h	s
32	0.015859	0.09	8.90	0.00009
50	0.015870	17.84	26.65	0.03555
100	0.015987	67.04	75.91	0.12777
150	0.016196	116.30	125.29	0.21226
200	0.016476	165.74	174.89	0.29046
300	0.017240	265.66	275.23	0.43205
400	0.018334	368.32	378.50	0.55970
695.5	0.034310	783.5	802.5	0.9732

continued

TABLE B1.4 (USCS UNITS) *continued*
Properties of Compressed Liquid Water
v in cubic feet per pound-mass
u and h in British thermal units per pound-mass
s in British thermal units per pound-mass and degree Rankine

Temp °F	Pressure = 3500 psia				Temp °F	Pressure = 4000 psia			
	v	u	h	s		v	u	h	s
32	0.015833	0.10	10.36	0.00009	32	0.015807	0.10	11.80	0.00005
50	0.015845	17.80	28.06	0.03545	50	0.015821	17.76	29.47	0.03534
100	0.015964	66.88	77.22	0.12746	100	0.015942	66.72	78.52	0.12714
150	0.016173	116.03	126.51	0.21181	150	0.016150	115.77	127.73	0.21136
200	0.016450	165.38	176.03	0.28988	200	0.016425	165.02	177.18	0.28931
300	0.017206	265.04	276.19	0.43121	300	0.017174	264.43	277.15	0.43038
400	0.018284	367.32	379.16	0.55851	400	0.018235	366.35	379.85	0.55734

Extracted from J. H. Keenan, F. G. Keyes, P. G. Hill, and J. G. Moore, "Steam Tables," Wiley, New York, 1979. Reprinted by permission of John Wiley & Sons, Inc.

TABLE B2.1 (USCS UNITS)
Properties of Saturated Refrigerant-12
Primary Argument: Temperature
v in cubic feet per pound-mass
u and h in British thermal units per pound-mass
s in British thermal units per pound-mass and degree Rankine

Temp. °F T	Pres. psia P	Specific Volume		Internal Energy		Enthalpy			Entropy	
		Sat. Liq. v_f	Sat. Vap. v_g	Sat Liq. u_f	Sat. Vap. u_g	Sat. Liq. h_f	h_{fg}	Sat. Vap. h_g	Sat. Liq. s_f	Sat. Vap. s_g
−40	9.305	0.01057	3.8868	−0.02	66.54	0.00	73.23	73.23	0.0000	0.1745
−30	11.999	0.01068	3.0684	2.07	67.54	2.10	72.25	74.35	0.0049	0.1731
−20	15.270	0.01796	2.4516	4.16	68.54	4.21	71.25	75.47	0.0098	0.1718
−15	17.150	0.01086	2.2011	5.24	69.03	5.28	70.74	76.02	0.0122	0.1713
−10	19.197	0.01092	1.9803	6.31	69.54	6.35	70.23	76.57	0.0146	0.1707
−5	21.437	0.01098	1.7871	7.38	70.04	7.42	69.71	77.13	0.0169	0.1702
0	23.863	0.01104	1.6157	8.45	70.54	8.50	69.18	77.68	0.0193	0.1698
5	26.505	0.01111	1.4649	9.53	71.04	9.58	68.64	78.22	0.0216	0.1693
10	29.356	0.01117	1.3303	10.61	71.54	10.68	68.09	78.77	0.0239	0.1689
20	35.765	0.01131	1.1045	12.80	72.54	12.87	66.98	79.85	0.0285	0.1682
25	39.341	0.01138	1.0093	13.90	73.03	13.98	66.40	80.38	0.0308	0.1678
30	43.182	0.01145	0.92401	15.01	73.53	15.10	65.82	80.92	0.0331	0.1675
40	51.705	0.01160	0.7784	17.24	74.52	17.35	64.62	81.97	0.0376	0.1669
50	61.432	0.01175	0.6598	19.49	75.50	19.63	63.38	83.00	0.0421	0.1664
60	72.462	0.01192	0.5625	21.77	76.47	21.93	62.08	84.01	0.0465	0.1660
70	84.90	0.01209	0.4820	24.08	77.42	24.27	60.73	85.00	0.0509	0.1656
80	98.58	0.01228	0.4150	26.41	78.38	26.63	59.32	85.96	0.0553	0.1652
85	106.64	0.01237	0.3868	27.59	78.78	27.83	58.58	86.42	0.0574	0.1650
90	114.42	0.01247	0.3586	28.77	79.28	29.03	57.85	86.88	0.0596	0.1649
95	123.07	0.01258	0.3349	29.96	79.70	30.25	57.07	87.32	0.0618	0.1647
100	131.72	0.01269	0.3111	31.16	80.18	31.47	56.30	87.77	0.0639	0.1645
105	141.30	0.01280	0.2908	32.37	80.58	32.70	55.49	88.19	0.0661	0.1644
110	150.87	0.01291	0.2706	33.58	81.05	33.94	54.67	88.61	0.0682	0.1642
115	161.43	0.01304	0.2533	34.80	81.53	35.19	53.91	89.10	0.0704	0.1640
120	171.99	0.01316	0.2360	36.03	81.89	36.45	52.95	89.40	0.0725	0.1639
140	220.63	0.01373	0.1803	41.05	83.43	41.61	49.19	90.80	0.0811	0.1631
160	278.74	0.01444	0.1380	46.26	84.73	47.00	44.85	91.86	0.0896	0.1620
180	347.61	0.01536	0.1050	51.75	85.66	52.74	39.68	92.42	0.0985	0.1605
200	428.84	0.01667	0.0781	57.75	85.94	59.08	33.06	92.14	0.1078	0.1580
233.2	598.3	0.0287	0.0287	76.22	76.22	79.40	0.00	79.40	0.1368	0.1368

Source: "1989 ASHRAE Handbook - Fundamentals," ASHRAE, Atlanta, GA, 1989. Reprinted with permission.

TABLE B2.2 (USCS UNITS)
Properties of Refrigerant-12 Vapor
Saturated Liquid and Vapor States are in Italics
v in cubic feet per pound-mass
u and h in British thermal units per pound-mass
s in British thermal units per pound-mass and degree Rankine

Temp °F	Pressure = 10 psia T$_{sat}$ = − 37.22° F			
	v	u	h	s
Liq	*0.0106*	*0.561*	*0.581*	*0.0014*
Vap	*3.6349*	*66.815*	*73.543*	*0.1741*
0	3.9890	71.176	78.559	0.1855
20	4.1761	73.585	81.314	0.1913
40	4.3615	76.039	84.112	0.1970
60	4.5457	78.541	86.955	0.2026
80	4.7288	81.090	89.843	0.2081
100	4.9114	83.684	92.774	0.2134
120	5.0932	86.322	95.749	0.2186
140	5.2743	89.005	98.767	0.2238
160	5.4552	91.733	101.83	0.2288

Temp °F	Pressure = 15 psia T$_{sat}$ = − 29.76° F			
	v	u	h	s
Liq	*0.0108*	*4.021*	*4.051*	*0.0094*
Vap	*2.4926*	*68.461*	*75.381*	*0.1719*
0	2.6281	70.952	78.248	0.1783
20	2.7562	73.391	81.043	0.1843
40	2.8827	75.872	83.875	0.1900
60	3.0078	78.393	86.744	0.1957
80	3.1319	80.958	89.653	0.2012
100	3.2553	83.565	92.603	0.2065
120	3.3779	86.215	95.593	0.2118
140	3.5001	88.907	98.624	0.2169
160	3.6217	91.635	101.690	0.2220

Temp °F	Pressure = 20 psia T$_{sat}$ = -8.15° F			
	v	u	h	s
Liq	*0.0109*	*6.701*	*6.741*	*0.0154*
Vap	*1.9060*	*69.722*	*76.778*	*0.1705*
20	2.0458	73.194	80.767	0.1791
40	2.1428	75.700	83.632	0.1850
60	2.2386	78.242	86.529	0.1907
80	2.3333	80.824	89.461	0.1962
100	2.4271	83.445	92.430	0.2016
120	2.5202	86.107	95.436	0.2069
140	2.6129	88.807	98.479	0.2120
160	2.7050	91.546	101.560	0.2171

Temp °F	Pressure = 30 psia T$_{sat}$ = 11.08° F			
	v	u	h	s
Liq	*0.0112*	*10.849*	*10.911*	*0.0244*
Vap	*1.3034*	*71.648*	*78.885*	*0.1688*
20	1.3344	72.784	80.193	0.1716
40	1.4023	75.345	83.132	0.1776
60	1.4688	77.933	86.089	0.1834
80	1.5342	80.550	89.069	0.1890
100	1.5986	83.199	92.076	0.1945
120	1.6623	85.885	95.115	0.1998
140	1.7254	88.605	98.186	0.2050
160	1.7881	91.361	101.290	0.2101

continued

TABLE B2.2 (USCS UNITS) *continued*
Properties of Refrigerant-12 Vapor
Saturated Liquid and Vapor States are in Italics
v in cubic feet per pound-mass
u and h in British thermal units per pound-mass
s in British thermal units per pound-mass and degree Rankine

Temp °F	Pressure = 40 psia T_{sat} = 25.88° F				Temp °F	Pressure = 50 psia T_{sat} = 38.01° F			
	v	u	h	s		v	u	h	s
Liq	*0.0114*	*14.096*	*14.180*	*0.0312*	*Liq*	*0.0116*	*16.813*	*16.920*	*0.0368*
Vap	*0.9935*	*73.123*	*80.478*	*0.1678*	*Vap*	*0.8037*	*74.332*	*81.770*	*0.1670*
40	1.0313	74.975	82.610	0.1721	40	0.8080	74.587	82.064	0.1676
60	1.0834	77.610	85.631	0.1780	60	0.8515	77.276	85.156	0.1737
80	1.1342	80.266	88.663	0.1837	80	0.8937	79.974	88.245	0.1795
100	1.1840	82.948	91.714	0.1893	100	0.9349	82.688	91.340	0.1852
120	1.2330	85.658	94.787	0.1947	120	0.9752	85.426	94.451	0.1906
140	1.2815	88.398	97.886	0.1999	140	1.0150	88.188	97.581	0.1959
160	1.3295	91.167	101.010	0.2051	160	1.0541	90.975	100.730	0.2011
180	1.3770	93.975	104.170	0.2101	180	1.0929	93.796	103.910	0.2061
200	1.4242	96.816	107.360	0.2150	200	1.1313	96.651	107.120	0.2111

Temp °F	Pressure = 60 psia T_{sat} = 48.61° F				Temp °F	Pressure = 80 psia T_{sat} = 66.20° F			
	v	u	h	s		v	u	h	s
Liq	*0.0117*	*19.176*	*19.306*	*0.0415*	*Liq*	*0.0120*	*84.626*	*23.375*	*0.0492*
Vap	*0.6750*	*75.362*	*82.858*	*0.1665*	*Vap*	*0.5109*	*77.061*	*84.626*	*0.1657*
60	0.6965	76.926	84.661	0.1700	80	0.5314	79.025	86.894	0.1700
80	0.7331	79.669	87.811	0.1759	100	0.5600	81.857	90.150	0.1759
100	0.7686	82.420	90.956	0.1817	120	0.5876	84.688	93.389	0.1816
120	0.8032	85.186	94.106	0.1872	140	0.6144	87.525	96.623	0.1871
140	0.8371	87.973	97.269	0.1926	160	0.6406	90.377	99.862	0.1924
160	0.8705	90.783	100.450	0.1978	180	0.6662	93.245	103.110	0.1975
180	0.9033	93.618	103.650	0.2029	200	0.6915	96.141	106.380	0.2026
200	0.9359	96.406	106.800	0.2078	220	0.7164	99.062	109.670	0.2075
					240	0.7411	101.996	112.970	0.2123

continued

TABLE B2.2 (USCS UNITS) *continued*
Properties of Refrigerant-12 Vapor
Saturated Liquid and Vapor States are in Italics
v in cubic feet per pound-mass
u and h in British thermal units per pound-mass
s in British thermal units per pound-mass and degree Rankine

Temp °F	Pressure = 100 psia $T_{sat} = 80.78°F$				Temp °F	Pressure = 120 psia $T_{sat} = 93.34°F$			
	v	u	h	s		v	u	h	s
Liq	*0.0123*	*26.591*	*26.818*	*0.0556*	*Liq*	*0.0125*	*29.562*	*29.841*	*0.0611*
Vap	*0.4102*	*78.437*	*86.029*	*0.1652*	*Vap*	*0.3419*	*79.587*	*87.181*	*0.1647*
100	0.4341	81.253	89.287	0.1711	100	0.3492	80.597	88.353	0.1669
120	0.4577	84.158	92.629	0.1770	120	0.3704	83.592	91.820	0.1729
140	0.4803	87.055	95.945	0.1826	140	0.3905	86.557	95.231	0.1787
160	0.5023	89.953	99.250	0.1880	160	0.4098	89.509	98.611	0.1843
180	0.5237	92.867	102.560	0.1933	180	0.4284	92.464	101.980	0.1896
200	0.5447	95.788	105.870	0.1984	200	0.4466	95.421	105.340	0.1948
220	0.5653	98.728	109.190	0.2033	220	0.4643	98.387	108.700	0.1998
240	0.5856	101.692	112.530	0.2082	240	0.4818	101.379	112.080	0.2047
260	0.6057	104.680	115.890	0.2129	260	0.4990	104.387	115.470	0.2095

Temp °F	Pressure = 140 psia $T_{sat} = 104.44°F$				Temp °F	Pressure = 160 psia $T_{sat} = 114.44°F$			
	v	u	h	s		v	u	h	s
Liq	*0.0128*	*32.228*	*32.559*	*0.0658*	*Liq*	*0.0130*	*34.661*	*35.047*	*0.0701*
Vap	*0.2923*	*80.573*	*88.147*	*0.1644*	*Vap*	*0.2546*	*81.427*	*88.967*	*0.1640*
120	0.3075	82.982	90.951	0.1693	120	0.2597	82.319	90.009	0.1659
140	0.3259	86.029	94.475	0.1753	140	0.2771	85.463	93.669	0.1721
160	0.3434	89.042	97.941	0.1809	160	0.2933	88.550	97.236	0.1779
180	0.3602	92.048	101.380	0.1864	180	0.3087	91.608	100.750	0.1835
200	0.3764	95.038	104.790	0.1917	200	0.3235	94.649	104.230	0.1888
220	0.3921	98.039	108.200	0.1968	220	0.3378	97.685	107.690	0.1940
240	0.4075	101.050	111.610	0.2017	240	0.3518	100.723	111.140	0.1990
260	0.4227	104.078	115.030	0.2065	260	0.3654	103.769	114.590	0.2039
					280	0.3788	106.833	118.050	0.2086
					300	0.3920	109.912	121.520	0.2133

continued

TABLE B2.2 (USCS UNITS) *continued*

Properties of Refrigerant-12 Vapor
Saturated Liquid and Vapor States are in Italics
v in cubic feet per pound-mass
u and h in British thermal units per pound-mass
s in British thermal units per pound-mass and degree Rankine

Temp °F	Pressure = 180 psia T$_{sat}$ = 123.56° F			
	v	u	h	s
Liq	*0.0133*	*36.909*	*37.351*	*0.0740*
Vap	*0.2249*	*82.175*	*89.668*	*0.1637*
140	0.2386	84.854	92.803	0.1690
160	0.2540	88.026	96.490	0.1751
180	0.2685	91.144	100.090	0.1808
200	0.2823	94.237	103.640	0.1863
220	0.2955	97.306	107.150	0.1915
240	0.3083	100.379	110.650	0.1966
260	0.3208	103.453	114.140	0.2015
280	0.3330	106.536	117.630	0.2063
300	0.3450	109.637	121.130	0.2109

Temp °F	Pressure = 200 psia T$_{sat}$ = 131.96° F			
	v	u	h	s
Liq	*0.0135*	*39.010*	*39.509*	*0.0776*
Vap	*0.2008*	*82.835*	*90.268*	*0.1634*
140	0.2073	84.191	91.864	0.1661
160	0.2223	87.468	95.697	0.1724
180	0.2361	90.661	99.401	0.1783
200	0.2491	93.809	103.030	0.1838
220	0.2615	96.920	106.600	0.1892
240	0.2734	100.028	110.150	0.1943
260	0.2850	103.130	113.680	0.1993
280	0.2963	106.242	117.210	0.2041
300	0.3073	109.353	120.730	0.2088

Temp °F	Pressure = 250 psia T$_{sat}$ = 150.54° F			
	v	u	h	s
Liq	*0.0141*	*43.634*	*44.416*	*0.0856*
Vap	*0.1567*	*84.153*	*91.404*	*0.1626*
160	0.1636	85.861	93.430	0.1659
180	0.1768	89.314	97.495	0.1723
200	0.1888	92.636	101.370	0.1783
220	0.1999	95.892	105.140	0.1839
240	0.2103	99.097	108.830	0.1893
260	0.2204	102.282	112.480	0.1944
280	0.2301	105.454	116.100	0.1994
300	0.2395	108.618	119.700	0.2042
320	0.2487	111.793	123.300	0.2089
340	0.2577	114.977	126.900	0.2134

Temp °F	Pressure = 300 psia T$_{sat}$ = 166.54° F			
	v	u	h	s
Liq	*0.0147*	*48.016*	*48.832*	*0.0925*
Vap	*0.1264*	*85.087*	*92.106*	*0.1616*
180	0.1356	87.687	95.217	0.1665
200	0.1475	91.296	99.486	0.1731
220	0.1581	94.731	103.510	0.1791
240	0.1678	98.081	107.400	0.1847
260	0.1770	101.363	111.190	0.1901
280	0.1857	104.619	114.930	0.1952
300	0.1941	107.853	118.630	0.2001
320	0.2022	111.073	122.300	0.2049
340	0.2101	114.294	125.960	0.2096

continued

TABLE B2.2 (USCS UNITS) *continued*
Properties of Refrigerant-12 Vapor
Saturated Liquid and Vapor States are in Italics
v in cubic feet per pound-mass
u and h in British thermal units per pound-mass
s in British thermal units per pound-mass and degree Rankine

Temp °F	Pressure = 350 psia $T_{sat} = 180.64°$ F				Temp °F	Pressure = 400 psia $T_{sat} = 193.27°$ F			
	v	u	h	s		v	u	h	s
Liq	*0.0154*	*51.936*	*52.933*	*0.0987*	*Liq*	*0.0162*	*55.651*	*56.850*	*0.1046*
Vap	*0.1041*	*85.681*	*92.425*	*0.1604*	*Vap*	*0.0867*	*85.945*	*92.364*	*0.1590*
200	0.1167	89.687	97.249	0.1679	200	0.0917	87.597	94.384	0.1621
220	0.1275	93.421	101.680	0.1745	220	0.1036	91.865	99.534	0.1697
240	0.1370	96.957	105.830	0.1805	240	0.1133	95.684	104.070	0.1763
260	0.1456	100.375	109.810	0.1861	260	0.1218	99.293	108.310	0.1823
280	0.1538	103.729	113.690	0.1914	280	0.1296	102.776	112.370	0.1879
300	0.1615	107.039	117.500	0.1965	300	0.1369	106.177	116.310	0.1931
320	0.1689	110.320	121.260	0.2014	320	0.1438	109.536	120.180	0.1982
340	0.1760	113.597	125.000	0.2061	340	0.1504	112.876	124.010	0.2030

Source: ASHRAE, "1989 Handbook - Fundamentals," ASHRAE, Atlanta, GA, 1989. Reprinted with permission.

TABLE B3.1 (USCS UNITS)
Properties of Saturated Ammonia
Primary Argument: Temperature
v in cubic feet per pound-mass
u and h in British thermal units per pound-mass
s in British thermal units per pound-mass and degree Rankine

Temp. °F T	Pres. psia P	Specific Volume		Enthalpy			Entropy	
		Sat. Liq. v_f	Sat. Vap. v_g	Sat. Liq. h_f	h_{fg}	Sat. Vap. h_g	Sat. Liq. s_f	Sat. Vap. s_g
−60	5.5	0.02277	44.96	−23.8	614.7	590.9	−0.0580	1.4799
−50	7.6	0.02299	33.18	−11.7	606.5	594.8	−0.0281	1.4524
−40	10.4	0.02321	24.90	0.0	598.7	598.7	0.0000	1.4265
−30	13.9	0.02345	18.99	11.2	591.1	602.4	0.0264	1.4022
−20	18.3	0.02369	14.68	22.2	583.7	605.9	0.0517	1.3792
−10	23.7	0.02393	11.50	33.1	576.2	609.3	0.0762	1.3575
0	30.4	0.02419	9.111	44.0	568.5	612.5	0.1000	1.3368
5	34.3	0.02432	8.143	49.4	564.6	614.1	0.1117	1.3269
10	38.5	0.02446	7.298	54.9	560.7	615.6	0.1234	1.3171
20	48.2	0.02473	5.903	65.8	552.6	618.4	0.1463	1.2984
30	59.80	0.02502	4.819	76.9	544.3	621.1	0.1690	1.2805
40	73.40	0.02533	3.966	88.0	535.6	623.6	0.1914	1.2633
50	89.20	0.02564	3.290	99.2	526.7	625.9	0.2134	1.2468
60	107.7	0.02597	2.748	110.5	517.4	627.9	0.2352	1.2309
70	128.9	0.02632	2.310	121.9	507.8	629.7	0.2568	1.2155
80	153.1	0.02668	1.953	133.4	497.8	631.3	0.2780	1.2005
90	180.7	0.02706	1.659	145.0	487.5	632.5	0.2990	1.1860
100	212.0	0.02747	1.417	156.7	476.8	633.4	0.3198	1.1717
110	247.1	0.02790	1.215	168.4	465.6	634.0	0.3404	1.1577
120	286.5	0.02836	1.046	180.4	453.9	634.3	0.3607	1.1438

Source: Generated from REFRIG program, Wiley Professional Software, Wiley, New York, 1985.

TABLE B3.2 (USCS UNITS)
Properties of Ammonia Vapor
v in cubic feet per pound-mass
h in British thermal units per pound-mass
s in British thermal units per pound-mass and degree Rankine

Temp °F	Pressure = 10 psia $T_{sat} = -41.3°$ F			Temp °F	Pressure = 20 psia $T_{sat} = -16.6°$ F		
	v	h	s		v	h	s
Liq	0.0232	−1.48	−0.0035	Liq	0.0238	25.93	0.0601
Vap	25.81	598.18	1.4298	Vap	13.50	607.06	1.3717
0	28.57	619.48	1.4783	0	14.08	616.13	1.3918
20	29.88	629.73	1.5002	20	14.77	626.92	1.4148
40	31.19	639.96	1.5210	40	15.44	637.57	1.4365
60	32.48	650.18	1.5411	60	16.11	648.13	1.4573
80	33.77	660.43	1.5604	80	16.77	658.65	1.4771
100	35.06	670.70	1.5791	100	17.42	669.14	1.4962
120	36.34	681.01	1.5972	120	18.07	679.64	1.5147
140	37.61	691.38	1.6148	140	18.72	690.16	1.5325
160	38.89	701.82	1.6320	160	19.36	700.72	1.5498
180	40.16	712.33	1.6486	180	20.01	711.33	1.5667
200	41.43	722.91	1.6649	200	20.65	722.00	1.5831
220	42.70	733.58	1.6809	220	21.29	732.74	1.5991

Temp °F	Pressure = 30 psia $T_{sat} = -0.6°$ F			Temp °F	Pressure = 40 psia $T_{sat} = 11.7°$ F		
	v	h	s		v	h	s
Liq	0.0242	43.38	0.0987	Liq	0.0245	56.70	0.1272
Vap	9.23	612.33	1.3379	Vap	7.04	616.06	1.3140
0	9.242	612.66	1.3386				
20	9.721	624.00	1.3628	20	7.196	620.99	1.3244
40	10.19	635.11	1.3855	40	7.560	632.58	1.3480
60	10.65	646.04	1.4069	60	7.914	643.90	1.3702
80	11.10	656.84	1.4273	80	8.260	655.00	1.3912
100	11.54	667.57	1.4469	100	8.601	665.97	1.4112
120	11.98	678.25	1.4656	120	8.937	676.85	1.4303
140	12.42	688.93	1.4837	140	9.270	687.68	1.4486
160	12.86	699.61	1.5012	160	9.601	698.49	1.4664
180	13.29	710.32	1.5183	180	9.929	709.31	1.4835
200	13.72	721.08	1.5348	200	10.26	720.16	1.5002
220	14.15	731.90	1.5510	220	10.58	731.05	1.5165

continued

TABLE B3.2 (USCS UNITS) *continued*
Properties of Ammonia Vapor
v in cubic feet per pound-mass
h in British thermal units per pound-mass
s in British thermal units per pound-mass and degree Rankine

Temp °F	Pressure = 50 psia T_{sat} = 21.7° F			Temp °F	Pressure = 75 psia T_{sat} = 41.1° F		
	v	h	s		v	h	s
Liq	0.0248	67.66	0.1501	Liq	0.0254	89.23	0.1938
Vap	5.705	618.90	1.2954	Vap	3.884	623.87	1.2614
40	5.981	629.98	1.3180				
60	6.273	641.69	1.3410	60	4.083	635.97	1.2852
80	6.558	653.12	1.3625	80	4.285	648.26	1.3084
100	6.836	664.34	1.3829	100	4.481	660.16	1.3300
120	7.110	675.42	1.4024	120	4.672	671.77	1.3504
140	7.380	686.41	1.4210	140	4.859	683.19	1.3698
160	7.648	697.36	1.4390	160	5.043	694.48	1.3883
180	7.914	708.29	1.4564	180	5.225	705.70	1.4061
200	8.178	719.23	1.4732	200	5.406	716.88	1.4233
220	8.440	730.20	1.4896	220	5.585	728.05	1.4400

Temp °F	Pressure = 100 psia T_{sat} = 56.0° F			Temp °F	Pressure = 125 psia T_{sat} = 68.3° F		
	v	h	s		v	h	s
Liq	0.0258	106.01	0.2266	Liq	0.0253	119.92	0.2530
Vap	2.950	627.14	1.2372	Vap	2.382	629.42	1.2182
60	2.983	629.88	1.2424				
80	3.146	643.14	1.2675	80	2.460	637.74	1.2337
100	3.301	655.79	1.2905	100	2.592	651.23	1.2582
120	3.451	667.99	1.3119	120	2.718	664.08	1.2808
140	3.597	679.88	1.3321	140	2.839	676.46	1.3018
160	3.740	691.54	1.3512	160	2.957	688.52	1.3216
180	3.881	703.06	1.3695	180	3.073	700.37	1.3404
200	4.019	714.49	1.3871	200	3.187	712.06	1.3584
220	4.156	725.87	1.4041	220	3.299	723.66	1.3757

continued

TABLE B3.2 (USCS UNITS) *continued*
Properties of Ammonia Vapor
v in cubic feet per pound-mass
h in British thermal units per pound-mass
s in British thermal units per pound-mass and degree Rankine

Temp °F	Pressure = 150 psia $T_{sat} = 78.8°F$			Temp °F	Pressure = 175 psia $T_{sat} = 88.1°F$		
	v	h	s		v	h	s
Liq	0.0266	132.01	0.2755	Liq	0.0270	142.68	0.2949
Vap	1.992	631.08	1.2023	Vap	1.714	632.27	1.1889
80	2.000	632.00	1.2040				
100	2.117	646.44	1.2303	100	1.776	641.41	1.2053
120	2.228	660.00	1.2541	125	1.900	659.21	1.2364
140	2.333	672.94	1.2761	150	2.017	675.84	1.2643
160	2.435	685.43	1.2966	175	2.128	691.70	1.2898
180	2.534	697.62	1.3159	200	2.235	707.07	1.3135
200	2.632	709.59	1.3344	225	2.339	722.13	1.3360
220	2.727	721.42	1.3520	250	2.441	737.00	1.3573

Temp °F	Pressure = 200 psia $T_{sat} = 96.3°F$			Temp °F	Pressure = 250 psia $T_{sat} = 110.8°F$		
	v	h	s		v	h	s
Liq	0.0273	152.34	0.3122	Liq	0.0279	169.32	0.3419
Vap	1.501	633.13	1.1769	Vap	1.203	169.32	1.1567
100	1.519	636.09	1.1822				
125	1.634	654.97	1.2153	125	1.258	645.92	1.1771
150	1.740	672.34	1.2443	150	1.351	664.99	1.2091
175	1.840	688.73	1.2707	175	1.437	682.57	1.2373
200	1.937	704.50	1.2951	200	1.519	699.22	1.2630
225	2.030	719.86	1.3179	225	1.597	715.24	1.2869
250	2.121	734.98	1.3396	250	1.673	730.88	1.3093
275	2.211	749.97	1.3604	275	1.747	746.28	1.3307
300	2.299	764.91	1.3804	300	1.819	761.56	1.3511

Source: Generated from program REFRIG, Wiley Professional Software, Wiley, New York, 1985.

TABLE B4 (USCS UNITS)
Equilibrium Properties of Water-Lithium Bromide Solution
Water Vapor Saturation Temperature (T_r) in °F
Water Vapor Saturation Pressure (P_r) in psia
Solution Enthalpy (h_s) in Btu/lb$_m$
Solution Specific Volume (v_s) in ft³/lb$_m$

Solution Temp		Concentration lbm-LiBr/lbm-solution						
°F		0.40	0.45	0.50	0.55	0.60	0.65	0.70
	v_s	0.01145	0.01098	0.01054	0.01008	0.009654	0.009210	0.008805
60	T_r	43.32	35.20					
	P_r	0.14	0.10					
	h_s	9.89	10.14					
65	T_r	48.15	39.80					
	P_r	0.17	0.12					
	h_s	12.78	12.88					
70	T_r	52.99	44.39	33.08				
	P_r	0.20	0.14	0.09				
	h_s	15.68	15.61	17.51				
75	T_r	57.82	48.99	37.55				
	P_r	0.24	0.17	0.11				
	h_s	18.58	18.35	20.12				
80	T_r	62.65	53.58	42.02				
	P_r	0.28	0.20	0.13				
	h_s	21.48	21.09	22.72				
90	T_r	72.32	62.78	50.97	37.23			
	P_r	0.39	0.28	0.18	0.11			
	h_s	27.30	26.59	27.95	33.37			
100	T_r	81.98	71.97	59.91	46.00			
	P_r	0.54	0.39	0.26	0.15			
	h_s	33.13	32.09	33.18	38.31			
110	T_r	91.65	81.16	68.86	54.78	39.55		
	P_r	0.74	0.53	0.35	0.21	0.12		
	h_s	38.97	37.60	38.41	43.25	52.54		
120	T_r	101.32	90.35	77.80	63.55	48.11	32.69	
	P_r	0.99	0.71	0.47	0.29	0.17	0.09	
	h_s	44.82	43.12	43.66	48.20	57.15	69.57	
130	T_r	110.98	99.55	86.75	72.33	56.68	40.91	
	P_r	1.31	0.94	0.63	0.39	0.23	0.13	
	h_s	50.68	48.65	48.90	53.14	61.76	73.85	
140	T_r	120.65	108.74	95.70	81.10	65.24	49.12	
	P_r	1.73	1.23	0.83	0.53	0.31	0.17	
	h_s	56.56	54.19	54.15	58.09	66.37	78.13	

continued

TABLE B4 (USCS UNITS)

Equilibrium Properties of Water-Lithium Bromide Solution
Water Vapor Saturation Temperature (T_r) in °F
Water Vapor Saturation Pressure (P_r) in psia
Solution Enthalpy (h_s) in Btu/lb$_m$
Solution Specific Volume (v_s) in ft^3/lb$_m$

Solution Temp		Concentration lb$_m$-LiBr/lb$_m$-solution						
°F		0.40	0.45	0.50	0.55	0.60	0.65	0.70
150	T_r	130.31	117.93	104.64	89.88	73.80	57.34	
	P_r	2.24	1.60	1.09	0.70	0.41	0.23	
	h_s	62.45	59.75	59.41	63.04	70.99	82.41	
175	T_r	154.48	140.91	127.00	111.82	95.22	77.89	
	P_r	4.15	2.96	2.05	1.35	0.82	0.47	
	h_s	77.22	73.67	72.58	75.42	82.52	93.11	
200	T_r	178.64	163.89	149.37	133.76	116.63	98.43	
	P_r	7.28	5.20	3.66	2.46	1.54	0.91	
	h_s	92.07	87.65	85.79	87.82	94.06	103.82	
225	T_r	202.81	186.88	171.73	155.70	138.04	118.98	99.67
	P_r	12.20	8.73	6.23	4.27	2.75	1.65	0.94
	h_s	107.00	101.70	99.04	100.23	105.60	114.52	125.45
250	T_r	226.97	209.86	194.09	177.64	159.45	139.52	118.86
	P_r	19.63	14.08	10.18	7.12	4.68	2.85	1.64
	h_s	122.00	115.81	112.32	112.66	117.15	125.23	135.51
275	T_r	251.14	232.84	216.46	199.57	180.86	160.07	138.06
	P_r	30.45	21.91	16.04	11.42	7.65	4.75	2.75
	h_s	137.08	129.99	125.63	125.10	128.70	135.95	145.56
300	T_r	275.31	255.82	238.82	221.51	202.28	180.61	157.26
	P_r	45.74	33.03	24.45	17.69	12.07	7.61	4.44
	h_s	152.24	144.23	138.98	137.55	140.26	146.66	155.61

Source: "ASHRAE Handbook Fundamentals," ASHRAE, New York, Atlanta, GA, 1989. Reprinted with permission.

Table B5 (USCS UNITS)

Temperature-Dependent Molar Constant Pressure Specific Heats

$\overline{c}_p^0 = a + bT + cT^2 + dT^3$

T in $°R$

$\overline{c}_p^0$ in Btu/lb_m-mole-$°R$

Substance		a	$b \times 10^2$	$c \times 10^5$	$d \times 10^9$	Temp. Range, $°R$	$\overline{c}_p^0$ 77° F
Acetylene	C_2H_2	5.21	1.2227	-0.4812	0.7457	491-2700	10.504
Air		6.713	0.02609	0.03540	-0.08052	491-3240	6.943
Ammonia	NH_3	6.5846	0.34028	0.073034	-0.27402	491-2700	8.580
Benzene	C_6H_6	-8.650	6.4322	-2.327	3.179	491-2700	19.673
Carbon Dioxide	CO_2	5.316	0.79361	-0.2581	0.3059	491-3240	8.881
Carbon Monoxide	CO	6.726	0.02222	0.03960	-0.09100	491-3240	6.945
Ethane	C_2H_6	1.648	2.291	-0.4722	0.2984	491-2740	12.635
Ethanol	C_2H_6O	4.75	2.781	-0.7651	0.821	491-2700	17.605
Ethylene	C_2H_4	0.944	2.075	-0.6151	0.7326	491-2740	10.426
Hydrogen	H_2	6.952	-0.02542	0.02952	-0.03565	491-3240	6.895
Hydrogen Chloride	HCl	7.244	-0.1011	0.09783	-0.1776	491-2740	6.956
i-Butane	C_4H_{10}	-1.890	5.520	-1.696	2.044	491-2740	23.178
Methane	CH_4	4.750	0.6666	0.09352	-0.4510	491-2740	8.529
Methanol	CH_4O	4.55	1.214	-0.0898	-0.329	491-1800	10.759
n-Butane	C_4H_{10}	0.945	4.929	-1.352	1.433	491-2740	23.737
n-Hexane	C_6H_{14}	1.657	7.328	-2.112	2.363	491-2740	35.284
n-Pentane	C_5H_{12}	1.618	6.028	-1.656	1.732	491-2740	29.481
Nitric Oxide	NO	7.008	-0.01247	0.07185	-0.1715	491-2700	7.122
Nitrogen	N_2	6.903	-0.02085	0.05957	-0.1176	491-3240	6.945
Nitrogen Dioxide	NO_2	5.48	0.7583	-0.260	0.322	491-2700	8.852
Nitrous Oxide	N_2O	5.758	0.7780	-0.2596	0.4331	491-2700	9.254
Oxygen	O_2	6.085	0.2017	-0.05275	0.05372	491-3240	7.024
Propane	C_3H_8	-0.966	4.044	-1.159	1.300	491-2740	17.609
Propylene	C_3H_6	0.753	3.162	-0.8981	1.008	491-2740	15.299
Sulfur	S_2	6.499	0.2943	-0.1200	0.1632	491-3240	7.759
Sulfur Dioxide	SO_2	6.157	0.7689	-0.2810	0.3527	491-3240	9.530
Sulfur Trioxide	SO_3	3.918	1.935	-0.8256	1.328	491-3240	12.134
Water Vapor	H_2O	7.700	0.02552	0.07781	-0.1472	491-3240	8.039
Monotonic Gases		$\overline{c}_v^0 = \frac{3}{2}\Re$		$\overline{c}_P^0 = \frac{5}{2}\Re$		All Temps.	

Source: B.G., Kyle, "Chemical and Process Thermodynamics, 2/c, 1992, p. 545. Adapted by permission of Prentice-Hall, Englewood Cliffs, NJ, 1984.

TABLE B6.1 (USCS UNITS)

Properties of Air as an Ideal Gas with Variable Specific Heats
Temperature in degrees Rankine
Specific enthalpy in British thermal units per pound-mass
Specific internal energy in British thermal units per pound-mass
s_0 in British thermal units per pound-mass and degree Rankin

T	h	u	s_0	P_r	v_r	T	h	u	s_0	P_r	v_r
360	86.04	61.36	0.50353	0.3347	398.5	2000	504.72	367.61	0.93189	173.10	4.281
380	90.83	64.78	0.51647	0.4042	348.3	2050	518.61	378.06	0.93875	191.30	3.970
400	95.62	68.19	0.52875	0.4835	306.5	2100	532.54	388.57	0.94546	211.0	3.687
420	100.40	71.61	0.54043	0.5733	271.4	2150	546.52	399.13	0.95204	232.2	3.430
440	105.19	75.03	0.55157	0.6745	241.7	2200	560.55	409.73	0.95850	255.2	3.194
460	109.98	78.45	0.56222	0.7878	216.31	2250	574.63	420.38	0.96482	279.8	2.979
480	114.78	81.87	0.57241	0.9142	194.52	2300	588.76	431.08	0.97103	306.4	2.781
500	119.57	85.29	0.58220	1.0544	175.68	2350	602.93	441.82	0.97713	334.8	2.600
520	124.36	88.71	0.59160	1.2094	159.29	2400	617.14	452.61	0.98311	365.4	2.433
540	129.16	92.14	0.60065	1.3801	144.96	2450	631.39	463.43	0.98899	398.1	2.280
560	133.96	95.57	0.60938	1.5675	132.35	2500	645.68	474.29	0.99476	433.1	2.1386
580	138.76	99.00	0.61781	1.7725	121.23	2550	660.01	485.20	1.00044	470.4	2.0081
600	143.57	102.44	0.62595	1.996	111.35	2600	674.38	496.13	1.00602	510.3	1.8875
620	148.38	105.88	0.63384	2.240	102.56	2650	688.78	507.11	1.01150	552.9	1.7758
640	153.20	109.32	0.64148	2.504	94.70	2700	703.22	518.12	1.01690	598.1	1.6723
660	158.01	112.77	0.64890	2.790	87.65	2750	717.69	529.16	1.02221	646.3	1.5764
680	162.84	116.22	0.65610	3.099	81.30	2800	732.19	540.23	1.02744	697.5	1.4872
700	167.67	119.68	0.66310	3.432	75.57	2850	746.72	551.33	1.03258	751.8	1.4043
720	172.51	123.14	0.66991	3.790	70.38	2900	761.28	562.47	1.03765	809.5	1.3272
740	177.35	126.62	0.67655	4.175	65.66	2950	775.87	573.63	1.04263	870.6	1.2553
760	182.20	130.10	0.68301	4.588	61.36	3000	790.49	584.83	1.04755	935.3	1.1883
780	187.06	133.58	0.68932	5.031	57.44	3050	805.14	596.05	1.05239	1003.8	1.1257
800	191.92	137.08	0.69548	5.504	53.85	3100	819.82	607.29	1.05716	1076.1	1.0672
820	196.79	140.58	0.70150	6.008	50.56	3150	834.52	618.56	1.06187	1152.6	1.0125
840	201.68	144.09	0.70738	6.547	47.53	3200	849.24	629.86	1.06651	1233.2	0.9613
860	206.57	147.61	0.71314	7.120	44.75	3250	863.99	641.18	1.07108	1318.3	0.9133
880	211.47	151.14	0.71877	7.730	42.18	3300	878.77	652.53	1.07559	1408.0	0.8683
900	216.38	154.68	0.72429	8.378	39.80	3350	893.56	663.90	1.08004	1502.4	0.8261
920	221.30	158.23	0.72970	9.066	37.60	3400	908.38	675.29	1.08443	1601.8	0.7864
940	226.23	161.79	0.73500	9.795	35.55	3450	923.22	686.70	1.08877	1706.3	0.7491
960	231.18	165.36	0.74020	10.567	33.66	3500	938.09	698.14	1.09304	1816.1	0.7140
980	236.13	168.95	0.74531	11.384	31.89	3550	952.97	709.60	1.09727	1931.5	0.6809
1000	241.10	172.54	0.75032	12.248	30.25	3600	967.88	721.07	1.10143	2052.6	0.6498
1040	251.06	179.76	0.76010	14.125	27.28	3650	982.80	732.57	1.10555	2179.7	0.6204
1080	261.08	187.03	0.76954	16.212	24.68	3700	997.74	744.09	1.10962	2312.9	0.5927

continued

TABLE B6.1 (USCS UNITS) *continued*
Properties of Air as an Ideal Gas with Variable Specific Heats
Temperature in degrees Rankine
Specific enthalpy in British thermal units per pound-mass
Specific internal energy in British thermal units per pound-mass
s_0 in British thermal units per pound-mass and degree Rankin

T	h	u	s_0	P_r	v_r	T	h	u	s_0	P_r	v_r
1120	271.14	194.35	0.77869	18.526	22.40	3750	1012.71	755.62	1.11363	2452.4	0.5665
1160	281.25	201.72	0.78756	21.09	20.381	3800	1027.69	767.18	1.11760	2598.6	0.5418
1200	291.41	209.15	0.79618	23.91	18.595	3850	1042.69	778.75	1.12153	2751.6	0.5184
1240	301.63	216.62	0.80455	27.01	17.006	3900	1057.71	790.34	1.12540	2911.6	0.4962
1280	311.89	224.14	0.81270	30.42	15.587	3950	1072.74	801.94	1.12923	3078.9	0.4753
1320	322.21	231.72	0.82063	34.16	14.317	4000	1087.79	813.57	1.13302	3254	0.4554
1360	332.58	239.34	0.82837	38.24	13.177	4050	1102.86	825.21	1.13676	3436	0.4366
1400	342.99	247.02	0.83592	42.69	12.150	4100	1117.95	836.87	1.14046	3627	0.4188
1440	353.46	254.74	0.84329	47.54	11.223	4150	1133.05	848.54	1.14412	3826	0.4018
1480	363.98	262.51	0.85050	52.80	10.384	4200	1148.17	860.23	1.14775	4034	0.3858
1520	374.54	270.34	0.85754	58.52	9.623	4300	1178.45	883.65	1.15487	4475	0.3560
1560	385.15	278.21	0.86443	64.70	8.932	4400	1208.78	907.14	1.16185	4955	0.3290
1600	395.81	286.12	0.87118	71.39	8.303	4500	1239.18	930.68	1.16868	5474	0.3046
1650	409.20	296.08	0.87942	80.51	7.593	4600	1269.63	954.27	1.17537	6035	0.2824
1700	422.66	306.11	0.88745	90.52	6.958	4700	1300.13	977.92	1.18193	6641	0.2622
1750	436.18	316.21	0.89529	101.49	6.388	4800	1330.69	1001.62	1.18836	7294	0.2438
1800	449.77	326.37	0.90294	113.48	5.877	4900	1361.29	1025.37	1.19467	7998	0.2270
1850	463.42	336.59	0.91043	126.56	5.415	5000	1391.94	1049.16	1.20086	8754	0.21161
1900	477.13	346.87	0.91774	140.81	4.999	5100	1422.64	1073.00	1.20694	9565	0.19753
1950	490.90	357.21	0.92489	156.29	4.622	5200	1453.38	1096.89	1.21291	10435	0.18461

Source: Abridged from Keenan, J. H., Chao, J., and Kaye, J., "Gas Tables," Wiley, New York, 1980. Reprinted by permission of John Wiley & Sons, Inc.

TABLE B6.2 (USCS UNITS)
Properties of Carbon Dioxide as an Ideal Gas
with Variable Specific Heats
Temperature in degrees Rankine
Specific enthalpy in British thermal units per pound-mass-mole
Specific internal energy in British thermal units per pound-mass-mole
$\bar{s}^0$ *in British thermal units per pound-mass-mole and degree Rankine*
M = 44.0098

T	$\bar{h}$	$\bar{u}$	$\bar{s}^0$	T	$\bar{h}$	$\bar{u}$	$\bar{s}^0$
360	2559.0	1844.1	47.736	1920	19952.2	16139.4	65.136
380	2714.8	1960.2	48.157	1960	20479.5	16587.2	65.408
400	2873.3	2079.0	48.563	2000	21008.7	17037.0	65.676
420	3034.4	2200.4	48.956	2040	21539.9	17488.7	65.938
440	3198.2	2324.4	49.337	2080	22074.0	17943.4	66.197
460	3364.6	2451.1	49.707	2120	22610.1	18400.0	66.451
480	3533.6	2580.4	50.066	2160	23147.1	18857.6	66.703
500	3705.1	2712.2	50.417	2200	23686.1	19317.2	66.951
520	3879.1	2846.5	50.758	2240	24227.0	19778.7	67.198
537	**4029.1**	**2962.7**	**51.041**	2280	24769.9	20242.2	67.438
540	4055.8	2983.4	51.091				
560	4234.7	3122.7	51.416	2320	25314.8	20707.6	67.672
580	4416.1	3264.3	51.735	2360	25860.6	21173.9	67.906
600	4599.8	3408.3	52.046	2400	26407.4	21641.3	68.137
620	4785.6	3554.4	52.351	2440	26956.1	22110.6	68.363
640	4973.9	3703.0	52.649	2480	27505.9	22580.9	68.588
660	5164.1	3853.4	52.942	2520	28057.5	23053.2	68.808
680	5356.5	4006.1	53.229	2560	28610.2	23526.4	69.025
700	5551.0	4160.9	53.511	2600	29164.8	24001.6	69.239
720	5747.4	4317.6	53.788	2640	29720.4	24477.7	69.451
740	5945.7	4476.2	54.060	2680	30276.9	24954.8	69.662
760	6146.1	4636.8	54.327	2720	30834.5	25432.9	69.869
780	6348.2	4799.2	54.589	2760	31393.0	25129.0	70.073
800	6552.1	4963.4	54.847	2800	31953.5	26393.1	70.272
820	6757.7	5129.3	55.101	2840	32513.9	26874.1	70.470
840	6965.2	5297.1	55.351	2880	33076.3	27357.1	70.667
860	7174.4	5466.5	55.597	2920	33639.7	27841.0	70.861
880	7385.1	5637.5	55.840	2960	34204.1	28326.0	71.052
900	7597.5	5810.2	56.078	3000	34768.5	28810.9	71.243
920	7811.5	5984.5	56.313	3040	35334.8	29297.8	71.429
940	8026.9	6160.2	56.545	3080	35901.1	29784.7	71.616
960	8243.8	6337.4	56.773	3120	36469.4	30273.5	71.799
980	8462.4	6516.2	56.999	3160	37037.7	30762.4	71.979
1000	8682.3	6696.4	57.221	3200	37606.0	31251.2	72.158
1040	9126.3	7061.0	57.656	3240	38176.2	31742.0	72.335
1080	9575.8	7431.1	58.080	3280	38746.4	32232.8	72.510

continued

TABLE B6.2 (USCS UNITS) *continued*
Properties of Carbon Dioxide as an Ideal Gas
with Variable Specific Heats
Temperature in degrees Rankine
Specific enthalpy in British thermal units per pound-mass-mole
Specific internal energy in British thermal units per pound-mass-mole
$\bar{s}^0$ *in British thermal units per pound-mass-mole and degree Rankine*
M = 44.0098

T	$\bar{h}$	$\bar{u}$	$\bar{s}^0$	T	$\bar{h}$	$\bar{u}$	$\bar{s}^0$
1120	10030.5	7806.3	58.494	3320	39318.6	32725.6	72.684
1160	10490.1	8186.5	58.897	3360	39890.8	33218.3	72.855
1200	10954.6	8571.6	59.290	3400	40463.0	33711.1	73.026
1240	11423.8	8961.3	59.675	3440	41037.1	34205.8	73.193
1280	11897.2	9355.3	60.051	3480	41611.3	34700.5	73.360
1320	12357.0	9753.7	60.418	3520	42186.4	35196.2	73.522
1360	12856.8	10156.0	60.778	3560	42762.5	35692.8	73.685
1400	13342.5	10562.3	61.130	3600	43338.6	36189.5	73.846
1440	13832.1	10972.5	61.475	3640	43915.6	36687.1	74.005
1480	14325.3	11386.2	61.812	3680	44492.7	37184.7	74.164
1520	14821.9	11803.4	62.144	3720	45070.7	37683.3	74.319
1560	15321.9	12224.0	62.468	3760	45648.8	38181.9	74.474
1600	15825.2	12647.8	62.787	3800	46228.8	38682.5	74.629
1640	16331.6	13074.8	63.099	3840	46807.8	39182.1	74.780
1680	16840.8	13504.5	63.406	3880	47388.8	39683.6	74.930
1720	17352.9	13937.2	63.707	3920	47969.7	40185.2	75.079
1760	17867.9	14372.8	64.003	3960	48550.7	40686.7	75.226
1800	18385.1	14810.5	64.294	4000	49132.7	41189.2	75.373
1840	18905.5	15251.5	64.580	4040	49714.6	41691.7	75.518
1880	19427.9	15694.5	64.860	4080	50296.6	42194.2	75.661

Abridged from Keenan, J. H., Chao, J., and Kaye, J., "Gas Tables," Wiley, New York, 1980. Reprinted by permission of John Wiley & Sons, Inc.

TABLE B6.3 (USCS UNITS)
Properties of Carbon Monoxide as an Ideal Gas
with Variable Specific Heats
Temperature in degrees Rankine
Specific enthalpy in British thermal units per pound-mass-mole
Specific internal energy in British thermal units per pound-mass-mole
$\bar{s}_0$ in British thermal units per pound-mass-mole and degree Rankine
M = 28.0104

T	$\bar{h}$	$\bar{u}$	$\bar{s}_0$	T	$\bar{h}$	$\bar{u}$	$\bar{s}_0$
360	2499.2	1784.3	44.399	1920	14007.7	10194.8	56.499
380	2638.3	1883.6	44.775	1960	14328.6	10436.3	56.664
400	2777.3	1983.0	45.132	2000	14650.7	10678.9	56.827
420	2916.4	2082.3	45.471	2040	14973.7	10922.5	56.987
440	3055.4	2181.7	45.794	2080	15297.7	11167.2	57.144
460	3194.5	2281.0	46.104	2120	15622.8	11412.7	57.299
480	3333.6	2380.4	46.400	2160	15948.7	11659.3	57.451
500	3472.7	2479.8	46.684	2200	16275.6	11906.7	57.601
520	3611.9	2579.3	46.956	2240	16603.3	12155.0	57.749
537	**3730.2**	**2663.8**	**47.180**	2280	16931.9	12404.1	57.894
540	3751.1	2678.7	47.219				
560	3890.3	2778.2	47.472	2320	17261.3	12654.1	58.038
580	4029.6	2877.8	47.717	2360	17591.5	12904.9	58.179
600	4169.0	2977.5	47.953	2400	17922.5	13156.5	58.318
620	4308.4	3077.2	48.182	2440	18254.3	13408.8	58.455
640	4448.0	3177.0	48.403	2480	18586.7	13661.8	58.590
660	4587.6	3276.9	48.618	2520	18919.9	13915.5	58.723
680	4727.4	3377.0	48.827	2560	19253.7	14169.9	58.855
700	4867.3	3477.2	49.029	2600	19588.2	14425.0	58.984
720	5007.4	3577.6	49.227	2640	19923.3	14680.7	59.112
740	5147.6	3678.1	49.419	2680	20259.1	14937.0	59.238
760	5288.0	3778.8	49.606	2720	20595.4	15193.9	59.363
780	5428.7	3879.7	49.789	2760	20932.4	15451.4	59.486
800	5569.5	3980.8	49.967	2800	21269.9	15709.5	59.607
820	5710.6	4082.2	50.141	2840	21607.9	15968.1	59.727
840	5852.0	4183.8	50.311	2880	21946.5	16227.2	59.846
860	5993.5	4285.7	50.478	2920	22285.6	16486.9	59.963
880	6135.4	4387.9	50.641	2960	22625.2	16747.0	60.078
900	6277.6	4490.3	50.801	3000	22965.3	17007.7	60.192
920	6420.1	4593.1	50.957	3040	23305.8	17268.8	60.305
940	6562.8	4696.1	51.111	3080	23646.8	17530.4	60.416
960	6705.9	4799.5	51.262	3120	23988.3	17792.4	60.527
980	6849.4	4903.2	51.409	3160	24330.2	18054.8	60.635
1000	6993.2	5007.3	51.555	3200	24672.5	18317.7	60.743
1040	7281.8	5216.5	51.838	3240	25015.2	18581.0	60.850
1080	7571.8	5427.1	52.111	3280	25358.3	18844.7	60.955

continued

TABLE B6.3 (USCS UNITS) *continued*

Properties of Carbon Monoxide as an Ideal Gas with Variable Specific Heats
Temperature in degrees Rankine
Specific enthalpy in British thermal units per pound-mass-mole
Specific internal energy in British thermal units per pound-mass-mole
$\bar{s}_0$ in British thermal units per pound-mass-mole and degree Rankine
M = 28.0104

T	$\bar{h}$	$\bar{u}$	$\bar{s}_0$	T	$\bar{h}$	$\bar{u}$	$\bar{s}_0$
1120	7863.4	5639.2	52.376	3320	25701.8	19108.8	61.059
1160	8156.5	5852.9	52.634	3360	26045.7	19373.2	61.162
1200	8451.2	6068.1	52.883	3400	26389.9	19638.0	61.264
1240	8747.4	6284.9	53.126	3440	26734.5	19903.2	61.364
1280	9045.2	6503.3	53.362	3480	27079.5	20168.7	61.464
1320	9344.5	6723.2	53.593	3520	27424.8	20434.5	61.563
1360	9645.4	6944.7	53.817	3560	27770.4	20700.7	61.660
1400	9947.9	7167.7	54.037	3600	28116.3	20967.2	61.757
1440	10251.9	7392.3	54.251	3640	28462.6	21234.0	61.853
1480	10557.4	7618.3	54.460	3680	28809.1	21501.2	61.947
1520	10864.3	7845.8	54.665	3720	29156.0	21768.6	62.041
1560	11172.7	8074.8	54.865	3760	29503.1	22036.3	62.134
1600	11482.5	8305.2	55.061	3800	29850.6	22304.3	62.226
1640	11793.7	8536.9	55.253	3840	30198.3	22572.6	62.317
1680	12106.3	8770.0	55.441	3880	30546.3	22841.1	62.407
1720	12420.1	9004.4	55.626	3920	30894.5	23110.0	62.496
1760	12735.2	9240.1	55.807	3960	31243.0	23379.0	62.585
1800	13051.6	9477.0	55.985	4000	31591.8	23648.3	62.672
1840	13369.1	9715.1	56.159	4040	31940.8	23917.9	62.759
1880	13687.8	9954.4	56.331	4080	32290.0	24187.7	62.845

Abridged from Keenan, J. H., Chao, J., and Kaye, J., "Gas Tables," Wiley, New York, 1980. Reprinted by permission of John Wiley & Sons, Inc.

TABLE B6.4 (USCS UNITS)

Properties of Hydrogen as an Ideal Gas with Variable Specific Heats
Temperature in degrees Rankine
Specific enthalpy in British thermal units per pound-mass-mole
Specific internal energy in British thermal units per pound-mass-mole
$\bar{s}^0$ *in British thermal units per pound-mass-mole and degree Rankine*
M=2.0158

T	$\bar{h}$	$\bar{u}$	$\bar{s}^0$	T	$\bar{h}$	$\bar{u}$	$\bar{s}^0$
360	2449.4	1734.4	28.501	1920	13400.4	9587.6	40.142
380	2580.3	1825.7	28.855	1960	13692.4	9800.1	40.293
400	2712.6	1918.3	29.194	2000	13984.3	10012.6	40.440
420	2846.1	2012.0	29.520	2040	14276.3	10225.1	40.585
440	2980.5	2106.8	29.832	2080	14570.2	10439.6	40.728
460	3115.9	2202.4	30.131	2120	14865.1	10655.0	40.869
480	3251.9	2298.7	30.421	2160	15160.9	10871.5	41.006
500	3388.5	2395.6	30.701	2200	15457.8	11088.9	41.141
520	3525.6	2492.9	30.971	2240	15754.6	11306.3	41.274
537	**3642.6**	**2576.1**	**31.192**	2280	16052.4	11524.6	41.407
540	3663.2	2590.9	31.230				
560	3801.1	2689.0	31.480	2320	16352.2	11745.0	41.536
580	3939.5	2787.7	31.724	2360	16652.9	11966.3	41.665
600	4078.0	2886.5	31.958	2400	16953.6	12187.6	41.794
620	4216.8	2985.5	32.187	2440	17255.4	12409.9	41.920
640	4355.7	3084.8	32.407	2480	17559.0	12634.1	42.043
660	4494.8	3184.1	32.620	2520	17862.7	12858.3	42.164
680	4633.9	3283.5	32.828	2560	18167.3	13083.5	42.283
700	4773.1	3383.0	33.031	2600	18473.9	13310.7	42.400
720	4912.6	3482.8	33.227	2640	18780.5	13537.9	42.517
740	5052.0	3582.5	33.418	2680	19088.1	13766.0	42.634
760	5191.5	3682.3	33.605	2720	19396.7	13995.1	42.750
780	5331.1	3782.1	33.785	2760	19707.2	14226.2	42.863
800	5470.6	3881.9	33.960	2800	20017.7	14457.3	42.974
820	5610.2	3981.8	34.135	2840	20328.2	14688.3	43.083
840	5749.9	4081.7	34.302	2880	20640.6	14921.3	43.192
860	5889.6	4181.8	34.467	2920	20954.0	15155.3	43.300
880	6029.3	4281.8	34.627	2960	21267.5	15389.3	43.407
900	6169.0	4381.8	34.784	3000	21582.9	15625.3	43.512
920	6308.8	4481.8	34.939	3040	21898.3	15861.2	43.617
940	6448.6	4581.9	35.090	3080	22214.6	16098.2	43.721
960	6588.5	4682.1	35.235	3120	22531.9	16336.1	43.824
980	6728.3	4782.2	35.380	3160	22850.3	16574.9	43.925
1000	6868.4	4882.5	35.521	3200	23168.6	16813.8	44.025
1040	7148.2	5082.9	35.795	3240	23488.8	17054.7	44.124
1080	7428.3	5283.6	36.059	3280	23810.1	17296.5	44.223

continued

TABLE B6.4 (USCS UNITS) *continued*

Properties of Hydrogen as an Ideal Gas with Variable Specific Heats
Temperature in degrees Rankine
Specific enthalpy in British thermal units per pound-mass-mole
Specific internal energy in British thermal units per pound-mass-mole
$\bar{s}^0$ in British thermal units per pound-mass-mole and degree Rankine
M=2.0158

T	h	u	$\bar{s}^0$	T	h	u	$\bar{s}^0$
1120	7708.6	5484.4	36.315	3320	24132.3	17539.3	44.320
1160	7989.0	5685.4	36.562	3360	24454.5	17782.0	44.416
1200	8269.7	5886.7	36.798	3400	24776.7	18024.8	44.511
1240	8550.7	6088.3	37.028	3440	25100.9	18269.6	44.606
1280	8832.0	6290.1	37.251	3480	25426.1	18515.3	44.702
1320	9113.6	6492.3	37.467	3520	25751.2	18761.0	44.795
1360	9395.6	6694.9	37.678	3560	26078.3	19008.7	44.886
1400	9678.0	6897.8	37.884	3600	26405.4	19256.3	44.978
1440	9960.5	7100.9	38.083	3640	26733.5	19505.0	45.069
1480	10243.7	7304.6	38.277	3680	27062.6	19754.6	45.158
1520	10527.8	7509.3	38.466	3720	27391.6	20004.2	45.248
1560	10811.9	7714.0	38.651	3760	27721.6	20254.8	45.335
1600	11097.1	7919.7	38.830	3800	28052.6	20506.4	45.423
1640	11382.2	8125.4	39.006	3840	28384.6	20758.9	45.508
1680	11669.2	8333.0	39.179	3880	28716.6	21011.5	45.595
1720	11956.3	8540.6	39.348	3920	29048.6	21264.0	45.681
1760	12243.4	8748.3	39.513	3960	29382.5	21518.5	45.766
1800	12531.4	8956.9	39.674	4000	29716.5	21773.0	45.850
1840	12820.4	9166.5	39.832	4040	30052.4	22029.5	45.933
1880	13110.4	9377.0	39.989	4080	30388.3	22285.9	46.014

Abridged from Keenan, J. H., Chao, J., and Kaye, J., "Gas Tables," Wiley, New York, 1980.
Reprinted by permission of John Wiley & Sons, Inc.

TABLE B6.5 (USCS UNITS)
Properties of Nitrogen as an Ideal Gas
with Variable Specific Heats
Temperature in degrees Rankine
Specific enthalpy in British thermal units per pound-mass-mole
Specific internal energy in British thermal units per pound-mass-mole
$\bar{s}^0$ *in British thermal units per pound-mass-mole and degree Rankine*
M=28.0134

T	$\bar{h}$	$\bar{u}$	$\bar{s}^0$	T	$\bar{h}$	$\bar{u}$	$\bar{s}^0$
360	2499.0	1784.1	42.954	1920	13897.1	10084.2	54.970
380	2638.1	1883.5	43.330	1960	14213.5	10321.2	55.133
400	2777.1	1982.8	43.687	2000	14531.1	10559.4	55.294
420	2916.2	2082.1	44.026	2040	14849.7	10798.6	55.451
440	3055.2	2181.5	44.350	2080	15169.4	11038.8	55.606
460	3194.3	2280.8	44.659	2120	15490.1	11280.1	55.759
480	3333.4	2380.2	44.955	2160	15811.8	11522.3	55.910
500	3472.5	2479.5	45.238	2200	16134.4	11765.6	56.058
520	3611.6	2578.9	45.511	2240	16458.0	12009.7	56.203
537	**3735.4**	**2669.0**	**48.966**	2280	16782.5	12254.7	56.347
540	3750.7	2678.3	45.774				
560	3889.8	2777.7	46.027	2320	17107.8	12500.6	56.488
580	4029.0	2877.2	46.271	2360	17434.0	12747.4	56.628
600	4168.2	2976.7	46.507	2400	17761.1	12995.0	56.765
620	4307.5	3076.2	46.735	2440	18088.9	13243.4	56.901
640	4446.8	3175.8	46.956	2480	18417.5	13492.6	57.034
660	4586.2	3275.5	47.171	2520	18746.9	13742.5	57.166
680	4725.6	3375.3	47.379	2560	19077.0	13993.2	57.296
700	4865.2	3475.1	47.581	2600	19407.8	14244.6	57.424
720	5004.9	3575.1	47.778	2640	19739.3	14496.6	57.551
740	5144.7	3675.1	47.970	2680	20071.5	14749.4	57.676
760	5284.6	3775.3	48.156	2720	20404.3	15002.8	57.799
780	5424.6	3875.6	48.338	2760	20737.8	15256.8	57.921
800	5564.8	3976.1	48.516	2800	21071.9	15511.5	58.041
820	5705.2	4076.8	48.689	2840	21406.5	15766.7	58.159
840	5845.8	4177.6	48.858	2880	21741.8	16022.5	58.277
860	5986.5	4278.7	49.024	2920	22077.6	16278.9	58.392
880	6127.5	4379.9	49.416	2960	22414.0	16535.8	58.507
900	6268.7	4481.4	49.344	3000	22750.9	16793.3	58.620
920	6410.1	4583.1	49.500	3040	23088.3	17051.3	58.732
940	6551.8	4685.1	49.652	3080	23426.2	17309.8	58.842
960	6693.7	4787.3	49.802	3120	23764.6	17568.7	58.951
980	6835.9	4889.8	49.948	3160	24103.5	17828.2	59.059
1000	6978.4	4992.5	50.092	3200	24442.9	18088.1	59.166
1040	7264.2	5198.9	50.372	3240	24782.7	18348.5	59.271
1080	7551.2	5406.5	50.643	3280	25122.9	18609.3	59.376

continued

TABLE B6.5 (USCS UNITS) *continued*

Properties of Nitrogen as an Ideal Gas with Variable Specific Heats
Temperature in degrees Rankine
Specific enthalpy in British thermal units per pound-mass-mole
Specific internal energy in British thermal units per pound-mass-mole
$\bar{s}^0$ *in British thermal units per pound-mass-mole and degree Rankine*
M=28.0134

T	$\bar{h}$	$\bar{u}$	$\bar{s}^0$	T	$\bar{h}$	$\bar{u}$	$\bar{s}^0$
1120	7839.5	5615.4	50.905	3320	25463.6	18870.5	59.479
1160	8129.2	5825.6	51.159	3360	25804.7	19132.2	59.581
1200	8420.2	6037.1	51.406	3400	26146.2	19394.2	59.682
1240	8712.6	6250.1	51.646	3440	26488.1	19656.7	59.782
1280	9006.4	6464.5	51.879	3480	26830.3	19919.5	59.881
1320	9301.7	6680.3	52.106	3520	27173.0	20182.7	59.979
1360	9598.4	6897.6	52.328	3560	27516.0	20446.3	60.076
1400	9896.5	7116.3	52.544	3600	27859.3	20710.3	60.172
1440	10196.1	7336.5	52.755	3640	28203.1	20974.5	60.267
1480	10497.1	7558.1	52.961	3680	28547.1	21239.2	60.361
1520	10799.6	7781.1	53.162	3720	28891.5	21504.1	60.454
1560	11103.4	8005.5	53.360	3760	29236.2	21769.4	60.546
1600	11408.6	8231.3	53.553	3800	29581.2	22035.0	60.637
1640	11715.2	8458.4	53.742	3840	29926.6	22300.9	60.728
1680	12023.1	8686.9	53.928	3880	30272.2	22567.1	60.817
1720	12332.3	8916.7	54.110	3920	30618.1	22833.6	60.906
1760	12642.8	9147.7	54.288	3960	30964.3	23100.3	60.994
1800	12954.6	9380.0	54.463	4000	31310.8	23367.4	61.081
1840	13267.6	9613.6	54.635	4040	31657.6	23634.7	61.167
1880	13581.7	9848.3	54.804	4080	32004.6	23902.3	61.253

Abridged from Keenan, J. H., Chao, J., and Kaye, J., "Gas Tables," Wiley, New York, 1980.
Reprinted by permission of John Wiley & Sons, Inc.

TABLE B6.6 (USCS UNITS)

Properties of Oxygen as an Ideal Gas with Variable Specific Heats
Temperature in degrees Rankine
Specific enthalpy in British thermal units per pound-mass-mole
Specific internal energy in British thermal units per pound-mass-mole
$\bar{s}^0$ *in British thermal units per pound-mass-mole and degree Rankine*
M=31.9988

T	$\bar{h}$	$\bar{u}$	$\bar{s}^0$	T	$\bar{h}$	$\bar{u}$	$\bar{s}^0$
360	2500.3	1785.4	46.176	1920	14497.3	10684.5	58.680
380	2639.4	1884.8	46.552	1960	14833.7	10941.4	58.854
400	2778.6	1984.3	46.909	2000	15170.9	11199.2	59.024
420	2917.9	2083.9	47.249	2040	15509.0	11457.8	59.192
440	3057.3	2183.5	47.573	2080	15847.9	11717.3	59.356
460	3196.8	2283.3	47.883	2120	16187.5	11977.5	59.518
480	3336.4	2383.2	48.180	2160	16527.8	12238.4	59.677
500	3476.2	2483.3	48.466	2200	16868.9	12500.0	59.833
520	3616.2	2583.6	48.740	2240	17210.7	12762.4	59.987
537	**3735.4**	**2669.0**	**48.966**	2280	17553.2	13025.5	60.139
540	3756.5	2684.1	49.005				
560	3897.0	2784.9	49.260	2320	17896.4	13289.2	60.288
580	4037.8	2886.0	49.508	2360	18240.1	13553.5	60.435
600	4178.9	2987.4	49.747	2400	18584.6	13818.5	60.580
620	4320.4	3089.2	49.979	2440	18929.6	14084.1	60.722
640	4462.3	3191.3	50.204	2480	19275.3	14350.3	60.863
660	4604.5	3293.8	50.423	2520	19621.5	14617.1	61.001
680	4747.2	3396.8	50.636	2560	19968.3	14884.5	61.138
700	4890.3	3500.2	50.843	2600	20315.7	15152.5	61.272
720	5033.9	3604.1	51.045	2640	20663.7	15421.0	61.405
740	5177.9	3708.4	51.243	2680	21012.2	15690.1	61.536
760	5322.5	3813.2	51.435	2720	21361.2	15959.7	61.666
780	5467.5	3918.6	51.624	2760	21710.8	16229.8	61.793
800	5613.1	4024.4	51.808	2800	22060.9	16500.5	61.919
820	5759.2	4130.8	51.988	2840	22411.6	16771.8	62.043
840	5905.8	4237.7	52.165	2880	22762.8	17043.5	62.166
860	6053.0	4345.2	52.338	2920	23114.5	17315.7	62.287
880	6200.7	4453.2	52.508	2960	23466.7	17588.5	62.407
900	6349.0	4561.7	52.675	3000	23819.4	17861.8	62.526
920	6497.8	4670.8	52.838	3040	24172.6	18135.6	62.643
940	6647.1	4780.4	52.999	3080	24526.3	18409.8	62.758
960	6797.0	4890.5	53.157	3120	24880.5	18684.6	62.872
980	6947.4	5001.2	53.312	3160	25235.2	18959.9	62.985
1000	7098.3	5112.5	53.464	3200	25590.4	19235.6	63.097
1040	7401.8	5336.5	53.762	3240	25946.1	19511.9	63.208
1080	7707.4	5562.6	54.050	3280	26302.2	19788.6	63.317

continued

TABLE B6.6 (USCS UNITS) *continued*

Properties of Oxygen as an Ideal Gas with Variable Specific Heats
Temperature in degrees Rankine
Specific enthalpy in British thermal units per pound-mass-mole
Specific internal energy in British thermal units per pound-mass-mole
$\bar{s}^0$ *in British thermal units per pound-mass-mole and degree Rankine*
M=31.9988

T	$\bar{h}$	$\bar{u}$	$\bar{s}^0$	T	$\bar{h}$	$\bar{u}$	$\bar{s}^0$
1120	8014.9	5790.8	54.330	3320	26658.9	20065.8	63.425
1160	8324.5	6020.9	54.601	3360	27016.0	20343.5	63.532
1200	8636.0	6252.9	54.865	3400	27373.6	20621.7	63.638
1240	8949.3	6486.8	55.122	3440	27731.7	20900.4	63.742
1280	9264.4	6722.5	55.372	3480	28090.3	21179.5	63.846
1320	9581.2	6959.8	55.616	3520	28449.4	21459.1	63.949
1360	9899.6	7198.8	55.853	3560	28808.9	21739.2	64.050
1400	10219.6	7439.4	56.085	3600	29168.9	22019.8	64.151
1440	10541.2	7681.5	56.312	3640	29529.3	22300.8	64.250
1480	10864.2	7925.1	56.533	3680	29890.3	22582.3	64.349
1520	11188.6	8170.0	56.749	3720	30251.7	22864.3	64.447
1560	11514.3	8416.3	56.961	3760	30613.5	23146.7	64.543
1600	11841.3	8663.9	57.168	3800	30975.9	23429.6	64.639
1640	12169.5	8912.7	57.370	3840	31338.7	23713.0	64.734
1680	12498.9	9162.6	57.569	3880	31701.9	23996.8	64.828
1720	12829.4	9413.7	57.763	3920	32065.7	24281.1	64.922
1760	13161.0	9665.9	57.954	3960	32429.8	24565.8	65.014
1800	13493.6	9919.1	58.141	4000	32794.5	24851.0	65.106
1840	13827.2	10173.3	58.324	4040	33159.6	25136.7	65.196
1880	14161.8	10428.4	58.504	4080	33525.1	25422.8	65.286

Abridged from Keenan, J. H., Chao, J., and Kaye, J., "Gas Tables," Wiley, New York, 1980. Reprinted by permission of John Wiley & Sons, Inc.

TABLE B6.7 (USCS UNITS)
Properties of Water Vapor as an Ideal Gas
with Variable Specific Heats
Temperature in degrees Rankine
Specific enthalpy in British thermal units per pound-mass-mole
Specific internal energy in British thermal units per pound-mass-mole
$\bar{s}^0$ in British thermal units per pound-mass-mole and degree Rankine
M = 18.0152

T	$\bar{h}$	$\bar{u}$	$\bar{s}^0$	T	$\bar{h}$	$\bar{u}$	$\bar{s}^0$
360	2846.9	2132.0	41.886	1920	16630.9	12818.1	56.203
380	3006.2	2251.6	42.317	1960	17034.7	13142.4	56.411
400	3165.7	2371.3	42.726	2000	17441.0	13469.3	56.616
420	3325.1	2491.1	43.115	2040	17850.0	13798.8	56.819
440	3484.7	2610.9	43.486	2080	18261.7	14131.1	57.019
460	3644.4	2730.9	43.841	2120	18676.0	14465.9	57.216
480	3804.1	2850.9	44.181	2160	19092.9	14803.4	57.411
500	3964.2	2971.3	44.507	2200	19512.3	15143.4	57.603
520	4124.3	3091.7	44.821	2240	19934.2	15485.8	57.793
537	**4260.5**	**3194.2**	**45.079**	2280	20358.7	15831.0	57.981
540	4284.6	3212.3	45.124				
560	4445.3	3333.2	45.416	2320	20785.6	16178.4	58.167
580	4606.2	3454.4	45.698	2360	21215.0	16528.4	58.350
600	4767.4	3575.9	45.971	2400	21646.8	16880.8	58.532
620	4928.8	3697.6	46.236	2440	22081.0	17235.5	58.711
640	5090.7	3819.7	46.493	2480	22517.6	17592.7	58.889
660	5252.9	3942.2	46.743	2520	22956.4	17952.0	59.064
680	5415.4	4065.1	46.985	2560	23397.5	18313.7	59.238
700	5578.5	4188.4	47.222	2600	23840.8	18677.6	59.410
720	5741.9	4312.0	47.452	2640	24286.5	19043.8	59.580
740	5905.8	4436.3	47.677	2680	24734.3	19412.2	59.748
760	6070.1	4560.9	47.896	2720	25184.1	19782.6	59.915
780	6235.1	4686.1	48.110	2760	25636.1	20155.1	60.080
800	6400.4	4811.7	48.319	2800	26090.1	20529.7	60.243
820	6566.3	4937.8	48.524	2840	26546.2	20906.4	60.405
840	6732.6	5064.5	48.724	2880	27004.3	21285.0	60.565
860	6899.5	5191.7	48.921	2920	27464.4	21665.6	60.724
880	7067.0	5319.4	49.113	2960	27926.3	22048.2	60.881
900	7234.9	5447.6	49.302	3000	28390.1	22432.5	61.036
920	7403.5	5576.5	49.487	3040	28855.9	22818.8	61.191
940	7572.7	5705.9	49.669	3080	29323.4	23206.9	61.343
960	7742.3	5835.8	49.848	3120	29792.6	23596.8	61.495
980	7912.5	5966.4	50.023	3160	30263.8	23988.4	61.645
1000	8083.4	6097.6	50.196	3200	30736.4	24381.7	61.793
1040	8426.8	6361.5	50.532	3240	31211.0	24776.8	61.941
1080	8772.7	6627.9	50.859	3280	31687.1	25173.5	62.087

continued

TABLE B6.7 (USCS UNITS) *continued*
Properties of Water Vapor as an Ideal Gas
with Variable Specific Heats
Temperature in degrees Rankine
Specific enthalpy in British thermal units per pound-mass-mole
Specific internal energy in British thermal units per pound-mass-mole
$\bar{s}^0$ *in British thermal units per pound-mass-mole and degree Rankine*
M = 18.0152

T	$\bar{h}$	$\bar{u}$	$\bar{s}^0$	T	$\bar{h}$	$\bar{u}$	$\bar{s}^0$
1120	9121.0	6896.8	51.175	3320	32164.9	25571.8	62.232
1160	9471.7	7168.1	51.483	3360	32644.1	25971.6	62.375
1200	9824.9	7441.8	51.782	3400	33125.0	26373.0	62.517
1240	10180.5	7718.0	52.074	3440	33607.4	26776.1	62.658
1280	10538.7	7996.8	52.358	3480	34091.4	27180.6	62.798
1320	10899.5	8278.2	52.636	3520	34576.8	27586.6	62.937
1360	11262.9	8562.2	52.907	3560	35063.6	27993.9	63.075
1400	11628.9	8848.7	53.172	3600	35551.9	28402.8	63.211
1440	11997.5	9137.9	53.432	3640	36041.5	28812.9	63.346
1480	12368.7	9429.7	53.686	3680	36532.6	29224.6	63.480
1520	12742.7	9724.2	53.935	3720	37024.9	29637.5	63.613
1560	13119.3	10021.4	54.180	3760	37518.6	30051.8	63.745
1600	13498.7	10321.3	54.420	3800	38013.5	30467.3	63.876
1640	13880.6	10623.8	54.656	3840	38509.9	30884.1	64.006
1680	14265.5	10929.2	54.888	3880	39007.3	31302.2	64.135
1720	14652.9	11237.2	55.116	3920	39506.0	31721.4	64.263
1760	15043.1	11548.0	55.340	3960	40005.9	32141.9	64.390
1800	15436.0	11861.4	55.561	4000	40507.0	32563.6	64.516
1840	15831.5	12177.6	55.778	4040	41009.2	32986.3	64.641
1880	16229.9	12496.5	55.992	4080	41512.5	33410.2	64.765

Abridged from Keenan, J. H., Chao, J., and Kaye, J., "Gas Tables," Wiley, New York, 1980. Reprinted by permission of John Wiley & Sons, Inc.

Table B7 (USCS UNITS)
Enthalpy and Gibbs Free Energy of Formation, and Standard Absolute Entropy of Selected Substances

$\bar{h}_f^0$ in British thermal units per pound-mass-mole
$\bar{g}_f^0$ in British thermal units per pound-mass-mole
$\bar{s}^0$ in British thermal units per pound-mass-mole and degree Rankine

Substance	Formula	M	$\bar{h}_f^0$	$\bar{g}_f^0$	$\bar{s}^0$
Carbon	$C(s)$	12.0112	0	0	1.36
Hydrogen	$H_2(g)$	2.0159	0	0	31.21
Nitrogen	$N_2(g)$	28.0134	0	0	45.77
Oxygen	$O_2(g)$	31.9988	0	0	49.00
Carbon Monoxide	$CO(g)$	28.0106	−47,540	−59,010	47.21
Carbon Dioxide	$CO_2(g)$	44.0100	−169,300	−169,680	51.07
Water	$H_2O(g)$	18.0153	−104,040	−98,350	45.11
Water	$H_2O(f)$	18.0153	−122,970	−102,040	16.71
Hydrogen Peroxide	$H_2O_2(g)$	34.0147	−58,640	−45,430	55.60
Ammonia	$NH_3(g)$	17.0306	−19,750	−7,140	45.97
Oxygen	$O(g)$	15.9994	107,210	99,710	38.47
Hydrogen	$H(g)$	1.0080	93,780	87,460	27.39
Nitrogen	$N(g)$	14.0067	203,340	195,970	36.61
Hydroxyl	$OH(g)$	17.0074	16,790	14,750	43.92
Methane	$CH_4(g)$	16.0112	−32,210	−21,860	44.49
Acetylene (Ethyne)	$C_2H_2(g)$	26.0382	97,540	87,990	48.00
Ethylene (Ethene)	$C_2H_4(g)$	28.0542	22,490	29,306	52.54
Ethane	$C_2H_6(g)$	30.0701	−36,420	−14,150	54.85
Propylene (Propene)	$C_3H_6(g)$	42.0813	8,790	26,980	63.80
Propane	$C_3H_8(g)$	44.0972	−44,680	−10,105	64.51
n-Butane	$C_4H_{10}(g)$	58.1243	−54,270	−6,760	74.11
n-Octane	$C_8H_{18}(g)$	114.2327	−89,680	7,110	111.55
n-Octane	$C_8H_{18}(f)$	114.2327	−107,530	2,840	86.23
Benzene	$C_6H_6(g)$	78.1147	35,680	55,780	64.34
Methyl Alcohol	$CH_3OH(g)$	32.0424	−86,540	−69,700	57.29
Methyl Alcohol	$CH_3OH(f)$	32.0424	−102,670	−71,570	30.30
Ethyl Alcohol	$C_2H_5OH(g)$	44.0536	−101,230	−72,520	67.54
Ethyl Alcohol	$C_2H_5OH(f)$	44.0536	−119,470	−75,240	38.40

Abridged from Wark, K., Jr., "Thermodynamics," McGraw-Hill, New York, 1988 as taken from JANAF, "Thermochemical Tables," Dow Chemical Company, Midland, MI, 1971; "Selected Values of Chemical Thermodynamic Properties," NBS Technical Note 270-3, 1968; and "API Research Project 44," Carnegie Press, 1953. Reproduced with permission of McGraw-Hill.

Thermodynamics Reference Charts

FIGURE C.1
Generalized compressibility charts.

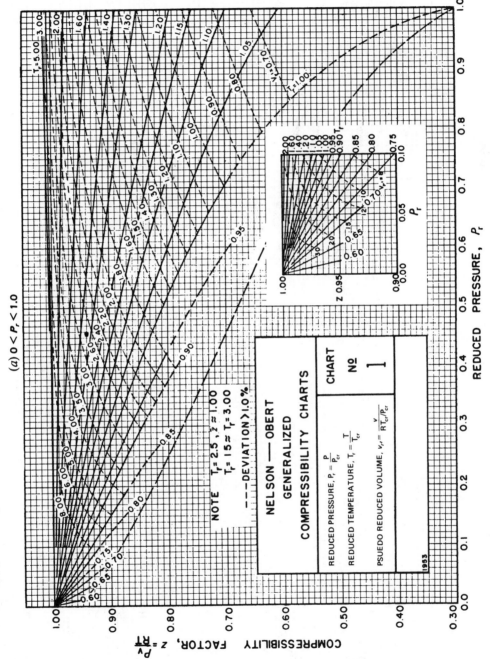

(a) $0 < P_r < 1.0$

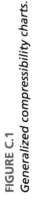

NOTE $T_r = 2.5, \; z \approx 1.00$
$T_r = 15 \approx T_r = 3.00$
---DEVIATION > 1.0%

NELSON — OBERT
GENERALIZED
COMPRESSIBILITY CHARTS

CHART No 1

REDUCED PRESSURE, $P_r = \dfrac{P}{P_{cr}}$

REDUCED TEMPERATURE, $T_r = \dfrac{T}{T_{cr}}$

PSUEDO REDUCED VOLUME, $v_r = \dfrac{v}{RT_{cr}/P_{cr}}$

1953

COMPRESSIBILITY FACTOR, $z = \dfrac{Pv}{RT}$

REDUCED PRESSURE, P_r

From John R. Howell and Richard O. Bukius, *Fundamentals of Engineering Thermodynamics*, Second edition, McGraw-Hill, New York, 1992. Used with permission.

FIGURE C.1

Generalized compressibility charts. (Continued)

(b) 0 < Pr < 7

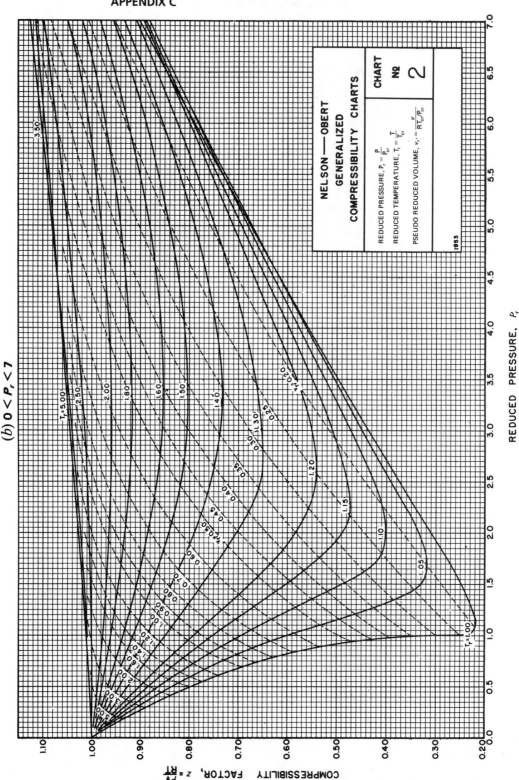

FIGURE C.1
Generalized compressibility charts. (Continued)

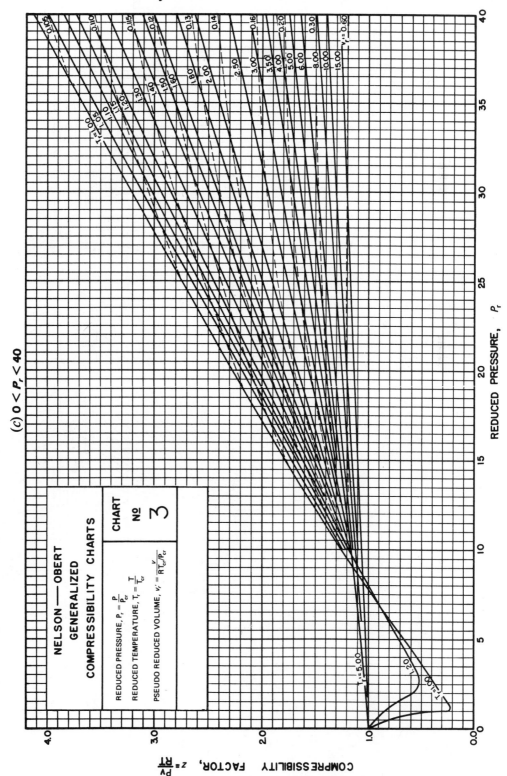

(c) $0 < P_r < 40$

NELSON — OBERT
GENERALIZED
COMPRESSIBILITY CHARTS

CHART
№ 3

REDUCED PRESSURE, $P_r = \dfrac{P}{P_{cr}}$

REDUCED TEMPERATURE, $T_r = \dfrac{T}{T_{cr}}$

PSEUDO REDUCED VOLUME, $v_r' = \dfrac{v}{RT_{cr}/P_{cr}}$

COMPRESSIBILITY FACTOR, $z = \dfrac{Pv}{RT}$

REDUCED PRESSURE, P_r

FIGURE C.2
Generalized enthalpy departure chart.

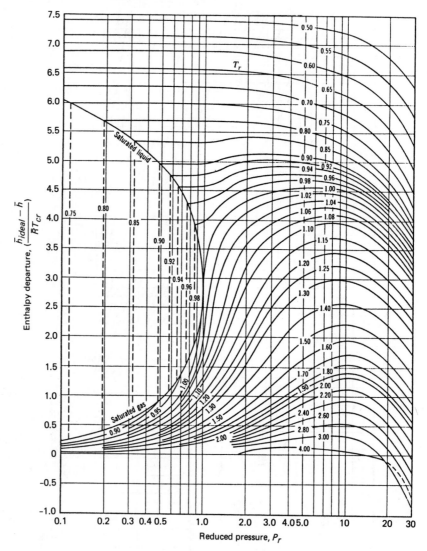

From John R. Howell and Richard O. Bukius, *Fundamentals of Engineering Thermodynamics,* Second edition, McGraw-Hill, New York, 1992. Used with permission.

FIGURE C.3
Generalized entropy departure chart.

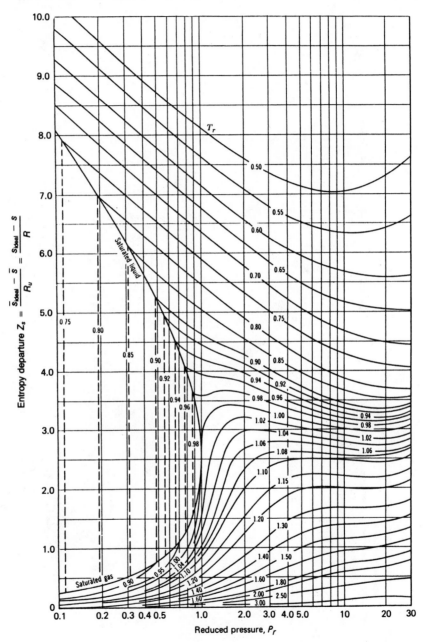

From John R. Howell and Richard O. Bukius, *Fundamentals of Engineering Thermodynamics,* SI Version, McGraw-Hill, New York, 1987. Used with permission.

FIGURE C.4.
Psychrometric Chart.

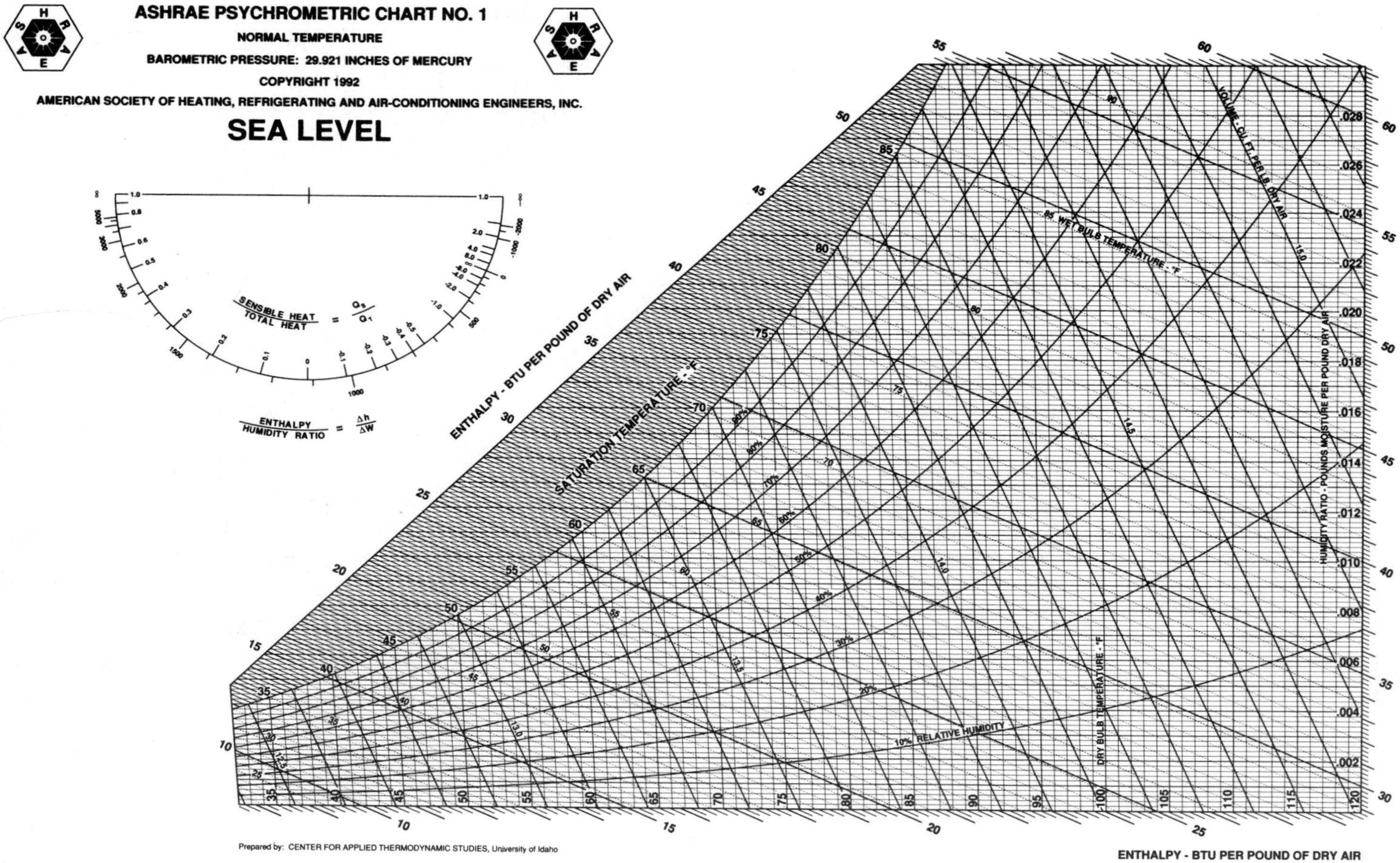

ASHRAE PSYCHROMETRIC CHART NO. 1

NORMAL TEMPERATURE

BAROMETRIC PRESSURE: 29.921 INCHES OF MERCURY

COPYRIGHT 1992

AMERICAN SOCIETY OF HEATING, REFRIGERATING AND AIR-CONDITIONING ENGINEERS, INC.

SEA LEVEL

Prepared by: CENTER FOR APPLIED THERMODYNAMIC STUDIES, University of Idaho

Reprinted by permission of the American Society of Heating, Refrigerating, and Air-Conditioning Engineers, Inc., Atlanta.

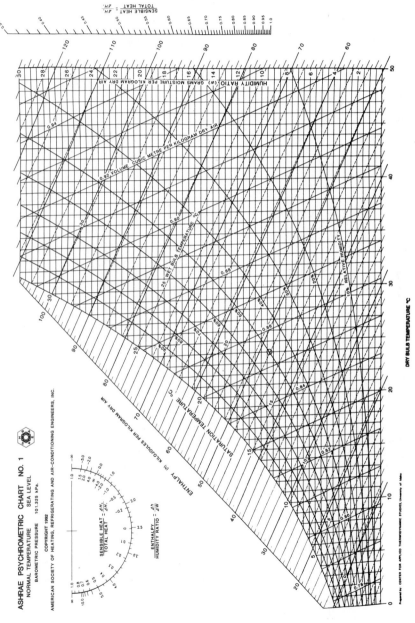

FIGURE C.5.
Psychrometric Chart.

Reprinted by permission of the American Society of Heating, Refrigerating, and Air-Conditioning Engineers, Inc., Atlanta.

Answers to Selected Problems

Chapter 1

1.7	Weight is an extensive property
1.10	Molar specific volume is an intensive property
1.13	This process is irreversible
1.19	$_1W_2 = -33.50$ kJ (cart)
	$_1W_2 = -67.00$ kJ (man)
1.23	$W = 1.867$ kJ
1.26	$w = 30.28$ lb$_f$
1.29	$w = 1920$ N
1.32	$e_k = 0.45$ kJ/kg
1.35	$E_p = 2.57$ Btu
1.38	$E_p = 19.6$ kJ
1.41	(a) $1500 \dfrac{\text{kN}}{\text{m}^2}$
	(b) $1{,}500{,}000 \dfrac{\text{kg}}{\text{m-s}^2}$
	(c) $1500 \dfrac{\text{kg}}{\text{km-s}^2}$
1.44	(a) $31{,}300 \dfrac{\text{lb}_f}{\text{ft}^2}$

Chapter 2

2.1	$W = 0.455$ kJ

2.4	$|\dot{Q}_h| = 17{,}820$ Btu/hr
2.7	$|\dot{W}| = 2.170$ kW
2.10	$\eta = 0.625$
2.13	$q_h = -85$ kJ $\quad q_l = 35$ kJ
2.16	$COP_{hp} = 5$
2.22	$\dot{Q}_h = 19{,}090$ Btu/hr
2.25	$|\dot{W}| = 0.8790$ kW
2.28	Claim is invalid
2.34	$\dot{Q}_l = 4$ kW $\quad \dot{S}_l = -0.02$ kJ/s-K
	$\dot{S}_h = 0.0351$ kJ/s-K
2.37	$\triangle S_h = -10$ Btu/$^\circ$ R
	$Q_h = -15{,}000$ Btu
2.40	$\triangle S_{everything} = -0.08337$ Btu/$^\circ$ R
2.43	$\dot{S}_h = 0.3401$ kJ/s-K
	$\dot{S}_l = -0.3664$ kJ/s-K
2.46	$T_l \geq 476^\circ$ R
2.49	$w/q_h = 0.7734$
2.52	$|w|q_l = 0.1099$
2.55	$COP_{ref,max} = 14.65$

Chapter 3

3.1	$M_{CH_4} = 16.033$ kg/kg-mole
	$M_{C_3H_8} = 44.067$ kg/kg-mole
3.4	M= 24.32 lb$_m$/lb$_m$-mole
3.7	$x_{CO_2} = 0.1806$

3.10 $\gamma = 58.18 \text{ lb}_f\text{ft}^3$

3.13 $\text{wf}_l = \text{x}_i$

3.16 $\Delta P = 1.44 \text{ lb}_f/\text{in}^2$

3.19 $P_3 = 110.0 \text{ psia}$

3.22 $P_2 = 392 \text{ psia}$

3.25 $P_1 = 14.86 \text{ psia}$

3.29 $T = 65.56^\circ \text{C}$

3.31 $T = 762^\circ \text{R}$

3.34 $T_1 = 1010^\circ \text{R} \quad P_2 = 0.9503 \text{ psia}$

3.37 $V_2 = 0.2542 \text{ m}^3$

3.40 $P_2 = 66.98 \text{ psia} \quad U_2 = 921 \text{ Btu}$

3.44 $u = 1076.3 \text{ Btu/lb}_m$

3.46 $U = 660.8 \text{ Btu} \quad S = 0.9812 \text{ Btu/}^\circ \text{R}$

3.49 $v = 0.001102 \text{ m}^3/\text{kg}$

3.52 $h = 301.36 \text{ kJ/kg}$

3.55 $T_2 = -30.1^\circ \text{C} \quad u_2 = 317.42 \text{ kJ/kg}$

3.58 $h_2 - h_1 = -806.9 \text{ kJ/kg}$

3.61 $T_2 \approx 220^\circ \text{C} \quad s_2 = 4.093 \text{ kJ/kg-K}$

3.64   Error = 0.0040

3.68 $\Delta V = 0.04338 \text{ m}^3$

3.73 $V = 4.154 \text{ m}^3$

3.76 $\Delta T = 466^\circ \text{F or }^\circ \text{R}$

3.79 $V_1 = 1.154 \text{ ft}^3$

3.82 $\Delta u = 46.69 \text{ kJ/kg} \quad \Delta h = 65.39 \text{ kJ/kg}$

3.85 $\Delta h = 45.95 \text{ kJ/kg}$

3.88 $v_2 - v_1 = -8.505 \text{ ft}^3/\text{lb}_m$

3.91 $\Delta u = \Delta h = 50 \text{ Btu/lb}_m$

         $\Delta u = 49.98 \text{ Btu/lb}_m$

         $\Delta h = 49.99 \text{ Btu/lb}_m$

         $\Delta u = 49.31 \text{ Btu/lb}_m$

         $\Delta h = 55.24 \text{ Btu/lb}_m$

Chapter 4

4.2 $_1W_2 = -5.136 \text{ Btu}$

4.5 $_1W_2 = 111.1 \text{ Btu}$

4.8 $_1W_2 = 192.4 \text{ Btu}$

4.11 $_1w_2 = 139.0 \text{ Btu/lb}_m$

4.17 $U_2 - U_1 = 15 \text{ Btu}$

4.20 $U_2 - U_1 = 0 \text{ Btu} \quad P_2 = 50 \text{ psia}$

4.23 $_1W_2 = 5.03(10^{-8})\text{J}$

4.26 $_1W_2 = -0.0214 \text{ Btu}$

4.29 $u_2 - u_1 = -100 \text{ kJ/kg}$

4.32 $_1Q_2 = 46.82 \text{ kJ}$

4.35 $_1q_2 = -2163 \text{ kJ/kg}$

4.38 $_1Q_2 = -83.66 \text{ Btu}$

4.41 $_1q_2 = -67.26 \text{ Btu/lb}_m$

         $_1w_2 = 0 \text{ Btu/lb}_m$

4.44   (a) $_1Q_2 = 1704 \text{ kJ}$

         (b) $_1Q_2 = 2387 \text{ kJ}$

4.47 $_1W_2 = 1500 \text{ kJ} \quad m = 13.94 \text{ kg}$

4.50 $_1W_2 = -7968 \text{ Btu} \quad _1Q_2 = -6405 \text{ Btu}$

4.53 $_1q_2 = -50 \text{ kJ/kg}$

4.56 $_1w_2 = -7.275 \text{ Btu/lb}_m$

         $_1q_2 = -56.30 \text{ Btu/lb}_m$

4.59 $T_2 = 10^\circ \text{F} \quad \Delta t = 16.43 \text{ hr}$

4.62 $_1q_3 = 341.0 \text{ kJ/kg}$

4.65 $_1Q_2 = 0 \text{ Btu} \quad _1W_2 = 15.59 \text{ Btu}$

4.68 $T_2 = 665^\circ \text{C} \quad _1w_2 = -888.4 \text{ kJ/kg}$

4.71 $s_2 - s_1 = -0.00285 \text{ Btu/lb}_m-^\circ \text{R}$

4.74   Helium

4.77 $s_2 - s_1 = -0.1044 \text{ Btu/lb}_m-^\circ \text{R}$

4.80 $_1w_2 = -76.74 \text{ Btu/lb}_m$

4.83 $T_2 = 579 \text{ K}$

4.89   Neon

4.91 $_1w_2 = -148.0 \text{ kJ/kg}$

         $_1q_2 = -103.6 \text{ kJ/kg}$

4.94 $T_{2,min} = 306 \text{ K}$

4.97   Process is possible

4.100 $V_{2,min} = 2.646 \text{ m}^3$

4.103   (a) $\Delta S_{R-12} = 0.00111 \text{ kJ/kg-K}$

         (b) $\Delta S_C = -0.0003437 \text{ kJ/kg-K}$

         (c) $\Delta S_{everything} = 0.0007663 \text{ kJ/kg-K}$

4.106 $_1w_2 = 132.4$ Btu/lb$_m$ $T_2 = 273.9°$ F

4.109 $_1W_2 = -4965$ kJ $T_2 = 830$ K

4.111 Air

4.115 Steam

4.118 0.3085

Chapter 5

5.2 $\dot{m} = 11.97$ kg/s $A_2 = 1.625$ m^2

5.5 $D = 3.634$ ft

5.8 $\dot{m}_3 = 10.2$ kg/s $V_3 = 14.67$ m/s

5.11 $\dot{m}_{ash} = 0.01$ lb$_m$/s $\dot{m}_{air} = 9.99$ lb$_m$/s

5.14 $w_{fw} = 0.1185$ Btu/lb$_m$

5.17 $\dot{m}_2 = 0.03580$ kg/s

   $\dot{m}_3 = 0.01534$ kg/s

   $\dot{w}_{fw} = 0$ kW

5.20 $V = 82.59$ ft/s $\dot{W}_{fw} = -0.421$ hp

5.23 $\dot{W} = -4.010$ kW

5.26 $\dot{m} = 3397$ lb$_m$/s

5.29 $\dot{W} = 6.498$ hp $A_2 = 32.73$ ft^2

5.32 $\dot{W} = 2128$ hp

5.35 $_1q_2 = 760.7$ Btu/lb$_m$

5.38 $\dot{W} = -0.0947$ kW

5.41 $V_1 = 515$ ft/s $V_2 = 941$ ft/s

5.44 $_1w_2 = -39.15$ Btu/lb$_m$

5.47 $V_2 = 755$ ft/s

5.50 $\dot{W} = 17,060$ hp

5.53 $\dot{W} = 5595$ kW

5.56 $V_2 = 1284$ ft/s

5.59 valid

5.62 $\dot{S}_{gen} = 181.8$ kJ/kg-K-hr

5.65 $T_2 = 1111°$ R $_1w_2 = -139.3$ Btu/lb$_m$

5.68 $_1q_2 = 848.8$ Btu/lb$_m$

5.74 $x_1 = 0.9562$

5.77 $_1q_3 = 341.0$ kJ/kg

5.80 $_1q_3 = 142.1$ kJ/kg

5.83 $_1w_3 = 0$ kJ/kg

5.86 $_1w_2 = -205.9$ kJ/kg

5.89 $s_{gen} = 0.09410$ kJ/kg-K

5.92 $\eta = 0.3878$

5.95 $\eta = 0.3927$

5.98 $V_2 = 253$ ft/s

5.100 $\eta = 0.96$

5.103 $\dot{W}_{min} = -1.518$ hp

5.107 $\eta_{2nd} = 0.88$

5.110 $\Delta\psi = 93.3$ kJ/kg $\eta_{2nd,open} = 1.00$

   $\eta_{2nd,part} = 0.9235$

Chapter 6

6.1 $\Delta t = 18.7$ min

6.4 $V_{max} = 5.03$ ft^3

6.7 $Q = -95,010$ kJ $W = -81,790$ kJ

6.10 $T_f = 240.4$ K

6.13 $T_f = 394$ K $m_f = 17,690$ kg

6.16 $T_f = 910°$ R $W = 370.2$ Btu

6.19 $Q = -81.75 \times 10^6$ Btu

   $W = -57.24 \times 10^6$ Btu

6.22 $V_f = 255$ m^3 $W = 18,000$ kJ

6.25 $\Delta s_\infty \geq (s_{in} - s)\Delta m$

6.28 $\Delta s \geq -4.995$ kJ/K

6.31 $W_{ptl} = -0.0045$ kJ

6.37 $W = -351.1 \times 10^6$ kJ

   $W_{ptl} = -277.3 \times 10^6$ kJ

Chapter 7

7.2 $\eta = 0.3$

7.5 $\dot{m} = 59.71$ lb$_m$/s

7.8 $\eta = 0.4294$

7.11 $\dot{Q}_{in} = 61.19 \times 10^6$ Btu/hr

   $\dot{Q}_{out} = -44.14 \times 10^6$ Btu/hr

   $\eta = 0.2786$

7.14 $T_2 = 293°$ C $T_4 = 284°$ C

   $\eta = 0.3348$

7.17	$\eta = 0.3777$
7.20	$\dot{m} = 0.1271 \text{ lb}_m/\text{s}$
7.23	$\dot{m} = 0.7756 \text{ lb}_m/\text{s}$
7.27	$\eta = 0.3153$
7.30	$P \approx 200 \text{ kPa}$
7.33	$\eta = 0.2945$
7.36	$\eta = 0.2812$
7.39	Boiler $w_{lost} = 428 \text{ Btu/lb}_m$
7.42	Boiler
7.45	COP= 6.026 $\dot{Q}_{in} = 343{,}100 \text{ Btu/hr}$
7.48	$\dot{m} = 0.014 \text{ lb}_m/\text{s}$ $\dot{W} = -1.675 \text{ kW}$
7.51	COP= 6.388
7.54	COP= 7.128
7.57	$\dot{W} = -4.659 \text{ kW}$
7.60	Condenser
7.63	$\dot{W} = -1.269 \text{ kW}$
7.66	Mixing point
7.69	$\dot{Q} = 0.2564 \times 10^6 \text{ Btu/hr}$
	$\dot{W} = -48.08 \text{ hp}$
7.72	Condenser $\dot{W}_{lost} = 26.84 \text{ Btu/s}$
7.75	COP $= 0.8279$
7.78	$\dot{Q}_{gen} = 11.33 \text{ Btu/s}$
	$\dot{Q}_{abs} = -15.50 \text{ Btu/s}$
	$\dot{Q}_{cond} = 10.39 \text{ Btu/s}$

Chapter 8

8.3	$T_h = 576 \text{ K}$ $r_P = 5.654$
	$r_v = 2.827$
8.6	$\eta_{max} = 0.6429$
8.9	$\eta = 0.4749$ $q_{in} = 247.4 \text{ Btu/lb}_m$
	$P_{mean} = 48.99 \text{ psia}$
8.12	$r_v = 7.20$
8.15	$\dot{W} = 23.91 \text{ hp}$
8.18	$q_{in} = 318.7 \text{ Btu/lb}_m$
	$q_{out} = 161.8 \text{ Btu/lb}_m$
	$\eta = 0.4923$

8.21	$P_{max} = 4829 \text{ kPa}$ $T_{max} = 1515 \text{ K}$
	$q_{in} = 554.5 \text{ kJ/kg}$ $\eta = 0.6460$
8.24	$w_{lost,net} = 27.50 \text{ Btu/lb}_m$
8.27	$\dot{W} = 24.83 \text{ kW}$
8.30	$w_{net} = 124.0 \text{ Btu/lb}_m$
	$q_{in} = 193.6 \text{ Btu/lb}_m$
	$\eta = 0.6405$
8.33	$T_{max} = 2693° \text{ R}$
	$\dot{Q}_{in} = 1.878 \times 10^6 \text{ Btu/hr}$
8.36	$\dot{W} = 131.1 \text{ hp}$
8.39	$W_{net} = 289.0 \text{ kJ}$
8.42	$T_h = 765° \text{ R}$ $m = 0.02647 \text{ lb}_m$
	$P_{max} = 125.0 \text{ psia}$
8.45	$q = 64.00 \text{ Btu/lb}_m$
8.48	$w_{lost,net} = 4.5 \text{ kJ/kg}$
8.51	$w_{net} = 9.032 \text{ kJ/kg}$ $\eta = 0.04687$
8.54	$\dot{w}_{net} = 266.3 \text{ kW}$
8.57	$\dot{Q}_{in} = 2339 \text{ Btu/s}$ $\dot{W}_{net} = 1405 \text{ Btu/s}$
	$\eta = 0.6007$
8.60	$\dot{W} = 360.7 \text{ hp}$ $\eta = 0.7213$
8.63	$w_{lost,net} = 180.1 \text{ kJ/kg}$
8.66	$\eta = 0.4479$
8.69	$\dot{Q}_{in} = 1760 \text{ Btu/s}$
8.72	$\eta = 0.3561$
8.75	$W_{turb} = 263.7 \text{ kW}$
	$\dot{W}_{comp} = -173.2 \text{ kW}$
	$\dot{Q}_{out} = 328.4 \text{ Btu/s}$
8.78	$\dot{Q}_{in} = 725.8 \text{ kW}$
8.81	$F = 8296 \text{ lb}_f$
8.84	$V = 530 \text{ m/s}$ $F = 166{,}000 \text{ N}$
8.87	$\dot{W}_{lost,net} = 5.161 \times 10^6 \text{ kW}$
8.90	$\eta = 0.5186$
8.93	$\text{COP}_r = 0.3655$
8.96	$\text{COP} = 1.165$ $\dot{m} = 0.3969 \text{ lb}_m/\text{s}$

Chapter 9

9.2 $\Delta h = 286.0$ kJ/kg

        $\Delta h = 281.0$ kJ/kg

9.4 $\Delta h = 200.6$ kJ/kg

9.6 $\Delta s = 0.1588$ kJ/kg-K

9.10 $\Delta u = 874.4$ kJ/kg

9.31 $\frac{v}{T} = f(P)$

9.34 $\mu \approx 0.012$ K/kPa

9.37 $P = 0$

9.45 $P \approx 99.4$ psia $P \approx 134.0$ psia

9.48 $V_2 = 0.02294$ m^3 $V_2 = 0.02301$ m^3

        $V_2 = 0.02303$ m^3

9.51 $P_2 = 34.95$ psia $P_2 = 35.02$ psia

        $P_2 = 35.08$ psia

9.54 $T_2 = 986°$ R $T_2 = 960°$ R

9.57 $v = 0.03804$ m^3/kg

        $v = 0.03521$ m^3/kg

        $v = 0.03521$ m^3/kg

        $v = 0.03576$ m^3/kg

9.60 $\Delta v = -0.01674$ m^3/kg

9.63 $T = 1549°$ R $T = 1565°$ R

        $T = 1551°$ R

9.67 $T = 560$ K

9.70 $\Delta h = -211.2$ Btu/lb$_m$

        $\Delta s = -0.0147$ Btu/lb$_m$-°R

        $\Delta h = -259.8$ Btu/lb$_m$

        $\Delta s = -0.0366$ Btu/lb$_m$-°R

9.73 $\Delta h = 911.3$ kJ/kg

        $\Delta s = 1.109$ kJ/kg-K

        $\Delta h = 1130$ kJ/kg

        $\Delta s = 1.544$ kJ/kg-K

9.76 $\Delta s = 0.8347$ kJ/kg-K

        $w = 675.2$ kJ/kg

        $\Delta s = 1.586$ kJ/kg-K

        $w = 1121$ kJ/kg

9.79 $w = -159.8$ kJ/kg

9.79 $w = -159.8$ kJ/kg

9.83 $w = -225.1$ Btu/lb$_m$

        $w = -223.0$ Btu/lb$_m$

9.86 $w = 151.5$ kJ/kg

        $w = 164.1$ kJ/kg

9.89 $w = -235.3$ Btu/lb$_m$

        $w = -214.9$ Btu/lb$_m$

9.91 $\eta = 0.5407$ $\eta = 0.5647$

9.94 $\eta = 0.7515$ $\eta = 0.7844$

Chapter 10

10.1 $M = 10.61$ kg/kg-mole

        $V = 3.761$ m^3

        $V_{O_2} = 0.07789$ m^3

        $P_{O_2} = 2.071$ kPa

10.4 $M = 10.61$ kg/kg-mole

10.7 $q = 326.6$ kJ/kg

10.10 $\Delta V = 32.11$ ft^3 $Q = 837.9$ Btu

10.13 $_1W_2 = 118.1$ kJ $_1Q_2 = 568.8$ kJ

10.16 $m = 4.727$ lb$_m$

10.19 $P_{C_2H_6} = 24.15$ kPa

10.22 $\dot{m} = 76,220$ lb$_m$/hr

        $\dot{V} = 254,500$ ft^3/hr

10.25 $\eta = 0.3865$

10.28 $\eta = 0.541$

10.31 $\eta = 0.3698$

10.34 $\eta_{2nd} = 0.852$ $w_{lost} = 38.8$ Btu/lb$_m$

10.37 $\dot{m} = 0.4088$ lb$_m$/s

        $\dot{m} = 0.05565$ lb$_m$/s

        $\dot{m} = 0.05565$ lb$_m$/s

        $\dot{V} = 0.05454$ ft^3/s

10.40 $_1w_2 = 58.19$ Btu/lb$_m$

10.43 $T_{wb} = 81.5°$ F $\omega = 0.0216$ lb$_m$/lb$_m$-da

        $h = 45.4$ Btu/lb$_m$-da $T_{dp} = 79.5°$ F

        $P_v = 0.489$ psia

10.46 $T \approx 85°$ F

10.49 $\phi = 62\%$ $\omega = 0.0190$ lb$_m$/lb$_m$-da
$h = 42.0$ Btu/lb$_m$-da $T_{wb} = 79°$ F
$P_v = 0.433$ psia

10.52 $T_{as} \approx 22°$ C

10.55 $\dot{m}_2/\dot{m}_{da} = 0.0204$ lb$_m$/lb$_m$-da
$q = -31.6$ Btu/lb$_m$-da

10.58 $\dot{m}_2/\dot{m}_{da} = 0.007$ kg/kg-da
$\dot{Q} = -89.5$ kW

10.61 $\dot{m}_2/\dot{m}_{da} = 0.0066$ lb$_m$/lb$_m$-da
$q = -0.94$ Btu/lb$_m$-da

10.64 $T \approx 24°$ C

10.67 $\dot{m}_{da} = 5.929$ kg/s

10.70 $\dot{m}_{da} = 3.118$ lb$_m$/s

10.73 $m = 20.00$ kg $Q = -57,350$ kJ

10.76 $T = 55°$ F

10.79 $q = 16.9$ Btu/lb$_m$-da

10.82 $\dot{m}_3 = 0.06698$ lb$_m$/s
$\dot{Q} = -16.16$ Btu/s

10.85 $T = 40°$ F $\phi = 30\%$

10.88 $m_{da,1}/m_{da,2} = 0.7533$ $T \approx= 22°$ C

10.91 $\phi = 8\%$

Chapter 11

11.2 $N_{CO_2} = 4$ kg-mole/kg-mole-fuel

11.5 $m_{H_2O}/m_{C_3H_8} = 1.634$
$m_{air}/m_{C_3H_8} = 15.63$

11.8 $m_{CO_2}/m_{C_4H_{10}} = 3.029$
$m_{air}/m_{C_4H_{10}} = 15.42$

11.11 $m_{H_2O}/m_{C_8H_{18}} = 1.420$

11.14 $M = 27.90$ kg/kg-mole

11.17 $m_{H_2O}/m_{C_8H_{18}} = 0.1578$

11.20 $A/F = 18.63$

11.23 $m_{H_2O}/m_{coal} = 0.5421$ $T_{dp} \approx 104°$ F

11.26 $T_{dp} = 51°$ C

11.29 $f = 0.511$

11.32 $T \approx 3080°$ F

11.32 $T \approx 3080°$ F

11.35 $q = -15,930$ Btu/lb$_m$

11.38 $T \approx 1417°$ F

11.41 $\bar{q} = -1,427,800$ kJ/kg-mole-fuel

11.44 $HHV = -13,960$ Btu/lb$_m$
$LHV = -13,420$ Btu/lb$_m$

11.47 $HHV = -12,890$ Btu/lb$_m$
$LHV = -12,320$ Btu/lb$_m$

11.50 $\bar{q} = -231,280$ Btu/lb$_m$-mole-fuel

11.53 $T \approx 2510°$ R $P = 68.7$ psia

11.56 $w_{ptl} = 18,190$ Btu/lb$_m$-fuel

11.59 $\delta\overline{\Phi} = -3,370,500$ kJ/kg-mole-fuel

11.62 $\overline{w}_{ptl} = 5,013,180$ kJ/kg-mole-fuel

11.65 $\overline{w}_{ptl} = -377,173$ Btu/lb$_m$-mole-fuel

Chapter 12

12.3 $g = -63.35$ kJ/kg

12.6 $m_f = 65.11$ kg

12.9 $P_v = 123.1$ kPa $P_v = 492.2$ kPa

12.12 $P_v = 0.3632$ psia

12.15 $y_g = 0.9714$ $y_f = 0.9675$

12.18 $\mu = 328.3$ Btu/lb$_m$-mole-N$_2$

12.22 $\ln K_p = -30.70$ $\ln K_p = -6.79$

12.25 $y = 0.4799$

12.28 $y = 0.2090$

12.31 $y_{CO_2} = 0.9562$ $y_{CO} = y_O = 0.0219$

12.34 $y_{CH_4} = 0.0059$ $y_{N_2} = 0.2529$
$y_C = 0.2471$ $y_{H_2} = 0.4941$

12.37 $y_{O_2} = 0.9973$ $y_O = 0.0268$

12.39 $\ln K_p = 32.294$

12.41 $\ln K_p = 12.368$

12.44 1

12.47 $y_{CO_2} = 0.1576$ $y_{H_2O} = 0.086$
$y_{O_2} = 0.0002$ $y_{2NO} = 0.0356$
$y_{N_2} = 0.7206$

12.50 $\bar{q} = -12,190$ kJ/kg-fuel

Index